de Gruyter Expositions in Mathematics 49

Applied Algebraic Dynamics

by

Vladimir Anashin and Andrei Khrennikov

Walter de Gruyter · Berlin · New York

Authors

Andrei Khrennikov
International Center for Mathematical Modeling
Växjö University
Vejdes plats 7
35195 Växjö, Sweden
E-mail: andrei.khrennikov@vxu.se

Vladimir Anashin
Institute for Information Security
Moscow State University
Leninskie Gory
119991 Moscow, Russia
E-mail: vladimir.anashin@u-picardie.fr

Mathematics Subject Classification 2000: 05B15, 11-02, 11B37, 11B50, 11B85, 11K41, 11K45, 12J25, 13M10, 20-02, 20E18, 22D40, 28D05, 30G06, 37-02, 37A05, 37A25, 37N20, 37N25, 37N30, 46S10, 60F20, 65C10, 68P25, 68Q99, 68N30, 81P99, 92C30, 92D20, 94A55, 94A60

Key words: Algebraic dynamical systems, p-adic numbers, measure-preserving transformations, ergodicity, profinite groups, automata, computer sciences, cryptography, p-adic probability, quantum theory, psychology, genetics, Latin squares, pseudorandom generators, stream ciphers.

∞ Printed on acid-free paper which falls within the guidelines of the ANSI to ensure permanence and durability.

ISSN 0938-6572
ISBN 978-3-11-020300-4

Bibliographic information published by the Deutsche Nationalbibliothek

The Deutsche Nationalbibliothek lists this publication in the Deutsche Nationalbibliografie; detailed bibliographic data are available in the Internet at http://dnb.d-nb.de.

Printing and binding: Hubert & Co. GmbH & Co. KG, Göttingen.
Cover design: Thomas Bonnie, Hamburg.

This book is dedicated to Kurt Hensel.

Preface

In this book, we develop methods of algebraic dynamics and apply them to concrete problems from computer science, cryptology, theoretical physics, cognitive science, psychology, neurophysiology, and genetics. Therefore this book is for pure mathematicians working in the theory of dynamical systems and related areas, as well as for applied scientists interested in the mentioned non-mathematical disciplines. Although all chapters of the book contain mathematical results, we tried to make 'applied' chapters somewhat independent from 'mathematical' chapters; that's why speaking on applied problems we introduce relevant mathematical notions and results more informally. However, in 'applied' chapters we make here and there proper references to 'mathematical' chapters for those applied scientists who are interested in the underlying mathematical theory. Also, in Chapter 1 we remind some notions and facts from algebra, number theory and p-adic analysis. A reader interested only in 'applied' chapters, may not read this chapter, since it is for references, and mainly serves as a sort of a glossary.

Now we make a brief outline of a general approach we mostly apply throughout the book. Recall that a (discrete, autonomous) *dynamical system* is just a pair $\langle \mathbb{S}, f \rangle$, where $f : \mathbb{S} \to \mathbb{S}$ is a map of a set $\mathbb{S}$ (configuration space) into itself. Dynamical system theory studies trajectories (orbits), i.e., sequences of iterations:

$$x_0,\ x_1 = f(x_0), \ldots, x_{i+1} = f(x_i) = f^{i+1}(x_0), \ldots .$$

Central questions are asymptotic behavior of these sequences, their distribution, etc. Often to obtain a rich model, one considers $\mathbb{S}$ which is endowed with a *metric* (or generally, a topology) and with a *measure*.

We speak about *algebraic* dynamics whenever we assume that the space $\mathbb{S}$ is endowed with a certain algebraic structure (a ring, a group, etc.), and that the map f somehow *agrees* with this algebraic structure; say, when f is either a polynomial over $\mathbb{S}$, or an automorphism of $\mathbb{S}$, or a composition of operations and endomorphisms, etc.

In real life settings we *never* deal with an infinite $\mathbb{S}$. Yet for a finite $\mathbb{S}$, every trajectory is eventually periodic, and so it is meaningless to speak of its asymptotic behavior. Unfortunately, in real life settings the set $\mathbb{S}$ is usually big; so big that we can not use computer simulations to answer the question where will be the point after N iterations for large N.

However, we can study behavior of trajectories on small $\mathbb{S}$ in order to understand what happens to trajectories when $\mathbb{S}$ becomes bigger and bigger. Thus, we have to study asymptotic behavior of trajectories when $\#\mathbb{S} \to \infty$ (here and throughout the book $\#\mathbb{S}$ denotes the number of elements in $\mathbb{S}$).

Obviously, we can say almost nothing nontrivial about this asymptotic behavior in a general case, for arbitrary maps of arbitrary finite sets. It turns out that we can say a lot about this behavior whenever $\mathbb{S}$ is endowed with an algebraic structure and f agrees with this structure. Say, when f is a polynomial, and finite algebraic systems $\mathbb{S}_n$ constitute a *projective spectrum*, which is also called an *inverse spectrum*:

$$\mathbb{S}_\infty \cdots \xrightarrow{\varphi_{n+1}} \mathbb{S}_n \xrightarrow{\varphi_n} \mathbb{S}_{n-1} \xrightarrow{\varphi_{n-1}} \cdots \xrightarrow{\varphi_1} \mathbb{S}_0.$$

Speaking loosely, a projective spectrum is a sequence of sets endowed with algebraic structures such that $\mathbb{S}_n$ can be "projected" to $\mathbb{S}_{n-1}$ – by the map φ_n – in such a way that the algebraic structure on $\mathbb{S}_n$ is "projected" on the algebraic structure of $\mathbb{S}_{n-1}$. This happens, for instance, when all $\mathbb{S}_n$ are algebraic systems of the same type (e.g., all are groups, or all are rings, etc.), and φ_n are epimorphisms. Given algebraic systems $\mathbb{S}_n$ and projections φ_n, the 'limit algebraic system' $\mathbb{S}_\infty$, which is called an *inverse limit* (or a *projective limit*) of the spectrum, can be rigorously defined. The very construction of the inverse limit of finite algebraic systems implies a natural metric (which is then necessarily a non-Archimedean metric), and a natural probabilistic measure on the algebraic system $\mathbb{S}_\infty$. This way one can lift[1] dynamics from $\mathbb{S}_n$ to dynamics on $\mathbb{S}_\infty$ and to study it there thus obtaining information about the dynamics on a finite $\mathbb{S}_n$.

An important class of such inverse limits is given by rings of p-adic integers[2] $\mathbb{Z}_p$ ($p > 1$ is a prime number), which are inverse limits of the residue class rings $\mathbb{Z}/p^n\mathbb{Z}$ modulo p^n (or briefly, of residue rings modulo p^n), $n = 1, 2, \ldots$. The corresponding projections φ_n are just reductions modulo p^n, which clearly are ring epimorphisms.

Although we can not apply directly inverse limits to obtain a field of p-adic numbers $\mathbb{Q}_p$, which is also one of the basic configuration spaces in this book, we remark that by suitable scalings $p^{-k}\mathbb{Z}_p, k = 1, 2, \ldots$, the ring $\mathbb{Z}_p$ can be 'extended' to the field $\mathbb{Q}_p$. As the ring $\mathbb{Z}_p$ is approximated by finite rings $\mathbb{Z}/p^n\mathbb{Z}$, in a precise algebraic meaning of the word[3], we may say that $\mathbb{Q}_p$ is 'approximated' by finite sets as well, up to the mentioned scalings.

---

[1] This is indeed a sort of Hensel's lift; the latter originates from Kurt Hensel's proof of his famous Lemma.

[2] Actually one of goals we pursue is to demonstrate that p-adic numbers, which appeared more than a century ago in Kurt Hensel's works as a pure mathematical construction, see e.g. [169], recently were recognized as a base for adequate descriptions of physical, biological, cognitive and information processing phenomena; to say nothing of the important role these numbers are playing in various mathematical sciences.

[3] In algebra they say that an algebraic system (i.e., a universal algebra) A is *approximated* by universal algebras of some class $\mathcal{A}$ whenever given $g, h \in A$, $g \neq h$, there exists a homomorphism φ of A into some algebra $B \in \mathcal{A}$ such that $\varphi(g) \neq \varphi(h)$.

Moreover, we will show in this book that *ergodic*[4] polynomial dynamics on finite commutative rings or on finite solvable (and *not* necessarily commutative!) groups with operators, can be described as 'projections' of corresponding p-adic ergodic dynamics.

Therefore there is tight connection between dynamics in finite sets and p-adic dynamics. Typically one can derive important features of dynamics in $\mathbb{Q}_p$ or $\mathbb{Z}_p$ from corresponding dynamics in "pre-limit" finite sets, residue rings modulo p^n, and vice versa. As said, such an approach is one of the main tools which will be used in this book, especially to study dynamical systems for applications in cryptology, automata theory, computer science, and pseudorandom number generation, see Chapters 8–11. In many other applications, especially to cognitive science, psychology, neurophysiology, genetics, see Chapters 14–17, finite sets $\mathbb{S}_n$ are given by rings of residue classes $(\bmod\, m^n)$, where $m > 1$ is an arbitrary natural number.

Although in real life settings we always deal with dynamics on a configuration space of finite order, this order varies from 'big' to 'very big'. Physics provides a good illustration for the latter case: In physics theoretical formalism was developed for dynamical systems in configuration spaces with coordinates from the *real continuum* (and not finite sets!). One of the reasons for this is an extremely big number of possible states for a physical system. Even for one dimensional particle, a fine description of its trajectory can be performed only in a space containing a huge number of points. In Newton's time, it was totally impossible to proceed with, e.g., difference equations. The model based on the real continuum became dominating in theoretical physics as well as in natural science, in general. The later development of computers and numerical methods provides a possibility to operate in finite (but extremely big) configuration spaces. However, the original (Newtonian) physical ideology was not changed. Discrete dynamics, e.g., given by difference equations, were considered as mathematical approximations of "real physical laws" given by differential equations – e.g., by second Newton law or by Maxwell equations. In the 1960s and, especially, 70s–80s, it was a good occasion to change this ideology.[5] Unfortunately, this chance was not used. A new attempt was done in the 1990s in connection with development of p-adic theoretical physics[6], Chapter 13. Unfortunately, neither of those approaches changed

[4]Recall that a dynamical system f on a configurations space $\mathbb{S}$ endowed with a probability measure μ is called ergodic whenever there is no (up to subsets of measure 0) f-invariant subsets other than the empty set and the whole set $\mathbb{S}$; this means, loosely speaking, that the probability the system falls into stationary states is 0.

[5]Say: "For any physical process, one can put limits of the precision of the numerical representation of data and introduce a configuration space containing a finite number of points. Only corresponding discrete dynamics are 'real', continuous dynamics in continuous (real) configuration spaces are only ideal mathematical constructions."

[6]First p-adic physical models were elaborated in the 1990s at Steklov Mathematical Institute of Russian Academy of Science by V. Vladimirov, I. Volovich, I. Aref'eva, E. Zelenov in collaboration with A. Khrennikov and B. Dragovich; important contributions to this domain were done by E. Witten, G. Parisi, P. Framton, Freund, Olson and others, see, e.g., monographs [201,214,407] and pioneer papers of Vladimirov and Volovich [404,405,408].

the general situation in physics.

On the other hand, in some areas, e.g., in computer science, cryptology, numerical analysis, etc., the dimension of a configuration space is much smaller; usually it is of order of a word bitlength of a computer. A trajectory in this case is a sequence of states, and the dynamics is often defined explicitly – by a state transition function. This function, which is a composition of basic instructions of a processor, can be regarded as a polynomial over a corresponding universal algebra.

For instance, in cryptology it is important to describe evolution of the initial state (which is usually a 'key'); that is, to describe the trajectory of a single particle, speaking in 'dynamical' terms. Knowledge that the number of 'bad keys' tends to zero as bitlength tends to infinity says nothing on whether the cipher is secure, being implemented as a program for a computer of a fixed word bitlength, which is normally rather small, 8, 16, 32, 64, or rarely 128, 512, 1024. Say, if we know only that the system is chaotic when the bitlength is infinite, this gives us almost nothing about the behavior of this system on a finite set: For instance, it is well known that the Bernoulli shift $x_0 + 2x_1 + 4x_2 + \cdots \mapsto x_1 + 2x_2 + 4x_3 + \cdots$ is a chaotic transformation on the space of 2-adic integers $\mathbb{Z}_2$. However, a counterpart of the Bernoulli shift on a finite configuration space $\{0, 1, 2, 3, \ldots, 2^n - 1\}$ of all n-bit numbers is a 1-bit shift towards less significant bits; this map obviously degenerates after at most n iterations, sending every number to 0. This is only one illustration from numerous others why the 'usual' real or complex dynamics approach does not match to describe evolutions of computer programs. Another illustration are numerical experiments with chaotic systems. They demonstrate that (we quote from [298]) "digital computers are absolutely incapable of showing true long-time dynamics of some chaotic systems, including the tent map, the Bernoulli shift map and their analogues, even in a high-precision floating-point arithmetic."

However, it turns out that basic computer instructions, both numerical ones (integer addition and multiplication) and logical ones (bit-by-bit logical OR, AND, XOR, NOT, ...) can be regarded as *continuous (1-Lipschitz) maps with respect to the 2-adic metric*; whence, all compositions of these instructions, i.e., corresponding *computer programs, are continuous* with respect to this metric as well. So in this case namely the 2-adic dynamics gives us a powerful tool to study behavior of these programs as their dynamics are essentially 2-adic, see Chapter 8.

Furthermore, if we consider an automaton whose input and output alphabets are the same m-letter set, a function this automaton evaluates – a transformation of input words to output ones – is again a 1-Lipschitz (whence, continuous) transformation on the space $\mathbb{Z}_m$ of m-adic integers. Note that automata are usual models for various information processes.

These remarks are a partial explanation of the fact that the algebraic dynamic approach turned out to be especially effective in application to various problems of *information processing* independently on where these problems arise; e.g., in computer science, cryptology, cognitive sciences, genetics or somewhere else.

However, we do not touch in this book other aspects of applied algebraic dynamics such as superstring theory, quantum mechanics and field theory (only a short review in Chapter 13), disordered systems (especially spin glasses), wavelets, theory of pseudo-differential operators, see, e.g., [201, 214, 407].

The theory of algebraic dynamical systems is intensively developing discipline on the boundary between various mathematical theories – dynamical systems, number theory, algebraic geometry, non-Archimedean analysis – and having numerous applications – cryptology, computer science, theoretical physics, cognitive science, genetics, and image analysis.

Traditionally dynamical systems were considered in the fields of real and complex numbers, $\mathbb{R}$ and $\mathbb{C}$. Later studies of dynamical systems in *finite fields and rings* were started. Number theory was widely used in these investigations. Theory of p-adic dynamical systems was developed as a natural generalization of dynamics in residue rings modulo p^n. It was generalized to arbitrary *non-Archimedean fields.*[7] This was the combination of *number theoretic and dynamical flows* towards algebraic dynamics. We can mention investigations of W. Narkiewicz, A. Batra, P. Morton and P. Patel, J. Silverman and G. Call, D. K. Arrowsmith, F. Vivaldi and Hatjispyros, J. Lubin, T. Pezda, H.-C. Li, L. Hsia, e.g., [40, 41, 45, 46, 82, 173, 174, 289–296, 302–304, 326–334, 334, 335, 338–342, 356–361, 401, 402], and recently J. A. G. Roberts and F. Vivaldi, W.-S. Chou and I. E. Shparlinski, A.-H. Fan, J. L. Chabert, Y. Fares, M.-T. Li and J.-Y. Yao, Y.-F. Wang, and D. Zhou, M. Misiurewicz, J. G. Stevens, and D. Thomas, A. Peinado, F. Montoya, J. Muñoz and A. J. Yuste, F. Durand and F. Paccaut, J. Kingsbery, A. Levin, A. Preygel, and C. E. Silva, see [83, 85, 110, 127–129, 131, 132, 261, 262, 319, 354, 372, 379].

This flow is closely related to the flow induced in *algebraic geometry.* In algebraic geometry fields of real and complex numbers, $\mathbb{R}$ and $\mathbb{C}$, do not play an exceptional role. All geometric structures can also be considered over non-Archimedean fields. Therefore, for people working in algebraic geometry, it was natural to try to generalize some mathematical structures to the non-Archimedean case, even if this structures did not directly belong to the domain of algebraic geometry; for example, dynamics in a non-Archimedean field $\mathbb{K}$. This (algebraic geometric) dynamical flow began with article of M. Herman and J. C. Yoccoz [170] on the problem of small divisors in non-Archimedean fields. It seems that this was the first publication on non-Archimedean dynamics. In further development of this dynamical flow the crucial role was played by J. Silverman, see, e.g., [380–382]. Investigations were continued by R. Benedetto, [52–61], J. Rivera-Letelier [366–369], C. Favre and J. Rivera-Letelier [133], F. Laubie and A. Movahhedi and A. Salinier [283], J.-P. Bézivin [64–67]. Finally, the fundamental book of J. Silverman [383] devoted to arithmetic problems in theory of dynamical systems was published.

---

[7] These are fields with absolute values for which the strong triangle inequality $|x + y| \leq \max(|x|, |y|)$ holds. We remark that fields of p-adic numbers $\mathbb{Q}_p$ are non-Archimedean.

Another flow towards algebraic dynamics has p-adic theoretical physics as its source. In 1989, Ruelle, Thiran, Verstegen, Weyers published the interesting article [373] on p-adic quantum mechanics and little bit later Thiran, Verstegen, Weyers published article [395] on p-adic dynamics, see also [400]. We also mention the earlier preprint [51] of Ben-Menahem. One of the authors of this book also used this pathway towards p-adic dynamical systems, from study of quantum models with $\mathbb{Q}_p$-valued functions, e.g., [201], to p-adic and more general non-Archimedean dynamical systems, e.g., [203, 214]. As the result, a strong research group on non-Archimedean dynamics was created at Växjö University, Sweden: Andrei Khrennikov, Karl-Olof Lindahl, Marcus Nilsson, Robert Nyqvist, and Per-Anders Svensson, [5, 256, 301, 347, 347, 348, 348, 392, 392]. Main efforts of this group were directed to study dependence of the number of cycles of a fixed length on the parameter p. Numerical simulations performed by Khrennikov and Nilsson for monomial dynamical systems, $x \mapsto x^n$, supported the conjecture on random dependence. Later they obtained rigorous mathematical results on corresponding probability distributions; in particular, averages and dispersions. These results are deeply coupled to classical results on the asymptotic distribution of the number of primes. Khrennikov, Nilsson, and Nyqvist [255] generalized these results to perturbations of monomial systems:

$$x \mapsto x^n + q(x),$$

where $q(x)$ is a polynomial which is 'small' comparing with the monomial part of the dynamics; smallness is defined as smallness of coefficients with respect to the p-adic absolute value. The degree of $q(x)$ does not play any role. Thus such dynamics can be extremely complex from the algebraic viewpoint. An attempt to find the distribution of the number of cycles of the fixed length for new classes of polynomials (which are not reducible to monomial in the sense of theory of perturbations) was done in [257]. In spite of the use of very advanced methods from number theory based on Chebotarev theorem, only a restricted class of new polynomial systems was investigated. The problem – *to find the probability distribution of the number of cycles of the fixed length L, say, e.g., $L = 6$, depending on p for an arbitrary polynomial dynamical system with rational coefficients* – has not yet been solved. Another domain of research of the Växjö group is dynamics in finite extensions of fields of p-adic numbers. The main problem under study is dependence (of course, random) of the number of cycles on p and the degree of extension. Strongest results in this direction were obtained by P.-A. Svensson [392, 393], see also Khrennikov and Svensson [258]. A. Khrennikov and K. O. Lindahl studied in [234, 301] the problem of linearization of p-adic and more general non-Archimedean dynamical systems, cf. M. Herman and J. C. Yoccoz [170]. K. O. Lindahl with his work [301] opened a new interesting domain of algebraic dynamics, namely, dynamics in non-Archimedean fields of prime characteristic.

We point out recent publications of Vladimir Arnold [37–39] devoted to chaotic aspects of arithmetic dynamics closely coupled to the problem of turbulence. A p-adic attack to this complicated problem was also done by S. Fishenko and E. Zenelov

[135]. However, the latter paper has no direct coupling to discrete dynamical systems.

In 1997, Andrei Khrennikov [214, 217] proposed to apply dynamical systems in rings $\mathbb{Z}_m$ for modeling of *cognitive processes,* especially in *psychology,* see Chapter 14. In applications to cognitive science the crucial role is played not by the algebraic structure of $\mathbb{Z}_m$, but by its *hierarchical structure* corresponding to the projective limit. We remark that the projective limit structure on $\mathbb{Z}_m$ can be geometrically realized as a *tree*. This treelike representation of $\mathbb{Z}_m$ gives a possibility to describe neuronal trees and production of mental information by such trees, see Chapter 15. Recently 4-adic and 2-adic dynamical systems were applied to genetics, Chapter 16. We also mention applications of m-adic numbers to image analysis – compression of information and image recognition, see Benois, Khrennikov, Kotovich, Borzystaya [62, 246, 247]. Unfortunately, mainly as a consequence of restriction to volume of the book, we were not able to present the latter domain of applied research in this book.

We also point out a flow towards algebraic dynamics which is extremely important for applications to computer science and cryptology, especially in connection with pseudorandom numbers and uniform distribution of sequences. This flow arose in 1992 starting with publications [21, 22] by one of the authors of the book, Vladimir Anashin; these works were succeeded by his works [23–26, 28, 29], see Chapters 8–11. Mainly this flow is motivated by the problem how to construct a computer program that produces random-looking sequence of numbers. To look any random, the sequence must be at least uniformly distributed in some precise meaning, it must also pass common statistical tests, and the performance of the corresponding program (or hardware device) must be sufficiently fast. To satisfy the latter condition, the program must be a not too complicated composition of basic computer instructions mentioned above (additions, multiplications, ORs, ANDs, XORs, etc.), which are, as said, continuous with respect to a 2-adic metric. Thus, to compile with the first condition, one may combine these instructions into a certain ergodic transformation f on $\mathbb{Z}_2$; then the corresponding sequence of iterations $x, f(x), f^2(x), \ldots$ will be necessarily uniformly distributed in $\mathbb{Z}_2$ and hence modulo 2^n, for all $n = 1, 2, \ldots$. This was a strong motivation to develop *p-adic ergodic theory*, see Chapter 4.

Programs that produce random-looking sequences of numbers, the *pseudorandom generators*, are needed for various applied purposes. For instance, pseudorandom numbers are used in computer experiments, modeling, various computer simulations, numerical analysis (recall quasi-Monte-Carlo methods), and cryptography; e.g., the so-called *stream ciphers* actually are cryptographically secure pseudorandom generators, see Chapter 10. That's why there is a huge number of works on pseudorandom numbers, both theoretical and practical. It is impossible to mention here even a small part of relevant papers, we only refer to volume 2 of the monograph by Donald Knuth 'The Art of Computer Programming' [267], to the monograph by Harald Niederreiter [344], and to the survey [126] by Graham Everest, Alf van der Poorten, Igor Shparlinsky, and Thomas Ward. For cryptographic applications of pseudorandom generators see books [315, 375] on practical cryptography. [8]

We note that currently there exists a variety of methods to construct pseudorandom numbers; these methods use different ideas and approaches from different branches of mathematics. Moreover, there exist pseudorandom generators whose theory is p-adic, and which nevertheless are based on approaches that are completely different from the approach presented in our book, see e.g. generators introduced by A. Klapper and M. Goresky [263], by D. Bosio and F. Vivaldi [74], see also [355, 403], and by C. Woodcock and N. Smart [412].

In Chapter 4, we develop p-adic ergodic theory for *1-Lipschitz* transformations on $\mathbb{Z}_p$; the latter theory leads to the theory of the so-called *congruential generators*, see Chapter 9, a sort of very popular and wide-spread pseudorandom generators. However, not all existing types of pseudorandom generators are congruential (e.g., the generators mentioned above are *not* congruential); thus, not all of them are covered by the p-adic ergodic theory from Chapter 4.

The most known congruential generators are *linear* congruential generators, which produce recurrence sequences whose law of recursion is $x_{i+1} = a \cdot x_i + b \pmod{N}$, where a, b are rational integers, and $N > 1$ is an integer. These generators are well studied (see e.g. [267]); however, they have immanent drawbacks due to their linearity, which leads either to cryptographic insecurity or to false results in some numerical simulations, see relevant discussions in [77, 267, 315, 375]. This fact stimulated since the late 1980s a huge search for new, non-linear types of congruential generators. The most important non-linear congruential generators are *polynomial generators*, which produce recurrence sequences whose law of recursion is $x_{i+1} = f(x_i) \pmod{N}$, where f is a polynomial with rational integer coefficients. The other types of congruential generators are *exponential*, when $x_{i+1} = a^{x_i} + b \pmod{N}$, *inversive*, when $x_{i+1} = (a \cdot x_i + b)^{-1} \pmod{N}$, and various combinations of these. We stress here that generators based on the so-called *T-functions*, which recently attracted significant attention in cryptography, are also congruential generators that correspond to the case when f is a composition of arithmetical (integer addition and multiplication) and logical (OR, AND, XOR, …) operations, and N is a power of 2.

One of the most important applications of the p-adic ergodic theory, whose development started in the early 1990s by works [21, 22] of one of the authors of the book, Vladimir Anashin, are namely congruential generators. Actually almost all results on periods of these generators, obtained earlier in different works by different authors, can be (and are) reproved and significantly generalized and strengthened by methods of p-adic ergodic theory, see Chapters 9 and 10. For instance, all mathematical results of papers [264, 265] by A. Klimov and A. Shamir, which initiated interest to T-functions in cryptographic community, either are contained among or immediately (and easily) follow from the results of works [21, 22] by Vladimir Anashin, who published them a decade prior to the mentioned publications of A. Klimov and A. Shamir,

[8] We note, however, that there are some highly questionable statements about these generators in these books, at our view.

see relevant examples in Chapters 9 and 10. Currently ideas and techniques of p-adic ergodic theory penetrated into cryptographic community: Several stream ciphers and cryptographic primitives are developed with these ideas, see e.g. relevant designs in [350], see also [28, 30, 273, 274]. We note that with the use of p-adic ergodic theory it became possible to establish certain crucial structural and distribution properties of sequences produced by congruential generators that doubtfully can be proved by other methods, see Chapter 11.

Another important application of p-adic ergodic theory, is computer science and automata theory, see Chapter 8. There we also reprove and/or generalize a number of known results and obtain new ones. For instance, we present new methods to construct fast algorithms to produce big quantities of large Latin squares; the latter are important for different applied areas, e.g., in experiment design, software testing, in communications, etc. In Subsection 11.1.2 we introduce a new measure of complexity of maps performed by automata; this measure clearly differs automata that use or do not use multiplication of variables; this in turn implies that for some crucial applications automata of the latter type are unacceptable, though they are faster. We expect in the near future new results in automata theory obtained by p-adic methods since every automaton, as said, can be considered as a 1-Lipschitz map of m-adic integers into themselves: Currently a research group from the Institute for Information Security at the Moscow State University is working at further applications of algebraic dynamics to various problems of computer science and cryptology.

It is worth noting here that methods of the p-adic ergodic theory developed in Chapter 4 turned out to be rather powerful from a theoretical point of view as well. We recall that the study of ergodicity of monomial dynamical systems, $x \mapsto x^n$, played an important role in the development of the p-adic dynamical theory. It was immediately observed that behavior of p-adic dynamical systems depends crucially on the prime parameter p. The main aim of investigations performed in papers of M. Gundlach, A. Khrennikov, and K.-O. Lindahl [160–162, 250, 300] was to find such a p-dependence for ergodicity, cf. Parry and Coelho [352], Bryk and Silva [80]. An interesting algebraic inter-relation between p and n guaranteing ergodicity was found. In [160–162, 250, 300] the problem of ergodicity of perturbed monomial dynamics was formulated: $x \mapsto x^n + q(x)$. This problem was announced at numerous international conferences and talks at many universities throughout the world. In the ergodic community it was recognized that this problem is rather complex; the problem has been unsolved until the end of 2005. Nevertheless, in 2005 Vladimir Anashin solved this problem in the most general case [27], for arbitrary 1-Lipschitz locally analytic dynamics, see Section 4.7.1.

To conclude with p-adic ergodic theory of 1-Lipschitz transformations on $\mathbb{Z}_p$, we remark that, for a special class of functions, namely, for 1-Lipschitz ergodic transformations on $\mathbb{Z}_p$ and for 1-Lipschitz measure-preserving transformations on $\mathbb{Z}_2$, it is possible to interpolate their iterations with respect to the discrete time, $t_n = 0, 1, \ldots$, to *continuous p-adic time* $t \in \mathbb{Z}_p$, see Subsection 4.8.1 of Chapter 4. This is a step to unification of p-adic discrete time dynamics with p-adic continuous time dynamics; the latter was considered by, e.g., B. Dwork, G. Gerotto, F. J. Sullivan, and P. Roba [112–115], see also A. Escassut, A. Khrennikov, N. Grande-Kimpe, L. Van Hamme [97,98,124,125].

Finally we concern another aspect of p-adic ergodic theory, the ergodic theory for profinite groups, see Part II. A mathematical part of this history started with a problem of P. Halmoš whether an automorphism of a locally compact but non-compact group can be an ergodic measure-preserving transformation, [167, p. 26]. The problem attracted notable attention and motivated a related study of affine ergodic transformations on a non-commutative groups G (that is, ergodic transformations of the form $x \mapsto gx^{\beta}$, where $g \in G$, and β is an automorphism of the group G), by B. Schreiber with co-workers, and by other authors, see e.g. [365] and references therein.[9] In the late 1960s the theory of polynomials over non-commutative algebraic structures, and especially over groups, emerged, see [286]; development of the latter naturally leaded then to the study of polynomial transformations on groups with operators, i.e., transformations of the form

$$x \mapsto g_1(x^{\omega_1})^{n_1} g_2(x^{\omega_2})^{n_2} \cdots g_k(x^{\omega_k})^{n_k} g_{k+1} = g(x^{\alpha_1})^{n_1}(x^{\alpha_2})^{n_2} \cdots (x^{\alpha_k})^{n_k},$$

where $g, g_1, \dots, g_{k+1} \in G$, $n_1, \dots, n_k$ are rational integers, and $\omega_1, \dots, \omega_k$ are operators, i.e., group endomorphisms, $\alpha_1, \dots, \alpha_k$ are endomorphisms of the group G. As any profinite group[10] can be endowed with a metric (which is called a profinite metric) and a measure, it is reasonable to ask what continuous with respect to the profinite metric transformations are measure-preserving or ergodic with respect to the mentioned measure. Recent works [261, 262] by J. Kingsbery, A. Levin, A. Preygel, and C. E. Silva give general sufficient and necessary conditions for measure-preservation and ergodicity of transformations in terms of actions of these transformations on all groups of the inverse spectrum; for instance, to determine whether a transformation is measure-preserving it is necessary to verify whether it induces a bijection on every group from the inverse limit, i.e., for infinite number of groups. Thus, it is reasonable to ask whether this verification can be done in a finite number of steps, and so to obtain explicit formulas for these transformations.

The latter setting is important for applications. Actually ergodic transformation on groups may be used to produce pseudorandom sequences of permutations in a manner ergodic transformations of p-adic integers are used to produce pseudorandom sequences of numbers. Pseudorandom sequences of permutations on finite sets are used in cryptography in construction of the so-called polyalphabetic substitution ciphers. A

[9] The mentioned problem is also connected with another flow in ergodic theory of actions (particularly $\mathbb{Z}^d$-actions) by group automorphisms on a compact metric group, see e.g. [111]. Although corresponding works deal with dynamical systems of algebraic nature, we note however that both the approach we develop in our book and the problems we study here have very little in common with this flow: actually the groups we consider in Part II have no ergodic automorphisms at all.

[10] a group that is an inverse limit of finite groups

well-known example of ciphers of this kind is produced by ENIGMA, an encryption machine used by Germany during World War II.

In Part II we consider a problem how to determine ergodic transformations on profinite groups with operators. We note that not all profinite groups admit polynomial ergodic transformation; however, using an earlier publication of Vladimir Anashin [19] that characterizes finite solvable groups having ergodic polynomials, we determine ergodic polynomial transformations on profinite groups with operators that are inverse limits of finite solvable groups. We emphasize that these dynamics on profinite groups can somehow be 'reduced to', or 'combined of', the p-adic dynamics on different spaces of p-adic integers.

These results may be considered, on the one hand, as a contribution to ergodic theory for non-commutative algebraic structures. In this connection, it is interesting to note that actually in Part II we mimic the approach from the p-adic ergodic theory, but with the use of a non-commutative differential calculus (instead of p-adic derivation), which originally arose in works of R. Fox on knot theory, see [94]. We believe that this approach can be expanded to develop ergodic theory on non-commutative algebraic systems other than groups with operators.

On the other hand, the ergodic theory for profinite groups, which we develop in Part II of the book, has applications to pseudorandom generators that are constructed not only with the use of arithmetical and logical instructions of a computer, but also with the use of flags, 1-bit registry operations that are used, e.g., to perform program jumps. Finally, basic ideas of this approach lead to new constructions of 'flexible' stream ciphers whose state update function and filter function are being modified during encryption, see Section 10.3

To conclude, we emphasize that all applied issues we touch in the book, which are looking so diverse by origin and nature, turned out to have a lot of common features that can be explained and understood by means of algebraic dynamics. So we hope that this book will be useful, not only for pure mathematicians (working in number theory, theory of dynamical systems, algebraic geometry, analysis, probability), but also for people (interested in mathematical modeling) working in cryptography, computer science, cognitive science, psychology, theoretical physics, and genetics.

Moscow/Växjö, 2004–2009

Vladimir Anashin
Andrei Khrennikov

Contents

Preface vii

1 Algebraic and number-theoretic background 1
1.1 Facts from number theory 1
1.1.1 Some useful equalities and congruences 1
1.1.2 Möbius and Euler functions, Legendre symbol 3
1.1.3 Distribution of prime numbers 5
1.2 Basic notions and facts from algebra 6
1.2.1 Universal algebras 6
1.2.2 Groups 9
1.2.3 Rings 14
1.3 Fields 17
1.3.1 Finite fields 17
1.3.2 Non-Archimedean fields 19
1.4 p-adic numbers 19
1.4.1 Canonical expansion of p-adic numbers 22
1.4.2 Tree-like structure of the p-adic numbers 24
1.5 Ultrametric spaces 24
1.6 The Haar measure 26
1.7 Non Archimedean rings, m-adic numbers 28
1.8 Extensions of the field of p-adic numbers 29
1.8.1 Finite extensions of $\mathbb{Q}_p$ 29
1.8.2 The algebraic closure of $\mathbb{Q}_p$ 32
1.8.3 Complex p-adic numbers 33
1.8.4 Krasner's lemma 33

I The Commutative Non-Archimedean Dynamics **35**

2 Dynamics on algebraic structures 37
2.1 Basic notions of dynamics 37
2.1.1 Ergodicity and uniform distribution of sequences 37

2.2 Dynamics on finite algebraic structures . . . 39
2.2.1 Hereditary dynamical properties and compatibility . . . 39
2.2.2 Ergodic polynomial transformations on finite Abelian groups with operators . . . 41
2.2.3 Ergodic polynomial transformations on finite commutative rings . . . 42

3 p-adic analysis . . . 48
3.1 Analysis in complete non-Archimedean fields . . . 48
3.2 Analytic functions . . . 51
3.3 Hensel's lemma . . . 52
3.4 Roots of unity . . . 54
3.5 Non-Archimedean normed spaces . . . 56
3.6 Multidimensional analysis . . . 57
3.7 The differentiability modulo p^k . . . 58
3.8 Compatible functions on $\mathbb{Z}_p$ . . . 62
3.8.1 Compatibility is equivalent to 1-Lipschitz . . . 63
3.8.2 Compatibility and differentiability . . . 66
3.9 Mahler expansion . . . 75
3.9.1 Identities modulo p^k . . . 76
3.9.2 Mahler expansions of compatible functions . . . 78
3.10 Special classes of locally analytic functions . . . 80
3.10.1 Class $\mathcal{C}$ . . . 80
3.10.2 Class $\mathcal{B}$ . . . 83
3.10.3 Class $\mathcal{A}$ . . . 87

4 p-adic ergodic theory . . . 90
4.1 Discrete dynamical systems . . . 90
4.2 Periodic points and their character . . . 90
4.3 Monomial dynamics . . . 93
4.3.1 Topologically transitive and minimality . . . 94
4.3.2 Unique ergodicity . . . 96
4.4 Measure-preserving and ergodic isometries on $\mathbb{Z}_p^n$ . . . 98
4.4.1 Measure-preserving isometries . . . 100
4.4.2 1-Lipschitz measure-preserving functions . . . 102
4.4.3 1-Lipschitz ergodic functions . . . 105
4.5 Ergodic 1-Lipschitz transformations on $\mathbb{Z}_p$ . . . 106
4.5.1 Ergodicity of affine mappings . . . 106
4.5.2 Ergodicity and measure-preservation in terms of coordinate functions . . . 108
4.5.3 Ergodicity and measure-preservation in terms of Mahler expansion . . . 111

4.6 Measure-preservation and ergodicity of uniformly differentiable functions on $\mathbb{Z}_p^n$ . . . 119
4.6.1 Conditions for measure-preservation . . . 119
4.6.2 No uniformly differentiable 1-Lipschitz ergodic transformations on $\mathbb{Z}_p^n$, $n \geq 2$ . . . 122
4.6.3 Differentiable ergodic transformations on $\mathbb{Z}_p$ . . . 125
4.6.4 Measure-preservation and ergodicity of $\mathcal{A}$-, $\mathcal{B}$-, and $\mathcal{C}$-functions . . . 132
4.7 Ergodic 1-Lipschitz transformations on p-adic spheres . . . 148
4.7.1 1-Lipschitz ergodic transformations on spheres . . . 148
4.7.2 Ergodicity of $\mathcal{B}$-functions and of analytic functions . . . 151
4.7.3 Ergodicity of perturbed monomial mappings . . . 153
4.7.4 Ergodicity of $\mathcal{A}$-functions on spheres . . . 155
4.8 Concluding remarks to p-adic ergodic theory . . . 156
4.8.1 Continuous p-adic dynamics . . . 156
4.8.2 Non-minimal dynamics. Non-compatible dynamics. Mixing 159

5 Asymptotic distribution of cycles . . . 162
5.1 Monomial systems in $\mathbb{C}_p$ and in finite extensions of $\mathbb{Q}_p$ . . . 163
5.2 Number of cycles of $x \mapsto x^n$ in $\mathbb{Q}_p$ . . . 166
5.3 Total number of cycles . . . 169
5.4 Possible values of the number of cycles . . . 171
5.5 Probability on the set of prime numbers . . . 172
5.6 Distribution of cycles . . . 174
5.7 Expectation value and dispersion . . . 176
5.8 Fuzzy cycles . . . 180

II The Non-Commutative Non-Archimedean Dynamics **197**

6 Basics of polynomial dynamics on groups . . . 199
6.1 Non-commutative differential calculus . . . 200
6.2 Bijective polynomials over finite groups . . . 204

7 Ergodic polynomials over groups with operators . . . 205
7.1 Basic properties of groups having ergodic polynomials . . . 206
7.2 Finite solvable groups having ergodic polynomials . . . 209
7.2.1 The multivariate case . . . 209
7.2.2 The univariate case: Nilpotent groups . . . 212
7.2.3 The univariate case: Solvable groups . . . 217
7.3 Ergodic theory for profinite groups . . . 232
7.3.1 Metric and measure on a profinite group . . . 233

7.3.2 Equations, the non-commutative Hensel's lemma, and measure-preserving polynomials over profinite groups 235
7.3.3 Ergodic polynomials over profinite groups 237

III Applications **243**

8 Automata, computers, combinatorics 245
8.1 Automata functions are continuous 245
8.2 Computers think 2-adically 252
8.3 Differentiable instructions and programs 259
8.4 Latin squares 262

9 Pseudorandom numbers 269
9.1 Pseudorandom generator is a dynamical system 271
9.1.1 What pseudorandom generators are good? 272
9.1.2 Why p-adic ergodic theory? 274
9.2 Congruential generators of the longest period 275
9.2.1 Types of congruential generators 277
9.2.2 Periods of congruential generators 279

10 Stream ciphers 305
10.1 How secure are congruential generators? 306
10.2 Wreath products 309
10.3 Counter-dependent generators 314
10.3.1 Special output functions 325
10.4 Generators based on multivariate functions 328
10.5 Security issues 334
10.5.1 The number of transitive compatible mappings 335
10.5.2 Key recovery and intractability 337

11 Structure of trajectories 340
11.1 Distribution in Euclidean space 340
11.1.1 Points falling on hyperplanes 341
11.1.2 Lacunas 347
11.2 Properties of coordinate sequences 358
11.2.1 Linear and 2-adic complexities 359
11.2.2 Structure of coordinate sequences 366
11.3 Distribution of k-tuples 371

12 p-adic probability theory 377
12.1 Historical remarks 377
12.2 Frequency probability theory 379

12.3 Ensemble probability 385
12.3.1 Ensembles of infinite volumes 386
12.3.2 The rules for working with p-adic probabilities 391
12.3.3 Negative probabilities and p-adic ensemble probabilities . . 396
12.4 Measures 396
12.5 p-adic probability space 400
12.6 p-adic probability measures on the space of binary sequences 402
12.7 Some technical p-adic results 403
12.8 p-adic tests for randomness 404
12.9 Some limit theorems 408
12.10 Recursive enumeration of the set of p-adic tests 410
12.11 No p-adic universal test 413

13 p-adic valued quantization 415
13.1 Toward quantum mechanics with p-adic valued wave functions . . . 415
13.2 Hilbert spaces 417
13.3 Groups of unitary isometric operators in the p-adic Hilbert space . . 419
13.4 Axiomatics of quantum mechanics with p-adic valued wave functions 421
13.5 Gaussian integral and spaces of square integrable functions 422
13.6 Gaussian representations of position and momentum operators . . . 425
13.7 One parameter groups generated by position and momentum operators 427
13.8 Operator calculus 427
13.9 Spectrum of p-adic position operator 428
13.10 Concluding remarks on p-adic quantization 431

14 m-adic modeling in cognitive science and psychology 433
14.1 On modeling of mental quantities 434
14.1.1 Representation of mental states by numbers 434
14.1.2 Encoding by branches of trees 437
14.1.3 Dynamical system approach, artificial intelligence 438
14.1.4 Unconscious and conscious dynamics – Freudian approach 439
14.1.5 Neuronal hierarchy 441
14.2 Mental space 442
14.3 Dynamical thinking in mental space 442
14.4 Associations and ideas 443
14.5 Neuronal realization 444
14.6 Model of cognitive psychology 446
14.7 Dynamics of associations and ideas 447
14.8 Advantages of dynamical processing of associations and ideas . . . 448
14.9 Transformation of unconscious mental flows into conscious flows . . 449
14.10 Hidden forbidden wishes, psychoanalysis 458
14.10.1 Hysteric reactions 460

14.10.2 Feedback control based on doubtful ideas 461
14.11 Neuro and mental cybernetic bases for the pleasure and reality principles 462
14.12 Consequences for psychology and neuropsychology 464
14.13 Consequences for psychoanalysis 465
14.14 Psycho-robots 467

15 Neuronal hierarchy behind the ultrametric mental space 468
15.1 Hierarchic neural pathways 469
15.2 Model: thinking on neuronal tree 470
15.2.1 Mental field on the brain 470
15.2.2 Probabilistic dynamics in the mental space 475
15.3 Diffusion model of dynamics of statistical mental state 478
15.3.1 Markovean body $\rightarrow$ mind fields 478
15.3.2 Thinking as m-adic diffusion 479
15.3.3 Discussion 481
15.4 Postulates 485

16 Gene expression from dynamics in the 2-adic space 487
16.1 Description of model 488
16.1.1 4-adic representation of nucleotides 488
16.1.2 DNA-reproduction and 4-adic dynamics 489
16.2 Genetic space 490
16.2.1 4-adic encoding of DNA and RNA 490
16.2.2 2-adic encoding 491
16.3 Dynamical model for degeneracy of the genetic code 492

17 Genetic code on the diadic plane 494
17.1 Vertebral mitochondrial and eucaryotic codes 495
17.2 Parametrization of the set of codons by the diadic plane 495
17.3 Genetic code on the diadic plane 498
17.4 Physical-chemical regularity of the genetic code 501

Bibliography 503

Notation 527

Index 529

Chapter 1

Algebraic and number-theoretic background

This chapter is to remind the reader some basic notions and results we use throughout our book.

1.1 Facts from number theory

In this section, we remind some important facts and useful formulas from number theory. We assume that the reader is familiar with residues modulo N and their basic properties.

1.1.1 Some useful equalities and congruences

Theorem 1.1 (Chinese Remainder Theorem). *Let $N \in \mathbb{N}$ be a natural number, $N > 1$. Represent $N = p_1^{e_1} p_2^{e_2} \cdots p_r^{e_r}$, where p_j, $1 \leqslant j \leqslant r$, are prime numbers, and r is the number of different primes in decomposition of N. Then, given arbitrary integers $a_j \in \{0, 1, \ldots, p_j^{e_j - 1} - 1\}$, $1 \leqslant j \leqslant r$, there exists an integer $a \in \{0, 1, \ldots, N-1\}$ such that $a \equiv a_j \pmod{p_j^{e_j}}$ for all $1 \leqslant j \leqslant r$.*

Note that the proof of Theorem 1.1 is constructive; that is, gives an algorithm to find this a explicitly, see any relevant book on number theory.

For $i \in \mathbb{N}_0$, $n \in \mathbb{N}$, the binomial coefficient is

$$\binom{n}{i} = \frac{n(n-1)\cdots(n-i+1)}{i!};$$

note that

$$\binom{n}{0} = 1,$$

by the definition. Note also that $\binom{n}{i} = 0$ for $i > n$.

Theorem 1.2 (Lucas' theorem). *Let $r = \sum_{i=0}^{N} r_i p^i$ and $n = \sum_{i=0}^{N} n_i p^i$ be base-p expansions of $r, n \in \mathbb{N}_0$: $r_i, n_i \in \{0, 1, \ldots, p-1\}$ $(i = 0, 1, 2, \ldots)$. Then the following* *congruence for binomial coefficients holds*:

$$\binom{r}{n} \equiv \binom{r_0}{n_0}\binom{r_1}{n_1}\cdots\binom{r_N}{n_N} \pmod{p}.$$

Proof. See e.g. [12]. □

Corollary 1.3. *Under the conditions of Theorem 1.2, let $\ell \geq 1$, $k \geq 1$, $n \leq p^k - 1$. Then*

$$\binom{p^k\ell - 1}{n} \equiv (-1)^n \pmod{p}.$$

Proof. Take $r = p^k\ell - 1$ in the statement of Theorem 1.2, then $r_i = p - 1$ for $i = 0, 1, \ldots, k-1$. Now Theorem 1.2 implies that

$$\binom{p^k\ell - 1}{n} \equiv \binom{p-1}{n_0}\binom{p-1}{n_1}\cdots\binom{p-1}{n_{k-1}} \equiv (-1)^n \pmod{p}$$

as obviously $\binom{p-1}{j} = \frac{(p-1)\cdot(p-2)\cdot\ldots\cdot(p-j)}{j!} \equiv (-1)^j \pmod{p}$ for all $j \in \{0, 1, \ldots, p-1\}$. □

Definition 1.4. A *difference* (with respect to variable x_i) of a function $f(x_1, \ldots, x_n)$ is

$$\Delta_i f(x_1, \ldots, x_n) = f(x_1, \ldots, x_{i-1}, x_i + 1, x_{i+1}, \ldots, x_n) - f(x_1, \ldots, x_n),$$

and the sth difference (with respect to variable x_i) of the function f is

$$\Delta_i^s f = \Delta_i^{s-1}(\Delta_i f), \qquad s = 1, 2, \ldots,$$

where $\Delta_i^0 f = f$ by the definition. We write $\Delta f(x)$ rather than $\Delta_1 f(x)$ for a univariate function f.

One verifies directly that

$$\Delta\binom{n}{i} = \binom{n+1}{i} - \binom{n}{i} = \binom{n}{i-1}. \tag{1.1}$$

Theorem 1.5 (Gregory–Newton formula). *The following identity holds for all $n \in \mathbb{N}$ and all functions g*:

$$g(y+n) = \sum_{i=0}^{\infty} \binom{n}{i} \Delta^i g(y).$$

Theorem 1.6 (Binomial inversion formula).

$$\alpha_m = \sum_{k=0}^{\infty} \binom{m}{k} \beta_k$$

if and only if

$$\beta_k = \sum_{m=0}^{\infty} (-1)^{m+k} \binom{k}{m} \alpha_m.$$

1.1.2 Möbius and Euler functions, Legendre symbol

Let us begin with the definition of the Möbius function.

Definition 1.7. Let $n \in \{1, 2, \ldots\}$. Then we can write

$$n = p_1^{e_1} p_2^{e_2} \cdots p_r^{e_r},$$

where p_j, $1 \leqslant j \leqslant r$, are prime numbers and r is the number of different primes. The function μ on $\mathbb{N}$ defined by $\mu(1) = 1$, $\mu(n) = 0$ if any $e_j > 1$ and $\mu(n) = (-1)^r$, if $e_1 = \cdots = e_r = 1$ is called the *Möbius function.*

The Möbius function has the following property, see for example [165] or [33],

$$\sum_{d|n} \mu(d) = \begin{cases} 1, & \text{if } n = 1, \\ 0, & \text{if } n > 1, \end{cases}$$

where d is a positive divisor of n. This property is used for proving the following classical result.

Theorem 1.8 (Möbius inversion formula). *Let f and g be functions defined for each $n \in \mathbb{N}$. Then,*

$$f(n) = \sum_{d|n} g(d) \tag{1.2}$$

if and only if

$$g(n) = \sum_{d|n} \mu(d) f(n/d). \tag{1.3}$$

We recall the definition of Euler's totient function and Euler's theorem.

Definition 1.9. Let n be a positive integer. Henceforth, we will denote by $\varphi(n)$ the number of natural numbers less than n which are relatively prime to n. The function φ is called *Euler's totient function.*

If p is a prime number then $\varphi(p^l) = p^{l-1}(p-1)$.

Theorem 1.10 (Euler's theorem). *If a is an integer relatively prime to b then* $a^{\varphi(b)} \equiv 1 \pmod{b}$.

For later use we also recall that

$$\varphi(n) = \sum_{d|n} \mu(d)\frac{n}{d}. \tag{1.4}$$

Theorem 1.11. *Let a, b and m be integers with m positive. If* $\gcd(a,m) \mid b$ *then the congruence*

$$ax \equiv b \pmod{m}$$

has exactly $\gcd(a,m)$ *solutions.*

Definition 1.12. Let p be an odd prime and let a be an integer. Suppose $p \nmid a$. If the congruence

$$x^2 \equiv a \pmod{p} \tag{1.5}$$

is solvable then a is called a *quadratic residue modulo p*, and if it has no solution, then a is called a *quadratic non residue modulo p*.

Definition 1.13. Let p be an odd prime and a an integer. Then define the function $a \mapsto \left(\frac{a}{p}\right)$, from $\mathbb{Z}$ to $\mathbb{Z}$, as

$$\left(\frac{a}{p}\right) = \begin{cases} 1, & \text{if } p \nmid a \text{ and } a \text{ is quadratic residue modulo } p, \\ 0, & \text{if } p \mid a, \\ -1, & \text{if } p \nmid a \text{ and } a \text{ is quadratic non residue modulo } p. \end{cases}$$

This function is called the *Legendre symbol.*

Denote the set of $\pmod{p}$-residue classes in $\mathbb{Z}$ by the symbol $\mathbb{F}_p$.

Theorem 1.14 (Lagrange). *If f is a polynomial of one-variable of degree n defined over $\mathbb{F}_p$ then it cannot have more than n roots, unless it is identically zero.*

Lagrange's theorem gives that the congruence (1.5) has exactly two solutions if $\left(\frac{a}{p}\right) = 1$. If $\left(\frac{a}{p}\right) = 0$, then the congruence (1.5) has the unique solutions $x = 0$. Hence, the congruence (1.5) has $\left(\frac{a}{p}\right) + 1$ solutions.

Theorem 1.15. *The Legendre symbol has the following properties*:

(1) $\left(\frac{ab}{p}\right) = \left(\frac{a}{p}\right)\left(\frac{b}{p}\right)$,

(2) *if* $a \equiv b \pmod{p}$ *then* $\left(\frac{a}{p}\right) = \left(\frac{b}{p}\right)$,

(3) $\left(\frac{a^2}{p}\right) = 1$ *and specially* $\left(\frac{1}{p}\right) = 1$,

(4) $\left(\frac{-1}{p}\right) = (-1)^{(p-1)/2}$,

(5) *if* $\gcd(a, p) = 1$ *then* $\left(\frac{a}{p}\right) = 1$ *if and only if* $a^{(p-1)/2} \equiv 1 \pmod p$ *(Euler's criterion).*

Corollary 1.16. *Let p be an odd prime. Then*

(1) $\left(\frac{p-1}{p}\right) = 1$ *if and only if* $p \equiv 1 \pmod 4$.

(2) $\left(\frac{p-1}{p}\right) = -1$ *if and only if* $p \equiv 3 \pmod 4$.

Proof. Because, $p - 1 \equiv -1 \pmod p$, Theorem 1.15 gives that $\left(\frac{p-1}{p}\right) = \left(\frac{-1}{p}\right) = (-1)^{(p-1)/2}$. We prove (1). Suppose that $(-1)^{(p-1)/2} = 1$, that is, $(p-1)/2 = 2k$ for some integer k. This is equivalent to $p = 4k + 1$, and (1) is proved. The proof of (2) is done with same method. □

Theorem 1.17. *Let p be a prime. The Diophantine equation*

$$x^2 + y^2 = p$$

is solvable in integers x and y if and only if $p = 2$ *or* $p \equiv 1 \pmod 4$.

1.1.3 Distribution of prime numbers

To be able to derive a formula for the number of cycles of some dynamical systems, we need to use some tools of number theory.

Let $x \in \mathbb{R}$, $x > 0$ and let $\pi(x)$ denote the number of primes not exceeding x. Since there are infinitely many primes, $\pi(x) \to \infty$, when $x \to \infty$. Legendre and Gauss conjectured at the end of the 18th century that

$$\lim_{x\to\infty} \frac{\pi(x)\log(x)}{x} = 1, \tag{1.6}$$

or in other words, $\pi(x)$ is asymptotic to $x/\log x$. This conjecture was proved in 1896 by Hadamard and de La Vallée Poussin [99, 163] and is known as *the prime number theorem*. They used the theory of analytic functions and properties of the Riemann zeta function

$$\zeta(s) = \sum_{n=1}^{\infty} \frac{1}{n^s}.$$

An elementary proof was presented in 1949 by Erdős and Selberg.

Let $\pi_{a,k}(x)$ be the number of primes not exceeding x in the arithmetic progression $nk + a$, $n = 0, 1, 2, \ldots$. Dirichlet proved that $\pi_{a,k}(x) \to \infty$ when $x \to \infty$ if and only if $(a, k) = 1$. This is known as *Dirichlet's theorem.* We also have a prime number theorem for arithmetic progressions:

$$\lim_{x\to\infty} \frac{\pi_{a,k}(x)\varphi(k)}{\pi(x)} = 1 \tag{1.7}$$

if $(a, k) = 1$. A proof can be found in [288].

1.2 Basic notions and facts from algebra

In this section, we remind some notions and facts about general universal algebras, as well as about concrete universal algebras we are dealing in our book most of all, rings and groups. Actually this section is mainly for making references and unifying terminology. Although we often start with very basic notions, such as a notion of a group, the reader is nevertheless assumed to be familiar with these beforehand, especially if he is going to read Part II on dynamics over non-commutative groups: Some proofs there involve various group-theoretic techniques, and the reader is better to have a certain (however, not too big) experience in group theory to understand details.

1.2.1 Universal algebras

We remind some basic concepts of universal algebra following mainly [286]. A *universal algebra* (or, briefly, an algebra, if this makes no confusion) is a non-empty set A endowed with a set of operations Ω (the latter set is often called a *signature* of the universal algebra). Every operation $\omega \in \Omega$ is a map from the nth Cartesian power A^n into A; the number n is called the *arity* of the operation ω. Two algebras A and B with operations Ω and Ψ, respectively, are said to *have the same type* (or to be *similar*) if there exists a one-to-one correspondence ι between Ω and Ψ that preserves arities; that is, the arity of $\omega \in \Omega$ is equal to the arity of $\iota(\omega) \in \Psi$. If A and B are algebras of the same type, we do not differ further ω from $\iota(\omega)$, if there is no fear of confusion. A subset $S \subset A$ is called a *subalgebra* if it is closed with respect to all operations from Ω: $\omega(a_1, \dots, a_n) \in S$ for all $a_1, \dots, a_n \in S$ and every (n-ary) operation $\omega \in \Omega$. An equivalence $\sim$ on A is called a *congruence* of the algebra A whenever $\sim$ agrees with all operations from Ω; that is, given an n-ary operation $\omega \in \Omega$ and elements $a_1, \dots, a_n, b_1, \dots, b_n \in A$ such that $a_i \sim b_i$ for all $i = 1, 2, \dots, n$, then necessarily $\omega(a_1, \dots, a_n) \sim \omega(b_1, \dots, b_n)$. If a class of equivalent elements contains more than one element, but not all elements of the algebra, a congruence is said to be proper. An algebra that has no proper congruences is sometimes called *simple*.

If A and B are algebras of the same type, the map $\varphi\colon A \to B$ is called a homomorphism whenever it agrees with all operations; that is, for every (n-ary) operation $\omega \in \Omega$ and every $a_1, \dots, a_n \in A$ we have that $\varphi(\omega(a_1, \dots, a_n)) = \omega(\varphi(a_1), \dots, \varphi(a_n))$. If φ is surjective (injective), it is called an epimorphism (monomorphism). If φ is simultaneously an epimorphism and a monomorphism, it is called an isomorphism. If $A = B$, the homomorphism φ is called an endomorphism, and if φ is an isomorphism, then φ is called an automorphism. Note that every homomorphism φ defines a congruence: $a \sim b$ if and only if $\varphi(a) = \varphi(b)$. Vice versa, every congruence defines an epimorphism of A onto algebra of classes of equivalent elements of A, the *factor algebra* of A with respect to the congruence $\sim$. The epimorphism ψ is said to be *natural* in this case. The congruence $\sim$ is sometimes called a *kernel* of φ.

Now we formulate one of the most important notions of the book, the compatibility. Loosely speaking, a map $F: A^k \to A^m$ is said to be compatible if it agrees with all congruences of A. Here is a formal definition:

Definition 1.18 (Compatibility). Let $F = (f_1, \ldots, f_m): A^k \to A^m$ be a map of the kth Cartesian power of the algebra A into its mth Cartesian power; that is, $f_j : A^k \to A$, for all $j = 1, 2, \ldots, m$. The map F is said to be *compatible*, if for every congruence $\sim$ of A and every elements $a_1, \ldots, a_n, b_1, \ldots, b_n \in A$ such that $a_i \sim b_i$ for all $i = 1, 2, \ldots, n$ we have that $f_j(a_1, \ldots, a_n) \sim f_j(b_1, \ldots, b_n)$, for all $j = 1, 2, \ldots, m$.

It is clear that every operation $\omega \in \Omega$ is compatible; whence, all compositions of operations from Ω as well. So we come to one more important notion, the notion of a polynomial over universal algebra. Loosely speaking, a polynomial is a composition of operations with variables and constants (the latter are elements of algebra A). Our formulation of this notion is somewhat different from the one of [286] and is a bit less formal. We do this to give the reader a right understanding of things that are clear in cases of concrete algebras, groups and rings, we mainly dealing with in this book, rather than to formulate this notion in the most general sense and full rigor. Otherwise we have also to formulate a notion of a variety of universal algebras, of a free products in varieties, etc. We refer the interested reader to the monograph [286] for these. We note, however, that as in our book we are more interested in *polynomial functions*, the maps induced by polynomials, rather than in polynomials themselves, the difference between these two notions of a polynomial over a universal algebra is not so significant since polynomial functions defined by polynomials in our sense and by the ones in the sense of the book [286] coincide.

Definition 1.19 (Polynomials over universal algebras). Let $X = \{x_1, x_2, \ldots\}$ be a set of variables, and let A be an algebra. Then

(1) Every variable x_i is a polynomial in variable x_i over the algebra A.

(2) Every element $a \in A$ is a polynomial on empty set of variables over the algebra A.

(3) If $w_1, \ldots, w_k$ are polynomials on sets of variables $X_1 \subset X, \ldots, X_k \subset X$, respectively, and if $\omega \in \Omega$ is a k-ary operation of A, then $\omega(w_1, \ldots, w_k)$ is a polynomial on the set of variables $X_1 \cup \cdots \cup X_k$ over the algebra A.[1]

[1] Note that a polynomial on empty set of variables is thus an element from A.

(4) No polynomials in variables X over the algebra A other than named in (1)–(3) do exist.

We define a notion of a polynomial in variables $X = \{x_1, \ldots, x_n\}$ in a similar manner; so further X is either a finite or countable set of variables.

Denote $A[X]$ the set of all polynomials in variables X over the algebra A; then $A[X]$ is the algebra of the same type as the algebra A: All operations from Ω are well defined on $A[X]$ (see (3) from Definition 1.19). Now we point out the difference between our definition of a polynomial and the classical one. For instance, let A be a field; then the polynomials x_1x_2 and x_2x_1 are equal in the classical meaning. However, according to Definition 1.19, these two polynomials are different. This is because the classical notion of a polynomial emerged as a polynomial over a commutative ring, so if we let variables to commute, we do not change the map defined by this polynomial. However, if we consider a non-commutative ring, the classical definition does not work any longer, since variables can not commute with each other, and with coefficients as well, without affecting the map defined by the polynomial. Actually to get rid off 'extra' polynomials, we must define a notion of polynomial over an algebra from a certain variety, see [286]. However, as said, these 'extra' polynomials imply no difference between two definitions if we consider polynomial maps.

From Definition 1.19 it immediately follows that every polynomial f in variables $Y \subset \{x_1, \ldots, x_n\}$ induces a map from A^n to A in an obvious way: Given $a_1, \ldots, a_n \in A$ we substitute a_j for x_j for all occurrences of x_j in f and all $j = 1, 2, \ldots, n$ and obtain an element $f(a_1, \ldots, a_n) \in A$ performing corresponding operations from Ω. This map is called a *polynomial map*, or a *polynomial function* induced by the polynomial f on A (for more rigorous definition of this notion see [286]). Definition 1.19 immediately implies that the following proposition is true:

Proposition 1.20. *Every polynomial function is compatible.*

An algebra A is called *n-polynomially complete*, or *n-functionally complete*, if every n-variate function on A valuated in A is a polynomial function, for a suitable n-variate polynomial over A. An algebra is called *polynomially complete* if it is n-polynomially complete, for all $n = 1, 2, 3, \ldots$. Comparing cardinalities of the set of polynomials in n variables and of the set of n-variate functions we immediately conclude that an n-polynomially complete algebra must be necessarily finite. Moreover, from Proposition 1.20 it is clear that an n-polynomially complete algebra must be simple.

One more notion from universal algebra that is especially important for the problems considered in our book is a notion of inverse limit of universal algebras. We say that a family $\{A_n : n = 0, 1, 2, \ldots\}$ of similar algebras form an *inverse spectrum*

$$\cdots \xrightarrow{\varphi_{n+1}} A_n \xrightarrow{\varphi_n} A_{n-1} \xrightarrow{\varphi_{n-1}} \cdots \xrightarrow{\varphi_1} A_0$$

whenever all φ_n, $n = 0, 1, 2, \ldots$, are epimorphisms. Denote A_∞ a set of all sequences of the form $(a_i) = \ldots, a_n, a_{n-1}, \ldots, a_0$ such that $a_i \in A_i$ and $\varphi_i(a_i) = a_{i-1}$, for all $i = 1, 2, 3, \ldots$. Given a k-ary operation $\omega \in \Omega$ and k sequences $(a_i^{(j)}) \in A_\infty$, $j = 1, 2, \ldots, k$, we define $\omega((a_i^{(1)}), \ldots, (a_i^{(k)})) = (\omega(a_i^{(1)}, \ldots, a_i^{(k)}))$. Thus, A_∞ is an algebra of the same type as the algebras A_n. The algebra A_∞ is called an *inverse limit* of algebras A_n and is denoted as

$$A_\infty = \lim_{\substack{\longleftarrow \\ n\to\infty}} A_n.$$

In this book, we mainly deal with a case when all A_n are finite (rings or groups). In this case the algebra A_∞ can be endowed also with a metric, which will be necessarily non-Archimedean, and with a probabilistic measure, the normalized Haar measure; namely this way we 'rise' polynomial dynamics from A_n to dynamics on A_∞. We will not go into further details here; we postpone these considerations until we study concrete inverse limits, the ring of p-adic integers $\mathbb{Z}_p$ in further sections and in Part I, or profinite groups in Part II.

1.2.2 Groups

This subsection is only to remind the reader some basic notions and facts from group theory; we mainly need these only in Part II of the book. We mainly follow the books [156, 164] in this subsection, to which the reader is referred for scrupulous texts on group theory.

A *semigroup* S is a universal algebra with a binary operation $\cdot$ (multiplication), which is associative: $a \cdot (b \cdot c) = (a \cdot b) \cdot c$, for all $a, b, c \in S$. A *group* G is a semigroup whose signature is extended by a 0-ary operation 1 (the identity of the group), and by a unary operation $()^{-1}$ (taking an inverse). All three operations are related by the identities $a \cdot 1 = 1 \cdot a = a$, $a \cdot a^{-1} = a^{-1} \cdot a = 1$, for all $a \in G$. As usual, we often omit the sign of multiplication $\cdot$ in group expressions. A group consisting only of 1 is called *trivial*.

The smallest number $n \in \mathbb{N}$ such that $g^n = 1$ is called the *order* of the element $g \in G$, if such a number exists. An element of order 2 is called an *involution*. According to the general definition of a subalgebra, a *subgroup* is a subset $H \subset G$ that contains 1 and is closed with respect to multiplication and inversion. A subgroup $H \subset G$ such that $H \neq \{1\}$ and $H \neq G$ is called *proper*. Given $c \in G$, the set $\{1, c^{\pm 1}, c^{\pm 2}, \ldots\}$ is a subgroup, a *cyclic subgroup generated by the element* c. It is obvious if c is an element of a finite order n, then the cyclic subgroup generated by c is merely a set $\{1, c, c^2, \ldots, c^{n-1}\}$.

Given a subgroup $H \subset G$, the set $aH = \{ah : h \in H\}$ is called a (left) *coset* of the group G with respect to the subgroup H. Right cosets are defined by an analogy. If a number of left (right) cosets with respect to H is finite, it is equal to the number of right (left) cosets and is called an *index* $|G : H|$ of the subgroup H in G. The number of elements of the (sub)group G (H) is called the order of the (sub)group; we denote the order by $\#G$ $(\#H)$. Lagrange's theorem yields: $\#G = |G : H| \cdot \#H$. The subgroup H is called *normal* (denoted as $H \lhd G$) if $gH = Hg$ for every $g \in G$. Normal subgroups define congruences on groups and vice versa: If $H \lhd G$, then cosets with respect to H are classes of equivalent elements with respect to the corresponding congruence. Thus, every normal subgroup defines a natural epimorphism φ onto a factor group G/H; H is called a *kernel* of φ and denoted by $\ker\varphi = H$. In other terms, a normal subgroup is a subgroup that is invariant with respect to every *inner automorphism* of the group; the latter automorphism is a conjugation by the element g: $x \mapsto x^g = g^{-1}xg$. A subgroup is said to be a *characteristic* subgroup if its invariant with respect to all automorphisms of a group. Finally, if a subgroup is invariant with respect to all endomorphisms of a group, it is called a *fully invariant* subgroup. The following theorem describes the structure of minimal (with respect to inclusion) normal subgroups in a finite group:

Theorem 1.21. *A minimal normal subgroup of a finite group is isomorphic to a direct power* (that is, to a Cartesian product of some isomorphic copies) *of a simple group.*

If H is a subgroup in G, then the unique maximal (with respect to inclusion) subgroup $N \supset H$ of G in which H is a normal subgroup is called a *normalizer* of the subgroup H, and is denoted by $N_G(H)$.

If $H \lhd G$, and if K is isomorphic to the factor group G/H (we denote this by $K \cong G/H$), we say that the group G is an *extension* of the group H by the group K. Given H and K, an extension of H by K is not unique. Among all extensions of H by K there always exist extensions of a special sort, *split extensions*, or *semidirect products*. These are defined as follows: Consider a group $\mathrm{Aut}\,(H)$ of all automorphisms of the group H (clearly, $\mathrm{Aut}\,(H)$ is a group with respect to composition of automorphisms), and consider a homomorphism $\theta\colon K \to \mathrm{Aut}\,(H)$. On the set of all ordered pairs $K \leftthreetimes H = \{(a,h) : a \in K, h \in H\}$ define multiplication as

$$(a_1, h_1) \cdot (a_2, h_2) = (a_1 a_2, h_1^{\theta(a_2)} h_2),$$

where $h^{\theta(a)}$ is the image of the element $h \in H$ under action of the automorphism $\theta(a) \in \mathrm{Aut}\,(H)$, $a \in K$. It could be verified that under the so defined multiplication the set $K \leftthreetimes H$ is a group, H is its normal subgroup, and the factor group with respect to H is isomorphic to K. Note that the definition of semidirect product depends on the homomorphism θ; for instance, when θ is a trivial homomorphism (that maps K onto a trivial subgroup), the semidirect product is merely a direct product.

Example 1.22. A symmetric group Sym(3) of degree 3 (that is, a group of all permutations on a set of three elements) is a semidirect product of a cyclic subgroup of order 3 (which is normal) by a cyclic subgroup of order 2. A symmetric group Sym(4) of degree 4 is a semidirect product of group K_4 of order 4, which is a direct product of two cyclic groups of order 2 each, by a symmetric group Sym(3). The group K_4 is called a *Klein group*.

A set $Z(G)$ of all elements of a group G that commute with all elements of G is called a *center* of the group G:

$$Z(G) = \{g \in G : gh = hg \text{ for all } h \in G\}.$$

It is clear that $Z(G)$ is a commutative subgroup of G (we recall that in group theory commutative groups are called *Abelian*). Moreover, $Z(G)$ is a characteristic (hence, normal) subgroup of G; however, not necessarily a fully invariant subgroup. Given a subset S in G, we denote $C_G(S) = \{g \in G : gs = sg \text{ for all } s \in S\}$ the *centralizer* of S in G. Thus, $Z(G) = C_G(G)$.

Given a group G, consider a canonical epimorphism $\varphi\colon G \to G/Z(G)$ and denote $Z_2(G) = \varphi^{-1}(Z(G/Z(G)))$. It is clear that $Z_2(G)$ is a characteristic subgroup of G, and that $Z_2(G) \supset Z_1(G) = Z(G)$. Proceeding this way, we obtain the so-called *upper central series* series $\cdots \supset Z_2(G) \supset Z_1(G) \supset \{1\}$ of subgroups in G. If the series reaches G (that is, if $Z_n(G) = G$ for some n), the group G is called *nilpotent*. The smallest n such that $Z_n(G) = G$ is called a *nilpotent class* of the nilpotent group G. Thus, Abelian groups are nilpotent groups of class 1. All subgroups and factor groups of nilpotent groups are also nilpotent. A counterpart of the upper central series is the *lower central series*, which are defined as follows: Recall that a *commutator of elements* $a, b \in G$ is the element $[a,b] = a^{-1}b^{-1}ab \in G$. Given subgroups $A, B \subset G$ we define their commutator $[A, B]$ as a subgroup generated by all commutators $[a,b]$, $a \in A$, $b \in B$. Then, terms of the lower central series are $L_1(G) = G$, $L_2(G) = [L_1(G), G]$, $L_3(G) = [L_2(G), G], \ldots$. It is clear that the series is descending, and that every $L_i(G)$ is a fully invariant subgroup in G, $i = 1, 2, \ldots$. A group G is nilpotent if and only if $L_m(G) = \{1\}$ for some $m \in \mathbb{N}$. If G is nilpotent of class n, then $L_i(G) \subset Z_{n-i+1}(G)$, for all $i = 1, 2, \ldots, n$.

An important example of finite nilpotent groups are p-groups; the latter are groups of orders p^n, for some n. A maximal p-subgroup of a finite group is called a *Sylow p-subgroup* of a group. Given p, all Sylow p-subgroups of a finite group G are conjugate in G, the order of every Sylow p-subgroup is equal to the maximum power of p that divides the order of G, and the number of all Sylow p-subgroups of G is congruent to 1 modulo p (Sylow theorem).

The following theorem completely characterizes finite nilpotent groups in terms of p-groups:

Theorem 1.23. *A finite group G is nilpotent if and only if for every $p \mid \#G$, a Sylow p-subgroup is normal* (thus, unique) *in G; the group G is then a direct product of all its Sylow p-subgroups, for all $p \mid \#G$.*

Example 1.24. It is not difficult to show that $\mathrm{Aut}\,(K_4) \cong \mathrm{Sym}(3)$: As the group K_4 is isomorphic to the additive group of a 2-dimensional vector space over a field $\mathbb{F}_2 = \mathbb{Z}/2\mathbb{Z}$ of two elements, $\mathrm{Aut}\,(K_4)$ is isomorphic to a group of all non-singular 2×2 matrices over $\mathbb{F}_2$. Now take arbitrary involution $\alpha \in \mathrm{Aut}\,(K_4)$ and consider the semidirect product D_2 of K_4 by a cyclic subgroup A (of order 2) generated by α: $D_2 = A \leftthreetimes K_4$. The group D_2 is of order 8; thus, nilpotent. The center of this group is of order 2; it is a cyclic group generated by the eigenvector of the matrix α. Moreover, $D_2/Z(D_2) \cong K_4$; thus, D_2 is a nilpotent group of class 2. The group D_2 is called a *dihedral group of order 8.*

A generalization of p-groups are *π-groups*, where π is a non-empty set of primes; finite π-groups are finite groups G such that $p \in \pi$ for every prime divisor $p \mid \#G$. Also, π'-groups are finite groups G such that $p \notin \pi$ for every prime divisor $p \mid \#G$. However, finite π-groups need not be necessarily nilpotent unless π is a one-element set. For instance, Sym(3) is a $\{2, 3\}$-group, and Sym(3) is not nilpotent.

Note that nilpotent groups can be obtained as sequential extensions of Abelian groups when the extended Abelian group lies in the center of the extension. These extensions are called *central.* If we consider non-central sequential extensions of Abelian groups, we obtain a solvable group. Namely, a group G is called *solvable* if it possesses a finite normal series

$$G = G_0 \rhd G_1 \rhd \cdots \rhd G_n \rhd G_{n+1} = \{1\} \tag{1.8}$$

all whose *factors* G_i/G_{i+1} are Abelian groups. We recall that series (1.8) is called *(sub)normal* whenever all G_i are normal subgroups in G (in G_{i-1}). Factors of subnormal series are also called *sections*; i.e., sections are merely factor groups of subgroups. Solvable groups are exactly those groups whose *derived series* ends with a trivial group: Recall that a derived (sub)group of group G is a subgroup G' generated by all commutators $[a, b] = a^{-1}b^{-1}ab$, $a, b \in G$. The second derived (sub)group G'' is $(G')'$, etc. It is not difficult to see that all these subgroups are fully invariant in G, and that $G' = L_2(G)$. The group G is solvable if and only if the nth derived group $G^{(n)}$ is trivial, for some n. The smallest n such that $G^n = \{1\}$ is called the *derived length* of the group G. Subnormal series (1.8) are called *chief* if G_{i+1} is a maximal normal subgroup of G_i, $i = 1, 2, \ldots, n$. A factor of chief series is called a *chief factor* of the group; all chief factors of a finite solvable groups are elementary Abelian, and vice versa. Recall that an *elementary Abelian p-group* is a finite Cartesian power of a cyclic group of prime order p. All subgroups and factor groups of solvable groups are also solvable.

Example 1.25. The symmetric group $G = \mathrm{Sym}(4)$ of all permutations of a set of four elements is solvable; its derived length is 3. Indeed, it is not difficult to verify that $G'' \cong K_4$ is a subgroup that consist of an identity permutation, and of permutations that are products of two disjoint cycles (there are 3 such permutations in $\mathrm{Sym}(G)$). The subgroup G' is the alternating subgroup Alt(4); it is a semidirect product of G'' by a subgroup of order 3, which is generated by a cycle of length 3.

Groups can be represented via *generators and relations*. Recall that a *free group* $F(x_1, \ldots, x_n)$ with free generators $x_1, \ldots, x_n$ is a set of all finite words of form

$x_{i_1}^{m_1} \cdots x_{i_k}^{m_k}$ where $i_j \in \{1, \dots, n\}, i_j \neq i_{j+1}, m_j \in \mathbb{Z} \setminus \{0\}, j = 1, \dots, n$. Multiplications is just a concatenation of words succeeded by reduction of terms: $x_i^m x_i^r = x_i^{m+r}$, $x_i^0 = 1$, 1 is the empty word. We write $F(x_1, \dots, x_n) = \operatorname{gp}(x_1, \dots, x_n \parallel \varnothing)$; that is, a free group is a group with empty set of relations. Now, given a group G generated by elements $g_1, \dots, g_n$, there exists a unique epimorphism $\tau\colon F(x_1, \dots, x_n) \to G$ such that $\tau(x_i) = g_i$, for all $i = 1, \dots, n$. Let $w_\ell(x_1, \dots, x_n) \in F(x_1, \dots, x_n)$, $\ell \in \{1, \dots, s\}$ be elements of the free group that generate $\ker \tau$ as a normal subgroup; that is, $\ker \tau$ is a minimal normal subgroup of $F(x_1, \dots, x_n)$ that contains all $w_\ell(x_1, \dots, x_n)$. We write then

$$G = \operatorname{gp}(g_1, \dots, g_n \parallel w_1(g_1, \dots, g_n) = 1, \dots, w_\ell(g_1, \dots, g_n) = 1),$$

a representation of the group G in generators $g_1, \dots, g_n$ and relations $w_\ell(g_1, \dots, g_n)$, $\ell = 1, \dots, s$.

Example 1.26. In Part II of the book we will need the following 2-groups represented by generators and relations:

- the *dihedral* group

$$\mathrm{D}_n = \operatorname{gp}(u, v \parallel u^2 = v^{2^n} = 1, v^u = v^{-1})$$

  of order 2^{n+1}, $n = 2, 3, 4, \dots$;
- the (generalized) *quaternion* group

$$\mathrm{Q}_n = \operatorname{gp}(u, v \parallel v^{2^n} = 1, v^u = v^{-1}, u^2 = v^{2^{n-1}})$$

  of order 2^{n+1}, $n = 2, 3, 4, \dots$;
- the *semidihedral* group

$$\mathrm{SD}_n = \operatorname{gp}(u, v \parallel u^2 = v^{2^n} = 1, v^u = v^{2^{n-1}-1})$$

  of order 2^{n+1}, $n = 3, 4, 5, \dots$.

All these groups D_n, Q_n, and SD_n are nilpotent of class n, their Frattini subgroups are generated by v^2 (thus, cyclic), and factor groups by Frattini subgroups are isomorphic to the Klein group K_4. Both D_n and SD_n are split extensions of a cyclic group of order 2^n (generated by v) by a cyclic group of order 2 (generated by u). However, the groups are not isomorphic one to another, since the action of u on a cyclic group generated by v is different in both cases. The group Q_n is also an extension of a cyclic group of order 2^n (generated by v) by a cyclic group of order 2; however, the extension is not split.

Further, if G is any of these groups D_n, SD_n, or Q_n, then G' is a cyclic subgroup generated by v^2, and thus $G'' = \{1\}$; so these groups are solvable, and their derived length is 2. In other words, all these groups are extensions of Abelian groups by

Abelian groups; such groups are called *metabelian*. However, all these groups D_n, SD_n, and Q_n are nilpotent of class n: $L_i(G)$ is a cyclic subgroup generated by $v^{2^{i-1}}$, $i = 2, 3, \ldots, n+1$; so $L_{n+1}(G) = \{1\}$ for either group $G \in \{D_n, SD_n, Q_n\}$.

The group $G[x_1, \ldots, x_n]$ of all polynomials in variables $x_1, \ldots, x_n$ over the group G is a *free product* of the group G by the group $F(x_1, \ldots, x_n)$; recall that a *free product* of groups A and B is a set of all words in the alphabet $A \setminus \{1\} \cup B \setminus \{1\}$, such that neighboring letters in a word are from different groups, multiplication of words is a concatenation succeeded by reduction of neighboring letters if they are in the same group (two neighboring letters from the same group must be replaced by a product of corresponding elements), 1 is the empty word. It is worth notice here that n-polynomially complete groups are exactly all finite simple non-Abelian groups, if $n > 1$, and also a group of order 2, if $n = 1$, see [286].

Non-generators of a group G are elements that can be removed from every set of generators of the group G such that the rest generators generate the whole group G. All non-generators of a group form a subgroup $\mathrm{Fr}(G)$, the *Frattini subgroup* of the group G; the subgroup $\mathrm{Fr}(G)$ is an intersection of all maximal subgroups of G. The Frattini subgroup is a characteristic subgroup in G, and it is nilpotent whenever G is finite. If G is a finite p-group, the factor group $G/\mathrm{Fr}(G)$ is an elementary Abelian group; that is, a Cartesian product of m cyclic groups of order p, and the number m is the number of generators in the smallest generating system of G. Actually, if the elements $g_1, \ldots, g_m \in G/\mathrm{Fr}(G)$ generate $G/\mathrm{Fr}(G)$, then every set $g'_1, \ldots, g'_m \in G$ such that $\varphi(g'_i) = g_i$, $i = 1, \ldots, m$, $\varphi\colon G \to G/\mathrm{Fr}(G)$ a canonical epimorphism, generates the whole group G (Burnside Basis Theorem). In particular, a factor group of a non-cyclic nilpotent group by its Frattini subgroup cannot be cyclic.

A notion of a *group with operators* is a generalization of a notion of a group. Actually a group G with a set of operators Ω is a group whose signature is extend by unary operations (that form Ω) such that every unary operation $\omega \in \Omega$ is an endomorphism of the group G: $(ab)^\omega = a^\omega b^\omega$, for all $a, b \in G$, $\omega \in \Omega$. Thus, every group can be considered as a group with empty set of operators. Further generalization is a notion of *groups with multioperators*; these are groups whose signatures are extended by a set of operations Ω, and Ω may consist of operations of various arities; however, if $w \in \Omega$ is an n-ary operation, then $w(1, \ldots, 1) = 1$. An important example of groups with multioperators are rings; they are considered in the next subsection.

1.2.3 Rings

In this subsection we remind some notions and facts from ring theory, mainly following [36, 314, 337, 343]. A *ring* R is a universal algebra with two operations $+$ (addition) and $\cdot$ multiplication, such that R with respect to $+$ is a commutative group (which is denoted as R^+) with neutral 0, which is called zero, and inverse $-$ (that is $-a$ is an additive inverse for $a \in R$, $a + (-a) = 0$), R is a semigroup with respect to $\cdot$, and $(a+b)\cdot c = (a\cdot c)+(b\cdot c)$, $c\cdot(a+b) = (c\cdot a)+(c\cdot b)$, for all $a, b, c \in R$. We mainly consider *commutative* rings in this book, that is, $a \cdot b = b \cdot a$, for all $a, b \in R$. As usual, we omit the sign of multiplication $\cdot$ in expressions, and we omit parenthesis according to the common rule: $a + bc = a + (b \cdot c)$. Whenever the ring R has an *identity*, that is, a multiplicative neutral element, we denote it as 1: $a \cdot 1 = 1 \cdot a$, for all $a \in R$. A ring having the identity is called a *ring with identity*. Further within this subsection 'ring' stands for 'commutative ring with identity'. The additive order of 1, that is, the smallest $n \in \mathbb{N}$ such that $n \cdot 1 = 0$, if such n exists, is called the *characteristic* of R, and is denoted by $\operatorname{char}(R)$. A ring is said to be of *zero characteristic* if no such n exists.

If an element $a \in R$ has a multiplicative inverse, it is denoted by a^{-1}: $a \cdot a^{-1} = a^{-1} \cdot a = 1$. All *invertible* elements (those having multiplicative inverses) are called *units*. They form a group R^* with respect to ring multiplication; this group is called a *unit group*, or a *multiplicative (sub)group* of the ring R. If $R^* = R \setminus \{0\}$, the ring R is called a *field*.

A non-zero element $a \in R$ is called a *zero divisor* whenever there exists an element $b \in R\setminus\{0\}$ such that $ab = 0$. An non-zero element $a \in R$ is called *nilpotent* whenever $a^n = 0$ for some $n \in \mathbb{N}$; the smallest such n is called the *nilpotency index of* a.

A ring R without zero divisors is called an *(integral) domain*. Every integral domain can be embedded into a field; the smallest one is called a *quotient field* of R and denoted as $Q(R)$. For instance, a ring $\mathbb{Z} = \{0, \pm 1, \pm 2, \ldots\}$ of all rational integers is an integral domain; its quotient field is $\mathbb{Q}$, the field of all rational numbers. An *integer-valued function* is a map $F : Q(R)^n \to Q(R)^m$ such that $F(R^n) \subset R^m$. We remind that any *integer-valued polynomial* f over $\mathbb{Q}$ in variable x can be expressed as

$$f(x) = \sum_{i=0}^{d} a_i \binom{x}{i},$$

where $a_i \in \mathbb{Z}$, $i = 0, 1, \ldots, d$, and vice versa, see a substantial monograph [81] on various aspects of integer-valued polynomials. Integer-valued functions on the field of p-adic numbers $\mathbb{Q}_p$ are the maps we are mostly focused at in our book.

A *module* over a ring R is a commutative group M with respect to operation $\oplus$, endowed with an 'external' operation of multiplication by elements of R: Given $r, s \in R$, $h, g \in M$, one defines this multiplication $r \cdot h \in M$ so that $(rs) \cdot h = r \cdot (s \cdot h)$ and $r \cdot (h \oplus g) = (r \cdot h) \oplus (r \cdot g)$. Vector spaces over fields are important example of modules; the other important example are ideals.

A non-empty subset $I \subset R$ is called an *ideal* whenever I is a subgroup with respect to ring addition $+$, and $ra \in I$ for all $r \in R$, $a \in I$. An ideal I is called *proper* whenever $I \neq R$ and $I \neq \{0\}$. An non-zero ideal is called *nilpotent* whenever $I^n = \{0\}$ for some $n \in \mathbb{N}$; that is, $a_1 \cdots a_n = 0$ for all $a_1, \ldots, a_n \in I$. The smallest n with this property is called the *nilpotency index of the ideal* I and denoted as $\operatorname{ind} I$. A unique maximal ideal $J \subset R$, $J \neq R$ (if it exists), is called a *radical* of the ring and denoted $J(R)$. A ring that has a radical is called a *local ring*. In particular, a field is a local ring whose radical is zero. Ideals are kernels of ring homomorphisms, and vice versa.

It is clear that given $a_1, \dots, a_n \in R$, the set $a_1 \cdot R + \cdots + a_n \cdot R$, which is a set of all sums $a_1 r_1 + \cdots + a_n r_n$, $r_1, \dots, r_n \in R$, is an ideal of R, the smallest ideal that contains $a_1, \dots, a_n$. This ideal is called an *ideal generated by elements* $a_1, \dots, a_n$. An ideal that is generated by a single element is called *principal*. A ring all whose ideals are principal, is called a *principal ideal ring*. It is clear that factor rings of principal ideal rings are again principal ideal rings.

Theorem 1.27. *A ring $R[x]$ of all polynomials in a variable x over a field R is a principal ideal ring.*

Now we remind some facts about finite rings; we need these mainly in Subsection 2.2.3. The following is true:

Proposition 1.28. *Every non-zero element of a finite ring is either a unit, or a zero divisor.*

Finite principal ideal rings can be constructed as Cartesian products (in ring theory they prefer the term 'direct sum') of fields and local rings.

Theorem 1.29. *Every finite principal ideal ring R is isomorphic to a direct sum of local principal ideal rings. Foremost, if R is local, then the radical J of R is nilpotent, and $\#R = (\#F)^{\operatorname{ind} J}$, where $F = R/J$ is a residue field of R.*

In applications to computer science and cryptology (see Part III) we mainly deal with residue rings $\mathbb{Z}/N\mathbb{Z}$ modulo N. For these rings, Theorem 1.29 yields:

Theorem 1.30 (Chinese Remainder Theorem, equivalent form). *Let N be a natural number, $N > 1$. Represent $N = p_1^{e_1} p_2^{e_2} \cdots p_r^{e_r}$, where p_j, $1 \leqslant j \leqslant r$, are prime numbers, and r is the number of different primes in decomposition of N. Then the residue ring $\mathbb{Z}/N\mathbb{Z}$ is a direct sum of residue rings $\mathbb{Z}/p_j^{e_j}\mathbb{Z}$, $1 \leqslant j \leqslant r$.*

For residue rings $\mathbb{Z}/N\mathbb{Z}$ there exists a simple way to determine whether a given element is invertible or a zero divisor, cf. Proposition 1.28:

Proposition 1.31 (Invertibility modulo N). *Let N be a natural number, $N > 1$. Represent $N = p_1^{e_1} p_2^{e_2} \cdots p_r^{e_r}$, where p_j, $1 \leqslant j \leqslant r$, are prime numbers, and r is the number of different primes in decomposition of N. Then the element a of the residue ring $\mathbb{Z}/N\mathbb{Z}$ is invertible if and only if $a \not\equiv 0 \pmod{p_j}$ for all $1 \leqslant j \leqslant r$.*

With the use of these results in combination with the following Proposition 1.32, it is easy to determine multiplicative subgroups of residue rings. Actually, this way we determine automorphism groups of finite cyclic groups.

Proposition 1.32. *Let p be a prime, let $k \in \mathbb{N}$. A group $(\mathbb{Z}/p^k\mathbb{Z})^*$ of all invertible elements of the residue ring $\mathbb{Z}/p^k\mathbb{Z}$ is a cyclic group of order $(p-1)\cdot p^{k-1}$ whenever p is odd. If $p = 2$ and $k > 2$ then $(\mathbb{Z}/2^k\mathbb{Z})^*$ is a direct product of a group of order 2 by a cyclic group of order 2^{k-2}. The group $(\mathbb{Z}/4\mathbb{Z})^*$ is a cyclic group of order 2, the group $(\mathbb{Z}/2\mathbb{Z})^*$ is trivial.*

The following theorem characterizes polynomially complete algebras in the class of all commutative rings.

Theorem 1.33 (Polynomial completeness of finite fields). *Let $n \in \mathbb{N}$. A commutative ring is n-polynomially complete if and only it is a finite field.*

Note that there are known explicit formulas that express a given map as a polynomial over a finite field, see Subsection 1.3.1.

In the sequel, we will need some more special types of rings, a ring of formal power series, and a (semi)group ring. Given a ring R and a variable x, consider all formal expressions $\sum_{i=0}^{\infty} a_i$, $a_i \in R$, $i = 0, 1, 2, \ldots$. We can define addition and multiplication of these sums by common rules for infinite series; as every coefficient of a sum or product is then a finite expression of finite number of coefficients of summands (respectively, factors), these operations are well defined. Thus we obtain a *ring* $R[[x]]$ *of formal power series*; its elements are called *formal power series* over R.

To construct a *(semi)group ring* RG we need a (semi)group G and a ring R. We then consider finite formal sums $a_1g_1 + \cdots + a_ng_n$, where all $a_j \in R$, $g_j \in G$, $g_j \neq g_i$ if $i \neq j$, $i, j \in \{1, \ldots, n\}$. Given $a, b \in R$, $g, h \in G$, we define addition $ag + bh = (a+b)h$ if $g = h$, multiplication $ag \cdot bh = (ab)(gh)$, and then expand these rules for addition and multiplication of the above formal sums in a standard way using the distributive law. We put $0g = 0$ for all $g \in G$; so 0 is an additive neutral of RG. We put $-ag = (-a)g$. Thus we obtain a ring, which is called a semigroup ring if G is a semigroup, and a group ring, if G is a group. The ring RG is commutative whenever both R and G are commutative, and which has an identity whenever both R and G have identities (multiplicative neutral elements).

1.3 Fields

In this section, we remind some facts (and related notions) about fields.

1.3.1 Finite fields

Finite fields have some special properties we use throughout the book. A characteristic char(F) of a finite field F is a prime number p, and $\#F = p^n$ for a suitable $n \in \mathbb{N}$. Given a prime p and a positive rational integer n, there exists (up to a ring isomorphism) a unique field of order p^n. We denote this unique field of p^n elements by $\mathbb{F}_{p^n}$. In particular, if $n = 1$, then $\mathbb{F}_p$ is isomorphic to the residue ring $\mathbb{Z}/p\mathbb{Z}$ modulo p.

A multiplicative subgroup $\mathbb{F}_{p^n}^*$ is a cyclic group of order $p^n - 1$; generators of this group are called *primitive elements* of the field $\mathbb{F}_{p^n}$. Thus, there are exactly $\varphi(p^n - 1)$ different primitive elements in $\mathbb{F}_{p^n}$, where φ is the Euler totient function.

As said (see Theorem 1.33), finite fields are polynomially complete rings. Given a map $\varphi\colon \mathbb{F}_q \to \mathbb{F}_q$, there exists a polynomial $f_\varphi(x) \in \mathbb{F}_q[x]$ such that $f_\varphi(z) = \varphi(z)$, for all $z \in \mathbb{F}_q$:

$$f_\varphi(x) = \sum_{z \in \mathbb{F}_q} \varphi(z) \cdot \frac{x^q - x}{z - x}. \tag{1.9}$$

We note that $f_\varphi(x)$ is indeed a polynomial over $\mathbb{F}_q$ as $x^q - x = \prod_{z \in \mathbb{F}_q}(x - z)$. Formula (1.9) holds since

$$\frac{x^q - x}{z - x} = \begin{cases} 1, & \text{whenever } x = z; \\ 0, & \text{otherwise.} \end{cases}$$

Using this method we can construct an interpolation polynomial for an arbitrary n-variate mapping from $\mathbb{F}_q^n$ to $\mathbb{F}_q^m$, as, e.g.,

$$\frac{x^q - x}{a - x} \cdot \frac{y^q - y}{b - y} = \begin{cases} 1, & \text{whenever } x = a \text{ and } y = b; \\ 0, & \text{otherwise,} \end{cases}$$

and henceforth. Moreover, we can interpolate simultaneously a mapping and its derivative, in the following way:

Proposition 1.34. *Given two mappings $\varphi\colon \mathbb{F}_q \to \mathbb{F}_q$ and $\psi\colon \mathbb{F}_q \to \mathbb{F}_q$, there exists a polynomial $f_{\varphi,\psi}(x) \in \mathbb{F}_p(x)$ such that*

- *$f_{\varphi,\psi}$ induces on $\mathbb{F}_q$ the mapping φ:*

$$f_{\varphi,\psi}(z) = \varphi(z) \quad \textit{for all } z \in \mathbb{F}_q,$$

- *a derivative $f'_{\varphi,\psi}(x)$ induces on $\mathbb{F}_q$ the mapping ψ:*

$$f'_{\varphi,\psi}(z) = \psi(z) \quad \textit{for all } z \in \mathbb{F}_q.$$

Proof. Given mappings φ and ψ, construct interpolation polynomials f_φ and f_ψ according to formula (1.9). Then

$$f_{\varphi,\psi}(x) = f_\varphi(x) - (x^q - x) \cdot (f'_\varphi(x) - f_\psi(x)).$$

Note that $z^q - z = 0$ for all $z \in \mathbb{F}_q$, that $(x^q - x)' = qx^{q-1} - 1$ is identically -1 on $\mathbb{F}_q$, and that $f'_\varphi(x)$ is a polynomial over $\mathbb{F}_q$ (as $f_\varphi(x)$ is a polynomial over $\mathbb{F}_q$). □

Note 1.35. This proposition can also be generalized to arbitrary mappings $\mathbb{F}_q^n$ to $\mathbb{F}_q^m$ with the use of interpolation formulas for n-variate mappings we mentioned above, as well as for higher order derivatives.

1.3.2 Non-Archimedean fields

Let K be a field. An *absolute value* on K is a function $|\cdot|\colon K \to \mathbb{R}$ such that

- $|x| \geqslant 0$, for all $x \in K$,
- $|x| = 0$ if and only if $x = 0$,
- $|xy| = |x||y|$, for all $x, y \in K$,
- $|x + y| \leqslant |x| + |y|$, for all $x, y \in K$.

If $|\cdot|$ in addition satisfies the *strong triangle inequality*

$$|x + y| \leqslant \max(|x|, |y|) \tag{1.10}$$

for all $x, y \in K$ then we say that $|\cdot|$ is *non-Archimedean*.

If $|x| = 1$ for all non-zero $x \in K$ we call $|\cdot|$ the *trivial absolute value*. It is easy to see that the trivial absolute value is non-Archimedean.

Proposition 1.36. *Let K be a field and let $|\cdot|$ be a non-Archimedean absolute value on K. Let $x, y \in K$ such that $|x| \neq |y|$. Then*

$$|x + y| = \max(|x|, |y|). \tag{1.11}$$

Proof. Assume that $|x| > |y|$. By the strong triangle inequality we have

$$|x| = |(x + y) - y| \leqslant \max(|x + y|, |y|).$$

The assumption $|x| > |y|$ implies $\max(|x + y|, |y|) = |x + y|$. Thus $|x| \leqslant x + y$. By the strong triangle inequality,

$$|x + y| \leqslant \max(|x|, |y|) = |x|.$$

We can conclude that $|x + y| = |x|$. □

1.4 p-adic numbers

In this section, we introduce a notion we are mostly dealing with in our book, the notion of a p-adic number.

Let p be a fixed prime number. By the fundamental theorem of arithmetics, each non-zero integer n can be written uniquely as

$$n = p^{\operatorname{ord}_p n} \hat{n},$$

where $\hat{n}$ is a non-zero integer, $p \nmid \hat{n}$, and $\operatorname{ord}_p n$ is a unique non-negative integer. The function $\operatorname{ord}_p\colon \mathbb{Z} \setminus \{0\} \to \mathbb{N}_0$ is called the *p-adic valuation*. If $a, b \in \mathbb{Z}^+$ then we define the p-adic valuation of $x = a/b$ as

$$\operatorname{ord}_p x = \operatorname{ord}_p a - \operatorname{ord}_p b. \tag{1.12}$$

One can easily show that the valuation is well defined. The valuation of x does not depend on the fractional representation of x. By using the p-adic valuation we will define a new absolute value on the field of rational numbers.

Definition 1.37. The *p-adic absolute value* of $x \in \mathbb{Q} \setminus \{0\}$ is given by

$$|x|_p = p^{-\operatorname{ord}_p x} \tag{1.13}$$

and $|0|_p = 0$.

Example 1.38. If $p = 2$ then $\operatorname{ord}_2 \frac{1}{2} = -1$ and $\left|\frac{1}{2}\right|_2 = 2$. Moreover $\operatorname{ord}_2 3 = 0$ and $|3|_2 = 1$. If $p = 3$ then $\operatorname{ord}_3 \frac{1}{2} = 0$, $\operatorname{ord}_3 3 = 1$, $\left|\frac{1}{2}\right|_3 = 1$ and $|3|_3 = \frac{1}{3}$.

Let X be a set and let ρ be a *metric* on X. Then ρ by definition has the following properties:

- For all $x, y \in X$, $\rho(x, y) \geqslant 0$ and $\rho(x, y) = 0$ if and only if $x = y$.
- For all $x, y \in X$, $\rho(x, y) = \rho(y, x)$.
- For all $x, y, z \in X$,
$$\rho(x, z) \leqslant \rho(x, y) + \rho(y, z)$$
(the triangle inequality).

We say that the pair (X, ρ) is a metric space.

The p-adic absolute value is non-Archimedean. It induces a metric

$$\rho(x, y) = |x - y|_p.$$

Two absolute values on a field K are said to be equivalent if they generate the same topology on K. Essentially there are only two types of non-trivial absolute values on $\mathbb{Q}$. This is the essence of the following theorem.

Theorem 1.39 (Ostrovski). *Every non-trivial absolute value on $\mathbb{Q}$ is either equivalent to the real absolute value or to one of the p-adic absolute values.*

For a proof of Ostrovski's theorem see, for example, [374] or [157].

Let ρ be a metric induced by the p-adic absolute value on $\mathbb{Q}$, $(\mathbb{Q}, \rho)$ is then a metric space. However, this space is not complete. There exist Cauchy sequences which do not converge to any element of $\mathbb{Q}$. We shall use the following result:

Theorem 1.40. *A sequence (x_j) in $\mathbb{Q}$ is a Cauchy sequence with respect to the p-adic absolute value if and only if*

$$\lim_{j \to \infty} |x_{j+1} - x_j|_p = 0. \tag{1.14}$$

Proof. If (x_j) is a Cauchy sequence then it is clear that $x_{j+1}-x_j \to 0$, when $j \to \infty$.

Assume now that (x_j) is a sequence that satisfies (1.14). Let $i > j$. Then there exists $k \in \mathbb{Z}^+$ such that $i = j + k$. We have

$$|x_i - x_j| \leqslant \max(|x_{j+k} - x_{j+k-1}|_p, |x_{j+k-1} - x_{j+k-2}|_p, \dots, |x_{j+1} - x_j|_p).$$

If $x_{j+1} - x_j \to 0$ when $j \to \infty$ it follows that $x_i - x_j \to 0$ when $i, j \to \infty$. Hence (x_j) is a Cauchy sequence. □

Example 1.41. There is no rational number x satisfying $x^2 = 7$. But since this equation has a solution modulo 3 ($x \equiv 1$) it is possible to construct a sequence $(x_j)_{j\geqslant 0}$ such that $x_j \equiv x_{j+1} \pmod{3^j}$ and $x_j^2 \equiv 7 \pmod{3^{j+1}}$. We have that (x_j) is a Cauchy sequence because

$$|x_j - x_{j+1}|_p \leqslant 3^{-(j+1)} \to 0,\ j \to \infty.$$

It is clear that the limit of this sequence must be a solution of $x^2 = 7$, since

$$|x_j^2 - 7|_p \leqslant 3^{-(j+1)} \to 0,\ j \to \infty.$$

Thus the limit does not belong to $\mathbb{Q}$. We have proved that $\mathbb{Q}$ endowed with the metric induced by the 3-adic absolute value is not complete.

In fact, we can generalize this example to any metric space $(\mathbb{Q}, \rho)$, where ρ is the metric induced by the p-adic absolute value, see [157]. The presence of such examples implies

Theorem 1.42. *The metric space $(\mathbb{Q}, \rho)$, where ρ is the metric induced by the p-adic absolute value is not complete.*

The completion of $\mathbb{Q}$ will be a field, the *field of p-adic numbers*, $\mathbb{Q}_p$. The p-adic absolute value is extended to $\mathbb{Q}_p$ and $\mathbb{Q}$ is dense in $\mathbb{Q}_p$. It is worth noting that

$$\{|x|_p : x \in \mathbb{Q}_p\} = \{|x|_p : x \in \mathbb{Q}\} = \{p^m : m \in \mathbb{Z}\} \cup \{0\}.$$

Finally, we mention some topological properties of fields of p-adic numbers. A topological space is locally compact if every point has a compact neighborhood. We recall that the space $\mathbb{Q}_p$ is locally compact. A field K endowed with a topology is said to be a *topological field* if the operations of addition, subtraction, multiplication and division are continuous. We also recall that the field of p-adic numbers is a topological field.

1.4.1 Canonical expansion of p-adic numbers

The set

$$B_1(0) = \{x \in \mathbb{Q}_p : |x|_p \leqslant 1\}$$

is called the set of *p-adic integers*. It is denoted by $\mathbb{Z}_p$. In fact, $\mathbb{Z}_p$ is a subring of $\mathbb{Q}_p$ and

$$B_1^-(0) = \{x \in \mathbb{Z}_p : |x|_p < 1\}$$

is a maximal ideal of $\mathbb{Z}_p$. The quotient ring

$$\mathbb{Z}_p / B_1^-(0)$$

is then a field, called the *residue class field* of $\mathbb{Q}_p$.

Theorem 1.43. *For each $x \in \mathbb{Z}_p$ there exists a sequence $(x_j)_{j \geqslant 0}$ such that*

$$x_j \in \mathbb{Z}, \quad 0 \leqslant x_j \leqslant p^{j+1} - 1, \quad x_{j+1} \equiv x_j \pmod{p^{j+1}}$$

for all $j \geqslant 0$ and $|x - x_j|_p \leqslant p^{-(j+1)}$.

Proof. Let $x \in \mathbb{Z}_p$. Because of the fact that $\mathbb{Q}$ is dense in $\mathbb{Q}_p$ we can find a rational number a/b such that $|x - a/b|_p \leqslant p^{-(j+1)}$ for every j. In fact, this number can be chosen to be an integer. Since

$$|a/b|_p \leqslant \max(|x|_p, |a/b - x|_p) \leqslant 1$$

it is clear that $p \nmid b$, so $\gcd(p^{j+1}, b) = 1$. Therefore there exist b' and p' such that $p' p^{j+1} + b'b = 1$ or equivalently $b'b \equiv 1 \pmod{p^{j+1}}$. We then have

$$|a/b - ab'|_p = |a/b|_p |1 - b'b|_p \leqslant p^{-(j+1)},$$

and $|x - ab'|_P \leqslant \max(|x - a/b|_p, |a/b - ab'|_p) \leqslant p^{-(j+1)}$. There is a unique integer x_j satisfying $0 \leqslant x_j \leqslant p^{j+1} - 1$ and $x_j \equiv ab' \pmod{j+1}$. It is clear that $|x_j - x|_p \leqslant p^{-(j+1)}$.

It remains to show that $x_{j+1} \equiv x_j \pmod{p^{j+1}}$. This follows from the fact that

$$|x_{j+1} - x_j|_p \leqslant \max(|x_{j+1} - x|_p, |x - x_j|_p) \leqslant \max(p^{-(j+2)}, p^{-(j+1)}) \leqslant p^{-(j+1)}.$$

□

Corollary 1.44. *The residue class field of $\mathbb{Q}_p$ is isomorphic to the finite field $\mathbb{F}_p$ of p elements.*

Proof. It follows from the theorem that the integers $\{0, 1, \ldots, p-1\}$ form a complete set of representatives of the cosets of $B_1^-(0)$. □

Theorem 1.45. *Every $x \in \mathbb{Z}_p$ can be expanded in the following way*

$$x = y_0 + y_1 p + y_2 p^2 + \cdots + y_j p^j + \cdots .$$

Proof. By expanding the elements of the sequence (x_j) from Theorem 1.43 in the base p we get

$$\begin{aligned} x_0 &= y_0, \quad 0 \leqslant y_0 \leqslant p-1, \\ x_1 &= y_0 + y_1 p, \quad 0 \leqslant y_1 \leqslant p-1, \\ x_2 &= y_0 + y_1 p + y_2 p^2, \quad 0 \leqslant y_2 \leqslant p-1, \\ &\vdots \\ x_j &= y_0 + y_1 p + \cdots + y_j p^j, \quad 0 \leqslant y_j \leqslant p-1. \end{aligned}$$

It is clear that the sum $\sum_{j \geqslant 0} y_j p^j$ converges. □

Note 1.46. In the sequel for $x \in \mathbb{Z}_p$ we use the notation $\delta_i(x) = y_i$, $i = 0, 1, 2, \ldots$. Thus $\delta_i(x) \in \{0, 1, \ldots, p-1\}$ for all $i = 0, 1, 2, \ldots$.

Note 1.47. A p-adic integer $x \in \mathbb{Z}_p$ is invertible in $\mathbb{Z}_p$ (that is, has a multiplicative inverse $x^{-1} \in \mathbb{Z}_p$, $x^{-1} \cdot x = 1$) if and only if $\delta_0(x) \neq 0$.

Corollary 1.48. *Every $x \in \mathbb{Q}_p$ can be expanded in the base p in the following way:*

$$x = \sum_{j \geqslant j_{\min}} y_j p^j, \tag{1.15}$$

where $j_{\min} = \operatorname{ord}_p x \in \mathbb{Z}$ and $0 \leqslant y_j \leqslant p-1$ for $j \geqslant j_{\min}$.

Proof. Let $x \in \mathbb{Q}_p$ and assume that $x \in \mathbb{Z}_p$. Let $y = p^{-\operatorname{ord}_p x} x$. Then

$$|p^{-\operatorname{ord}_p x} x|_p = p^{\operatorname{ord}_p x} \cdot p^{-\operatorname{ord}_p x} = 1.$$

Thus $y \in \mathbb{Z}_p$. That is, every $x \notin \mathbb{Q}_p$ can be written as $x = y \cdot p^{-m}$ for some positive integer m and $y \in \mathbb{Z}_p$. By Theorem 1.45 we obtain an expansion of y. If we then divide it by p^m we get (1.15). □

For each positive integer $m \geqslant 2$ we can expand a real number r with respect to the base m in the following way:

$$r = \sum_{i \leqslant i_{\max}} r_i m^i, \tag{1.16}$$

for some integer $i_{\max}$. A real number r can have infinitely many negative powers in this expansion, but a p-adic number can have infinitely many positive powers in the expansion (1.15).

Example 1.49. For every prime p we have the following expansion of -1,

$$-1 = (p-1) + (p-1)p + (p-1)p^2 + \cdots,$$

since $1 + (p-1) + (p-1)p + (p-1)p^2 + \cdots = 0$.

Example 1.50. In $\mathbb{Q}_2$, the rational number $1/3$ has the expansion

$$1/3 = 1 + 1 \cdot 2 + 0 \cdot 2^2 + 1 \cdot 2^3 + 0 \cdot 2^4 + \cdots.$$

1.4.2 Tree-like structure of the p-adic numbers

Rings of p-adic numbers have a simple geometric structure. These are homogeneous trees with p branches leaving each vertex and one incoming branch.

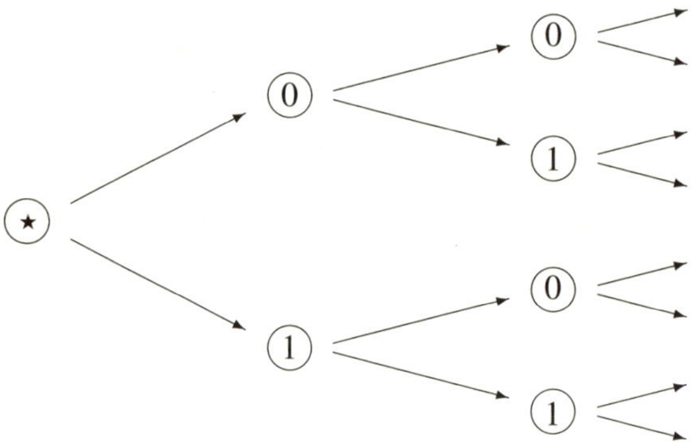

Figure 1.1. The 2-adic tree

1.5 Ultrametric spaces

Let (X, ρ) be a metric space. If ρ also has the property that

$$\rho(x, z) \leqslant \max(\rho(x, y), \rho(y, z)) \tag{1.17}$$

(the strong triangle inequality) then ρ is said to be an *ultrametric*. A set endowed with an ultrametric is called an *ultrametric space*.

Proposition 1.51. *In an ultrametric space all triangles are isosceles. More precise, if X is an ultrametric space with metric ρ and $a, b, c \in X$ such that $\rho(a, b) \neq \rho(b, c)$ then $\rho(a, c) = \max(\rho(a, b), \rho(b, c))$.*

Proof. Assume that $\rho(a,b) < \rho(b,c)$. We then have

$$\rho(a,c) \leqslant \max(\rho(a,b), \rho(b,c)) = \rho(b,c)$$

and

$$\rho(b,c) \leqslant \max(\rho(a,b), \rho(a,c)) = \rho(a,c)$$

since $\rho(a,b) < \rho(b,c)$. □

It is impossible to embed an ultrametric space of more than three points in a plane. But it is possible to use other frameworks for visualizing an ultrametric space, for example trees.

Let (X,ρ) be a metric space. Let $a \in X$ and let $r \in \mathbb{R}^+$. The *open ball of radius r with center a* is the set

$$B_r^-(a) = \{x \in X : \rho(a,x) < r\}.$$

The *closed ball of radius r with center a* is the set

$$B_r(a) = \{x \in X : \rho(a,x) \leqslant r\}.$$

The set

$$S_r(a) = \{x \in X : \rho(a,x) = r\}$$

is called the *sphere of radius r with center a*.

In further considerations it is sometimes important to underline in which metric space a ball or a sphere is taken. We then use the symbols $B_r^-(a,X)$, $B_r(a,X)$ and $S_r(a,X)$. Proposition 1.51 has some remarkable consequences for the balls in X.

Proposition 1.52. *Every element of a ball can be regarded as a center of it.*

Proof. We prove the proposition in the case of an open ball $B_r^-(a) \subset X$. Let $b \in B_r^-(a)$. We want to prove that $B_r^-(b) = B_r^-(a)$. Take $x \in B_r^-(b)$ then

$$\rho(x,a) \leqslant \max(\rho(x,b), \rho(b,a)) < r$$

so $B_r^-(b) \subseteq B_r^-(a)$. In the same way we obtain $B_r^-(a) \subseteq B_r^-(b)$. Thus $B_r^-(a) = B_r^-(b)$. □

Proposition 1.53. *Each open ball is both open and closed.*

Proof. It is trivial that an open ball is an open set. We prove that each ball $B_r^-(a)$ is closed. Let b be a limit point of $B_r^-(a)$. Let $s \leqslant r$. Then

$$B_s^-(b) \cap B_r^-(a) \neq \varnothing$$

since b is a limit point. Let $c \in B_s^-(b) \cap B_r^-(a)$. By the strong triangle inequality we have

$$\rho(b,a) \leqslant \max(\rho(b,c), \rho(c,a))$$

so $b \in B_r^-(a)$. That is, $B_r^-(a)$ contains all its limit points and it is therefore closed. □

Proposition 1.54. *Each closed ball of positive radius is both open and closed.*

Proof. We will prove that the ball $B_r(a)$, $r > 0$ is open. Let $b \in B_r(a)$ and let $s \in \mathbb{R}$ such that $0 < s < r$. We then have $B_s^-(b) \subseteq B_r(a)$ since if $x \in B_s^-(b)$ then

$$\rho(x, a) \leqslant \max(\rho(x, b), \rho(b, a)).$$

The proof that a closed ball is closed is similar to the proof that the open ball is closed. □

Proposition 1.55. *Let B_1 and B_2 be balls in X. Then either B_1 and B_2 are ordered by inclusion ($B_1 \subseteq B_2$ or $B_2 \subseteq B_1$) or B_1 and B_2 are disjoint.*

Proof. We will prove this for two open balls; the proofs of the other cases are identical. Let $a, b \in X$ and let $r, s \in \mathbb{R}^+$ such that $r \geqslant s > 0$. Assume that $B_s^-(b) \cap B_r^-(a) \neq \varnothing$. Then there is $c \in B_s^-(b) \cap B_r^-(a)$ such that $B_r^-(c) = B_r^-(a)$ and $B_s^-(c) = B_s^-(b)$. Of course, $B_s^-(c) \subseteq B_r^-(c)$ so $B_s^-(b) \subseteq B_r^-(a)$ and the proposition is proved. □

Definition 1.56. A topological space X is *connected* if it cannot be represented as a union of two disjoint non-empty open sets. A connected subspace of X which is not properly contained in a larger connected subspace of X is called a *connected component* of X.

Definition 1.57. A topological space X is said to be *totally disconnected* if we for each pair $a, b \in X$ can find open subsets A, B of X such that $a \in A$, $b \in B$, $A \cap B = \varnothing$ and $A \cup B = X$.

It is easy to prove that the components of a totally disconnected space are the singleton sets $\{x\}$, for $x \in X$. Since any ball in an ultrametric space is open and closed, we obtain the following simple, but very important result:

Theorem 1.58. *An ultrametric space is totally disconnected.*

Every non-Archimedean field can be regarded as an ultrametric space with the metric $\rho(x, y) = |x - y|$ induced by the absolute value.

1.6 The Haar measure

On $\mathbb{Q}_p$ (as on any locally compact group) there exists the Haar measure, i.e., a positive measure dx invariant under shifts, $d(x + a) = dx$, and normalized by the equality

$$\int_{|x|_p \leq 1} dx = 1.$$

The invariant measure dx on the field $\mathbb{Q}_p$ is extended to an invariant measure $d^n x = dx_1 \cdots dx_n$ on $\mathbb{Q}_p^n$ in the standard way.

We set $B_\gamma \equiv B_{p^\gamma}(0)$, $\gamma \in \mathbb{Z}$ and $S_\gamma \equiv S_{p^\gamma}(0)$. We have (see [407])

$$\begin{aligned} \int_{B_\gamma} dx &= p^\gamma, \\ \int_{S_\gamma} dx &= p^\gamma\Big(1-\frac{1}{p}\Big), \quad \gamma \in \mathbb{Z}. \end{aligned} \tag{1.18}$$

If f is an integrable function on $\mathbb{Q}_p$, then ([407])

$$\begin{aligned} \int_{B_N} f(x)\,dx &= \sum_{\gamma=-\infty}^{N} \int_{S_\gamma} f(x)\,dx, \\ \int_{S_\gamma} f(x)\,dx &= \int_{B_\gamma} f(x)\,dx - \int_{B_{\gamma-1}} f(x)\,dx. \end{aligned} \tag{1.19}$$

Let A be a measurable subset in $\mathbb{Q}_p^n$. Denote by $L_\rho(A)$ the set of all functions $f(x)$ such that

$$\int_A |f(x)|^\rho\, d^n x < \infty \quad (\rho \geq 1).$$

We also have a formula for the change of variables ([407]):

$$\int_{\mathbb{Q}_p} f(x)\,dx = \int_{\mathbb{Q}_p} f\Big(\frac{1}{\xi}\Big)\frac{1}{|\xi|_p^2}\,d\xi. \tag{1.20}$$

Since the Haar measure is a countably additive measure on the σ-algebra of Borel subsets, we have the ordinary Lebesgue dominated convergence theorem:

Theorem 1.59. *If a sequence of functions $f_k \in L_1(\mathbb{Q}_p^n)$, $k \to \infty$, converges almost everywhere in $\mathbb{Q}_p^n$ (with respect to the measure $d^n x$) to a function f, i.e.,*

$$f_k(x) \to f(x), \quad k \to \infty, \quad x \in \mathbb{Q}_p^n, \quad a.e.,$$

and there exists a function $\psi \in L_1(\mathbb{Q}_p^n)$ such that

$$|f_k(x)| \leq \psi(x), \quad k \in \mathbb{N}, \quad x \in \mathbb{Q}_p^n, \quad a.e.,$$

then the following equality holds:

$$\lim_{k\to\infty} \int_{\mathbb{Q}_p^n} f_k(x)\,d^n x = \int_{\mathbb{Q}_p^n} f(x)\,d^n x.$$

1.7 Non-Archimedean rings, m-adic numbers

Let F be a ring[2]. Recall that a *norm* is a mapping $|\cdot|: F \to \mathbb{R}^+$ satisfying the following conditions:

$$|x| = 0 \Longleftrightarrow x = 0 \quad \text{and} \quad |1| = 1, \tag{1.21}$$

$$|xy| \leqslant |x||y|, \tag{1.22}$$

$$|x + y| \leq |x| + |y|. \tag{1.23}$$

The ring F with the norm $|\cdot|$ is called a *normed ring.*[3] Set

$$|F| = \{r \in \mathbb{R}^+ : r = |x|, x \in F\}.$$

The inequality (1.23) is the well-known triangle axiom. A norm is said to be non-Archimedean if the *strong triangle axiom* is valid, i.e., $|x + y| \leq \max(|x|, |y|)$. A ring F with a non-Archimedean norm is said to be a *non-Archimedean ring.* We shall use the following property of a non-Archimedean norm: $|x+y| = \max(|x|, |y|)$, if $|x| \neq |y|$, cf. Section 1.3.2. If a norm $|\cdot|$ has the property $|xy| = |x||y|$, then it is called *absolute value.* This definition matches with the definition of the absolute value on a field.

Denote by $\mathbb{Z}(F)$ the ring generated in F by its unity element. If F has zero characteristic (i.e., $n \cdot 1 = 1 + \cdots + 1 \neq 0$ for any $n = 1, 2, \ldots$), then $\mathbb{Z}(F)$ is isomorphic to the ring of integers $\mathbb{Z}$. Therefore in this case we can consider $\mathbb{Z}$ as a subring of F. In what follows we consider only normed rings F which have zero characteristic.

Let $|\cdot|$ be a norm on a ring F. Then the function $\rho(x, y) = |x - y|$ is a metric on F. It is a translation invariant metric, i.e. $\rho(x + h, y + h) = \rho(x, y)$. Let $|\cdot|$ be a non-Archimedean norm. Then the corresponding metric ρ satisfies the strong triangle inequality: $\rho(x, y) \leqslant \max[\rho(x, z), \rho(z, y)]$. Thus it is an *ultrametric.*

If we repeat considerations of Section 1.4 for an arbitrary natural number $m > 1$, we construct the system of the so called m-adic numbers $\mathbb{Q}_m$ (by completing $\mathbb{Q}$ with respect to the m-adic metric $\rho(x, y) = |x - y|_m$). However, this system is not in general a field. There exist in general divisors of zero in $\mathbb{Q}_m$, thus $\mathbb{Q}_m$ is only a ring. It is important for our further considerations to remark that m-adic numbers have canonical expansions of the form (1.15) (with m instead of p).

For instance, any m-adic integer $x \in \mathbb{Z}_m$ has a canonical m-adic expansion of the form

$$x = y_0 + y_1 m + y_2 m^2 + \cdots + y_j m^j + \cdots,$$

[2]Within this section, by a ring we always mean a commutative ring with identity 1.

[3]Later, in Section 3.5, we introduce the notion of a normed linear space. One should be careful, since in the latter case one has inequality (instead of equality) in the analog of (1.22). Moreover, in Subsection 1.8.1 the notion of norm will appear in totally different context. In particular, it will be $\mathbb{Q}_p$-valued. We hope that such operating with "norm" in various contexts will not disturb readers. It is impossible to do anything, since these are traditional terminologies.

where $y_0, y_1, \ldots \in \{0, 1, \ldots, m-1\}$; $|x|_m = m^{-i}$, where i is the smallest non-negative rational integer such that $y_i \neq 0$, or $|x|_m = 0$ (that is, $x = 0$) if no such i exists.

1.8 Extensions of the field of p-adic numbers

This section is quite complicated from the algebraic viewpoint. At the same time results of this section are not important for the main part of this book. In principle, it is sufficient to know that, in contrast to the real case, finite extensions of $\mathbb{Q}_p$ are not reduced to a single quadratic extension. We remind that all finite extensions of $\mathbb{R}$ coincide with the quadratic extension $\mathbb{C} \equiv \mathbb{R}(\sqrt{-1})$. The latter is algebraically closed. In the p-adic case already quadratic extensions can be non-isomorphic to each other. The same is valid for extensions of higher orders. Non of finite extensions is algebraically closed. Thus by starting with any polynomial and by extending $\mathbb{Q}_p$ with roots of this polynomial (which do not belong to $\mathbb{Q}_p$) we obtain an extension of $\mathbb{Q}_p$, say L, such that one can find another polynomial with coefficients from $\mathbb{Q}_p$ whose roots do not belong to L. Algebraic closure of $\mathbb{Q}_p$ has infinite dimension as a linear space over $\mathbb{Q}_p$. It is not complete – as a metric space – with respect to a natural extension of the p-adic absolute value. By completing it we obtain the algebraically closed field which is a complete metric space. This is the field of complex p-adic numbers $\mathbb{C}_p$. In principle, the reader can proceed on the basis of this brief description of the structure of algebraic extensions of $\mathbb{Q}_p$ and omit coming sections.

1.8.1 Finite extensions of $\mathbb{Q}_p$

Everywhere below we denote by K a finite extension of the p-adic numbers. Let $m = [K : \mathbb{Q}_p]$ denote the dimension of K as a vector space over $\mathbb{Q}_p$. The p-adic absolute value $|\cdot|_p$ can be extended to K, in the unique way. See [157], [374] or [371] for detail.

Suppose that L and K are two finite extensions of $\mathbb{Q}_p$ which form a tower $\mathbb{Q}_p \subset K \subset L$. Let $|\cdot|_K$ be the unique extension of the p-adic valuation on K, and let $|\cdot|_L$ be the unique extension of the p-adic valuation on L. The restriction of $|\cdot|_L$ to elements of K is a non-Archimedean valuation on K and therefore, by uniqueness, $|x|_K = |x|_L$ for every $x \in K$. Hence, the valuation of x does not depend on the context.

Still, we know that there exists a unique extension of the p-adic valuation, but how can we evaluate the p-adic valuation on elements in K? To be able to evaluate the p-adic valuation on elements in $K \setminus \mathbb{Q}_p$, we need a function

$$\mathbf{N}_{K/\mathbb{Q}_p} : K \to \mathbb{Q}_p,$$

which satisfies the equality

$$\mathbf{N}_{K/\mathbb{Q}_p}(xy) = \mathbf{N}_{K/\mathbb{Q}_p}(x)\,\mathbf{N}_{K/\mathbb{Q}_p}(y).$$

This function is called the *norm* from K to $\mathbb{Q}_p$. There exist several ways to define $\mathbf{N}_{K/\mathbb{Q}_p}$, all equivalent. Below, three of them are listed.

(1) Let $\alpha \in K$ and consider K as a finite-dimensional $\mathbb{Q}_p$-vector space. The map from K to K defined by multiplication by α is a $\mathbb{Q}_p$-linear map. Since it is linear it corresponds to a matrix. Then define $\mathbf{N}_{K/\mathbb{Q}_p}$ to be the determinant of this matrix.

(2) Let $\alpha \in K$ and consider the subfield $\mathbb{Q}_p(\alpha)$. Let $r = [K : \mathbb{Q}_p(\alpha)]$, $T(\alpha, \mathbb{Q}_p)$ be the minimal polynomial of α over $\mathbb{Q}_p$ and let $n = \deg(T(\alpha, \mathbb{Q}_p))$. Then the norm is defined as

$$\mathbf{N}_{K/\mathbb{Q}_p}(\alpha) = (-1)^{nr} a_0^r,$$

where $T(\alpha, \mathbb{Q}_p) = a_n x^n + a_{n-1}x^{n-1} + \cdots + a_1 x + a_0$.

(3) Suppose that K is a normal extension of $\mathbb{Q}_p$. Let $G(K/\mathbb{Q}_p)$ be the Galois group of this extension. Then, for $\alpha \in K$, the norm is defined as

$$\mathbf{N}_{K/\mathbb{Q}_p}(\alpha) = \prod \sigma(\alpha), \quad \text{for all } \sigma \in G(K/\mathbb{Q}_p).$$

Observe that $|G(K/\mathbb{Q}_p)| = [K : \mathbb{Q}_p]$, because K is a normal extension of $\mathbb{Q}_p$ and $\mathbb{Q}_p$ is of characteristic zero.

Example 1.60. Let ε be an element in $\mathbb{Q}_p$ such that $\sqrt{\varepsilon} \notin \mathbb{Q}_p$. Consider the quadratic extension $K = \mathbb{Q}_p(\sqrt{\varepsilon})$. Then $[K : \mathbb{Q}_p] = 2$ and $\{1, \sqrt{\varepsilon}\}$ is a basis for K over $\mathbb{Q}_p$, that is, each element in K can be written in the form $a + b\sqrt{\varepsilon}$, where $a, b \in \mathbb{Q}_p$.

(1) The linear map $x \mapsto (a + b\sqrt{\varepsilon})x$ maps 1 to $a + b\sqrt{\varepsilon}$, and $\sqrt{\varepsilon}$ to $\varepsilon b + a\sqrt{\varepsilon}$, so its matrix with respect to the basis $\{1, \sqrt{\varepsilon}\}$ is

$$\mathsf{M} = \begin{pmatrix} a & \varepsilon b \\ b & a \end{pmatrix}.$$

Therefore, $\mathbf{N}_{K/\mathbb{Q}_p}(a + b\sqrt{\varepsilon}) = \det(\mathsf{M}) = a^2 - \varepsilon b^2$.

(2) If $\alpha = a + b\sqrt{\varepsilon}$ then $r = 1$, and if $\alpha = a$ then $r = 2$. In the case $r = 2$ we have $T(\alpha, \mathbb{Q}_p) = x - a$, and the norm is $(-1)^{1 \cdot 2} a^2 = a^2$. In the case $r = 1$, the irreducible polynomial for α over $\mathbb{Q}_p$ must be of degree two. Since $(a + b\sqrt{\varepsilon})^2 = a^2 + \varepsilon b^2 + 2ab\sqrt{\varepsilon}$ is equivalent with

$$(a + b\sqrt{\varepsilon})^2 - 2a(a + b\sqrt{\varepsilon}) + (a^2 - \varepsilon b^2) = 0,$$

we must have that $T(\alpha, \mathbb{Q}_p) = x^2 - 2ax + (a^2 - \varepsilon b^2)$, and the norm is equal to $(-1)^{2 \cdot 1}(a^2 - \varepsilon b^2)^1 = a^2 - \varepsilon b^2$. Hence $\mathbf{N}_{K/\mathbb{Q}_p}(a + b\sqrt{\varepsilon}) = a^2 - \varepsilon b^2$, either if b is equal to zero or not.

(3) Since $|G(K/\mathbb{Q}_p)| = [K : \mathbb{Q}_p] = 2$, there exist two $\mathbb{Q}_p$-automorphisms:

$$\iota : a + b\sqrt{\varepsilon} \mapsto a + b\sqrt{\varepsilon} \quad \text{and} \quad \sigma : a + b\sqrt{\varepsilon} \mapsto a - b\sqrt{\varepsilon},$$

and $\mathbf{N}_{K/\mathbb{Q}_p}(a + b\sqrt{\varepsilon}) = \iota(a + b\sqrt{\varepsilon})\sigma(a + b\sqrt{\varepsilon}) = a^2 - \varepsilon b^2$.

Theorem 1.61. *Let K be a finite extension of $\mathbb{Q}_p$ and $n = [K:\mathbb{Q}_p]$. Then the function $|\cdot|\colon K \to \mathbb{R}^+$ defined by*

$$|x| = \sqrt[n]{|N_{K/\mathbb{Q}_p}(x)|_p}$$

is a non-Archimedean valuation on K that extends $|\cdot|_p$.

Since $|\cdot|$ is unique, $|\cdot|_p$ can also be used to denote the extended p-adic valuation. From algebra we know that for each finite extension K of $\mathbb{Q}_p$ there exists a finite normal extension of $\mathbb{Q}_p$ which contains K. The smallest such normal extension of $\mathbb{Q}_p$ is called the *normal closure* of $\mathbb{Q}_p$ over K. If K is not a normal extension of $\mathbb{Q}_p$ and we want to define a norm by using $\mathbb{Q}_p$-automorphisms, then we consider the normal closure of $\mathbb{Q}_p$ over K and use the third definition of the norm.

Let $x \in K$ and let $|x|_p = p^{-t}$. We set $\operatorname{ord}_p x = t$. Thus by definition:

$$|x|_p = p^{-\operatorname{ord}_p x}.$$

Let K be a finite field extension of $\mathbb{Q}_p$ and $n = [K:\mathbb{Q}_p]$. For $x \in K$ set $y = \mathbf{N}_{K/\mathbb{Q}_p}(x)$. Then we have by Theorem 1.61 that

$$|x|_p = \sqrt[n]{|y|_p} = \sqrt[n]{p^{-\operatorname{ord}_p y}} = p^{-\operatorname{ord}_p y/n} = p^{-\operatorname{ord}_p x},$$

where $\operatorname{ord}_p x = \operatorname{ord}_p y/n$, that is, $\operatorname{ord}_p x \in \frac{1}{n}\mathbb{Z}$, because $\operatorname{ord}_p y \in \mathbb{Z}$.

If $a, b \in K$ then $\operatorname{ord}_p ab = \operatorname{ord}_p a + \operatorname{ord}_p b$. This gives that ord_p is a homomorphism from the multiplicative group $K^\times$ to the additive group $\mathbb{Q}$. Then the image $\operatorname{Im}(\operatorname{ord}_p)$ is an additive subgroup of $\mathbb{Q}$, and $\operatorname{Im}(\operatorname{ord}_p) \subseteq \frac{1}{n}\mathbb{Z}$. Let d/e be in $\operatorname{Im}(\operatorname{ord}_p)$, where d and e are relatively prime, chosen so that the denominator e is the largest possible. This choice can be done because e has to be a divisor of n, and the set of possible divisors is bounded. Since d and e are relatively prime, there must be a multiple of d which is congruent to 1 modulo e, that is, we can find r and s such that $rd = 1 + se$. But then

$$r\frac{d}{e} = \frac{1+se}{e} = \frac{1}{e} + s$$

is in $\operatorname{Im}(\operatorname{ord}_p)$. Since $s \in \mathbb{Z} \subset \frac{1}{n}\mathbb{Z}$, it follows that $1/e \in \operatorname{Im}(\operatorname{ord}_p)$. Since e was chosen to be the largest possible denominator in $\operatorname{Im}(\operatorname{ord}_p)$, it follows that $\operatorname{Im}(\operatorname{ord}_p) = \frac{1}{e}\mathbb{Z}$. This unique positive integer e is called the *ramification index* of K over $\mathbb{Q}_p$. The extension K over $\mathbb{Q}_p$ is called *unramified* if $e = 1$, *ramified* if $e > 1$ and *totally ramified* if $e = n$.

Definition 1.62. We say that an element $\pi \in K$ is a uniformizer if $\operatorname{ord}_p \pi = 1/e$.

We call the set

$$\mathcal{O}_K = \{x \in K : |x| \leqslant 1\}$$

the *valuation ring* of K. The set

$$\mathcal{P}_K = \{x \in K : |x| < 1\}$$

is its maximal ideal. Since $\mathcal{O}_K$ is a local ring (this means that it has a unique maximal ideal) all the elements of $\mathcal{O}_K \setminus \mathcal{P}_K$ are units (invertible elements) of $\mathcal{O}_K$. The quotient ring $\mathcal{O}_K/\mathcal{P}_K$ is a field (because $\mathcal{P}_K$ was maximal). We call it the *residue class field of* K.

The set of units in $\mathcal{O}_K$ are denoted by $\mathcal{O}_K^\times$ and it is equal to the unit sphere (in K) with center in zero $S_1(0, K)$:

$$S_1(0, K) = \mathcal{O}_K^\times.$$

The *valuation group* is

$$\mathcal{V}_K = \{|x|_p : x \in K \setminus \{0\}\}.$$

We state a few facts about the extension K:

- K is locally compact and complete.
- Each $x \in K$ can be written as $x = u\pi^{v_\pi(x)}$, where $u \in \mathcal{O}_K^\times$ and $v_\pi(x) = \frac{\operatorname{ord}_p x}{e}$.
- The degree of $\mathbb{K}$ as a field extension of $\mathbb{F}_p$ (the residue class field of $\mathbb{Q}_p$ is isomorphic to $\mathbb{F}_p$) is $f = m/e$. Hence $\mathbb{K} = \mathbb{F}_{p^f}$.
- The multiplicative group $\mathbb{K}^*$ is cyclic and it has $p^f - 1$ elements.
- Let $C = \{c_0, c_1, \ldots, c_{p^f-1}\}$ be a fixed complete set of representatives of the cosets of $\mathcal{P}_K$ in $\mathcal{O}_K$. Then every $x \in K$ has a unique π-adic expansion of the form
$$x = \sum_{i \geqslant i_0} a_i \pi^i,$$
where $i_0 \in \mathbb{Z}$ and $a_i \in C$ for every $i \geqslant i_0$.

1.8.2 The algebraic closure of $\mathbb{Q}_p$

We now want to construct a field that contains all zeros of all polynomials over $\mathbb{Q}_p$.

Definition 1.63. Let K be a field. If every polynomial in $K[x]$ has a zero in K then K is said to be *algebraically closed*. If K is a field extension of L and K is algebraically closed then K is said to be an *algebraic closure* of L: $K = \bar{L}$.

Let U be the union of all finite extensions of $\mathbb{Q}_p$. It can be proven that it is an algebraic closure of $\mathbb{Q}_p$, that is $U = \overline{\mathbb{Q}_p}$. If $x \in \overline{\mathbb{Q}_p}$ then x belongs to the finite extension $\mathbb{Q}_p(x)$. We can define $|x|$ by using the unique extension of the p-adic absolute value to $\mathbb{Q}_p(x)$. It can be shown that the absolute value does not depend on the field we take it in. Therefore, it makes sense to say that it is the absolute value of $x \in \overline{\mathbb{Q}_p}$. So, we have extended the p-adic absolute value to $\overline{\mathbb{Q}_p}$. The image of $\overline{\mathbb{Q}_p} \setminus \{0\}$ under the extended p-adic valuation is $\mathbb{Q}$. In other words, the possible positive absolute values are p^r, where $r \in \mathbb{Q}$.

The algebraic closure $\overline{\mathbb{Q}}_p$ of $\mathbb{Q}_p$ is an infinite extension, this follows from the fact that there exist irreducible polynomials of any degree over $\mathbb{Q}_p$. See [157] or [371] for details.

1.8.3 Complex p-adic numbers

Unfortunately, $\overline{\mathbb{Q}}_p$ is not complete with the metric induced by the extended p-adic absolute value. We complete $\overline{\mathbb{Q}}_p$ and obtain a new field $\mathbb{C}_p$ which is algebraically closed. The latter fact is Krasner's theorem. We are lucky that in the p-adic case by completing the algebraic closure we again obtain an algebraically closed field. In principle it might occur that the completion is not algebraically closed. So the process "algebraic closure $\to$ completion $\to$ algebraic closure $\to$ completion $\to \ldots$" might have many (or even infinitely many) steps. But by Krasner's theorem this process has only one step.

We call $\mathbb{C}_p$ the *complex p-adic numbers*. We sum up some more facts about $\mathbb{C}_p$:

- The possible positive absolute values of the elements of $\mathbb{C}_p$ is p^r, where $r \in \mathbb{Q}$.
- The field $\mathbb{C}_p$ is algebraically closed (Krasner's theorem).
- The field $\mathbb{C}_p$ is not locally compact.

As we can see, there is a great difference between the real and the p-adic case. The algebraic closure of $\mathbb{R}$ is $\mathbb{C}$, that is, an extension of degree 2. The field $\mathbb{C}$ is complete with respect to the ordinary absolute value. The algebraic closure of $\mathbb{Q}_p$ is an infinite extension of $\mathbb{Q}_p$, that is, not complete.

1.8.4 Krasner's lemma

The following theorem gives us some information about the internal structure of an algebraically closed non-Archimedean field.

Theorem 1.64 (Krasner's lemma). *Let K be a complete non-Archimedean field of characteristic zero. Let x and y be elements in the algebraic closure of K and let $x_1, x_2, \ldots, x_n$ be the conjugates of x (different from x) over K. If*

$$|x - y|_p < |x - x_i|_p \quad \text{for } 1 \leqslant i \leqslant n,$$

then $K(x) \subseteq K(y)$.

Part I

The Commutative Non-Archimedean Dynamics

Chapter 2

Dynamics on algebraic structures

In this chapter we consider dynamics on commutative algebraic structures, groups and rings, and explain how these dynamics relate to p-adic dynamics.

2.1 Basic notions of dynamics

Usually a *dynamical system on a measurable space* $\mathbb{S}$ is understood as a triple $(\mathbb{S}; \mu; f)$, where $\mathbb{S}$ is a set endowed with a measure μ, and

$$f : \mathbb{S} \to \mathbb{S}$$

is a *measurable function*; that is, an f-preimage of any measurable subset is a measurable subset. Basic definitions from dynamical system theory, as well as the ones from the theory of uniform distribution of sequences, can be found in [276]; see also [183] as a comprehensive monograph on various aspects of dynamical systems theory.

A *trajectory* of the dynamical system is a sequence

$$x_0, x_1 = f(x_0), \dots, x_i = f(x_{i-1}) = f^i(x_0), \dots$$

of points of the space $\mathbb{S}$, x_0 is called an *initial* point of the trajectory. If $F : \mathbb{S} \to \mathbb{T}$ is a measurable mapping to some other measurable space $\mathbb{T}$ with a measure ν (that is, if an F-preimage of any ν-measurable subset of $\mathbb{T}$ is a μ-measurable subset of X), the sequence $F(x_0), F(x_1), F(x_2), \dots$ is called an *observable*.

A mapping $F : \mathbb{S} \to \mathbb{Y}$ of a measurable space $\mathbb{S}$ into a measurable space $\mathbb{Y}$ endowed with probabilistic measure μ and ν, respectively, is said to be *measure-preserving* whenever $\mu(F^{-1}(S)) = \nu(S)$ for each measurable subset $S \subset \mathbb{Y}$. In case $\mathbb{S} = \mathbb{Y}$ and $\mu = \nu$, a measure preserving mapping F is said to be *ergodic* whenever for each measurable subset S such that $F^{-1}(S) = S$ holds either $\mu(S) = 1$ or $\mu(S) = 0$.

2.1.1 Ergodicity and uniform distribution of sequences

Let $\mathbb{A}$ be a compact topological group[1], and let μ be its Haar measure. We assume that the Haar measure is normalized, so that it takes values in a real interval [0, 1]. Thus, the Haar measure μ is a natural probabilistic measure on $\mathbb{A}$.

[1] a group endowed with a topology where all group operations are continuous

Let, further, $\{a_n\}_{n=0}^{\infty}$ be a sequence of elements of $\mathbb{A}$, let N be a nonnegative rational integer and let U be a subset of $\mathbb{A}$. Put $\nu_N(U) = \sum_{n=0}^{N-1} \chi_U(a_n)$, where χ_U is a characteristic function of the subset U; that is, $\chi_U(a) = 1$ if and only if $a \in U$, and $\chi_U(a) = 0$ otherwise. In other words, $\nu_N(U)$ is the number of terms of a finite subsequence $(a_n)_{n=0}^{N-1}$ that lie in U.

Definition 2.1 ([276]). The sequence $(a_n)_{n=0}^{\infty}$ is called *uniformly distributed* (with respect to the measure μ) whenever

$$\liminf_{N\to\infty} \frac{\nu_N(U)}{N} \geq \mu(U)$$

for all open subsets $U \subseteq \mathbb{A}$ (equivalently, if

$$\limsup_{N\to\infty} \frac{\nu_N(U)}{N} \leq \mu(U)$$

for all closed subsets $U \subseteq \mathbb{A}$.) An equivalent form of the definition yields:

$$\lim_{N\to\infty} \frac{\nu_N(U)}{N} = \mu(U)$$

for all Borel sets $U \subseteq \mathbb{A}$ such that $\mu(\mathrm{cl}(U) \setminus \mathrm{Int}\,(U)) = \varnothing$, where $\mathrm{Int}\,(U)$ is the union of all open subsets of U, and $\mathrm{cl}(U)$ is the closure of U.

For instance, a sequence $(s_i)_{i=0}^{\infty}$ of p-adic integers is uniformly distributed (with respect to the normalized Haar measure μ_p on $\mathbb{Z}_p$) if and only if it is *uniformly distributed modulo* p^k for all $k = 1, 2, \ldots$. That is, for every $a \in \{0, 1, \ldots, p^k - 1\}$ relative numbers of occurrences of a in the initial segment of length N in the sequence $(s_i \bmod p^k)_{i=0}^{\infty}$ of residues modulo p^k are asymptotically equal; i.e.,

$$\lim_{N\to\infty} \frac{\nu_N(a)}{N} = \frac{1}{p^k},$$

where $\nu_N(a) = \#\{s_i \equiv a \pmod{p^k} : i < N\}$, see [276] for details. Note that $\nu_N(a) = \nu_N(a + p^k\mathbb{Z}_p)$, the number of occurrences of elements of the ball $a + p^k\mathbb{Z}_p$ among the first N terms of the sequence $(s_i)_{i=0}^{\infty}$. Obviously, in the definition of m-dimensional uniformly distributed sequences $(a_n \in \mathbb{Z}_p^m)_{n=0}^{\infty}$ the above equation should be replaced by

$$\lim_{N\to\infty} \frac{\nu_N(a + p^k\mathbb{Z}_p^m)}{N} = p^{-km}.$$

In the sequel, measure-preserving and ergodic mappings will serve us as a tool to construct uniformly distributed sequences for various applied purposes, see e.g. Chapter 9. In these applications we actually use the following basic result of ergodic theory (see e.g. [276, Chapter 3: Definition 1.1, Exercise 1.10, Lemma 2.2]).

Proposition 2.2. *Let $\mathbb{S}$ and $\mathbb{T}$ be compact topological groups, let $f : \mathbb{S} \to \mathbb{T}$ be a mapping that is continuous and measurable with respect to the Haar measure. If $(a_n)_{n=0}^{\infty}$ is a uniformly distributed sequence over $\mathbb{S}$ and f is measure-preserving, then the sequence $(f(a_n))_{n=0}^{\infty}$ is uniformly distributed over $\mathbb{T}$.*

If additionally $\mathbb{S} = \mathbb{T}$, f is ergodic, and $\mathbb{S}$ is separable[2]*, then the sequence $(f^n(a))_{n=0}^{\infty}$ is uniformly distributed for almost all $a \in S$.*

2.2 Dynamics on finite algebraic structures

Actually in real life settings we usually deal with dynamical systems on finite sets; that is, when the order $\#\mathbb{A}$ of the group $\mathbb{A}$ is finite. Then every subset U of $\mathbb{A}$ is open and closed simultaneously, and $\mu(U) = \#U \cdot (\#\mathbb{A})^{-1}$. The uniform distribution of a sequence $\{a_n\}_{n=0}^{\infty}$ in this particular case implies that

$$\lim_{N\to\infty} \frac{\nu_N(U)}{N} = \frac{\#U}{\#\mathbb{A}}$$

for each subset $U \subseteq \mathbb{A}$. Moreover, if groups $\mathbb{A}$ and $\mathbb{B}$ are of finite order, then the mapping $f : \mathbb{A} \to \mathbb{B}$ is measure-preserving if and only if $\#f^{-1}(a) = \#f^{-1}(b)$ for all $a, b \in A$. Such mappings are called *balanced*. Obviously, the mapping $f : \mathbb{A} \to \mathbb{A}$ preserves measure μ if and only if it is bijective; that is, f is a permutation on $\mathbb{A}$. Finally, f is ergodic if and only if this permutation has only one cycle of length $\#\mathbb{A}$. In the latter case we say that f is *transitive* on $\mathbb{A}$. Note that whenever f is transitive, the corresponding trajectory is just a periodic sequence, and its shortest period is of length $\#\mathbb{A}$; that is, every element from $\mathbb{A}$ occurs at the period exactly once. We call these sequences *strictly* uniformly distributed.

2.2.1 Hereditary dynamical properties and compatibility

Let A be a universal algebra (e.g., a group, or a ring), let $f : A \to A$ be a compatible mapping. Let $\varphi\colon A \to B$ be any epimorphism of the universal algebra A onto a universal algebra B of the same kind, and let $x, y \in A$ be arbitrary elements of A such that their φ-images coincide, $\varphi(x) = \varphi(y)$. Then $\varphi(f(x)) = \varphi(f(y))$ since f is compatible. Thus, the mapping $f\varphi\colon B \to B$ defined as $(f\varphi)(b) = \varphi(f(a))$ for $b \in B$, $a \in \varphi^{-1}(b)$, is well defined. So each compatible transformation on A defines a unique transformation on each epimorphic image of A. As each epimorphism of A defines a unique congruence of A and vice versa, we say that f *possesses some property* $\mathfrak{P}$ *modulo congruence* η if the mapping induced by f on the corresponding epimorphic image possesses $\mathfrak{P}$. The following easy proposition holds:

[2] that is, contains a countable dense subset

Proposition 2.3. *Let A be a finite group, let η be a congruence of A, and let $F: A^n \to A^m$* (where $m \leq n$) *be a balanced* (*resp., bijective, transitive*) *compatible mapping of the nth Cartesian power A^n onto the mth Cartesian power A^m of the group A. Then F is balanced* (*resp., bijective, transitive*) *modulo η. If H is a kernel of the congruence η, $k = |A : H|$, then the mapping $F: A^n \to A^n$ is transitive if and only if F is transitive modulo η and the iterated mapping $F^{kn}: H^n \to H^n$ is transitive on H^n.*

Moreover, if A is a direct product of groups B and C, $A = B \times C$, then F is balanced on A if and only if F is balanced both on B and C, i.e., modulo each congruence corresponding to a projection onto a direct factor. Finally, the mapping $F: A \to A$ is transitive if and only if it is transitive both on B and C and orders $\#B$ and $\#C$ are coprime.

Proof. Since H is a kernel of the congruence η, H is a normal subgroup which is a kernel of a canonical epimorphism of A onto a factor-group $A/\eta = A/H$. Denote by $+$ the group operation of the group A (which needs not be necessarily commutative). Choose an arbitrary element $c \in A^m$ and consider the following inclusion:

$$F(x_1 + H, \dots, x_n + H) \subseteq c + H^m. \tag{2.1}$$

Choose an arbitrary system $\mathcal{S} \subseteq H^{(n)}$ of elements which contains one and only one element of each coset $h + H^n$. Let t be a number of elements of $\mathcal{S}$ which satisfy (2.1). Consider an inclusion

$$F(a_1, \dots, a_n) \in c + H^m. \tag{2.2}$$

If $\mathbf{x} = (x_1, \dots, x_n) \in S$ and if $\mathbf{x}$ satisfies (2.1), then each element $(a_1, \dots, a_n)$ which lies in the coset $(x_1, \dots, x_n) + H^n$, satisfies (2.2) since F is compatible. Thus, the number of elements of A^n that satisfy (2.2) is exactly $t \cdot \#H^n$.

On the other hand, let F be balanced. Then for each $d \in c + H^m$ the equation $F(a_1, \dots, a_n) = d$ has exactly $\#A^{n-m}$ solutions in A^n and consequently there exist exactly $\#A^{n-m} \cdot \#H^m$ elements of A^n that satisfy (2.2). In view of the argument above this implies that $\#A^{n-m} \cdot \#H^m = t \cdot \#H^n$. Hence, $t = \#(A/H)^{n-m}$. Thus, t does not depend on the choice of c and, consequently, F induces a balanced mapping of a factor-group $(A/H)^n$ onto a factor-group $(A/H)^m$. The rest of the proof is quite obvious and we omit it. □

Surprisingly, it turns out that to describe dynamics on a finite set we often have to study dynamics on infinite spaces; for instance, there exist deep connections between measure-preservation and ergodicity on $\mathbb{Z}_p$ on the one hand, and measure preservation and ergodicity *modulo* p^k on the other hand. Loosely speaking, certain dynamics on the space $\mathbb{Z}_p$, which is a continuum, is totally determined by dynamics on finite residue rings $\mathbb{Z}/p^k\mathbb{Z}$, and vice versa. We postpone these considerations as well as exact statements till Section 4.4.

The most "natural" compatible transformation of a universal algebra is a polynomial transformation. However, ergodic polynomials (i.e, polynomials that induce ergodic transformations on the universal algebra) exist not over every universal algebra. Actually, the existence of ergodic polynomial imposes strict limitations on the structure of a universal algebra. As ergodicity is the leading theme of the book, we first introduce some important examples of universal algebras *having ergodic polynomials*; i.e., of algebras such that there exist polynomials over these algebras that induce ergodic transformations on these algebras. In this section, we consider only finite universal algebras; now we describe finite Abelian groups with operators and finite commutative rings that admit of ergodic (whence, transitive) polynomials. A similar problem for finite non-Abelian groups is much more complicated, and we postpone it until Part II.

2.2.2 Ergodic polynomial transformations on finite Abelian groups with operators

Let G be a finite Abelian group with operation $+$ written additively, let Ω be a set of operators on G; that is, every element $\omega \in \Omega$ induces an endomorphism of the group G: $(a+b)^{\omega} = a^{\omega} + b^{\omega}$ for all $a, b \in G$. It is clear that as the group G is Abelian, any ergodic (i.e., transitive) polynomial transformation must be of the form $x \mapsto a + x^{\alpha}$, where α lies in the ring $\operatorname{Env}\Omega$ generated by endomorphisms of G induced by operators from Ω, and moreover, that α must be an automorphism of G. Recall that as G is Abelian, all its endomorphisms form a ring with respect to addition and multiplication (i.e., composition) of endomorphisms. That is, finite Abelian groups having ergodic polynomials are exactly finite Abelian groups having transitive *affine* transformations $x \mapsto a + x^{\alpha}$. Groups having transitive affine transformations were studied in [179], under the name of *single orbit groups*. We summarize results from [179] concerning Abelian groups (with operators) that have transitive polynomials, in the following theorem:

Theorem 2.4. *A finite Abelian group G with a set of operators Ω has ergodic polynomials if and only if G is isomorphic to one of the following groups*:

(1) *A cyclic group $\mathrm{C}(m)$, $m = 1, 2, \ldots$, with arbitrary set of operators Ω.*

(2) *The Klein group K_4 with $\Omega \ni \omega$ inducing a non-identity involution on K_4.*

(3) *A direct product of a group of type 2 by a group of type 1 of odd order.*

Note 2.5. As the Klein group K_4 is isomorphic to the additive group of a 2-dimensional vector space over $\mathbb{F}_2$, it is not difficult to prove that the affine transformation $x \mapsto a + x^{\psi}$ on the Klein group K_4 ($a \in \mathrm{K}_4$, $\psi \in \operatorname{End}(\mathrm{K}_4)$) is transitive on K_4 if and only if ψ is a non-identity automorphism whose square $\psi^2 = \psi \circ \psi$ is an identity automorphism, and $a^{\psi} \neq a$.

As every endomorphism of the cyclic group $\mathrm{C}(n)$ (written additively) is a multiplication by m, all affine transformations of $\mathrm{C}(n)$ are in fact transformations of the form $x \mapsto (a + mx) \bmod n$ of the residue ring $\mathbb{Z}/n\mathbb{Z}$ modulo n. Thus, in view of the

Chinese Remainder Theorem 1.1 and Proposition 2.3 to characterize transitive transformations of this form, it suffices to consider only the case when n is a power of a prime. Theorem 4.36 (and Lemma 4.37) actually completely describe transitive affine transformations of residue rings $\mathbb{Z}/p^k\mathbb{Z}$, p prime, in force of Theorem 4.23.

All these results, in view of Proposition 2.3, give us a complete description of all finite Abelian groups (with operators) having transitive polynomials, as well as transitive polynomial transformations themselves, in explicit forms. Starting at this point, we can try to expand these considerations in two directions: First, to the case of non-Abelian groups, and second, to the case of other commutative universal algebras; the most important of the latter are commutative rings. We deal with ergodic polynomial transformations on non-Abelian groups in Part II of the book; we consider commutative rings having transitive polynomials in the next subsection. As we shall see, in both cases the problem of description of corresponding ergodic transformations will inevitably lead us to the non-Archimedean dynamics.

2.2.3 Ergodic polynomial transformations on finite commutative rings

Now we are going to demonstrate that residue rings and finite fields are, loosely speaking, the only 'interesting' finite commutative rings that have polynomial ergodic transformations; that is, for most applied areas we restrict ourselves to dynamics on residue rings or finite fields rather than on more exotic rings. However, polynomial dynamics on residue rings can be naturally 'raised' to dynamics on the ring $\mathbb{Z}_p$ of p-adic integers as the latter ring is an inverse limit of residue rings $\mathbb{Z}/p^n\mathbb{Z}$, $n = 1, 2, \ldots$.

Let R be a finite commutative ring with identity 1 (i.e., 1 is a multiplicative neutral element of R). Existence of univariate transitive polynomials over R significantly restricts the structure of R:

Proposition 2.6. *Whenever R has transitive polynomials, R is a principal ideal ring.*

Proof. Indeed, let I be a non-zero ideal in R of index n (i.e., $n = \#(R/I)$), and let $f(x) \in R[x]$ be a transitive polynomial over R. Then, as the transformation $z \mapsto f^n(z)$ is transitive on I, every element z from I can be represented as $z = f^{kn}(0)$ for a suitable $k \in \mathbb{N}_0$. That is, z is a linear combination (with coefficients from R) of powers of the element $f^n(0)$. Hence, $I = f^n(0) \cdot R$; i.e., I is generated by the constant term of the polynomial $f^n(x)$. □

Proposition 2.6 shows that whenever R has a transitive polynomial, R is a direct sum of local principal ideal rings, see Subsection 1.2.3. That is, every direct summand is either a field or a ring that has a unique maximal non-zero ideal, a radical of the ring. By Proposition 2.3, the ring R has a transitive polynomial if and only if every direct summand has a transitive polynomial, and orders of direct summands are pairwise coprime. From Subsection 1.2.3 we know that every finite field is polynomially complete; in particular, every finite field has transitive polynomials. Thus, to characterize finite commutative rings that have transitive polynomials it suffices to restrict ourselves to finite local rings whose radicals are non-zero.

Theorem 2.7 ([19]). *A local ring R has transitive polynomials if and only if one of the following alternatives holds*[3]:

(1) $R = \mathbb{F}_{p^n}$, *a field of p^n elements, $n = 1, 2, \ldots$;*
(2) $R = \mathbb{Z}/p^n\mathbb{Z}$, *a residue ring modulo p^n, p prime, $n = 1, 2, \ldots$;*
(3) $R = \mathbb{F}_p[x]/x^2 \cdot \mathbb{F}_p[x]$, *$p$ prime;*
(4) $R = \mathbb{F}_p[x]/x^3 \cdot \mathbb{F}_p[x]$, *$p \in \{2, 3\}$;*
(5) $R = \mathbb{Z}[x]/p^2 \cdot \mathbb{Z}[x] + x^3 \cdot \mathbb{Z}[x] + (x^2 - p) \cdot \mathbb{Z}[x]$, *$p \in \{2, 3\}$;*
(6) $R = \mathbb{Z}[x]/9 \cdot \mathbb{Z}[x] + x^3 \cdot \mathbb{Z}[x] + (x^2 + 3) \cdot \mathbb{Z}[x]$.

Note 2.8. It is obvious that the ring $R = \mathbb{Z}[x]/p^2 \cdot \mathbb{Z}[x] + x^3 \cdot \mathbb{Z}[x] + (x^2 - p) \cdot \mathbb{Z}[x]$ is a factor ring of the ring of polynomials in variable x over the residue ring $\mathbb{Z}/p^2\mathbb{Z}$, modulo the ideal generated by two polynomials, x^3 and $x^2 - p$. That is, the order of this ring R is p^3.

In a similar manner, it is easy to demonstrate that the ring $R = \mathbb{Z}[x]/9 \cdot \mathbb{Z}[x] + x^3 \cdot \mathbb{Z}[x] + (x^2 + 3) \cdot \mathbb{Z}[x]$ is a factor ring of the ring of polynomials in variable x over the residue ring $\mathbb{Z}/9\mathbb{Z}$, modulo the ideal generated by two polynomials, x^3 and $x^2 + 3$. That is, the order of this ring R is 27.

To prove Theorem 2.7, we need the following lemma.

Lemma 2.9. *Let a finite local ring R have transitive polynomials; let I be an ideal of R, and let the nilpotent index*[4] $\operatorname{ind} I$ *of I be 2. Then the additive subgroup I^+ of I is isomorphic either to a cyclic p-group for some prime p, or to the Klein group* $\mathrm{K}_4 = \mathrm{C}(2) \times \mathrm{C}(2)$ *of order 4.*

Proof. Let $f(x) \in R[x]$ be a transitive polynomial on R. As f induces a compatible transformation on R, f maps every coset with respect to some ideal onto a coset with respect to the same ideal; in particular, $f(a + I) = f(a) + I$ for all $a \in R$. From here it follows that if $k = \#R/I$, then the kth iterate $f^k(x)$ of the polynomial f induces a transitive transformation on I. As $I^2 = \{0\}$, then the mentioned transformation (which is itself a polynomial over R) must be of the form $z \mapsto a + bz$, for suitable $a \in I$, $b \in R$. As a multiplication by b is an endomorphism of the additive group I^+, the group I^+ satisfies the conditions of Theorem 2.4. However, $\#I^+ \mid \#R$ and $\#R = \#F^{\operatorname{ind} J(R)}$, where $J(R)$ is a radical (i.e., a unique maximal ideal) of R, and $F = R/J(R)$ is a residue field of R, see Subsection 1.2.3. Hence, I^+ is a p-group, where $p = \operatorname{char} F$, and the conclusion follows. □

[3] We characterize rings up to isomorphisms.

[4] That is, the smallest $k \in \mathbb{N}$ such that $I^k = \{0\}$; recall that we have by definition $I^k = \{a_1 \cdots a_k : a_1, \ldots, a_k \in R\}$.

Proof of Theorem 2.7. We start with a proof that the conditions of the theorem are necessary. Let $f(x) \in R[x]$ be an ergodic polynomial over a local ring R. Denote $J = J(R)$, a radical of R. According to the note that precedes the statement of Theorem 2.7, we may assume that R is a local ring with a non-zero radical J. In this case, the following claim is true:

Claim 1: The residue field $F = R/J(R)$ is prime; i.e., $F = \mathbb{F}_p$ for some prime p.

To prove the claim, we may assume that $\operatorname{ind} J = 2$; otherwise consider a factor-ring $\bar{R} = R/J^2$, which has the same residue field as R, has ergodic polynomial by Proposition 2.3, and $\operatorname{ind} J(\bar{R}) = 2$. Under this assumption, we can consider J as a module over F, whence, as a vector space over the field F. By Proposition 2.6, the ideal J is principal; whence, the dimension of this vector space is 1. That is, $J = \{r\xi : r \in F\}$. However, there exists a transitive transformation on J of the form $z \mapsto a+bz$, $a \in J, b \in R$ (see the proof of Lemma 2.9). As $a = \bar{a}\xi$, $z = u\xi$, $bz = \bar{b}u\xi$ for some $\bar{a}, \bar{b}, u \in F$, the transformation $\tau : u \mapsto \bar{a}+\bar{b}u$ must be transitive on F. Note that then $\bar{a} \neq 0$. Moreover, it is clear that $\tau^i(0) = \bar{a} \cdot (1+\bar{b}+\cdots+\bar{b}^{i-1})$, for all $i = 1, 2, \ldots$. Now, assuming that $\bar{b} \neq 1$, we have that $\tau^i(0) = \bar{a} \cdot (\bar{b}-1)^{-1} \cdot (\bar{b}^i - 1)$, for all $i = 0, 1, 2, \ldots$. From here, putting $i = q = \#F$, we conclude that $\tau^q(0) = \bar{a} \neq 0$ since $\bar{b}^q = \bar{b}$, see Subsection 1.3.1. However, this contradicts the transitivity of τ, as the latter obviously implies that $\tau^q(0) = 0$. So, necessarily $\bar{b} = 1$; but then $\tau^l(0) = i\bar{a}$, and thus $\tau^p(0) = 0$, where $p = \operatorname{char} F$. That is, necessarily $q = p$ since τ is transitive on F.

Claim 2: If $p = \operatorname{char} F$ is odd and if $\operatorname{ind} J \geq 4$, then the additive group $(J^2)^+$ of the ideal J^2 is cyclic.

We shall prove the claim by induction on $\operatorname{ind} J$. If $\operatorname{ind} J = 4$ then $\operatorname{ind} J^2 = 2$, so $(J^2)^+$ is cyclic by Lemma 2.9. Now let the claim be true if $\operatorname{ind} J < n$; let us prove that then it is true if $\operatorname{ind} J = n$, $n > 4$.

Assume that $(J^2)^+$ is not a cyclic group. This assumption implies that then $(J^2)^+$ is a direct sum of two cyclic groups: of the group J^{n-1} of order p, and of the cyclic group of order p^{n-3}. Indeed, in view of Claim 1, $\#J^{n-1} = p$ (see Subsection 1.2.3), so $(J^{n-1})^+$ is a cyclic group. The group $(J^2/J^{n-1})^+$ is a cyclic group by induction hypothesis, and $\#(J^2/J^{n-1})^+ = p^{n-3}$, as it easily follows from Claim 1 and relevant results mentioned in Subsection 1.2.3. Now take $a \in J^2$ so that the coset $a + J^{n-1}$ is a generator of the cyclic group $(J^2/J^{n-1})^+$. Then the additive order of a must be p^{n-3}; otherwise, if this order is greater than p^{n-3}, the group $(J^2)^+$ is cyclic as $\#(J^2)^+ = p^{n-2}$. So the additive cyclic group A generated by a have a zero intersection with $(J^{n-1})^+$, $A \cap (J^{n-1})^+ = \{0\}$, since otherwise $A \supset (J^{n-1})^+$ and whence $(J^2)^+$ is cyclic. Thus, $(J^2)^+$ is a direct product of A and of $(J^{n-1})^+$.

On the other hand, by Lemma 2.9 every group $(J^k)^+$ must be cyclic whenever $k \geq \frac{n}{2}$. Then, as it follows from Claim 1 in combination with relevant results mentioned in Subsection 1.2.3, the order of this group $(J^k)^+$ is p^{n-k}, $(J^k)^+ \supset (J^{n-1})^+$, and the latter inclusion is strict for $k < n-1$. However, the direct product $A \times (J^{n-1})^+$ contains no cyclic subgroups of order greater than p that contain $(J^{n-1})^+$ as a proper subgroup. Thus, the assumption that $(J^2)^+$ is not a cyclic group leads to a contradiction.

Claim 3: If char $F = 2$*, and if* ind $J \geq 6$*, then the additive group* $(J^3)^+$ *of the ideal* J^3 *is cyclic.*

This can be proved by a group-theoretic argument similar to that from the proof of Claim 2. We leave details to the reader.

Claim 4: If for some n the group $(J^n)^+$ *is cyclic, then either R is isomorphic to the residue ring* $\mathbb{Z}/p^k\mathbb{Z}$*, p prime, or* ind $J \leq n+1$.

Recall that we denote by 1 the identity (a unique multiplicative neutral element) of R; thus $p \cdot 1 \in R$ is a sum of p of identities 1, and we denote $p \cdot 1$ via $p \in R$. As $R/J = \mathbb{F}_p$, then $p \in J$.

Let $p \in J \setminus J^2$. Then R is isomorphic to $\mathbb{Z}/p^k\mathbb{Z}$, where $k = \operatorname{ind} J$. This can be proved by induction on ind J with the use of a standard ring-theoretic argument. Indeed, if ind $J = 2$ then $p \cdot 1 \neq 0$ since otherwise elements $0, 1, 2 \cdot 1, \ldots, (p-1) \cdot 1$ form a subfield F isomorphic to $\mathbb{F}_p$, and so R is a direct sum of F and of J; thus, R is not a local ring. Assuming the claim is true for ind $J < k$, we see that if ind $J = k$, then R/J^{k-1} is isomorphic to $\mathbb{Z}/p^{k-1}\mathbb{Z}$, and so the smallest non-zero power of p that is zero in R is at least p^{k-1}. If $p^{k-1} = 0$ in R then R is isomorphic to a direct sum of $\mathbb{Z}/p^{k-1}\mathbb{Z}$ and of J^{k-1}; whence, R is not a local ring. Thus, $p^{k-1} \neq 0$ in R; then k is the additive order of $p \in R$, so R is isomorphic to $\mathbb{Z}/p^k\mathbb{Z}$ as $\#R = p^k$, see Subsection 1.2.3.

Now let $p \in J^2$. We will show that the assumption ind $J \geq n+2$ leads to a contradiction in this case. For this purpose it suffices to assume that ind $J = n+2$ since otherwise we consider the factor ring R/J^{n+2} instead of R. But then $pJ^n \subset J^2 \cdot J^n = J^{n+2} = \{0\}$; so, as $(J^n)^+$ is cyclic by our assumption, the order of the group $(J^n)^+$ must be p. From here it follows that $(J^n)^+$ can not include (as a proper subgroup) the cyclic group $(J^{n+1})^+$, which is also of order p since $J^{n+2} = \{0\}$ and $R/J = \mathbb{F}_p$. The contradiction proves that ind $J \leq n+1$.

Finally from Claims 1–4 we deduce that *if the local ring R with a non-zero radical J(R) has transitive polynomials, then either R is isomorphic to the residue ring* $\mathbb{Z}/p^k\mathbb{Z}$*,* $k =$ ind $J(R)$*, or* ind $J(R) \leq 3$ *whenever p is odd, or* ind $J(R) \leq 5$ *whenever* $p = 2$. In other words, either R is a residue ring, or R is "small": $\#R \leq p^3$ for p odd, $\#R \leq 32$ for $p = 2$. So to conclude the proof that the conditions of Theorem 2.7 are necessary it suffices to describe the latter "small" local rings explicitly. To do this, we will use results on characterization of finite local principal ideal rings from [36, 337].

We start with the case p odd. Let ind $J = 2$, then $\#R = p^2$, and thus either R is isomorphic to $\mathbb{Z}/p^2\mathbb{Z}$, or char $R = p$. In the latter case R is isomorphic to the factor ring $\mathbb{F}_p[x]/x^2 \cdot \mathbb{F}_p[x]$ of the ring $\mathbb{F}_p[x]$ of univariate polynomials over the field $\mathbb{F}_p$ modulo the ideal generated by x^2, see [36, Theorem 3]. Thus, R is a ring of type 3 from the statement of Theorem 2.7.

Further, if $\operatorname{ind} J = 3$, then $\#R = p^3$, and thus R is either isomorphic to the residue ring $\mathbb{Z}/p^3\mathbb{Z}$ (whenever char $R = 3$), or char $R \mid p^2$. In the latter case, by the argument similar to that from the proof of Lemma 2.9 it can be shown that there exist $a_0, a_1, a_2 \in R$ such that the mapping $\gamma\colon z \mapsto a_0 + a_1 z + a_2 z^2$ is transitive on J. Then, as the mapping $\bar{\gamma}\colon z \mapsto a_0 + a_1 z$ must be transitive on J/J^2, by the argument similar to that at the end of the proof of Claim 1 it can be demonstrated that $a_1 = 1 + b$ for a suitable $b \in J$. Now by direct calculations we obtain

$$\gamma^p(0) = a_0 \cdot (p + b \cdot (1 + 2 + \cdots + (p-1)) + a_0 a_2 \cdot (1^2 + 2^2 + \cdots + (p-1)^2)). \quad (2.3)$$

From here it follows that $\gamma^p(0) = 0$ if $p > 3$: Indeed, as 2 and 6 have multiplicative inverses 2^{-1} and 6^{-1} in R in the latter case, from (2.3) we deduce that

$$\gamma^p(0) = a_0 \cdot (p + b \cdot 2^{-1} p(p-1) + a_0 a_2 \cdot 6^{-1} p(p-1)(2p-1)). \quad (2.4)$$

The equality (2.4) immediately implies that $\gamma^p(0) = 0$ in the case char $R = p$. However, in the case char $R = p^2$ necessarily $p \cdot 1 \in J^2$ (recall that 1 is the identity of R), and thus $a_0 p = 0$ as $a_0 \in J$; hence, $\gamma^p(0) = 0$ in this case as well.

But on the other hand, γ^p must be transitive on $J^2 \neq \{0\}$; so $\gamma^p(0)$ can not be 0. The contradiction shows that the only possibility remains under our restrictions, $p = 3$. In this case, if char $R = 3$ from [36, Theorem 3] we deduce that R is isomorphic to the ring $\mathbb{F}_3[x]/x^3\mathbb{Z}_3[x]$ of type 4 from the statement of the theorem we are proving. In the case when char $R = 9$ two types of rings are possible, of type 5 and 6. Indeed, as R is a principal ideal ring by Proposition 2.6, the ideal J is generated by some $a \in R$, so R is generated by a over a subring generated by 1, and the latter subring is isomorphic to $\mathbb{Z}/9\mathbb{Z}$. Then, $a^3 = 0$ as $\operatorname{ind} J = 3$; thus $a^2 \in J^2$, $a^2 \neq 0$; now as $3 \in J^2$ (since char $R = 9$), so the equality $a^2 = \pm 3$ must hold in R. That is, R is either of type 5 or of type 6, depending on the sign in the latter equality.

The remaining case when $p = 2$ and $\operatorname{ind} J \le 5$ can be studied in a similar way. If $\operatorname{ind} J = 2$ then $\#R = 4$, so R is isomorphic either to $\mathbb{Z}/4\mathbb{Z}$ (if char $R = 4$) or to $\mathbb{F}_2[x]/x^2\mathbb{F}_2[x]$ (if char $R = 2$). If $\operatorname{ind} J = 3$ then $\#R = 8$, so R is isomorphic either to $\mathbb{Z}/8\mathbb{Z}$ (if char $R = 8$) or to the ring of type 4 or 5, by [36, Theorem 3]. Now we will show that whenever $\operatorname{ind} J \in \{4, 5\}$ then necessarily R is isomorphic to the residue ring $\mathbb{Z}/2^{\operatorname{ind} J}\mathbb{Z}$.

Let first $\operatorname{ind} J = 4$. Assume that R is not isomorphic to $\mathbb{Z}/16\mathbb{Z}$; i.e., that char $R \mid 8$. Then, in a way similar to that from the proof of Lemma 2.9 we conclude that there exists a polynomial $u(y) = a_0 + a_1 y + a_2 y^2 + a_3 y^3 \in R[y]$ that is transitive on J. Then necessarily $a_0 \in J$, and the polynomial $u^2(y)$ must be transitive on J^2. From here by direct calculations we obtain that $u^2(z) = u^2(0) + a_1^2 z$ for all $z \in J^2$. However, $a_1 = 1 + b$ for a suitable $b \in J$ (this can be shown in a way similar to that from the end of the proof of Claim 1); so $u^2(z) = u^2(0) + z$ for all $z \in J^2$. Now from the transitivity of the latter mapping u^2 on J^2 it follows that the group $(J^2)^+$ must be cyclic. But then Claim 4 implies that $\operatorname{ind} J \le 3$, a contradiction.

Now consider the final case, ind $J = 5$. Assume that char $R \mid 16$; then by Proposition 2.3 we see that the factor ring R/J^4 has transitive polynomials, and so the argument of the preceding case implies that R/J^4 must be isomorphic to the residue ring $\mathbb{Z}/16\mathbb{Z}$. However, by Proposition 2.6 $J^4 = b\mathbb{R}$ for a suitable non-zero $b \in R$; but then the set $\{0, 8 \cdot 1, 8 \cdot 1 + b, b\}$ is a non-principal ideal of the ring R, a contradiction to Proposition 2.6. This concludes the proof that the conditions of Theorem 2.7 are necessary.

To prove that these conditions are sufficient, we just present transitive polynomials for rings of type 3–6, as by Theorem 1.33 finite fields are polynomially complete and thus have transitive polynomials, and the polynomial $1 + y$ in variable y is obviously transitive on the residue ring $\mathbb{Z}/p^k\mathbb{Z}$.

Let us show that the polynomial $f(y) = 1+y+xy^p \in R[y]$ is transitive on the ring $R = F_p[x]/x^2 \cdot \mathbb{F}_p[x]$ (we take x as a representative of the coset $x + x^2 \cdot \mathbb{F}_p[x] \in R$). Indeed, this polynomial f is transitive on the factor ring of the ring R modulo the ideal xR as the latter factor ring is isomorphic to $\mathbb{F}_p$ and $f(z) = 1 + z$ for all $z \in R/xR$. It is easy to see that $f^i(z) = f^i(0) + (f^i)'(0)z$ for all $z \in xR$, $i = 0, 1, 2, \ldots$, where $'$ stands for derivation. Now direct calculations show that $f^p(z) = -x + z$ for all $z \in xR$. As $(xR)^+$ is a cyclic group of order p, then f^p is transitive on xR. Thus we finally conclude that the polynomial f is transitive on R.

A similar argument (or direct verification) shows that if R is a ring of type 4 or 5 with $p = 2$, then the polynomial $f(y) = 1+y+xy^3$ is transitive on R. Finally, if R is a ring of type 4–6 with $p = 3$, then the polynomial $f(y) = 1+y+y^2(y^3-y)^2+xy^2$ is transitive on R. The latter can be proved by the argument similar to that in the case of rings of type 3: As the polynomial $y^2(y^3 - y)^2$ is identically 0 on the factor ring R/xR, and this ring is a ring of type 3, the polynomial f is transitive on R/xR. Then direct calculations show that $f^9(z) = x^2 + z$ for all $z \in x^2R$, whence f^9 is transitive on x^2R. □

Chapter 3

p-adic analysis

In this chapter we develop tools and techniques of p-adic analysis that will be necessary to study p-adic dynamics in further chapters.

3.1 Analysis in complete non-Archimedean fields

Let K be a complete non-Archimedean field or integral domain. For example K can be $\mathbb{Q}_p$, or $\mathbb{Z}_p$, or a finite extension of $\mathbb{Q}_p$ or $\mathbb{C}_p$. The concepts of convergence, continuity and derivative are defined in K in the same way as in $\mathbb{R}$. A sequence (x_n) in K converges to $x \in K$ if $\lim_{n\to\infty} |x_n - x| = 0$.

Definition 3.1. Let $O \subseteq K$ be an open set and let $x \in O$. A function $f : O \to K$ is said to be continuous at x if for every $\varepsilon > 0$ there exists $\delta > 0$ such that, for every $y \in O$, $|f(y) - f(x)| < \varepsilon$ whenever $|y - x| < \delta$.

Definition 3.2. Let $O \subseteq K$ be an open set, let $f : O \to K$ be a function and let $x \in O$. We say that f is differentiable at x if the limit[1]

$$f'(x) = \lim_{h\to 0} \frac{f(x+h) - f(x)}{h}$$

exists. If $f'(x)$ exists for every $x \in O$ we say that f is differentiable in O and we call $x \mapsto f'(x)$ the derivative of f.

Let us now state some remarkable results of the analysis in K. First we can extend Theorem 1.40 to a general non-Archimedean field:

Theorem 3.3. *A sequence (x_n) in K is Cauchy if and only if*

$$\lim_{n\to\infty} |x_{n+1} - x_n| = 0.$$

[1]Note that in contrast to the limit in the definition of a convergent sequence, which is a limit with respect to metric in $\mathbb{R}$, the limit we use in the definition of a derivative is a limit with respect to a non-Archimedean metric in K. We use the same symbol lim for both limits when there is no risk of misunderstanding; otherwise we use $\lim^p$ for a p-adic limit, and lim for a limit in $\mathbb{R}$.

Theorem 3.4. *If a sequence* (x_n) *in* K *converges to a non-zero element* $x \in K$ *then we have* $|x_n| = |x|$ *for sufficiently large* n.

Theorem 3.5. *Let* (x_n) *be a sequence in* K. *The series* $\sum_{n=0}^{\infty} x_n$ *converges if and only if* $\lim_{n\to\infty} x_n = 0$.

Proof. Let $s_n = \sum_{j=0}^{n} x_j$. The sequence converges if and only if s_n is a Cauchy sequence, since K is complete. By Theorem 3.3 s_n is a Cauchy sequence if and only if

$$|s_{n+1} - s_n| \to \infty, \quad n \to \infty.$$

Since $|a_n| = |s_{n+1} - s_n|$ we are done. □

In the sequel we will need the following classical result of Legendre, see, e.g., [11, Corollary 3.2.2], [268, Chapter 1, Section 2, Exercise 13], [214].

Lemma 3.6 (Valuation of a factorial). *Let a natural number* n *be written in the canonical representation* $n = a_0 + a_1 p + \cdots + a_m p^m$. *Denote*

$$\mathrm{wt}_p\, n = \sum_{k=0}^{m} a_k,$$

the p*-adic weight of* n. *Then*

$$\mathrm{ord}_p\, n! = \frac{n - \mathrm{wt}_p\, n}{p-1}.$$

Corollary 3.7 (Valuation of a binomial coefficient). *For all* $i, k \in \mathbb{N}_0$,

$$\mathrm{ord}_p \binom{i+k}{i} = \frac{1}{p-1}(\mathrm{wt}_p\, i + \mathrm{wt}_p\, k - \mathrm{wt}_p(i+k)).$$

Example 3.8. Let $a_n = n$, $b_n = n!$ and $c_n = p^n$. Since $|a_{n+1} - a_n|_p = 1$ it follows that (a_n) is not a Cauchy sequence and hence it is not convergent. From Lemma 3.6 it follows that the number of factors of p in $n!$ is $\frac{n-\mathrm{wt}_p n}{p-1}$, where $\mathrm{wt}_p\, n = a_0 + a_1 + \cdots + a_N$ if $n = a_0 + a_1 p + \cdots + a_N p^N$. If $k+1$ is the number of digits in n then $\mathrm{wt}_p\, n \leqslant (k+1)(p-1)$. We also have $p^k \leqslant n < p^{k+1}$ so $k \leqslant \log_p n < k+1$. This implies that

$$\lim_{n\to\infty} -\frac{n - \mathrm{wt}_p\, n}{p-1} \leqslant \lim_{n\to\infty} \frac{-n + (\log_p n + 1)(p-1)}{p-1} = -\infty,$$

hence $|b_n|_p = |n!|_p \to 0$ as $n \to \infty$. Since $|p^n|_p = p^{-n}$ it is clear that $c_n \to 0$ as $n \to \infty$.

Example 3.9. Since $n! \to 0$ and $p^n \to 0$ as $n \to \infty$ it is clear that $\sum_{n=0}^{\infty} n!$ and $\sum_{n=0}^{\infty} p^n$ converge.

We point to an interesting number theoretic conjecture related to the factorial series: "For any p, its sum is a rational number (depending on p)." Numerous numerical simulations performed by Wim Schikhof strongly supported this conjecture. However, no rigorous prove has been provided. Of course, one cannot exclude that, in spite of numerical simulations, for some class of prime numbers sums are p-adically irrational.

Example 3.10. In $\mathbb{Q}_p$ a differentiable function may have zero derivative everywhere but still not being locally constant. The function $f : \mathbb{Q}_p \to \mathbb{Q}_p$ is defined by

$$f(x) = \begin{cases} 1, & |x|_p \geqslant 1, \\ p^{2n}, & 1/p^n \leqslant |x|_p < 1/p^{n-1}, \\ 0, & x = 0. \end{cases}$$

Then f is not locally constant around $x = 0$, but still $f'(0) = 0$. In fact

$$\lim_{h \to 0} \frac{f(0+h) - f(0)}{h} = \lim_{h \to 0} \frac{f(h)}{h}$$

and if $1/p^n \leqslant |h|_p < 1/p^{n-1}$ then

$$\frac{f(h)}{h} \lessapprox \frac{1/p^{2n}}{1/p^n} = \frac{1}{p^n} \to 0$$

as $n \to \infty$ ($h \to 0$).

Example 3.11. There exists a function $g : \mathbb{Z}_p \to \mathbb{Z}_p$ such that $g' = 0$ and g is injective. Let $x \in \mathbb{Z}_p$. Then $x = \sum_{j=0}^{\infty} a_j p^j$, where $a_j \in \{0, 1, \dots, p-1\}$ for all $j \geqslant 0$. We define

$$g(x) = \sum_{j=0}^{\infty} a_j p^{2j}.$$

First we prove that g is injective. Let $x = \sum_{j=0}^{\infty} a_j p^j \in \mathbb{Z}_p$, $y = \sum_{j=0}^{\infty} b_j p^j$ and assume that $x \neq y$. Then we can find an integer $n \geqslant 0$ such that $|x - y|_p = p^{-n}$, $a_n \neq b_n$ but $a_j = b_j$ for $0 \leqslant j \leqslant n-1$. If $g(x) = g(y)$ then

$$0 = |g(x) - g(y)| = p^{-2n}.$$

This is impossible. Hence $x = y$ and g is injective.

Let us now prove that $g' = 0$. Let x and y be as above. We can find $h \in \mathbb{Z}_p$ such that $y = x + h$. We have

$$|g(x) - g(x+h)|_p = p^{-2n} = |x - (x+h)|_p^2 = |h|_p^2$$

and

$$\lim_{h \to 0} \frac{|g(x) - g(x+h)|_p}{|h|_p} = \lim_{h \to 0} |h|_p = 0.$$

We have proved that $g'(x) = 0$ for all $x \in \mathbb{Z}_p$.

3.2 Analytic functions

Let K be a complete non-Archimedean field and let (a_n) be a sequence in K. We say that $f(x) = \sum a_n x^n$ is a formal power series. It defines a continuous function on the open ball of radius $\rho = 1/\limsup |a_n|^{1/n}$. The function can be extended to the closed ball of radius ρ if $|a_n|\rho^n \to 0$. As in the classical case we call ρ the radius of convergence. In contrary to what happens in the classical case the power series converges for all or none of the points of the sphere of radius ρ.

Theorem 3.12. *Functions defined by power series are differentiable.*

As in the complex case, functions defined by power series are called *analytic functions*.

Theorem 3.13 (Maximum principle). *Let $K = \mathbb{C}_p$ and $f : B_r(a) \to \mathbb{C}_p$ be an analytic function having the power series expansion*

$$f(x) = \sum_{n=0}^{\infty} b_n (x-a)^n.$$

Then

$$\sup_{B_r(a)} |f(x)|_p = \sup_{S_r(a)} |f(x)|_p = \max_n |b_n|_p r^n.$$

The proof can be found in [371] and in [374]. It is based on the fact that $\mathbb{C}_p$ is not locally compact. The maximum principle is not true for locally compact spaces such as $\mathbb{Q}_p$ and its finite extensions.

Example 3.14. We define the p-adic exponential function by the standard power series

$$e^x = \sum_{j=0}^{\infty} \frac{x^j}{j!},$$

where in general $x \in \mathbb{C}_p$.

What about radius of convergence of the exponential function? This series converges if and only if $|x|_p < p^{-1/(p-1)}$. If $x \in \mathbb{Q}_p$, $p \neq 2$, then it converges if and only if $|x|_p \leqslant 1/p$. If $x \in \mathbb{Q}_2$ then the series converges if and only if $|x|_2 \leqslant 1/4$.

In the same way, i.e., by considering corresponding power series, we can introduce p-adic trigonometric functions:

$$\sin x = \sum_{j=0}^{\infty} \frac{(-1)^j x^{2j+1}}{(2j+1)!}, \quad \cos x = \sum_{j=0}^{\infty} \frac{(-1)^j x^{2j}}{(2j)!}.$$

They have the same domains of definition (in $\mathbb{C}_p$ and $\mathbb{Q}_p$) as the exponential function.

We shall also use the p-adic logarithmic function, see, for example, [374]. We restrict our considerations to the case of $\mathbb{Q}_p$. Let $u \in B_{p^{-1}}(1)$. Then the p-adic logarithmic function $u \mapsto \ln_p u$ (inverse to the exponential function) is well defined. For $u = 1 + x$ with $|x|_p \leqslant 1/p$, we have

$$\ln_p u = \sum_{k=1}^{\infty} \frac{(-1)^{k+1} x^k}{k}. \tag{3.1}$$

By using (3.1) we can obtain that

$$\ln_p : B_{p^{-1}}(1) \to B_{p^{-1}}(0)$$

is an isometry.

3.3 Hensel's lemma

Let K be a finite extension of $\mathbb{Q}_p$, $\mathcal{O}_K = \{x : |x|_p \leqslant 1\}$ and let π be a uniformizer, see Subsection 1.8.1. We remark that for $K = \mathbb{Q}_p$, $\pi = p$. Those who proceeded without reading Section 1.8 can consider just the latter (simplest) case through this section.

Let $\alpha, \beta \in \mathcal{O}_K$. We say that $\alpha = \beta \pmod{\pi^\gamma}$ if α and β belongs to the same coset in $\mathcal{O}_K/\pi^\gamma \mathcal{O}_K$ or that $|\alpha - \beta|_p \leqslant |\pi|_p^\gamma$.

Theorem 3.15. *Let $F(x)$ be a polynomial over $\mathcal{O}_K$. Assume that there exists $\alpha_0 \in \mathcal{O}_K$ and $\gamma \in \mathbb{N}$ such that*

$$F(\alpha_0) \equiv 0 \pmod{\pi^{2\gamma+1}},$$
$$F'(\alpha_0) \equiv 0 \pmod{\pi^{\gamma}},$$
$$F'(\alpha_0) \not\equiv 0 \pmod{\pi^{\gamma+1}}.$$

Then there exists $\alpha \in \mathcal{O}_K$ such that $F(\alpha) = 0$ and $\alpha \equiv \alpha_0 \pmod{\pi^{\gamma+1}}$.

Proof. Assume that we have constructed a sequence $(\alpha_n) \in \mathcal{O}_K$ such that

$$F(\alpha_n) \equiv 0 \pmod{\pi^{2\gamma+1+n}}, n \geqslant 0, \tag{3.2}$$

$$\alpha_n \equiv \alpha_{n-1} \pmod{\pi^{\gamma+n}}, n \geqslant 1. \tag{3.3}$$

In the first part of this proof we will show that under this assumption the theorem is true.

It is easy to see that (α_n) is a Cauchy sequence in K. In fact

$$|\alpha_n - \alpha_{n-1}|_p \leqslant |\pi|_p^{\gamma+n} \to 0,$$

when $n \to \infty$ since $|\pi|_p < 1$.

Let α be the limit of (α_n). This limit exists, since K is a complete field (it is a finite-dimensional vector space over a complete field). It is clear that $\alpha \in \mathcal{O}_K$. Let us prove that $F(\alpha) = 0$. For every $n \in \mathbb{N}$ we have

$$|F(\alpha) - 0|_p \leqslant \max\{|F(\alpha_n)|_p, |F(\alpha_n) - F(\alpha)|_p\}.$$

By (3.2), $|F(\alpha_n)|_p \to 0$, when $n \to \infty$, and by the continuity of F, $|F(\alpha_n) - F(\alpha)|_p \to 0$. Hence $|F(\alpha)|_p = 0$ and therefore $F(\alpha) = 0$.

We have to show that $\alpha \equiv \alpha_0 \pmod{\pi^{\gamma+1}}$. Since (α_n) converges we can find a natural number n such that $|\alpha - \alpha_n|_p \leqslant |\pi|_p^{\gamma+1}$. For such n we have

$$|\alpha_n - \alpha_0|_p \leqslant \max\{|\alpha_0 - \alpha_1|_p, \dots, |\alpha_{n-1} - \alpha_n|_p\} \leqslant |\pi|_p^{\gamma+1}$$

and

$$|\alpha_0 - \alpha|_p \leqslant \max\{|\alpha_0 - \alpha_n|_p, |\alpha_n - \alpha|_p\} \leqslant |\pi|_p^{\gamma+1}.$$

In other words $\alpha \equiv \alpha_0 \pmod{\pi^{\gamma+1}}$.

We have left to construct the sequence (α_n). Let

$$\alpha_n = \alpha_{n-1} - \frac{F(\alpha_{n-1})}{F'(\alpha_{n-1})}$$

for $n \geqslant 1$. We will prove by induction that (α_n) satisfies (3.2) and (3.3). For $n = 0$ the congruence (3.2) holds by the assumptions. Let us now assume that (3.2) and (3.3) hold for a fixed n. We will now prove that they hold for $n + 1$.

By the hypothesis we have $\alpha_n \equiv \alpha_0 \pmod{\pi^{\gamma+1}}$ and therefore $\alpha_n = \alpha_0 + \beta_n \pi^{\gamma+1}$ for some $\beta_n \in \mathcal{O}_K$. Since $F'(\alpha_0) \equiv 0 \pmod{\pi^{\gamma}}$ and $F'(\alpha_0) \not\equiv 0 \pmod{\pi^{\gamma+1}}$, we have $F'(\alpha_0) = \beta_0 \pi^{\gamma}$, where $|\beta_0| = 1$, or $\beta_0 \in \mathcal{O}_K^{\times}$, the set of units (the unit sphere with center at zero), see Subsection 1.8.1.

By formal differentiation we obtain

$$F'(\alpha_n) = F'(\alpha_0) + \beta\pi^{\gamma+1} = (\beta_0 + \pi\beta)\pi^{\gamma}$$

and therefore we can write $F'(\alpha_n) = \epsilon_n \pi^{\gamma}$ for some ϵ_n such that $|\epsilon_n|_p = 1$. By the induction hypothesis we have $F(\alpha_n) = \epsilon_n \pi^{2\gamma+1+n}$ for some ϵ_n such that $|\epsilon_n|_p \leqslant 1$. Therefore

$$\alpha_{n+1} = \alpha_n + \frac{\epsilon_n}{\epsilon_n}\pi^{\gamma+1+n},$$

and hence $\alpha_{n+1} \in \mathcal{O}_K$ and $\alpha_{n+1} \equiv \alpha_n \pmod{\pi^{\gamma+1+n}}$. We have to prove that $F(\alpha_{n+1}) \equiv 0 \pmod{\pi^{2\gamma+2+n}}$. A formal Taylor series expansion of F at α_n is

$$F(x) = F(\alpha_n) + F'(\alpha_n)(x - \alpha_n) + G(x)(x - \alpha_n^2),$$

where $G(x)$ is a polynomial over $\mathcal{O}_K$. Hence

$$F(\alpha_{n+1}) = \left(\frac{F(\alpha_n)}{F'(\alpha_n)}\right)^2 G(\alpha_{n+1}) = \left(\frac{\epsilon_n}{\epsilon_n}\pi^{\gamma+1+n}\right)^2 G(\alpha_{n+1})$$

and therefore

$$F(\alpha_{n+1}) \equiv 0 \pmod{\pi^{2\gamma+2+n}}.$$

Thus, we have constructed the sequence and the proof is finished. □

In particular, for $\gamma = 0$ we have:

Corollary 3.16 (Hensel's lemma). *Let $F \in \mathcal{O}_K[x]$ and suppose that there exists $\alpha_0 \in \mathcal{O}_K$ such that $F(\alpha_0) \equiv 0 \pmod{\pi}$ and $F'(\alpha_0) \not\equiv 0 \pmod{\pi}$. Then there exists $\alpha \in \mathcal{O}_K$ such that $F(\alpha) = 0$ and $\alpha \equiv \alpha_0 \pmod{\pi}$.*

We have a more general form of Hensel's lemma.

Theorem 3.17 (General form of Hensel's lemma). *Let K be a complete non-Archimedean field and let $\mathcal{O}_K = \{x \in K : |x| \leqslant 1\}$. Let f be a polynomial with coefficients in $\mathcal{O}_K$. If $x \in \mathcal{O}_K$ and*

$$|f(x)| < |f'(x)|^2$$

then there exists a root $y \in \mathcal{O}_K$ of f such that

$$|y - x| = |f(x)/f'(x)| < |f'(x)|.$$

Moreover, this is the only root of f in the open ball of center x and radius $|f'(x)|$.

A proof of this theorem can be found in [371].

3.4 Roots of unity

Let K be a finite extension of $\mathbb{Q}_p$ and let $\mathbb{K}$ be the residue class field. The multiplicative group $\mathbb{K}^*$ is cyclic and has $p^f - 1$ elements. Since a cyclic group has a cyclic subgroup of order d for each divisor d of $p^f - 1$, for every $d \mid p^f - 1$ there exists $x \in \mathbb{K}$ that generates the subgroup of d elements and we also have $x^d = 1$. We say that x is a primitive root of unity. It generates a group of d roots to the polynomial $x^d - 1$ in $\mathbb{K}$.

Let us denote the d roots $x_1, \dots, x_d$. Take now d elements $y_1, \dots, y_d$ of $\mathcal{O}_K^\times$ such that $y_j \in x_j$. Here $\mathcal{O}_K^\times$ is the set of units (the unit sphere with center at zero), see Subsection 1.8.1.

Then there are d approximate roots of $F(x) = x^d - 1 = 0$ in $\mathcal{O}_K^\times$ because $F(y_j) \equiv 0 \pmod{\pi}$ and $F'(y_j) \not\equiv 0 \pmod{\pi}$.

Of course, the d different y_j are located in d different cosets of $\mathcal{P}_K$. Hence they are noncongruent modulo π. By Hensel's lemma, for each $d \mid p^f - 1$, the equation $x^d - 1 = 0$ has d solutions in K. Thus we proved the following result which will be useful in our further considerations:

Proposition 3.18. *$\mathcal{O}_K^\times$ contains the $(p^f - 1)$-roots of unity.*

Proposition 3.19. *Let n be an integer that is relatively prime to $p^f - 1$. Let $x^n = 1$. Then $x \equiv 1 \pmod{\pi}$ or in other words $x \in B_1^-(1)$.*

Proof. It is clear that x belongs to an element of $\mathbb{K}^*$ (since $|x|_p = 1$). Since m is relatively prime to the order of the group $\mathbb{K}^*$, the only possibility is that $x \in 1$ in $\mathbb{K}^*$. [There are no groups of order m in $\mathbb{K}^*$ since the order of the subgroup must divide the order of the group (Lagrange's theorem).] □

Lemma 3.20. *If $x \equiv 1 \pmod{\pi}$ then $x^p \equiv 1 \pmod{\pi^2}$ and $x^{p^r} \equiv 1 \pmod{\pi^{r-1}}$.*

Proof. We first prove that $x^p \equiv 1 \pmod{\pi^2}$. There exists $y \in \mathcal{P}_K$ such that $x = 1 + y$. We then have

$$x^p = (1+y)^p = 1 + py + y^2 \sum_{j=2}^{p} \binom{p}{j} y^{j-2}.$$

Since $p \in \mathcal{P}_K$ we have that $x \equiv 1 \pmod{\pi^2}$. We will now prove that $x^{p^r} \equiv 1 \pmod{\pi^{r-1}}$ by induction over r. If we assume that $x^{p^{r-1}} \equiv 1 \pmod{\pi^{r-2}}$ then there is $y \in \pi^{r-1}\mathcal{O}_K$ such that $x^{p^{r-1}} = 1 + y$. Then

$$x^{p^r} = (1+y)^p = 1 + py + y^2 \sum_{j=2}^{p} \binom{p}{j} y^{j-2}$$

and hence $x^{p^r} \equiv 1 \pmod{\pi^{r-1}}$. □

Proposition 3.21. *If $x \in B_1^-(1)$ such that $x^n = 1$ then n is divisible by a power of p and x is a root of unity for that power of p.*

Proof. Assume that $p \nmid n$, then there exists r such that $p^r \equiv 1 \pmod{n}$. Since $x \equiv 1 \pmod{\pi}$ it follows from the lemma that

$$x = x^{p^r} \equiv 1 \pmod{\pi^{r-1}}.$$

If we replace r by a multiple of r then we see that x is congruent to 1 for an arbitrary large power of π. We can draw the conclusion that $x = 1$. If $n = n'p^\eta$, for some $\eta \in \mathbb{N}$ and $p \nmid n'$, then $x^n = (x^{p^\eta})^{n'} = 1$. It also follows that $x^{p^\eta} = 1$. Hence, x is a root of unity for some power of p. □

Theorem 3.22. *Let ζ be a p^tth root of unity in K. Then $|\zeta - 1|_p = |p|_p^{1/\varphi(p^t)}$, where $\varphi(p^t) = p^{t-1}(p-1)$* (Euler's totient function).

See [371] for a proof.

Corollary 3.23. *Let e be the ramification index of K. Then the number of roots of unity whose order is a power of p is less than or equal to $e/(1 - 1/p)$.*

Theorem 3.24. *Let $n \in \mathbb{N}$, $n \geqslant 2$ and $p \nmid n$. Then the equation $x^n - 1 = 0$ has $(n, p^f - 1)$ different solutions in $\mathcal{O}_K^\times$.*

Proof. For such n, $\mathcal{O}_K^\times$ contains only roots of $x^n - 1 = 0$, that is $(p^f - 1)$-roots of unity. Hence the equation has $(n, p^f - 1)$ different solutions. □

3.5 Non-Archimedean normed spaces

Essentials of non-Archimedean functional analysis can be found, e.g., in books of Monna [322], van Rooji [399], or Schikhof [374].

Let E be a linear space over a non-Archimedean field K. The latter has the absolute value $|\cdot|$. A *non-Archimedean norm* on E is a map

$$\|\cdot\| : E \to \mathbb{R}_+$$

satisfying the following conditions:

(a) $\|x\| = 0 \iff x = 0$;

(b) $\|\lambda x\| = |\lambda| \, \|x\|, \quad \lambda \in K$;

(c) $\|x + y\| \leq \max(\|x\|, \|y\|)$.

The latter inequality is the *strong triangle inequality* for the norm. A linear space E endowed with a norm is called a *normed space*.

We remark that the definition of the norm on a linear space differs from the definition of the norm on a ring, see Section 1.7: instead of equality (b), one has an inequality. In principle, one can consider more complex algebraic objects, namely, normed modules over normed rings. For such objects, equality (b) should be modified to inequality to match with the definition of the norm on a ring. Finally, we point out that the definition of the norm on a linear space matches well with the definition of the absolute value on a field.

As usual, we define a non-Archimedean Banach space E as a complete normed space over K. The metric

$$\rho(x, y) = \|x - y\|$$

is ultrametric, see Section 1.5 for details. Hence every non-Archimedean Banach space is totally disconnected. All balls $B_r(a) = \{x \in E : \|x - a\| \leqslant r\}$ are clopen.

The *dual space* E' is defined as space of continuous K-linear functionals $l : E \to K$. Let us introduce the standard norm on E':

$$\|l\| = \sup_{x \neq 0} |l(x)|_K / \|x\|.$$

The space E' endowed with this norm is a *Banach space*.

The simplest example of a non-Archimedean Banach space is the space

$$K^n = K \times \cdots \times K \quad (n \text{ times})$$

with the non-Archimedean (canonical) norm

$$\|x\| = \max_{1 \leqslant j \leqslant n} |x_j|.$$

More interesting examples are infinite-dimensional non-Archimedean Banach spaces realized as spaces of sequences. Set

$$c_0 \equiv c_0(K) = \{x = (x_n)_{n=1}^{\infty} \in K^{\infty} : \lim_{n \to \infty} x_n = 0\}$$

and $\|x\| = \max_n |x_n|$.

To simplify notation, for the finite-dimensional space K^n, the canonical norm will be simply denoted by the same symbol as the absolute value on K: $|x|$ and in the p-adic case

$$|x|_p = \max_{1 \leqslant j \leqslant n} |x_j|_p,$$

i.e., by the same symbol as the p-adic absolute value. We hope that such notations will not induce misunderstanding.

3.6 Multidimensional analysis

All considerations of Sections 3.1, 3.5 are easily generalized to Cartesian products of non Archimedean fields or rings. Such Cartesian products are endowed with max-norms, see Section 3.5. As in the real and complex cases, multidimensional analogues are generated by using norms, instead of absolute values.

In what follows we mostly consider n-variate functions defined on $\mathbb{Q}_p^n$ (or on $\mathbb{Z}_p^n$) and valuated in $\mathbb{Q}_p^m$ (or in $\mathbb{Z}_p^m$). For the reader's convenience, in the sequel we reformulate (or remind) basic notions of p-adic analysis for considered cases, when needed.

Definition 3.25. A function $F : \mathbb{Q}_p^n \to \mathbb{Q}_p^m$ is said to be *uniformly continuous* if and only if for every $M \in \mathbb{N}_0$ there exists $N \in \mathbb{N}_0$ such that $|f(x) - f(y)|_p \le p^{-N}$ whenever $|x - y|_p \le p^{-M}$.

The function F is said to satisfy the *Lipschitz condition* with a constant $\alpha = p^t$, $t \in \mathbb{Z}$, (to be an α-*Lipschitz*, for short) whenever

$$|f(x) - f(y)|_p \le \alpha \cdot |x - y|_p. \tag{3.4}$$

The function F is said to be *asymptotically* α-*Lipschitz* whenever (3.4) holds uniformly for all points x, y that are sufficiently close to each other, that is, there exists $K \in \mathbb{N}_0$ such that (3.4) holds whenever $|x - y|_p \le p^{-K}$.

The definition can be re-stated for F defined on an open subset of $\mathbb{Q}_p^n$ (e.g., on $\mathbb{Z}_p^n$) in an obvious manner.

Definition 3.26 (Differentiable function). A function $F\colon \mathbb{Q}_p^n \to \mathbb{Q}_p^m$ is said to be *differentiable* at the point $\mathbf{u} = (u_1, \dots, u_n) \in \mathbb{Q}_p^n$ if there exists a positive $n \times m$ matrix $F'_k(\mathbf{u})$ over $\mathbb{Q}_p$ (called *the Jacobi matrix* of the function F at the point $\mathbf{u}$) such that for all sufficiently small $\mathbf{h}$ the function F can be represented in the form

$$F(\mathbf{u}+\mathbf{h}) = F(\mathbf{u}) + \mathbf{h} \cdot F'_k(\mathbf{u}) + \alpha(\mathbf{u}, \mathbf{h}),$$

where

$$\lim_{\mathbf{h}\to\mathbf{0}} \frac{|\alpha(\mathbf{u}, \mathbf{h})|_p}{|\mathbf{h}|_p} = 0.$$

The function F is said to be *uniformly differentiable* on $\mathbb{Q}_p^n$ whenever there exists $K \in \mathbb{N}$ such that F can be represented in the above form for all $\mathbf{u} \in \mathbb{Q}_p^n$ and all $\mathbf{h}$ with a norm not greater than p^{-K}, $|\mathbf{h}|_p \le p^{-K}$.

The definition of a uniformly differentiable function can also be re-stated for F defined on an open subset of $\mathbb{Q}_p^n$ (e.g., for F defined on $\mathbb{Z}_p^n$) in an obvious manner.

3.7 The differentiability modulo p^k

In this section, we introduce a concept of the *derivative modulo* p^k, which is very important in further studies of p-adic dynamics. This concept was originally introduced in the beginning of the 1990s by Vladimir Anashin, see [21, 22].

Let $s \in \mathbb{N}$, and let $\mathbf{a} = (a_1, \dots, a_n)$ and $\mathbf{b} = (b_1, \dots, b_n)$ be arbitrary points of $\mathbb{Q}_p^n$. We write $\mathbf{a} \equiv \mathbf{b} \pmod{p^s}$ if and only if $|a_i - b_i|_p \le p^{-s}$ (or, which is the same, if and only if $a_i = b_i + c_i p^s$ for suitable $c_i \in \mathbb{Z}_p$, $i = 1, 2, \dots, n$). In other words, we use sometimes for better convenience $\mathbf{a} \equiv \mathbf{b} \pmod{p^s}$ rather than $|\mathbf{a} - \mathbf{b}|_p \le p^{-s}$ meaning that both $\mathbf{a}$ and $\mathbf{b}$ lie in some ball of radius p^{-s} of the space $\mathbb{Q}_p^n$.

Note that for all $s \in \mathbb{N}$ the binary relation $\cdot \equiv \cdot \pmod{p^s}$ is a congruence of $\mathbb{Q}_p$ whenever $\mathbb{Q}_p$ is considered as a module over a ring $\mathbb{Z}_p$, see Subsection 1.2.1 for a general definition of a congruence on a universal algebra. In other words, we can work with the relation $\cdot \equiv \cdot \pmod{p^s}$ in a usual manner; e.g., multiply both parts by a p-adic integer, add congruences partwise, etc.

Now we generalize the main notion of Calculus, a derivative.

Definition 3.27. A function

$$F = (f_1, \dots, f_m)\colon \mathbb{Z}_p^n \to \mathbb{Z}_p^m$$

is said to be *differentiable modulo* p^k at the point $\mathbf{u} = (u_1, \dots, u_n) \in \mathbb{Z}_p^n$ if there exists a positive integer rational N and an $n \times m$ matrix $F'_k(\mathbf{u})$ over $\mathbb{Q}_p$ (called *the*

Jacobi matrix modulo p^k of the function F at the point $\mathbf{u}$) such that for every positive rational integer $K \geq N$ and every $\mathbf{h} = (h_1, \ldots, h_n) \in \mathbb{Z}_p^n$ the congruence

$$F(\mathbf{u} + \mathbf{h}) \equiv F(\mathbf{u}) + \mathbf{h} \cdot F'_k(\mathbf{u}) \pmod{p^{k+K}} \tag{3.5}$$

holds whenever $|\mathbf{h}|_p \leq p^{-K}$. In the case $m = 1$ the Jacobi matrix modulo p^k is called a *differential modulo* p^k. In the case $m = n$ a determinant of the Jacobi matrix modulo p^k is called a *Jacobian modulo* p^k. Entries of the Jacobi matrix modulo p^k are called *partial derivatives modulo* p^k of the function F at the point $\mathbf{u}$.

A partial derivative (respectively, a differential) modulo p^k we sometimes denote by $\frac{\partial_k f_i(\mathbf{u})}{\partial_k x_j}$ (respectively, by $d_k F(\mathbf{u}) = \sum_{i=1}^n \frac{\partial_k F(\mathbf{u})}{\partial_k x_i} d_k x_i$).

Note that congruence (3.5) holds if and only if the function $F(u + h)$ can be represented in the form

$$F(\mathbf{u} + \mathbf{h}) = F(u) + \mathbf{h} \cdot F'_k(\mathbf{u}) + \alpha(\mathbf{u}, \mathbf{h}) \tag{3.6}$$

for sufficiently small $\mathbf{h}$ (that is, when $|\mathbf{h}|_p \leq p^{-K}$ for some $K \in \mathbb{N}$), where

$$\frac{|\alpha(\mathbf{u}, \mathbf{h})|_p}{|\mathbf{h}|_p} \leq p^{-k}. \tag{3.7}$$

The notion of a function that is differentiable modulo p^k is of high importance for applications, see Chapters 8 and 9, and especially Section 8.3 for 'natural' examples of these functions. So we briefly discuss this notion here.

Compared to differentiability (cf. Definition 3.26), the differentiability modulo p^k is a weaker restriction. Speaking loosely, in a univariate case ($m = n = 1$), Definition 3.27 just yields that

$$\frac{F(\mathbf{u} + \mathbf{h}) - F(\mathbf{u})}{\mathbf{h}} \approx F'_k(\mathbf{u}).$$

Note that whenever $\approx$ ('approximately') stands for an '*arbitrarily high* precision' one obtains a common definition of differentiability of a p-adic function: For arbitrary $k \in \mathbb{N}$ there exists $K \in \mathbb{N}$ such that (3.6) and (3.7) hold whenever $|h|_p \leq p^{-K}$.

However, if $\approx$ stands for a 'precision that is *not worse than* p^{-k}', one obtains the differentiability modulo p^k: In this case k in (3.7) is fixed, and both (3.6) and (3.7) hold for sufficiently small h. Note that the notion of a derivative modulo p^k is a sort of a mathematical rigorism for an ill-defined notion of a 'derivative up to k digits after a point', which often is used in common speech.

Obviously, whenever a function is differentiable in a classical meaning, and if its derivative is a p-adic integer, then the function is differentiable modulo p^k for all $k = 1, 2, \ldots$. In this case the derivative modulo p^k of the function is just a reduction modulo p^k of its classical derivative: Note that according to Definition 3.27 partial derivatives modulo p^k are determined up to a summand that is 0 modulo p^k. The converse is also true: If a function is differentiable modulo p^k for all sufficiently large k then it is differentiable (in the classical meaning).

In cases when all partial derivatives modulo p^k at all points of $\mathbb{Z}_p^n$ are p-adic integers we say that the function F has *integer-valued derivative modulo* p^k. In these cases we can associate to each partial derivative modulo p^k a unique element of the ring $\mathbb{Z}/p^k\mathbb{Z}$; a Jacobi matrix modulo p^k at each point $\mathbf{u} \in \mathbb{Z}_p^n$ thus can be considered as a matrix over the ring $\mathbb{Z}/p^k\mathbb{Z}$.

Functions that have integer-valued derivatives are important in further considerations: In Section 3.8 we will demonstrate that a 1-Lipschitz function has integer-valued derivatives (modulo some p^k) whenever the function is differentiable (modulo some p^k).

The following definition is an analog of the classical one:

Definition 3.28. A function $F : \mathbb{Z}_p^n \to \mathbb{Z}_p^m$ is said to be *uniformly differentiable modulo* p^k *on* $\mathbb{Z}_p^n$ if and only if there exists $K \in \mathbb{N}$ such that congruence (3.5) holds simultaneously for all $\mathbf{u} \in \mathbb{Z}_p^n$ whenever $|\mathbf{h}|_p \le p^{-K}$. The smallest of these K is denoted by $N_k(F)$.

Note 3.29. The number $N_k(F)$ plays an important role in further considerations.

The 'rules of derivation modulo p^k' of functions that have integer-valued derivatives modulo p^k are similar to the ones in the classical case. The only difference is that these rules are congruences modulo p^k, and not equalities.

Proposition 3.30. *Let* $G : \mathbb{Z}_p^s \to \mathbb{Z}_p^n$ *and* $F : \mathbb{Z}_p^n \to \mathbb{Z}_p^m$ *be differentiable modulo* p^k *at the points* $\mathbf{v} = (v_1, \dots, v_s)$ *and* $\mathbf{u} = G(\mathbf{v})$*, respectively, and let all partial derivatives modulo* p^k *of the functions* G *and* F *at the points, respectively,* $\mathbf{v}$ *and* $\mathbf{u}$ *are* p*-adic integers. Then the composition* $F \circ G : \mathbb{Z}_p^s \to \mathbb{Z}_p^m$ *is uniformly differentiable modulo* p^k *at the point* $\mathbf{v}$*, all its partial derivatives modulo* p^k *at this point are* p*-adic integers, and*

$$(F \circ G)'_k(\mathbf{v}) \equiv G'_k(\mathbf{v})F'_k(\mathbf{u}) \pmod{p^k}.$$

In particular, if functions $f, g : \mathbb{Z}_p \to \mathbb{Z}_p$ *are differentiable modulo* p^k *at the point* $u \in \mathbb{Z}_p$*, and if their derivatives modulo* p^k *at this point are integer-valued, then*

$$(f + g)'_k(u) \equiv f'_k(u) + g'_k(u) \pmod{p^k};$$

$$(f \cdot g)'_k(u) \equiv f'_k(u)g(u) + f(\mathbf{u})g'_k(u) \pmod{p^k}.$$

If, moreover, there exists an open ball $U \ni u$ *such that* $g(r) \not\equiv 0 \pmod{p}$ *at every point* $r \in U$*, then the function*

$$\frac{f}{g} : U \to \mathbb{Z}_p$$

is differentiable modulo p^k at the point u, has integer-valued derivative modulo p^k at this point, and

$$\left(\frac{f}{g}\right)'_k(u) = \frac{f'_k(u)g(u) - f(u)g'_k(u)}{g(u)^2}.$$

If additionally the functions F, G, f, g are uniformly differentiable modulo p^k, and if their derivatives modulo p^k are integer-valued everywhere on $\mathbb{Z}_p$, then the same is true for the functions $F \circ G$, $f + g$, and $f \cdot g$.

Finally, if $g(v) \not\equiv 0 \pmod{p}$ for all $v \in \mathbb{Z}_p$, then the function $\frac{f}{g}$ is integer-valued and uniformly differentiable modulo p^k everywhere on $\mathbb{Z}_p$, and its partial derivative modulo p^k is integer-valued at all points of $\mathbb{Z}_p$.

Sketch proof. A proof of this proposition, with minor changes due to the non-Archimedean metric, follows (up to the use of congruences modulo p^n rather than equations) the one of the classical Calculus. The argument is still valid since a congruence modulo p^n is a congruence relation on the ring $\mathbb{Z}_p$; whence we can, for instance, multiply both parts of some congruence modulo p^n by a p-adic unit (i.e., by a p-adic integer with a norm 1) without affecting the validity of this congruence. □

Note 3.31. Proposition 3.30 does not hold for functions whose derivatives modulo p^k are not integer-valued. However, both a sum of (uniformly) differentiable modulo p^k functions and a product of such function by a p-adic integer are still (uniformly) differentiable modulo p^k, since a congruence modulo p^n is a congruence relation on $\mathbb{Q}_p$ when $\mathbb{Q}_p$ is considered as module over the ring $\mathbb{Z}_p$.

Proposition 3.32. *If the function $F = (f_1, \ldots, f_m)\colon \mathbb{Z}_p^n \to \mathbb{Z}_p^m$ is uniformly differentiable modulo p^k, then each of its derivatives modulo p^k is a periodic function, and the length of the period is $p^{N_k(F)}$*(cf. Definition 3.28).

Proof. The proof can obviously be restricted to the case $m = n - 1$. According to Definitions 3.27 and 3.28, if $|h|_p \le p^{-K}$ then for all $u \in \mathbb{Z}_p$ and $K \ge N_k(F)$ the following congruence holds:

$$F(u+h) - F(u) \equiv h\frac{\partial_k F(u)}{\partial_k x} \pmod{p^{k+K}}. \tag{3.8}$$

Taking $|h_1|_p \le |h|_p$ and substituting $u = u_1 + h_1$ into (3.8), represent

$$F(u+h) - F(u) = F(u_1 + h_1 + h) - F(u_1) - (F(u_1 + h_1) - F(u_1)). \tag{3.9}$$

Now applying (3.8) to (3.9) we obtain that

$$F(u+h) - F(u) \equiv (h_1 + h)\frac{\partial_k F(u_1)}{\partial_k x} - h_1\frac{\partial_k F(u_1)}{\partial_k x} \pmod{p^{k+K}},$$

and conclude that

$$F(u+h) - F(u) \equiv h \frac{\partial_k F(u_1)}{\partial_k x} \pmod{p^{k+K}} \tag{3.10}$$

since a congruence modulo p^{k+K} is a congruence relation of the module $\mathbb{Q}_p$ over the ring $\mathbb{Z}_p$, see Note 3.31. Now comparing (3.8) and (3.10), and taking $h = p^K$ we obtain that

$$\frac{\partial_k F(u_1)}{\partial_k x} \equiv \frac{\partial_k F(u)}{\partial_k x} \pmod{p^k}$$

whenever $|u_1 - u|_p \le p^{-N_k(F)}$. □

Note 3.33. Nowhere in the proof we demand that the derivatives modulo p^k must be integer-valued!

In other words, Proposition 3.32 implies that *each partial derivative modulo p^k can be considered as a function defined on (and valuated in) the residue ring $\mathbb{Z}/p^{N_k(F)}\mathbb{Z}$.* Moreover, if a continuation $\tilde{F}$ of the function $F = (f_1, \dots, f_m)\colon \mathbb{N}_0^n \to \mathbb{N}_0^m$ to the space $\mathbb{Z}_p^n$ is uniformly differentiable modulo p^k on $\mathbb{Z}_p^n$, then one can simultaneously continue the function F together with all its (partial) derivatives modulo p^k to the whole space $\mathbb{Z}_p^n$. Consequently, we may study if necessary (partial) derivatives modulo p^k of the function $\tilde{F}$ rather than those of F, and vice versa. For example, a partial derivative $\frac{\partial_k f_i(\mathbf{u})}{\partial_k x_j}$ modulo p^k vanishes modulo p^k at no point of $\mathbb{Z}_p^n$ (that is, $\frac{\partial_k f_i(\mathbf{u})}{\partial_k x_j} \not\equiv 0 \pmod{p^k}$ for all $u \in \mathbb{Z}_p^n$, or equivalently, $\left|\frac{\partial_k f_i(\mathbf{u})}{\partial_k x_j}\right|_p > p^{-k}$ everywhere on $\mathbb{Z}_p^n$) if and only if $\frac{\partial_k f_i(\mathbf{u})}{\partial_k x_j} \not\equiv 0 \pmod{p^k}$ for all $u \in \{0, 1, \dots, p^{N_k(F)} - 1\}$.

3.8 Compatible functions on $\mathbb{Z}_p$

In this section we consider compatible mappings of the ring $\mathbb{Z}_p$ as they are important in various applications, e.g., to computer science and cryptology, since basic microchip instructions can be viewed as compatible mappings of the ring of 2-adic integers. We mainly follow the works [21, 22] in this section.

Since the only congruences of the ring $\mathbb{Z}_p$ (that is, binary equivalence relations that agree with addition and multiplication of $\mathbb{Z}_p$, cf. Definition 1.18) are congruences modulo p^k, $k \in \mathbb{N}$, we state the following

Definition 3.34. A function $F = (f_1, \dots, f_m)\colon \mathbb{Z}_p^n \to \mathbb{Z}_p^m$ is called (*asymptotically*) *compatible* if (there exists a nonnegative rational integer N such that for each $k \ge N$) the congruence $u \equiv v \pmod{p^k}$ implies $F(u) \equiv F(v) \pmod{p^k}$, for every pair $u, v \in \mathbb{Z}_p^n$.

Since every class of congruent elements from $\mathbb{Z}_p$ with respect to a congruence modulo p^k is a coset with respect to ideal $p^k\mathbb{Z}_p$ of the ring $\mathbb{Z}_p$, and every such coset is a ball of radius p^{-k} in the metric space $\mathbb{Z}_p$, and vice versa, it is clear that (asymptotically) compatible functions map (sufficiently small) balls into balls, and vice versa, all mappings that map (sufficiently small) balls into balls are (asymptotically) compatible.

3.8.1 Compatibility is equivalent to 1-Lipschitz

Let F be (asymptotically) compatible, and let $|u - v|_p = p^{-\ell}(\le p^{-N})$; i.e., $u \equiv b \pmod{p^\ell}$. According to Definition 3.34 we conclude that $F(u) \equiv F(b) \pmod{p^\ell}$; that is, $|F(u) - F(v)|_p \le p^{-\ell} = |u - v|_P$.

In other words, asymptotically compatible functions are precisely all those functions that satisfy the uniform Lipschitz condition

$$|F(u) - F(v)|_p \le |u - v|_p \tag{3.11}$$

for each pair of points (u, v) which are sufficiently close one to another, i.e. such points that $|u - v|_p \le p^{-N}$; compatible functions satisfy this condition for all pairs $u, v \in \mathbb{Z}_p^n$. Since (asymptotically) compatible functions satisfy the Lipschitz condition, they are continuous and, consequently, uniformly continuous on $\mathbb{Z}_p$.

We conclude:

> *Compatible mappings of the ring $\mathbb{Z}_p$ into itself are 1-Lipschitz functions, and vice versa. Whence, compatible mappings of the ring $\mathbb{Z}_p$ into itself are uniformly continuous transformations on the metric space $\mathbb{Z}_p$.*

So we further use the term 'compatible functions' along with a term '1-Lipschitz function' in this book. We reserve the notation $\mathcal{L}_1$ for the class of 1-Lipschitz functions, and $\bar{\mathcal{L}}_1$ for asymptotically compatible functions.

We already mentioned that compatible mappings are important in various applications, see Chapters 8 and 9 for details: As basic microchip instructions are compatible mappings of the ring of 2-adic integers, these instructions (as well as their compositions, i.e., computer programs) are uniformly continuous functions on $\mathbb{Z}_2$. This observation hints to a possibility to apply the non-Archimedean analysis and the non-Archimedean dynamics to various problems of computer science. This is why we are particularly focused at dynamical properties of 1-Lipschitz functions in this book.

Now we characterize compatible functions in terms of the so-called *coordinate functions*; the latter are functions $\delta_i(f(x_1, \dots, x_n))$ defined on $\mathbb{Z}_p^n$ and valuated in $\{0, 1, \dots, p-1\}$: The ith coordinate function is merely a value of coefficient of the ith term in a canonical p-adic expansion of $f(x_1, \dots, x_n)$, see Note 1.46.

Proposition 3.35. *A function $f : \mathbb{Z}_p^n \to \mathbb{Z}_p$ is compatible if and only if for every $i = 1, 2, \dots$ the ith coordinate function $\delta_i(f(x_1, \dots, x_n))$ does not depend on $\delta_{i+k}(x_s)$, for all $s = 1, 2, \dots, n$ and $k = 1, 2, \dots$.*

Proof. Let the function $\delta_i(f(x_1, \dots, x_n))$ depend on $\delta_{i+k}(x_s)$ for some i, s, k; i.e., let there exist $(u_1, \dots, u_n)$ and $(v_1, \dots, v_n)$ in $\mathbb{Z}_p^n$ such that $u_j = v_j$ for $j = 1, 2, \dots, n$; $j \neq s$, and

$$\delta_{i+k}(u_s) \neq \delta_{i+k}(v_s), \quad \delta_r(u_s) = \delta_r(v_s)$$

for all $r = 0, 1, 2, \dots$; $r \neq i + k$, and

$$\delta_i(f(u_1, \dots, u_n)) \neq \delta_i(f(v_1, \dots, v_n)). \tag{3.12}$$

This means that $(u_1, \dots, u_n) \equiv (v_1, \dots, v_n) \pmod{p^{i+k}}$, i.e., in particular

$$(u_1, \dots, u_n) \equiv (v_1, \dots, v_n) \pmod{p^{i+1}},$$

whereas in view of (3.12)

$$f(u_1, \dots, u_n) \not\equiv f(v_1, \dots, v_n) \pmod{p^{i+1}},$$

a contradiction to the compatibility of f. □

Note 3.36. From the proof of Proposition 3.35 it immediately follows that *a function $f: \mathbb{Z}_p^n \to \mathbb{Z}_p$ is asymptotically compatible if and only if there exists $N \in \mathbb{N}_0$ such that for every $i = N, N+1, N+2, \dots$ the ith coordinate function $\delta_i(f(x_1, \dots, x_n))$ does not depend on $\delta_{i+k}(x_s)$, for all $s = 1, 2, \dots, n$ and $k = 1, 2, \dots$.*

Proposition 3.35 demonstrates that a compatible function $F: \mathbb{Z}_p^n \to \mathbb{Z}_p^m$ is just a triangular function from a p-valued logic, and vice versa, every triangular function defines a compatible function $F: \mathbb{Z}_p^n \to \mathbb{Z}_p^m$.

Definition 3.37. Recall that an n-variate *triangular function* (of a p-valued logic) is a mapping

$$\Phi\colon (\alpha_0^{\downarrow}, \alpha_1^{\downarrow}, \alpha_2^{\downarrow}, \dots) \mapsto (\Phi_0^{\downarrow}(\alpha_0^{\downarrow}), \Phi_1^{\downarrow}(\alpha_0^{\downarrow}, \alpha_1^{\downarrow}), \Phi_2^{\downarrow}(\alpha_0^{\downarrow}, \alpha_1^{\downarrow}, \alpha_2^{\downarrow}), \dots),$$

where $\alpha_i^{\downarrow} \in \mathbb{B}_p^n$ is an n-dimensional columnar vector; $\mathbb{B}_p = \{0, 1, \dots, p-1\}$, and the mapping $\Phi_i^{\downarrow}: (\mathbb{B}_p^n)^{i+1} \to \mathbb{B}_p^m$ maps n-dimensional vectors $\alpha_0^{\downarrow}, \dots, \alpha_i^{\downarrow}$ to an m-dimensional vector $\Phi_i^{\downarrow}(\alpha_0^{\downarrow}, \dots, \alpha_i^{\downarrow}) \in \mathbb{B}_p^m$. Accordingly, a univariate triangular function f is a mapping

$$(\chi_0, \chi_1, \chi_2, \dots) \stackrel{f}{\mapsto} (\psi_0(\chi_0); \psi_1(\chi_0, \chi_1); \psi_2(\chi_0, \chi_1, \chi_2); \dots),$$

where $\chi_j \in \{0, 1, \dots, p-1\}$, and each $\psi_j(\chi_0, \dots, \chi_j) \in \{0, 1, \dots, p-1\}$ is a function in variables $\chi_0, \dots, \chi_j$ of a p-valued logic.

Triangular functions define p-adic functions in an obvious manner: e.g., a univariate triangular function f sends a p-adic integer

$$\chi_0 + \chi_1 \cdot p + \chi_2 \cdot p^2 + \cdots$$

to the p-adic integer

$$\psi_0(\chi_0) + \psi_1(\chi_0, \chi_1) \cdot p + \psi_2(\chi_0, \chi_1, \chi_2) \cdot p^2 + \cdots .$$

Seemingly the triangular functions originate from automata theory: Actually, every automaton on p symbols (with n inputs and m outputs) defines a triangular function Φ, and vice versa, see Chapter 8 for details. Note that in automata theory triangular functions are also known under the name of *determined functions*, as well as of *automata functions*, see e.g. [413]. In cryptology, triangular functions are usually considered only for $p = 2$ and are called *T-functions* by some authors in this case, see Chapter 9.

In further study we need one more characterization of compatible p-adic functions.

Proposition 3.38. *A continuous function $f : \mathbb{Z}_p^n \to \mathbb{Z}_p$ is compatible if and only if every function $\frac{1}{i}\Delta_j^i f$ (where $j = 1, 2, \ldots, n$; $i = 1, 2, \ldots$) is integer-valued on $\mathbb{Z}_p$* (i.e., all its values on $\mathbb{Z}_p$ are p-adic integers).

Proof. In view of (3.11) we conclude that $f : \mathbb{Z}_p^n \to \mathbb{Z}_p$ is compatible if and only if

$$|f(x_1, \ldots, x_{i-1}, x_i + h, x_{i+1}, \ldots, x_n) - f(x_1, \ldots, x_n)|_p \le |h|_p \tag{3.13}$$

for all $x_1, \ldots, x_n, h \in \mathbb{Z}_p$ and all $i = 1, 2, \ldots, n$; or, equivalently, if and only if the p-adic number

$$\alpha_h = \frac{1}{h}(f(x_1, \ldots, x_{i-1}, x_i + h, x_{i+1}, \ldots, x_n) - f(x_1, \ldots, x_n)) \tag{3.14}$$

is a p-adic integer for all $h \in \mathbb{Z}_p \setminus \{0\}$ and all $x_1, \ldots, x_n \in \mathbb{Z}_p$.

As $f(x_1, \ldots, x_n)$ is continuous, (3.13) holds for all $h \in \mathbb{Z}_p$ if and only if it holds for all $h \in \mathbb{N}$, since $\mathbb{N}$ is a dense subset in $\mathbb{Z}_p$. Thus, a continuous function f is compatible if and only if α_h is a p-adic integer for each positive rational integer h. Now applying the Gregory–Newton formula (Theorem 1.5), we conclude that for a positive rational integer h the p-adic number α_h can be expressed as

$$\alpha_h = \frac{1}{h}\sum_{j=1}^{h}\binom{h}{j}\Delta_i^j f(x_1, \ldots, x_n) = \sum_{j=1}^{h}\binom{h-1}{j-1}\frac{1}{j}\Delta_i^j f(x_1, \ldots, x_n).$$

Thus, the function f is compatible if and only if each p-adic number

$$\alpha_m = \sum_{k=0}^{m-1}\binom{m}{k}\frac{1}{k+1}\Delta_i^{k+1} f(x_1, \ldots, x_n) \tag{3.15}$$

is a p-adic integer for $m = 1, 2, 3, \ldots$. Now applying combinatorial relations of Theorem 1.6 we express $\frac{1}{k+1}\Delta_i^{k+1} f(x_1, \ldots, x_n)$ from (3.15) via the numbers α_m:

$$\frac{1}{k+1}\Delta_i^{k+1} f(x_1, \ldots, x_n) = \sum_{m=0}^{k} (-1)^{m+k} \binom{k}{m} \alpha_{m+1}, \tag{3.16}$$

where $k = 0, 1, 2, \ldots$. Now (3.16) implies that all fractions $\frac{1}{k+1}\Delta_i^{k+1} f(x_1, \ldots, x_n)$ are p-adic integers whenever all α_n for $n = 0, 1, 2, 3, \ldots$ are p-adic integers; whereas (3.15) implies the converse. Whence, all α_m for $m = 0, 1, 2, \ldots$ are p-adic integers if and only if all fractions $\frac{1}{k+1}\Delta_i^{k+1} f(x_1, \ldots, x_n)$ for $k = 0, 1, 2, \ldots$ are p-adic integers. □

3.8.2 Compatibility and differentiability

The following theorem demonstrates that 1-Lipschitz functions are tightly related to functions that are uniformly differentiable (or at least are uniformly differentiable modulo some p^k) and have integer-valued derivatives.

Theorem 3.39. *Let a function $F = (f_1, \ldots, f_m) \colon \mathbb{Z}_p^n \to \mathbb{Z}_p^m$ be uniformly differentiable modulo p, and let it have integer-valued derivatives modulo p at all points of $\mathbb{Z}_p^n$. Then*

$$F(x_1, \ldots, x_n) = P(x_1, \ldots, x_n) + C(x_1, \ldots, x_n)$$

where P is a periodic function with a period of length $p^{N_1(F)}$, and C is a compatible function. Consequently, F is asymptotically compatible, and C is uniformly differentiable modulo p.

Proof. Put

$$P(x_1, \ldots, x_n) = (f_1(x_1, \ldots, x_n) \bmod p^{N_1(F)}, \ldots, f_m(x_1, \ldots, x_n) \bmod p^{N_1(F)}),$$
$$C(x_1, \ldots, x_n) = F(x_1, \ldots, x_n) - P(x_1, \ldots, x_n).$$

For $l \geq N_1(F)$ and all $s_1, \ldots, s_n \in \mathbb{Z}_p$ Definition 3.27 implies that

$$F(x_1 + s_1 p^l, \ldots, x_n + s_n p^l) \equiv F(x_1, \ldots, x_n) \pmod{p^l} \tag{3.17}$$

since $F_1'(x_1, \ldots, x_n)$ is a matrix over $\mathbb{Z}/p\mathbb{Z}$, and consequently

$$(s_1 p^l, \ldots, s_n p^l) F_1'(x_1, \ldots, x_n) \equiv (0, \ldots, 0) \pmod{p^l}.$$

In particular, (3.17) implies that F is asymptotically compatible. This in turn means that for $i \geq N_1(F)$ the function $\delta_i(f_j(x_1, \ldots, x_n))$ depends only on

$$\delta_0(x_1), \ldots, \delta_0(x_n), \ldots, \delta_i(x_1), \ldots, \delta_i(x_n);$$

i.e., this is a periodic function with a period of length p^{i+1}. Hence C is compatible.

On the other hand, (3.17) implies that if $i < N_1(F)$ then $\delta_i(f_j(x_1,\dots,x_n))$ does not depend on $\delta_r(x_t)$ for $r = N_1(F), N_1(F)+1,\dots$ and $t = 1,2,\dots,n$; that is, for all $i = 0,1,\dots,N_1(F)-1$ and all $j = 1,2,\dots,m$ the function $\delta_i(f_j(x_1,\dots,x_n))$ is periodic with a period of length $p^{N_1(F)}$. Hence the function $P(x_1,\dots,x_n)$ is periodic with a period of length $p^{N_1(F)}$ since

$$f_j(x_1,\dots,x_n) \bmod p^{N_1(F)} = \sum_{i=0}^{N_1(F)-1} \delta_i(f_j(x_1,\dots,x_n))p^i$$

for $j = 1,2,\dots,m$. Thus $P(x_1,\dots,x_n)$ is a pseudo-constant, whence has zero derivatives. We conclude finally that the function $C = F - P$ is uniformly differentiable modulo p, and that the corresponding partial derivatives of C and F modulo p pairwise coincide. □

Note 3.40. From the proof of Theorem 3.39 it easily follows that *any asymptotically compatible function is a sum of a compatible function and of a periodic function with a period of length p^K for some $K \in \mathbb{N}_0$, and vice versa, any such sum is asymptotically compatible* since the congruence (3.17) of the proof of Theorem 3.39 is equivalent to the asymptotic compatibility of F. Moreover, this K is equal to N from the statement of Note 3.36: Actually from the proof of Theorem 3.39, as well as from the proof of Proposition 3.35, it can be easily deduced that *a function $f: \mathbb{Z}_p^n \to \mathbb{Z}_p$ is asymptotically compatible if and only if there exists $N \in \mathbb{N}_0$ such that $f(x_1,\dots,x_n) = g(x_1,\dots,x_n) + c(x_1,\dots,x_n)$, where $c: \mathbb{Z}_p^n \to \mathbb{Z}_p$ is a compatible function* (which is identically 0 modulo p^N) *and $g: \mathbb{Z}_p^n \to \{0,1,\dots,p^N-1\}$ is a periodic function with a period of length p^N.*

Indeed, from the proof of Theorem 3.39, as well as from the proof of Proposition 3.35, it follows that $g(x_1,\dots,x_n) = f(x_1,\dots,x_n) \bmod p^N$ and $c(x_1,\dots,x_n) = f(x_1,\dots,x_n) - g(x_1,\dots,x_n)$, where the mapping $\cdot \bmod p^N : \mathbb{Z}_p \to \{0,1,\dots, p^N-1\}$ is just a reduction modulo p^N of a p-adic integer:

$$z \bmod p^N = \delta_0(z) + \delta_1(z)\cdot p + \dots + \delta_{N-1}(z)\cdot p^{N-1}.$$

Thus, the most essential component of any asymptotically compatible function is a compatible function: for instance, the function f is differentiable if and only if its compatible summand c is differentiable since every periodic function with a period whose length is a power of p is differentiable everywhere on $\mathbb{Z}_p$ and its derivative is 0. So in the sequel we focus our study on compatible functions making remarks about asymptotically compatible ones whenever it is reasonable.

From Subsection 1.2.1 we know that polynomial mappings of a universal algebra are compatible; thus, all polynomials with p-adic integer coefficients are 1-Lipschitz. Since a derivative of this polynomial is also a polynomial with p-adic integer coefficients, the derivative is integer-valued. Integer-valued functions that have integer-valued derivatives are sometimes called *twice integer-valued*.

Polynomials over $\mathbb{Z}_p$ are important examples of twice integer-valued functions. Yet there exists a much wider class of twice integer-valued p-adic functions. The following easy proposition holds:

Proposition 3.41. *Let a compatible function* $F = (f_1, \dots, f_m)\colon \mathbb{Z}_p^n \to \mathbb{Z}_p^m$ *be uniformly differentiable modulo* p^k *at the point* $\mathbf{u} \in \mathbb{Z}_p^n$. *Then* $\left|\frac{\partial_k f_i(\mathbf{u})}{\partial_k x_j}\right|_p \le 1$, *i.e., F has integer-valued derivatives modulo* p^k.

Proof. In view of Definition 3.27 it is sufficient to prove the proposition for $m = n = 1$. Now let a compatible mapping $f\colon \mathbb{Z}_p \to \mathbb{Z}_p$ be uniformly differentiable modulo p^k at the point $x \in \mathbb{Z}_p$; that is, $f(x + p^t s) \equiv f(x) + p^t s f_k'(x) \pmod{p^{k+K}}$ for all $t \ge K, s \in \mathbb{Z}_p$, and K sufficiently large. In particular, $f(x+p^K) \equiv f(x)+p^K f_k'(x) \pmod{p^{k+K}}$. Since the compatibility of f implies that $f(x + p^K) - f(x) = rp^K$ for a suitable $r \in \mathbb{Z}_p$, the latter congruence implies that $rp^K = p^K f_k'(x) + zp^{k+K}$ for suitable $z \in \mathbb{Z}_p$. We conclude finally that $f_k'(x) \in \mathbb{Z}_p$. □

Note 3.42. Obviously, Proposition 3.41 remains true for asymptotically compatible functions as well.

Now we state a criterion for a differentiability modulo p of a compatible univariate function and find a formula for a derivative modulo p.

Theorem 3.43. *A compatible function* $f\colon \mathbb{Z}_p \to \mathbb{Z}_p$ *is differentiable modulo p at the point* $u \in \mathbb{Z}_p$ *if and only if*

$$\frac{\Delta^i f(u)}{i} \equiv 0 \pmod{p}$$

for all sufficiently large i. If this condition is satisfied, the derivative $f_1'(u)$ modulo p of the function f at the point u is

$$f_1'(u) \equiv \sum_{i=1}^{\infty}(-1)^{i-1}\frac{\Delta^i f(u)}{i} \equiv \sum_{t=0}^{\infty}\sum_{k=1}^{p-1}(-1)^{k-1}\frac{\Delta^{kp^t} f(u)}{kp^t} \pmod{p}.$$

Note 3.44. Since f is compatible, the fraction $\frac{\Delta^j f(u)}{j}$ is a p-adic integer for all $j \in \mathbb{N}$, see Proposition 3.38.

To prove the theorem, we need some technical lemmas.

Lemma 3.45. *Let* $f\colon \mathbb{Z}_p \to \mathbb{Z}_p$ *be a compatible function, let* $u \in \mathbb{Z}_p$, *and let a base-p expansion of i contain more than one nonzero digits* (i.e., $i \ne p^\alpha l$ for $\alpha \in \{0, 1, 2, \dots\}, l \in \{1, 2, \dots, p-1\}$). *Then* $\frac{1}{i}\Delta^i f(u) \equiv 0 \pmod{p}$.

Proof. Since

$$\frac{1}{i}\Delta^i f(u) = \sum_{j=1}^{i}(-1)^{i+j}\binom{i-1}{j-1}\frac{1}{j}(f(u+j)-f(u)),$$

see (3.16) and (3.14) of Proposition 3.38, it is sufficient to demonstrate that

$$S(i) = \sum_{j=1}^{\infty}(-1)^j\binom{i-1}{j-1}\frac{1}{j}(f(u+j)-f(u)) \equiv 0 \pmod p$$

whenever $i \neq lp^\alpha$, where $l \in \{1, 2, \dots, p-1\}$ and $\alpha \in \mathbb{N}_0$. Note that all fractions $\frac{1}{j}(f(u+j)-f(u))$ are p-adic integers since f is compatible.

Represent $j \in \mathbb{N}$ as $j = p^r l + p^{r+1}t$ where $r = \mathrm{ord}_p\, j$, $l \in \{1, 2, \dots, p-1\}$, $t \in \mathbb{N}_0$. We have then

$$S(i) = \sum_{r=0}^{\infty}\sum_{l=1}^{p-1}\sum_{t=0}^{\infty}(-1)^{p^r l + p^{r+1}t}\binom{i-1}{p^r l + p^{r+1}t - 1}\frac{f(u+p^r l + p^{r+1}t) - f(u)}{p^r l + p^{r+1}t}. \tag{3.18}$$

The compatibility of f implies that

$$f(u + p^r l + p^{r+1}t) = f(u + p^r l) + p^{r+1}\xi$$

for a suitable $\xi \in \mathbb{Z}_p$; hence

$$\frac{f(u+p^r l + p^{r+1}t) - f(u)}{p^r l + p^{r+1}t} \equiv \frac{f(u+p^r l) - f(u)}{p^r l + p^{r+1}t} \pmod p \tag{3.19}$$

since $l + pt$ is a unit in $\mathbb{Z}_p$. Whence

$$\frac{f(u+p^r l) - f(u)}{p^r l + p^{r+1}t} \equiv \frac{f(u+p^r l) - f(u)}{p^r l} \pmod p \tag{3.20}$$

since $(l+pt)^{-1} \equiv l^{-1} \pmod p$. Now from (3.18) in view of (3.19) and of (3.20) we conclude that

$$S(i) \equiv \sum_{r=0}^{\infty}\sum_{l=1}^{p-1}\frac{f(u+p^r l) - f(u)}{p^r l}\sum_{t=0}^{\infty}(-1)^{l+t}\binom{i-1}{p^r l + p^{r+1}t - 1} \pmod p, \tag{3.21}$$

since $(-1)^p \equiv -1 \pmod p$ for every prime p. Denote

$$\sigma_r(i) = \sum_{t=0}^{\infty}(-1)^{l+t}\binom{i-1}{p^r l + p^{r+1}t - 1}. \tag{3.22}$$

Note that whenever s is a p-adic integer, $\operatorname{ord}_p s = k$, then the jth digit $\delta_j(s-1)$ of the base-p expansion of $s-1$ is $p-1$ for $j < k$. With this in mind, we consider cases $\operatorname{ord}_p i < r$, $\operatorname{ord}_p i > r$, and $\operatorname{ord}_p i = r$ separately.

Case 1: $\operatorname{ord}_p i < r$. The above note in view of Lucas' Theorem 1.2 implies that

$$\binom{i-1}{p^r l + p^{r+1}t - 1} \equiv 0 \pmod p$$

whenever $\operatorname{ord}_p i < r$, and consequently, that $\sigma_r(i) \equiv 0 \pmod p$ in this case.

Case 2: $\operatorname{ord}_p i > r$. In this case Lucas' Theorem 1.2 implies that

$$\binom{i-1}{p^r l + p^{r+1}t - 1} \equiv \binom{p-1}{l-1}\binom{\nu(i,r)}{t} \pmod p, \tag{3.23}$$

where $\nu(i,r) = \lfloor \frac{i-1}{p^{r+1}} \rfloor$, the integral part of $\frac{i-1}{p^{r+1}}$. Now, since

$$\binom{p-1}{l-1} \equiv (-1)^{l-1} \pmod p,$$

combining (3.23) and (3.22) we conclude that

$$\sigma_r(i) \equiv -\sum_{t=0}^{\infty} (-1)^t \binom{\nu(i,r)}{t} \pmod p. \tag{3.24}$$

Further, since

$$\sum_{\ell=0}^{\infty} (-1)^\ell \binom{m}{\ell} = \begin{cases} 1, & \text{if } m = 0, \\ 0, & \text{otherwise,} \end{cases}$$

the right hand part of (3.24) is zero modulo p whenever $\nu(i,r) \neq 0$, that is, whenever $i > p^{r+1}$. However, we are considering the case $\operatorname{ord}_p i > r$; thus, since the conditions $\operatorname{ord}_p i > r$ and $i \leq p^{r+1}$ hold simultaneously only if $i = p^{r+1}$, the condition $\sigma_r(i) \not\equiv 0 \pmod p$ necessarily implies that $i = p^{r+1}$ in the case under consideration.

Case 3: $\operatorname{ord}_p i = r$. In a manner similar to the one of case 2 we prove that

$$\sigma_r(i) \equiv \sum_{t=0}^{\infty} (-1)^{l+t} \binom{\delta_r(i)-1}{l-1}\binom{\nu(i,r)}{t} \pmod p,$$

and that the sum in the right hand part of this congruence may not vanish modulo p only if the following two conditions $\delta_r(i) \geq l$ and $\nu(i,r) = 0$ hold simultaneously. But these two conditions hold simultaneously only if $i = p^r \delta_r(i)$. This in view of (3.21) and (3.22) finishes the proof of Lemma 3.45. □

Lemma 3.46. *Let $f : \mathbb{Z}_p \to \mathbb{Z}_p$ be a compatible function, and let $u, h \in \mathbb{Z}_p$. Then the following congruence holds*:

$$f(u+h) \equiv f(u) + h\Bigg(\tilde{f}_m(u) + \sum_{i=2}^{p-1} \frac{\Delta^{ip^m} f(u)}{ip^m} \binom{h-1}{ip^m - 1}\Bigg) \pmod{p^{m+1}},$$

where $m = \operatorname{ord}_p h$ and

$$\tilde{f}_m(u) \equiv \sum_{t=0}^{m-1} \sum_{l=1}^{p-1} (-1)^{l-1} \frac{\Delta^{lp^t} f(u)}{lp^t} + \frac{\Delta^{p^m} f(u)}{p^m} \pmod{p}.$$

In particular, if $p = 2$ then

$$f(u+h) \equiv f(u) + h \sum_{i=0}^{m} \frac{\Delta^{2^i} f(u)}{2^i} \pmod{2^{m+1}}.$$

Proof. In view of the compatibility of f it is sufficient to prove the lemma under assumption that $h = p^m\chi$, $\chi \in \{1, 2, \dots, p-1\}$. Applying the Gregory–Newton formula of Theorem 1.5, we see that

$$f(u + p^m\chi) = \sum_{i=0}^{p^m\chi} \binom{p^m\chi}{i} \Delta^i f(u);$$

thus

$$f(u + p^m\chi) = f(u) + p^m\chi \sum_{i=1}^{p^m\chi} \binom{p^m\chi - 1}{i-1} \frac{\Delta^i f(u)}{i}$$

since

$$\ell\binom{\ell-1}{j-1} = j\binom{\ell}{j}.$$

Now Lemma 3.45 implies that

$$f(u + p^m\chi) \equiv f(u) + p^m\chi\Bigg(\sum_{t=0}^{m-1}\sum_{l=1}^{p-1} \binom{p^m\chi - 1}{p^t l - 1} \frac{\Delta^{lp^t} f(u)}{lp^t} + \sum_{j=1}^{\chi} \binom{p^m\chi - 1}{p^m j - 1} \frac{\Delta^{jp^m} f(u)}{jp^m}\Bigg) \pmod{p^{m+1}}.$$

From here, combining the congruence

$$\binom{\chi-1}{j-1} \equiv \binom{p^m\chi - 1}{p^m j - 1} \pmod{p}, \tag{3.25}$$

which follows immediately from Lucas' Theorem 1.2, and an obvious congruence

$$\binom{p-1}{k} \equiv (-1)^k \pmod{p},$$

we deduce that

$$f(u + p^m\chi) \equiv f(u) + p^m\chi\Bigg(\sum_{t=0}^{m-1}\sum_{l=1}^{p-1}(-1)^{l-1}\frac{\Delta^{lp^t} f(u)}{lp^t} + \sum_{j=1}^{\chi}\binom{\chi-1}{j-1}\frac{\Delta^{jp^m} f(u)}{jp^m}\Bigg) \pmod{p^{m+1}}.$$

The latter congruence implies that

$$f(u + p^m\chi) \equiv f(u) + p^m\chi\Bigg(\tilde{f}_m(u) + \sum_{j=2}^{p-1}\binom{\chi-1}{j-1}\frac{\Delta^{jp^m} f(u)}{jp^m}\Bigg) \pmod{p^{m+1}},$$

since $\chi \in \{1, 2, \dots, p-1\}$. This in view of (3.25) proves Lemma 3.46. □

Proof of Theorem 3.43. If $\frac{\Delta^i f(u)}{i} \equiv 0 \pmod{p}$ for all $i \geq N$ then in view of Lemma 3.46 the following congruences hold:

$$f(u+h) \equiv f(u) + h\tilde{f}_m(u) \pmod{p^{m+1}},$$
$$\tilde{f}_m(u) \equiv \tilde{f}_{m+1}(u) \equiv \cdots \pmod{p}$$

for all sufficiently small $h \in \mathbb{Z}_p$ (i.e. for all h with $|h|_p = p^{-m}$, where m sufficiently large). Consequently, f is differentiable modulo p at the point $u \in \mathbb{Z}_p$.

Vice versa, let the function f be differentiable modulo p at the point u, i.e. let there exist $N \in \mathbb{N}$ and $c \in \mathbb{Q}_p$ such that

$$f(u+h) \equiv f(u) + hc \pmod{p^{m+1}}, \tag{3.26}$$

where $|h|_p = p^{-m}, m \geq N$. From (3.26) in view of Lemma 3.46 we deduce that

$$\tilde{f}_m(u) + \sum_{j=2}^{p-1}\binom{h-1}{jp^m-1}\frac{\Delta^{jp^m} f(u)}{jp^m} \equiv c \pmod{p} \tag{3.27}$$

for all $m \geq N$. In the case $p = 2$ the sum in the left hand part of congruence (3.27) vanishes, so suppose for a moment that $p \neq 2$. According to Lucas' Theorem 1.2 we then have

$$\sum_{j=2}^{p-1}\binom{h-1}{jp^m-1}\frac{\Delta^{jp^m} f(u)}{jp^m} \equiv \sum_{j=2}^{p-1}\binom{h_m-1}{j-1}\frac{\Delta^{jp^m} f(u)}{jp^m} \pmod{p},$$

where $h_m = \delta_m(h)$, the mth p-adic digit of h. So in view of (3.27) the function $\Psi_u(h_m)$ defined by the equation

$$\Psi_u(h_m) = \delta_0\left(\sum_{j=2}^{p-1}\binom{h_m - 1}{j-1}\frac{\Delta^{jp^m} f(u)}{jp^m}\right)$$

is a constant whenever $|h|_p = p^{-m}, m \geq N$. In particular, $\Psi_u(h_m) = \Psi_u(1) = 0$, and this implies that for all $m \geq N$ the following system of congruences modulo p holds:

$$\sum_{j=2}^{p-1}\binom{k-1}{j-1}\frac{\Delta^{jp^m} f(u)}{jp^m} \equiv 0 \pmod{p} \quad (k = 2, 3, \dots, p-1). \tag{3.28}$$

System (3.28) of congruences is triangular, so necessarily

$$\frac{\Delta^{jp^m} f(u)}{jp^m} \equiv 0 \pmod{p} \quad (j = 2, 3, \dots, p-1) \tag{3.29}$$

for all $m \geq N$. Now from (3.27) in view of (3.29) and Lemma 3.46, we deduce that for each prime p the following congruence holds:

$$\sum_{t=0}^{N-1}\sum_{l=1}^{p-1}(-1)^{l-1}\frac{\Delta^{lp^t} f(u)}{lp^t} + \sum_{s=N}^{m}\frac{\Delta^{p^s} f(u)}{p^s} \equiv c \pmod{p}, \tag{3.30}$$

where c does not depend on m. Hence

$$\frac{\Delta^{p^s} f(u)}{p^s} \equiv 0 \pmod{p} \tag{3.31}$$

for all $s \geq N + 1$. Finally combining (3.29) and (3.31) with Lemma 3.45, we obtain that

$$\frac{\Delta^i f(u)}{i} \equiv 0 \pmod{p}$$

for all $i \geq p^N + 1$. The second statement of Theorem 3.43 follows from (3.30) in view of Lemma 3.45 since $c \equiv f_1'(u) \pmod{p}$, see (3.26). □

Now it is worth comparing here notions of differentiability and of differentiability modulo p^k once again. As for differentiability of a function $f : \mathbb{Z}_p \to \mathbb{Q}_p$ at the point $u \in \mathbb{Z}_p$, the following result is known (see e.g. [308, Chapter 13, Theorem 1]):

Theorem 3.47. *A function $f : \mathbb{Z}_p \to \mathbb{Q}_p$ is differentiable at the point $u \in \mathbb{Z}_p$ if and only if*

$$\lim_{i\to\infty}^{p} \frac{\Delta^i f(u)}{i} = 0.$$

If this condition is satisfied, the derivative $f'(u)$ of the function f at the point u is

$$f'(u) = \sum_{i=1}^{\infty}(-1)^{i-1}\frac{\Delta^i f(u)}{i}.$$

Comparing Theorem 3.47 to Theorem 3.43 it is reasonable to suppose that a similar result should hold for differentiability modulo p^k, $k \geq 2$. Note that the case $k = 2$ is of highest importance in view of Theorem 4.55 on ergodicity. Thus we set the following problem:

Open Question 3.48. Is it true that a compatible function $f : \mathbb{Z}_p \to \mathbb{Z}_p$ is differentiable modulo p^k ($k \geq 2$) at the point $u \in \mathbb{Z}_p$ if and only if

$$\frac{\Delta^i f(u)}{i} \equiv 0 \pmod{p^k}$$

for all sufficiently large i?

Note that anyway a formula from Theorem 3.47 holds for a derivative modulo p^k as well, in the following sense:

Proposition 3.49. *If the function $f : \mathbb{Z}_p \to \mathbb{Z}_p$ is differentiable modulo p^k at the point $u \in \mathbb{Z}_p$, then*

$$f'_k(u) \equiv \lim_{\ell\to\infty}\left(\sum_{i=1}^{z_\ell}(-1)^{i-1}\frac{\Delta^i f(u)}{i}\right) \bmod p^k$$

for every sequence $\{z_\ell \in \mathbb{N}_0\}_{\ell=0}^{\infty}$ that converges to 0 with respect to the p-adic metric.

Proof. Applying the Gregory–Newton formula of Theorem 1.5, we see that

$$f(u + z_\ell) = \sum_{i=0}^{z_\ell}\binom{z_\ell}{i}\Delta^i f(u);$$

thus

$$f(u + z_\ell) = f(u) + z_\ell \cdot \sum_{i=1}^{z_\ell}\binom{z_\ell - 1}{i - 1}\frac{\Delta^i f(u)}{i}$$

since

$$z_\ell\binom{z_\ell - 1}{j - 1} = j\binom{z_\ell}{j}.$$

However, as $\binom{z}{j}$ is a continuous function on $\mathbb{Z}_p$, $\lim^p_{z\to 0}\binom{z-1}{j} = (-1)^j$, so

$$\frac{f(u + z_\ell) - f(u)}{z_\ell} = \sum_{i=1}^{z_\ell}\binom{z_\ell - 1}{i - 1}\frac{\Delta^i f(u)}{i} \equiv \sum_{i=1}^{z_\ell}(-1)^{i-1}\frac{\Delta^i f(u)}{i} \pmod{p^k}$$

for all sufficiently large ℓ. As $\lim^{p}_{\ell\to\infty}\left(\frac{f(u+z_\ell)-f(u)}{z_\ell} \bmod p^k\right) = f'_k(u)$ by the definition of a derivative modulo p^k, the conclusion follows. □

In other words, Proposition 3.49 claims that the function

$$S(z) = \sum_{i=1}^{z}(-1)^{i-1}\frac{\Delta^i f(u)}{i} \bmod p^k$$

of variable z is constant on a sufficiently small ball $p^N\mathbb{Z}_p$: $S(z) = f'_k(u)$ for all $z \in p^N\mathbb{Z}_p$. That is, differentiability modulo p^k implies that all sums

$$\sum_{i=p^N t+1}^{p^N(t+1)} (-1)^{i-1}\frac{\Delta^i f(u)}{i}$$

are 0 modulo p^k, for all $t = 1, 2, \ldots$ and sufficiently large N, and our Question 3.48 asks whether differentiability modulo p^k implies that all terms of these sums are 0 modulo p^k. Now we only know that the answer is affirmative for $k = 1$ (see Theorem 3.43); for $k > 1$ the problem is still open.

3.9 Mahler expansion

In this section, we introduce Mahler expansion, a useful technique which we will need in further chapters to study dynamics produced by a compatible (i.e., 1-Lipschitz) mapping. We characterize p-adic 1-Lipschitz functions in terms of Mahler expansion in this section as well. We follow works [21, 22, 24] in further considerations in the section.

Every function $f : \mathbb{N}_0 \to \mathbb{Z}_p$ (or, respectively, $f : \mathbb{N}_0 \to \mathbb{Z}$) has the only *Mahler expansion*, that is, has a unique representation via the so-called *Mahler interpolation series*

$$f(x) = \sum_{i=0}^{\infty} a_i \binom{x}{i}, \tag{3.32}$$

where $a_i \in \mathbb{Z}_p$ (respectively, $a_i \in \mathbb{Z}$), $i = 0, 1, 2, \ldots$, and

$$\binom{x}{i} = \frac{x(x-1)\cdots(x-i+1)}{i!}$$

for $i = 1, 2, \ldots$;

$$\binom{x}{0} = 1,$$

by the definition.

Various properties of the function $f : \mathbb{Z}_p \to \mathbb{Z}_p$ can be expressed via properties of coefficients of its Mahler expansion. We recall some basic facts about Mahler series, referring to [308] or [374] for their proofs.

If f is uniformly continuous on $\mathbb{N}_0$ with respect to the p-adic metric, it can be uniquely expanded to a uniformly continuous function on $\mathbb{Z}_p$. Hence the interpolation series for f converges uniformly on $\mathbb{Z}_p$. The following is true: The series (3.32) converges uniformly on $\mathbb{Z}_p$ if and only if

$$\overset{p}{\lim_{i\to\infty}} a_i = 0. \tag{3.33}$$

Hence a uniformly convergent series defines a uniformly continuous function on $\mathbb{Z}_p$. The function f represented by the interpolation series (3.32) is (uniformly) differentiable everywhere on $\mathbb{Z}_p$ if and only if

$$\overset{p}{\lim_{i\to\infty}} \frac{a_{i+n}}{i} = 0 \tag{3.34}$$

for all $n \in \mathbb{N}_0$; in this case the following formula for the derivative holds:

$$f'(x) = \sum_{i=1}^{\infty} (-1)^{i+1} \frac{\Delta^i f(x)}{i}. \tag{3.35}$$

The function f is analytic on $\mathbb{Z}_p$ if and only if

$$\overset{p}{\lim_{i\to\infty}} \frac{a_i}{i!} = 0. \tag{3.36}$$

To represent functions of several variables we use interpolation series of the following form:

$$f(x_1, \dots, x_n) = \sum_{(i_1,\dots,i_n)\in\mathbb{N}_0^n} a_{i_1,\dots,i_n} \binom{x_1}{i_1}\binom{x_2}{i_2}\cdots\binom{x_n}{i_n}; \tag{3.37}$$

here $a_{i_1,\dots,i_n} \in \mathbb{Z}_p$.

Open Question 3.50. Find an analog of condition (3.34) for uniform differentiability modulo p^k on $\mathbb{Z}_p$.

3.9.1 Identities modulo p^k

This is an auxiliary subsection; we describe here a special class of functions, which are, loosely speaking, sufficiently small with respect to a p-adic metric, but not too small.

Definition 3.51. A function $F : \mathbb{Z}_p^n \to \mathbb{Z}_p^m$ is called *an identity modulo* p^k if for every $\mathbf{u} \in \mathbb{Z}_p^n$ the following congruence holds:

$$F(\mathbf{u}) \equiv (0, \dots, 0) \pmod{p^k}.$$

In other words, F is an identity modulo p^k if and only if $|F(\mathbf{u})|_p \le p^{-k}$ for all $\mathbf{u} \in \mathbb{Z}_p^n$.

We need to characterize identities modulo p^k in order to study the behavior of compatible functions modulo some p^k since it is clear that two compatible functions coincide modulo p^k if and only if their difference is an identity modulo p^k. The following easy proposition characterizes identities modulo p^k in terms of Mahler expansion.

Proposition 3.52. *A function $f : \mathbb{Z}_p^n \to \mathbb{Z}_p$ is an identity modulo p^k if and only if all coefficients of its Mahler expansion* (3.37) *are 0 modulo p^k:*

$$|a_{i_1,\dots,i_n}|_p \le p^{-k}$$

for all $(i_1, \dots, i_n) \in \mathbb{N}_0^n$.

Proof. Induction on n. Let $n = 1$. As f is a continuous function, and as $\mathbb{N}_0$ is a dense subset in $\mathbb{Z}_p$, f is an identity modulo some p^k if and only if

$$\sum_{i=0}^{s} a_i \binom{s}{i} \equiv 0 \pmod{p^k} \tag{3.38}$$

for all $s = 0, 1, 2, \dots$. However, a triangular system of congruences (3.38) has a unique solution

$$0 \equiv a_0 \equiv a_1 \equiv a_2 \equiv \cdots \pmod{p^k}; \tag{3.39}$$

hence for $n = 1$ the proposition is true. As

$$f(x_1, \dots, x_{n-1}, s) = \sum_{i=0}^{s} g_i(x_1, \dots, x_{n-1}) \binom{x}{i}$$

for every $s \in \mathbb{N}_0$, then by a similar argument we conclude that $f(x_1, \dots, x_n)$ is an identity modulo p^k if and only if

$$g_i(x_1, \dots, x_{n-1}) \equiv 0 \pmod{p^k}$$

for all $x_1, \dots, x_{n-1} \in \mathbb{Z}_p$ and all $i = 0, 1, 2, \dots$. By the induction, in view of (3.37) the latter condition holds if and only if the congruences

$$a_{i_1,\dots,i_{n-1},i} \equiv 0 \pmod{p^k}$$

hold for all $i = 0, 1, 2, \dots$. □

3.9.2 Mahler expansions of compatible functions

In this subsection we characterize compatible functions in terms of Mahler expansions. Recall that $\lfloor\alpha\rfloor$ for a real α denotes the integral part of α, that is, the nearest to α rational integer which does not exceed α. Note that

$$\lfloor\log_p \alpha\rfloor = (\text{a number of digits in a base-}p \text{ expansion for } \alpha) - 1.$$

So to unify notation we assume further that $\lfloor\log_p 0\rfloor = 0$, by the definition.

Theorem 3.53. *A function* $f : \mathbb{Z}_p^n \to \mathbb{Z}_p$ *represented by Mahler expansion* (3.37) *is compatible if and only if*

$$|a_{i_1,\ldots,i_n}|_p \leq p^{-\mu(i_1,\ldots,i_n)},$$

where $\mu(i_1,\ldots,i_n) = \max\{\lfloor\log_p i_k\rfloor : k = 1,2,\ldots,n\}$.

In particular, a univariate function $f : \mathbb{Z}_p \to \mathbb{Z}_p$ *represented by Mahler expansion* (3.32) *is compatible if and only if*

$$|a_i|_p \leq p^{-\lfloor\log_p i\rfloor}$$

for all $i = 1,2,\ldots$.

Proof. Induction on n. Let $n = 1$. According to Proposition 3.38, the function f is compatible if and only if $\frac{\Delta^i f(x)}{i}$ is a p-adic integer for all $x \in \mathbb{Z}_p$, $i = 1,2,\ldots$. Yet

$$\frac{\Delta^i f(x)}{i} = \frac{1}{i}\sum_{j=i}^{\infty} a_j \binom{x}{j-i} \tag{3.40}$$

in view of (1.1). Now (3.40) implies that $\frac{\Delta^i f(x)}{i}$ is a p-adic integer if and only if

$$\sum_{j=i}^{\infty} a_j \binom{x}{j-i}$$

is an identity modulo $p^{\operatorname{ord}_p i}$. Proposition 3.52 implies now that $\frac{\Delta^i f(x)}{i}$ is a p-adic integer if and only if the following congruences hold simultaneously for all $j \geq i$:

$$a_j \equiv 0 \pmod{p^{\operatorname{ord}_p i}}. \tag{3.41}$$

Thus, f is compatible if and only if congruences (3.41) hold simultaneously for all $i = 1,2,\ldots$ and all $j \geq i$. This means (since $\lfloor\log_p j\rfloor = \max\{\operatorname{ord}_p i : i = 1,2,3,\ldots,j\}$) that the following congruences hold simultaneously:

$$a_j \equiv 0 \pmod{p^{\lfloor\log_p j\rfloor}} \quad (j = 1,2,3,\ldots).$$

This proves Theorem 3.53 for $n = 1$.

Now let the statement of the theorem be true for all r-variate functions that satisfy the conditions of the theorem, $r < n$. Represent

$$f(x_1, \dots, x_n) = \sum_{j=0}^{\infty} g_j(x_1, \dots, x_{n-1}) \binom{x_n}{j},$$

where all functions g_j are uniformly continuous on $\mathbb{Z}_p^{n-1}$, for all $j = 1, 2, \dots$:

$$g_j(x_1, \dots, x_{n-1}) = \sum_{(i_1,\dots,i_{n-1})\in\mathbb{N}_0^{n-1}} a_{i_1,\dots,i_{n-1},j} \binom{x_1}{i_1}\binom{x_2}{i_2}\cdots\binom{x_{n-1}}{i_{n-1}}.$$

According to Proposition 3.38, the function $f(x_1, \dots, x_n)$ is compatible if and only if all fractions $\frac{1}{i}\Delta_s^i f(x_1, \dots, x_n)$ are p-adic integers, for all $i = 1, 2, \dots$, all $s = 1, 2, \dots, n$, and all $x_1, \dots, x_n \in \mathbb{Z}_p$. Using an argument similar to that of the case $n = 1$ we conclude that

$$\frac{1}{i}\Delta_s^i f(x_1, \dots, x_n) = \begin{cases} \sum_{j=i}^{\infty} \frac{1}{i} g_j(x_1, \dots, x_{n-1})\binom{x_n}{j-i}, & \text{if } s = n, \\ \sum_{j=0}^{\infty} \frac{1}{i}\Delta_s^i g_j(x_1, \dots, x_{n-1})\binom{x_n}{j}, & \text{otherwise.} \end{cases} \tag{3.42}$$

If $s \neq n$, all functions $\frac{1}{i}\Delta_s^i f(x_1, \dots, x_n)$ $(i = 1, 2, \dots)$ are simultaneously integer-valued if and only if all functions $\frac{1}{i}\Delta_s^i g_j(x_1, \dots, x_{n-1})$ are simultaneously integer-valued, for all $j = 0, 1, 2, \dots$ and all $i = 1, 2, \dots$. This in force of Proposition 3.38 implies that every function $g_j(x_1, \dots, x_n)$ $(j = 0, 1, 2, \dots)$ is compatible. By induction hypothesis, the latter holds if and only if the following inequalities hold simultaneously:

$$|a_{i_1,\dots,i_{n-1},j}|_p \le p^{-\mu(i_1,\dots,i_{n-1})} \quad (j, i_1, \dots, i_n \in \mathbb{N}_0). \tag{3.43}$$

If $s = n$, then by an argument similar to that of the case $n = 1$ from (3.42) we deduce that all functions $\frac{1}{i}\Delta_n^i f(x_1, \dots, x_n)$ $(i = 1, 2, \dots)$ are integer-valued if and only if the following inequalities hold simultaneously for all $j = 1, 2, \dots$ and all $x_1, \dots, x_{n-1} \in \mathbb{Z}_p$:

$$|g_j(x_1, \dots, x_{n-1})|_p \le p^{-\lfloor \log_p j \rfloor}. \tag{3.44}$$

But these conditions imply that every function $g_j(x_1, \dots, x_{n-1})$ is an identity modulo $p^{\lfloor \log_p j \rfloor}$; whence, in view of Proposition 3.52, the following conditions hold simultaneously for all $i_1, \dots, i_{n-1} \in \mathbb{N}_0$ and all $j \in \mathbb{N}$:

$$|a_{i_1,\dots,i_{n-1},j}|_p \le p^{-\lfloor \log_p j \rfloor}. \tag{3.45}$$

Now combining (3.43) with (3.45) we finish the proof of Theorem 3.53. □

Corollary 3.54 (cf. [166]). *An integer-valued polynomial* $f(x) \in \mathbb{Q}[x]$ *is compatible as a mapping of the ring* $\mathbb{Z}$ *into* $\mathbb{Z}$ (that is, according to Definition 1.18, a congruence $a \equiv b \pmod{m}$ implies a congruence $f(a) \equiv f(b) \pmod{m}$, for all $m \in \mathbb{N}\setminus\{1\}$ and all $a, b \in \mathbb{Z}$) *if and only if* f *can be represented in the following form*:

$$f(x) = a_0 + \sum_{i=1}^{d} a_i \cdot \operatorname{lcm}(1, 2, \ldots, i) \cdot \binom{x}{i},$$

where $a_0, a_1, \ldots \in \mathbb{Z}$*, and* $\operatorname{lcm}(k, l, m, \ldots)$ *for* $k, l, m, \ldots \in \mathbb{N}$ *is the least common multiple of* $k, l, m, \ldots$.

Proof. The result follows immediately from Theorem 3.53: The compatibility of f on the ring $\mathbb{Z}$ is obviously equivalent to the compatibility of f on all rings $\mathbb{Z}_p$, for each prime p; now just note that $p^{\lfloor \log_p i \rfloor}$ is the greatest power of p which does not exceed i. □

3.10 Special classes of locally analytic functions

In this section we study some important classes of locally analytic functions on $\mathbb{Z}_p$, which were originally introduced in [24].

3.10.1 Class $\mathcal{C}$

Note 3.55. According to Section 3.2, the power series $\sum_{i=0}^{\infty} c_i x^i$, where $c_i \in \mathbb{Q}_p$ for $i = 0, 1, 2 \ldots$, converges everywhere on $\mathbb{Z}_p$ if and only if $\lim^p_{i\to\infty} c_i = 0$; under the latter condition the series defines a continuous function on $\mathbb{Z}_p$.

Of course, in general a function defined by this series may not be integer-valued, not speaking about compatibility. Consider, however, a special case when all coefficients c_i are p-adic integers. Namely, in the ring $\mathbb{Z}_p[[x]]$ of all formal power series in one variable x over the ring $\mathbb{Z}_p$ consider a set $\mathcal{C}(x)$ of all series

$$s(x) = \sum_{i=0}^{\infty} c_i x^i \quad (c_i \in \mathbb{Z}_p,\ i = 0, 1, 2, \ldots) \tag{3.46}$$

that converge everywhere on $\mathbb{Z}_p$. In other words, $s(x) \in \mathcal{C}(x)$ if and only if $\lim^p_{i\to\infty} c_i = 0$. Under these assumptions the series $s(x) \in \mathcal{C}(x)$ defines on $\mathbb{Z}_p$ an integer-valued function $s \colon \mathbb{Z}_p \to \mathbb{Z}_p$, which is called a *$\mathcal{C}$-function*.

Proposition 3.56. *Every $\mathcal{C}$-function $s \colon \mathbb{Z}_p \to \mathbb{Z}_p$ is uniformly differentiable on $\mathbb{Z}_p$; its derivative is integer-valued everywhere on $\mathbb{Z}_p$.*

Proof. From Theorem 3.12 we already know that the function s is differentiable. Consider a formal derivative $s'(x) \in \mathbb{Z}_p[[x]]$ of the series $s(x)$:

$$s'(x) = \sum_{i=1}^{\infty} i c_i x^{i-1}.$$

Since $0 \leq |ic_i|_p = |i|_p|c_i|_p \leq |c_i|_p$, and $\lim^p_{i\to\infty} c_i = 0$, we conclude that $\lim^p_{i\to\infty} ic_i = 0$, and hence that $s'(x) \in \mathcal{C}(x)$. We assert that the function $s' : \mathbb{Z}_p \to \mathbb{Z}_p$ is a derivative of the function $s : \mathbb{Z}_p \to \mathbb{Z}_p$ with respect to the p-adic metric.

Indeed, it is known that in the ring $\mathbb{Z}_p[[x, y]]$ of all formal power series in variables x, y over $\mathbb{Z}_p$ the following equality holds:

$$s(x + y) = \sum_{i=0}^{\infty} \frac{s^{(i)}(x)}{i!} y^i,$$

where $s^{(i)}(x) \in \mathbb{Z}_p[[x]]$ $(i = 1, 2, \ldots)$ is the ith formal derivative of the series $s(x)$, and $s^{(0)}(x) = s(x)$. By the assertion just proved, $s^{(i)}(x) \in \mathcal{C}(x)$ for all $i = 0, 1, 2, \ldots$. Thus,

$$\frac{s^{(i)}(u)}{i!} = \sum_{j=i}^{\infty} c_j \binom{j}{i} u^{j-i} \in \mathbb{Z}_p \tag{3.47}$$

for every $u \in \mathbb{Z}_p$. However,

$$\left|\frac{s^{(i)}(u)}{i!}\right|_p = \left|\sum_{j=i}^{\infty} c_j \binom{j}{i} u^{j-i}\right|_p \leq \max\{|c_j|_p : j = i, i + 1, \ldots\},$$

and consequently,

$$\lim^p_{i\to\infty} \frac{s^{(i)}(u)}{i!} = 0, \tag{3.48}$$

since $\lim^p_{i\to\infty} c_i = 0$. Thus, for every $u \in \mathbb{Z}_p$ we conclude that

$$s(u + y) = \sum_{i=0}^{\infty} \frac{s^{(i)}(u)}{i!} y^i \in \mathcal{C}(y). \tag{3.49}$$

Finally, if $s(x) \in \mathcal{C}(x)$, then the Taylor series (3.49) at the point $u \in \mathbb{Z}_p$ converges to s everywhere on $\mathbb{Z}_p$. In particular, for $h \in \mathbb{Z}_p$ we obtain that

$$s(u + h) = s(u) + s'(u)h + \alpha(u, h),$$

where

$$\lim^p_{h\to 0} \frac{\alpha(u, h)}{h} = \lim^p_{h\to 0} h \sum_{i=2}^{\infty} \frac{s^{(i)}(u)}{i!} h^{i-2} = 0,$$

since $\sum_{i=2}^{\infty} \frac{s^{(i)}(u)}{i!} h^{i-2} \in \mathbb{Z}_p$ in view of (3.47), (3.48) and of Note 3.55. Moreover,

$$\left|\frac{\alpha(u,h)}{h}\right|_p = \left|h \sum_{i=2}^{\infty} \frac{s^{(i)}(u)}{i!} h^{i-2}\right|_p \leq |h|_p$$

for all $u, h \in \mathbb{Z}_p$. Whence, s is uniformly differentiable on $\mathbb{Z}_p$, and s' is a derivative of the function s. □

From this proposition we immediately deduce the following

Corollary 3.57. *A class $\mathcal{C}$ of all $\mathcal{C}$-functions is closed with respect to derivations; all $\mathcal{C}$-functions are infinitely many times differentiable.*

Now consider Mahler expansions for functions defined by series from $\mathcal{C}(x)$: Let

$$s(x) = \sum_{i=0}^{\infty} s_i \binom{x}{i} \tag{3.50}$$

be an interpolation series for the function $s(x) \in \mathcal{C}(x)$ defined by convergent power series (3.46). We note:

Proposition 3.58. *All fractions $\frac{s_i}{i!}$ are p-adic integers, for all $i = 0, 1, 2, \ldots$.*

Proof. Indeed,

$$s(x) = \sum_{k=0}^{\infty} c_k x^k = \sum_{k=0}^{\infty} c_k \sum_{i=0}^{k} S(k,i) i! \binom{x}{i} = \sum_{i=0}^{\infty} i! \binom{x}{i} \sum_{k=i}^{\infty} S(k,i) c_k, \tag{3.51}$$

where $S(k,i)$ is a Stirling number of the second kind; that is, $S(k,i)$ the number of ways to partition a set of k elements into i nonempty subsets, see e.g. [158] for definitions and useful formulas.

Further, since all Stirling numbers $S(k,i)$ are rational integers, $|S(k,i)|_p \leq 1$; whence, as the power series (3.46) is convergent, $\lim_{i\to\infty}^{p} c_i = 0$, and thus $\lim_{k\to\infty}^{p} S(k,i)c_k = 0$. Consequently, the series $\sum_{k=i}^{\infty} S(k,i)c_k$ converges to some $A_i \in \mathbb{Z}_p$, for all $i = 0, 1, 2, \ldots$. This proves our assertion since

$$s_i = i! \sum_{k=i}^{\infty} S(k,i) c_k = i! A_i \quad (i = 0, 1, 2, \ldots), \tag{3.52}$$

see (3.51). □

In other words, Proposition 3.58 shows that any function defined by a series from $\mathcal{C}(x)$ can be represented as falling factorial series $s(x) = \sum_{i=0}^{\infty} b_i x^{\underline{i}}$ over $\mathbb{Z}_p$ (i.e., $b_i = \frac{s_i}{i!} \in \mathbb{Z}_p$ for all $i = 0, 1, 2, \ldots$) where $x^{\underline{0}} = 1$, $x^{\underline{i}} = x(x-1)\cdots(x-i+1)$ by the definition.

3.10.2 Class $\mathcal{B}$

We now consider a wider class $\mathcal{B}(x)$ of falling factorial series with p-adic integer coefficients; that is, $f(x) \in \mathcal{B}(x)$ if and only if $f(x) = \sum_{i=0}^{\infty} b_i x^{\underline{i}}$ ($b_i \in \mathbb{Z}_p$). In other words,

$$\mathcal{B}(x) = \left\{ \sum_{i=0}^{\infty} a_i \binom{x}{i} : \frac{a_i}{i!} \in \mathbb{Z}_p;\ i = 0, 1, 2, \ldots \right\}. \tag{3.53}$$

In force of a criterion for convergence of Mahler interpolation series (see (3.33)) series from $\mathcal{B}(x)$ are uniformly convergent on $\mathbb{Z}_p$ and thus define uniformly continuous functions on $\mathbb{Z}_p$, which we call *$\mathcal{B}$-functions*.

Denote by $\mathcal{B}$ a class of all functions defined by series from $\mathcal{B}(x)$. Note that any two distinct series from $\mathcal{B}(x)$ (respectively, from $\mathcal{C}(x)$) define two distinct functions on $\mathbb{Z}_p$: For functions defined by series from $\mathcal{B}(x)$ the assertion follows from the definition of $\mathcal{B}$-functions in view of Proposition 3.52. As for functions defined by series from $\mathcal{C}(x)$, we note that the above mentioned interpolation series (3.50) for $s(x) \in \mathcal{C}(x)$ defines a function, which is identically 0 on $\mathbb{Z}_p$ if and only if all coefficients s_i are 0. Whence, $A_i = 0$ for $i = 0, 1, 2, \ldots$, see (3.52). However, $A_i = \sum_{k=i}^{\infty} S(k,i)c_k$, thus $c_i = \sum_{k=i}^{\infty} s(k,i)A_k = 0$, where $s(k,i)$ are Stirling numbers of the first kind, and the assertion follows. So in the sequel we do not differ series from functions they define.

The class $\mathcal{B}$ is endowed with a non-Archimedean metric

$$D_p(f, g) = \max_{z \in \mathbb{Z}_p} |f(z) - g(z)|_p,$$

in other words, the distance between two $\mathcal{B}$-functions f and g is p^{-N} whenever N is the largest natural integer such that these functions are congruent to each other modulo p^N. The following is true:

Proposition 3.59. *The class $\mathcal{B}$ is a complete* (with respect to the metric D_p) *metric space of 1-Lipschitz functions that are differentiable everywhere on $\mathbb{Z}_p$. The class $\mathcal{B}$ is closed with respect to additions, multiplications, derivations, and compositions of functions. A countable set $\mathcal{P}$ of all polynomials with non-negative rational integer coefficients is a dense subset of $\mathcal{B}$. The class $\mathcal{C}$ is a proper subclass of $\mathcal{B}$: $\mathcal{C} \subset \mathcal{B}$, $\mathcal{C} \neq \mathcal{B}$.*

Proof. Combining Theorem 3.53 with Lemma 3.6 it is not difficult to demonstrate that a $\mathcal{B}$-function is compatible (that is, 1-Lipschitz), with the use of the obvious inequality $\mathrm{wt}_p\, i \le (p-1)(\lfloor \log_p i \rfloor + 1)$, which holds for all $i = 1, 2, \ldots$ and each prime p.

Now we prove that $\mathcal{B}$-functions are uniformly differentiable on $\mathbb{Z}_p$, and that $\mathcal{B}$ is closed with respect to derivations: If $f \in \mathcal{B}$, then $f' \in \mathcal{B}$. Recall that a uniformly continuous function $f : \mathbb{Z}_p \to \mathbb{Z}_p$ that is represented by the interpolation series (3.32) is uniformly differentiable on $\mathbb{Z}_p$ if an only if (3.34) holds for all $n \in \mathbb{N}_0$. Yet the latter condition is obviously true for $f \in \mathcal{B}$ since $\operatorname{ord}_p a_i \geqslant \operatorname{ord}_p i! = \frac{1}{p-1}(i - \operatorname{wt}_p i)$ (see Lemma 3.6), and $\lfloor \log_p i \rfloor \geqslant \operatorname{ord}_p i$ for all $i = 0, 1, 2, \ldots$. Thus, the derivative f' of the function f is defined everywhere on $\mathbb{Z}_p$, and

$$f'(x) = \sum_{i=1}^{\infty}(-1)^{i+1}\frac{\Delta^i f(x)}{i},$$

see (3.35). However, $\frac{\Delta^i f(x)}{i} = \frac{1}{i}\sum_{j=i}^{\infty} a_j \binom{x}{j-i}$; consequently,

$$\sum_{i=1}^{\infty}(-1)^{i+1}\frac{\Delta^i f(x)}{i} = \sum_{k=0}^{\infty}\binom{x}{k}\sum_{i=1}^{\infty}(-1)^{i+1}\frac{a_{k+i}}{i}. \tag{3.54}$$

Since (3.34) holds, the series $\sum_{i=1}^{\infty}(-1)^{i+1}\frac{a_{k+i}}{i}$ converges for every $k \in \mathbb{N}_0$ to some $S_k \in \mathbb{Q}_p$. Moreover,

$$\begin{aligned}
\operatorname{ord}_p \frac{a_{k+i}}{i} &= \operatorname{ord}_p a_{k+i} - \operatorname{ord}_p i \ \geq\ \operatorname{ord}_p (k+i)! - \lfloor \log_p i \rfloor \\
&= \frac{1}{p-1}(i + k - \operatorname{wt}_p(i+k)) - \lfloor \log_p i \rfloor \\
&= \frac{1}{p-1}(i - \operatorname{wt}_p i) - \lfloor \log_p i \rfloor + \frac{1}{p-1}(k - \operatorname{wt}_p k) \\
&\qquad + \frac{1}{p-1}(\operatorname{wt}_p k - \operatorname{wt}_p(i+k) + \operatorname{wt}_p i) \\
&\geq \frac{1}{p-1}(k - \operatorname{wt}_p k) \ =\ \operatorname{ord}_p k!,
\end{aligned}$$

where the latter inequality holds since $\frac{1}{p-1}(i - \operatorname{wt}_p i) \geq \lfloor \log_p i \rfloor$ and $\frac{1}{p-1}(\operatorname{wt}_p k - \operatorname{wt}_p(i+k) + \operatorname{wt}_p i) = \operatorname{ord}_p \binom{i+k}{i} \geq 0$, see Lemma 3.6 and Corollary 3.7. Thus, $\frac{S_k}{k!} \in \mathbb{Z}_p$ for all $k \in \mathbb{N}_0$; whence $f' \in \mathcal{B}$.

Now we prove that $\mathcal{B}$ is a closure (with respect to the metric D_p) of the class of all functions induced by polynomials with non-negative rational integer coefficients. Since every polynomial from $\mathbb{Z}_p[x]$ is congruent modulo p^k to some polynomial with non-negative rational integer coefficients, it suffices to prove that $\mathcal{B}$ is a closure of $\mathbb{Z}_p[x]$ with respect to the metric D_p.

From the definition of the class $\mathcal{B}$ it easily follows that every function $f \in \mathcal{B}$ can be uniformly approximated by polynomials over $\mathbb{Z}_p$: For each $n \in \mathbb{N}$ there exists a polynomial $f_n(x) \in \mathbb{Z}_p[x]$ such that $f(z) \equiv f_n(z) \pmod{p^n}$ for all $z \in \mathbb{Z}_p$. Actually, the series $\sum_{j=0}^{\infty} r_j \binom{x}{j}$ defines a function that is identically 0 modulo p^n if and only if all $r_j \equiv 0 \pmod{p^n}$, see Proposition 3.52. So in view of Lemma 3.6 we may put $f_n(x) = \sum_{i=0}^{\omega(n)} a_i \binom{x}{i}$, where $\omega(n) = \max\{j \in \mathbb{N}_0 : \frac{1}{p-1}(j - \operatorname{wt}_p j) < n\}$.

The inverse assertion is also true: Suppose a function $f : \mathbb{Z}_p \to \mathbb{Z}_p$ can be uniformly approximated by polynomials over $\mathbb{Z}_p$ in the above mentioned sense; then $f \in \mathcal{B}$. To prove this assertion assume that $f(z) \equiv f_i(z) \pmod{p^i}$ for all $z \in \mathbb{Z}_p$, where $f_i(x) \in \mathbb{Z}_p[x]$, $i = 1, 2, \ldots$. Every polynomial $f_i(x)$ of degree d_i has one and only one Mahler expansion (3.32): $f_i(x) = \sum_{j=0}^{d_i} a_{ij}\binom{x}{j}$, where $a_{ij} \in \mathbb{Z}_p$ and $\operatorname{ord}_p a_{ij} \geq \operatorname{ord}_p(j!)$ in view of (3.52), since $f_i \in \mathcal{C} \subset \mathcal{B}$. Given a function f, every polynomial $f_i(x)$ is uniquely determined up to a summand that is 0 modulo p^i everywhere on $\mathbb{Z}_p$. So we may assume that $d_i = \omega(i)$; then coefficients of the polynomial $f_i(x)$ are determined uniquely up to summands whose p-adic norms do not exceed p^{-i}. This implies that $a_{i+1,j} \equiv a_{ij} \pmod{p^i}$ (we assume $a_{ij} = 0$ for $j > \omega(i)$). Hence, $\lim^p_{i\to\infty} a_{ij} = a_j \in \mathbb{Z}_p$, and $\frac{a_j}{j!} \in \mathbb{Z}_p$. Consequently, a series $\sum_{i=0}^{\infty} a_i\binom{x}{i}$ defines a function $\tilde{f} \in \mathcal{B}$, which is uniformly continuous on $\mathbb{Z}_p$. The function $\tilde{f}$ is equal to f since $f(z) \equiv f_i(z) \equiv \tilde{f}(z) \pmod{p^i}$ for all $z \in \mathbb{Z}_p$ and all $i = 1, 2, \ldots$.

Actually we have proved that $\mathcal{B}$ is a complete metric space with respect to the metric D_p; from here it follows that $\mathcal{B}$ is closed with respect to additions, multiplications and compositions of functions: If $f, g \in \mathcal{B}$ then $f + g, f \cdot g, f(g) \in \mathcal{B}$. Indeed, let g be uniformly approximated by a sequence $\{g_n(x) \in \mathbb{Z}_p[x] : n = 1, 2, \ldots\}$, that is, $g_n(z) \equiv g(z) \pmod{p^n}$ for all $z \in \mathbb{Z}_p$. Now compatibility of the function f implies that $D_p(f(g), f(g_n)) \leq p^{-n}$, i.e., that the sequence $f(g_n)$ converges to $f(g)$ with respect to the distance D_p as $n \to \infty$. Yet $f(g_n) \in \mathcal{B}$ for every $n = 1, 2, \ldots$: If f is uniformly approximated by a sequence $\{f_m(x) \in \mathbb{Z}_p[x] : m = 1, 2, \ldots\}$, then $f_m(g_n(z)) \equiv f(g_n(z)) \pmod{p^m}$ for all $z \in \mathbb{Z}_p$. Hence, the sequence $\{f_m(g_n(x)) \in \mathbb{Z}_p[x] : m = 1, 2, \ldots\}$ converges to the function $f(g_n)$ with respect to the distance D_p, and $f_m(g_n) \in \mathcal{B}$, since $f_m(g_n)$ is a polynomial over $\mathbb{Z}_p$. Consequently, $f(g) \in \mathcal{B}$ in view of completeness of $\mathcal{B}$ with respect to D_p.

Finally, the inclusion $\mathcal{B} \supset \mathcal{C}$ is strict. A function $\sum_{i=0}^{\infty} i!\binom{x}{i} = \sum_{i=0}^{\infty} x^{\underline{i}}$ lies in $\mathcal{B}$, yet $f(x) \notin \mathcal{C}$: f is not analytic on $\mathbb{Z}_p$ in view of (3.36). □

Although a $\mathcal{B}$-function is not necessarily analytic on $\mathbb{Z}_p$, it is analytic on all balls of radii less than 1 (these functions are called *locally analytic of order 1* in [374]). We re-state the definition for functions defined on (and valuated in) the ball $\mathbb{Z}_p$.

Definition 3.60. A function $f : \mathbb{Z}_p \to \mathbb{Z}_p$ is said to be *locally analytic of order r* ($r = 1, 2, \ldots$) whenever

$$f(a + p^r h) = \sum_{i=0}^{\infty} p^{ir} h^i \frac{f^{(i)}(a)}{i!}$$

for all $a, h \in \mathbb{Z}_p$. Here, as usual, $f^{(i)}(a)$ stands for the ith derivative of the function f at the point $a \in \mathbb{Z}_p$.

The following result was proved by Y. Amice [16, Chapter III, Section 10, Theorem 3, Corollary 1(c)]:

Proposition 3.61 (Amice). *A function* $f(x) = \sum_{i=0}^{\infty} a_i x^{\underline{i}}$ $(a_i \in \mathbb{Q}_p)$ *is locally analytic of order* r *on* $\mathbb{Z}_p$ *if and only if*

$$\lim_{i\to\infty}\left(\frac{i}{p-1}\cdot\left(1-\frac{1}{p^r}\right)-\log_p|a_i|_p\right)=+\infty.$$

Now we are able to prove a Taylor theorem for $\mathcal{B}$-functions:

Theorem 3.62 (Taylor theorem for $\mathcal{B}$-functions). *For every* $f \in \mathcal{B}$, $a, h \in \mathbb{Z}_p$ *and* $k = 1, 2, 3, \ldots$ *the following equality holds:*

$$f(a+p^k h) = f(a) + f'(a)\cdot p^k h + \frac{f''(a)}{2!}\cdot p^{2k}h^2 + \frac{f'''(a)}{3!}\cdot p^{3k}h^3 + \cdots. \quad (3.55)$$

Moreover, all $\frac{f^{(j)}(a)}{j!}$ *are* p*-adic integers,* $j = 0, 1, 2, \ldots$.

Proof. The first claim of Theorem 3.62 immediately follows from Proposition 3.61 which obviously holds with $r = 1$ for any $\mathcal{B}$-function f in force of definition of the class $\mathcal{B}$, see (3.53). To prove the second claim of the theorem we note that

$$f^{(n)}(x) = \sum_{k=0}^{\infty}\binom{x}{k}\sum_{i_1,i_2,\ldots,i_n\geq 1}\frac{a_{k+i_1+i_2+\cdots+i_n}}{i_1\cdot i_2\cdot\ldots\cdot i_n}(-1)^{n+i_1+i_2+\cdots+i_n}.$$

This equation can be easily proved by induction on n in view of (3.35) and (3.54). However,

$$\sum_{i_1,i_2,\ldots,i_n\geq 1}\frac{a_{k+i_1+i_2+\cdots+i_n}}{i_1\cdot i_2\cdot\ldots\cdot i_n}(-1)^{n+i_1+i_2+\cdots+i_n}$$

$$=\sum_{s=n}^{\infty}\sum_{\substack{i_1,i_2,\ldots,i_n\geq 1\\ i_1+i_2+\cdots+i_n=s}}\frac{a_{k+s}}{i_1\cdot i_2\cdot\ldots\cdot i_n}(-1)^{n+s}, \quad (3.56)$$

and $\frac{a_{k+s}}{i_1\cdot i_2\cdot\ldots\cdot i_n} = \frac{a_{k+s}}{s!}\frac{s!}{i_1\cdot i_2\cdot\ldots\cdot i_n} \in \mathbb{Z}_p$ since both $\frac{(i_1+i_2+\cdots+i_n)!}{i_1\cdot i_2\cdot\ldots\cdot i_n} \in \mathbb{Z}$ and $\frac{a_{k+s}}{(k+s)!} \in \mathbb{Z}_p$, see the definition of a $\mathcal{B}$-function (3.53) for the latter. Thus, the sum

$$\sigma_s = \sum_{\substack{i_1,i_2,\ldots,i_n\geq 1\\ i_1+i_2+\cdots+i_n=s}}\frac{a_{k+s}}{i_1\cdot i_2\cdot\ldots\cdot i_n}(-1)^{n+s}$$

in the right-hand side of (3.56) is a p-adic integer. Moreover, as $\frac{a_{k+s}}{i_1\cdot i_2\cdot\ldots\cdot i_n} = \frac{a_{k+s}}{j_1\cdot j_2\cdot\ldots\cdot j_n}$ whenever $j_1, j_2, \ldots, j_n$ is a permutation of $i_1, i_2, \ldots, i_n$, the sum σ_s is a multiple of $n!$, i.e., $\frac{\sigma_s}{n!} \in \mathbb{Z}_p$. This proves the theorem. □

3.10.3 Class $\mathcal{A}$

Some important functions (for instance, some compatible integer-valued polynomials over $\mathbb{Q}_p$; i.e., polynomials that not necessarily have integer p-adic coefficients yet map $\mathbb{Z}_p$ into itself and satisfy the Lipschitz condition with constant 1 everywhere on $\mathbb{Z}_p$) do not lie in $\mathcal{B}$, see examples further. However, they lie in a wider class $\mathcal{A}$:

Definition 3.63. A function $f : \mathbb{Z}_p \to \mathbb{Z}_p$ lies in $\mathcal{A}$ (and is said to be an $\mathcal{A}$-*function*) if and only if f is compatible (i.e., satisfies the Lipschitz condition with constant 1), and $p^n f \in \mathcal{B}$ for some non-negative rational integer n.

Now, since $f = \frac{1}{p^n} g$ for a suitable $\mathcal{B}$-function g and suitable non-negative rational integer n, from Theorem 3.62 we immediately conclude that the Taylor theorem for every $\mathcal{A}$-function f holds in the following form:

Theorem 3.64 (Taylor theorem for $\mathcal{A}$-functions). *For every $f \in \mathcal{A}$, $a, h \in \mathbb{Z}_p$ and $k = 1, 2, 3, \ldots$ the function $f(a + p^k h)$ in variable h can be represented via convergent Taylor series*:

$$f(a+p^k h) = f(a) + f'(a) \cdot p^k h + \frac{f''(a)}{2!} \cdot p^{2k} h^2 + \frac{f'''(a)}{3!} \cdot p^{3k} h^3 + \cdots . \quad (3.57)$$

Note that $\frac{f^{(j)}(a)}{j!}$ are *not* necessarily p-adic integers now; however, in view of the second claim of Theorem 3.62, $\left|\frac{f^{(j)}(a)}{j!}\right|_p \le p^n$ for all $j = 1, 2, \ldots$. Moreover, $f'(a)$ is a p-adic integer in view of Proposition 3.41.

Concluding the section we consider some examples of $\mathcal{A}$-, $\mathcal{B}$-, and $\mathcal{C}$-functions, which are important for some applications (e.g. for inversive and exponential pseudorandom generators), see Chapter 9 for details.

It is obvious that a p-adic logarithm $\ln_p(1 + px) = \sum_{i=1}^{\infty} (-1)^{i+1} \frac{p^i}{i} x^i$ lies in $\mathcal{C}$ since $\operatorname{ord}_p i \le \lfloor \log_p i \rfloor$ and thus $\frac{p^i}{i} \in \mathbb{Z}_p$ for all $i = 1, 2, \ldots$ and $\lim^p_{i\to\infty} \frac{p^i}{i} = 0$.

A rational function over $\mathbb{Z}_p$, i.e. a function $f(x) = \frac{u(x)}{v(x)}$, where $u(x), v(x)$ are polynomials with p-adic integer coefficients, lies in $\mathcal{B}$ providing the denominator vanishes modulo p nowhere on $\mathbb{Z}_p$. Indeed, once $v(z) \not\equiv 0 \pmod p$ for every $z \in \mathbb{Z}_p$, there exists a multiplicative inverse for $v(z)$ in the residue ring $\mathbb{Z}/p^n\mathbb{Z}$, for every $n = 1, 2, \ldots$. Thus $\frac{u(z)}{v(z)} \equiv u(z)v(z)^{\varphi(p^n)-1} \pmod{p^n}$, where φ is Euler's totient function. Hence, the function f can be uniformly approximated (with respect to the metric D_p) by polynomials $u(x)v(x)^{\varphi(p^n)-1} \in \mathbb{Z}_p[x]$, $n = 1, 2, \ldots$; so $f \in \mathcal{B}$ in force of Proposition 3.59.

Another type of $\mathcal{B}$-functions are exponential ones. For instance, consider a function a^x with $a \equiv 1 \pmod p$ (hence, $a = 1 + pr$ for a suitable $r \in \mathbb{Z}_p$). Then $a^x = \sum_{i=0}^{\infty} p^i r^i \binom{x}{i}$; it is well known (see e.g. [308, Chapter 14, Section 5]) that for $p \ne 2$ this function is analytic on $\mathbb{Z}_p$ (whence, lies in $\mathcal{C}$). If $p = 2$ and r is odd, then a^x is not analytic on $\mathbb{Z}_2$, thus not in $\mathcal{C}$. Nevertheless, in the latter case a^x is in $\mathcal{B}$ since

$\mathrm{ord}_2\, i! = i - \mathrm{wt}_2\, i$ and thus $(1+2r)^x = \sum_{i=0}^{\infty} 2^i r^i \binom{x}{i} \in \mathcal{B}$. It is not difficult to see that the function $(1+4r)^x$ is in $\mathcal{C}$. So, summing all these considerations we conclude that if $a \in \mathbb{Z}_p$, $a \equiv 1 \pmod p$ then the function a^x is in $\mathcal{B}$.

Exponential functions of the considered type are special cases of functions of more general form u^v, where $u(z) \equiv 1 \pmod p$ for all $z \in \mathbb{Z}_p$.

Proposition 3.65. *Let $u, v\colon \mathbb{Z}_p \to \mathbb{Z}_p$ be compatible* (that is, 1-Lipschitz) *functions and let $u(z) \equiv 1 \pmod p$ for all $z \in \mathbb{Z}_p$. Then the function $f(z) = u(z)^{v(z)}$ is well defined for all $z \in \mathbb{Z}_p$, integer-valued and compatible. Moreover, if $w, v \in \mathcal{B}$, $u(z) = 1 + p \cdot w(z)$, then $f \in \mathcal{B}$.*

Proof. From the above argument considering a function a^x it immediately follows that the function f is well defined on $\mathbb{Z}_p$ and that it is integer-valued. To prove the compatibility of f note that for arbitrary $b, c, d \in \mathbb{Z}_p$ and $n = 1, 2, \ldots$ one has $(a + p^n b)^{c + p^n d} = (a + p^n b)^c ((a + p^n b)^{p^n})^d$, since elementary properties of powers are of the same form both in real and p-adic cases, see e.g. [308, Chapter 14, Section 5]. As both u and v are compatible functions, for arbitrary $z, r \in \mathbb{Z}_p$ there exist $s, t \in \mathbb{Z}_p$ such that $(u(z + p^n r))^{v(z + p^n r)} = (u(z) + p^n t)^{v(z) + p^n s}$; hence $(u(z + p^n r))^{v(z + p^n r)} = (u(z) + p^n t)^{v(z)}((u(z) + p^n t)^{p^n})^s \equiv (u(z) + p^n t)^{v(z)} \pmod{p^n}$ since $(u(z) + p^n t)^{p^n} \equiv 1 \pmod{p^n}$. Here is a proof of the latter congruence: As $u(z) \equiv 1 \pmod p$, for a suitable $k \in \mathbb{Z}_p$ we have $u(z) + p^n t = 1 + pk$; yet $(1+pk)^{p^n} = \sum_{i=0}^{p^n} k^i p^i \binom{p^n}{i} = \sum_{i=0}^{p^n} k^i \frac{p^i}{i!} (p^n)^{\underline{i}} \equiv 1 \pmod{p^n}$ since $\frac{p^i}{i!} \in \mathbb{Z}_p$ in view of Lemma 3.6. Finally denoting by $\overline{v(z)} = v(z) \bmod p^n$ the least nonnegative residue of $v(z)$ modulo p^n, for a suitable $h \in \mathbb{Z}_p$ we obtain

$$
\begin{aligned}
f(z + p^n r) &\equiv (u(z) + p^n t)^{v(z)} = (u(z) + p^n t)^{\overline{v(z)}} (u(z) + p^n t)^{p^n h} \\
&\equiv (u(z) + p^n t)^{\overline{v(z)}} = \sum_{i=0}^{\overline{v(z)}} u(z)^{\overline{v(z)} - i} p^{ni} t^i \binom{\overline{v(z)}}{i} \\
&\equiv (u(z))^{\overline{v(z)}} \\
&\equiv (u(z))^{\overline{v(z)}} (u(z))^{p^n h} = (u(z))^{v(z)},
\end{aligned}
$$

where $\cdot \equiv \cdot$ stands for $\cdot \equiv \cdot \pmod{p^n}$. So f is compatible.

To prove the rest of the proposition note that for every $z \in \mathbb{Z}_p$ and every $n = 1, 2, \ldots$ the congruence $(u(z))^{v(z)} \equiv \sum_{i=0}^{n-1} (u(z) - 1)^i \binom{v(z)}{i} \pmod{p^n}$ holds since $|u(z) - 1|_p \le \frac{1}{p}$. This implies that

- in view of Proposition 3.59, all functions $f_n = \sum_{i=0}^{n-1} \frac{p^i}{i!} v^{\underline{i}} w^i$ are in $\mathcal{B}$ since all fractions $\frac{p^i}{i!}$ are p-adic integers, see Lemma 3.6;
- the sequence $(f_n)_{n=1}^{\infty}$ converges to f with respect to the metric D_p.

From here it follows that $f \in \mathcal{B}$ in force of Proposition 3.59. □

A natural (and important) example of an $\mathcal{A}$-function, which is not necessarily a $\mathcal{B}$-function, is an integer-valued polynomial over $\mathbb{Q}_p$ of degree d that satisfies Lipschitz condition with a constant 1, i.e., a function $f(x) = \sum_{i=0}^{d} a_i p^{\lfloor \log_p i \rfloor} \binom{x}{i}$, where $a_i \in \mathbb{Z}_p, i = 0, 1, 2, \ldots$. This example stresses the importance of $\mathcal{A}$-functions: In view of Theorem 3.53 and Proposition 3.52, every 1-Lipschitz function can be approximated (with respect to the metric D_p) by $\mathcal{A}$-functions.

Chapter 4

p-adic ergodic theory

This is one of the main chapters of the book. Here we develop p-adic ergodic theory, mostly for 1-Lipschitz dynamics on $\mathbb{Z}_p^n$.

4.1 Discrete dynamical systems

This chapter and Chapter 5 are devoted to discrete non-Archimedean dynamical systems, namely iterations of the type

$$x_{n+1} = f(x_n), \tag{4.1}$$

where $f : X \to X$ and in further considerations we will let X be $\mathbb{Q}_p$, a finite extension of $\mathbb{Q}_p$, or $\mathbb{C}_p$, or $\mathbb{Z}_p$ as well as cartesian products of such fields and rings. Below, we will sometimes write "the dynamical system $f(x)$" when referring to the dynamical system that is described by iterations of f.

4.2 Periodic points and their character

We recall once again that for a given point x_0 the set of points $\{f^m(x_0) : m \in \mathbb{N}\}$ is called the *trajectory* or *orbit* through x_0. Some orbits of a dynamical system are of particular interest:

Definition 4.1. A point $x_0 \in X$ is said to be a *periodic point* if there exists $r \in \mathbb{N}$ such that $f^r(x_0) = x_0$. The least r with this property is called the length of *period* of x_0. If x_0 has period r, it is called an *r-periodic point.* A 1-periodic point is called a *fixed point*. The orbit of an r-periodic point x_0 is

$$\{x_0, x_1, \ldots, x_{r-1}\},$$

where $x_j = f^j(x_0)$, $0 \leqslant j \leqslant r-1$. This orbit is called an *r-cycle*.

An r-cycle consists of r different r-periodic points. Each element of the cycle has the cycle as its orbit. As a simple consequence we have that the number of r-periodic point of a discrete dynamical system is always divisible by r.

To study the long-time behavior of a dynamical system, we have to introduce a metric on X. Let K be a complete non-Archimedean field (in the same way we can proceed in the multidimensional case). We consider the dynamical system

$$f : B \to K, \quad x \mapsto f(x), \tag{4.2}$$

where $B = B_R(a)$, for some $R \in \mathbb{R}^+$ and some $a \in K$, or $B = K$ and $f : B \to B$ is an analytic function.

Definition 4.2. Let x_0 be an r-periodic point and let $g(x) = f^r(x)$. If there exists a ball $B_\rho^-(x_0)$ such that for every $x \in B_\rho^-(x_0)$ we have

$$\lim_{s\to\infty} g^s(x) = x_0$$

then we say that x_0 is an *attractor*. The set

$$A(x_0) = \{x \in x : \lim_{s\to\infty} g^s(x) = x_0\}$$

is called the *basin of attraction* of x_0.

Definition 4.3. Let x_0 be an r-periodic point. If there exists a ball $B_\rho^-(x_0)$ such that $|x - x_0| < |g(x) - x_0|$ for every $x \in B_\rho^-(x_0)$, $x \neq x_0$ then x_0 is said to be a *repeller*.

Definition 4.4 (see [214]). Let x_0 be a r-periodic point. If there exists an open ball $B_\rho^-(x_0)$ such that for every $\rho' < \rho$ the spheres $S_{\rho'}(x_0)$ are invariant under the map $g = f^r$ then $B_\rho^-(x_0)$ is said to be a *Siegel disk* and x_0 is said to be a center of a Siegel disk. The union of all Siegel disks with center x_0 is the *Siegel disk of maximal radius* of x_0. It is denoted by $\mathrm{SI}(x_0)$.

Definition 4.5. An r-periodic point x_0 is said to be *attractive* if $|g'(x_0)| < 1$, *indifferent* if $|g'(x_0)| = 1$ and *repelling* if $|g'(x_0)| > 1$.

The essence of this definition is clarified in Theorem 4.7.

The following lemma and theorem and their proofs are taken from [214].

Lemma 4.6. *Let $f : B \to K$ be an analytic function and let $a \in B$ and $f'(a) \neq 0$. Then there exists $r > 0$ such that*

$$s = \max_{2\leqslant n<\infty} \left|\frac{1}{n!}\frac{d^n f}{dx^n}(a)\right|_K r^{n-1} < |f'(a)|. \tag{4.3}$$

If $r > 0$ satisfies this inequality and $B_r(a) \subset B$ then

$$|f(x) - f(y)| = |f'(a)||x - y| \tag{4.4}$$

for all $x, y \in B_r(a)$.

Proof. We consider the case $B = B_R(a)$. We have $f(x) - f(y) = [f'(a) + T(x, y, a)](x - y)$ with

$$T(x, y, a) = \sum_{n=2}^{\infty} \frac{1}{n!} \frac{d^n f}{dx^n}(a)[(x-a)^{n-1} + (y-a)(x-a)^{n-2} + \cdots + (y-a)^{n-1}]. \quad (4.5)$$

Denote the expression in the square brackets by $U_n(x, y, a)$. Let $x, y \in B_r(a), r \leqslant R$. By the strong triangle inequality we obtain: $|U_n(x, y, a)|_K \leqslant r^{n-1}$. Set

$$\sigma(\rho) = \max_{2 \leqslant n < \infty} \left| \frac{1}{n!} \frac{d^n f}{dx^n}(a) \right|_K \rho^{n-2}, \quad \rho > 0.$$

By the analyticity of f on $B_R(a)$ we have $\sigma(R) \leqslant \|f\|_R / R^2 < \infty$. As $\sigma(r) \leqslant \sigma(R)$ for any $r \leqslant R$, we obtain

$$\sup_{x, y \in B_r(a)} |T(x, y, a)|_K \leqslant r\sigma(R) \to 0, \quad r \to 0. \quad (4.6)$$

Hence, if $f'(a) \neq 0$ then there exists $r > 0$ satisfying (4.3). We obtain (4.4) for such an r. □

Theorem 4.7. *Let a be a fixed point of the analytic function $f : B \to K$. Then*:

(i) *If a is an attracting point of f then it is an attractor of the dynamical system* (4.2). *If $r > 0$ satisfies the inequality*

$$q = \max_{1 \leqslant n < \infty} \left| \frac{1}{n!} \frac{d^n f}{dx^n}(a) \right|_K r^{n-1} < 1, \quad (4.7)$$

and $B_r(a) \subset B$ then $B_r(a) \subset A(a)$.

(ii) *If a is an indifferent point of f then it is the center of a Siegel disk. If $r > 0$ satisfies the inequality* (4.3) *and $B_r(a) \subset B$ then $B_r(a) \subset \mathrm{SI}(a)$.*

(iii) *If a is a repelling point of f then a is a repeller of the dynamical system* (4.2).

Proof. If $f'(a) \neq 0$ and $r > 0$ satisfies (4.3) (with $B_r(a) \subset B$), then it suffices to use the previous lemma.

If a is an arbitrary attracting point then again by (4.6) there exists $r > 0$ satisfying (4.7). Thus we have $|f(x) - f(y)|_K < q|x - y|_K$, $q < 1$, for all $x, y \in B_r(a)$. Consequently a is an attractor of (2.1) and $B_r(a) \subset A(a)$. □

For stronger results on the basin of attraction and the maximal Siegel disk, see [253]. The following lemma follows directly from the chain rule:

Lemma 4.8. *Let x_0 be an r-periodic point and let $g(x) = f^r(x)$. Then*

$$\frac{dg}{dx}(x_0) = \prod_{j=0}^{r} f'(x_j), \tag{4.8}$$

where $x_j = f^j(x_0)$.

Theorem 4.9. *If one r-periodic point of an r-cycle is an attractor (repeller, center of a Siegel disc) then all the r-periodic points of that cycle are attractors (repellers, centers of Siegel discs).*

Proof. It is easy to see that all $\frac{dg}{dx}(x_j)$ for $0 \leqslant j \leqslant r-1$ are equal. It is just a matter of reordering the factors in the product of (4.8). From Theorem 4.7 it follows that they all have the same character. □

In view of this theorem, it makes sense to speak about the basin of attraction of a cycle.

Definition 4.10. Let γ be an r-cycle $\{x_0, x_1, \ldots, x_{r-1}\}$. The basin of attraction of γ is defined as $A(\gamma) = \bigcup_{x\in\gamma} A(x)$, where $A(x)$ is the basin of attraction of x.

4.3 Monomial dynamics

By a monomial dynamical system in $\mathbb{Q}_p$ we mean a discrete dynamical system that is described by iterations of

$$f(x) = x^n, \quad n \in \mathbb{N}, n \geqslant 2. \tag{4.9}$$

In this section we study in detail ergodic behavior of p-adic monomial dynamical systems. Behavior of p-adic dynamical systems depends crucially on the prime parameter p. The main aim of investigations performed in the papers [160–162, 250, 300] was to find such a p-dependence for ergodicity, cf. [80, 352].

We recall that the study of ergodicity of monomial dynamical systems on p-adic spheres was important for development of p-adic dynamical systems theory. Results of [160–162, 250, 300] presented in this section were essentially generalized in [27], for arbitrary 1-Lipschitz locally analytic dynamical systems on p-adic spheres, see Section 4.7. From the viewpoint of applications, pseudorandom number generation provides the main motivation to study ergodicity of p-adic dynamical systems, see Chapter 9: p-adic ergodic dynamical systems give a huge class of excellent pseudorandom generators which are so important in cryptography, as well as in other applied ares, such as numerical analysis, quasi-Monte Carlo methods, and computer simulations. Of course, study of p-adic ergodicity is very important from purely mathematical viewpoint. It is a natural generalization of real and complex ergodicity, cf. [409].

Let ψ_n be a (monomial) mapping on $\mathbb{Z}_p$ taking x to x^n. Then all spheres $S_{p^{-l}}(1)$ are ψ_n-invariant if and only if n is a multiplicative unit, i.e., $(n, p) = 1$.

In particular ψ_n is an isometry on $S_{p^{-l}}(1)$ if and only if $(n, p) = 1$. Therefore we will henceforth assume that n is a unit. Also note that, as a consequence, $S_{p^{-l}}(1)$ is not a group under multiplication. Thus our investigations are *not about the dynamics on a compact (Abelian) group.* Hence, extended theory of ergodic systems which was developed for locally compact groups cannot be applied to our problem.

We remark that monomial mappings, $x \mapsto x^n$, are topologically transitive and ergodic with respect to Haar measure on the unit circle in the complex plane. We obtained [160–162, 250, 300] an analogous result for monomial dynamical systems over p-adic numbers. The process is, however, not straightforward. The result will depend on the natural number n. Moreover, in the p-adic case we never have ergodicity on the unit circle, but on the circles around the point 1.

4.3.1 Topologically transitive and minimality

The fields of p-adic numbers $\mathbb{Q}_p$ are interesting topological structures. Therefore it is useful to start not directly with the study of ergodicity (which assumes the presence of a measure), but with the study of topological transitivity and minimality of p-adic dynamical systems, cf. [409]. Moreover, applications to pseudorandom generators in Chapter 9 are, in fact, based on topological transitivity and minimality.

Let us consider the dynamical system

$$x \mapsto x^n$$

on spheres $S_{p^{-l}}(1)$. The result depends crucially on the following well-known result from group theory. We set

$$\langle n \rangle = \{n^N : N = 0, 1, 2, \ldots\}$$

for a natural number n. The following lemma is actually a restatement of Proposition 1.32:

Lemma 4.11. *Let $p > 2$ and l be any natural number, then the natural number n is a generator of $(\mathbb{Z}/p^\ell\mathbb{Z})^*$ if and only if n is a generator of $(\mathbb{Z}/p^2\mathbb{Z})^*$. The group $(\mathbb{Z}/2^\ell\mathbb{Z})^*$ is noncyclic for $l \geqslant 3$.*

Recall that a dynamical system given by a continuous transformation ψ on a compact metric space X is called *topologically transitive* if there exists a dense orbit $\{\psi^n(x) : n \in \mathbb{N}\}$ in X, and (one-sided) *minimal*, if all orbits for ψ in X are dense. For the case of monomial systems $x \mapsto x^n$ on spheres $S_{p^{-l}}(1)$ topological transitivity means the existence of an $x \in S_{p^{-l}}(1)$ such that each $y \in S_{p^{-l}}(1)$ is a limit point in the orbit of x, i.e. can be represented as

$$y = \lim_{k\to\infty} x^{n^{N_k}}, \tag{4.10}$$

for some sequence $\{N_k\}$, while minimality means that such a property holds for any $x \in S_{p^{-l}}(1)$. Our investigations are based on the following theorem.

Theorem 4.12. *For $p \neq 2$ the set $\langle n \rangle$ is dense in $S_1(0)$ if and only if n is a generator of $(\mathbb{Z}/p^2\mathbb{Z})^*$.*

Proof. We have to show that for every $\epsilon > 0$ and every $x \in S_1(0)$ there is a $y \in \langle n \rangle$ such that $|x - y|_p < \epsilon$. Let $\epsilon > 0$ and $x \in S_1(0)$ be arbitrary. Because of the discreteness of the p-adic metric we can assume that $\epsilon = p^{-k}$ for some natural number k. But (according to Lemma 4.11) if n is a generator of $(\mathbb{Z}/p^2\mathbb{Z})^*$, then n is also a generator of $(\mathbb{Z}/p^\ell\mathbb{Z})^*$ for every natural number l (and $p \neq 2$) and especially for $l = k$. Consequently there is an N such that $n^N = x \bmod p^k$. From the definition of the p-adic metric we see that $|x - y|_p < p^{-k}$ if and only if x equals to $y \bmod p^k$. Hence we have that $\left|x - n^N\right|_p < p^{-k}$. □

Let us consider $p \neq 2$ and for $x \in B_{p^{-1}}(1)$ the p-adic exponential function $t \mapsto x^t$, see, for example [374]. This function is well defined and continuous as a map from $\mathbb{Z}_p$ to $\mathbb{Z}_p$. In particular, for each $a \in \mathbb{Z}_p$, we have

$$x^a = \lim_{k \to a} x^k, \quad k \in \mathbb{N}. \tag{4.11}$$

We shall also use properties of the p-adic logarithmic function, see Section 3.2. We recall that $\ln_p : B_{p^{-1}}(1) \to B_{p^{-1}}(0)$ is an isometry:

$$|\ln_p x_1 - \ln_p x_2|_p = |x_1 - x_2|_p, \quad x_1, x_2 \in B_{1/p}(1)\,. \tag{4.12}$$

Lemma 4.13. *Let $x \in B_{p^{-1}}(1), x \neq 1, a \in \mathbb{Z}_p$ and let $\{m_k\}$ be a sequence of natural numbers. If $x^{m_k} \to x^a, k \to \infty$, then $m_k \to a$ as $k \to \infty$, in $\mathbb{Z}_p$.*

This is a consequence of the isometric property of $\ln_p$.

Theorem 4.14. *Let $p \neq 2$ and $l \geq 1$. Then the monomial dynamical system $x \mapsto x^n$ is minimal on the circle $S_{p^{-l}}(1)$ if and only if n is a generator of $\mathbb{F}^*_{p^2}$.*

Proof. Let $x \in S_{p^{-l}}(1)$. Consider the equation $x^a = y$. What are the possible values of a for $y \in S_{p^{-l}}(1)$? We prove that a can take an arbitrary value from the sphere $S_1(0)$. We have that $a = \frac{\ln_p x}{\ln_p y}$. As $\ln_p : B_{p^{-1}}(1) \to B_{p^{-1}}(0)$ is an isometry, we have $\ln_p(S_{p^{-l}}(1)) = S_{p^{-l}}(1)$. Thus $a = \frac{\ln_p x}{\ln_p y} \in S_1(0)$ and moreover, each $a \in S_1(0)$ can be represented as $\frac{\ln_p x}{\ln_p y}$ for some $y \in S_{p^{-l}}(1)$.

Let y be an arbitrary element of $S_{p^{-l}}(1)$ and let $x^a = y$ for some $a \in S_1(0)$. By Theorem 4.12 if n is a generator of $(\mathbb{Z}/p^2\mathbb{Z})^*$, then each $a \in S_1(0)$ is a limit point of the sequence (n^N). Thus $a = \lim_{k\to\infty} n^{N_k}$ for some subsequence $\{N_k\}$. By using the continuity of the exponential function we obtain (4.10).

Suppose now that, for some n, $x^{n^{N_k}} \to x^a$. By Lemma 4.13 we obtain that $n^{N_k} \to a$ as $k \to \infty$. If we have (4.10) for all $y \in S_{p^{-l}}(1)$, then each $a \in S_1(0)$ can be approximated by elements n^N. In particular, all elements $\{1, 2, \ldots, p-1, p+1, \ldots, p^2-1\}$ can be approximated with respect to modp^2. Thus n is a generator of $(\mathbb{Z}/p^2\mathbb{Z})^*$. □

Example 4.15. In the case $p = 3$ we have that ψ_n is minimal if $n = 2$, 2 is a generator of $(\mathbb{Z}/9\mathbb{Z})^* = \{1, 2, 4, 5, 7, 8\}$. But for $n = 4$ it is not; $\langle 4 \rangle \bmod 3^2 = \{1, 4, 7\}$. We can also see this by noting that $S_{1/3}(1) = B_{1/3}(4) \cup B_{1/3}(7)$ and that $B_{1/3}(4)$ is invariant under ψ_4.

Corollary 4.16. *If a is a fixed point of the monomial dynamical system $x \mapsto x^n$, then this is minimal on $S_{p^{-l}}(a)$ if and only if n is a generator of $(\mathbb{Z}/p^2\mathbb{Z})^*$.*

Example 4.17. Let $p = 17$ and $n = 3$. In $\mathbb{Q}_{17}$ there is a primitive 3rd root of unity. Moreover, 3 is also a generator of $(\mathbb{Z}/17^2\mathbb{Z})^*$. Therefore there exist nth roots of unity different from 1 around which the dynamics is minimal.

4.3.2 Unique ergodicity

In the following we will show that the minimality of the monomial dynamical system $\psi_n : x \mapsto x^n$ on the sphere $S_{p^{-l}}(1)$ is equivalent to its *unique ergodicity*. The latter property means that there exists a unique probability measure on $S_{p^{-l}}(1)$ and its Borel σ-algebra which is invariant under ψ_n. We will see that this measure is in fact the normalized restriction of the Haar measure on $\mathbb{Z}_p$. Moreover, we will also see that the ergodicity of ψ_n with respect to Haar measure is also equivalent to its unique ergodicity. We should point out that – though many results are analogous to the case of the (irrational) rotation on the circle, our situation is quite different, in particular as we do not deal with dynamics on topological subgroups.

Lemma 4.18. *Assume that ψ_n is minimal. Then the Haar measure m is the unique ψ_n-invariant measure on $S_{p^{-l}}(1)$.*

Proof. First note that minimality of ψ_n implies that $(n, p) = 1$ and hence that ψ_n is an isometry on $S_{p^{-l}}(1)$. Then, as a consequence of Theorem 27.5 in [374], it follows that $\psi_n(B_r(a)) = B_r(\psi_n(a))$ for each ball $B_r(a) \subset S_{p^{-l}}(1)$. Consequently, for every open set $U \neq \varnothing$ we have $S_{p^{-l}}(1) = \bigcup_{N=0}^{\infty} \psi_n^N(U)$. It follows for a ψ_n-invariant measure μ that $\mu(U) > 0$.

Moreover we can split $S_{p^{-l}}(1)$ into disjoint balls of radii $p^{-(l+k)}$, $k \geq 1$, on which ψ_n acts as a permutation. In fact, for each $k \geq 1$, $S_{p^{-l}}(1)$ is the union,

$$S_{p^{-l}}(1) = \bigcup B_{p^{-(l+k)}}(1 + b_l p^l + \cdots + b_{l+k-1} p^{l+k-1}), \tag{4.13}$$

where $b_i \in \{0, 1, \ldots, p-1\}$ and $b_l \neq 0$.

We now show that ψ_n is a permutation on the partition (4.13). Recall that every element of a p-adic ball is the center of that ball, and as pointed out above $\psi_n(B_r(a)) = B_r(\psi_n(a))$. Consequently we have for all positive integers k, $\psi_n^k(a) \in B_r(a) \Rightarrow \psi_n^k(B_r(a)) = B_r(\psi_n^k(a)) = B_r(a)$ so that $\psi_n^{Nk}(a) \in B_r(a)$ for every natural number N. Hence, for a minimal ψ_n a point of a ball B of the partition (4.13) must move to another ball in the partition.

Furthermore the minimality of ψ_n shows indeed that ψ_n acts as a permutation on balls. By invariance of μ all balls must have the same positive measure. As this holds for any k, μ must be the restriction of Haar measure m. □

The arguments of the proof of Lemma 4.18 also show that Haar measure is always ψ_n-invariant. Thus if ψ_n is uniquely ergodic, the unique invariant measure must be the Haar measure m. Under these circumstances it is known [409] that ψ_n must be minimal.

Theorem 4.19. *The monomial dynamical system $\psi_n : x \mapsto x^n$ on $S_{p^{-l}}(1)$ is minimal if and only if it is uniquely ergodic in which case the unique invariant measure is the Haar measure.*

Let us mention that unique ergodicity yields in particular the ergodicity of the unique invariant measure, i.e., the Haar measure m, which means that

$$\frac{1}{N}\sum_{i=0}^{N-1} f(x^{n^i}) \to \int f\,dm \quad \text{for all } x \in S_{p^{-l}}(1), \tag{4.14}$$

and all continuous functions $f : S_{p^{-l}}(1) \to \mathbb{R}$.

On the other hand the arguments of the proof of Lemma 4.18, i.e., the fact that ψ_n acts as a permutation on each partition of $S_{p^{-l}}(1)$ into disjoint balls if and only if $\langle n \rangle = (\mathbb{Z}/p^2\mathbb{Z})^*$, proves that if n is not a generator of $(\mathbb{Z}/p^2\mathbb{Z})^*$ then the system is not ergodic with respect to Haar measure. Consequently, if ψ_n is ergodic then $\langle n \rangle = (\mathbb{Z}/p^2\mathbb{Z})^*$ so that the system is minimal by Theorem 4.14, and hence even uniquely ergodic by Theorem 4.19. Since unique ergodicity implies ergodicity one has the following.

Theorem 4.20. *The monomial dynamical system $\psi_n : x \mapsto x^n$ on $S_{p^{-l}}(1)$ is ergodic with respect to Haar measure if and only if it is uniquely ergodic.*

Even if the monomial dynamical system $\psi_n : x \mapsto x^n$ on $S_{p^{-l}}(1)$ is ergodic, it never can be mixing, especially not weak-mixing. This can be seen from the fact that an abstract dynamical system is weak-mixing if and only if the product of such two systems is ergodic. If we choose a function f on $S_{p^{-l}}(1)$ and define a function F on $S_{p^{-l}}(1) \times S_{p^{-l}}(1)$ by $F(x, y) := f(\ln_p x / \ln_p y)$ (which is well defined as $\ln_p$ does not vanish on $S_{p^{-l}}(1)$), we obtain a non-constant function satisfying $F(\psi_n(x), \psi_n(y)) = F(x, y)$. This shows, see [409], that $\psi_n \times \psi_n$ is not ergodic, and hence ψ_n is not weak-mixing with respect to any invariant measure, in particular the restriction of Haar measure.

Let us consider the ergodicity of a perturbed system

$$\psi_q = x^n + q(x), \tag{4.15}$$

for some polynomial q such that $q(x)$ equals to 0 mod p^{l+1}, $|q(x)|_p < p^{-(l+1)}$. This condition is necessary in order to guarantee that the sphere $S_{p^{-l}}(1)$ is invariant. For such a system to be ergodic it is necessary that n is a generator of $(\mathbb{Z}/p^2\mathbb{Z})^*$. This follows from the fact that for each $x = 1 + a_l p^l + \cdots$ on $S_{p^{-l}}(1)$ (so that $a_l \neq 0$) the condition on q gives

$$\psi_q^N(x) \equiv 1 + n^N a_l \pmod{p^{l+1}}.$$

Now ψ_q acts as a permutation on the $p-1$ balls of radius $p^{-(l+1)}$ if and only if $\langle n \rangle = (\mathbb{Z}/p^2\mathbb{Z})^*$. Consequently, a perturbation (4.15) cannot make a nonergodic system ergodic.

In [160–162, 250, 300] the problem of ergodicity of perturbed monomial dynamics on p-adic spheres was formulated, it was announced at numerous international conferences and talks at many universities throughout the world. Nevertheless, it remained unsolved until 2005, when Vladimir Anashin solved it in the most general case [27], for 1-Lipschitz locally analytic dynamical systems, see Subsection 4.7.1.

4.4 Measure-preserving and ergodic isometries on $\mathbb{Z}_p^n$

The main goal of this section is to establish connections between the dynamics produced by isometries on a continuum phase space $\mathbb{Z}_p^n$ with the dynamics on finite phase spaces $(\mathbb{Z}/p^k\mathbb{Z})^n$. It turns out that any 1-Lipschitz (i.e., compatible) measure-preserving (respectively, ergodic) transformation on $\mathbb{Z}_p$ is an isometry which induces permutations (respectively, permutations with a single cycle) on all residue rings $\mathbb{Z}/p^k\mathbb{Z}$, $k = 1, 2, \ldots$, and vice versa. Now we describe this more formally.

For every $k = 1, 2, \ldots$, a mapping $\cdot \bmod p^k : \mathbb{Z}_p \to \mathbb{Z}/p^k\mathbb{Z}$

$$z \mapsto z \bmod p^k = \left(\sum_{i=0}^{\infty} \delta_i(z) \cdot p^i\right) \bmod p^k = \sum_{i=0}^{k-1} \delta_i(z) \cdot p^i \tag{4.16}$$

is an epimorphism of the ring $\mathbb{Z}_p$ onto the residue ring $\mathbb{Z}/p^k\mathbb{Z}$: Recall that $\delta_i(z)$ is a coefficient of the ith term in a canonical p-adic expansion of $x \in \mathbb{Z}_p$, see Note 1.46, so the sum in the right-hand part of (4.16) can be considered as an element of the residue ring $\mathbb{Z}/p^k\mathbb{Z}$.

Given a 1-Lipschitz (whence, compatible, see Subsection 3.8.1) function $f : \mathbb{Z}_p \to \mathbb{Z}_p$, a mapping $f \bmod p^k : r \mapsto f(r) \bmod p^k$, where $r = \sum_{i=0}^{k-1} \delta_i(r) \cdot p^i \in \mathbb{Z}/p^k\mathbb{Z}$, is a well-defined mapping of the residue ring $\mathbb{Z}/p^k\mathbb{Z}$ into itself, see Subsection 2.2.1. We call this mapping an *induced function modulo* p^k. We can expand the mapping $\cdot \bmod p^k$ to Cartesian powers $\mathbb{Z}_p^n$; we denote the corresponding mapping from $\mathbb{Z}_p^n$ onto $(\mathbb{Z}/p^k\mathbb{Z})^n$ by the same symbol $\cdot \bmod p^k$ and now, given a 1-Lipschitz function $F\colon \mathbb{Z}_p^n \to \mathbb{Z}_p^m$ we define in an obvious manner $F \bmod p^k\colon (\mathbb{Z}/p^k\mathbb{Z})^n \to \mathbb{Z}/p^k\mathbb{Z})^m$, the induced function modulo p^k.

Definition 4.21 (cf. Section 2.2). A 1-Lipschitz function $F\colon \mathbb{Z}_p^n \to \mathbb{Z}_p^m$ is said to be *balanced modulo* p^k (respectively, *bijective, transitive modulo* p^k) whenever the induced function $F \bmod p^k\colon (\mathbb{Z}/p^k\mathbb{Z})^n \to \mathbb{Z}/p^k\mathbb{Z})^m$ is balanced (respectively, bijective, transitive).

Note 4.22. Definition 4.21 can be re-stated for an asymptotically compatible function F (see Definition 3.34) in an obvious manner: The only difference is that for an asymptotically compatible function the induced function is well defined modulo p^k for all sufficiently large k.

A central result of this section is the following theorem, which was announced in [24] and proved in [27]:

Theorem 4.23. *For $m = n = 1$, a 1-Lipschitz function $F\colon \mathbb{Z}_p^n \to \mathbb{Z}_p^m$ is measure-preserving (or, accordingly, ergodic) if and only if it is bijective (accordingly, transitive) modulo p^k for all $k = 1, 2, 3, \ldots$.*

For $n \geq m$, the function F is measure-preserving if and only if it is balanced modulo p^k, for all $k = 1, 2, 3, \ldots$.

The theorem follows directly from Propositions 4.33, 4.34 and 4.35 below.

Note 4.24. As it can be seen from the proofs of Propositions 4.33, 4.34 and 4.35 below, Theorem 4.23 remains true whenever in the statement we change 'all k' to 'all sufficiently large k'. Moreover, in this form, *Theorem 4.23 holds for asymptotically compatible functions as well* (that is, for functions from $\bar{\mathcal{L}}_1$ rather than from $\mathcal{L}_1$, see Subsection 3.8.1): For an asymptotically compatible function F we just take $k \geq N$, where $N \in \mathbb{N}$ is a number from the statement of Note 3.36, see also Note 3.40; proofs of all results of Section 4.4 can be easily modified for this case.

Note that with respect to minimality and unique ergodicity compatible (i.e., 1-Lipschitz) transformations on $\mathbb{Z}_p$ behave similarly to monomial maps, see Section 4.3; recently F. Durand and F. Paccaut proved the following, see [110, Theorem 6]:

Theorem 4.25 ([110]). *Let $f\colon \mathbb{Z}_p \to \mathbb{Z}_p$ be an onto compatible map. The following propositions are equivalent:*

- *f is minimal;*

- *f is conjugate to the translation $t(x) = x + 1$ on $\mathbb{Z}_p$;*
- *f is uniquely ergodic;*
- *f is ergodic.*

4.4.1 Measure-preserving isometries

First we prove that a 1-Lipschitz function $F : \mathbb{Z}_p^n \to \mathbb{Z}_p^n$ preserves measure if and only if it is bijective modulo p^k, for all $k = 1, 2, \ldots$. We consider the case $n = 1$ just to simplify notation; the statements of Propositions 4.26 and 4.28 as well as of Notes 4.27, 4.30 and of Corollary 4.29 remain true for arbitrary $n \in \mathbb{N}$, the respective proofs are quite similar to those for the case $n = 1$. It is worth noting here that Proposition 4.26 can be deduced also from a more general result stated in Subsection 4.4.2. However, we present a separate proof for this proposition to obtain some extra information on the functions of the considered type.

Proposition 4.26. *A 1-Lipschitz measure-preserving function $f : \mathbb{Z}_p \to \mathbb{Z}_p$ is a bijection of $\mathbb{Z}_p$ onto itself.*

Proof. We prove that f is both injective and surjective.

Claim 1: Under the conditions of Proposition 4.26 the function f is injective.

Indeed, if there exist $a, b \in \mathbb{Z}_p$ ($a \neq b$) such that $f(a) = f(b) = z$ then for some k the balls $a + p^k\mathbb{Z}_p$ and $b + p^k\mathbb{Z}_p$ are disjoint, whereas $f(a + p^k\mathbb{Z}_p), f(a + p^k\mathbb{Z}_p) \subset z + p^k\mathbb{Z}_p$. Hence $\mu_p(f^{-1}(z + p^k\mathbb{Z}_p)) \geq 2 \cdot p^{-k}$ since $f^{-1}(z + p^k\mathbb{Z}_p) \supset f^{-1}(a + p^k\mathbb{Z}_p), f^{-1}(b + p^k\mathbb{Z}_p)$; so f does not preserve μ_p.

Claim 2: Under the conditions of Proposition 4.26 the function f is bijective modulo p^k for all $k = 1, 2, \ldots$.

Otherwise for suitable $a, b \in \mathbb{Z}_p$ ($a \neq b$) and k, the balls $a + p^k\mathbb{Z}_p$ and $b + p^k\mathbb{Z}_p$ are disjoint, whereas $f(a + p^k\mathbb{Z}_p), f(a + p^k\mathbb{Z}_p) \subset z + p^k\mathbb{Z}_p$. Yet this leads to a contradiction, see Claim 1.

Claim 3: Under the conditions of Proposition 4.26 the function f is surjective.

Take arbitrary $z \in \mathbb{Z}_p$. Then in view of Claim 2 there exists exactly one $x_1 \in \mathbb{Z}/p\mathbb{Z}$ such that $f(x_1) \equiv z \pmod{p}$ (here and further we identify elements of the residue ring $\mathbb{Z}/p^k\mathbb{Z}$ with non-negative rational integers $0, 1, \ldots, p^k - 1$ in an obvious way). Similarly, there exists exactly one $x_2 \in \mathbb{Z}/p^2\mathbb{Z}$ such that $f(x_2) \equiv z \pmod{p^2}$; whence necessarily $x_2 \equiv x_1 \pmod{p}$, etc.

So we obtain a sequence $x_2, x_2, \ldots$ such that $|f(x_i) - z|_p \leq p^{-i}$ and $|x_{i+1} - x_i|_p \leq p^{-i}$ for $i = 1, 2, \ldots$. It is an exercise to show now that the sequence $x_2, x_2, \ldots$ is a Cauchy sequence (which hence converges to some $x \in \mathbb{Z}_p$), and that $f(x) = z$. □

Note 4.27. As a bonus we have that whenever a 1-Lipschitz function $g : \mathbb{Z}_p \to \mathbb{Z}_p$ is bijective modulo p^k for all $k = 1, 2, \ldots$, it is a bijection of $\mathbb{Z}_p$ onto $\mathbb{Z}_p$, see proofs of Claims 2 and 3 above.

Proposition 4.28. *Let a 1-Lipschitz function $g\colon \mathbb{Z}_p \to \mathbb{Z}_p$ be bijective modulo p^k for all $k = 1, 2, \dots$. Then g preserves measure.*

Proof. In view of Note 4.27 the function g is a bijection of $\mathbb{Z}_p$ onto $\mathbb{Z}_p$; whence, there exists an inverse function $f = g^{-1}$, which is also a bijection of $\mathbb{Z}_p$ onto $\mathbb{Z}_p$. Moreover, f is continuous since g is continuous.

Claim 1: f is 1-Lipschitz.

If there are $a, b \in \mathbb{Z}_p$ such that $a \equiv b \pmod{p^k}$ and $f(a) \not\equiv f(b) \pmod{p^k}$ then assuming $a = g(u)$, $b = g(v)$ for uniquely defined $u, v \in \mathbb{Z}_p$ we have $g(u) \equiv g(v) \pmod{p^k}$ and $f(g(u)) \not\equiv f(g(v)) \pmod{p^k}$; that is, $g(u) \equiv g(v) \pmod{p^k}$ and $u \not\equiv v \pmod{p^k}$. The latter contradicts the conditions of Proposition 4.28.

Claim 2: $f(a + p^k\mathbb{Z}_p) = f(a) + p^k\mathbb{Z}_p$ for every $a \in \mathbb{Z}_p$ and every $k = 1, 2, \dots$.

In view of Claim 1, $f(a + p^k\mathbb{Z}_p) \subset f(a) + p^k\mathbb{Z}_p$. To prove the inverse inclusion, denote $f(a) = b$; then $g(b) = a$. Since g is 1-Lipschitz, $g(b + p^k\mathbb{Z}_p) \subset g(b) + p^k\mathbb{Z}_p$. Applying a bijection f to the both sides of this inclusion, one obtains $b + p^k\mathbb{Z}_p \subset f(g(b) + p^k\mathbb{Z}_p)$, since f is 1-Lipschitz (see Claim 1); that is, $f(a) + p^k\mathbb{Z}_p \subset f(a + p^k\mathbb{Z}_p)$, the needed inverse inclusion.

Claim 3: f is bijective modulo p^k for all $k = 1, 2, \dots$.

Assuming there exist $u, v \in \mathbb{Z}_p$ and $k \in \{1, 2, \dots\}$ such that $u \equiv v \pmod{p^k}$ and $f(u) \not\equiv f(v) \pmod{p^k}$ one obtains that $u + p^k\mathbb{Z}_p = v + p^k\mathbb{Z}_p$, yet $f(u) + p^k\mathbb{Z}_p \neq f(v) + p^k\mathbb{Z}_p$, a contradiction in view of Claim 2.

Claim 4: f satisfies the conditions of Proposition 4.28.

See Claims 1 and 3.

Claim 5: $g(a + p^k\mathbb{Z}_p) = g(a) + p^k\mathbb{Z}_p$ for every $a \in \mathbb{Z}_p$ and every $k = 1, 2, \dots$.

See Claim 4.

Claim 6: $\mu_p(g(M)) = \mu_p(M)$, for every measurable $M \subset \mathbb{Z}_p$.

Since M is measurable, then

$$\mu_p(M) = \inf\{\mu_p(V)\colon V \supset M, V \text{ is open in } \mathbb{Z}_p\}.$$

Since V is open, it is a disjoint union of a countable number of balls V_j of non-zero radius each: $V = \bigcup_{j\in J} V_j$. Then $g(V) = \bigcup_{j \in J} g(V_j)$, since g is a bijection. Note that in view of Claim 5, each $g(V_j)$ is a ball of a radius that is equal to the one of the ball V_j; that is, $\mu_p(g(V_j)) = \mu_p(V_j)$, for all $j \in J$. Moreover, the balls are disjoint: $g(V_i) \cap g(V_j) = \varnothing$ whenever $i \neq j$ (since $f(g(V_i) \cap g(V_j)) = V_i \cap V_j$ in view of Claim 2). This implies that $\mu_p(g(V)) = \mu_p(V)$. Note that $g(V)$ is open since g is a continuous bijection. Hence,

$$\mu_p(g(M)) \le \inf\{\mu_p(g(V))\colon V \supset M, V \text{ is open in } \mathbb{Z}_p\} = \mu_p(M).$$

In view of Claim 4, one has then $\mu_p(f(R)) \le \mu_p(R)$, for every measurable $R \subset \mathbb{Z}_p$. Now we take $R = g(M)$ (whence $f(R) = M$) and obtain $\mu_p(M) \le \mu_p(g(M))$, thus proving the proposition. □

Corollary 4.29. *A 1-Lipschitz function $f: \mathbb{Z}_p \to \mathbb{Z}_p$ preserves measure if and only if it is bijective modulo p^k for all $k = 1, 2, \ldots$.*

Proof. Necessity of the conditions is proved by Claim 2 of Proposition 4.26, whereas their sufficiency is proved by Proposition 4.28. □

Note 4.30. As a bonus we have that *every 1-Lipschitz measure-preserving function $f: \mathbb{Z}_p \to \mathbb{Z}_p$ is an isometry*: A distance between two points is just a radius of the smallest ball that contains them both; however, as it was shown, a measure-preserving 1-Lipschitz mapping is a bijection that merely permutes balls of pairwise equal radii.

4.4.2 1-Lipschitz measure-preserving functions

Now we prove that a 1-Lipschitz function $F: \mathbb{Z}_p^n \to \mathbb{Z}_p^m$, $m \le n$, preserves measure if and only if it is balanced modulo p^k, for all $k = 1, 2, \ldots$. We need the following lemma.

Lemma 4.31. *Let a 1-Lipschitz function $F: \mathbb{Z}_p^n \to \mathbb{Z}_p^m$, $m \le n$, be balanced modulo p^k, for all $k = 1, 2, \ldots$. Then for every $b \in \mathbb{Z}_p^m$ a full preimage $F^{-1}(b + p^s\mathbb{Z}_p^m)$ is a union of $p^{s(n-m)}$ pairwise disjoint balls $a_j + p^s\mathbb{Z}_p^n$, $j = 1, 2, \ldots, p^{s(n-m)}$.*

Proof. We start with proving the lemma 'modulo p^k'.

Claim 1: For every $\bar{b}_k \in (\mathbb{Z}/p^k)^m$, a full preimage $\bar{F}_k^{-1}(\bar{b}_k + p^s(\mathbb{Z}/p^k\mathbb{Z})^m)$ of the coset $\bar{b}_k + p^s(\mathbb{Z}/p^k\mathbb{Z})^m \subset (\mathbb{Z}/p^k\mathbb{Z})^m$ (modulo the ideal $p^k(\mathbb{Z}/p^k\mathbb{Z})^m$ of the ring $(\mathbb{Z}/p^k\mathbb{Z})^m$) is a disjoint union of $p^{s(n-m)}$ suitable pairwise disjoint cosets (modulo the ideal $p^s(\mathbb{Z}/p^k\mathbb{Z})^n$ of the ring $(\mathbb{Z}/p^k\mathbb{Z})^n$):

$$\bar{F}_k^{-1}(\bar{b}_k + p^s(\mathbb{Z}/p^k\mathbb{Z})^m) = \bigcup_{j=1}^{p^{s(n-m)}} (\bar{a}_{k,j} + p^s(\mathbb{Z}/p^k\mathbb{Z})^n).$$

Here and further we assume that $s \le k$. In this case

$$\#(\bar{b}_k + p^s(\mathbb{Z}/p^k\mathbb{Z})^m) = p^{m(k-s)},$$

and since F is balanced modulo p^k, then

$$\#\bar{F}_k^{-1}(\bar{b}_k + p^s(\mathbb{Z}/p^k\mathbb{Z})^m) = p^{k(n-m)} \cdot p^{m(k-s)} = p^{kn-ms}. \tag{4.17}$$

Further, since F is balanced modulo p^s, then $\#\bar{F}_s^{-1}(\bar{b}_s) = p^{s(n-m)}$, for every $\bar{b}_s \in \{0, 1, \ldots, p^s - 1\}^m = (\mathbb{Z}/p^s\mathbb{Z})^m$. Take $\bar{b}_s \equiv \bar{b}_k \pmod{p^s}$ and let

$$\bar{F}_s^{-1}(\bar{b}_s) = \{\bar{a}_{s,1}, \ldots, \bar{a}_{s,p^{s(n-m)}}\} \subset (\mathbb{Z}/p^s\mathbb{Z})^n = \{0, 1, \ldots, p^s - 1\}^n.$$

For $j = 1, 2, \ldots, p^{s(n-m)}$ choose (and fix) $\bar{a}_{k,j} \in (\mathbb{Z}/p^k\mathbb{Z})^n$ so that $\bar{a}_{k,j} \equiv \bar{a}_{s,j} \pmod{p^s}$. Note that the latter congruence, in accordance with what has been agreed at the beginning of Section 3.7, just means that $|\bar{a}_{k,j} - \bar{a}_{s,j}|_p \le p^{-s}$; that is $\bar{a}_{k,j}^{(i)} \equiv \bar{a}_{s,j}^{(i)} \pmod{p^s}$ for each ith component $\bar{a}_{k,j}^{(i)}$ of $\bar{a}_{k,j} \in (\mathbb{Z}/p^k\mathbb{Z})^n = \{0, 1, \ldots, p^k - 1\}^n$, $i = 1, 2, \ldots, n$.

Now for $j = 1, 2, \ldots, p^{s(n-m)}$ take $\hat{a}_{k,j} \in (\mathbb{Z}/p^k\mathbb{Z})^n$ so that $\hat{a}_{k,j} \equiv \bar{a}_{s,j} \pmod{p^s}$; that is, $\hat{a}_{k,j} \in \bar{a}_{k,j} + p^s(\mathbb{Z}/p^k\mathbb{Z})^n$, and vice versa. Since F is 1-Lipschitz, $\bar{F}_k(\hat{a}_{k,j}) \equiv \bar{b}_s \pmod{p^s}$; thus, $\bar{F}_k(\hat{a}_{k,j}) \in \bar{b}_k + p^s(\mathbb{Z}/p^k\mathbb{Z})^m$ (recall that $\bar{b}_s \equiv \bar{b}_k \pmod{p^s}$ by our choice). So every $\hat{a}_{k,j}$ is an $\bar{F}_k$-preimage of a certain element of the coset $b_k + p^s(\mathbb{Z}/p^k\mathbb{Z})^m$, and there are exactly $p^{s(n-m)} \cdot p^{n(k-s)} = p^{nk-ms}$ these elements $\hat{a}_{k,j}$. Comparing this number with what is given by equation (4.17), we conclude that all these $\hat{a}_{k,j}$ constitute the full preimage $\bar{F}_k^{-1}(\bar{b}_k + p^s(\mathbb{Z}/p^k\mathbb{Z})^m)$, which is then just the union of cosets $\bar{a}_{k,j} + p^s(\mathbb{Z}/p^k\mathbb{Z})^n$ over $j \in \{1, \ldots, p^{s(n-m)}\}$. These cosets are disjoint since all $\bar{a}_{k,j}$ are different modulo p^s.

Claim 2: For $j = 1, 2, \ldots, p^{s(n-m)}$ fix $a_j \in \mathbb{Z}_p^n$ such that $a_j \equiv \bar{a}_{s,j} \pmod{p^s}$, where $\bar{a}_{s,j}$ are defined as above for $\bar{b}_k \equiv b \pmod{p^k}$. Then

$$F^{-1}(b + p^s\mathbb{Z}_p^m) = \bigcup_{j=1}^{p^{s(n-m)}} (a_j + p^s\mathbb{Z}_p^n).$$

First note that in this setting the definition of $\bar{a}_{s,j}$ (whence, of a_j) does not depend on k, only on b and s, since for $\bar{b}_k \equiv b \pmod{p^k}$ the set $\{\bar{a}_{s,1}, \ldots, \bar{a}_{s,p^{s(n-m)}}\}$ is just a full $\bar{F}_s$-preimage of $(b \bmod p^s)$; here $(b \bmod p^s)$ is a unique non negative rational integer that lays at the distance p^{-s} from the point b; an approximation of b by a non-negative rational integer with precision p^{-s} with respect to a p-adic metric. In other words, given $b \in \mathbb{Z}_p^m$, we put $\bar{b}_s \equiv b \pmod{p^s}$, where $\bar{b}_s \in \{1, 2, \ldots, p^s - 1\}^m$, then take all solutions $\bar{a}_{s,j} \in \{1, 2, \ldots, p^s - 1\}^n$ of the congruence $\bar{F}_s(x) = \bar{b}_s \pmod{p^s}$ in indeterminate x, and after that, for each of these $p^{s(n-m)}$ solutions $\bar{a}_{s,j}$, we choose an arbitrary $a_j \in \mathbb{Z}_p^n$ so that $a_j \equiv \bar{a}_{s,j} \pmod{p^s}$.

From the definition of $\bar{a}_j$ it follows immediately that for every $h \in (\mathbb{Z}_p)^n$, $F(a_j + p^s \cdot h) \equiv b \pmod{p^s}$ since F is 1-Lipschitz; whence $F^{-1}(b + p^s\mathbb{Z}_p^m) \supset \bigcup_{j=1}^{p^{s(n-m)}} (a_j + p^s\mathbb{Z}_p^n)$. Thus, we must prove the inverse inclusion only.

Given $c \in b + p^s\mathbb{Z}_p^m$, for every $k \ge s$ it follows from Claim 1 that $F^{-1}(c) \in \bar{F}_k^{-1}(c \bmod p^k) + p^k\mathbb{Z}_p^n$, where $\bar{F}_k^{-1}(c \bmod p^k)$ is a subset of the finite set $\bigcup_{j=1}^{p^{s(n-m)}} (\bar{a}_{k,j} + p^s \cdot \{0, 1, \ldots, p^{k-s} - 1\}^n)$.

Thus, applying Claim 1 we obtain:

$$\begin{aligned}
F^{-1}(c) &\in \bigcap_{k=s}^{\infty}(\bar{F}_k^{-1}(c \bmod p^k) + p^k\mathbb{Z}_p^n)\\
&\subset \bigcap_{k=s}^{\infty}\bigcup_{j=1}^{p^{s(n-m)}}(\bar{a}_{k,j} + p^s\cdot\{0,1,\dots,p^{k-s}-1\}^n + p^k\mathbb{Z}_p^n)\\
&= \bigcup_{j=1}^{p^{s(n-m)}}\bigcap_{k=s}^{\infty}(\bar{a}_{k,j} + p^s\cdot\{0,1,\dots,p^{k-s}-1\}^n + p^k\mathbb{Z}_p^n)\\
&= \bigcup_{j=1}^{p^{s(n-m)}}\bigcap_{k=s}^{\infty}(\bar{a}_{s,j} + p^s\cdot\{0,1,\dots,p^{k-s}-1\}^n + p^k\mathbb{Z}_p^n)\\
&= \bigcup_{j=1}^{p^{s(n-m)}}(\bar{a}_{s,j} + p^s\mathbb{Z}_p^n) = \bigcup_{j=1}^{p^{s(n-m)}}(a_j + p^s\mathbb{Z}_p^n).
\end{aligned}$$

This finishes the proof of Lemma 4.31. □

Corollary 4.32. $\mu_p(F^{-1}(b + p^s\mathbb{Z}_p^m)) = \sum_{j=1}^{p^{s(n-m)}} \mu_p(a_j + p^s\mathbb{Z}_p^n) = p^{s(n-m)} \cdot p^{-sn} = p^{-sm} = \mu_p(b + p^s\mathbb{Z}_p^m))$.

Proposition 4.33. *Under the conditions of Lemma 4.31, the function F preserves measure.*

Proof. Balls of the form $b + p^s\mathbb{Z}_p^m$ constitute a base of a σ-ring of all measurable sets of the space $\mathbb{Z}_p^m$. In view of Corollary 4.32, F is then a measurable mapping; that is, any preimage of a measurable set is measurable. Now let's find $\mu_p(F^{-1}(M)$ for a measurable $M \subset \mathbb{Z}_p^m$.

Any open measurable subset $A \subset \mathbb{Z}_p^m$ is a disjoint union of such balls; hence, $F^{-1}(A)$ is open measurable subset of $\mathbb{Z}_p^n$, and $\mu_p(F^{-1}(A)) = \mu_p(A)$ in view of Corollary 4.32. Further, for a measurable M one has $\mu_p(M) = \inf\{\mu_p(V) \colon V \supset M, V \text{ is open in } \mathbb{Z}_p^m\}$; thus,

$$\mu_p(F^{-1}(M)) \le \inf\{\mu_p(F^{-1}(V)) : V \supset M, V \text{ is open in } \mathbb{Z}_p^m\} = \mu_p(M).$$

On the other hand, $\mu_p(M) = \sup\{\mu_p(W) : W \subset M, W \text{ is closed in } \mathbb{Z}_p^m\}$. Since each ball $b + p^s\mathbb{Z}_p^m$ is closed in $\mathbb{Z}_p^m$, each closed subset $W \subset \mathbb{Z}_p^m$ is a countable union of such balls (and, maybe, points); hence, the union is disjoint, whence $\mu_p(F^{-1}(W))$ is a closed subset of $\mathbb{Z}_p^n$, and $\mu_p(F^{-1}(W)) = \mu_p(W)$ in view of Corollary 4.32. Thus,

$$\mu_p(F^{-1}(M)) \ge \sup\{\mu_p(F^{-1}(W)) : W \subset M, W \text{ is closed in } \mathbb{Z}_p^m\} = \mu_p(M).$$

Finally we get $\mu_p(F^{-1}(M)) = \mu_p(M)$, thus proving the proposition. □

We now prove the inverse statement.

Proposition 4.34. *Any 1-Lipschitz measure-preserving function $F : \mathbb{Z}_p^n \to \mathbb{Z}_p^m$ is balanced modulo p^k, for all $k = 1, 2, \ldots$.*

Proof. Assume that for some k there exist $\bar{x}, \bar{y} \in (\mathbb{Z}/p^k\mathbb{Z})^m = \{0, 1, \ldots, p^k - 1\}^m$ such that $\#\bar{F}_k^{-1}(\bar{x}) \neq \#\bar{F}_k^{-1}(\bar{y})$; note that both $F_k^{-1}(\bar{x})$ and $F_k^{-1}(\bar{y})$ lie in a finite set $(\mathbb{Z}/p^k\mathbb{Z})^n = \{0, 1, \ldots, p^k - 1\}^n$. Consider two balls $\bar{x} + p^k\mathbb{Z}_p^m$ and $\bar{y} + p^k\mathbb{Z}_p^m$ in $\mathbb{Z}_p^m$. Then

$$F^{-1}(\bar{x} + p^k\mathbb{Z}_p^m) = \bigcup_{z \in \bar{F}_k^{-1}(\bar{x})} (z + p^k\mathbb{Z}_p^n),$$

$$F^{-1}(\bar{y} + p^k\mathbb{Z}_p^m) = \bigcup_{z \in \bar{F}_k^{-1}(\bar{y})} (z + p^k\mathbb{Z}_p^n).$$

Thus, $\mu_p(F^{-1}(\bar{x} + p^k\mathbb{Z}_p^m)) \neq \mu_p(F^{-1}(\bar{y} + p^k\mathbb{Z}_p^m))$; a contradiction. □

4.4.3 1-Lipschitz ergodic functions

We finally characterize ergodic functions among all 1-Lipschitz functions $F : \mathbb{Z}_p^n \to \mathbb{Z}_p^n$.

Proposition 4.35. *A 1-Lipschitz function $F : \mathbb{Z}_p^n \to \mathbb{Z}_p^n$ is ergodic if and only if F is transitive modulo p^k, for all $k = 1, 2, \ldots$.*

Proof. We start with the 'if' part of the statement. By the definition, the function F is ergodic whenever $F^{-1}(A) = A$ implies either $\mu_p(A) = 1$ or $\mu_p(A) = 0$, for any measurable $A \subset \mathbb{Z}_p^n$. Let F be transitive modulo p^k for every $k = 1, 2, \ldots$, yet let F be not ergodic. That is, let there exist a measurable non-empty $A \subset \mathbb{Z}_p^n$ such that $0 < \mu_p(A) < 1$ and $F^{-1}(A) = A$ (whence $F(A) = A$, since F is a bijection, see Corollary 4.29 and Proposition 4.26).

We claim that then there exists a closed F-invariant subset $\overline{C} \subset A$ (that is, $F^{-1}(\overline{C}) = \overline{C}$) such that $1 > \mu_p(\overline{C}) > 0$. Moreover, this closed subset $\overline{C}$ is a union of some finite number of balls of pairwise equal radii.

Indeed, as any open subset of $\mathbb{Z}_p^n$ is a countable union of balls, and since a complement of a ball of a positive radius r is a union of a finite number of balls of this radius r, every closed subset of $\mathbb{Z}_p^n$ is a countable union of balls, some of which are, maybe, of zero radius (i.e., points). However,

$$\mu_p(A) = \sup\{\mu_p(S) : S \subset A, S \text{ is closed in } \mathbb{Z}_p^n\},$$

since μ_p is a regular measure. Thus, there exists a closed subset $B \subset A$ such that $\mu_p(B) > 0$ since $\mu_p(A) > 0$. Hence, there exists a subset $C \subset B$, which is a ball of a positive radius r; thus, $\mu_p(C) > 0$. Since by Corollary 4.29 and Proposition 4.26 the mapping F is a 1-Lipschitz and measure-preserving bijection, both $F^{-1}(C)$ and $F(C)$ are balls of the same radius r. Thus, the set $\overline{C} = \bigcup_{s=-\infty}^{\infty} F^s(C)$ is an F-invariant subset of A: $F^{-1}(\overline{C}) = \overline{C}$, and $\overline{C} \subset A$. As the union $\bigcup_{s=-\infty}^{\infty} F^s(C)$ is a union of balls of the same radius r, then $\overline{C}$ is a union of a finite number of balls of radius r, since there are only finitely many balls of the radius r. Obviously, $\mu_p(\overline{C}) < 1$ since $\mu_p(A) < 1$ by our assumption. Also, $\mu_p(\overline{C}) \geq \mu_p(C) > 0$.

Now, to prove the 'if' part of the proposition we may additionally suggest that A is either a ball (of radius, say, $1 > p^k > 0$), or A is not a ball, yet a union of a finite number of balls of radius $r = p^k > 0$ each. In all cases the mapping $\bar{F}_k$ is not transitive since it has a proper invariant subset, which consists of all images modulo p^k of these balls. Yet this contradicts our assumption that F is transitive modulo p^k for all $k = 1, 2, \ldots$.

Now we prove the 'only if' part of the proposition. Let F be ergodic. Then F preserves measure, so in view of Corollary 4.29 for each $k = 1, 2, \ldots$ the mapping $\bar{F}_k$ is a permutation of the elements of the ring $(\mathbb{Z}/p^k\mathbb{Z})^n$. In case for some k the permutation $\bar{F}_k$ has more than one cycle, we have that there exists a proper subset $\bar{A} \subset (\mathbb{Z}/p^k\mathbb{Z})^n = \{0, 1, \ldots, p^k - 1\}^n$ such that $\bar{F}_k(\bar{A}) = \bar{A}$. This implies that $F(\bar{A} + p^k\mathbb{Z}_p^n) = \bar{A} + \mathbb{Z}_p^n$, i.e. $F^{-1}(\bar{A} + p^k\mathbb{Z}_p^n) = \bar{A} + p^k\mathbb{Z}_p^n$, since F is a bijection, see Proposition 4.26. Yet $\mu_p(\bar{A} + p^k\mathbb{Z}_p^n) = (\#\bar{A}) \cdot p^{-kn}$, and $0 < (\#\bar{A}) \cdot p^{-kn} < 1$, since $\bar{A}$ is a proper subset in $\{0, 1, \ldots, p^k - 1\}^n$. This contradicts to our assumption that F is ergodic. □

4.5 Ergodic 1-Lipschitz transformations on $\mathbb{Z}_p$

In this section we obtain various results on ergodicity (and measure-preservation) for 1-Lipschitz maps from $\mathbb{Z}_p^n$ to $\mathbb{Z}_p^n$. We mainly follow [21, 24].

4.5.1 Ergodicity of affine mappings

In this subsection we obtain explicit conditions for ergodicity of affine mappings from $\mathbb{Z}_p^n$ onto $\mathbb{Z}_p^n$, i.e., of mappings $F = (f_1, \ldots, f_n)\colon \mathbb{Z}_p^n \to \mathbb{Z}_p^n$, where every function $f_j(x_1, \ldots, x_n)$ is of the form

$$f_j(x_1, \ldots, x_n) = a_{j,0} + a_{j,1} \cdot x_1 + \cdots + a_{j,n} \cdot x_n,$$

$a_{j,0}, a_{j,1}, \ldots, a_{j,n} \in \mathbb{Z}_p$. Actually in this subsection we restrict our study with the case $n = 1$ only, since *no affine ergodic transformation on* $\mathbb{Z}_p^n$ *exists for* $n > 1$; the latter claim follows from a much more general result which we prove in Subsection 4.6.2, see Theorem 4.51 there.

So we consider a transformation $f(x) = ax + b$ on the space $\mathbb{Z}_p$, where $a, b \in \mathbb{Z}_p$. This case serves as a base for further considerations; also, it is important for applications: Transformations of this sort give rise to a class of random number generators, the so-called linear congruential generators, see Chapter 9 for details. Generators of this kind are well studied, see e.g. [267, Subsection 3.2.1] and references therein. Now we will actually just reproduce corresponding results after re-stating them in dynamical terms.

In view of Theorem 4.23 it is clear that f is measure-preserving if and only if a has a multiplicative inverse modulo p^k for all $k = 1, 2, \dots$ (that is, a is a unit in $\mathbb{Z}_p$); in other words, if and only if $a \not\equiv 0 \pmod p$.

Theorem 4.36. *The function $f(x) = ax + b$, where $a, b \in \mathbb{Z}_p$, is an ergodic transformation on $\mathbb{Z}_p$ if and only if following conditions hold simultaneously*:

$$b \not\equiv 0 \pmod p; \tag{4.18}$$

$$a \equiv 1 \pmod p, \text{ for } p \text{ odd}; \tag{4.19}$$

$$a \equiv 1 \pmod 4, \text{ for } p = 2. \tag{4.20}$$

Proof. In view of Theorem 4.23 we must prove that f is transitive modulo p^k if and only if the conditions of Theorem 4.36 hold. We prove this by induction on k, and we state a base of induction as a lemma:

Lemma 4.37. *The function $f(x) = ax + b$ is transitive modulo p if and only if $b \not\equiv 0 \pmod p$ and $a \equiv 1 \pmod p$.*

Proof. It is clear that $a \not\equiv 0 \pmod p$ (otherwise f is a constant) and that $b \not\equiv 0 \pmod p$ (otherwise 0 is a fixed point of f). Now, as for every $i = 1, 2, \dots$

$$f^i(x) = a^i x + b(a^{i-1} + a^{i-2} + \dots + a + 1) \tag{4.21}$$

we conclude that if $a \not\equiv 1 \pmod p$ then $f^p(x) = a^p x + b(a^p - 1)(a - 1)^{-1}$ where $(a - 1)^{-1}$ is a multiplicative inverse of $(a - 1)$ modulo p. Thus, as $z^p \equiv z \pmod p$ for every $z \in \mathbb{Z}$, we have $f^p(x) = ax + b \pmod p$, i.e., $f^p(x) \equiv f(x) \pmod p$. However, if f is transitive modulo p then $f^p(x) \equiv x \pmod p$. This contradiction proves that $a \equiv 1 \pmod p$.

The converse statement of Lemma 4.37 is obvious: If $a \equiv 1 \pmod p$ then (4.21) implies that $f^i(x) \equiv x + bi \pmod p$, i.e., given $x, y \in \{0, 1, \dots, p - 1\}$ from the congruence $x+bi \equiv y \pmod p$ one finds $i \in \{0, 1, \dots, p-1\}$ (since $b \not\equiv 0 \pmod p$) such that $f^i(x) \equiv y \pmod p$. □

Now we assume that the conditions of Theorem 4.36 imply transitivity of f modulo p^k; we claim that then f is transitive modulo p^{k+1}.

As f is measure-preserving, f is bijective modulo p^{k+1}; thus, as f is transitive modulo p^k, it is clear that f is transitive modulo p^{k+1} whenever $f^i(0) \equiv 0 \pmod{p^{k+1}}$ implies $i \equiv 0 \pmod{p^{k+1}}$. Note that $f^i(0) \equiv 0 \pmod{p^{k+1}}$ implies $i \equiv 0 \pmod{p^k}$ since f is transitive modulo p^k. Now we just calculate $f^i(0) \bmod p^{k+1}$ for $i = p^k \ell$.

As $a = 1 + pr$ for a suitable $r \in \mathbb{Z}_p$, from (4.21) we get

$$f^i(0) = b \cdot \frac{(1+pr)^i - 1}{pr} = b \sum_{j=1}^{i} p^{j-1} r^{j-1} \binom{i}{j}. \tag{4.22}$$

Now represent

$$\binom{i}{j} = \frac{1}{j!} i(i-1)\cdots(i-j+1) = \frac{i}{j} \cdot \left(\frac{i}{1} - 1\right) \cdot \left(\frac{i}{2} - 1\right) \cdots \left(\frac{i}{j-1} - 1\right).$$

As $\frac{i}{t} \in \mathbb{Z}_p$ for $i = p^k \ell, t = 1, 2, \ldots, i-1$ we conclude that $\operatorname{ord}_p \binom{p^k \ell}{j} \geq k - \operatorname{ord}_p j$, so from (4.22) it follows that

$$f^{p^k \ell}(0) \equiv b \cdot p^k \ell \pmod{p^{k+1}} \text{ for } p \text{ odd, and} \tag{4.23}$$

$$f^{2^k \ell}(0) \equiv b \cdot \left(2^k \ell + 2r \binom{2^k \ell}{2}\right) \pmod{2^{k+1}}, \tag{4.24}$$

since $j - \operatorname{ord}_p j < 2$ if and only if either $j = 1$, or $p = 2$ and $j = 2$.

Whenever p is odd, from (4.24) it follows that $f^{p^k \ell}(0) = 0 \pmod{p)^{k+1}}$ if and only if $\ell \equiv 0 \pmod{p}$, thus proving our claim for odd p. For $p = 2$, however, (4.24) implies that $f^{2^k \ell}(0) \equiv 0 \pmod{2^{k+1}}$ ether when ℓ is even, or when both ℓ and r are odd. Yet the latter case does not hold since $a \equiv 1 \pmod{4}$.

We conclude finally that the conditions of Theorem 4.36 are sufficient. In view of Theorem 4.23, the above argument shows that these conditions are also necessary. □

We stress a leading idea of the proof:

Note 4.38. Given a 1-Lipschitz (that is, a compatible) measure-preserving function $f : \mathbb{Z}_p \to \mathbb{Z}_p$, which is transitive modulo p^k, the function f is transitive modulo p^{k+1} if and only if $f^{p^k \ell}(z) \equiv z \pmod{p^{k+1}}$ implies $\ell \equiv 0 \pmod{p}$ for some (or, equivalently, every) $z \in \mathbb{Z}_p$.

In the sequel, we exploit this observation frequently. Note also that the statement of Note 4.38 holds for asymptotically compatible functions as well, once k is sufficiently large.

4.5.2 Ergodicity and measure-preservation in terms of coordinate functions

In this subsection we prove criteria of measure-preservation and of ergodicity for 1-Lipschitz functions $f : \mathbb{Z}_2 \to \mathbb{Z}_2$ in terms of coordinate functions, which were defined in Subsection 3.8.1. Recall that according to Proposition 3.35 every 1-Lipschitz function $f : \mathbb{Z}_2 \to \mathbb{Z}_2$ can be represented in a form

$$f\left(\sum_{i=0}^{\infty} \chi_i \cdot 2^i\right) = \sum_{j=0}^{\infty} \psi_i(\chi_0, \dots, \chi_i) \cdot 2^i \tag{4.25}$$

where $\chi_i \in \{0, 1\}$, and each ith coordinate function $\psi_i(\chi_0, \dots, \chi_i) = \delta_i(f(x))$ is a Boolean function in Boolean variables $\chi_0, \dots, \chi_i$; that is, $\psi_i : \{0, 1\}^{i+1} \to \{0, 1\}$; $i = 0, 1, 2, \dots$.

The following Theorem 4.39 is just a re-statement in dynamical terms of a known (at least since the mid 1970s) result from the theory of Boolean functions, the so-called bijectivity/transitivity criterion for triangular Boolean mappings. Although the criterion was cited in the literature (see e.g. [21, Lemma 4.8]), its author is not known.

Recall that an *algebraic normal form*, the ANF, of the Boolean function $\psi_i(\chi_0, \dots, \chi_i)$ is a representation of this function via $\oplus$ (addition modulo 2, that is, logical 'exclusive or') and $\cdot$ (multiplication modulo 2, that is, logical 'and', or conjunction). In other words, the ANF of the Boolean function ψ is its representation in the form

$$\psi(\chi_0, \dots, \chi_j) = \beta \oplus \beta_0\chi_0 \oplus \beta_1\chi_1 \oplus \dots \oplus \beta_{0,1}\chi_0\chi_1 \oplus \cdots,$$

where $\beta, \beta_0, \dots \in \{0, 1\}$ and $\chi_0, \dots, \chi_j$ are Boolean variables. The ANF is sometimes called a *Boolean polynomial* since obviously an ANF $\psi(\chi_0, \dots, \chi_j)$ can be considered as an element of a factor-ring of the ring of $(j + 1)$-variate polynomials $(\mathbb{Z}/2\mathbb{Z})[x_0, \dots, x_j]$, with coefficients from the residue ring $\mathbb{Z}/2\mathbb{Z}$, modulo an ideal generated by all polynomials $x_i^2 - x_i$, $i = 0, 1, \dots, j$.

Recall that the *weight* of the Boolean function ψ in $(j + 1)$ variables is the number of $(j + 1)$-bit words that *satisfy* ψ; that is, the weight is a cardinality of the truth set of ψ, and the *truth set* of ψ is the set all points from $\{0, 1\}^{j+1}$ where ψ takes value 1.

Theorem 4.39 (folklore). *The function f defined by equation* (4.25) *is measure-preserving if and only if for every $i = 0, 1, \dots$ the ANF of the ith coordinate function is*

$$\psi_i(\chi_0, \dots, \chi_i) = \chi_i \oplus \varphi_i(\chi_0, \dots, \chi_{i-1}),$$

where φ_i is an ANF of a Boolean function in Boolean variables $\chi_0, \dots, \chi_{i-1}$, and φ_0 is a constant from $\{0, 1\}$.

The function f is ergodic if and only if, additionally, $\varphi_0 = 1$, and every Boolean function φ_i is of odd weight, that is, takes value 1 exactly at an odd number of points from $\{0, 1\}^i$ for $i = 1, 2, \dots$. The latter takes place if and only if a degree of the ANF of φ_i for $i \geq 1$ is exactly i, that is, if and only if the ANF of φ_i contains a monomial $\chi_0 \cdots \chi_{i-1}$.

Proof. Collecting together all terms of the ANF that do not contain a variable χ_j we write the function ψ_i in the following form:

$$\psi_i(\chi_0, \dots, \chi_i) = \chi_i \cdot \omega_i(\chi_0, \dots, \chi_{i-1}) \oplus \varphi_i(\chi_0, \dots, \chi_{i-1}),$$

where both $\omega_i(\chi_0, \ldots, \chi_{i-1})$ and $\varphi_i(\chi_0, \ldots, \chi_{i-1})$ are Boolean functions in Boolean variables $\chi_0, \ldots, \chi_{i-1}$.

Obviously, whenever all $\omega_i(\chi_0, \ldots, \chi_{i-1})$ are identically 1, the function f is measure-preserving in view of Theorem 4.23 since f is bijective modulo 2^{k+1} for every $k = 0, 1, 2, \ldots$: To find a preimage of the mapping $f \bmod 2^k$ one must solve a system of Boolean equations

$$\begin{cases} \chi_0 \oplus \varphi_0 & = \alpha_0, \\ \chi_1 \oplus \varphi_1(\chi_1) & = \alpha_1, \\ & \vdots \\ \chi_k \oplus \varphi_k(\chi_0, \ldots, \chi_{k-1}) & = \alpha_k, \end{cases}$$

which has a unique solution given any $\alpha_0, \ldots, \alpha_k \in \{0, 1\}$.

Conversely, let i be the smallest number such that $\omega_i(\chi_0, \ldots, \chi_{i-1}) = 0$ for a certain vector $(\varepsilon_0, \ldots, \varepsilon_{i-1})$ of zeros and ones. Then

$$f(\varepsilon_0 + \varepsilon_1 \cdot 2 + \cdots + \varepsilon_{i-1} \cdot 2^{i-1} + 0 \cdot 2^i) \equiv f(\varepsilon_0 + \varepsilon_1 \cdot 2 + \cdots + \varepsilon_{i-1} \cdot 2^{i-1} + 1 \cdot 2^i) \pmod{2^{i+1}}.$$

Whence f is not bijective modulo 2^{i+1}, thus not measure-preserving in view of Theorem 4.23.

Now, to prove the ergodicity part of the statement we first note that f is transitive modulo 2 if and only if $\psi_0(\chi_0) = \chi_0 \oplus 1$. Further, if f is transitive modulo 2^{k+1}, then f is transitive modulo 2^j for all $j = 1, 2, \ldots, k$; so the ith coordinate function $\delta_i(f^{2^k})(x)$ of the 2^kth iterate of the function f is

$$\delta_i(f^{2^k}(\chi_0 + \chi_1 \cdot 2 + \chi_2 \cdot 4 + \cdots)) = \begin{cases} \chi_i, & \text{if } i < k; \\ \chi_k \oplus \sigma, & \text{if } i = k, \end{cases} \tag{4.26}$$

where σ is a sum modulo 2 of all values of the Boolean function φ_k at all points from $\{0, 1\}^k$; that is, σ is the weight modulo 2 of the function φ_k. From (4.26) it follows then that the transitivity of the function f modulo 2^{k+1} implies $\sigma = 1$; otherwise $f^{2^k}(x) \equiv x \pmod{2^{k+1}}$ for every $x \in \mathbb{Z}_2$. Thus, a weight of the function φ_k must be odd.

The rest of the statement of the theorem is a well-known result from the theory of Boolean functions: *A weight of a Boolean function is odd if and only if its ANF is of maximum degree.*

To prove this claim consider a Boolean function $\psi(\chi_0, \ldots, \chi_j)$ in Boolean variables $\chi_0, \ldots, \chi_j$. For $\alpha, \beta \in \{0, 1\}$ define $\alpha^\beta = 1$ whenever $\alpha = \beta$ and $\alpha^\beta = 0$, otherwise. Then we can write the Boolean function ψ in the form

$$\psi(\chi_0, \ldots, \chi_j) = \bigoplus_{(\beta_0, \ldots, \beta_j) \in T(\psi)} \chi_0^{\beta_0} \cdots \chi_j^{\beta_j}, \tag{4.27}$$

where $T(\psi) \subset \{0,1\}^{j+1}$ is a truth set of the Boolean function ψ. To obtain ANF from representation (4.27) we substitute $\chi^\beta = \chi \oplus \beta \oplus 1$ and perform all multiplications and additions modulo 2; it is obvious then that the coefficient $\mathrm{Coef}_{\chi_0 \cdots \chi_j}\, \psi$ of the term $\chi_0 \cdots \chi_j$ (of degree $j+1$, which is a maximum degree of any Boolean function in $j+1$ variables) in the ANF of the Boolean function ψ is $\#T(\psi) \bmod 2$. □

4.5.3 Ergodicity and measure-preservation in terms of Mahler expansion

Recall that every function $f: \mathbb{Z}_p \to \mathbb{Z}_p$ can be expressed via the Mahler interpolation series (3.32)

$$f(x) = \sum_{i=0}^{\infty} a_i \binom{x}{i},$$

where $a_i \in \mathbb{Z}_p$, $i = 0, 1, 2, \ldots$. We now are going to describe how one can determine from the coefficients a_i whether f is measure-preserving or, respectively, ergodic. A central result of this subsection is the following

Theorem 4.40. *The function f defines a 1-Lipschitz measure-preserving transformation on $\mathbb{Z}_p$ whenever the following conditions hold simultaneously*:

$$a_1 \not\equiv 0 \pmod{p}; \tag{4.28}$$

$$a_i \equiv 0 \pmod{p^{\lfloor \log_p i \rfloor + 1}}, \quad i = 2, 3, \ldots. \tag{4.29}$$

The function f defines a 1-Lipschitz ergodic transformation on $\mathbb{Z}_p$ whenever the following conditions hold simultaneously:

$$a_0 \not\equiv 0 \pmod{p}; \tag{4.30}$$

$$a_1 \equiv 1 \pmod{p}, \quad \textit{for } p \textit{ odd}; \tag{4.31}$$

$$a_1 \equiv 1 \pmod{4}, \quad \textit{for } p = 2; \tag{4.32}$$

$$a_i \equiv 0 \pmod{p^{\lfloor \log_p (i+1) \rfloor + 1}}, \quad i = 2, 3, \ldots. \tag{4.33}$$

Moreover, in the case $p = 2$ these conditions are necessary: Namely, if f is 1-Lipschitz and measure-preserving then conditions (4.28) *and* (4.29) *hold simultaneously; if f is 1-Lipschitz and ergodic then conditions* (4.30), (4.32) *and* (4.33) *hold simultaneously.*

Thus, Theorem 4.40 gives a complete description of 1-Lipschitz measure-preserving (respectively, of 1-Lipschitz ergodic) transformations on $\mathbb{Z}_p$ for $p = 2$ in terms of Mahler expansion. We also show in this subsection that $p = 2$ is the only case when the conditions of Theorem 4.40 are necessary.

To prove the theorem we need some extra results, which are of interest by their own.

Lemma 4.41. *Given a 1-Lipschitz function* $v\colon \mathbb{Z}_p \to \mathbb{Z}_p$ *and p-adic integers* c, d, $c \not\equiv 0 \pmod{p}$, *the function* $g(x) = d + cx + p \cdot v(x)$ *preserves measure, and the function* $h(x) = c + x + p \cdot \Delta v(x)$ *is ergodic.* (Recall that Δ is a difference operator: $\Delta v(x) = v(x+1) - v(x)$ by the definition.)

Proof. In view of Theorem 4.23 we must show that the function g (respectively, h) is bijective (respectively, transitive) modulo p^k for all $k = 1, 2, 3, \ldots$.

First we prove by induction on k that g is bijective modulo p^k for all $k = 1, 2, 3, \ldots$. The assertion is obviously true for $k = 1$.

Assume our claim is true for $k = 1, 2, \ldots, n-1$. Let us prove that it holds for $k = n$. Let $g(a) \equiv g(b) \pmod{p^n}$ for some p-adic integers a, b. Then $a \equiv b \pmod{p^{n-1}}$ by induction hypothesis. Hence $p \cdot v(a) \equiv p \cdot v(b) \pmod{p^n}$ since v is 1-Lipschitz. Further, the congruence $g(a) \equiv g(b) \pmod{p^n}$ implies the congruence $c \cdot a + p \cdot v(a) \equiv c \cdot b + p \cdot v(b) \pmod{p^n}$, and consequently, $c \cdot a \equiv c \cdot b \pmod{p^n}$. Since $c \not\equiv 0 \pmod{p}$, the latter congruence implies that $a \equiv b \pmod{p^n}$ thus proving the first assertion of Lemma 4.41.

To prove the remaining part of the statement we note that the assertion we just proved implies that the function h preserves measure. To prove transitivity of h modulo p^k for all $k = 1, 2, 3, \ldots$ we use induction on k.

From Lemma 4.37 it follows that h is transitive modulo p. Assume that h is transitive modulo p^{k-1} and pursue as in Note 4.38.

We calculate successively

$$h^1(x) = c + x + p \cdot v(x+1) - p \cdot v(x),$$

$$\vdots$$

$$\begin{aligned} h^j(x) &= h(h^{j-1}(x)) = cj + h^{j-1}(x) + p \cdot v(h^{j-1}(x) + 1) - p \cdot v(h^{j-1}(x)) \\ &= cj + x + p \sum_{i=0}^{j-1} v(h^i(x) + 1) - p \sum_{i=0}^{j-1} v(h^i(x)), \end{aligned}$$

and so on. We recall that $h^0(x) = x$ by the definition. Thus

$$h^{p^{k-1}\ell}(x) = cp^{k-1}\ell + x + p \sum_{i=0}^{p^{k-1}\ell-1} v(h^i(x) + 1) - p \sum_{i=0}^{p^{k-1}\ell-1} v(h^i(x)). \quad (4.34)$$

However, as h is transitive modulo p^{k-1} and 1-Lipschitz, we see that

$$\sum_{i=0}^{p^{k-1}\ell-1} v(h^i(x) + 1) \equiv \sum_{i=0}^{p^{k-1}\ell-1} v(h^i(x)) \equiv \ell \sum_{z=0}^{p^{k-1}-1} v(z) \pmod{p^{k-1}},$$

so (4.34) implies that $h^{p^{k-1}\ell}(x) \equiv cp^{k-1}\ell + x \pmod{p^k}$. Yet $c \not\equiv 0 \pmod{p}$; thus if $h^{p^{k-1}\ell}(0) \equiv 0 \pmod{p^k}$ then necessarily $\ell \equiv 0 \pmod{p}$. This proves Lemma 4.41 in view of Note 4.38. □

Corollary 4.42. *Under the assumptions of Lemma 4.41 let $r \equiv 1 \pmod p$ if p is odd, and let $r \equiv 1 \pmod 4$ if $p = 2$. Then the function $c + rx + p\cdot\Delta v(x)$ is ergodic.*

Proof. We have $r = 1 + ps$ for odd p (respectively $r = 1 + 4s$ for $p = 2$) where $s \in \mathbb{Z}_p$. Now, since p is odd, the function $u(x) = s\binom{x}{2}$ (respectively, $u(x) = 2s\binom{x}{2}$) is a polynomial over $\mathbb{Z}_p$, thus, 1-Lipschitz. Consequently, the function $v_1(x) = u(x) + v(x)$ is 1-Lipschitz either. Since $\Delta v_1(x) = sx + \Delta v(x)$ for odd p (respectively, $\Delta v_1(x) = 2sx + \Delta v(x)$ for $p = 2$), the proof is finished in view of Lemma 4.41. □

Proof of Theorem 4.40. Recall that according to Theorem 3.53, a function $f : \mathbb{Z}_p \to \mathbb{Z}_p$ is 1-Lipschitz if and only if it can be represented in the form

$$f(x) = b_0 + \sum_{i=1}^{\infty} b_i p^{\lfloor \log_p i \rfloor} \binom{x}{i}, \tag{4.35}$$

where $b_i \in \mathbb{Z}_p$, $i = 0, 1, 2, \ldots$. As $\lfloor \log_p i \rfloor = \lfloor \log_p (i+1) \rfloor$ for all $i = 1, 2, \ldots$ but $i = p^t - 1$ $(t = 1, 2, 3, \ldots)$, and as

$$\Delta v(x) = \sum_{i=1}^{\infty} b_i p^{\lfloor \log_p i \rfloor} \binom{x}{i-1},$$

(see (1.1)) sufficiency of the conditions of Theorem 4.40 follows now from Lemma 4.41 and Corollary 4.42.

To prove that for $p = 2$ the conditions of Theorem 4.40 are necessary we will express coefficients of algebraic normal forms of coordinate functions (see Subsection 4.5.2) via coefficients of Mahler expansion (4.35) and then apply Theorem 4.39.

During the proof we denote $\chi_i = \delta_i(x) \in \{0, 1\}$. Then for arbitrary $n \in \mathbb{N}$ and $x \in \mathbb{Z}_2$ Lemma 3.46 implies that

$$f(x) \equiv f(\chi_0 + \chi_1 2 + \cdots + \chi_{n-1} 2^{n-1}) + 2^n \chi_n \tilde{f}_n(x) \pmod{2^{n+1}}, \tag{4.36}$$

where

$$\tilde{f}_n(x) \equiv \sum_{s=0}^{n} \frac{\Delta^{2^s} f(x)}{2^s} \pmod 2. \tag{4.37}$$

From (3.40) (see proof of Theorem 3.53) we conclude that

$$\frac{\Delta^{2^s} f(x)}{2^s} = \frac{1}{2^s} \sum_{i=2^s}^{\infty} a_i \binom{x}{i-2^s} = \sum_{i=2^s}^{\infty} b_i 2^{\lfloor \log_2 i \rfloor - s} \binom{x}{i-2^s}.$$

This, in view of Lucas' Theorem 1.2, implies that the following congruences modulo 2 hold:

$$\begin{aligned}\delta_0\left(\frac{\Delta^{2^s} f(x)}{2^s}\right) &\equiv \sum_{i=2^s}^{2^{s+1}-1} \delta_0(b_i)\binom{x}{i-2^s} \\ &\equiv \sum_{j=0}^{2^s-1} \delta_0(b_{2^s+j})\binom{x_0}{\delta_0(j)}\cdots\binom{x_{s-1}}{\delta_{s-1}(j)} \pmod 2. \end{aligned} \tag{4.38}$$

From (4.38) it follows that $\delta_0\left(\frac{\Delta^{2^s} f(x)}{2^s}\right)$ does not depend on $\chi_s, \chi_{s+1}, \dots$ and that $\delta_0(\Delta f(x)) \equiv b_1 \equiv a_1 \pmod 2$. Now the latter congruence in view of (4.37) implies

$$\tilde{f}_n(x) \equiv \tilde{f}_n(\bar{x}_n) \equiv \sum_{s=1}^{n} \frac{\Delta^{2^s} f(\bar{x}_s)}{2^s} + a_1 \pmod 2 \tag{4.39}$$

where here and in the following $\bar{x}_k$ stands for $x \bmod 2^k = \chi_0 + \chi_1 2 + \cdots + \chi_{k-1} 2^{k-1}$ $(k = 1, 2, \dots)$.

Theorem 4.39 implies now that f preserves measure if and only if the following two conditions hold:

- f is bijective modulo 2, (4.40)
- $\tilde{f}_n(x) \equiv 1 \pmod 2$ for all $n = 1, 2, \dots$ and all $x \in \mathbb{Z}_2$. (4.41)

As $f(x) \equiv a_0 + a_1 x \pmod 2$ then condition (4.40) is equivalent to the following condition:

$$a_1 \equiv 1 \pmod 2. \tag{4.42}$$

Now, in view of (4.39) and (4.42), condition (4.41) holds if and only if the following condition

$$\frac{\Delta^{2^s} f(\bar{x}_s)}{2^s} \equiv 0 \pmod 2 \tag{4.43}$$

holds for all $s = 1, 2, 3, \dots$ and all $x \in \mathbb{Z}_2$. However, in view of (4.38), condition (4.43) holds for all $s = 1, 2, 3, \dots$ and all $x \in \mathbb{Z}_2$ if and only if the condition

$$b_i \equiv 0 \pmod 2 \tag{4.44}$$

holds for all $i = 2, 3, \dots$. As $a_i = b_i 2^{\lfloor \log_2 i \rfloor}$ for $i = 1, 2, \dots$, then (4.42) and (4.44) imply necessity of conditions (4.28) and (4.29) when $p = 2$.

Further, as an ergodic function f preserves measure, from Theorem 4.39 in view of (4.36) and condition (4.41) we conclude that the ANF of the Boolean function $\delta_i(f(x)) = \psi_i(\chi_0, \dots, \chi_i)$ is of the following form:

$$\psi_i(\chi_0, \dots, \chi_i) = \varphi_i(\chi_0, \dots, \chi_{i-1}) \oplus x_i \tag{4.45}$$

where $\varphi_i(\chi_1, \dots, \chi_{i-1}) = \delta_i(f(\chi_0 + \cdots + \chi_{i-1}2^{i-1}))$ and φ_0 is a constant. Now from Theorem 4.39 it follows that once the function f is ergodic, $\varphi_0 = 1$, and the coefficient $\mathrm{Coef}_{\chi_0 \cdots \chi_{i-1}} \varphi_i$ of the monomial $\chi_0 \cdots \chi_{i-1}$ in the ANF φ_i must be 1 for all $i = 1, 2, \dots$.

Since obviously $\varphi_0 \equiv a_0 \pmod 2$, we conclude now that f is a 1-Lipschitz ergodic function if and only if the following conditions (4.46)–(4.49) hold simultaneously:

$$a_0 \equiv 1 \pmod 2; \tag{4.46}$$

$$a_1 \equiv 1 \pmod 2; \tag{4.47}$$

$$a_j \equiv 0 \pmod{2^{\lfloor \log_2 j \rfloor + 1}}, \text{ for all } j = 2, 3, \dots; \tag{4.48}$$

$$C_i = 1, \quad \text{for all } i = 1, 2, \dots, \tag{4.49}$$

where $C_i = \mathrm{Coef}_{\chi_0 \cdots \chi_{i-1}} \varphi_i$. To finish the proof, we use the following recursive formula for $\mathrm{Coef}_{\chi_0 \cdots \chi_{i-1}} \varphi_i$:

Lemma 4.43. *If a 1-Lipschitz function f preserves measure, then*

$$\mathrm{Coef}_{x_0 \cdots x_n} \varphi_{n+1} \equiv \delta_1(b_{2^{n+1}-1}) + \mathrm{Coef}_{x_0 \cdots x_{n-1}} \varphi_n \pmod 2$$

for all $n = 1, 2, \dots$.

Proof. We begin as in the proof of Lemma 3.46: Using the Gregory–Newton formula from Theorem 1.5 and taking into account that $\chi_n \in \{0, 1\}$, we conclude that

$$\begin{aligned} f(\bar{x}_{n+1}) &= f(\bar{x}_n) + \chi_n \sum_{i=1}^{2^n} \Delta^i f(\bar{x}_n) \binom{2^n}{i} \\ &= f(\bar{x}_n) + 2^n \chi_n \sum_{k=0}^{2^n - 1} \frac{\Delta^{k+1} f(\bar{x}_n)}{k+1} \binom{2^n - 1}{k}. \end{aligned}$$

Hence,

$$\delta_{n+1}(f(\bar{x}_{n+1})) \equiv \delta_{n+1}(f(\bar{x}_n)) + \delta_1(\chi_n S_n) + \delta_n(f(\bar{x}_n))\delta_0(\chi_n S_n) \pmod 2, \tag{4.50}$$

where

$$S_n = \sum_{k=0}^{2^n - 1} \frac{\Delta^{k+1} f(\bar{x}_n)}{k+1} \binom{2^n - 1}{k}.$$

As by Lucas' Theorem 1.2, $\binom{2^n - 1}{k} \equiv 1 \pmod 2$ for all $k \in \{0, 1, \dots, 2^n - 1\}$, then, combining together Lemma 3.45 and Lemma 3.46, we conclude that

$$S_n \equiv \sum_{s=0}^{n} \frac{\Delta^{2^s} f(\bar{x}_n)}{2^s} \equiv \tilde{f}_n(\bar{x}_n) \pmod 2. \tag{4.51}$$

However, $\tilde{f}_n(\bar{x}_n) \equiv 1 \pmod 2$ since f preserves measure, see (4.41). Then (4.51) implies that

$$\delta_0(S_n) = 1. \tag{4.52}$$

This, in view of (4.50), implies that

$$\begin{aligned} &\mathrm{Coef}_{\chi_0\cdots\chi_n}\ \delta_{n+1}(f(\bar{x}_{n+1})) \\ &\qquad \equiv \mathrm{Coef}_{\chi_0\cdots\chi_n}\ \delta_1(\chi_n S_n) + \mathrm{Coef}_{\chi_0\cdots\chi_{n-1}}\ \delta_n(f(\bar{x}_n)) \pmod 2. \end{aligned} \tag{4.53}$$

As $\delta_1(\chi_n S_n) = \chi_n \delta_1(S_n)$ then

$$\mathrm{Coef}_{\chi_0\cdots\chi_n}\ \delta_1(\chi_n S_n) = \mathrm{Coef}_{\chi_0\cdots\chi_{n-1}}\ \delta_1(S_n). \tag{4.54}$$

Now we must calculate $\delta_1(S_n)$. From 'school-textbook' algorithms of addition and multiplication of 2-adic integers $u_k, v_k \in \mathbb{Z}_2$; $u_k = \delta_0(u_k) + \delta_1(u_k) \cdot 2 + \delta_2(u_k) \cdot 2^2 + \cdots$ and $v_k = \delta_0(v_k) + \delta_1(v_k) \cdot 2 + \delta_2(v_k) \cdot 2^2 + \cdots$, it follows that

$$\begin{aligned} &\delta_1\left(\sum_{k=0}^{m} u_k v_k\right) \\ &\equiv \sum_{k=0}^{m} \delta_0(u_k)\delta_1(v_k) + \sum_{k=0}^{m} \delta_1(u_k)\delta_0(v_k) + \delta_1\left(\sum_{k=0}^{m} \delta_0(u_k v_k)\right) \pmod 2. \end{aligned} \tag{4.55}$$

For $\epsilon_k \in \{0,1\}, k = 0,1,2,\ldots,m$, denote $\Xi(\epsilon_0,\ldots,\epsilon_m) = \delta_1(\sum_{k=0}^{m} \epsilon_k)$, then clearly

$$\Xi(\epsilon_0,\ldots,\epsilon_m) \equiv \left\lfloor \frac{1}{2} \mathrm{Wt}(\epsilon_0,\ldots,\epsilon_m) \right\rfloor \pmod 2,$$

where $\mathrm{Wt}(\epsilon_0,\ldots,\epsilon_m)$ is the number of nonzero coordinates of a binary vector $(\epsilon_0,\ldots,\epsilon_m)$. Now assuming $m = 2^n - 1, u_k = \frac{\Delta^{k+1} f(\bar{x}_n)}{k+1}, v_k = \binom{2^n-1}{k}$ we apply (4.55) to calculate $\delta_1(S_n)$.

From Lucas' Theorem 1.2 it follows that

$$\delta_0(v_k) = \delta_0\left(\binom{2^n - 1}{k}\right) = 1$$

for all $k = 0,1,\ldots,2^n - 1$. Hence,

$$\delta_1\left(\sum_{k=0}^{m} \delta_0(u_k v_k)\right) = \delta_1\left(\sum_{k=0}^{m} \delta_0(u_k)\delta_0(v_k)\right) = \Xi(\delta_0(u_0),\ldots,\delta_0(u_m)).$$

Further, from Lemma 3.45 it follows that for all $k \neq 2^r - 1$

$$\delta_0(u_k) = \delta_0\left(\frac{\Delta^{k+1} f(\bar{x}_n)}{k+1}\right) = 0. \tag{4.56}$$

As f preserves measure, (4.43) holds for all $s = 1, 2, \dots$ and all $x \in \mathbb{Z}_2$, so from (4.56) it follows that $\delta_0(u_1) = \cdots = \delta_0(u_m) = 0$, whence the function

$$\Xi(\delta_0(u_0), \dots, \delta_0(u_m)) = \delta_1\Big(\sum_{k=0}^{m} \delta_0(u_k v_k)\Big)$$

in the right hand part of (4.55) vanishes. Finally applying (4.56) and (4.43) to (4.55) we conclude that

$$\delta_1(S_n) \equiv \delta_0(\Delta f(\bar{x}_n)) + \sum_{k=0}^{2^n-1} \delta_1\left(\frac{\Delta^{k+1} f(\bar{x}_n)}{k+1}\right) \pmod 2. \tag{4.57}$$

As f preserves measure, then (4.42) and (4.44) hold; thus, the coefficients b_i of the Mahler expansion (4.35) satisfy the following conditions:

$$\begin{cases} b_1 \equiv 1 \pmod 2; \\ b_i = 2c_i, \text{ for appropriate } c_i \in \mathbb{Z}_2;\ i = 2, 3, \dots. \end{cases} \tag{4.58}$$

Hence for every $s \geq 2$, $s = \hat{s} \cdot 2^{\operatorname{ord}_2 s}$, $\hat{s}$ odd, we have

$$\frac{\Delta^s f(x)}{s} = \frac{2}{\hat{s}} \sum_{i=s}^{\infty} c_i 2^{\lfloor \log_2 i \rfloor - \operatorname{ord}_2 s} \binom{x}{i-s}, \tag{4.59}$$

in view of (1.1) (we note that $\hat{s}$ is a unit of $\mathbb{Z}_2$, thus $\hat{s}$ has a multiplicative inverse $\frac{1}{\hat{s}} \in \mathbb{Z}_2$). Consequently, (4.59) implies that

$$\delta_1\left(\frac{\Delta^s f(x)}{s}\right) \equiv \sum_{i=s}^{\infty} c_i 2^{\lfloor \log_2 i \rfloor - \operatorname{ord}_2 s} \binom{x}{i-s} \pmod 2. \tag{4.60}$$

Since we have that either $\lfloor \log_2 i \rfloor > \operatorname{ord}_2 s$ or $s \leq i$ hold in all cases except the case when $s = 2^r, 2^r \leq i \leq 2^{r+1} - 1$, congruence (4.60) implies that

$$\delta_1\left(\frac{\Delta^s f(x)}{s}\right) \equiv \begin{cases} \sum_{i=2^r}^{2^{r+1}-1} c_i \binom{x}{i-2^r} \pmod 2, & \text{if } s = 2^r \text{ for } r = 1, 2, \dots; \\ 0 \pmod 2, & \text{otherwise.} \end{cases} \tag{4.61}$$

Further, from (4.35) in view of (4.58) and (1.1) we derive that

$$\Delta f(x) \equiv b_1 \pmod 4. \tag{4.62}$$

Now from (4.57) in view of (4.58), (4.61) and (4.62) it follows that

$$\delta_1(S_n) \equiv 1 + \delta_1(b_1) + \sum_{s=1}^{n} \sum_{j=0}^{2^s-1} c_{j+2^s} \binom{\bar{x}_n}{j} \pmod 2. \tag{4.63}$$

From here with the use of Lucas' Theorem 1.2 we deduce that

$$\mathrm{Coef}_{\chi_0\cdots\chi_{n-1}}\,\delta_1(S_n) \equiv c_{2^{n+1}-1} \pmod 2.$$

The latter congruence in view of (4.58), (4.53) and (4.54) finishes the proof of Lemma 4.43. □

Now we can finish our proof of Theorem 4.40. Lemma 4.43 implies that

$$\mathrm{Coef}_{\chi_0\cdots\chi_{i-1}}\,\delta_1(f(\bar{x}_i)) \equiv \sum_{r=2}^{i}\delta_1(b_{2^r-1}) + \mathrm{Coef}_{\chi_0}\,\delta_1(f(\chi_0)) \pmod 2. \tag{4.64}$$

From (4.35) we have $f(\chi_0) = b_0 + b_1\chi_0$, so taking into account (4.58) we conclude that

$$\delta_1(f(\chi_0)) \equiv \delta_1(b_0) + \chi_0(\delta_1(b_1) + \delta_0(b_0)) \pmod 2.$$

Thus, (4.64) in view of (4.46) implies that

$$\mathrm{Coef}_{\chi_0\cdots\chi_{i-1}}\,\delta_1(f(\bar{x}_i)) \equiv 1 + \sum_{r=1}^{i}\delta_1(b_{2^r-1}) \pmod 2,$$

since $a_0 = b_0$. This means that the condition (4.49) is equivalent to the following condition

$$\sum_{r=1}^{i}\delta_1(b_{2^r-1}) \equiv 0 \pmod 2; \quad i = 1, 2, 3, \ldots,$$

or, equivalently, to the condition

$$\delta_1(b_{2^r-1}) = 0 \quad (r = 1, 2, 3, \ldots). \tag{4.65}$$

As $a_j = b_j 2^{\lfloor \log_2 j \rfloor}$, then, combining together (4.46), (4.47), (4.48) and (4.65), we finish the proof of Theorem 4.40. □

We conclude the section with a useful theorem that enables one to construct 1-Lipschitz measure-preserving and ergodic transformations on $\mathbb{Z}_2$ from an arbitrary 1-Lipschitz function $v\colon \mathbb{Z}_2 \to \mathbb{Z}_2$:

Theorem 4.44. *A function $f : \mathbb{Z}_2 \to \mathbb{Z}_2$ is 1-Lipschitz and measure-preserving (respectively, is 1-Lipschitz and ergodic) if and only if it can be represented in the form $f(x) = c + x + 2\cdot v(x)$ (respectively, in the form $f(x) = 1 + x + 2\cdot \Delta v(x)$), where $c \in \mathbb{Z}_2$ and v is a 1-Lipschitz function.*

Proof. Follows immediately from Theorem 4.40 in view of Theorem 3.53 and formula (1.1). □

4.6 Measure-preservation and ergodicity of uniformly differentiable functions on $\mathbb{Z}_p^n$

In this section we study (following [21, 24]) ergodicity and/or measure-preservations of functions $F\colon \mathbb{Z}_p^n \to \mathbb{Z}_p^m$ that are uniformly differentiable (modulo p) and have integer-valued derivatives (modulo p). Recall that in view of Theorem 3.39 all these functions are asymptotically compatible, that is, $\bar{\mathcal{L}}_1$-functions, i.e. functions that are 1-Lipschitz on all sufficiently small balls. So for these functions F the induced functions $F \bmod p^k\colon (\mathbb{Z}_p/p^k\mathbb{Z})^n \to (\mathbb{Z}_p/p^k\mathbb{Z})^m$ are well defined whenever k is sufficiently large, say, $k \geq N_1(F)$. Thus, we can apply Theorem 4.23 to study measure-preservation and ergodicity of F, see Note 4.24. As 1-Lipschitz uniformly differentiable functions are a special case of functions under consideration, the theory that follows can be applied to various important classes of functions, e.g., for analytic functions on $\mathbb{Z}_p$ ($\mathcal{C}$-functions), $\mathcal{B}$-functions, $\mathcal{A}$-functions (in particular, for twice integer-valued polynomials over $\mathbb{Q}_p$), etc. Also, the theory works for a number of problems arising in computer science, numerical simulations, cryptology, see Chapters 8 and 9.

4.6.1 Conditions for measure-preservation

In this subsection we study a question when a uniformly differentiable (modulo p) function is measure-preserving providing that all derivatives (modulo p) of this function are integer-valued.

Theorem 4.45. *Let the function $F\colon \mathbb{Z}_p^n \to \mathbb{Z}_p^m$, $m \leq n$, be uniformly differentiable modulo p, and let all partial derivatives modulo p of the function F be integer-valued. Then F is measure preserving whenever it is balanced modulo p^k for some $k \geq N_1(F)$, and the rank* $\operatorname{rk} F_1'(\mathbf{y})$ *of its Jacobi matrix $F_1'(\mathbf{y})$ modulo p is m at all points $\mathbf{y} \in \mathbb{Z}_p^n$.*

Moreover, in the case $m = n$ these conditions are also necessary: If $F\colon \mathbb{Z}_p^n \to \mathbb{Z}_p^n$ is measure-preserving then F is bijective modulo p^k for all $k \geq N_1(F)$, and $\det F_1'(\mathbf{y}) \not\equiv 0 \pmod p$ *for all $\mathbf{y} \in \mathbb{Z}_p^n$.*

Finally, the function $F\colon \mathbb{Z}_p^n \to \mathbb{Z}_p^n$ is measure-preserving if and only if F is bijective modulo p^k for some $k \geq N_1(f) + 1$.

Proof. During the proof we consider elements of a ring $(\mathbb{Z}/p^r\mathbb{Z})^\ell$ as ordered strings of ℓ numbers from $\{0, 1, \dots, p^r - 1\}$. With this in mind, for $\mathbf{w} \in (\mathbb{Z}/p^s\mathbb{Z})^m$ denote $F_s^{-1}(\mathbf{w}) = \{\mathbf{v} \in (\mathbb{Z}/p^s\mathbb{Z})^n : F(\mathbf{v}) \equiv \mathbf{w} \pmod{p^s}\}$, a preimage of $\mathbf{w}$ with respect to the function $F \bmod p^s\colon (\mathbb{Z}/p^s\mathbb{Z})^n \to (\mathbb{Z}/p^s\mathbb{Z})^m$. Let $s \geq k \geq N_1(F)$. Since F is asymptotically compatible, F is a sum of a compatible function and a periodic function with a period of length $p^{N_1(F)}$ (see Theorem 3.39); so we conclude that if $\mathbf{u} \in F_{s+1}^{-1}(\mathbf{w})$, then $\bar{\mathbf{u}} \in F_s^{-1}(\bar{\mathbf{w}})$. Here and further in the proof $\bar{\mathbf{a}} = (\bar{a}_1, \dots, \bar{a}_m) \in (\mathbb{Z}/p^s\mathbb{Z})^m$ stands for $\mathbf{a} \bmod p^s = (a_1 \bmod p^s, \dots, a_m \bmod p^s)$, where $\mathbf{a} = (a_1, \dots, a_m) \in (\mathbb{Z}/p^{s+1}\mathbb{Z})^m$. Put $\mathbf{z} = \bar{\mathbf{u}} + p^s\mathbf{h} \in (\mathbb{Z}/p^{s+1}\mathbb{Z})^n$, where $\mathbf{h} \in (\mathbb{Z}/p\mathbb{Z})^n$. In view of uniform differentiability of the function F modulo p (see Definition 3.27), we have

$$F(\mathbf{z}) \equiv F(\bar{\mathbf{u}}) + p^s\mathbf{h} \cdot F_1'(\bar{\mathbf{u}}) \pmod{p^{s+1}}. \tag{4.66}$$

Since $F(\bar{\mathbf{u}}) \equiv \bar{\mathbf{w}} + p^s\mathbf{b} \pmod{p^{s+1}}$ and $\mathbf{w} = \bar{\mathbf{w}} + p^s\mathbf{c}$ for suitable $\mathbf{b}, \mathbf{c} \in (\mathbb{Z}/p\mathbb{Z})^m$, in view of (4.66) we conclude that $\mathbf{z} \in F_{s+1}^{-1}(\mathbf{w})$ if and only if $\bar{\mathbf{z}} \in F_s^{-1}(\mathbf{w})$ (i.e., $\bar{\mathbf{u}} \in F_s^{-1}(\mathbf{w})$) and $\mathbf{h}$ satisfies the following system of linear equations over a field $\mathbb{Z}/p\mathbb{Z}$:

$$\mathbf{b} + \mathbf{h} \cdot F_1'(\bar{\mathbf{u}}) = \mathbf{c}. \tag{4.67}$$

Thus, if columns of the matrix $F_1'(\bar{\mathbf{u}})$ are linearly independent over $\mathbb{Z}/p\mathbb{Z}$, then the linear system (4.67) has exactly p^{n-m} pairwise distinct solutions $\mathbf{h} \in (\mathbb{Z}/p\mathbb{Z})^n$ given arbitrary $\mathbf{b}, \mathbf{c} \in (\mathbb{Z}/p\mathbb{Z})^m$. From here it follows that

$$\#F_{s+1}^{-1}(\mathbf{w}) = (\#F_s^{-1}(\bar{\mathbf{w}})) \cdot p^{n-m}. \tag{4.68}$$

Hence, if F is balanced modulo p^s (i.e., if $\#F_s^{-1}(\bar{\mathbf{w}})$ does not depend on $\bar{\mathbf{w}}$) and if a rank of the matrix $F_1'(\bar{\mathbf{w}})$ is m, for all $\bar{\mathbf{w}} \in (\mathbb{Z}/p^s\mathbb{Z})^n$, then (4.68) implies that F is balanced modulo p^{s+1}. However, in view of Proposition 3.32, the matrix $F_1'(\bar{\mathbf{w}})$ depends only on $\bar{\mathbf{w}} \bmod p^{N_1(F)}$. This in view of Note 4.24 proves the first claim of Theorem 4.45.

To prove the second claim, take $m = n$ and suppose that $F : \mathbb{Z}_p^n \to \mathbb{Z}_p^n$ is a measure-preserving function. In view of Note 4.24 this implies that F is bijective modulo p^k for all $k \geq N_1(F)$. Definition 3.27 of uniform differentiability modulo p implies that

$$F(\mathbf{u} + p^k\mathbf{h}) \equiv F(\mathbf{u}) + p^k\mathbf{h} \cdot F_1'(\mathbf{u}) \pmod{p^{k+1}} \tag{4.69}$$

for all $\mathbf{u}, \mathbf{h} \in \mathbb{Z}_p$. Here $F_1'(\mathbf{u})$ is an $n \times n$ matrix over a field $\mathbb{Z}/p\mathbb{Z}$. If $\det F_1'(\mathbf{u}) \equiv 0 \pmod{p}$ for some $\mathbf{u} \in \mathbb{Z}_p^n$ (or, equivalently, for some $\mathbf{u} \in \{0, 1, \dots, p^{N_1(F)} - 1\}^n$ in view of periodicity of partial derivatives modulo p, see Proposition 3.32), then there exists $\mathbf{h} \in \{0, 1, \dots, p-1\}^n, \mathbf{h} \not\equiv (0, \dots, 0) \pmod{p}$ such that $\mathbf{h} \cdot F_1'(\mathbf{u}) \equiv (0, \dots, 0) \pmod{p}$. However, then (4.69) implies that $F(\mathbf{u} + p^k\mathbf{h}) \equiv F(\mathbf{u}) \pmod{p^{k+1}}$, in contradiction to bijectivity modulo p^{k+1} of the function F, since $\mathbf{u} + p^k\mathbf{h} \not\equiv \mathbf{u} \pmod{p^{k+1}}$.

Finally, if F is bijective modulo some $k \geq N_1(F) + 1$ then F is bijective modulo p^{k-1} due to compatibility of F (see Proposition 2.3) and $\det F_1'(\mathbf{u}) \equiv 0 \pmod{p}$ nowhere on $\mathbb{Z}_p$ since otherwise the above argument implies that F is not bijective modulo p^k. Thus, F is measure-preserving in force of the first claim of Theorem 4.45. □

Note 4.46. The bound given by Theorem 4.45 is sharp: That is, there exists a function $f : \mathbb{Z}_p \to \mathbb{Z}_p$ such that

- f is uniformly differentiable modulo p,
- a derivative f_1' is integer-valued,
- f is bijective modulo $p^{N_1(f)}$,
- f is not bijective modulo $p^{N_1(f)+1}$, and
- f is not measure-preserving.

For instance, a polynomial $f(x) = 1 + x^p$ is bijective modulo p, $N_1(f) = 1$; however, the polynomial f is not measure-preserving since $f'(z) \equiv 0 \pmod{p}$ for all $z \in \mathbb{Z}_p$.

We also stress the following note since it is important for applications, e.g. in computer science and cryptology, see Chapters 9 and 10.

Note 4.47. Due to periodicity of partial derivatives modulo p, see Proposition 3.32, *in order to verify whether the condition* rk $F_1'(\mathbf{y}) = m$ *from the statement of Theorem 4.45 (or, respectively the condition* $\det F_1'(\mathbf{y}) \not\equiv (0, \dots, 0) \pmod{p}$) *holds for all* $\mathbf{y} \in \mathbb{Z}_p^n$, *it is sufficient to verify these conditions only for* $\mathbf{y} \in \{0, 1, \dots, p^{N_1(F)} - 1\}^n$.

The following obvious corollary of Theorem 4.45 holds:

Corollary 4.48. *Under the assumptions of Theorem 4.45 let $m = 1$. Then F if measure-preserving whenever F is balanced modulo p^k for some $k \geq N_1(F)$, and all partial derivatives modulo p of the function F vanish simultaneously at no point of $(\mathbb{Z}/p^k\mathbb{Z})^n$. If additionally $n = 1$, then F is measure-preserving if and only if it is bijective modulo $p^{N_1(F)}$ and its derivative modulo p vanishes at no point of $\{0, 1, \dots, p^{N_1(F)} - 1\}$. Equivalently, if $m = n = 1$ then F is measure-preserving if and only if F is bijective modulo $p^{N_1(F)+1}$.*

Corollary 4.48 immediately implies that a polynomial from $\mathbb{Z}_p[x_1, \dots, x_n]$ is measure-preserving whenever it is balanced modulo p and all its partial derivatives vanish modulo p simultaneously at no point from $(\mathbb{Z}/p\mathbb{Z})^n = \{0, 1, \dots, p-1\}^n$; in particular, a polynomial from $\mathbb{Z}_p[x]$ is measure-preserving if and only it is bijective modulo p and its derivative vanishes modulo p nowhere (moreover, a polynomial from $\mathbb{Z}_p[x]$ is measure-preserving if and only it is bijective modulo p^2). It is worth noting here that these results about polynomials over $\mathbb{Z}_p$ (as well as analogs of these results for polynomials over commutative rings) are well known in the theory of polynomials over universal algebras, see e.g. [286]; however, it turns out that these results remain true for a class of functions that is much wider than polynomials, namely, they hold for $\mathcal{B}$-functions also. We postpone a proof of these results, see Corollary 4.70; now we discuss the question whether the mentioned sufficient conditions of measure-preservation for polynomials are necessary. Unfortunately, the answer is negative: The following counter-example is based on ideas from [180].

Example 4.49. Consider a polynomial $f(x,y) = 2x + y^3$ over $\mathbb{Z}_2$, in variables x, y. As $\frac{\partial f(x,y)}{\partial x} = 2$, $\frac{\partial f(x,y)}{\partial y} = 3y^2$, both partial derivatives are 0 modulo 2 whenever $y \equiv 0 \pmod 2$. Nevertheless, f is a measure-preserving mapping from $\mathbb{Z}_2^2$ onto $\mathbb{Z}_2$. Here is a proof.

By induction on ℓ we prove that f is balanced modulo p^ℓ for all $k = 1, 2, \dots$. The claim follows then from Theorem 4.23. For $\ell = 1$ we have that $f(x,y) \equiv y \pmod 2$, that is, f is balanced modulo 2.

Let $\ell > 1$. We will show that for every $z \in \mathbb{Z}/2^\ell\mathbb{Z}$ there exist exactly 2^ℓ pairs (x,y) such that $f(x,y) \equiv z \pmod{2^\ell}$ and $(x,y) \in \{0, 1, \dots, 2^\ell - 1\}^2$.

Indeed, if $z = 1 + 2r$ for some $r \in \{0, 1, \dots, 2^{\ell-1} - 1\}$, then it follows that $y = 1 + 2k$ for some $k \in \{0, 1, \dots, 2^{\ell-1} - 1\}$. So $2x + (1 + 2k)^3 \equiv 1 + 2r \pmod{2^\ell}$ implies $x + 3k + 6k^2 + 4k^3 \equiv r \pmod{2^{\ell-1}}$. The left hand part of the latter congruence is a polynomial $g(x,k)$ in x, k. The polynomial $g(x,k)$ is measure-preserving in view of Theorem 4.45. This implies that the congruence $g(x,k) \equiv r \pmod{2^{\ell-1}}$ in unknowns x, k has exactly $2^{\ell-1}$ solutions in $\{0, 1, \dots, 2^{\ell-1} - 1\}^2$.

If $z = 2r$ for some $r \in \{0, 1, \dots, 2^{\ell-1} - 1\}$, then it follows that $y = 2k$ for some $k \in \{0, 1, \dots, 2^{\ell-1} - 1\}$; consequently, the congruence $f(x,y) \equiv z \pmod{2^\ell}$ implies the congruence $x + 4k^3 \equiv r \pmod{2^{\ell-1}}$. The polynomial $d(x,k) = x + 4k^3$ is measure-preserving in view of Theorem 4.45. Now using an argument similar to that of the case $z = 1 + 2r$ we conclude that the congruence $f(x,y) \equiv 2r \pmod{2^\ell}$ in unknowns x, y has exactly 2^ℓ solutions in $\{0, 1, \dots, 2^\ell - 1\}^2$. This proves that f is measure-preserving.

Theorem 4.45 together with Example 4.49 gives rise to the following problem, which is important both for theory and for various applications (e.g., in computer science and cryptology, see Chapters 9 and 10); however, the problem is not solved even in the case F is a polynomial over $\mathbb{Z}_p$ (or over $\mathbb{Z}$).

Open Question 4.50. Find necessary and sufficient conditions of measure-preservation for the function $F\colon \mathbb{Z}_p^n \to \mathbb{Z}_p^m$, $m < n$, from the statement of Theorem 4.45.

4.6.2 No uniformly differentiable 1-Lipschitz ergodic transformations on $\mathbb{Z}_p^n$, $n \geq 2$

Now we start studying conditions for ergodicity of functions that are uniformly differentiable modulo p and have integer-valued derivatives modulo p. This class of functions contains all asymptotically compatible (in particular, 1-Lipschitz) functions that are uniformly differentiable modulo p, see Proposition 3.41. It turns out that among these functions, ergodic ones exist only in dimension 1; namely:

Theorem 4.51. *Let an ergodic function $F\colon \mathbb{Z}_p^n \to \mathbb{Z}_p^n$ be uniformly differentiable modulo p, and let all its partial derivatives modulo p be integer-valued. Then $n = 1$.*

To prove Theorem 4.51, we need two lemmas. Recall that we call an identity modulo p^k a function that is 0 modulo p^k everywhere, see Definition 3.51.

Lemma 4.52. *Let a function $f: \mathbb{Z}_p^n \to \mathbb{Z}_p$ be uniformly differentiable modulo p, let it have integer-valued derivatives modulo p, and let f be an identity modulo p^k for some $k > N_1(f)$. Then every partial derivative modulo p of the function f is an identity modulo p.*

Proof. Fix arbitrary $x_0, x_1, \ldots, x_{i-1}, x_{i+1}, \ldots, x_n \in \mathbb{Z}_p$ and consider a function $g_i(x_0, x_1, \ldots, x_n) = x_i + x_0 f(x_1, \ldots, x_n)$ of variate x_i. It is clear that g_i is uniformly differentiable modulo p^k, its derivative modulo p^k is integer-valued, and g_i is bijective modulo p^k. As $k > N_1(f)$, in view of Theorem 4.45, g_i is measure-preserving, so its derivative modulo p is not zero modulo p everywhere on $\mathbb{Z}_p$, i.e.,

$$\frac{\partial_1}{\partial_1 x_i} g_i(u_0, \ldots, u_n) = 1 + u_0 \cdot \frac{\partial_1}{\partial_1 x_i} f(u_1, \ldots, u_n) \not\equiv 0 \pmod p \tag{4.70}$$

for all $u_0, \ldots, u_n \in \mathbb{Z}_p$. If

$$\frac{\partial_1}{\partial_1 x_i} f(u_1, \ldots, u_n) \equiv d \not\equiv 0 \pmod p$$

for some $u_1, \ldots, u_n \in \mathbb{Z}_p$, then taking u_0 such that $u_0 d \equiv -1 \pmod p$ we obtain a contradiction to (4.70). This proves Lemma 4.52. □

Lemma 4.53. *Let a function $H: \mathbb{Z}_p^n \to \mathbb{Z}_p^n$ be uniformly differentiable modulo p, and let H has integer-valued derivatives modulo p. If H is bijective modulo p^k and if H induces a trivial permutation modulo p^{k-1}* (i.e., an identity transformation on $(\mathbb{Z}/p^{k-1}\mathbb{Z})^n$) *for some $k > N_1(H) + 1$, then H induces on $(\mathbb{Z}/p^k\mathbb{Z})^n$ either a trivial permutation, or a permutation of multiplicative order p* (that is, either this permutation is a unit element of a finite symmetric group $\mathrm{Sym}(p^{kn})$ on p^{kn} elements, or an order of this permutation, as an element from $\mathrm{Sym}(p^{kn})$, is p.)

Proof. Let G be an arbitrary function that satisfies the conditions of Lemma 4.53, and let $N_1(G) = N_1(H)$. Represent both H and G in the following form:

$$H(x_1, \ldots, x_n) = (x_1, \ldots, x_n) + U(x_1, \ldots, x_n);$$
$$G(x_1, \ldots, x_n) = (x_1, \ldots, x_n) + V(x_1, \ldots, x_n).$$

Then both U and V are uniformly differentiable modulo p, have integer-valued derivatives modulo p, and $N_1(U) = N_1(V) = N_1(H)$. Moreover, both U and V are identities modulo p^{k-1} whenever $k - 1 > N_1(H)$. Then Lemma 4.52 implies that $U_1' = V_1' = 0$ everywhere on $\mathbb{Z}_p^n$. As $|U|_p \le p^{-k+1}$ and $|V|_p \le p^{-k+1}$ everywhere on $\mathbb{Z}_p^n$, and as both U and V are uniformly differentiable modulo p, from (3.5) we deduce that

$$\begin{aligned} H(G(h_1,\dots,h_n)) &= H((h_1,\dots,h_n)+V(h_1,\dots,h_n)) \\ &\equiv H(h_1,\dots,h_n)+V(h_1,\dots,h_n)\cdot H_1'(h_1,\dots,h_n) \\ &\equiv H(h_1,\dots,h_n)+V(h_1,\dots,h_n)+V(h_1,\dots,h_n)\cdot U_1'(h_1,\dots,h_n) \\ &\equiv (h_1,\dots,h_n)+U(h_1,\dots,h_n)+V(h_1,\dots,h_n) \pmod{p^k} \end{aligned}$$

for all $h_1,\dots,h_n \in \mathbb{Z}_p$. This implies, in particular, that for all $s \in \mathbb{N}$ the following congruence for iterates of H holds:

$$H^s(h_1,\dots,h_n) \equiv (h_1,\dots,h_n)+s\cdot U(h_1,\dots,h_n) \pmod{p^k}.$$

As U is an identity modulo p^{k-1}, the latter congruence implies that $H^p(h_1,\dots,h_n) \equiv (h_1,\dots,h_n) \pmod{p^k}$ for all $h_1,\dots,h_n \in \mathbb{Z}_p$. This proves Lemma 4.53 since in view of Theorem 4.45 the function H is measure-preserving and thus in view of Theorem 4.23 induces a permutation of elements of $(\mathbb{Z}/p^k\mathbb{Z})^n$. □

Proof of Theorem 4.51. As F is an ergodic $\bar{\mathcal{L}}_1$-function, in view of Theorem 4.23 and Note 4.24 there exists $k > N_1(F)+1$ such that F is transitive modulo p^n for all $n \ge k-1$. The function F then permutes elements of $(\mathbb{Z}/p^k\mathbb{Z})^n$; we denote the corresponding permutation by $\sigma_k(F)$. Consider a permutation $\sigma = \sigma_k(F)^{p^{(k-1)n}}$. As F is transitive modulo p^k, the multiplicative order of the permutation σ is p^n; hence σ is not a trivial permutation (not a unit element of a group $\mathrm{Sym}(p^{kn})$).

On the other hand, $\sigma = \sigma_k(F^{p^{(k-1)n}})$. But $F^{p^{(k-1)n}}$ is bijective modulo p^k and induces a trivial permutation modulo p^{k-1} (the latter claim follows from the transitivity of F modulo p^{k-1}). Since σ is not trivial, in view of Lemma 4.53 a multiplicative order of permutation σ must be p. However, according to the preceding argument, the multiplicative order of σ is p^n, so necessarily $n = 1$. □

Of course, there exist non-differentiable 1-Lipschitz ergodic transformations on $\mathbb{Z}_p^n$ for every $n > 1$. Actually, given a 1-Lipschitz ergodic transformation f on $\mathbb{Z}_p$, one can construct a 1-Lipschitz ergodic transformation on $\mathbb{Z}_p^n$ for every $n > 1$ in the following way.

Consider a bijection $B\colon \mathbb{Z}_p^n \to \mathbb{Z}_p$ defined by the rule $\delta_k(B(x_0,\dots,x_{n-1})) = \delta_\ell(x_r)$, where $r \in \{0,1,\dots,n-1\}$ is the least non-negative residue of $k \in \{0,1,2,\dots\}$ modulo n, $k = \ell\cdot n + r$, $(x_0,\dots,x_{n-1}) \in \mathbb{Z}_p^n$. Loosely speaking, we consider an element of $\mathbb{Z}_p^n$ as an entry of a table of n one-side infinite rows (say, stretching from left to right) of symbols from $\{0,1,\dots,p-1\}$, and to this table we put into a correspondence an infinite string of symbols from $\{0,1,\dots,p-1\}$ (that is, an element from $\mathbb{Z}_p$) obtained by reading successively elements of each column of the table, from top to bottom and from left to right.

Now take a 1-Lipschitz transformation $H : \mathbb{Z}_p \to \mathbb{Z}_p$ and a conjugate transformation $H^B(x_0, \dots, x_{n-1}) = B^{-1}(H(B(x_0, \dots, x_{n-1})))$

$$H^B(x_0, \dots, x_{n-1}) = (f_0(x_0, \dots, x_{n-1}), \dots, f_{n-1}(x_0, \dots, x_{n-1})) \colon \mathbb{Z}_p^n \to \mathbb{Z}_p^n.$$

Obviously, by Theorem 4.23, the conjugate mapping H^B is 1-Lipschitz and ergodic whenever the mapping H is ergodic: Given a univariate triangular mapping H (see Subsection 3.8.1 about these)

$$x = \sum_{i=0}^{\infty} \chi_i \cdot p^i = (\chi_0, \chi_1, \chi_2, \dots) \overset{H}{\mapsto} (\psi_0(\chi_0); \psi_1(\chi_0, \chi_1); \psi_2(\chi_0, \chi_1, \chi_2); \dots),$$

we just construct an n-variate triangular mapping

$$\begin{array}{ccccccccc}
\chi_0 & \chi_n & \chi_{2n} & \cdots & \overset{f_0}{\mapsto} & \psi_0(x) & \psi_n(x) & \psi_{2n}(x) & \cdots \\
\chi_1 & \chi_{n+1} & \chi_{2n+1} & \cdots & \overset{f_1}{\mapsto} & \psi_1(x) & \psi_{n+1}(x) & \psi_{2n+1}(x) & \cdots \\
 & & & & \vdots & & & & \\
\chi_{n-1} & \chi_{2n-1} & \chi_{3n-1} & \cdots & \overset{f_{n-1}}{\mapsto} & \psi_{n-1}(x) & \psi_{2n-1}(x) & \psi_{3n-1}(x) & \cdots
\end{array}$$

where $\chi_0, \chi_1, \dots \in \{0, 1, \dots, p-1\}$, $\psi_m(x) = \psi_m(\chi_0, \dots, \chi_m) \in \{0, 1, \dots, p-1\}$, $m = 0, 1, 2, \dots$. Now assuming that the rows in the left-hand part are new variables, $x_j = (\chi_j, \chi_{n+j}, \chi_{2n+j}, \dots) = \sum_{i=0}^{\infty} \chi_{in+j} \cdot p^i$ $(j = 0, 1, \dots, n-1)$, we see that the n-variate mapping $H^B = (f_0, f_1, \dots, f_{n-1})$, where $f_j(x_0, \dots, x_{n-1}) = \sum_{i=0}^{\infty} \psi_{in+j}(x) \cdot p^i$ for $j = 0, 1, \dots, n-1$, is transitive modulo p^k for all $k = 1, 2, \dots$ whenever H is transitive modulo p^k for all $k = 1, 2, \dots$.

This easy construction of multivariate ergodic transformation is of some importance in computer science. However, it would be highly desirable to characterize multivariate 1-Lipschitz ergodic transformations of $\mathbb{Z}_p^n$ that can not be reduced in this sense to univariate ergodic transformations. Thus we state:

Open Question 4.54. Characterize 1-Lipschitz ergodic transformations on $\mathbb{Z}_p^n$, $n > 1$.

4.6.3 Differentiable ergodic transformations on $\mathbb{Z}_p$

In this subsection we study conditions for ergodicity of differentiable transformations on $\mathbb{Z}_p$. A central result of this subsection is Theorem 4.55, which gives sufficient and necessary conditions of ergodicity for functions that are uniformly differentiable modulo p^2. We note that to prove Theorem 4.55 we use a wide generalization of a method of M. V. Larin from the proof of his criterion of transitivity modulo p^n of a polynomial with rational integer coefficients, [282].[1]

Theorem 4.55. *Let a function $f : \mathbb{Z}_p \to \mathbb{Z}_p$ be uniformly differentiable modulo p^2, and let a derivative modulo p^2 of the function f be integer-valued. Then f is ergodic if and only if it is transitive modulo p^n for some (equivalently, for every) $n \geq N_2(f) + 1$ whenever p is odd or, respectively, for some (equivalently, for every) $n \geq N_2(f) + 2$ whenever $p = 2$.*

To prove the theorem, we need a lemma.

Lemma 4.56. *Let the function $f : \mathbb{Z}_p \to \mathbb{Z}_p$ be uniformly differentiable modulo p, let its derivative modulo p be integer-valued, and let the function f be transitive modulo p^k for some $k \geq N_1(f)+1$. Then f induces on $\mathbb{Z}/p^{k+1}\mathbb{Z}$ a permutation that is either a single cycle of length p^{k+1} or a product of p pairwise disjoint cycles of length p^k each.*

Proof. A general idea of the proof is as follows: As f is transitive (whence, bijective) modulo p^k for some $k \geq N_1(f)+1$, then in view of Theorem 4.45 f is bijective modulo p^{k+1}. The corresponding permutation of elements of the residue ring $\mathbb{Z}/p^{k+1}\mathbb{Z}$ is a product of disjoint cycles, and a reduction modulo p^k maps every this cycle on the whole residue ring $\mathbb{Z}/p^k\mathbb{Z}$ since f is transitive modulo p^k. Thus, a length of a cycle must be a multiple of p^k. Further, as f is asymptotically compatible (see the very beginning of Section 4.6), f maps balls (of radii less than $p^{-N_1(f)}$) into balls; thus, as p-adic ball are cosets in the ring $\mathbb{Z}_p$ with respect to ideals generated by powers of p, the iterate $f^{p^k} \bmod p^{k+1}$ permutes cosets of the ring $\mathbb{Z}/p^{k+1}\mathbb{Z}$ with respect to ideal generated by p^k. Moreover, as $f^{p^k} \bmod p^k$ is an identity transformation on $\mathbb{Z}/p^k\mathbb{Z}$, every this coset must be invariant with respect to action of $f^{p^k} \bmod p^{k+1}$. Now it is clear that whenever this action is transitive on the coset, then f is transitive on $\mathbb{Z}/p^{k+1}\mathbb{Z}$. However, it turns out that $f^{p^k} \bmod p^{k+1}$ acts on the coset by an affine transformation; that is, the action is conjugate to an affine transformation on the finite field of p elements. Here Lemma 4.37 comes into play.

With all this in mind, we start a proof. For $x \in \mathbb{Z}_p$ denote $\chi_i = \delta_i(x) \in \{0, 1, \dots, p-1\}$, a coefficient of the ith term in a p-adic canonical expansion of x; $i = 0, 1, 2, \dots$ (see Theorem 1.45 and Note 1.46). Now Definition 3.28 of uniform differentiability modulo p^k implies that for an arbitrary $x \in \mathbb{Z}_p$ and $s \geq N_1(f) = N$ the following congruence holds:

$$f(\chi_0 + \chi_1 \cdot p + \dots + \chi_{s-1} \cdot p^{s-1} + \chi_s \cdot p^s) \equiv f(\chi_0 + \chi_1 \cdot p + \dots + \chi_{s-1} \cdot p^{s-1}) + \chi_s \cdot p^s \cdot f_1'(\chi_0 + \chi_1 \cdot p + \dots + \chi_{s-1} \cdot p^{s-1}) \pmod{p^{s+1}}. \quad (4.71)$$

[1]Although Larin's criterion of transitivity modulo p^n for polynomials over $\mathbb{Z}$ was cited since the beginning of the 1990s in different papers, see e.g. [21–23], it was first published in 2002, see [282]; for odd p the criterion was also obtained by D. L. Desjardins and M. E. Zieve, see [101].

The latter congruence implies that the sth coordinate function $\delta_s(f(x))$ of the function f is of the following form:

$$\delta_s(f(x)) \equiv \Phi_s(\chi_0, \dots, \chi_{s-1}) + \chi_s \cdot f_1'(x) \pmod{p}, \tag{4.72}$$

where $\Phi_s(\chi_0, \dots, \chi_{s-1}) = \delta_s(f(\chi_0 + \chi_1 \cdot p + \dots + \chi_{s-1} \cdot p^{s-1}))$. As a derivative $f_1'(x)$ modulo p is a periodic function with a period of length p^N (see Proposition 3.32), $f_1'(x)$ depends only on $\chi_0, \dots, \chi_{N-1}$; so we can rewrite (4.72) in the form

$$\delta_s(f(x)) \equiv \Phi_s(\chi_0, \dots, \chi_{s-1}) + \chi_s \cdot \Psi(\chi_0, \dots, \chi_{N-1}) \pmod{p}, \tag{4.73}$$

where $\Psi(\chi_0, \dots, \chi_{N-1}) = f_1'(x)$.

Further, as a chain rule holds for derivatives modulo p as well (see Proposition 3.30), we conclude that for a derivative modulo p of the rth iterate of f ($r = 1, 2, \dots$) the following congruence holds:

$$(f^r(x))_1' \equiv \prod_{j=0}^{r-1} f_1'(f^j(x)) \pmod{p}. \tag{4.74}$$

As f is uniformly differentiable modulo p, f is asymptotically compatible (see the very beginning of Section 4.6); so transitivity of f modulo p^k for some $k \ge N$ implies transitivity of f modulo p^n for all $k \ge n \ge N$, see Proposition 2.3. However, as f_1' depends only on $\chi_0, \dots, \chi_{N-1}$, and as f is transitive modulo p^N, from (4.74) we deduce that

$$(f^{p^n}(x))_1' \equiv \left(\prod_{\gamma_0, \dots, \gamma_{N-1}=0}^{p-1} \Psi(\gamma_0, \dots, \gamma_{N-1}) \right)^{p^{n-N}} \pmod{p}. \tag{4.75}$$

Denote a product in brackets in the right hand part of (4.75) by Π. Then, as the function f^{p^n} is uniformly differentiable modulo p and its derivative modulo p is integer-valued, from (4.73) and (4.75) we conclude that

$$\delta_n(f^{p^n}(x)) \equiv \Xi_n(\chi_0, \dots, \chi_{n-1}) + \chi_n \cdot \Pi^{p^{n-N}} \pmod{p}, \tag{4.76}$$

where $\Xi_n(\chi_0, \dots, \chi_{n-1}) = \delta_n(f^{p^n}(\chi_0 + x_1 \cdot p + \dots + \chi_{n-1} \cdot p^{n-1}))$. As an asymptotically compatible function f is transitive modulo p^{n+1} for $k \ge n \ge N$, the function f^{p^n}, on the one hand, induces a trivial permutation modulo p^n, and on the other hand, induces on each coset $a + p^n \cdot (\mathbb{Z}/p^{n+1}\mathbb{Z})$ of the residue ring $\mathbb{Z}/p^{n+1}\mathbb{Z}$ a permutation that is a cycle of length p. This in particular implies that the right hand part of (4.76), considered as a function in a variable x_n, must be a permutation; moreover, it must be a cycle of length p on $\{0, 1, \dots, p-1\}$. However, as this function is an affine transformation on a finite field $\mathbb{Z}/p\mathbb{Z}$, from Lemma 4.37 it follows that $\Pi^{p^{n-N}} \equiv 1$

(mod p); whence $\Pi \equiv 1 \pmod p$ (since $z^p \equiv z \pmod p$, see Subsection 1.3.1). Finally we conclude that

$$f^{p^k}(x) \equiv f^{p^k}(\chi_0 + \chi_1 \cdot p + \cdots + \chi_k \cdot p^k)$$
$$\equiv \chi_0 + \chi_1 \cdot p + \cdots + \chi_{k-1} \cdot p^{k-1} + p^k \cdot (\Xi_k(\chi_0, \ldots, \chi_{k-1}) + \chi_k) \pmod{p^{k+1}}. \quad (4.77)$$

The latter congruence implies that f induces a permutation modulo p^{k+1}, which we denote as σ. We claim that if

$$\Xi_k(\chi_0, \ldots, \chi_{k-1}) \not\equiv 0 \pmod p$$

for some (equivalently, all) $\chi_0, \ldots, \chi_{k-1} \in \{0, 1, \ldots, p-1\}$, then f is transitive modulo p^{k+1}; otherwise the permutation σ is a product of p disjoint cycles of length p^k each.

To prove the latter claim, take arbitrary $\gamma_0, \ldots, \gamma_k \in \{0, 1, \ldots, p-1\}$ and denote C a cycle of the permutation σ that contains a point $\gamma_0+\gamma_1 \cdot p+\cdots+\gamma_{k-1} \cdot p^{k-1}+\chi_k \cdot p^k \in \mathbb{Z}/p^{k+1}\mathbb{Z}$. As f is transitive modulo p^k, then $C \bmod p^k = \mathbb{Z}/p^k\mathbb{Z}$; thus, p^k is a factor of $\#C$, the length of the cycle C.

Now, if $\Xi_k(\gamma_0, \ldots, \gamma_{k-1}) \not\equiv 0 \pmod p$, then (4.77) implies that

$$f^{p^k}(\gamma_0 + \gamma_1 \cdot p + \cdots + \gamma_{k-1} \cdot p^{k-1} + \chi_k \cdot p^k)$$
$$\not\equiv \gamma_0 + \gamma_1 \cdot p + \cdots + \gamma_{k-1} \cdot p^{k-1} + \chi_k \cdot p^k \pmod{p^{k+1}}, \quad (4.78)$$

i.e., that $\#C > p^k$. On the other hand, (4.77) implies that $\#C$ is a factor of p^{k+1}. Finally we conclude that in this case $\#C = p^{k+1}$; that is, f is transitive modulo p^{k+1}.

Now let the congruence $\Xi_k(\gamma_0, \ldots, \gamma_{k-1}) \equiv 0 \pmod p$ hold for some $\gamma_0, \ldots, \gamma_k \in \{0, 1, \ldots, p-1\}$. Then this congruence must hold for all $\gamma_0, \ldots, \gamma_k \in \{0, 1, \ldots, p-1\}$, since otherwise in view of the preceding argument the function f is transitive modulo p^{k+1}, so (4.78) holds for all $\gamma_0, \ldots, \gamma_k \in \{0, 1, \ldots, p-1\}$; this in view of (4.77) implies that $\Xi_k(\gamma_0, \ldots, \gamma_{k-1}) \not\equiv 0 \pmod p$ for all $\gamma_0, \ldots, \gamma_k \in \{0, 1, \ldots, p-1\}$, a contradiction. Thus, in the case under consideration (4.77) implies that σ^{p^k} is an identity permutation; hence, $\#C = p^k$ as p^k is a factor of $\#C$. This finally proves Lemma 4.56. □

Note 4.57. During the proof of Lemma 4.56 we have shown that whenever the function f is transitive modulo $p^{N_1(f)+1}$ (in particular, whenever f is ergodic) then necessarily $\prod_{i=0}^{p^{N_1(f)}-1} f_1'(f^i(x)) \equiv 1 \pmod p$ for every $x \in \mathbb{Z}_p$.

Proof of Theorem 4.55. During the proof of Lemma 4.56 we have established that if f is transitive modulo p^k for some $k \geq N_1(f)$ then f is transitive modulo p^n for all $k \ge n \ge N_1(f)$. This in view of Theorem 4.23 and Note 4.24 proves the 'only if' part of the statement of Theorem 4.55 since f is asymptotically compatible and $N_2(f) + 1 > N_1(f)$.

To prove the 'if' part of the statement, in view of Theorem 4.23 and Note 4.24 it is sufficient to prove that if $n \ge N_2(f) + 1$ (resp., if $n \ge N_2(f) + 2$ for $p = 2$) and if f is transitive modulo p^n, then it is transitive modulo p^{n+1}. In turn, to prove the latter claim, in view of Lemma 4.56 it is sufficient to prove that not every element of the residue ring $\mathbb{Z}/p^{n+1}\mathbb{Z}$ is a fixed point of the transformation $f^{p^n} \bmod p^{n+1}$:

$$f^{p^n}(x) \not\equiv x \pmod{p^{n+1}} \tag{4.79}$$

for some $x \in \mathbb{Z}_p$.

As transitivity of f modulo p^n implies transitivity of f modulo p^{n-1}, then

$$f^{p^{n-1}}(x) = x + p^{n-1}\xi(x), \tag{4.80}$$

where $\xi \colon \mathbb{Z}_p \to \mathbb{Z}_p$; note that $\xi(x) \not\equiv 0 \pmod p$ for all $x \in \mathbb{Z}_p$ since otherwise Lemma 4.56 implies that f is not transitive modulo p^n. Further, as f is uniformly differentiable modulo p^2 and its derivative modulo p^2 is integer-valued, the rth iterate f^r is uniformly differentiable modulo $p^2$and its derivative modulo p^2 is integer-valued, for all $r = 1, 2, \dots$; moreover, $(f^r(x))'_2 \equiv \prod_{j=0}^{r-1} f'_2(f^j(x)) \pmod{p^2}$, cf. (4.74). Now, as $n - 1 \ge N_2(f)$, then using a chain rule for derivatives modulo p^2 and an obvious equality $f^{sp^{n-1}}(x) = f^{(s-1)p^{n-1}}(x + p^{n-1}\xi(x))$ $(s = 1, 2, \dots)$, which follows from (4.80), we successively calculate

$$\begin{aligned} f^{p^n}(x) &\equiv f^{(p-1)p^{n-1}}(x) + p^{n-1}\xi(x) \prod_{j=0}^{(p-1)p^{n-1}-1} f'_2(f^j(x)) \\ &\equiv f^{(p-2)p^{n-1}}(x) + p^{n-1}\xi(x) \left(\prod_{j=0}^{(p-2)p^{n-1}-1} f'_2(f^j(x)) + \prod_{j=0}^{(p-1)p^{n-1}-1} f'_2(f^j(x)) \right) \\ &\equiv \cdots \equiv x + p^{n-1}\xi(x) \left(1 + \sum_{i=1}^{p-1} \prod_{j=0}^{(p-i)p^{n-1}-1} f'_2(f^j(x)) \right) \pmod{p^{n+1}}. \end{aligned} \tag{4.81}$$

As f'_2 is a periodic function with a period of length $p^{N_2(f)}$ (see Proposition 3.32) and f is transitive modulo p^{n-1}, we conclude that for arbitrary $i, j \in \mathbb{N}$ the following congruence holds:

$$f'_2(f^j(x)) \equiv f'_2(f^{j+ip^{n-1}}(x)) \pmod{p^2}.$$

In view of the transitivity of f modulo p^{n-1}, the latter congruence implies that

$$\prod_{j=0}^{(p-i)p^{n-1}-1} f'_2(f^j(x)) \equiv \alpha(x)^{p-i} \pmod{p^2},$$

where

$$\alpha(x) = \prod_{j=0}^{p^{n-1}-1} f_2'(f^j(x)).$$

In view of (4.81) we now conclude that

$$f^{p^n}(x) \equiv x + p^{n-1}\xi(x)\left(1 + \sum_{i=1}^{p-1} \alpha(x)^i\right) \pmod{p^{n+1}}. \tag{4.82}$$

Again, as f_2' is a periodic function with a period of length $p^{N_2(f)}$, and as f is transitive modulo p^{n-1} for $n-1 \geq N_2(f)$, then $\alpha(x) \bmod p^2$ does not depend on x; namely

$$\alpha(x) = \prod_{j=0}^{p^{n-1}-1} f_2'(f^j(x)) \equiv \left(\prod_{z=0}^{p^{N_2(f)}-1} f_2'(z)\right)^{p^{n-1-N_2(f)}} \pmod{p^2}. \tag{4.83}$$

We claim that $\alpha(x) \equiv 1 \pmod{p}$. Indeed, during the proof of Lemma 4.56 we have already established that if $k \geq N_1(f)$ and if f is transitive modulo p^k, then

$$\prod_{j=0}^{p^{N_1(f)}-1} f_1'(f^j(x)) \equiv 1 \pmod{p} \tag{4.84}$$

for all $x \in \mathbb{Z}_p$, see the proof of (4.77). From Definition 3.27 of a derivative modulo some p^ℓ it follows that $f_2'(x) \equiv f_1'(x) \pmod{p}$; consequently,

$$\alpha(x) \equiv 1 + p\beta \pmod{p^2} \tag{4.85}$$

for some $\beta \in \mathbb{N}_0$. In view of (4.84) and (4.85), from (4.82) we deduce now that

$$f^{p^n}(x) \equiv x + p^{n-1} \cdot \xi(x)\left(p + p\beta \sum_{i=1}^{p-1} i\right) \pmod{p^{n+1}}; \tag{4.86}$$

so for $p \neq 2$ we conclude that

$$f^{p^n}(x) \equiv x + p^n \cdot \xi(x) \pmod{p^{n+1}}.$$

This, in view of Lemma 4.56, proves Theorem 4.55 in the case $p \neq 2$ since $\xi(x) \not\equiv 0 \pmod{p}$, see (4.80) and the text thereafter.

For the case $p = 2$, congruence (4.86) implies that

$$f^{2^n}(x) \equiv x + 2^n(1+\beta) \pmod{2^{n+1}}; \tag{4.87}$$

so to finish the proof it is sufficient to show that β is even.

For $n \geq N_2(f) + 2$ the transitivity of f modulo 2^n implies that f is transitive modulo $2^{N_2(f)+2}$, so in view of the definition of a derivative modulo p^2 we have that

$$f^{2^N}(x + 2^N\xi) \equiv f^{2^N}(x) + 2^N\xi \prod_{j=0}^{2^N-1} f_2'(f^j(x)) \pmod{2^{N+2}} \tag{4.88}$$

where $N = N_2(f)$, $\xi \in \mathbb{Z}_2$. As f is transitive modulo 2^{N+2}, we conclude that for every $x \in \{0, 1, \dots, 2^N - 1\}$ the mapping

$$\varphi_x : \xi \mapsto \delta_N(f^{2^N}(x + 2^N\xi)) + 2 \cdot \delta_{N+1}(f^{2^N}(x + 2^N\xi)) \quad (\xi \in \{0, 1, 2, 3\})$$

is a cycle of length 4 on the residue ring $\mathbb{Z}/4\mathbb{Z}$. From (4.85) and (4.83) we now conclude that

$$\prod_{j=0}^{2^N-1} f_2'(f^j(x)) \equiv 1 + 2\beta \pmod 4;$$

thus, (4.88) implies now that

$$\varphi_x(\xi) \equiv c(x) + \xi(1 + 2\beta) \pmod 4, \tag{4.89}$$

where $c(x) = \delta_N(f^{2^N}(x)) + 2\delta_{N+1}(f^{2^N}(x))$. However, for every x the mapping φ_x is transitive modulo 4, so (4.89) in view of Theorem 4.36 implies that $\beta \equiv 0 \pmod 2$. This ends the proof of Theorem 4.55. □

Note 4.58. The bound given by Theorem 4.55 is sharp: e.g., for odd p there exists a function $f : \mathbb{Z}_p \to \mathbb{Z}_p$ such that

- f is uniformly differentiable modulo p^2,
- a derivative f_2' is integer-valued,
- f is transitive modulo $p^{N_2(f)}$,
- f is not transitive modulo $p^{N_2(f)+1}$,
- f is not ergodic.

A 1-Lipschitz function $f(x) = \delta_0(x + 1)$ serves as a respective example: The function f is uniformly differentiable, its derivative is 0 everywhere on $\mathbb{Z}_p$, and $N_2(f) = 1$, f is transitive modulo p. However, f is not even bijective (not speaking of transitivity) modulo p^2; thus, f is not ergodic in view of Theorem 4.23.

Note 4.59. *A straightforward analog of Theorem 4.55 for functions that are uniformly differentiable modulo p is not true.* Namely, for every $n \in \mathbb{N}$ there exists a 1-Lipschitz function $f : \mathbb{Z}_2 \to \mathbb{Z}_2$ such that f is uniformly differentiable modulo 2, $f_1' = 1$ everywhere on $\mathbb{Z}_2$, $N_1(f) = 1$, f is transitive modulo 2^k for $k = 1, 2, \dots, n$, and f is not transitive modulo 2^k for all $k > n$. By the argument similar to that which follows, one can construct a counterexample for $p \neq 2$ as well.

Indeed, for $x \in \mathbb{Z}_2$ consider its canonical 2-adic expansion $x = \chi_0 + \chi_1 \cdot 2 + \chi_2 \cdot 2^2 + \cdots$, where $\chi_0, \chi_1, \chi_2, \ldots \in \{0,1\}$. Consider a function

$$f(x) = \sum_{i=0}^{\infty} \psi_i(\chi_0, \ldots, \chi_i) \cdot 2^i,$$

where every $\psi_i(x_0, \ldots, x_i)$ is a Boolean function linear with respect to the Boolean variable x_i; that is, the algebraic normal form (ANF) of the function $\psi_i(x_0, \ldots, x_i)$ is

$$\psi_i(\chi_0, \ldots, \chi_i) = \varphi_i(\chi_0, \ldots, \chi_{i-1}) \oplus \chi_i,$$

see Subsection 4.5.2. The function f is 1-Lipschitz. Moreover, direct calculations show that for arbitrary $s \in \mathbb{N}$ and $h \in \mathbb{Z}_2$ there holds a congruence $f(x + 2^s h) \equiv f(x) + 2^s h \pmod{2^{s+1}}$; whence, the function f is uniformly differentiable modulo 2, $f_1'(x) = 1$ for all $x \in \mathbb{Z}_2$, and $N_1(f) = 1$.

Now, given $n \in \mathbb{N}$, take a function f such that $\varphi_0 = 1$, all Boolean functions $\varphi_i(\chi_0, \ldots, \chi_{i-1})$ are of odd weight for all $i = 1, 2, \ldots$ but $i = n$, and $\varphi_n(\chi_0, \ldots, \chi_{n-1})$ is of even weight. Then, according to Theorem 4.39, f is transitive modulo 2^k for $k = 1, 2, \ldots, n$, but f is not transitive modulo 2^{n+1}; thus, f is not ergodic.

Note, however, that in contrast to Theorem 4.55, the essential part of it, Lemma 4.56, holds for functions that are uniformly differentiable modulo p, and not necessarily modulo p^2. As in applications some important functions are differentiable modulo p, and not modulo p^2 (e.g., a function XOR, see Example 8.11), it is highly desirable to find necessary and sufficient conditions of ergodicity for functions that are uniformly differentiable modulo p, and not modulo p^2. So we set the following problem:

Open Question 4.60. Find necessary and sufficient conditions of ergodicity for 1-Lipschitz functions $f : \mathbb{Z}_p \to \mathbb{Z}_p$ that are uniformly differentiable modulo p.

4.6.4 Measure-preservation and ergodicity of $\mathcal{A}$-, $\mathcal{B}$-, and $\mathcal{C}$-functions

Theorems 4.45 and 4.55 exhibit a 'Hensel's-lemma-like' phenomenon that often occurs in p-adic dynamics: A behavior of a dynamical system on the whole continuum space is determined by its 'behavior modulo p^k', i.e. on a finite space (cf. Hensel's lemma, Corollary 3.16). Actually this phenomenon is of 'ultrametric nature' rather than of 'p-adic nature' since it holds for ultrametric (and not necessarily p-adic) spaces, see examples in Part II, e.g. in Subsection 7.3.3. This phenomenon is important in applications: e.g., to determine whether a dynamical system is ergodic (that is, transitive) on a large finite space, it is sufficient to determine whether it is ergodic on a relatively small finite space; for a smaller space one may use computers, whereas for a larger space this is not possible.

Thus, it is important to estimate $N_1(f)$ and $N_2(f)$ with the highest possible precision to reduce computational costs. Moreover, although both Theorems 4.45 and

4.55 give sharp bounds for cardinality of these smaller spaces where one must verify measure preservation (respectively, ergodicity) of a dynamical system, see Notes 4.46 and 4.58, these bounds are sharp only in the class of all functions that are uniformly differentiable modulo p (respectively, modulo p^2). However, for narrower classes of functions these bounds can obviously be sharpened; e.g., for affine functions: Theorem 4.36 together with Lemma 4.37 implies that an affine function $f(x) = ax + b$ is ergodic if and only if it is transitive modulo p whenever p is odd, or modulo 4 whenever $p = 2$, and *not* modulo p^2 and modulo 8, respectively, as follows from Theorem 4.55. In this subsection we show that for some important classes of functions the said bounds can be significantly reduced. Moreover, we calculate these bounds explicitly in contrast to those given by Theorems 4.45 and 4.55: It might be not an easy problem to find $N_1(f)$ and $N_2(f)$ given an arbitrary f.

We start with $\mathcal{A}$-functions. Let $f \in \mathcal{A}$, then, according to Definition 3.63 of $\mathcal{A}$-functions, $p^n \cdot f \in \mathcal{B}$ for a suitable $n \in \mathbb{N}_0$. Given $f \in \mathcal{A}$, denote $\rho(f) = \min\{n \in \mathbb{N}_0 : p^n \cdot f \in \mathcal{B}\}$; put

$$\lambda(f) = \min\left\{k \in \mathbb{N} : 2 \cdot \frac{p^k - 1}{p - 1} - k > \rho(f)\right\}.$$

The following theorem is true.

Theorem 4.61. *Let $f \in \mathcal{A}$, and let p be an odd prime. The function f is measure-preserving if and only if it is bijective modulo $p^{\lambda(f)+1}$. The function f is ergodic if and only if it is transitive modulo $p^{\lambda(f)+1}$ whenever $p \notin \{2,3\}$, or modulo $p^{\lambda(f)+2}$ whenever $p \in \{2,3\}$.*

Basically, our proof of Theorem 4.61 will follow lines of the proof of Theorem 4.55; however, we will need more than 2 terms in decomposition of the function $f(x + p^k h)$ modulo some power of p, cf. (4.71). According to Theorem 3.64, we can develop any $\mathcal{A}$-function f into Taylor series; unfortunately, coefficients $\frac{f^{(j)}(x)}{j!}$ of terms in the series are not necessarily p-adic integers if $j > 1$. So we are going to develop more delicate techniques to calculate $f(x + p^k h)$ modulo some power of p. We start with some technical results.

Lemma 4.62. *The sequence $\varkappa(i) = \operatorname{ord}_p i! - \lfloor \log_p i \rfloor$ $(i = 1, 2, 3, \dots)$ is monotone nondecreasing.*

Proof. Obviously, $\operatorname{ord}_p i! \ge \operatorname{ord}_p (i-1)!$; so if $\lfloor \log_p i \rfloor = \lfloor \log_p (i-1) \rfloor$ then $\varkappa(i-1) \le \varkappa(i)$. Assume now that $\lfloor \log_p j \rfloor > \lfloor \log_p (j-1) \rfloor$ for some positive rational integer j. Evidently, $\lfloor \log_p j \rfloor + 1$ is a number of digits in a base-p expansion of j. Hence, our assumption holds if and only if $j - 1 = (p-1) + (p-1)p + \cdots + (p-1)p^n = p^{n+1} - 1$ for some $n \in \mathbb{N}_0$. But then $\operatorname{ord}_p j! = \operatorname{ord}_p (j-1)! + n$, $\lfloor \log_p (j-1) \rfloor = n$, $\lfloor \log_p j \rfloor = n + 1$, and thus $\varkappa(j) > \varkappa(j-1)$. □

As f is 1-Lipschitz, in view of Theorem 3.53 it can be represented in the following form:

$$f(x) = b_0 + \sum_{i=1}^{\infty} b_i p^{\lfloor \log_p i \rfloor} \binom{x}{i}, \tag{4.90}$$

where $b_j \in \mathbb{Z}_p$, $j = 0, 1, 2, \ldots$. Everywhere during the proof of Theorem 4.61 we assume that f is represented in this form. In the following we denote $\lambda(f)$ by λ, and $\rho(f)$ by ρ.

Lemma 4.63. *Under the assumptions of Theorem 4.61, let p be an odd prime; then the following is true*:

$$b_i \equiv \begin{cases} 0 \pmod{p}, & \text{for } i \geq 2p^\lambda; \\ 0 \pmod{p^2}, & \text{for } i \geq 3p^\lambda. \end{cases}$$

Proof. Represent f as

$$f(x) = b_0 + \sum_{i=1}^{\infty} \frac{1}{i!} b_i p^{\lfloor \log_p i \rfloor} x^{\underline{i}},$$

where, we recall, $x^{\underline{i}} = x(x-1)\cdots(x-i+1)$ is the ith falling factorial power of x, $x^{\underline{0}} = 1$. As $f \in \mathcal{A}$, i.e., as $\left|b_i p^{\lfloor \log_p i \rfloor}\right|_p \leq p^\rho |i!|_p$ we conclude that

$$\operatorname{ord}_p b_i \geq \operatorname{ord}_p i! - \lfloor \log_p i \rfloor - \rho, \tag{4.91}$$

for all $i = 1, 2, \ldots$.

In view of Lemma 4.62, to finish the proof of Lemma 4.63 it is sufficient to show only that $\varkappa(2p^\lambda) - \rho \geq 1$ and $\varkappa(3p^\lambda) - \rho \geq 2$. We recall that $\operatorname{ord}_p i! = \frac{1}{p-1}(i - \operatorname{wt}_p i)$, see Lemma 3.6.

As $p \neq 2$, we conclude that $\varkappa(2p^\lambda) - \rho = \frac{1}{p-1}(2p^\lambda - 2) - \lambda - \rho \geq 1$ in view of the definition of $\lambda = \lambda(f)$. Hence, if $p \neq 3$, then

$$\varkappa(3p^\lambda) - \rho = \frac{1}{p-1}(3p^\lambda - 3) - \lambda - \rho = \varkappa(2p^\lambda) + \frac{1}{p-1}(p^\lambda - 1) - \rho \geq 2.$$

This proves Lemma 4.63 for $p \neq 3$.

Finally, let $p = 3$. Then

$$\varkappa(3p^\lambda) - \rho = \varkappa(3^{\lambda+1}) - \rho = \frac{1}{2}(3^{\lambda+1} - 1) - \lambda - 1 - \rho \geq 2,$$

otherwise in view of the inequality $3^\lambda - 1 - \lambda > \rho$, which follows directly from the definition of λ, we conclude that

$$\frac{1}{2}(3^{\lambda+1} - 1) - \lambda - 1 - 3^\lambda + 1 + \lambda < 1,$$

i.e., that $3^\lambda - 1 < 2$; so $\lambda < 1$, a contradiction. This finishes the proof of Lemma 4.63. □

Corollary 4.64. *Under the assumptions of Theorem 4.61, let p be an odd prime; then for every $i \in \mathbb{N}$ the following is true*:

$$\frac{\Delta^i f(x)}{i} \equiv \begin{cases} 0 \pmod{p^2}, & \text{if } i \geq 2p^{\lambda}+1; \\ 0 \pmod{p}, & \text{if } i \geq p^{\lambda}+1. \end{cases}$$

Proof. As $\Delta^j\binom{x}{i} = \binom{x}{i-j}$ if $i \geq j$ and $\Delta^j\binom{x}{i} = 0$ if $i < j$ (see (1.1)) then

$$\frac{\Delta^i f(x)}{i} = \frac{1}{\hat{\imath}} \sum_{j=i}^{\infty} b_j p^{\lfloor \log_p j \rfloor - \operatorname{ord}_p j} \binom{x}{j-i},$$

where $\hat{\imath} = i p^{-\operatorname{ord}_p i} \in \mathbb{Z}_p$, $\operatorname{ord}_p \hat{\imath} = 0$. Now the result is obvious in view of Lemma 4.63. □

Recall that every $\mathcal{A}$-function is infinitely many times differentiable on $\mathbb{Z}_p$, and its derivative f' is integer-valued, see Subsection 3.10.3.

Proposition 4.65. *Under the assumptions of Theorem 4.61, let p be an odd prime; then $N_1(f) \leq \lambda$, $N_2(f) \leq \lambda + 1$, and*

$$f'(x) \equiv \sum_{i=1}^{p^{\lambda}} (-1)^{i-1} \frac{\Delta^i f(x)}{i} \pmod{p},$$

$$f'(x) \equiv \sum_{i=1}^{2p^{\lambda}} (-1)^{i-1} \frac{\Delta^i f(x)}{i} \pmod{p^2},$$

where $\lambda = \lambda(f)$.

Proof. To prove Proposition 4.65 we show that for all $x, h \in \mathbb{Z}_p$

$$f(x + p^m h) \equiv f(x) + p^m h \cdot f'(x) \pmod{p^{m+2}} \tag{4.92}$$

whenever $m \geq \lambda + 1$, and that

$$f(x + p^m h) \equiv f(x) + p^m h \cdot f'(x) \pmod{p^{m+1}} \tag{4.93}$$

whenever $m \geq \lambda$.

Since f is 1-Lipschitz, it is sufficient to prove congruences (4.92) and (4.93) only for $h \in \{1, 2, \dots, p^2 - 1\}$. By the Gregory–Newton formula (see Theorem 1.5),

$$f(x+n) = \sum_{i=0}^{n} \binom{n}{i} \Delta^i f(x);$$

thus, for $n = p^m h$ we obtain that

$$f(x + p^m h) = f(x) + p^m h \varphi_m(x, h), \tag{4.94}$$

where

$$\varphi_m(x, h) = \sum_{i=1}^{p^m h} \binom{p^m h - 1}{i - 1} \frac{\Delta^i f(x)}{i}. \tag{4.95}$$

Now from Corollary 4.64 we deduce that

$$\varphi_m(x, h) \equiv \sum_{i=1}^{p^\lambda} \binom{p^m h - 1}{i - 1} \frac{\Delta^i f(x)}{i} \pmod p \tag{4.96}$$

whenever $m \geq \lambda$ and that

$$\varphi_m(x, h) \equiv \sum_{i=1}^{2p^\lambda} \binom{p^m h - 1}{i - 1} \frac{\Delta^i f(x)}{i} \pmod {p^2} \tag{4.97}$$

whenever $m \geq \lambda + 1$.

In view of Corollary 1.3 from (4.96) it follows that

$$\varphi_m(x, h) \equiv \sum_{i=1}^{p^\lambda} (-1)^{i-1} \frac{\Delta^i f(x)}{i} \pmod p,$$

for $m \geq \lambda$ thus proving the assertion of Proposition 4.65 that deals with estimates of $\mathrm{N}_1(f)$ and with the residue $f'(x) \bmod p$.

To prove the remaining part of the statement of Proposition 4.65 we first note that for $i = 1, 2, \ldots, 2p^\lambda$ the following obvious equality holds:

$$\binom{p^m h - 1}{i - 1} = \prod_{k=0}^{i-2} \frac{p^m h - (k + 1)}{k + 1} = \prod_{j=1}^{i-1} \left(\frac{h}{\hat{j}} p^{m - \operatorname{ord}_p j} - 1 \right), \tag{4.98}$$

where $\hat{j} = j p^{-\operatorname{ord}_p j}$ is a unit of $\mathbb{Z}_p$, (i.e., $\hat{j}$ has a multiplicative inverse $\frac{1}{\hat{j}}$ in $\mathbb{Z}_p$); hence, every term of the product in the right hand part of (4.98) is a p-adic integer.

If $i \leq p^\lambda$ then $m - \operatorname{ord}_p j \geq 2$ for all $j = 1, 2, \ldots, i - 1$; so (4.98) implies that

$$\binom{p^m h - 1}{i - 1} \equiv (-1)^{i-1} \pmod {p^2}. \tag{4.99}$$

If $p^\lambda + 1 \leq i \leq 2p^\lambda$ and $j \in \{1, 2, \ldots, i - 1\}$ then $m - \operatorname{ord}_p j = 1$ only in the case when simultaneously $j = p^\lambda$ and $m = \lambda + 1$; otherwise $m - \operatorname{ord}_p j \geq 2$. However, if $m - \operatorname{ord}_p j = 1$ then

$$\frac{\Delta^i f(x)}{i} \equiv 0 \pmod p$$

(see Corollary 4.64); hence in both cases we have that

$$\left(\frac{h}{\hat{j}}p^{m-\operatorname{ord}_p j}-1\right)\frac{\Delta^i f(x)}{i} \equiv -\frac{\Delta^i f(x)}{i} \pmod{p^2}.$$

From here in view of (4.98) we deduce that

$$\binom{p^m h-1}{i-1}\frac{\Delta^i f(x)}{i} \equiv (-1)^{i-1}\frac{\Delta^i f(x)}{i} \pmod{p^2} \tag{4.100}$$

for all $i = 1, 2, \ldots, 2p^\lambda$. Now combining together (4.97), (4.99), and (4.100) we conclude that

$$\varphi_m(x,h) \equiv \sum_{i=1}^{2p^\lambda}(-1)^{i-1}\frac{\Delta^i f(x)}{i} \pmod{p^2}.$$

This in view of (4.94), (4.95), and (4.97) completes the proof of Proposition 4.65. □

Lemma 4.66. *Under the assumptions of Theorem 4.61, let p be an odd prime; then the function*

$$\theta(x) = \sum_{j=2}^{p-1}(-1)^j\sum_{i=1}^{j-1}\frac{1}{i}\cdot\frac{\Delta^{jp^{\lambda-1}}f(x)}{jp^{\lambda-1}} + \sum_{k=1}^{p-1}(-1)^{k-1}\frac{\Delta^{kp^{\lambda-1}+p^\lambda}f(x)}{kp^\lambda} + \frac{\Delta^{2p^\lambda}f(x)}{2p^{\lambda+1}}.$$

is integer-valued, $\theta(a) = \theta(b) \pmod p$ whenever $a \equiv b \pmod{p^\lambda}$, and

$$f(x+p^\lambda h) \equiv f(x) + p^\lambda h \cdot f'(x) + p^{\lambda+1}h^2\cdot\theta(x) \pmod{p^{\lambda+2}}$$

for all $x, h \in \mathbb{Z}_p$.

Proof. First we prove that θ is integer-valued, i.e., maps $\mathbb{Z}_p$ into $\mathbb{Z}_p$. As f is 1-Lipschitz, every fraction $\frac{\Delta^s f(x)}{s}$ $(s = 1, 2, 3, \ldots)$ is a p-adic integer (see Proposition 3.38); so it is sufficient to show only that the following functions $\alpha(x)$ and $\beta_k(x)$ are integer-valued

$$\alpha(x) = \frac{\Delta^{2p^\lambda}f(x)}{2p^{\lambda+1}};$$

$$\beta_k(x) = \frac{\Delta^{kp^{\lambda-1}+p^\lambda}f(x)}{kp^\lambda},$$

for all $k \in \{1, 2, \ldots, p-1\}$. As

$$\Delta^i f(x) = \sum_{j=i}^{\infty} b_j p^{\lfloor \log_p j \rfloor}\binom{x}{j-i} \tag{4.101}$$

for $i = 1, 2, 3, \ldots$, and as

$$b_j p^{\lfloor \log_p j \rfloor} \equiv 0 \pmod{p^{\lambda+1}} \tag{4.102}$$

for all rational integers $j \geq 2p^\lambda$, by (4.90) and Lemma 4.63, then $\alpha(x) \in \mathbb{Z}_p$ for all $x \in \mathbb{Z}_p$. If $j \geq kp^{\lambda-1} + p^\lambda$ then $\lfloor \log_p j \rfloor \geq \lambda$; so (4.101) implies that $\beta_k(x) \in \mathbb{Z}_p$ for all $x \in \mathbb{Z}_p$.

Now we prove that for all $a, b \in \mathbb{Z}_p$ the congruence $a \equiv b \pmod{p^\lambda}$ implies a congruence $\theta(a) \equiv \theta(b) \pmod{p}$. From (4.101) it follows that

$$\alpha(x) \equiv \frac{1}{2} \sum_{j=2p^\lambda}^{3p^\lambda - 1} \frac{1}{p} \cdot b_j \binom{x}{j - 2p^\lambda} \pmod{p}. \tag{4.103}$$

Note that in (4.103) every fraction $\frac{b_j}{p}$ is a p-adic integer by Lemma 4.66. Now, as $a \equiv b \pmod{p^\lambda}$, then Lucas' Theorem 1.2 implies that for all $j = 2p^\lambda, 2p^\lambda + 1, \ldots, 3p^\lambda - 1$ the following congruence holds:

$$\binom{a}{j - 2p^\lambda} \equiv \binom{b}{j - 2p^\lambda} \pmod{p}.$$

Thus, (4.103) implies that

$$\alpha(a) \equiv \alpha(b) \pmod{p}. \tag{4.104}$$

Further, combining (4.101) with Lemma 4.63 we conclude that the following congruence holds for all $k = 1, 2, \ldots, p-1$:

$$\beta_k(x) \equiv \frac{1}{k} \sum_{j=kp^{\lambda-1}+p^\lambda}^{2p^\lambda - 1} b_j \binom{x}{j - kp^{\lambda-1} - p^\lambda} \pmod{p}.$$

From this congruence, by Lucas' Theorem 1.2 it follows that

$$\beta_k(a) \equiv \beta_k(b) \pmod{p} \tag{4.105}$$

whenever $a \equiv b \pmod{p^\lambda}$.

Further, denote

$$\gamma_k(x) = \frac{\Delta^{kp^{\lambda-1}} f(x)}{kp^{\lambda-1}};$$

then in view of (4.101) we conclude that

$$\gamma_k(x) \equiv \frac{1}{k} \sum_{j=kp^{\lambda-1}}^{p^\lambda - 1} b_j \binom{x}{j - kp^{\lambda-1}} \pmod{p}$$

for all $k = 1, 2, \ldots, p-1$. Now applying Lucas' Theorem 1.2 once again we conclude that

$$\gamma_k(a) \equiv \gamma_k(b) \pmod p \tag{4.106}$$

whenever $a \equiv b \pmod{p^\lambda}$. Now from (4.104)–(4.106) it follows that the congruence $a \equiv b \pmod{p^\lambda}$ implies the congruence $\theta(a) \equiv \theta(b) \pmod p$.

Now we prove the final assertion of Lemma 4.66. Our proof will follow the lines of the proof of Proposition 4.65; however, now we are considering the case $m = \lambda$ rather than $m \geq \lambda + 1$. Actually we will derive a congruence for $f(x + p^\lambda h)$ modulo $p^{\lambda+2}$ from equality (4.94) with $m = \lambda$. In order to do this, we must find a residue of $\varphi_\lambda(x, h)$ (see (4.95)) modulo p^2. Again, as f is 1-Lipschitz, during the proof we may assume that $h \in \mathbb{N}$.

In view of Lemma 3.45, from (4.98) it follows that if $i \in \{1, 2, \ldots, 2p^\lambda\}$ and either $i \leq p^{\lambda-1}$, or $p^{\lambda-1} < i < p^\lambda$, $p^{\lambda-1}$ is not a factor of i, then

$$\binom{p^\lambda h - 1}{i - 1} \frac{\Delta^i f(x)}{i} \equiv (-1)^{i-1} \frac{\Delta^i f(x)}{i} \pmod{p^2}. \tag{4.107}$$

Let now $i = kp^{\lambda-1}$ for $k \in \{2, 3, \ldots, p-1\}$; then (4.98) implies that

$$\binom{p^\lambda h - 1}{i - 1} \equiv (-1)^{kp^{\lambda-1}-1} + (-1)^k ph \sum_{j=1}^{k-1} \frac{1}{j} \pmod{p^2}. \tag{4.108}$$

Further, if $p^\lambda \leq i < 2p^\lambda$ and $\operatorname{ord}_p i \neq \lambda, \lambda - 1$ then combining (4.101) together with (4.102) we see that

$$\frac{\Delta^i f(x)}{i} = 0 \pmod{p^2}. \tag{4.109}$$

Now we find residues modulo p^2 of terms $\binom{p^\lambda h-1}{i-1} \frac{\Delta^i f(x)}{i}$ of the function $\varphi_\lambda(u, h)$ (see (4.95)) in two remaining cases, when $i = \nu p^\lambda$, $\nu \in \{1, 2\}$, and, respectively, when $i = kp^{\lambda-1} + p^\lambda$, $k \in \{1, 2, \ldots, p-1\}$. In the latter case in view of Corollary 4.64 and (4.98) the following congruence holds:

$$\binom{p^\lambda h - 1}{i - 1} \frac{\Delta^i f(x)}{i} \equiv (-1)^{i-1} \frac{\Delta^i f(x)}{i} + (-1)^{k-1} h \frac{\Delta^i f(x)}{i} \pmod{p^2}. \tag{4.110}$$

It is obvious that for all $k = 1, 2, \ldots, p-1$ the following trivial equality holds:

$$\left(1 + \frac{p}{k}\right) \frac{\Delta^{kp^{\lambda-1}+p^\lambda} f(x)}{kp^{\lambda-1} + p^\lambda} = \frac{\Delta^{kp^{\lambda-1}+p^\lambda} f(x)}{kp^{\lambda-1}}. \tag{4.111}$$

As, in view of Corollary 4.64,

$$\frac{\Delta^{kp^{\lambda-1}+p^\lambda} f(x)}{kp^{\lambda-1} + p^\lambda} \equiv 0 \pmod p,$$

then, since $\frac{p}{k} \in \mathbb{Z}_p$ and $\operatorname{ord}_p \frac{p}{k} = 1$, the equality (4.111) implies that

$$\frac{\Delta^{kp^{\lambda-1}+p^{\lambda}} f(x)}{kp^{\lambda-1}+p^{\lambda}} \equiv \frac{\Delta^{kp^{\lambda-1}+p^{\lambda}} f(x)}{kp^{\lambda-1}} \pmod{p^2}.$$

From here, substituting $i = kp^{\lambda-1} + p^{\lambda}$ to (4.110), we deduce that

$$\binom{p^{\lambda}h-1}{kp^{\lambda-1}+p^{\lambda}-1} \frac{\Delta^{kp^{\lambda-1}+p^{\lambda}} f(x)}{kp^{\lambda-1}+p^{\lambda}} \equiv (-1)^{kp^{\lambda-1}+p^{\lambda}-1} \frac{\Delta^{kp^{\lambda-1}+p^{\lambda}} f(x)}{kp^{\lambda-1}+p^{\lambda}} + (-1)^{k-1} ph \cdot \beta_k(x) \pmod{p^2}. \quad (4.112)$$

In the case $i = p^{\lambda}$, the equality (4.98) implies that

$$\binom{p^{\lambda}h-1}{p^{\lambda}-1} \equiv (-1)^{p^{\lambda}-1} - ph \sum_{j=1}^{p-1} \frac{1}{j} \equiv (-1)^{p^{\lambda}-1} \pmod{p^2} \quad (4.113)$$

since $\sum_{j=1}^{p-1} \frac{1}{j} \equiv \sum_{j=1}^{p-1} j \equiv 0 \pmod{p}$ for $p \neq 2$. Finally for $i = 2p^{\lambda}$ from (4.98) in view of Corollary 4.64 we conclude that

$$\begin{aligned} \binom{p^{\lambda}h-1}{2p^{\lambda}-1} \frac{\Delta^{2p^{\lambda}} f(x)}{2p^{\lambda}} &\equiv (-1)^{2p^{\lambda}-1} \frac{\Delta^{2p^{\lambda}} f(x)}{2p^{\lambda}} + h \frac{\Delta^{2p^{\lambda}} f(x)}{2p^{\lambda}} \\ &\equiv (-1)^{2p^{\lambda}-1} \frac{\Delta^{2p^{\lambda}} f(x)}{2p^{\lambda}} + hp \cdot \alpha(x) \pmod{p^2}. \quad (4.114) \end{aligned}$$

Now collecting together (4.107), (4.109), and (4.112)–(4.114), we finish the proof of Lemma 4.66 in the same way as in Proposition 4.65. □

Lemma 4.67. *Under the assumptions of Theorem 4.61, let p be an odd prime; then for all $x, h \in \mathbb{Z}_p$ the following congruence holds*:

$$f'(x + p^{\lambda}h) \equiv f'(x) + 2ph \cdot \theta(x) \pmod{p^2}.$$

Here θ is the function defined in the statement of Lemma 4.66.

Proof. From Proposition 4.65 it follows that

$$f'(x + p^{\lambda}h) \equiv \sum_{i=1}^{2p^{\lambda}} (-1)^{i-1} \frac{\Delta^i f(x + p^{\lambda}h)}{i} \pmod{p^2}. \quad (4.115)$$

For $i = 1, 2, \ldots, 2p^\lambda$ Lemma 4.66 implies that

$$\frac{\Delta^i f(x+p^\lambda h)}{i} \equiv \frac{\Delta^i f(x)}{i} + hp^{\lambda-\mathrm{ord}_p} \cdot \frac{\Delta^i f'(x)}{\hat{\imath}} + h^2 p^{\lambda+1-\mathrm{ord}_p i} \cdot \frac{\Delta^i \theta(x)}{\hat{\imath}} \pmod{p^2}, \quad (4.116)$$

where $\hat{\imath} = ip^{-\mathrm{ord}_p i}$ is a unit in $\mathbb{Z}_p$; that is, $\hat{\imath}$ has a multiplicative inverse $\frac{1}{\hat{\imath}} \in \mathbb{Z}_p$.

We show now that a term of order 2 (with respect to h) in (4.116) is 0 modulo p^2. If this term is not 0 modulo p^2, then necessarily $i \in \{p^\lambda, 2p^\lambda\}$. However, from (4.101) it follows that in this case

$$\begin{aligned}\frac{\Delta^{i+kp^{\lambda-1}} f(x)}{kp^{\lambda-1}} &\equiv 0 \pmod{p},\\ \frac{\Delta^{i+kp^{\lambda-1}+p^\lambda} f(x)}{kp^\lambda} &\equiv 0 \pmod{p},\\ \frac{\Delta^{i+2p^\lambda} f(x)}{2p^\lambda} &\equiv 0 \pmod{p},\end{aligned} \quad (4.117)$$

for all $k \in \{1, 2, \ldots, p-1\}$. Now, by the definition of θ, from (4.117) it follows that $\frac{\Delta^i \theta(x)}{\hat{\imath}} \equiv 0 \pmod{p}$ for $i \in \{p^\lambda, 2p^\lambda\}$, and thus

$$h^2 p^{\lambda+1-\mathrm{ord}_p i} \cdot \frac{\Delta^i \theta(x)}{\hat{\imath}} \equiv 0 \pmod{p^2}; \qquad i = 1, 2, \ldots, 2p^\lambda. \quad (4.118)$$

Now we consider a term of order 1 in (4.116). If this term is not 0 modulo p^2 then necessarily $i \in \{1, 2, \ldots, 2p^\lambda\}$ and $\mathrm{ord}_p i \geq \lambda - 1$; that is, $i \in \{p^\lambda, 2p^\lambda, kp^{\lambda-1}, kp^{\lambda-1} + p^\lambda : k = 1, 2, \ldots, p-1\}$. Combining together Corollary 4.64, Proposition 4.65, and Lemma 3.45 we conclude that

$$f'(x) \equiv \frac{\Delta^{p^\lambda} f(x)}{p^\lambda} + \sum_{t=0}^{\lambda-1}\sum_{\tau=1}^{p-1} (-1)^{\tau-1} \frac{\Delta^{\tau p^t} f(x)}{\tau p^t} \pmod{p}; \quad (4.119)$$

whence,

$$\Delta^i f'(x) \equiv \frac{\Delta^{i+p^\lambda} f(x)}{p^\lambda} + \sum_{t=0}^{\lambda-1}\sum_{\tau=1}^{p-1} (-1)^{\tau-1} \frac{\Delta^{i+\tau p^t} f(x)}{\tau p^t} \pmod{p}. \quad (4.120)$$

The latter congruence in force of (4.101) and Lemma 4.63 implies that $\Delta^i f'(x) \equiv 0 \pmod{p}$ when $i \in \{kp^{\lambda-1} + p^\lambda : k = 1, 2, \ldots, p-1\}$; consequently,

$$hp \cdot \frac{\Delta^{kp^{\lambda-1}+p^\lambda} f'(x)}{k+p} \equiv 0 \pmod{p^2}; \qquad k = 1, 2, \ldots, p-1, \quad (4.121)$$

since a multiplicative inverse $\frac{1}{k+p}$ of $k+p$ is in $\mathbb{Z}_p$ for $k = 1, 2, \ldots, p-1$.

If $i \in \{kp^{\lambda-1} : k = 1, 2, \dots, p-1\}$ then in view of Lemma 4.63 from (4.101) and (4.120) we deduce that

$$\Delta^{kp^{\lambda-1}} f'(x) \equiv \frac{\Delta^{kp^{\lambda-1}+p^\lambda} f(x)}{p^\lambda} + \sum_{\tau=1}^{p-k-1} (-1)^{\tau-1} \frac{\Delta^{(\tau+k)p^{\lambda-1}} f(x)}{\tau p^{\lambda-1}} \pmod{p}. \tag{4.122}$$

If $i = 2p^\lambda$ then Proposition 4.65 implies that

$$\Delta^{2p^\lambda} f'(x) \equiv \sum_{j=1}^{2p^\lambda} (-1)^{j-1} \frac{\Delta^{j+2p^\lambda} f(x)}{j} \pmod{p^2}.$$

This, in view of (4.101) and Lemma 4.63 implies that

$$\Delta^{2p^\lambda} f'(x) \equiv 0 \pmod{p^2}. \tag{4.123}$$

Now we consider the case $i = p^\lambda$. Proposition 4.65 implies that

$$\Delta^{p^\lambda} f'(x) \equiv \sum_{j=1}^{1+p^\lambda} (-1)^{j-1} \frac{\Delta^{j+p^\lambda} f(x)}{j} \pmod{p^2}, \tag{4.124}$$

since for $j = p^\lambda + 1, \dots, 2p^\lambda$ from (4.101) in view of Lemma 4.63 it follows that

$$\frac{\Delta^{j+p^\lambda} f(x)}{j} \equiv 0 \pmod{p^2}.$$

Moreover, (4.101) implies that the latter congruence holds also for all $j \le p^\lambda - 1$ such that $j \ne kp^{\lambda-1}$, where $k = 1, 2, \dots, p-1$. Thus, from (4.124) we deduce that

$$\Delta^{p^\lambda} f'(x) \equiv \frac{\Delta^{2p^\lambda} f(x)}{p^\lambda} + \sum_{k=1}^{p-1} (-1)^{k-1} \frac{\Delta^{kp^{\lambda-1}+p^\lambda} f(x)}{kp^{\lambda-1}} \pmod{p^2}. \tag{4.125}$$

Now, substituting (4.118), (4.121), (4.122), (4.123), (4.125) to (4.116) and summing up all obtained congruences for i ranging from 1 to $2p^\lambda$, in view of (4.115) and Proposition 4.65 we conclude that

$$\begin{aligned} f'(x + p^\lambda h) \equiv f'(x) + hp \cdot \Bigg(& \sum_{k=1}^{p-1} \frac{(-1)^{k-1}}{k} \sum_{\tau=1}^{p-k-1} (-1)^{\tau-1} \frac{\Delta^{(\tau+k)p^{\lambda-1}} f(x)}{\tau p^{\lambda-1}} \\ & + \sum_{k=1}^{p-1} (-1)^{k-1} \frac{\Delta^{kp^{\lambda-1}+p^\lambda} f(x)}{kp^\lambda} \Bigg) \\ & + h \sum_{k=1}^{p-1} (-1)^{k-1} \frac{\Delta^{kp^{\lambda-1}+p^\lambda} f(x)}{kp^{\lambda-1}} + h \cdot \frac{\Delta^{2p^\lambda} f(x)}{p^\lambda} \pmod{p^2}. \end{aligned} \tag{4.126}$$

Easy calculations in $\mathbb{Q}_p$ prove that the following equality for $k, \tau \in \{1, 2, \dots, p-1\}$ is true:

$$\sum_{k+\tau=m} \frac{1}{k\tau} = \sum_{k+\tau=m} \frac{1}{(m-\tau)\tau} = \frac{1}{m} \sum_{k+\tau=m} \left(\frac{1}{\tau} + \frac{1}{m-\tau} \right) = \frac{2}{m} \sum_{\tau=1}^{m-1} \frac{1}{\tau}.$$

From here it follows that

$$\begin{aligned}
&\sum_{k=1}^{p-1} \frac{(-1)^{k-1}}{k} \sum_{\tau=1}^{p-k-1} (-1)^{\tau-1} \frac{\Delta^{(\tau+k)p^{\lambda-1}} f(x)}{\tau p^{\lambda-1}} \\
&= \sum_{m=1}^{p-1} (-1)^m \sum_{k+\tau=m} \frac{1}{k\tau} \cdot \frac{\Delta^{mp^{\lambda-1}} f(x)}{p^{\lambda-1}} = 2 \sum_{m=1}^{p-1} (-1)^m \sum_{\tau=1}^{m-1} \frac{1}{\tau} \cdot \frac{\Delta^{mp^{\lambda-1}} f(x)}{mp^{\lambda-1}}.
\end{aligned} \tag{4.127}$$

As it was shown in the proof of Lemma 4.66, both $\alpha(x)$ and $\beta_k(x)$ are p-adic integers for $k = 1, 2, \dots, p-1$ and $x \in \mathbb{Z}_p$; thus

$$\begin{aligned}
2hp \cdot \alpha(x) &= h \cdot \frac{\Delta^{2p^\lambda} f(x)}{p^\lambda}, \\
hp \cdot \beta_k(x) &= h \cdot \frac{\Delta^{kp^{\lambda-1}+p^\lambda} f(x)}{kp^{\lambda-1}},
\end{aligned} \tag{4.128}$$

and the fractions in the right-hand part are p-adic integers. Finally, the assertion of Lemma 4.67 follows from (4.126), (4.127), (4.128), and from the definition of the function θ. □

Proof of Theorem 4.61. For $p = 2$, Theorem 4.61 follows from Theorem 4.40 in view of Lemma 4.62. Indeed, under the conditions of Theorem 4.61, the coefficients a_j of the Mahler expansion $f(x) = \sum_{j=0}^{\infty} a_j \binom{x}{j}$ of the function f satisfy the following congruence: $2^\rho a_i \equiv 0 \pmod{2^{\mathrm{ord}_2(i!)}}$. However, from the definition of λ in view of Lemma 4.62 it follows that $\mathrm{ord}_2(i!) - \lfloor \log_2 i \rfloor \geq \rho + 1$ for all $i \geq 2^{\lambda+1}$, as $\mathrm{ord}_2(2^{\lambda+1}!) = 2^{\lambda+1} - 1$, see Lemma 3.6. That is, $a_i \equiv 0 \pmod{2^{\lfloor \log_2 i \rfloor + 1}}$ for all $i \geq 2^{\lambda+1}$. A similar argument proves that $a_i \equiv 0 \pmod{2^{\lfloor \log_2 (i+1) \rfloor + 1}}$ for all $i \geq 2^{\lambda+2}$. In view of Theorem 4.40, this proves Theorem 4.61 in the case $p = 2$.

Now let $p \neq 2$. The first assertion of Theorem 4.61 in this case immediately follows from Theorem 4.45 and Proposition 4.65. Further, if $p = 3$, then, as $N_2(f) \leq \lambda + 1$ according to Proposition 4.65, the second assertion of Theorem 4.61 follows from Theorem 4.55.

Thus, we only must prove the second assertion of Theorem 4.61 for $p \notin \{2, 3\}$. As $N_2(f) \leq \lambda + 1$ according to Proposition 4.65, by Theorem 4.55 it is sufficient to show that f is transitive modulo $p^{\lambda+2}$ whenever f is transitive modulo $p^{\lambda+1}$. For this purpose, in view of Lemma 4.56 it is sufficient only to prove that $f^{p^{\lambda+1}}(x) \not\equiv x \pmod{p^{\lambda+2}}$ for at least one $x \in \mathbb{Z}_p$. Now we merely calculate $f^{p^{\lambda+1}}(x) \bmod p^{\lambda+2}$.

Under our assumptions, f is transitive modulo p^λ since f is 1-Lipschitz. Then by Lemma 4.56 we conclude that

$$f^{p^\lambda}(x) = x + p^\lambda \cdot \xi(x), \quad \xi(x) \not\equiv 0 \pmod{p}, \tag{4.129}$$

for all $x \in \mathbb{Z}_p$; here $\xi\colon \mathbb{Z}_p \to \mathbb{Z}_p$ is a function defined everywhere on $\mathbb{Z}_p$.

We claim that for all $i = 0, 1, 2, \dots$ the following congruence holds:

$$\begin{aligned} f^{p^\lambda+i}(x) &\equiv f^i(x) + p^\lambda \cdot \xi(x) \prod_{j=0}^{i-1} f'(f^j(x)) \\ &+ p^{\lambda+1} \cdot \xi(x)^2 \prod_{j=0}^{i-1} f'(f^j(x)) \sum_{k=0}^{i-1} \frac{\theta(f^k(x))}{f'(f^k(x))} \prod_{\tau=0}^{k-1} f'(f^\tau(x)) \pmod{p^{\lambda+2}}. \end{aligned} \tag{4.130}$$

Recall that a sum (respectively, a product) over an empty set of indices is assumed to be 0 (respectively, 1). Note also that as f is transitive modulo $p^{\lambda+1}$, f is bijective modulo $p^{\lambda+1}$. Then, however, as $\lambda + 1 \geq N_1(f) + 1$ in force of Proposition 4.65, Corollary 4.48 implies that f is measure-preserving, and that $f'(z) \not\equiv 0 \pmod{p}$ for all $z \in \mathbb{Z}_p$. Thus, denominators of all fractions in (4.130) have multiplicative inverses in $\mathbb{Z}_p$; so during the proof of (4.130) and further on, we assume that all calculations are performed in $\mathbb{Z}_p$.

To prove (4.130) we note that according to the chain rule, $\prod_{j=0}^{i-1} f'(f^j(x)) = (f^i(x))'$, (4.130) can be rewritten in the form

$$\begin{aligned} f^{p^\lambda+i}(x) &\equiv f^i(x) + p^\lambda \cdot \xi(x) \cdot (f^i(x))' \\ &+ p^{\lambda+1} \cdot \xi(x)^2 \cdot (f^i(x))' \sum_{k=0}^{i-1} \frac{(f^k(x))'}{f'(f^k(x))} \cdot \theta(f^k(x)) \pmod{p^{\lambda+2}} \end{aligned}$$

and then proved by induction on i. Indeed, for $i = 0$ our claim trivially follows from (4.129). Now we substitute the above expression for $f^{p^\lambda+i}(x) \bmod p^{\lambda+2}$ into the equation $f^{p^\lambda+i+1}(x) = f(f^{p^\lambda+i}(x))$ and with the use of Lemma 4.66 and obvious direct calculations we prove the demanded congruence for $f^{p^\lambda+i+1}(x)$. We omit details.

Now we apply (4.130) to calculate $f^{p^{\lambda+1}}(x) \bmod p^{\lambda+2}$. We have

$$\begin{aligned} f^{p^\lambda+i}(x) &\equiv f^i(x) + p^\lambda \cdot \xi(x) \cdot A_i(x) \\ &+ p^{\lambda+1} \cdot \xi(x)^2 \cdot B_i(x) \pmod{p^{\lambda+2}}, \end{aligned} \tag{4.131}$$

where

$$A_i(x) = (f^i(x))' = \prod_{j=0}^{i-1} f'(f^j(x));$$

$$B_i(x) = (f^i(x))' \sum_{k=0}^{i-1} \frac{(f^k(x))'}{f'(f^k(x))} \theta(f^k(x))$$

$$= \left(\prod_{j=0}^{i-1} f'(f^j(x))\right) \cdot \left(\sum_{k=0}^{i-1} \frac{\theta(f^k(x))}{f'(f^k(x))^2} \prod_{\tau=0}^{k} f'(f^\tau(x))\right).$$

Lemma 4.67 implies that $f'(a + p^\lambda h) \equiv f'(a) \pmod{p}$. From here we deduce that $f'(f^k(x)) \equiv f'(f^r(x)) \pmod{p}$ whenever $k \equiv r \pmod{p^\lambda}$, as f is transitive modulo p^λ. By the latter reason, $\theta(f^k(x)) \equiv \theta(f^r(x)) \pmod{p}$ whenever $k \equiv r \pmod{p^\lambda}$, in view of Lemma 4.66. Further, $N_1(f) \le \lambda$ by Proposition 4.65, and f is transitive modulo $p^{\lambda+1}$ by our assumption, so necessarily

$$\prod_{\tau=0}^{p^\lambda-1} f'(f^\tau(x)) \equiv 1 \pmod{p}, \tag{4.132}$$

see the proof of Lemma 4.56; consequently, $\prod_{\tau=0}^{k} f'(f^\tau(x)) \equiv \prod_{\tau=0}^{r} f'(f^\tau(x)) \pmod{p}$ whenever $k \equiv r \pmod{p^\lambda}$. Finally we conclude that

$$B_{tp^\lambda}(x) \equiv t \sum_{\tau=0}^{p^\lambda-1} \frac{\theta(f^\tau(x))}{f'(f^\tau(x))^2} \prod_{\nu=0}^{\tau} f'(f^\nu(x)) \equiv t \cdot B_{p^\lambda}(x) \pmod{p}, \tag{4.133}$$

for every $t \in \mathbb{N}$.

Now we calculate $A_{tp^\lambda}(x) \bmod p^2$ for $t \in \mathbb{N}$. Congruence (4.131) in view of (4.132) implies that

$$f^{kp^\lambda+\tau}(x) \equiv f^\tau(x) + kp^\lambda \cdot \xi(x) \prod_{j=0}^{\tau-1} f'(f^j(x)) \pmod{p^{\lambda+1}}, \tag{4.134}$$

for all $k \in \mathbb{N}$ and all $\tau \in \{0, 1, \dots, p^\lambda - 1\}$. As Lemma 4.67 implies that $f'(u) \equiv f'(v) \pmod{p^2}$ whenever $u \equiv v \pmod{p^{\lambda+1}}$, and as

$$A_{tp^\lambda}(x) = \prod_{k=0}^{t-1} \prod_{\tau=0}^{p^\lambda-1} f'(f^{kp^\lambda+\tau}(x)),$$

we conclude in view of congruence (4.134) that

$$A_{tp^\lambda}(x) = \prod_{k=0}^{t-1} \prod_{\tau=0}^{p^\lambda-1} f'\left(f^\tau(x) + kp^\lambda \cdot \xi(x) \prod_{j=0}^{\tau-1} f'(f^j(x))\right) \pmod{p^2}.$$

This implies in view of Lemma 4.67, that

$$A_{tp^\lambda}(x) = \prod_{k=0}^{t-1}\prod_{\tau=0}^{p^\lambda-1}\left(f'(f^\tau(x)) + 2kp\cdot\xi(x)\cdot\theta(f^\tau(x))\prod_{j=0}^{\tau-1} f_2'(f^j(x))\right)$$

$$\equiv \prod_{k=0}^{t-1}\left(\prod_{\tau=0}^{p^\lambda-1} f'(f^\tau(x)) + 2kp\cdot\xi(x)\cdot B_{p^\lambda}(x)\right) \pmod{p^2}. \quad (4.135)$$

According to (4.132),

$$\prod_{j=0}^{p^\lambda-1} f'(f^j(x)) = 1 + p\varepsilon$$

for a suitable $\varepsilon \in \mathbb{Z}_p$; consequently, (4.135) implies that

$$A_{tp^\lambda}(x) \equiv \prod_{k=0}^{t-1}\left(1 + p\varepsilon + 2kp\cdot\xi(x)\cdot B_{p^\lambda}(x)\right)$$

$$\equiv 1 + tp\varepsilon + pt(t-1)\cdot\xi(x)\cdot B_{p^\lambda}(x) \pmod{p^2}. \quad (4.136)$$

Now combining together (4.131), (4.133), and (4.136) we conclude that

$$f^{(t+1)p^\lambda}(x) = f^{p^\lambda+tp^\lambda}(x)$$

$$\equiv f^{tp^\lambda}(x) + p^\lambda\cdot\xi(x) + \varepsilon tp^{\lambda+1}\cdot\xi(x) + p^{\lambda+1}t^2\cdot\xi(x)^2\cdot B_{p^\lambda}(x) \pmod{p^{\lambda+2}}. \quad (4.137)$$

Finally, by obvious induction on n, from (4.137) and (4.129) we deduce that

$$f^{np^\lambda}(x) \equiv x + np^\lambda\cdot\xi(x) + \varepsilon p^{\lambda+1}\cdot\xi(x)\cdot\frac{n(n-1)}{2}$$

$$+ p^{\lambda+1}\cdot\xi(x)^2\cdot B_{p^\lambda}(x)\cdot\frac{n(n-1)(2n-1)}{6} \pmod{p^{\lambda+2}}.$$

From here it follows in particular that $f^{p^{\lambda+1}}(x) \equiv x + p^{\lambda+1}\cdot\xi(x) \pmod{p^{\lambda+2}}$ since $p \neq 2, 3$. However, the latter congruence in view of (4.129) implies that $f^{p^{\lambda+1}}(x) \not\equiv x \pmod{p^{\lambda+2}}$. This finally proves Theorem 4.61. □

Note 4.68. With the use of Theorem 4.61 we can determine *whether a given integer-valued and compatible polynomial $f(x) \in \mathbb{Q}_p[x]$ is ergodic*. Represent $f(x)$ in the form $f(x) = \frac{g(x)}{r}$, where $r \in \mathbb{Z}_p$, $g(x) \in \mathbb{Z}_p[x]$, and at least one coefficient of $g(x)$ is coprime with p. Actually, r is a least common denominator of all coefficients of $f(x)$ represented as irreducible fractions: We assume that $f(x)$ is represented in a falling factorial basis $x^{\underline{0}} = 1, x^{\underline{1}} = x, x^{\underline{2}} = x(x-1), \dots$, or in a standard basis $1, x, x^2, \dots$. Then $\rho(f) = \operatorname{ord}_p r$; note that r does not depend on a choice of a basis. Now we easily find $\lambda(f)$ and determine (e.g., by direct calculations) whether f is transitive modulo $p^{\lambda(f)+1}$ in the case $p \neq 2, 3$ or, respectively, modulo $p^{\lambda(f)+2}$ whenever $p = 2$ or $p = 3$.

Actually one can determine whether a polynomial $f(x) \in \mathbb{Q}_p[x]$ induces a 1-Lipschitz measure-preserving (respectively, ergodic) transformation on $\mathbb{Z}_p$ by evaluating f at $\approx p^3 \cdot \deg f$ points:

Proposition 4.69. *A polynomial $f(x) \in \mathbb{Q}_p[x]$ induces a 1-Lipschitz measure-preserving (respectively, ergodic) transformation on $\mathbb{Z}_p$ if and only if the mapping $z \mapsto f(z) \bmod p^{\lfloor \log_p(\deg f)\rfloor+3}$ is a compatible and bijective (respectively, transitive) transformation on the residue ring $\mathbb{Z}/p^{\lfloor \log_p(\deg f)\rfloor+3}\mathbb{Z}$.*

Proof. We prove only the ergodicity claim; a proof of the measure-preservation claim goes along similar lines and thus is omitted.

The coefficients $a_i \in \mathbb{Q}_p$ $(i = 0, 1, \dots, d)$ in the Mahler expansion of the polynomial $f(x)$ of degree d are completely determined by the values of $f(x)$ at the points $0, 1, \dots, d$. In particular, all values $f(0), f(1), \dots, f(d)$ are p-adic integers if and only if all coefficients $a_i \in \mathbb{Q}_p$ $(i = 0, 1, \dots, d)$ are p-adic integers, i.e., if and only if the polynomial $f(x)$ is integer-valued.

As $\Delta^i f(x) = 0$ for $i > \deg f = d$, in view of Theorem 3.53 from the proof of Proposition 3.38 it follows that f is a 1-Lipschitz transformation on $\mathbb{Z}_p$ if and only if f induces a compatible transformation on the residue ring $\mathbb{Z}/p^k\mathbb{Z}$ for some (arbitrarily fixed) $k \geq \lfloor \log_p d \rfloor + 1$.

By Theorem 4.61, an integer-valued polynomial $f(x) \in \mathbb{Q}_p[x]$ that induces a 1-Lipschitz transformation on $\mathbb{Z}_p$ is ergodic (on $\mathbb{Z}_p$) if and only if f is transitive modulo p^k for any (arbitrarily fixed) $k \geq \lambda(f) + 2$. Considering Mahler expansion (4.90) for $f(x)$, $f(x) = b_0 + \sum_{i=1}^{d} b_i p^{\lfloor \log_p i \rfloor}\binom{x}{i}$, where $b_j \in \mathbb{Z}_p$ for $j = 0, 1, 2, \dots$, we conclude that $\rho(f)$ is the least nonnegative rational integer that is not smaller than any of $\operatorname{ord}_p(i!) - \lfloor \log_p i \rfloor - \operatorname{ord}_p b_i$ $(i = 1, 2, \dots, d)$. Thus, since the function $\operatorname{ord}_p(i!) - \lfloor \log_p i \rfloor$ is monotone nondecreasing by Lemma 4.62, every $k \in \mathbb{N}$ that satisfies the inequality $2 \cdot \frac{p^k-1}{p-1} - k > \operatorname{ord}_p(d!) - \lfloor \log_p d \rfloor$ can not be smaller than $\lambda(f)$. However, $\operatorname{ord}_p(d!) = \frac{1}{p-1} \cdot (d - \operatorname{wt}_p d)$ by Lemma 3.6; so taking an arbitrary $k \in \mathbb{N}$ that satisfies the inequality

$$2 \cdot \frac{p^k - 1}{p - 1} - k > \frac{d}{p-1}, \tag{4.138}$$

we conclude that $k \geq \lambda(f)$. Elementary considerations show that $k = \lfloor \log_p d \rfloor + 1$ satisfies inequality (4.138) thus ending the proof. □

It is obvious that in some cases the conditions of Theorem 4.61 and of Proposition 4.69 can be relaxed; e.g., it is obvious that whenever $p > 3$, the proposition remains true after replacing $p^{\lfloor \log_p(\deg f)\rfloor+3}$ by $p^{\lfloor \log_p(\deg f)\rfloor+2}$. However, the point is that for some important classes of functions these bounds can be tightened significantly, so that the conditions depend only on the whole class rather than on a concrete function from the class:

Corollary 4.70. *A $\mathcal{B}$-function* (and thus a $\mathcal{C}$-function) *f is measure-preserving if and only if f is bijective modulo p^2. The function f is ergodic if and only if f is transitive modulo p^2 whenever $p \notin \{2,3\}$, or modulo p^3 whenever $p \in \{2,3\}$.*

Proof. By the definition of the class $\mathcal{B}$, $\rho(f) = 0$ for every $f \in \mathcal{B}$; whence, $\lambda(f) = 1$, and the conclusion follows from Theorem 4.61. □

From here we immediately deduce

Corollary 4.71 (cf. [101, 282]). *A polynomial $f \in \mathbb{Z}_p[x]$ is ergodic if and only if f is transitive modulo p^2 whenever $p \notin \{2,3\}$, or modulo p^3 whenever $p \in \{2,3\}$.*

Note 4.72. The bounds given by Corollary 4.71(and therefore by Corollary 4.70) are sharp: A polynomial $2x^3 + 3x + 5$ is transitive modulo 4, and is not transitive modulo 8 (whence, is not ergodic on $\mathbb{Z}_2$); a polynomial

$$1 + x - x(x-1)(x-2)(x-3)(x-4)(x-6)(x-7)$$

is transitive modulo 9, and is not transitive modulo 27 (whence, is not ergodic on $\mathbb{Z}_3$); a polynomial $1 + x^p$ is transitive modulo p, and is not transitive (even is not bijective in view of Corollary 4.48) modulo p^2; whence, is not measure-preserving on $\mathbb{Z}_p$. The first two examples are taken from [282].

4.7 Ergodic 1-Lipschitz transformations on p-adic spheres

In this section we study 1-Lipschitz ergodic transformations on spheres centered at $y \in \mathbb{Z}_p$. Main results of this section are Theorems 4.79 and 4.84, which give complete characterizations of $\mathcal{B}$-functions and of $\mathcal{A}$-functions that are ergodic on a p-adic sphere, as well as Proposition 4.83, which solves a similar problem for perturbed monomial mappings. Results of this section were obtained in [27].

4.7.1 1-Lipschitz ergodic transformations on spheres

Let $S_{p^{-r}}(y)$ be a sphere of radius $\frac{1}{p^r}$, $r \geq 1$, with a center at $y \in \mathbb{Z}_p$; that is

$$S_{p^{-r}}(y) = \left\{ z \in \mathbb{Z}_p : |z - y|_p = \frac{1}{p^r} \right\}.$$

Note that the sphere is a disjoint union of balls of radius $\frac{1}{p^{r+1}}$ each,

$$S_{p^{-r}}(y) = \bigcup_{s=1}^{p-1} (y + p^r s + p^{r+1}\mathbb{Z}_p), \tag{4.139}$$

since $S_{p^{-r}}(y)$ is a set-theoretic complement of the ball $y + p^{r+1}\mathbb{Z}_p$ in the ball $y + p^r\mathbb{Z}_p$. So $S_{p^{-r}}(y)$ is a closed and simultaneously an open (whence, a measurable) subset of $\mathbb{Z}_p$. We consider a measure $\hat{\mu}_p$ induced on $S_{p^{-r}}(y)$ by the Haar measure μ_p on the whole space $\mathbb{Z}_p$; we assume that $\hat{\mu}_p$ is normalized so that $\hat{\mu}_p(S_{p^{-r}}(y)) = 1$. Now, if $f \in \mathcal{L}_1$ is a 1-Lipschitz mapping of $\mathbb{Z}_p$ into $\mathbb{Z}_p$ such that the sphere $S_{p^{-r}}(y)$ is invariant under action of f (that is, $f(S_{p^{-r}}(y)) \subset S_{p^{-r}}(y)$), we can consider a restriction of f (which we denote by the same symbol f) on the sphere $S_{p^{-r}}(y)$ and study ergodicity of the restriction f with respect to the measure $\hat{\mu}_p$. We say then that *f is ergodic on the sphere $S_{p^{-r}}(y)$* whenever $S_{p^{-r}}(y)$ is invariant under action of f, and the action is ergodic with respect to $\hat{\mu}_p$, in the above mentioned meaning.

The following easy proposition holds:

Proposition 4.73. *Whenever $S_{p^{-r}}(y)$ is invariant under action of $f \in \mathcal{L}_1$, $f(y) \equiv y \pmod{p^r}$.*

Proof. Since $S_{p^{-r}}(y)$ is invariant, and since f maps balls into balls, $f(y + p^r s + p^{r+1}\mathbb{Z}_p) \subset y + p^r \hat{s} + p^{r+1}\mathbb{Z}_p$ for a suitable $\hat{s} \in \{1, 2, \dots, p-1\}$ (see (4.139)). However, $f(y + p^r s) \equiv f(y) \pmod{p^r}$ since $f \in \mathcal{L}_1$, and the result follows. □

From this proposition we immediately derive the following

Corollary 4.74. *Let all spheres around $y \in \mathbb{Z}_p$ of radii less than $\varepsilon > 0$ be invariant under action of $f \in \mathcal{L}_1$. Then $f(y) = y$.*

Further, as it can be seen from the respective proofs, all results of Section 4.4 hold not only for the whole space $\mathbb{Z}_p$, but (up to a proper re-statement) for any finite disjoint union of balls of pairwise equal radii as well. Moreover, following the lines of these proofs, corresponding results can be proved for any arbitrary measurable subset of $\mathbb{Z}_p$ of a positive measure rather than for the whole space $\mathbb{Z}_p$ only. We summarize this as the following important note:

Note 4.75. A 1-Lipschitz mapping $f : \mathbb{Z}_p \to \mathbb{Z}_p$ is ergodic on the sphere $S_{p^{-r}}(y)$ if and only if it induces on the residue ring $\mathbb{Z}/p^{k+1}\mathbb{Z}$ a mapping which is transitive on all subsets

$$S_{p^{-r}}(y) \bmod p^{k+1} = \{y + p^r s + p^{r+1}\mathbb{Z} : s = 1, 2, \dots, p-1\} \subset \mathbb{Z}/p^{k+1}\mathbb{Z},$$

$k = r, r+1, \dots$ (that is, permutates cyclically elements of every subset $S_{p^{-r}}(y) \bmod p^{k+1}$, see Section 2.2).

It is worth noting also that whenever a 1-Lipschitz mapping f is ergodic on the sphere $S_{p^{-r}}(y)$, f is a bijection of this sphere onto itself; moreover, it is an isometry on this sphere, see Notes 4.27 and 4.30. The same holds for balls.

From these notices we deduce the following lemma:

Lemma 4.76. *A 1-Lipschitz mapping $f: \mathbb{Z}_p \to \mathbb{Z}_p$ is ergodic on the sphere $S_{p^{-r}}(y)$ if and only if the following two conditions hold simultaneously:*

(1) *the mapping $z \mapsto f(z) \bmod p^{r+1}$ is transitive on the set*

$$S_{p^{-r}}(y) \bmod p^{r+1} = \{y + p^r s : s = 1, 2, \dots, p-1\} \subset \mathbb{Z}/p^{r+1}\mathbb{Z};$$

(2) *the mapping $z \mapsto f^{p-1}(z) \bmod p^{r+t+1}$ is transitive on the set*[2]

$$B_{p^{-(r+1)}}(y + p^r s) \bmod p^{r+t+1} = \{y + p^r s + p^{r+1} S : S = 0, 1, 2, \dots, p^t - 1\},$$

for all $t = 1, 2, \dots$ and some (equivalently, all) $s \in \{1, 2, \dots, p-1\}$.

Condition (2) holds if and only if f^{p-1} is an ergodic transformation on the ball $B_{p^{-(r+1)}}(y + p^r s) = y + p^r s + p^{r+1}\mathbb{Z}_p$ of radius $\frac{1}{p^{r+1}}$ centered at $y + p^r s$, for some (equivalently, all) $s \in \{1, 2, \dots, p-1\}$.

Proof. As every 1-Lipschitz ergodic transformation f of the sphere is bijective on this sphere, and f is an isometry on this sphere as well (see above notions), $f(a + p^k\mathbb{Z}_p) = f(a) + p^k\mathbb{Z}_p$, for all $a \in \mathbb{Z}_p$ and all $k = 1, 2, \dots$. Thus, the mapping $z \mapsto f(z) \bmod p^{k+1}$ $(k > r)$ permutes cyclically elements of the set

$$S_{p^{-r}}(y) \bmod p^{k+1} = \{y + p^r s + p^{r+1} S : s = 1, 2, \dots, p-1; S = 0, 1, 2, \dots, p^{k-r} - 1\}$$

if and only if conditions (1) and (2) hold simultaneously for $t = k - r$. This proves the first part of the statement of the lemma, in view of Note 4.75. The second part of the statement is just an analogue of Note 4.75 for balls rather than for spheres. □

Note 4.77. Obviously, Lemma 4.76 holds for spheres of radii 1 as well, in the following form: *A 1-Lipschitz transformation $f: S_1(y) \to S_1(y)$ on the sphere*

$$S_1(y) = \bigcup_{s \in \{0, \dots, p-1\} \setminus \{y\}} s + p\mathbb{Z}_p$$

is ergodic if and only if $f \bmod p$ is transitive on the set $\{0, \dots, p-1\} \setminus \{y\}$ and f^{p-1} is an ergodic transformation on every (equivalently, some) ball $B_{p^{-1}}(s)$, $s \in \{0, \dots, p-1\} \setminus \{y\}$.

Note 4.78. It is clear that both Lemma 4.76 and Note 4.75 hold for 1-Lipschitz mapping with domain $S_{p^{-r}}(y)$ rather than with domain $\mathbb{Z}_p$; that is, f may be defined only on the sphere $S_{p^{-r}}(y)$ rather than on the whole space $\mathbb{Z}_p$.

[2] That is, the sets $S_{p^{-r}}(y) \bmod p^{r+1}$ and $B_{p^{-(r+1)}}(y + p^r s) \bmod p^{r+t+1}$ are fuzzy cycles, see Definition 5.42 further.

4.7.2 Ergodicity of $\mathcal{B}$-functions and of analytic functions

We say that $z \in \mathbb{Z}_p$ is *primitive modulo* p^k whenever $z \bmod p^k$ generates the whole group $(\mathbb{Z}/p^k\mathbb{Z})^*$ of invertible elements of the residue ring $\mathbb{Z}/p^k\mathbb{Z}$. Note that whenever $k > 2$ we speak on primitivity modulo p^k only for odd p, see Proposition 1.32.

Theorem 4.79. *Let the function f lie in $\mathcal{B}$. The function f is ergodic on the sphere $S_{p^{-r}}(y)$ of sufficiently small radius p^{-r} if and only if one of the following alternatives holds:*

(1) *Whenever p is odd, then simultaneously*

- $f(y) \equiv y \pmod{p^{r+1}}$,
- *$f'(y)$ is primitive modulo p^2.*

(2) *Whenever $p = 2$, then simultaneously*

- $f(y) \equiv y \pmod{2^{r+1}}$,
- $f(y) \not\equiv y \pmod{2^{r+2}}$,
- $f'(y) \equiv 1 \pmod 4$.

Note 4.80. Within the context of the theorem, 'sufficiently small' means that $r \geq 2$ if $p > 3$, or $r \geq 3$ if $p \leq 3$.

Proof. As it immediately follows from Theorem 3.62, for every $g \in \mathcal{B}$ and all $k \in \mathbb{Z}_p$, $k = 1, 2, 3, \ldots$ the equality

$$g(a + p^k h) = g(a) + g'(a) \cdot p^k h + p^{2k} h^2 \cdot \hat{g}(h) \tag{4.140}$$

holds for a suitable $\mathcal{C}$-function $\hat{g}$ of variable h.[3] Since $f(y) = y + p^r z$ for a suitable $z \in \mathbb{Z}_p$ in view of Proposition 4.73, from (4.140) we deduce the following equality

$$\begin{aligned} f(y+p^r s+p^{r+1}S) &= f(y)+(p^r s+p^{r+1}S)\cdot f'(y)+p^{2r}\cdot(s+pS)^2\cdot\hat{w}(s+pS) \\ &= y + p^r z + p^r s \cdot f'(y) + p^{r+1} S \cdot f'(y) + p^{2r} \cdot v(s) + p^{2r+1} \cdot w(S), \end{aligned} \tag{4.141}$$

where v, $\hat{w}$ and w are $\mathcal{C}$-functions in respective variables and $r \geq 1$ (note that we have used (4.140) twice; with $g = f$, $a = y$, $p^k h = p^r s + p^{r+1} S$, for the first time, and with $g = w$, $a = s$, $p^k h = p^S$, for the second time). Note that w depends also on s, yet this is of no importance in the following argument.

[3] Of course, coefficients of series (3.53) that represents the function $p^{2k} \cdot g \in \mathcal{B}$ depend also on a and k, but this is of no importance at the moment.

Iterating (4.141) we conclude that

$$\begin{aligned} f^{p-1}(y+p^r s + p^{r+1} S) \\ &= y + p^r z \sum_{i=0}^{p-2} (f'(y))^i + p^r s \cdot (f'(y))^{p-1} \\ &\quad + p^{r+1} S \cdot (f'(y))^{p-1} + p^{2r} \cdot \breve{v}(s) + p^{2r+1} \cdot \breve{w}(S) \end{aligned} \tag{4.142}$$

for suitable $\breve{v}$ and $\breve{w}$, which are $\mathcal{B}$-functions (as compositions of $\mathcal{C}$-functions).

Now, to satisfy condition (2) of Lemma 4.76, the ball $y + p^r s + p^{r+1}\mathbb{Z}_p$ must be invariant under action of f^{p-1}, and f^{p-1} must act ergodically on this ball. However, (4.142) implies that the ball is invariant if and only if

$$\sigma(z,s) = z \sum_{i=0}^{p-2} (f'(y))^i + s \cdot (f'(y))^{p-1} \equiv s \pmod p. \tag{4.143}$$

Assuming the ball is invariant, we conclude that $\sigma(z,s) = s + p \cdot \gamma(z,s)$ for a suitable p-adic integer $\gamma(z,s)$. So, having s fixed, from (4.142) we see that under this assumption the following equality holds:

$$\begin{aligned} f^{p-1}(y + p^r s + p^{r+1} S) \\ = y + p^r s + p^{r+1} \cdot (\gamma(z,s) + S \cdot (f'(y))^{p-1} + p^{r-1} \cdot \breve{v}(s) + p^r \cdot \breve{w}(S)). \end{aligned}$$

Thus, to satisfy condition (2) of Lemma 4.76, the following $\mathcal{B}$-function

$$G_{z,s}(S) = \gamma(z,s) + S \cdot (f'(y))^{p-1} + p^{r-1} \cdot \breve{v}(s) + p^r \cdot \breve{w}(S) \tag{4.144}$$

in variable S must be ergodic on $\mathbb{Z}_p$.

Now, whenever $r > 1$ and $p > 3$, or whenever $r > 2$ and $p \le 3$, from Corollary 4.70 we deduce that the $\mathcal{B}$-function $G_{z,s}(S)$ from (4.144) is ergodic on $\mathbb{Z}_p$ if and only if the polynomial

$$L_{z,s}(S) = \gamma(z,s) + p^{r-1} \cdot \breve{v}(s) + S \cdot (f'(y))^{p-1} \tag{4.145}$$

of degree 1 in variable S is transitive modulo p^2 for $p > 3$, or modulo p^3 for $p \le 3$. But this in view of Theorem 4.36 and (4.145) implies that $f'(y) \not\equiv 0 \pmod p$. Now, as $f(y) = y + p^r z$, from (4.141) it follows that to satisfy condition (1) of Lemma 4.76, the mapping $s \mapsto z + s \cdot f'(y) \pmod p$ must be transitive on the multiplicative group (i.e., on the whole group of units) $(\mathbb{Z}/p\mathbb{Z})^*$ of the field $\mathbb{Z}/p\mathbb{Z}$. Hence, $z \equiv 0 \pmod p$ (that is, $f(y) \equiv y \pmod{p^{r+1}}$) since otherwise $s \mapsto 0 \pmod p$ for $s \equiv -\frac{z}{f'(y)} \pmod p$. From this moment we consider cases $p = 2$ and $p > 2$ separately.

Case 1: $p > 2$. In this case the mapping $s \mapsto s \cdot f'(y) \pmod p$ is transitive on $(\mathbb{Z}/p\mathbb{Z})^*$ if and only if $f'(y)$ is a primitive element of the field $\mathbb{Z}_p$ (that is, $f'(y)$ generates the cyclic group $(\mathbb{Z}/p\mathbb{Z})^*$).

Whenever this holds, every ball $y+p^r s+p^{r+1}\mathbb{Z}_p$, $s \in \{1,2,\dots,p-1\}$ is invariant under action of f^{p-1} in view of (4.143). Moreover, since $z \equiv 0 \pmod p$, in the case when $f'(y)$ is primitive modulo p we have that $\sigma(z,s) \equiv s \cdot (f'(y))^{p-1} \pmod{p^2}$ and whence $\gamma(z,s) \equiv bs \pmod p$, where $(f'(y))^{p-1} = 1+pb$, $b \in \mathbb{Z}_p$ (see (4.143) and the text thereafter for the definition of $\sigma(z,s)$ and $\gamma(z,s)$).

Now, the polynomial $L_{z,s}(S)$ (see (4.145)) in variable S is ergodic on $\mathbb{Z}_p$ (and thus condition (2) of Lemma 4.76 is satisfied) if and only if $b \not\equiv 0 \pmod p$, see Theorem 4.36. Yet this means that $f'(y)$ must be a generator of the multiplicative group $(\mathbb{Z}/p^2\mathbb{Z})^*$.

Case 2: $p = 2$. In this case the sphere $S_{2^{-r}}(y) = y + 2^r + 2^{r+1}\mathbb{Z}_2$ is a ball, see (4.139). Moreover, the above condition $f'(y) \not\equiv 0 \pmod p$ means that $f'(y) \equiv 1 \pmod 2$, and so the condition that the mapping $s \mapsto s \cdot f'(y) \pmod p$ is transitive on the multiplicative group $(\mathbb{Z}/p\mathbb{Z})^*$, which just means that $z + f'(y) \equiv 1 \pmod 2$ in this case, is automatically satisfied since we have already proved that $z \equiv 0 \pmod p$, (i.e., that $z = pc$ for suitable $c \in \mathbb{Z}_p$) for any p.

Further, if the polynomial $L_{z,s}(S)$ in variable S is transitive modulo p^3 then $f'(y) \equiv 1 \pmod 4$, see (4.145) and Theorem 4.36. That is, $f'(y) = 1 + 4b$ for some $b \in \mathbb{Z}_2$. Hence $\gamma(z,s) = c + 2b$ (see (4.143) and the text thereafter), so in view of (4.145) and Theorem 4.36, if $L_{z,s}(S)$ is transitive modulo 8, then $c \equiv 1 \pmod 2$; that is, $f(y) = y+2^r z = y+2^{r+1}c \not\equiv y \pmod{2^{r+2}}$. This proves Theorem 4.79. □

Corollary 4.81. *Let $y \in \mathbb{Z}_p$ be a fixed point of the function $f \in \mathcal{B}$, and let p be odd. Then, f is ergodic on all spheres around y of sufficiently small radii if and only if f is ergodic on some sphere around y of a sufficiently small radius.*

From Theorem 4.79 we immediately derive a complete characterization of $\mathcal{C}$-functions that are ergodic on p-adic spheres.

Theorem 4.82. *Let f be a $\mathcal{C}$-function. Whenever p is odd, the mapping $z \mapsto f(z)$ is an ergodic transformation on every sufficiently small sphere centered at $y \in \mathbb{Z}_p$ if and only if the following two conditions hold simultaneously:*

- *$f(y) = y$, and*
- *the derivative $f'(y)$ of the function f at the point $y \in \mathbb{Z}_p$ is primitive modulo p^2.*

In the case $p = 2$ no $\mathcal{C}$-function exists such that the mapping $z \mapsto f(z)$ is ergodic on all spheres around $y \in \mathbb{Z}_2$ of radii less than ε, whatever $\varepsilon > 0$ is taken.

Proof. This is obvious in view of Theorem 4.79 and Corollary 4.74. □

4.7.3 Ergodicity of perturbed monomial mappings

The following important consequence of Theorem 4.79 serves as a characterization of ergodic perturbed monomial transformation on spheres (cf. Section 4.3):

Proposition 4.83. *The perturbed monomial mapping* $f: x \mapsto x^\ell + q(x)$, *where* $q(x) = p^{r+1}u(x)$ *for some function* $u \in \mathcal{B}$ (e.g., for a polynomial $u(x) \in \mathbb{Z}_p[x]$) *is ergodic on the sphere* $S_{p^{-r}}(1)$ (where $r > 1$) *if and only if* ℓ *is primitive modulo* p^2.

Proof. Immediately follows from Theorem 4.79 with the only exception of the case $p = 3$ and $r = 2$. To handle this case, some extra efforts must be made. Namely, for $p = 3$ by Theorem 3.62 we conclude that

$$f^2(1 + 3^r s + 3^{r+1}S) = f^2(1) + (3^r s + 3^{r+1}S) \cdot f'(f(1)) \cdot f'(1)$$
$$+ \frac{1}{2}(3^r s + 3^{r+1}S)^2 \cdot (f''(f(1)) \cdot (f'(1))^2 + f'(f(1)) \cdot f''(1)) + 3^{3r+1} \cdot \hat{w}(S), \tag{4.146}$$

where $\hat{w}(S)$ is a $\mathcal{B}$-function in variable S. Now taking $f(x) = x^\ell + 3^{r+1}q(x)$, from (4.146) we derive that

$$f^2(1 + 3^r s + 3^{r+1}S) = 1 + (\ell + 1)3^{r+1}u(1) + (3^r s + 3^{r+1}S) \cdot \ell^2$$
$$+ \frac{1}{2}(3^r s + 3^{r+1}S)^2 \cdot \ell^2(\ell - 1)(\ell + 1) + 3^{2r+1}v(s) + 3^{2r+2}w(S), \tag{4.147}$$

where v and w are $\mathcal{B}$-functions in variables s and S, respectively. However, ℓ must be primitive modulo 3 (see case 2 of the proof of Theorem 4.79); so $\ell \equiv 2 \pmod 3$. Hence, $\ell^2 = 1 + 3b$ for a suitable $b \in \mathbb{Z}$. Also, $\ell(\ell - 1)(\ell + 1)$ is a multiple of 3; combining this altogether with (4.147) we conclude that

$$f^2(1 + 3^r s + 3^{r+1}S) = 1 + 3^r s + 3^{r+1} \cdot (b + (\ell + 1) \cdot u(1) + S\ell^2$$
$$+ 3^r \cdot \breve{v}(s) + 3^{r+1} \cdot \breve{w}(S)), \tag{4.148}$$

for suitable $\mathcal{B}$-functions $\breve{v}$ and $\breve{w}$. Now we must check whether the $\mathcal{B}$-function

$$L(S) = b + (\ell + 1) \cdot u(1) + S\ell^2 + 3^r \cdot \breve{v}(s) + 3^{r+1} \cdot \breve{w}(S)$$

is ergodic on $\mathbb{Z}_3$; cf. (4.144) where the residue term is $p^r \cdot \breve{w}(S)$ rather than $3^{r+1} \cdot \breve{w}(S)$ as in the case under consideration. The reason for this is that now an extra factor 3 in the fourth term of (4.147) arises because of the multiplier $\ell(\ell - 1)(\ell + 1)$.

Applying Corollary 4.70 and Theorem 4.36 to the $\mathcal{B}$-function L in variable S we see that L is ergodic on $\mathbb{Z}_p$ if and only if $b \not\equiv 0 \pmod 3$ (since $(\ell + 1)q(1) \equiv 0 \pmod 3$; we remind that $\ell \equiv 2 \pmod 3$). Thus, we finally conclude that ℓ must be primitive modulo p^2. □

Some known results on ergodicity of polynomial mappings also follow from Theorem 4.79. For instance, [80] concerns ergodicity of simple polynomial mappings $M_{a,\ell}: z \mapsto az^\ell$ on spheres, where $\ell > 0$ is rational integer, $a \in \mathbb{Z}_p$. From Hensel's lemma it follows that whenever $\ell \not\equiv 1 \pmod{p}$ and $a \in B_{p^{-1}}(1)$, the mapping $M_{a,\ell}$ has a unique fixed point $x_0 \in B_{p^{-1}}(1)$ (see [80, Lemma 8.2]). Under these assumptions, from Theorem 4.79 it immediately follows that $M_{a,\ell}$ is ergodic on $S_{p^{-r}}(x_0)$ (for p odd) if and only if $a \cdot \ell$ is primitive modulo p^2, that is, *if and only if ℓ is primitive modulo p^2* since $a \equiv 1 \pmod{p}$ by the assumption; cf. [80, Theorem 8.4]. Similarly, the translation $T_{a,b}\colon z \mapsto az + b$, with $a, b \in \mathbb{Z}_p$, has a fixed point $y_0 = \frac{b}{1-a} \in \mathbb{Q}_p$ whenever $a \neq 1$. In case $y \in \mathbb{Z}_p$, Theorem 4.79 yields $T_{a,b}$ is ergodic on $S_{p^{-r}}(y)$ *if and only if a is primitive modulo p^2*, cf. [80, Theorem 7.3].[4]

In view of Theorem 4.79 it is obvious that these results remain true in a 'perturbed form', that is, for mappings $z \mapsto M_{a,\ell}(z) + p^{r+1}v(z)$ and $z \mapsto T_{a,b} + p^{r+1}v(z)$, where v is an arbitrary polynomial over $\mathbb{Z}_p$ (or even a $\mathcal{B}$-function), despite in this case x_0 (respectively, y_0) are not necessarily fixed points of the corresponding mappings.

4.7.4 Ergodicity of $\mathcal{A}$-functions on spheres

Some important functions (for instance, some compatible integer-valued polynomials over $\mathbb{Q}_p$; i.e., those polynomials, which have not necessarily integer p-adic coefficients, that map $\mathbb{Z}_p$ into itself, and that satisfy Lipschitz condition with the constant 1 everywhere on $\mathbb{Z}_p$) do not lie in $\mathcal{B}$. However, they lie in a wider class $\mathcal{A}$, see Subsections 3.10.2 and 3.10.3. Fortunately we can determine whether an $\mathcal{A}$-function is ergodic on a p-adic sphere as well.

Theorem 4.84. *The statement of Theorem 4.79 remains true for $f \in \mathcal{A}$.*

Proof. The definition of an $\mathcal{A}$-function implies that $f = \frac{1}{p^n}\bar{f}$ for a suitable $\mathcal{B}$-function $\bar{f}$ and a suitable non-negative rational integer n, see Section 3.10. Then with the use of Theorem 3.64 we can re-write the key equation (4.140) of Theorem 4.79 in the following form:

$$g(a + p^k h) = g(a) + g'(a) \cdot p^k h + p^{2k-n} \cdot h^2 \cdot \hat{g}(h), \tag{4.149}$$

where $g \in \mathcal{A}$, $p^n g \in \mathcal{B}$, $\hat{g} \in \mathcal{C}$, and k is sufficiently large (so that $2k - n$ is positive). Then from (4.141) we obtain (for a sufficiently large r) that

$$\begin{aligned} f(y + p^r s + p^{r+1} S) &= f(y) + (p^r s + p^{r+1} S) \cdot f'(y) + p^{2r-n} \cdot (s + pS)^2 \cdot \hat{w}(s + pS) \\ &= y + p^r z + p^r s \cdot f'(y) + p^{r+1} S \cdot f'(y) + p^{2r-n} \cdot v(s) + p^{2r+1-n} \cdot w(S), \end{aligned} \tag{4.150}$$

where v, $\hat{w}$ and w are $\mathcal{C}$-functions in the respective variables. Now we assume that r is so large that the inequality $2r - n \geq r + 3$ holds, and finish the proof in a manner similar to that of Theorem 4.79. □

[4] We note however that we prove not exactly the same results as in [80] since we impose conditions that are slightly different from the ones in [80].

Note 4.85. In contrast to Theorem 4.79, within the conditions of Theorem 4.84 it depends also on n (i.e., on the function f) how small the sphere $S_{p^{-r}}(y)$ must be to satisfy the theorem.

Now we make some conclusions to Section 4.7. With the use of Theorem 4.79 we immediately obtain a number of examples of various functions that are ergodic on a p-adic sphere: For instance, whenever a positive rational integer ℓ generates modulo p^2 the whole group of units of the residue ring $\mathbb{Z}/p^2\mathbb{Z}$, the functions $1 + \ell \cdot (-1 + x + p^2 \cdot v(x))$ and $\ell \cdot (ax + a^x - 2a) + 1$ are ergodic on all (sufficiently small) spheres around 1, for every $a \in 1 + p^2\mathbb{Z}_p$ and every $\mathcal{B}$-function v (say, for a polynomial v over $\mathbb{Z}_p$); accordingly, the functions $\ell \cdot x + \ln_p(1 + p^2x)$ and $\frac{\ell \cdot x}{1+p^2x}$ are ergodic on all (sufficiently small) spheres around 0 (here $\ln_p$ stands for the p-adic logarithm: $\ln_p(1 + pz) = \sum_{i=1}^{\infty}(-1)^{i+1}\frac{p^i z^i}{i}$).

It is worth noting here that by virtue of Theorem 4.79 perturbed monomial mappings on spheres are ergodic whenever the perturbations are 'p-adically small' $\mathcal{B}$-functions (and even $\mathcal{A}$-functions), and not only 'p-adically small' polynomials over $\mathbb{Z}_p$: e.g., a perturbed monomial $x^\ell + \frac{1}{p}(x^p - x)^2$ is an integer-valued polynomial over $\mathbb{Q}_p$ (and *not* a polynomial over $\mathbb{Z}_p$) which is ergodic on sufficiently small spheres.

Here are examples of $\mathcal{A}$-functions (which are *not* $\mathcal{B}$-functions) that are ergodic on all sufficiently small spheres around 0:

$$\ell \cdot x + \ln_p(1 + p^2x) + \frac{1}{p}(x^p - x)^2; \quad \frac{\ell \cdot x}{1 + p^2x} + \frac{1}{p}(x^p - x)^2.$$

Note that our proofs of main results of the section use that $\mathcal{A}$-functions (whence, $\mathcal{B}$-functions) are locally analytic of order 1, in terminology of [374]. Within this context it would be interesting to answer the following question.

Open Question 4.86. Is it possible to expand Theorem 4.79 to the class of all 1-Lipschitz functions that are locally analytic of order n, $n = 1, 2, \ldots$?

4.8 Concluding remarks to p-adic ergodic theory

In this section, we make some comments and conclusions about questions that naturally arose in connection with presented p-adic ergodic theory for 1-Lipschitz transformations on $\mathbb{Z}_p$. These questions mainly concern dynamics with a continuous time, dynamics outside the class of 1-Lipschitz maps, and the non-minimal dynamics.

4.8.1 Continuous p-adic dynamics

In this subsection, we demonstrate that *every discrete 1-Lipschitz ergodic dynamical system f on $\mathbb{Z}_p$ can be extended to a dynamical system with a continuous p-adic time*. In other words, whenever $f\colon \mathbb{Z}_p \to \mathbb{Z}_p$ is 1-Lipschitz and ergodic, the function

$f^n(x), n \in \mathbb{Z}_0$, can be expanded to the function $f^t(x)$, $(t,x) \in \mathbb{Z}_p^2$, which is continuous as a 2-variate function (actually, it is 1-Lipschitz). Moreover, in the case $p = 2$ we show that given an arbitrary 1-Lipschitz measure-preserving function $f: \mathbb{Z}_2 \to \mathbb{Z}_2$, which is *not* necessarily ergodic, the function $f^n(x)$, $n \in \mathbb{N}_0$, can be expanded to the function $f^t(x)$, $(t,x) \in \mathbb{Z}_2^2$, which is continuous as a 2-variate function. Thus, we stress that the p-adic time arises very naturally in the p-adic ergodic theory, although currently we are not aware whether this concept has a physical or other applied meaning.

Let $f: \mathbb{Z}_p \to \mathbb{Z}_p$ be a 1-Lipschitz ergodic transformation on $\mathbb{Z}_p$. For every $n \in \mathbb{N}_0$ the nth iteration $f^n(x)$ is well defined. We assert that given $t \in \mathbb{Z}_p$, there exists a limit (with respect to the p-adic metric)

$$\overset{p}{\lim_{n_j \to t}} f^t(x),$$

where $(n_j \in \mathbb{N}_0)_{j=0}^{\infty}$ is an arbitrary sequence that tends p-adically to $t \in \mathbb{Z}_p$.

Indeed, let $n = m + p^k\ell, m, n, \ell \in \mathbb{N}_0, k \in \mathbb{N}$. Then $f^n(x) \equiv f^m(x) \pmod{p^k}$ as $f(x)$ is transitive modulo p^k for all $k \in \mathbb{N}$, by Theorem 4.23. That is, $|f^n(x) - f^m(x)|_p \le p^{-k}$ whenever $|n - m|_p \le p^{-k}$. This proves our assertion as $\mathbb{N}_0$ is dense in $\mathbb{Z}_p$.

Thus, the 2-variate function $f^t(x)\colon \mathbb{Z}_p^2 \to \mathbb{Z}_p$, $t, p \in \mathbb{Z}_p$, is well defined. Note that the argument above implies that $f^t(x)$ is 1-Lipschitz with respect to the variable $t \in \mathbb{Z}_p$. We claim that the function $f^t(x)$ is continuous as a 2-variate function of $t, x \in \mathbb{Z}_p$. Indeed, given $x, x', t, t' \in \mathbb{Z}_p$ such that $|x - x'|_p \le p^{-n}$, $|t - t'|_p \le p^{-m}$ we see that $|f^t(x) - f^{t'}(x')|_p \le p^{-k}$ for $k = \min\{n, m\}$ since in this case $t \equiv t' = r \pmod{p^k}$ and $x \equiv x' \equiv z \pmod{p^k}$ for suitable $r, z \in \{0, 1, \dots, p^k - 1\}$; so $f^t(x) \equiv f^{t'}(x') \equiv f^r(z) \pmod{p^k}$ as f is transitive modulo p^k by Theorem 4.23. Thus we have proved the following proposition:

Proposition 4.87. *Given a 1-Lipschitz ergodic transformation f on $\mathbb{Z}_p$, the function $f^t(x)$ is a 1-Lipschitz function defined for all $(t, z) \in \mathbb{Z}_p^2$ and valuated in $\mathbb{Z}_p$.*

Foremost, for every $x \in \mathbb{Z}_p$ *the function $f^t(x)$ is measure-preserving as a function of variable $t \in \mathbb{Z}_p$.* Indeed, given $n, m \in \mathbb{N}_0, n \ne m$, take $k \in \mathbb{N}$ such that $p^k > n, m$. Then $f^n(x) \not\equiv f^m(x) \pmod{p^k}$ for every $x \in \mathbb{Z}_p$ since f is transitive modulo p^k. This proves that given $x \in \mathbb{Z}_p$, the function $f^t(x)$ of variable $t \in \mathbb{Z}_p$ is bijective modulo p^k for all $k \in \mathbb{N}$; thus this function $f^t(x)$ is measure-preserving by Theorem 4.23. Thus the following proposition is true:

Proposition 4.88. *The 2-variate function $f^t(x)$ from Proposition 4.87 is measure-preserving with respect to variable $t \in \mathbb{Z}_p$.*

Example 4.89. Given an ergodic affine transformation $f(x) = ax + b$ on $\mathbb{Z}_p$, the 2-variate function from Proposition 4.87 is of the form $f^t(x) = bt + x$ if $a = 1$, and

$$f^t(x) = b \cdot \frac{a^t - 1}{a - 1} + a^t x,$$

if $a \neq 1$. Note that by Theorem 4.36, $p \nmid b$ and $a \equiv 1 \pmod p$.

Indeed, if $a = 1$ then $f^n(x) = bn + x$ for all $n \in \mathbb{N}_0$; so given $t \in \mathbb{Z}_p$, we have that $\lim^p_{n\to t} f^n(x) = bt + x$.

Let now $a \neq 1$. Then by Theorem 4.36, $a - 1 = pz$ if $p \neq 2$, and $a - 1 = 4z$ if $p = 2$, for a suitable $z \in \mathbb{Z}_p$. Given $n \in \mathbb{N}$, we have then that $f^n(x) = b \cdot (a^{n-1} + a^{n-2} + \cdots + 1) + a^n x$. Let $k = \operatorname{ord}_p z$, i.e., $z = p^k \hat{z}$ where $p \nmid \hat{z}$, then

$$a^{n-1} + a^{n-2} + \cdots + 1 = \frac{a^n - 1}{a - 1} = \begin{cases} \sum_{i=1}^n p^{(k+1)(i-1)} \hat{z}^{i-1} \binom{n}{i}, & \text{if } p > 2; \\ \sum_{i=1}^n 2^{(k+2)(i-1)} \hat{z}^{i-1} \binom{n}{i}, & \text{if } p = 2, \end{cases}$$

is a p-adic integer. It is well known (see e.g. [308, Chapter 14, Section 5]) that under the above restrictions on a, the function a^t is analytic on $\mathbb{Z}_p$; so we see that $\lim^p_{n\to t} a^n = a^t$, and the conclusion follows.

Now we consider the 2-adic case. Let $f: \mathbb{Z}_2 \to \mathbb{Z}_2$ be a 1-Lipschitz measure-preserving function. Thus, by Theorem 4.23, f is bijective modulo 2^n, for all $n \in \mathbb{N}$; so every map $f \bmod 2^n: x \mapsto f(x) \bmod 2^n$ of the residue ring $\mathbb{Z}/2^n\mathbb{Z}$ into itself is a permutation on $\{0, 1, \ldots, 2^n - 1\}$. We claim that *every cycle of this permutation has the length* 2^ℓ, *for a suitable* $\ell \in \mathbb{N}_0$. We proceed by induction on n.

For $n = 1$ the claim is obvious since $f \bmod 2$ is either the identity map (whose cycles are all of length $2^0 = 1$) or the map $x \mapsto x + 1$, which consist of the only cycle of length 2.

Let the claim be true for all $1 \le n < k$; let us prove it for $n = k$. Given $x \in \mathbb{Z}/2^k\mathbb{Z}$, denote $\chi_i^j = \delta_i(f^j(x))$, the ith digit in a base-2 expansion of the jth iterate of f. Note that $\chi_i^0 = \delta_i(x)$, $i = 0, 1, 2, \ldots$. From Theorem 4.39 it follows that

$$\chi_i^1 = \chi_i^0 \oplus \varphi_i(\chi_0^0, \ldots, \chi_{i-1}^0),$$

where ψ_i is a Boolean function in i Boolean variables; iterating this equality, we obtain that

$$\chi_i^j = \chi_i^0 \oplus \sum_{\ell=0}^{j-1} \varphi_i(\chi_0^\ell, \ldots, \chi_{i-1}^\ell), \tag{4.151}$$

for all $i = 0, 1, 2, \ldots$, where the sum $\sigma_i^j = \sum_{\ell=0}^{j-1} \varphi_i(\chi_0^\ell, \ldots, \chi_{i-1}^\ell)$ in the right-hand side is taken modulo 2; so $\sigma_i^j \in \{0, 1, \}$. If $f^r(x) \equiv x \pmod{2^k}$, then $f^r(x) \equiv x \pmod{2^{k-1}}$, so by induction hypothesis, the smallest $r \in \mathbb{N}$ that satisfies the latter congruence is a power of 2: $r = 2^s$, for a suitable $s \in \mathbb{N}_0$. Hence, either $f^{2^s}(x) \equiv x$

(mod 2^k), or $f^{2^s}(x) \not\equiv x \pmod{2^k}$, and in the latter case $\chi_i^{2^s} = \chi_i^0$, $\chi_{k-1}^{2^s} \equiv \chi_{k-1}^0 + 1 \pmod 2$, $i = 0, 1, 2, \ldots, k-2$. Thus, in the latter case $\sigma_{k-1}^{2^s} \equiv 1 \pmod 2$ in view of congruence (4.151). But then necessarily $\chi_{k-1}^{2^{s+1}} \equiv \chi_{k-1}^0 + \sigma_{k-1}^{2^{s+1}} \equiv \chi_{k-1}^{2^{s+1}} \pmod 2$ since $\sigma_{k-1}^{2^{s+1}} = 2 \cdot \sigma_{k-1}^{2^s} \equiv 0 \pmod 2$; we just note that $\sigma_{k-1}^{2^s}$ is a sum modulo 2 of all values of the Boolean function φ_{k-1} when the number $\chi_0 + \chi_1 \cdot 2 + \cdots + \chi_{k-2} \cdot 2^{k-2}$ runs through the cycle of the permutation $f \bmod 2^{k-1}$ that contains the residue $x \bmod 2^{k-1}$. So in this case the length of the cycle of the permutation $f \bmod 2^k$ that contains the residue $x \bmod 2^k$ is 2^{s+1}.

Now everything is ready to prove the following proposition:

Proposition 4.90. *For every 1-Lipschitz measure-preserving function $f: \mathbb{Z}_2 \to \mathbb{Z}_2$, the function $f^n(x)$, $n \in \mathbb{N}_0$, can be expanded to the function $f^t(x)$, $(t, x) \in \mathbb{Z}_2^2$, which is a 1-Lipschitz* (thus, continuous) *2-variate function defined for every $(t, x) \in \mathbb{Z}_2^2$ and valuated in $\mathbb{Z}_2$.*

Proof. We mimic the proof of Proposition 4.87. Let $n = m + 2^k\ell$, $m, n, \ell \in \mathbb{N}_0$, $k \in \mathbb{N}$. Then $f^n(x) \equiv f^m(x) \pmod{2^k}$ as the residue $x \bmod 2^k$ lies on some cycle of length 2^ℓ, $\ell \le k$. That is, $|f^n(x) - f^m(x)|_2 \le 2^{-k}$ whenever $|n - m|_2 \le 2^{-k}$. This proves that given $t \in \mathbb{Z}_2$, the limit $\lim^2_{n \to t} f^n(x)$ exists.

Now, given $x, x', t, t' \in \mathbb{Z}_2$ such that $|x - x'|_2 \le 2^{-n}$, $|t - t'|_2 \le 2^{-m}$ we see that $|f^t(x) - f^{t'}(x')|_2 \le 2^{-k}$ for $k = \min\{n, m\}$ since in this case $t \equiv t' \equiv r \pmod{2^k}$ and $x \equiv x' \equiv z \pmod{2^k}$ for suitable $r, z \in \{0, 1, \ldots, 2^k - 1\}$; so $f^t(x) \equiv f^{t'}(x') \equiv f^r(z) \pmod{2^k}$ as z lies on a cycle of length 2^ℓ, $\ell \le k$, of the permutation $f \bmod 2^k$. □

To conclude this subsection, we note in connection with Example 4.89 that for applications to e.g. numerical analysis (and computer modeling) it is important in some cases to have explicit expressions $f^t(x)$, for not to make all the iterations from the very first point but immediately start with the point at the moment t, for a certain t. So we formulate (somewhat informally) an open question:

Open Question 4.91. Find explicit representations for $f^t(x)$ via continuous p-adic time t for 1-Lipschitz ergodic transformations f on $\mathbb{Z}_p$ other than affine ones.

4.8.2 Non-minimal dynamics. Non-compatible dynamics. Mixing

Non-minimal 1-Lipschitz dynamics

In this chapter we were mainly interested in ergodicity of 1-Lipschitz transformations on $\mathbb{Z}_p$; recall that for 1-Lipschitz measure-preserving transformations on $\mathbb{Z}_p$ ergodicity is equivalent to minimality, see Theorem 4.25. We focused on ergodicity since it is important theoretically, as well for numerous applications in computer science and cryptology; however, non-minimal 1-Lipschitz transformations are also interesting both from theoretical and applied viewpoints. It would be highly interesting to determine ergodic components of a non-ergodic 1-Lipschitz transformation on $\mathbb{Z}_p$. Loosely speaking, this problem for non-minimal (and even non-measure preserving) 1-Lipschitz transformations on $\mathbb{Z}_p$ is equivalent to the question how to determine behavior of an arbitrary 1-Lipschitz transformation modulo p^n and to study how this behavior changes as n goes to infinity. This turned out to be a complicated question, no answer for a general case is known at the time being. A work in this direction was started in [101]; we also note recent works [130, 131] and references therein.

p^n-Lipschitz dynamics and mixing

In connection with the study of ergodicity of 1-Lipschitz transformations on $\mathbb{Z}_p$ in this chapter, it is reasonable to put a question on *mixing*. Recall that a μ-preserving map $f: \mathbb{S} \to \mathbb{S}$ on a measurable space with a measure μ is called mixing, see [276], whenever given two measurable subsets A, $B \subset \mathbb{S}$, $\lim_{n\to\infty} \mu(f^{-n}(A) \cap B) = \mu(A)\mu(B)$. A mixing map is necessarily ergodic.

None of the 1-Lipschitz ergodic maps $f: \mathbb{Z}_p \to \mathbb{Z}_p$ are mixing; moreover, their entropy is always 0 since they are conjugate to a translation $x \mapsto x+1$ on $\mathbb{Z}_p$, see Theorem 4.25. However, among p-Lipschitz maps mixing ones clearly exist; e.g., the Bernoulli shift $s: x \mapsto \lfloor \frac{x}{p} \rfloor$, $x \in \mathbb{Z}_p$, see [262] on mixing transformations on $\mathbb{Z}_p$. However, not every mixing map $f: \mathbb{Z}_p \to \mathbb{Z}_p$ is good for applications for, e.g., pseudorandom generation: If we apply the p-Bernoulli shift to an element from a finite residue ring $\mathbb{Z}_p/p^n\mathbb{Z}$, the corresponding trajectory becomes 0 after at most n iterations; so by no meaning the corresponding sequence of iterates $x, s(x), s^2(x), \dots$ on $\mathbb{Z}/p^n\mathbb{Z}$ can be considered as random-looking. Actually in applications to pseudorandom generation we need only those maps $f: \mathbb{Z}_p \to \mathbb{Z}_p$ that induce on every sufficiently large ring $\mathbb{Z}/p^n\mathbb{Z}$ a transformation with a long cycle, so long that a probability to fall onto a short cycle is negligible. Here by induced map $f \bmod p^n : \mathbb{Z}/p^n\mathbb{Z} \to \mathbb{Z}/p^n\mathbb{Z}$ we meaning a map $(f \bmod p^n)(x) = f(x) \bmod p^n$ when x runs over the numbers $0, 1, \dots, p^n - 1$. Note that as now we do not assume that f is compatible, cases when simultaneously $f(x) \not\equiv f(y) \pmod{p^n}$ and $x \equiv y \pmod{p^n}$, $x, y \in \mathbb{Z}_p$, may occur. We say temporarily that the map $f \bmod p^n$ (though non-compatible) is bijective modulo p^n whenever $x \mapsto f(x) \bmod p^n$ is a permutation of $0, 1, \dots, p^n - 1$. A natural question arises, what are (non-compatible) maps $f: \mathbb{Z}_p \to \mathbb{Z}_p$ that are bijective modulo p^n (in the above meaning) for all $n = 1, 2, \dots$. The following result was obtained by I. Yurov in [418]:

Theorem 4.92. *A non-compatible uniformly differentiable map $f: \mathbb{Z}_p \to \mathbb{Z}_p$ is bijective modulo p^n for all $n = 1, 2, \dots$, if and only if $p = 2$ and $f(x) = g\big(\frac{x\cdot(x+1)}{2}\big)$, where $g: \mathbb{Z}_2 \to \mathbb{Z}_2$ is a 1-Lipschitz measure-preserving transformation on $\mathbb{Z}_2$.*

Note that by Theorem 4.92 all non-compatible (thus, non 1-Lipschitz) maps f that are bijective modulo p^n for all n are then necessarily 2-Lipschitz. Note also that some properties of the pseudorandom generator with the recursion law $x_{i+1} = \frac{x_i \cdot (x_i+1)}{2} \bmod 2^n$ were studied in [412].[5]

In connection with Theorem 4.92, the following question naturally arises:

Open Question 4.93. Does there exist a polynomial g over $\mathbb{Z}_2$ such that the composition $f(x) = g(\frac{x\cdot(x+1)}{2})$ is transitive modulo 2^n, for all n?[6]

For applied purposes, g may be not necessarily a polynomial, but a (not too complicated) analytic function, or $\mathcal{A}$-function as well. Numerical experiments show that such a polynomial g exists for $n \leq 20$.

[5] Actually authors studied a generator with the recursion law $x_{i+1} = \frac{x_i \cdot (x_i-1)}{2} \bmod 2^n$ on numbers $1, 2, \ldots, 2^n - 1, 2^n$, assuming $2^n \bmod 2^n = 2^n$, so there is no much difference.

[6] That is, the permutation $x \mapsto g(\frac{x\cdot(x+1)}{2}) \bmod 2^n$ of numbers $0, 1, \ldots, 2^n - 1$ consists of the only cycle of length 2^n.

Chapter 5

Asymptotic distribution of cycles

As was pointed out, the presence of the parameter p – taking prime values $p = 2, 3, 5, \ldots, 1997, 1999, \ldots$ – is one of the most distinguishing features of the theory of p-adic dynamical systems. As we have seen, the ergodic behavior of such systems depends crucially on this parameter. In this chapter we shall study the dependence of the number of cycles of the fixed length on p. This behavior is characterized by a high degree of stochasticity. Therefore one may expect to obtain definite values only in average with respect to p – by using Dirichlet's mean value (which is well known in number theory). We shall also study in detail the structure of the set of cyclic points and their character for the fixed field of p-adic numbers. The structure of cycles plays an important role in, e.g., applications to cognitive science and genetics, see Chapters 14, 16. Cycles can be used for encoding of ideas in models of thinking on p-adic (and more general m-adic) trees. It is interesting to know dependence of the structure of cycles (a special class of ideas) on the parameter p which can be used, e.g., as the basis of the frequency encoding of information.

We again consider monomial dynamical systems. These systems have been studied in [214, 216, 254–257, 345, 346, 385] and corresponding random dynamical systems (random combination of iterations of various monomial systems) in [5] and [256].

We shall also point out recent publications of Vladimir Arnold [37–39] devoted to chaotic aspects of theory of dynamical systems in finite fields and rings. These publications attracted a lot of attention, see, e.g., [379] on critical analysis of Arnold's papers. One of the problems studied by Arnold has some relation to our studies of monomial dynamical systems. He studied the following problem in the residue ring $\mathbb{Z}/l\mathbb{Z}$ modulo l and made a number of conjectures about the length of the orbits.

Take an integer $g \geqslant 2$ with $\gcd(g, l) = 1$. Arnold studied the dynamical properties of the residue $g^m \bmod l$. Denote by $t_q(l)$ the multiplicative order of g modulo l. It was suggested [39] that for $g = 2$ the average multiplicative order

$$T_g(l) = \frac{1}{L} \sum_{l=1:(g,l)=1}^{L} t_g(l)$$

grows as

$$T_g(l) \sim c(g) \frac{L}{\log L} \tag{5.1}$$

with some constant $c(g)$ depending only on g.

However, Shparlinski noticed [379] that the classical result of Hooley on Artin's conjecture implies, under the *Extended Riemann Hypothesis, that the conjecture* (5.1) *is wrong* and in fact

$$T_g(l) \geqslant c(g)\frac{L}{\log L}\exp C(g)(\log\log\log L)^{3/2} \tag{5.2}$$

with some constant $C(g)$ depending only on g.

In [379] one can find extended bibliography related to this problem. We do not go deeper into details, since we study another type of average, namely, the average of the number of cycles of a *fixed length* r. This average is not unbounded and it has the definite limit for $L \to \infty$.

5.1 Monomial systems in $\mathbb{C}_p$ and in finite extensions of $\mathbb{Q}_p$

We shall consider some results for the dynamical system $p(x) = x^n$, $n = 2, 3, \ldots$, in $\mathbb{C}_p$. Recall that $\mathbb{C}_p$ is the completion of the algebraic closure of $\mathbb{Q}_p$. To find the fixed points we have to solve the equation $p(x) = x$. It is easy to see that 0 is a fixed point to $p(x)$ and $A(0, \mathbb{C}_p) = B_1^-(0, \mathbb{C}_p)$. Further, $A(\infty, \mathbb{C}_p) = \mathbb{C}_p \setminus B_1(0, \mathbb{C}_p)$. So the other fixed points are elements in $S_1(0, \mathbb{C}_p)$ and are roots of unity.

We denote by $\Gamma^{(n)}$ the set of all nth roots of unity in $\mathbb{C}_p$ and define the following subsets in $\mathbb{C}_p$,

$$\Gamma_n = \bigcup_{j=1}^{\infty} \Gamma^{(n^j)} \quad \text{and} \quad \Gamma_u = \bigcup_{(n,p)=1} \Gamma_n.$$

Each $\Gamma^{(n)}$ contains a primitive nth root of unity, since each $\Gamma^{(n)}$ is a cyclic group under multiplication. The set Γ_n contains therefore an infinite number of primitive roots of unity which are not elements of $\mathbb{Q}_p$. So $\mathbb{Q}_p(\Gamma_n)$ must be an infinite field extension of $\mathbb{Q}_p$. If E is a finite field extension of $\mathbb{Q}_p$ then $\Gamma_n \setminus E \neq \varnothing$.

Lemma 5.1. *If $x, y \in \Gamma_u$, $x \neq y$, then $|x - y|_p = 1$.*

Proof. Let $\xi \in \Gamma_u \cap B_1^-(1, \mathbb{C}_p)$ be an nth root of unity, $\gcd(n, p) = 1$. Then it exists an element $\gamma \in B_1^-(0, \mathbb{C}_p)$ such that $\xi = 1 + \gamma$. Hence, from $1 = \xi^n = (1+\gamma)^n = 1 + \binom{n}{1}\gamma + \binom{n}{2}\gamma^2 + \cdots + \binom{n}{n}\gamma^n$ it follows that

$$|\gamma|_p |\tbinom{n}{1} + \tbinom{n}{2}\gamma + \cdots + \tbinom{n}{n}\gamma^{n-1}|_p = 0.$$

But $|\binom{n}{1}|_p = 1$ and $|\binom{n}{2}\gamma + \cdots + \binom{n}{n}\gamma^{n-1}|_p < 1$, so by the isosceles triangle principle $|\binom{n}{1} + \binom{n}{2}\gamma + \cdots + \binom{n}{n}\gamma^{n-1}|_p = 1$. Thus, $\gamma = 0$, that is, $\xi = 1$ and therefore is $\Gamma_u \cap B_1^-(1, \mathbb{C}_p) = \{1\}$. This proves that if $x \in \Gamma_u$, $x \neq 1$, then $|1 - x|_p = 1$, because $|1 - x|_p \leqslant \max\{|1|_p, |x|_p\} = 1$. Let $x, y \in \Gamma_u$, $x \neq y$. Then there exist positive integers m and n such that $x^m = 1$, $y^n = 1$, $\gcd(m, p) = 1$ and $\gcd(n, p) = 1$. Since $\gcd(mn, p) = 1$ we have that $y/x \in \Gamma_u$ and therefore $|x - y|_p = |x|_p \cdot |1 - y/x|_p = 1$. □

It is clear that $B_1^-(1, \mathbb{C}_p) \subset S_1(0, \mathbb{C}_p)$. Lemma 5.1 says that if $x, y \in \Gamma_u$ then the open balls $B_1^-(x, \mathbb{C}_p)$ and $B_1^-(y, \mathbb{C}_p)$ are disjoint. It can be shown (see Schikhof [374], Lemma 33.2, p. 103) that each coset of $B_1^-(0, \mathbb{C}_p)$ in $S_1(0, \mathbb{C}_p)$ contains exactly one element of Γ_u.

Let E be a finite field extension of $\mathbb{Q}_p$ and $\xi \in \Gamma_u$. To prove that $B_1^-(\xi, \mathbb{C}_p) \cap E = \emptyset$, we use the *Teichmüller character*, which is defined as

$$\omega_p \colon S_1(0, \mathbb{C}_p) \to \Gamma_u \quad \text{where} \quad \omega_p(x) = \lim_{n \to \infty} x^{p^{n!}}.$$

The Teichmüller character ω_p maps an element $x \in S_1(0, \mathbb{C}_p)$ into the unique element $\xi \in \Gamma_u$ for which $|\xi - x|_p < 1$ (see Schikhof [374], pp. 103–104). Let $x \in S_1(0, \mathbb{C}_p)$. Then, the sequence $x, x^p, x^{p^{2!}}, x^{p^{3!}}, \ldots$ is a Cauchy sequence.

Lemma 5.2. *Let E be finite field extension of $\mathbb{Q}_p$ and $\xi \in \Gamma_u \setminus E$. Then*

$$B_1^-(\xi, \mathbb{C}_p) \cap E = \emptyset.$$

Proof. Suppose $B_1^-(\xi, \mathbb{C}_p) \cap E \neq \emptyset$ and let $x \in B_1^-(\xi, \mathbb{C}_p) \cap E$. Since E is a field we have that $x^{p^{n!}} \in E$ for all positive integers n and therefore $\omega_p(x) \in E$, since E is complete. But $\omega_p(x) = \xi$, so we have a contradiction. □

There are two main categories of the dynamical systems $x \mapsto x^n$ in $\mathbb{C}_p$; $p \mid n$ and $p \nmid n$. First, let us consider the case when $p \nmid n$. In [214] we find the following theorem.

Theorem 5.3. *Suppose that $p \nmid n$. Then, the dynamical system $p(x) = x^n$ has $n - 1$ fixed points $\xi_{j,n-1}$, $j = 1, 2, \ldots, n - 1$, on the sphere $S_1(0, \mathbb{C}_p)$ and all these points are centers of Siegel disks. Moreover, $\mathrm{SI}(\xi_{j,n-1}) = B_1^-(\xi_{j,n-1})$. If $n - 1 = p^l$ for some positive integer l then $\mathrm{SI}(\xi_{j,n-1}) = \mathrm{SI}(1, \mathbb{C}_p)$ for all j, $1 \leqslant j \leqslant n - 1$. If instead $p \nmid n - 1$ then $\xi_{j,n-1} \in S_1(1)$ and $\mathrm{SI}(\xi_{j,n-1}) \cap \mathrm{SI}(\xi_{i,n-1}) = \emptyset$ if $j \neq i$.*

Let us now consider the case when $p \mid n$. The next two theorems are proved in [214].

Theorem 5.4. *The dynamical system $p(x) = x^n$ has $n - 1$ fixed points $\xi_{j,n-1}$, $j = 1, 2, \ldots, n - 1$, on the sphere $S_1(0, \mathbb{C}_p)$. These points are attractors and $B_1^-(\xi_{j,n-1}, \mathbb{C}_p) \subset A(\xi_{j,n-1}, \mathbb{C}_p)$. For any $k = 2, 3, \ldots$, all k-cycles are also attractors and open unit balls are contained in basins of attraction.*

Theorem 5.5. *For the dynamical system $p(x) = x^n$, where $n = mp^k$, $\gcd(m, p) = 1$ and $k \geqslant 1$, the basin of attraction of 1 is*

$$A(1, \mathbb{C}_p) = \bigcup_{\xi} B_1^-(\xi, \mathbb{C}_p), \quad \xi \in \Gamma_m.$$

The open balls $B_1^-(\xi, \mathbb{C}_p)$ have empty intersection for different points ξ.

Corollary 5.6. *Let E be a finite field extension of $\mathbb{Q}_p$ and e the ramification index of E over $\mathbb{Q}_p$. For the dynamical system $p(x) = x^n$, where $n = mp^k$, $\gcd(m, p) = 1$ and $k \geqslant 1$, the basin of attraction of 1 is*

$$A(1, E) = \bigcup B_{p^{-1/e}}(\xi, E), \quad \xi \in \Gamma_m \cap E.$$

Proof. It is a direct consequence of Lemma 5.2 and Theorem 5.5. □

From now on, let E be a finite field extension of $\mathbb{Q}_p$ and e the ramification index of E over $\mathbb{Q}_p$. The image of ord_p is the set

$$\left\{0, \pm\frac{1}{e}, \pm\frac{2}{e}, \dots, \pm\frac{e-1}{e}, \pm 1, \pm\frac{e+1}{e}, \dots\right\}.$$

Let $x \in S_1(0, E)$ and $\gamma \in B_{p^{-1/e}}(0, E)$. Lemma 3.6 implies that

$$\mathrm{ord}_p\, k! \leqslant k - 1 \tag{5.3}$$

with strict inequality for $p > 2$. Thus

$$\left|\frac{1}{k!}\right|_p = p^{\mathrm{ord}_p k!} \leqslant p^{k-1}.$$

Since $|\gamma|_p \leqslant p^{-1/e}$, it follows that

$$\left|\frac{\gamma^{k-1}}{k!}\right|_p \leqslant p^{-(k-1)/e} p^{k-1} = p^{(k-1)(e-1)/e}.$$

Then for $1 \leqslant k \leqslant n$

$$\begin{aligned}
\left|\tbinom{n}{k}\right|_p |\gamma|_p^k &= |n(n-1)\cdots(n-k+1)|_p |\gamma|_p \left|\frac{\gamma^{k-1}}{k!}\right|_p \\
&\leqslant p^{(k-1)(e-1)/e} |n(n-1)\cdots(n-k+1)|_p |\gamma|_p \\
&\leqslant p^{(k-1)(e-1)/e} |n|_p |\gamma|_p.
\end{aligned}$$

Especially, if $e = 1$, $p = 2$ and n is an odd integer then we have that $\left|\binom{n}{k}\right|_2 |\gamma|_2 < |n|_2|\gamma|_p$, since $|n|_2 = 1$ and $|n-1|_2 < 1$. Finally,

$$|(x+\gamma)^n - x^n|_p = \left|\sum_{k=1}^{n} \binom{n}{k} x^{n-k}\gamma^k\right|_p$$
$$\leqslant \max_{1\leqslant k\leqslant n} \left\{\left|\binom{n}{k}\right|_p |\gamma|_p^k\right\} \leqslant p^{(n-1)(e-1)/e}|n|_p|\gamma|_p.$$

If $e = 1$, that is, E is an unramified field extension of $\mathbb{Q}_p$, and if $p > 2$ or if $p = 2$ when n is an odd integer then we have equality, by the isosceles triangle principle and (5.3). If $e > 1$ then we have strict inequality for all p. But this is not a good estimate of $(x+\gamma)^n - x^n$ when E is a ramified field extension of $\mathbb{Q}_p$.

Lemma 5.7. *Let $x \in S_1(0, E)$ and $\gamma \in E$. Then*

$$|(x+\gamma)^n - x^n|_p \leqslant |\gamma|_p \max\{|n|_p, |\gamma|_p\}. \tag{5.4}$$

If E is an unramified field extension of $\mathbb{Q}_p$ and $\gamma \in B_{p^{-1/e}}(0, E)$ then

$$|(x+\gamma)^n - x^n|_p \leqslant |n|_p|\gamma|_p,$$

with equality for $p > 2$ or for $p = 2$ when n is an odd integer.

Proof. It remains to show inequality (5.4). We have that

$$|(x+\gamma)^n - x^n|_p = \left|\binom{n}{1}x^{n-1}\gamma + \binom{n}{2}x^{n-2}\gamma^2 + \cdots + \binom{n}{n}\gamma^n\right|_p$$
$$= |\gamma|_p \left|\binom{n}{1}x^{n-1} + \gamma\left[\binom{n}{2}x^{n-2} + \cdots + \binom{n}{n}\gamma^{n-2}\right]\right|_p.$$

Moreover, $\left|\binom{n}{1}x^{n-1}\right|_p = |n|_p$, $|\gamma|_p \left|\binom{n}{2}x^{n-2} + \cdots + \binom{n}{n}\gamma^{n-2}\right|_p \leqslant |\gamma|_p$ and by the strong triangle inequality, inequality (5.4) is proved. □

5.2 Number of cycles of $x \mapsto x^n$ in $\mathbb{Q}_p$

In this section we will study the dynamical system (4.9) over $\mathbb{Q}_p$. From the former section we know that 0 and ∞ are attractive fixed points, $A(0) = B_1(0, \mathbb{Q}_p)$ and $A(\infty) = \mathbb{Q}_p \setminus B_1(0, \mathbb{Q}_p)$. All other periodic points are located on $S_1(0, \mathbb{Q}_p)$.

Fixed points of (4.9) on $S_1(0)$ are solutions of the equation $x^{n-1} = 1$, hence they are $(n-1)$th roots of unity. Periodic points, of period r, are solutions of the equation

$$x^{n^r-1} = 1 \tag{5.5}$$

and are therefore $(n^r - 1)$th roots of unity. It follows directly from the definition of the periodic points that the set of solutions of equation (5.5) not only contains the periodic points of period r but also the periodic points with periods that divide r.

We use (m, n) to denote the greatest common divisor of two positive integers m and n. The following fact follows directly from theorems of Section 3.4 in Chapter 3.

Theorem 5.8. *The equation $x^l = 1$ has $(l, p-1)$ solutions in $\mathbb{Q}_p$ for $p > 2$. If $p = 2$ then $x^l = 1$ has two solutions ($x = 1$ and $x = -1$) if l is even and one solution ($x = 1$) if l is odd.*

Corollary 5.9. *The only roots of unity in $\mathbb{Q}_p$ are the $(p-1)$th roots of unity.*

We also mention some other facts about the roots of unity in $\mathbb{Q}_p$.

Lemma 5.10. *If $p \nmid n$ and x and y are nth roots of unity, $x \neq y$, then $|x - y|_p = 1$.*

Proof. Since $|x|_p = |y|_p = 1$, it is clear that $|x - y|_p \leqslant 1$. Assume that $|x - y|_p < 1$. Then there is z such that $|z|_p < 1$ and $x = y + z$. We have

$$0 = |x^n - y^n|_p = |(y+z)^n - y^n|_p = \left| \sum_{j=1}^{n} \binom{n}{j} y^{n-j} z^j \right|_p$$

$$= |z|_p \left| n y^{n-1} + z \sum_{j=2}^{n} \binom{n}{j} y^{n-j} z^{j-2} \right|_p .$$

Because of the fact that $|ny^{n-1}|_p = 1$ and that $|\sum_{j=2}^{n} \binom{n}{j} y^{n-j} z^{j-2}|_p \leqslant 1$, we have that

$$\left| n y^{n-1} + z \sum_{j=2}^{n} \binom{n}{j} y^{n-j} z^{j-2} \right|_p = 1$$

from Theorem 1.36. We must then have $|z|_p = 0$ so $z = 0$. This implies that $x = y$, which is a contradiction. This gives us $|x - y|_p = 1$ and the theorem is proved. □

Corollary 5.11. *If $p \nmid n$, $x \neq 1$ and $x^n = 1$ then $|x - 1|_p = 1$. Thus $x \in S_1(1)$.*

Proof. Just set $y = 1$ in the theorem above. □

Theorem 5.12. *Let x and y be two nth roots of unity in $\mathbb{Q}_p$ and let $x \neq y$. If $p > 2$ then $|x - y|_p = 1$. If $p = 2$ then $|x - y|_2 = 1/2$.*

Proof. If $p > 2$ then any nth root of unity in $\mathbb{Q}_p$ is a $(p-1)$th root of unity, see Corollary 5.9. Since $p \nmid p - 1$ it follows from Lemma 5.10 that $|x - y|_p = 1$. If $p = 2$ the only possibility that $x \neq y$ is that $x = 1$ and $y = -1$ (or vice versa). Hence $|1 - (-1)|_2 = |2|_2 = 1/2$. □

Let $N(n, r, p)$ denote the number of periodic points of period r of (4.9) on $S_1(0) \subseteq \mathbb{Q}_p$. We know that each r-cycle contains r r-periodic points. If we denote by $\mathcal{N}(n,r,p)$ the number of r-cycles in $S_1(0) \subseteq \mathbb{Q}_p$, then

$$\mathcal{N}(n, r, p) = N(n, r, p)/r. \tag{5.6}$$

In [214] we find the following theorem about the existence of r-cycles.

Theorem 5.13. *Let $p > 2$ and let $m_j = (n^j - 1, p - 1)$. The dynamical system $f(x) = x^n$ has r-cycles ($r \geqslant 2$) in $\mathbb{Q}_p$ if and only if m_r does not divide any m_j, $1 \leqslant j \leqslant r - 1$.*

Proof. Let us assume that $m_r \nmid m_j$ for $1 \leqslant j \leqslant r - 1$. Consider the equation

$$x^{n^r - 1} = 1. \tag{5.7}$$

According to Theorem 5.8 this equation has m_r roots in $\mathbb{Q}_p$. Hence, all solutions of (5.7) are solutions of

$$x^{m_r} = 1.$$

Let $a_1 = \xi_{m_r}$ be a m_rth primitive root of unity. The sequence

$$(a_1, a_1^n, a_1^{n^2}, \ldots, a_1^{n^{r-1}}) \tag{5.8}$$

is a cycle whose length divides r. We now prove that the length of the sequence in (5.8) is actually r. Suppose that this is a cycle of length s, where $s < r$ (and $s \mid r$). We then have $a_1^{n^s} = a_1$ and $a_1^{n^s - 1} = 1$. The equation $x^{n^s - 1} = 1$ has m_s roots in $\mathbb{Q}_p$ and these roots satisfy $x^{m_s} = 1$. Since a_1 is a primitive m_rth root of unity we must have $m_r \mid m_s$, but this is a contradiction to our assumption.

Let us now assume that m_r divides some m_j, $1 \leqslant j \leqslant r - 1$. We want to prove that there are no cycles of length r. Suppose that there exists $b \in S_1(0)$ such that $b^{n^r - 1} = 1$. This equation has m_r solutions in $\mathbb{Q}_p$, therefore $b^{m^r} = 1$. The fact that m_r divides m_j implies that $b^{m_j} = 1$ and that $b^{n^j - 1} = 1$, since $m_j \mid b^{n^j - 1}$. We can make the conclusion that there are no cycles of length r. □

We have the following relation between m_j, $N(n, j, p)$ and $\mathcal{N}(n, j, p)$

$$m_j = \sum_{i \mid j} N(n, i, p) = \sum_{i \mid j} i \mathcal{N}(n, i, p). \tag{5.9}$$

When considering the phenomena involving p-adic numbers, the case $p = 2$ is often the odd man out. Let us consider this case.

Theorem 5.14. *The dynamical system $f(x) = x^n$ over $\mathbb{Q}_2$ has no cycles of order $r \geqslant 2$.*

Proof. If n is even then it follows from Theorem 5.8 that (4.9) has only one fixed point in $\mathbb{Q}_2$. It also follows that n^r is even for all $r \geqslant 2$ and this implies that $f^r(x) = x^{n^r}$ only has one fixed point in $\mathbb{Q}_2$ which also is the fixed point of $f(x) = x^n$. Hence f has no periodic points of period r. The case when n is odd is studied in a similar way. □

We are now ready to derive a formula for the number of periodic points of the monomial system (4.9). Observe that according to Theorem 5.8 we have for $p > 2$ that $(n^r - 1, p - 1)$ gives the number of periodic points of period r and periods that divide r. We have the following theorem.

Theorem 5.15. *Assume that $p > 2$. Then the number of r-periodic points of (4.9) in $S_1(0)$ is given by*

$$N(n, r, p) = \sum_{d|r} \mu(d)(n^{r/d} - 1, p - 1). \tag{5.10}$$

Proof. The theorem follows directly from the Möbius inversion formula and (5.9). □

The number of cycles of length r of (4.9) is given by

$$\mathcal{N}(n, r, p) = \frac{N(n, r, p)}{r} = \frac{1}{r} \sum_{d|r} \mu(d)(n^{r/d} - 1, p - 1). \tag{5.11}$$

Remark 5.16. If we assume that $r \geqslant 2$ then by Theorem 5.14, $N(n, r, 2) = 0$. If $p = 2$ in (5.10) we get that $N(n, r, 2) = 0$. Hence, we can use formula (5.10) also for $p = 2$ if $r \geqslant 2$.

Remark 5.17. Formula (5.11) implies the following result which may be interesting in number theory: For every natural number $n \geqslant 2$ and prime number $p > 2$ the number $\sum_{d|r} \mu(d)(n^{r/d} - 1, p - 1)$ is divisible by r.

5.3 Total number of cycles

In this section we will determine the total number of cycles of a monomial dynamical system in $\mathbb{Q}_p$ for a fixed p. Let $n \geqslant 2$ be a natural number. Denote by $p^*(n)$ the number we obtain if we remove the factors dividing n from the factorization of $p - 1$. That is, $p^*(n)$ is the largest divisor of $p - 1$ which is relatively prime to n.

Lemma 5.18. *We have for each $r \in \mathbb{N}$*

$$(n^r - 1, p - 1) = (n^r - 1, p^*(n)). \tag{5.12}$$

Proof. Since $n^r - 1 \equiv -1 \pmod q$ if $q \mid n$, we can remove the prime factors from $p - 1$ that divide n without changing the value of $(n^r - 1, p - 1)$. □

Lemma 5.19. *Let $(q, n) = 1$. Then there exists a least positive integer $\bar{r}$ such that $n^{\bar{r}} \equiv 1 \pmod q$ and if $n^r \equiv 1 \pmod q$ then $\bar{r} \mid r$.*

Proof. Since $(q, n) = 1$ it follows from Theorem 1.10 that $n^{\varphi(q)} \equiv 1 \pmod{q}$. It is clear that there exists a least $\bar{r}$ such that $n^{\bar{r}} \equiv 1 \pmod{q}$ and $\bar{r} \leqslant \varphi(q)$. There are numbers a and b, such that $r = a\bar{r} + b$, and $b < \bar{r}$. If we assume that $n^r \equiv 1 \pmod{q}$, we have the following relation

$$1 \equiv n^r \equiv n^{a\bar{r}+b} \equiv n^b.$$

Since $\bar{r}$ was the least positive integer such that $n^{\bar{r}} \equiv 1 \pmod{q}$ we have $b = 0$ and hence $\bar{r} \mid r$. □

Lemma 5.20. *There is a least integer $\hat{r}(n)$, such that*

$$(n^{\hat{r}(n)} - 1, p^*(n)) = p^*(n).$$

Proof. By Lemma 5.19 there is a least integer $\hat{r}(n)$ such that $n^{\hat{r}(n)} \equiv 1 \pmod{p^*(n)}$. Hence $p^*(n) \mid n^{\hat{r}(n)} - 1$ and the theorem is proved. □

Theorem 5.21. *Let $p > 2$ be a fixed prime number, let $n \geqslant 2$ be a natural number. If $R \geqslant \hat{r}(n)$ then*

$$\sum_{r=1}^{R} N(n, r, p) = p^*(n). \tag{5.13}$$

Proof. We first prove that $N(n, r, p) = 0$ if $r > \hat{r}(n)$. Since $(n^{\hat{r}(n)} - 1, p - 1) = p^*(n)$ and every $m_r = (n^r - 1, p - 1) \mid p^*(n)$, $r > \hat{r}(n)$, by Theorem 5.13 $N(n, r, p) = 0$.

Next we want to prove that if $r \nmid \hat{r}(n)$ then $N(n, r, p) = 0$. Let l_1 be a divisor of $p^*(n)$. Let q be the least integer such that $n^q - 1 \equiv 0 \pmod{l_1}$. Since $n^{\hat{r}(n)} \equiv 1 \pmod{p^*(n)}$ we have $n^{\hat{r}(n)} \equiv 1 \pmod{l_1}$. By Lemma 5.19 we obtain $q \mid \hat{r}(n)$.

The only possible values of $(n^r - 1, p - 1)$ are the divisors of $p^*(n)$. In the above paragraph we have shown that the least number q for which we have $(n^q - 1, p - 1) = l_1$ and $l_1 \mid p^*(n)$, must be a divisor of $\hat{r}(n)$. Hence if $r \nmid \hat{r}(n)$ then $N(n, r, p) = 0$.

So far we have proved that

$$\sum_{r=1}^{R} N(n, r, p) = \sum_{r \mid \hat{r}(n)} N(n, r, p).$$

It remains to prove that

$$\sum_{r \mid \hat{r}(n)} N(n, r, p) = p^*(n).$$

From (5.9) we know that

$$(n^r - 1, p^*(n)) = \sum_{d \mid r} N(n, d, p).$$

By setting $r = \hat{r}(n)$ we finish the proof of the theorem. □

Corollary 5.22. *Let $p > 2$. The dynamical system* (4.9) *has $p^*(n)$ periodic points on $S_1(0) \subseteq \mathbb{Q}_p$.*

Theorem 5.23. *Let $p > 2$. The total number, $\mathcal{N}_{\mathrm{Tot}}(n, p)$, of cycles of* (4.9) *on $S_1(0) \subseteq \mathbb{Q}_p$ is given by*

$$\mathcal{N}_{\mathrm{Tot}}(n, p) = \sum_{r|\hat{r}} \mathcal{N}(n, r, p) = \sum_{r|\hat{r}} \frac{1}{r} \sum_{d|r} \mu(d)(n^{r/d} - 1, p - 1). \tag{5.14}$$

Proof. From the proof of Theorem 5.21 we know that there are only cycles of lengths that divide $\hat{r}(n)$. The rest follows from (5.11). □

Example 5.24. Let us consider the monomial system $f(x) = x^2$ $(n = 2)$. If $p = 137$ then by Corollary 5.22 the dynamical system has $p^*(2) = 17$ periodic points. By Theorem 5.23 it has $K_{\mathrm{Tot}}(2, 137) = 3$ cycles. In fact, the monomial system $f(x) = x^2$ has one cycle of length 1 (one fixed point) and two cycles of length 8.

If we consider the same system, for $p = 1999$, then the total number of periodic points is $p^*(2) = 999$ and the total number of cycles is $K_{\mathrm{Tot}}(2, 1999) = 31$. In fact, the system has one cycle of length 1, 2, 6 and 18 and also 27 cycles of length 36.

Example 5.25. Let us now consider the dynamical system $f(x) = x^3$. If $p = 137$ then there are 136 periodic points and 13 cycles. In fact, there are two fixed points, three cycles of length 2 and 8 cycles of length 16. If $p = 1999$ then there are two fixed points and four cycles of length 18, so there are 74 periodic points and six cycles.

5.4 Possible values of the number of cycles

In this chapter we use probabilistic methods to study the behavior of cycles in $\mathbb{Q}_p$ for $p \to \infty$. By calculating the average $p \to \infty$ we obtain some number theoretical relations. The result presented in this section can also be obtained by algebraic methods, see [257].

Let n and r be given integers $n, r \geqslant 2$. Let $s(n, r, p) = (n^r - 1, p - 1)$. It is clear that the values $s(n, r, p)$ can attain are divisors of $n^r - 1$. The number of possible values of $s(n, r, p)$ is, of course, less or equal to the number of positive divisors of $n^r - 1$. Henceforth we will denote by $\tau(m)$, the number of positive divisors of m.

Lemma 5.26. *If $d \mid r$ then $n^{r/d} - 1 \mid n^r - 1$.*

Proof. Let $k = r/d$, then we can write $n^r - 1 = n^{dk} - 1$. Since

$$(n^k - 1) \sum_{j=0}^{d-1} n^{kj} = n^k \sum_{j=0}^{d-1} n^{kj} - \sum_{j=0}^{d-1} n^{kj} = \sum_{j=1}^{d} n^{kj} - \sum_{j=0}^{d-1} n^{kj} = n^{dk} - 1$$

we have $n^k - 1 \mid n^r - 1$. We have proved the lemma. □

Theorem 5.27. *For fixed n and r it is possible to express $\mathcal{N}(n, r, p)$ as a function η of $s(n, r, p)$. In fact,*

$$\mathcal{N}(n, r, p) = \eta(s(n, r, p)) = \frac{1}{r} \sum_{d|r} \mu(d)(n^{r/d} - 1, s(n, r, p)). \tag{5.15}$$

Proof. Lemma 5.26 implies that

$$(n^{r/d} - 1, p - 1) = (n^{r/d} - 1, s(n, r, p))$$

and the theorem follows. □

Of course, the number of possible values of $\mathcal{N}(n, r, p)$ for fixed n and r is finite.

Example 5.28. Let $n = 3$ and $r = 6$. We have $n^r - 1 = 728 = 2^3 \cdot 7 \cdot 13$. Table 5.1 shows the possible values of $s(3, 6, p)$ and $\mathcal{N}(3, 6, p)$. The divisors 7, 13 and 91 of 728 are not possible values of $s(3, 6, p)$, because $p - 1$ is divisible by 2 for every prime $p > 2$. $s(3, 6, p)$ takes value 1 only for $p = 2$.

$s(3, 6, p)$	$\mathcal{N}(3, 6, p)$
1	0
2	0
4	0
14	2
28	4
56	8
26	0
52	4
104	12
182	26
336	56
728	116

Table 5.1. Values of $s(3, 6, p)$ and $\mathcal{N}(3, 6, p)$ for $n = 3$ and $r = 6$.

Example 5.29. Let $n = 2$ and $r = 12$. We then have $n^r - 1 = 4095 = 3^2 \cdot 5 \cdot 7 \cdot 13$. Table 5.2 shows the possible values of $s(2, 12, p)$ and $\mathcal{N}(2, 12, p)$. In this case all the divisors of $n^r - 1$ are possible values of $s(n, r, p)$.

5.5 Probability on the set of prime numbers

In this section we will define an analogue of a probability measure on the set of prime numbers. Let us first recall the definition of a Kolmogorov probability space, see for

$s(2, 12, p)$	$\mathcal{N}(2, 12, p)$
1	0
3	0
5	0
7	0
9	0
13	1
15	0
21	0
35	2
39	3
45	2
63	0

$s(2, 12, p)$	$\mathcal{N}(2, 12, p)$
65	5
91	7
105	6
117	9
195	15
273	21
315	20
455	37
585	47
819	63
1365	111
4095	335

Table 5.2. Values of $s(2, 12, p)$ and $\mathcal{N}(2, 12, p)$ for the case $n = 2$ and $r = 12$.

example [378]. A probability space is a triple $(\Omega, \mathcal{G}, \mathbf{P})$ where Ω is any set and $\mathcal{G}$ is a σ-algebra of subsets of Ω and $\mathbf{P}$ is a σ-additive measure on $\mathcal{G}$ with values in $[0, 1]$.

Let Ω_{prime} denote the set of prime numbers and let P_M be the set of the first M prime numbers. It is natural to define the "probability" of a set $A \in \Omega_{\text{prime}}$ by

$$\mathbf{P}(A) = \lim_{M\to\infty} \frac{|A \cap P_M|}{M}. \tag{5.16}$$

Let $\mathcal{F}$ be the family of subsets $A \subseteq \Omega_{\text{prime}}$ such that the limit in (5.16) exists. The problem is now that if $A, B \in \mathcal{F}$ it is not necessary that $A \cup B \in \mathcal{F}$. Hence $\mathcal{F}$ is not an algebra of sets and definitely not a σ-algebra, see [349] and [242]. Instead we consider the generalized probability space $(\Omega_{\text{prime}}, \mathcal{F}, \mathbf{P})$, see [242] for the general theory.

The absence of the conventional probability measure induces some difficulties. However, some "probabilistic features" are preserved, see the following propositions whose proofs can be found in [242].

Proposition 5.30. *If $A, B \in \mathcal{F}$ and $A \cap B = \varnothing$ then $A \cup B \in \mathcal{F}$ and*

$$\mathbf{P}(A \cup B) = \mathbf{P}(A) + \mathbf{P}(B).$$

Proposition 5.31. *Let $A, B \in \mathcal{F}$. Then the following properties are equivalent:* 1) $A \cup B \in \mathcal{F}$, 2) $A \cap B \in \mathcal{F}$, 3) $A \setminus B \in \mathcal{F}$, *and* 4) $B \setminus A \in \mathcal{F}$. *We also have the following relations*:

$$\mathbf{P}(A \cup B) = \mathbf{P}(A) + \mathbf{P}(B) - \mathbf{P}(A \cap B)$$

and

$$\mathbf{P}(A \setminus B) = \mathbf{P}(A) - \mathbf{P}(A \cap B).$$

Another problem is to define an analogue of a random variable in the case of generalized probability space. We will define it only in a special case, see [242] for the general theory. We first recall that a random variable, see for example [378], on a probability space $(\Omega, \mathcal{G}, \mathbf{P})$ is a measurable function $\xi\colon (\Omega, \mathcal{G}) \to (\mathbb{R}, \mathcal{B})$, where $\mathcal{B}$ is the Borel σ-algebra of $\mathbb{R}$. Let ξ be a mapping from Ω_{prime} to a finite subset $F \in \mathbb{N}$. If $\xi^{-1}(\{x\}) \in \mathcal{F}$ for every $x \in F$, we will call ξ a *random variable*. If ξ is a random variable, then we define the probability that $\xi = x$ as $\mathbf{P}(\xi^{-1}(\{x\}))$. We define the expectation of ξ as

$$\mathbf{E}\xi = \sum_{x \in F} x\mathbf{P}(\xi^{-1}(\{x\})), \tag{5.17}$$

and the variance of ξ as

$$\mathbf{V}\xi = \sum_{x \in F} x^2\mathbf{P}(\xi^{-1}(\{x\})) - (\mathbf{E}\xi)^2. \tag{5.18}$$

It is easy to show that

$$\mathbf{E}\xi = \lim_{M\to\infty} \frac{1}{M} \sum_{p \in P_M} \xi(p) \tag{5.19}$$

and

$$\mathbf{V}\xi = \lim_{M\to\infty} \frac{1}{M} \sum_{p \in P_M} \xi(p)^2 - (\mathbf{E}\xi)^2. \tag{5.20}$$

5.6 Distribution of cycles

For fixed n and r, we consider $\mathcal{N}(n, r, p)$ as a random variable (in the sense of the previous section), $\xi(p)$, on Ω_{prime}. Let us also consider $s(n, r, p)$, for fixed n and r as a random variable, $\zeta(p)$, on Ω_{prime}.

From Section 5.4 we know that ξ only takes a finite number, say γ, of values. Let us denote them by ξ_j, where $1 \leqslant j \leqslant \gamma$. In this section we will compute the probability for ξ having the value ξ_j. Denote the number of prime numbers in P_M such that $d \mid p - 1$ by the symbol $\pi(d, M)$.

Lemma 5.32. *Let n and r be fixed numbers ($n \geqslant 2$ and $r \geqslant 2$). If $A(t, M)$ is the number of primes $p \in P_M$ such that $(n^r - 1, p - 1) = t$ then*

$$A(t, M) = \sum_{k \mid \frac{n^r - 1}{t}} \mu(k)\pi(kt, M). \tag{5.21}$$

Proof. Let $m = n^r - 1$. It is easy to see that

$$\pi(t, M) = \sum_{r \mid \frac{m}{t}} A(rt, M).$$

Since

$$\pi(kt, M) = \sum_{r \mid \frac{m}{kt}} A(rkt, M),$$

the right-hand side of (5.21) can be written

$$\sum_{k \mid \frac{m}{t}} \sum_{r \mid \frac{m}{kt}} \mu(k) A(rkt, M).$$

If $k' = rk$ then

$$\sum_{k \mid \frac{m}{t}} \mu(k)\pi(kt, M) = \sum_{k' \mid \frac{m}{t}} A(k't, M) \sum_{k \mid k'} \mu(k) = A(t, m)$$

by the properties of the Möbius function. □

Theorem 5.33. *Let s_j, $1 \leqslant j \leqslant \tau(n^r - 1)$ be a positive divisor of $n^r - 1$. Then the probability, $\omega(s_j)$, that $\zeta(p) = s_j$ is given by*

$$\omega(s_j) = \sum_{k \mid \frac{n^r - 1}{s_j}} \mu(k) \frac{1}{\varphi(ks_j)}.$$

Proof. Let $A(s_j, M)$ denote the number of prime numbers, $p \leqslant p_M$ such that $\zeta(p) = s_j$. By Lemma 5.32

$$A(s_j, M) = \sum_{k \mid \frac{n^r - 1}{s_j}} \mu(k)\pi(s_j k, M).$$

The probability that $\zeta(p) = s_j$ is given by limit

$$\lim_{M \to \infty} \frac{A(s_j, M)}{M} = \sum_{k \mid \frac{n^r - 1}{s_j}} \mu(k) \lim_{M \to \infty} \frac{\pi(s_j k, M)}{M}.$$

By the prime number theorem for primes in arithmetic progressions, see (1.7),

$$\lim_{M \to \infty} \frac{A(s_j, M)}{M} = \sum_{k \mid \frac{n^r - 1}{s_j}} \mu(k) \frac{1}{\varphi(ks_j)}$$

and the theorem is proved. □

Theorem 5.34. *The probability of $\xi(p) = \xi_i$ is given by*

$$\nu(\xi_i) = \sum_{s_j \in \mathcal{S}_i} \omega(s_j), \tag{5.22}$$

where $\mathcal{S}_i$ is the set of positive divisors x of $n^r - 1$ such that $\eta(x) = \xi_i$.

Proof. The theorem follows directly from Theorem 5.33 and Theorem 5.27. □

Example 5.35. Let $n = 3$ and $r = 6$ then the probabilities of the possible values of $\xi(p)$ is shown in Table 5.3.

ξ_j	$\nu(\xi_j)$
0	$\frac{230}{288}$
2	$\frac{22}{288}$
4	$\frac{16}{288}$
8	$\frac{11}{288}$
12	$\frac{5}{288}$
26	$\frac{2}{288}$
56	$\frac{1}{288}$
116	$\frac{1}{288}$

Table 5.3. Probabilities for $n = 3$ and $r = 6$.

Example 5.36. Let $n = 2$ and $r = 12$. In Table 5.4 we can see the probabilities of the possible values of ξ.

ξ_j	$\nu(\xi_j)$
0	$\frac{1463}{1728}$
1	$\frac{45}{1728}$
2	$\frac{88}{1728}$
3	$\frac{30}{1728}$
5	$\frac{15}{1728}$
6	$\frac{22}{1728}$
7	$\frac{9}{1728}$
9	$\frac{15}{1728}$

ξ_j	$\nu(\xi_j)$
15	$\frac{10}{1728}$
20	$\frac{11}{1728}$
21	$\frac{6}{1728}$
37	$\frac{3}{1728}$
47	$\frac{5}{1728}$
63	$\frac{3}{1728}$
111	$\frac{2}{1728}$
335	$\frac{1}{1728}$

Table 5.4. Probabilities for $n = 2$ and $r = 12$.

5.7 Expectation value and dispersion

In this section we will calculate expectation and variance of ξ. First, we will do this calculations for ζ. The cornerstone of these calculations is the following theorem.

Theorem 5.37. *Let $m \in \mathbb{Z}^+$. Then*

$$\lim_{M\to\infty} \frac{1}{M} \sum_{p\in P_M} (m, p-1) = \tau(m).$$

Proof. With the notations of Lemma 5.32 we have

$$\sum_{p \in P_M} (m, p-1) = \sum_{d \mid m} d A(d, M).$$

According to Lemma 5.32 we have

$$A(d, M) = \sum_{k \mid \frac{m}{d}} \mu(k) \pi(kd, M).$$

This gives us

$$\sum_{p \in P_M} (m, p-1) = \sum_{d \mid m} \sum_{k \mid \frac{m}{d}} d \mu(k) \pi(kd, M)$$

and if we set $t = kd$ then

$$\sum_{p \in P_M} (m, p-1) = \sum_{t \mid m} \pi(t, M) \sum_{k \mid t} \frac{t}{k} \mu(k) = \sum_{t \mid M} \pi(t, M) \varphi(t),$$

according to (1.4). From (1.7) we obtain

$$\lim_{M \to \infty} \frac{1}{M} \sum_{p \in P_M} (m, p-1) = \sum_{t \mid m} \lim_{M \to \infty} \frac{\pi(t, M)\varphi(t)}{M} = \tau(m).$$

□

We set $m = n^r - 1$. By (5.19) we get $\mathbf{E}\zeta = \tau(n^r - 1)$. We are now ready to calculate the expectation value of ξ.

Theorem 5.38. *We have*

$$\mathbf{E}\xi = \lim_{M \to \infty} \frac{1}{M} \sum_{p \in P_M} \xi(p) = \frac{1}{r} \sum_{d \mid r} \mu(d) \tau(n^{r/d} - 1). \tag{5.23}$$

The proof follows immediately from (5.19) and Theorem 5.37 and the fact that

$$\xi(p) = \frac{1}{r} \sum_{d \mid r} \mu(d)(n^{r/d} - 1, p - 1).$$

Example 5.39 (Computer simulation). Let $f(x) = x^2$. We are interested in the number of cycles of length 12 of this system for different primes p. We can use formula (5.10) and plot the number of cycles of length 12 as a function of p. In this way we obtain a graph with a high degree of randomness, see [254, 256]: the number of cycles of this length fluctuates essentially when p increases. However, the asymptotical inclination of the graph can be found numerically and it coincides with the expectation $\frac{1}{12} \sum_{d \mid 12} \mu(d) \tau(2^1 2 - 1)$ given by (5.23).

We calculate the variance of ξ. As in the calculation of $\mathbf{E}\xi$ we first calculate the variance of ζ. In fact, we have the following theorem that is a generalization of Theorem 5.37.

Theorem 5.40. *If m and n are non-negative integers then*

$$\lim_{M\to\infty}\frac{1}{M}\sum_{p\in P_M}(m,p-1)(n,p-1)=\sum_{a|m}\sum_{b|n}\frac{\varphi(a)\varphi(b)}{\varphi(\mathrm{lcm}(a,b))}. \tag{5.24}$$

Proof. We start with some notations. We set

$$B(n,m,M)=\frac{1}{M}\sum_{p\in P_M}(m,p-1)(n,p-1).$$

If $d \mid m$ and $k \mid n$ then $A(d,k,M)$ denotes the number of prime numbers $p\in P_M$ such that $(m,p-1)=d$ and $(n,p-1)=k$. It is easy to see that

$$B(n,m,M)=\sum_{d|m}\sum_{k|n}dkA(d,k,M).$$

Let $\pi(d,k,M)$ be the number of prime numbers $p\in P_M$ such that $d \mid p-1$ and $k \mid p-1$. We have the following relation between π and A:

$$\pi(d,k,M)=\sum_{r|\frac{m}{d}}\sum_{s|\frac{n}{k}}A(dr,ks,M). \tag{5.25}$$

We will now prove that

$$A(d,k,M)=\sum_{r|\frac{m}{d}}\sum_{s|\frac{n}{k}}\mu(r)\mu(s)\pi(dr,ks,M). \tag{5.26}$$

By (5.25)

$$\pi(dr,ks,M)=\sum_{r_1|\frac{m}{dr}}\sum_{s_1|\frac{n}{ks}}A(drr_1,kss_1,M).$$

We can now write the right-hand side of (5.26) as

$$\sum_{\hat r|\frac{m}{d}}\sum_{\hat s|\frac{n}{k}}\sum_{r|\hat r}\sum_{s|\hat s}\mu(r)\mu(s)A(d\hat r,k\hat s,M),$$

where $\hat r = rr_1$ and $\hat s = ss_1$. By the properties of the Möbius function we obtain that the right-hand side of (5.26) is equal to $A(d,k,M)$ which completes the proof of (5.26).

By (5.26) we obtain

$$B(m,n,M)=\sum_{d|m}\sum_{r|\frac{m}{d}}d\mu(r)\sum_{k|n}\sum_{s|\frac{n}{k}}k\mu(s)\pi(dr,ks,M). \tag{5.27}$$

Let $a = dr$ and $b = ks$. Then

$$B(m, n, M) = \sum_{a|m}\sum_{b|n} \pi(a, b, \mathrm{lcm}(a, b, M)) \sum_{r|b} \frac{a}{r}\mu(r) \sum_{s|b} \frac{b}{s}\mu(s)$$
$$= \sum_{a|m}\sum_{b|n} \pi(a, b, \mathrm{lcm}(a, b, M))\varphi(a)\varphi(b).$$

For a positive integer x, $\pi(x, M)$ denotes the number of prime numbers $p \in P_M$ such that $x \mid p - 1$. It is easily seen that $\pi(a, b, M) = \pi(\mathrm{lcm}(a, b), M)$.

We are now ready to calculate the limit $\lim_{M\to\infty} B(m, n, M)/M$. We have

$$\lim_{M\to\infty} \frac{1}{M} B(n, m, M) = \sum_{a|m}\sum_{b|n} \varphi(a)\varphi(b) \lim_{M\to\infty} \frac{\pi(\mathrm{lcm}(a, b), M)}{M}$$
$$= \sum_{a|m}\sum_{b|n} \frac{\varphi(a)\varphi(b)}{\varphi(\mathrm{lcm}(a, b))},$$

where the last equality follows from (1.7). □

It follows from the theorem above and (5.20) that

$$\mathbf{V}\zeta(p) = \sum_{a,b|n^r-1} \frac{\varphi(a)\varphi(b)}{\mathrm{lcm}(a, b)} - \tau(n^r - 1)^2. \tag{5.28}$$

Corollary 5.41. *Let ξ be as above. Then*

$$\mathbf{E}\xi^2(p) = \frac{1}{r^2} \sum_{d|r}\sum_{k|r} \mu(d)\mu(k) \sum_{a|n^{(r/d)}-1}\ \sum_{b|n^{(r/k)}-1} \frac{\varphi(a)\varphi(b)}{\varphi(\mathrm{lcm}(a, b))}. \tag{5.29}$$

Proof. We have

$$\mathbf{E}\xi^2(p) = \lim_{M\to\infty} \frac{1}{M} \sum_{p\in P_M} \frac{1}{r^2} \sum_{d|r}\sum_{k|r} \mu(r)\mu(k)(n^{(r/d)} - 1, p - 1)(n^{(r/k)} - 1)$$
$$= \frac{1}{r^2} \sum_{d|r}\sum_{k|r} \lim_{M\to\infty} \mu(r)\mu(k) \frac{1}{M} \sum_{p\in P_M} (n^{(r/d)} - 1, p - 1)(n^{(r/k)} - 1).$$

The corollary now follows from the theorem. □

The variance of ξ is according to Corollary 5.41 and (5.20) given by

$$\mathbf{V}\xi(p) = \frac{1}{r^2} \sum_{d|r}\sum_{k|r} \mu(d)\mu(k) \sum_{a|n^{(r/d)}-1}\ \sum_{b|n^{(r/k)}-1} \frac{\varphi(a)\varphi(b)}{\varphi(\mathrm{lcm}(a, b))}$$
$$- \left(\frac{1}{r} \sum_{d|r} \mu(d)\tau(n^{(r/d)} - 1)\right)^2.$$

5.8 Fuzzy cycles

To describe the dynamics outside the cycles on $S_1(0)$ we introduce the concept of *fuzzy cycles*, see Khrennikov [214].

Definition 5.42. A set of m different balls of radius $r = 1/p^l$ in $\mathbb{Q}_p$

$$\{B_r(a_0), B_r(a_1), \dots, B_r(a_{m-1})\}$$

is said to be a *fuzzy cycle* of order l and length m if

$$f(B_r(a_i)) \subseteq B_r(a_{i+1 \pmod m})$$

for $0 \leqslant i \leqslant m-1$.

There is a one-to-one correspondence between the fuzzy cycles of order 1 and the cycles in $\mathbb{Q}_p$, Proposition 4.3, p. 296, Khrennikov [214]. However, the structure of fuzzy cycle of orders $l \geqslant 2$ is not trivial. Some numerical experiments to clarify the structure were performed in Khrennikov's book [214] and especially in the paper of Khrennikov and Nilsson [254]. In this chapter the structure of fuzzy cycles is investigated by analytic methods, see [256] for more details.

Global dynamics

We begin this section with two simple propositions on monomial functions that will be useful in the description of the dynamics.

Proposition 5.43. *Let $x, y \in S_1(0) \subset \mathbb{Q}_p$ and suppose that $|x-y|_p < 1$. Then for all natural numbers n,*

$$|x^n - y^n|_p = |n|_p |x-y|_p$$

for $p > 2$.

To prove it, it suffices to note that $x \mapsto x^n$ is 1-Lipschitz, thus $|f(x) - f(y)|_p \leq |f'(z)|_p \cdot |x-y|_p$.

The next proposition can be found in Khrennikov [232].

Proposition 5.44. *The image, under $f(x) = x^n$, of a ball in $B_1(0) \setminus \{0\}$ is again a ball in $B_1(0) \setminus \{0\}$. Moreover, if $a \in B_1(0) \setminus \{0\}$ and ρ is such that $B_\rho(a) \subseteq B_1(0) \setminus \{0\}$ then $f(B_\rho(a)) = B_s(f(a))$, where $s = \rho |n|_p |a|_p^{n-1}$.*

Proof. Let $B_\rho(a) \subseteq B_1(0) \setminus \{0\}$, where $\rho = 1/p^m$ for some positive integer m. Since $0 \notin B_\rho(a)$, we have $|a|_p > \rho$.

By using Lemma 4.6 one can prove that if $a, \xi \in B_1(0)$ and $|a|_p > |\xi|_p$, then

$$|(a+\xi)^n - a^n|_p \leqslant |n|_p |\xi|_p |a|_p^{n-1} \tag{5.30}$$

for all positive integers n. From (5.30) we can easily conclude that $f(B_\rho(a)) \subseteq B_s(f(a))$.

We are now going to prove that $f(B_\rho(a)) = B_s(f(a))$. Let $y \in B_s(a^n)$. Hence, $y = a^n + \beta$, where $|\beta|_p \leqslant s$. To prove that $f(B_\rho(a)) = B_s(f(a))$ we must find ξ, such that $|\xi|_p \leqslant \rho$ and $(a+\xi)^n = a^n + \beta$. The last equation is equivalent to $(1+\xi/a)^n = 1 + \beta/a^n$, which has the formal solution

$$\xi = a((1+\beta/a^n)^{1/n} - 1).$$

The p-adic binomial $(1+x)^{1/n}$, see [374], is analytic over $\mathbb{Q}_p$ for $|x|_p \leqslant |n|_p/p$. Since

$$|\beta/a^n|_p \leqslant \rho|n|_p/|a|_p \leqslant |n|_p/p,$$

it follows that $\xi \in \mathbb{Q}_p$. It remains to be shown that $|\xi|_p \leqslant \rho$. We know from [374] that for $|x|_p \leqslant |n|_p/p$,

$$(1+x)^{1/n} = \sum_{j=0}^{\infty} \binom{1/n}{j} x^j,$$

where $\binom{1/n}{j} = (1/n)(1/n-1)\cdots(1/n-j+1)/j!$. From, e.g., [374] we also have the estimate $|j!|_p \leqslant p^{(j-1)/(1-p)}$.

We get

$$|\xi|_p \leqslant |a|_p \max_{1\leqslant j<\infty} \frac{|\beta|_p^j}{|a^n|_p^j |j!|_p} \leqslant \rho \max_{1\leqslant j<\infty} \left(\frac{\rho p^{1/(p-1)}}{|a|_p}\right)^{j-1} \leqslant \rho.$$

□

Corollary 5.45. *Let $f(x) = x^n$. Then the image of the ball $B_{1/p}(j)$, $1 \leqslant j \leqslant p-1$ is equal to the ball $B_{1/p}(k)$, where $k \equiv j^n \pmod p$, $1 \leqslant k \leqslant p-1$.*

Proof. From Proposition 5.44 it follows that $B_{1/p}(j)$ is mapped onto

$$B_{|n|_p/p}(f(j)) \subseteq B_{1/p}(f(j)).$$

Since $k \in B_{1/p}(f(j))$ we have $B_{1/p}(f(j)) = B_{1/p}(k)$. □

Observe that if $p \nmid n$ then $f(B_{1/p}(j)) = B_{1/p}(k)$ but if $p \mid n$ then $f(B_{1/p}(j)) \subset B_{1/p}(k)$.

Theorem 5.46 (see [214], p. 296). *All the elements of a ball of radius $1/p$ that does not contain periodic points are after a number of iterations of f mapped into a ball (of radius $1/p$) that contains a periodic point.*

Proof. Follows directly from the fact that there is a one-to-one correspondence between the fuzzy cycles of order 1 and the cycles in $\mathbb{Q}_p$. □

In the rest of this section we will study the dynamics of the balls of radius $1/p$ in $S_1(0)$. We do this by identifying each ball with an element of $\mathbb{F}_p^* \simeq (\mathbb{Z}/p\mathbb{Z})^*$. Each ball in $S_1(0)$ of radius $1/p$ can be written as $B_{1/p}(j)$, where $1 \leqslant j \leqslant p-1$. Identify this ball with $\bar{j}$, the residue class in $(\mathbb{Z}/p\mathbb{Z})^*$ containing j.

We know that there is a one-to-one correspondence between the periodic points of f over $\mathbb{F}_p$ and over $\mathbb{Q}_p$.

Definition 5.47. Let G_P denote the set of periodic points of $f(x)$ over $\mathbb{F}_p^*$. Let G_A denote the set of points in $\mathbb{F}_p^*$ that are attracted to 1.

Theorem 5.48. *The set G_P is a cyclic subgroup of $\mathbb{F}_p^*$. An element $x \in G_P$ is a generator of G_P if and only if x is an $\hat{r}(p)$-periodic point.*

Proof. We begin to show that G_P is a subgroup of $\mathbb{F}_p^*$. Let $x, y \in G_P$. Then there are least integers s and t such that $x^{n^s} = x$, $y^{n^t} = y$, $m = sm'$ and $m = tm''$. Let now m be the least common multiple of s and t then

$$xy^{n^m} = x^{n^m} y^{n^m} = x^{n^{sm'}} y^{n^{tm''}} = x^{(n^s)^{m'}} y^{(n^t)^{m''}} = xy.$$

Hence, $xy \in G_P$ since it is a m-periodic point. Let x^{-1} be the inverse of x in $\mathbb{F}_p^*$. We must show that $x^{-1} \in G_P$. We have

$$(x^{-1})^{n^s-1} = (x^{-1})^{n^s-1} x^{n^s-1} = (x^{-1}x)^{n^s-1} = 1^{n^s-1} = 1$$

so $x^{-1} \in G_P$. That is, G_P is a subgroup of $\mathbb{F}_p^*$. Since $\mathbb{F}_p^*$ itself is cyclic it follows that G_P is cyclic.

We now show that if g is a generator of G_P then it is a $\hat{r}(p)$-periodic point. Remember that $\hat{r}(p)$ was the least positive number such that $n^{\hat{r}(p)} - 1$ was divisible by $p^*(n)$. Assume that there is a number d such that $d \mid \hat{r}(p)$ and $g^{n^d-1} = 1$. Since g is a generator of G_P and the order of G_P is $p^*(n)$ we must have $p^*(n) \mid n^d - 1$ and hence $d = \hat{r}(p)$. We also know that G_P has $\varphi(p^*(n))$ generators.

Since $x^{n^{\hat{r}(p)}-1} = 1$ has $(n^{\hat{r}(p)} - 1, p-1)$ solutions and $\varphi((n^{\hat{r}(p)} - 1, p-1))$ primitive solutions, there are $\varphi((n^{\hat{r}(p)} - 1, p-1))$ $\hat{r}(p)$-periodic points in $\mathbb{F}_p^*$. Since $(n^{\hat{r}(p)} - 1, p-1) = p^*(n)$ there is exactly the same number of $\hat{r}(p)$-periodic points and generators of G_P. Every generator is an $\hat{r}(p)$-periodic point. Thus every $\hat{r}(p)$-periodic point is a generator of G_P. □

Theorem 5.49. *The set G_A is a cyclic subgroup of $\mathbb{F}_p^*$.*

Proof. We can describe G_A in the following way

$$G_A = \{x \in \mathbb{F}_p^* : x^{n^m} = 1 \text{ for some } m \in \mathbb{Z}^+\}.$$

Let $x, y \in G_A$ then there are m_1 and m_2 such that $x^{n^{m_1}} = 1$ and $y^{n^{m_2}} = 1$. Let m be the least common multiplier of m_1 and m_2 then $(xy)^m = x^{n^m} y^{n^m} = 1$, so $xy \in G_A$. Let x^{-1} be the inverse of x in $\mathbb{F}_p^*$. Then

$$(x^{-1})^{n^{m_1}} = (x^{-1})^{n^{m_1}} x^{n^{m_1}} = (x^{-1}x)^{n^{m_1}} = 1$$

and therefore $x^{-1} \in G_A$. We have proved that G_A is a subgroup of $\mathbb{F}_p^*$. Since $\mathbb{F}_p^*$ is cyclic it follows that G_A is cyclic. □

Definition 5.50. We call G_P the periodic group of the dynamical system and G_A the attractor group.

It might seem strange to call G_A the attractor group of the whole system, since it only contains points that are attracted to the fixed point 1. But, we will see that G_A determines completely the dynamics outside of balls containing periodic points.

Theorem 5.51. *$\mathbb{F}_p^*/G_A \simeq G_P$ and for $|G_A| = (p-1)/p^*(n)$.*

Proof. Let $\psi : \mathbb{F}_p^* \to, \psi(x) = x^{n^{p-1}}$. Let $x, y \in \mathbb{F}_p^*$ then

$$\psi(xy) = (xy)^{n^{p-1}} = x^{n^{p-1}} y^{n^{p-1}} = \psi(x)\psi(y),$$

so ψ is a homomorphism. After at most $p-1$ iterations every $x \in \mathbb{F}_p^*$ is mapped onto a periodic point. Hence $\operatorname{Im}\psi \subseteq G_P$. Let $y \in G_P$ and assume that y has period r. Let now m be such that $m + p - 1 \equiv 0 \pmod r$ then

$$\psi(y^{n^m}) = (y^{n^m})^{n^{p-1}} = y^{n^{m+p-1}} = y.$$

This proves that $\operatorname{Im}\psi = G_P$. We also have that $\ker\psi = G_A$. By the fundamental homomorphism theorem $\mathbb{F}_p^*/G_A \simeq G_P$. Since $|G_P| = p^*(n)$ we obtain that $|G_P| = (p-1)/p^*(n)$. □

Definition 5.52. Let $x \in G_P$. For $j \geqslant 1$ we denote by $A_j(x)$ the set of points in $\mathbb{F}_p^*$ that are mapped into x at first time after j iterations of f without passing any other periodic point on its way. We call $A_j(x)$ the jth attractor set of x. Observe that the pre-image of x is an element in $A_1(x)$.

We can now make a partition of the attractor group G_A in the following way,

$$G_A = \bigcup_{j \geqslant 1} A_j(1). \tag{5.31}$$

Definition 5.53. Let $x \in G_P$. By $G_A(x)$ we denote the set of points of $\mathbb{F}_p^*$ that are mapped onto x without passing any other periodic point on the way.

We have the following partition of $G_A(x)$,

$$G_A(x) = \bigcup_{j\geqslant 1} A_j(x).$$

Of course, $G_A(1) = G_A$, the attractor group.

Let us now study the cosets of G_A. Let $y \in G_P$ and assume that y is r-periodic then

$$yG_A = \bigcup_{j\geqslant 1}\{ys : s \in A_j(1)\}.$$

Since $(ys)^{n^j} = y^{n^j}$ for every $s \in A_j(1)$ we have

$$\{ys : s \in A_j(1)\} = A_j(y^{n^j \pmod r})$$

and hence

$$yG_A = \bigcup_{j\geqslant 1} A_j(y^{n^j \pmod r}).$$

We also have

$$A_j(y) = y^{n^{r-j} \pmod r} A_j(1)$$

so

$$G_A(y) = \bigcup_{j\geqslant 1} y^{n^{r-j} \pmod r} A_j(1).$$

There is a one-to-one correspondence between the sets $A_j(1)$ and $A_j(y)$. We therefore have

$$|G_A(y)| = |G_A| = (p-1)/p^*(n). \tag{5.32}$$

We are now going to show that the structure of G_A also inherits to $G_A(y)$. Remember that G_A was the set of points in $\mathbb{F}_p^*$ that were attracted to $1 \in \mathbb{F}_p^*$. Let $b_1 \in A_j(1)$ and take $a_1 \in f^{-1}(\{b_1\})$ arbitrary. Of course $a_1 \in A_{j+1}(1)$. Let b_y be the corresponding element to b_1 in $A_j(y)$ (that is $b_y = y^{n^{r-j} \pmod r} b_1$). The question is now: Will the corresponding elements, a_y, in $A_{j+1}(y)$ be mapped onto b_y? The answer is yes, because

$$(a_y)^n = \left(y^{n^{r-j} \pmod r} a_1\right)^n = y^{n^{r-j} \pmod r} b_1 = b_y.$$

Local dynamics

Let us now investigate the dynamics on the balls of radius $1/p$ on $S_1(0)$ that contain a periodic point.

Definition 5.54. Let a be an r-periodic point of f and let $l \in \mathbb{Z}^+$. The sphere

$$S_{p^l}(a) = \{x : |x - a|_p = 1/p^l\}$$

is called the l-sphere of a. Let $A = \{a_0, a_1, \ldots, a_{r-1}\}$ be a cycle of length r. Then by the l-sphere of A we mean the union of the l-spheres of the periodic points contained in A.

If $p \nmid n$ then the maximal Siegel disk of a periodic point x_0 is $\mathrm{SI}(x_0) = B_{1/p}(x_0)$ and the Siegel annulus of an r-cycle $\{x_0, \ldots, x_{r-1}\}$ is

$$\mathrm{SI}(\{x_0, \ldots, x_{r-1}\}) = \bigcup_j B_{1/p}(x_j).$$

We can find out more about the dynamics by using the notion of the l-sphere.

Theorem 5.55. *Let a be an indifferent r-periodic point. If x belongs to the l-sphere of a then $f(x)$ belongs to the l-sphere of $f(a)$.*

Proof. Let x be a point in the l-sphere of a. Then $|x - a|_p = 1/p^l$. We are going to show that $|f(x) - f(a)|_p = 1/p^l$. Since a is indifferent, $p \nmid n$. Therefore, by Lemma 5.43,

$$|f(x) - f(a)|_p = |x^n - a^n|_p = |x - a|_p = 1/p^l.$$

□

See Figure 5.1.

Theorem 5.56. *Let a be an attractive r-periodic point and let $n = p^k n'$, where $p \nmid n'$. If x belongs to the l-sphere of a then $f(x)$ belongs to the $l+k$-sphere of $f(a)$. Moreover, $f(S_{1/p^l}(a)) = S_{1/p^{l+k}}(f(a))$.*

Proof. Take x in the l-sphere of a arbitrary, then $|x - a| = 1/p^l$. Since $|n| = 1/p^k$ it follows from Theorem 5.43 that

$$|f(x) - f(a)|_p = |x^n - a^n|_p = |n|_p |x - a|_p = 1/p^k \cdot 1/p^l = 1/p^{l+k}.$$

To prove the second part, we observe that $f(B_{1/p^l}(a)) = B_{1/p^{l+k}}(f(a))$ and that $f(B_{1/p^{l+1}}(a)) = B_{1/p^{l+k+1}}(f(a))$. Together with the first part we now get the identity $f(S_{1/p^l}(a)) = S_{1/p^{l+k}}(f(a))$. □

Corollary 5.57. *If a is an attractive r-periodic point of $f(x) = x^n$, $n = p^k n'$ where $p \nmid n'$ and x belongs to the l-sphere with center at a then $f^r(x)$ belongs to the $l + rk$-sphere with center at a. Moreover, $f(S_{1/p^l}(a)) = S_{1/p^{l+rk}}(a)$.*

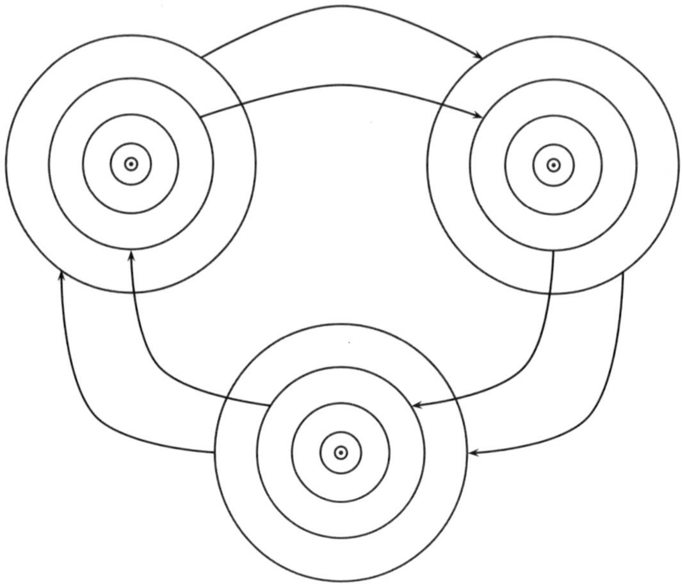

Figure 5.1. The l-sphere dynamics around a 3-cycle, where the periodic points are centers of Siegel disks.

Proof. Apply the theorem r times. □

See Figure 5.2. It follows from the discussion above that the basin of attraction of an r-cycle $\{x_0, \ldots, x_{r-1}\}$ is

$$A(\{x_0, \ldots, x_{r-1}\}) = \bigcup_{0 \leqslant j \leqslant r-1} \bigcup_{y \in \bar{x}_j G_A} B_{1/p}(y),$$

where $\bar{x_j} G_A$ are cosets of the attractor group.

Dynamics around neutral points

We will start to investigate fuzzy cycles in the spheres around an indifferent fixed point $a \in S_1(0)$. Let $l \geqslant 1$ and consider the l-sphere of a. Let $t \geqslant 0$, t will play the role of the depth parameter in the l-sphere. Let

$$I_t = \{i_0, i_1, \ldots, i_t\},$$

where $1 \leqslant i_0 \leqslant p-1$ and $0 \leqslant i_j \leqslant p-1$ for $1 \leqslant j \leqslant t$. We set

$$b(l, I_t) = a + i_0 p^l + i_1 p^{l+1} + \cdots + i_t p^{l+t}.$$

We are interested in fuzzy cycles inside of the l-sphere of a. The balls in the l-sphere of a at depth t are $B_{1/p^{l+t+1}}(b(l, I_t))$. Our aim is to determine the fuzzy cycles of

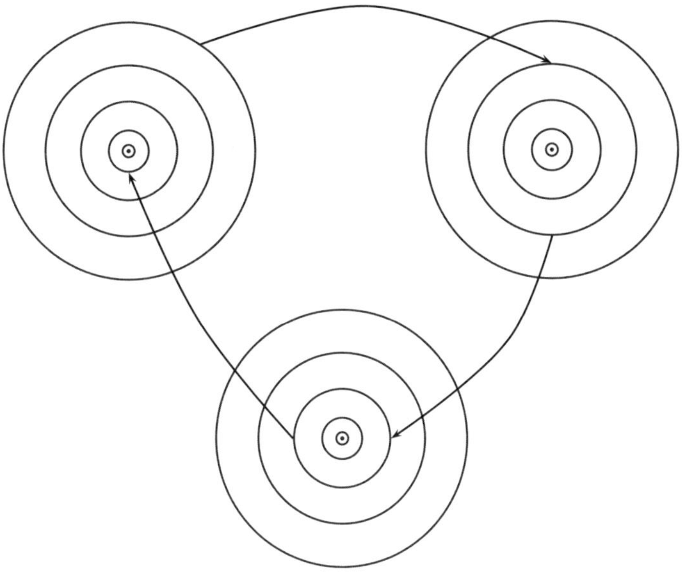

Figure 5.2. The l-sphere dynamics around a 3-cycle, where n and p are such that $p \mid n$ but $p^2 \nmid n$.

order $l + t + 1$. So we are interested in finding the least positive number m such that

$$f^m(B_{1/p^{l+t+1}}(b(l, I_t))) \subseteq B_{1/p^{l+t+1}}(b(l, I_t)).$$

In fact we can prove equality.

Lemma 5.58. *Let m_0 be the order of $\bar{n}$ (the canonical image of n) in $\mathbb{F}_p^*$. The least m for which*

$$f^m(B_{1/p^{l+1}}(b(l, I_0))) = B_{1/p^{l+1}}(b(l, I_0))$$

is equal to m_0.

Proof. First, we prove that $f^m(B_{1/p^{l+1}}(b(l, I_0))) \subseteq B_{1/p^{l+1}}(b(l, I_0))$. We have

$$\begin{aligned}|f^m(b(l, I_0)) - b(l, I_0)|_p &= |(a + i_0 p^l)^{n^m} - (a + i_0 p^l)|_p \\ &= \left| a^{n^m} - a + n^m i_0 p^l a^{n^m - 1} - i_0 p^l + \sum_{k=2}^{n^m} \binom{n^m}{k} a^{n^m - k} (i_0 p^l)^k \right|_p \\ &\leqslant |i_0 p^l (n^m - 1)|_p,\end{aligned}$$

since $lk \geqslant l + 1$ for every $k \geqslant 2$. This is less than or equal to $1/p^{l+1}$ if and only if $n^m \equiv 1 \pmod{p}$. Hence, the least m, satisfying

$$f^m(B_{1/p^{l+1}}(b(l, I_0))) \subseteq B_{1/p^{l+1}}(b(l, I_0))$$

is $m = m_0$, the order of $\bar{n}$ in $\mathbb{F}_p^*$. By Theorem 5.44 f^m maps $B_{1/p^{l+1}}(b(l, I_0))$ onto a ball of radius $1/p^{l+1}$ and this ball must be $B_{1/p^{l+1}}(b(l, I_0))$, so we have proved the equality. □

The number m_0 will play a large role in the future analysis of the dynamics. Let $s_0 \geqslant 0$ be the unique number satisfying $n^{m_0} = 1 + n' p^{s_0}$, where $p \nmid n'$. Like m_0, s_0 will also be crucial for the dynamics on the l-spheres. This we will see in the following theorem.

Theorem 5.59. *Let m_0 be as in the lemma above and let*

$$m_j = \begin{cases} 1, & 1 \leqslant j < s_0, \\ p, & j \geqslant s_0. \end{cases} \tag{5.33}$$

The least positive integer m for which

$$f^m(B_{1/p^{l+t+1}}(b(l, I_t))) = B_{1/p^{l+t+1}}(b(l, I_t))$$

is equal to $\prod_{j=0}^t m_j$. Moreover the unique number s_t, $t \geqslant 1$, defined by

$$n^{\prod_{j=0}^t m_j} = 1 + n'_t p^{s_t}, \quad p \nmid n'_t,$$

is given by

$$s_t = \begin{cases} s_0, & t < s_0, \\ t+1, & t \geqslant s_0. \end{cases}$$

Proof. We will prove this theorem by induction. By Lemma 5.58 the theorem is true for $t = 0$. We assume that the theorem is true for t and prove that it is then also true for $t + 1$. First, we find the least positive integer m such that

$$|f^m(b(l, I_{t+1})) - b(l, I_{t+1})| \leqslant 1/p^{l+t+2}.$$

(That $f^m(B_{1/p^{l+t+1}}(b(l, I_t))) = B_{1/p^{l+t+1}}(b(l, I_t))$ will follow in the same way as in the proof of Lemma 5.58.) Of course, m must be a multiple of $\prod_{j=0}^t m_j$. Set $m = m_{t+1} \prod_{j=0}^t m_j$ and let $N = n^m$. We have to prove that $m_{t+1} = 1$ if $t + 1 < s_0$ and that $m_{t+1} = p$ if $t + 1 \geqslant s_0$. We have

$$\begin{aligned} f^m(b(l, I_t)) &= (b(l, I_t))^N \\ &= a + N(i_0 p^l + \cdots + i_{t+1} p^{l+t+1}) \\ &\quad + \sum_{k=2}^{N} \binom{N}{k} a^{N-k} (i_0 p^l + \cdots + i_{t+1} pl + t + 1)^k. \end{aligned}$$

We will show that the sum in the last term has an absolute value that is less than or equal to $1/p^{l+t+2}$, that is, each term in the sum contains at least $l+t+2$ factors of p.

Consider the binomial coefficient for $k \geqslant 2$

$$\binom{N}{k} = \frac{N(N-1)}{(k-1)k} \cdot \frac{(N-1-1)\cdots(N-1-(k-2))}{1\cdots(k-2)}. \tag{5.34}$$

By the induction hypothesis we know that we can write

$$N-1 = (1+n'_t p^{s_t})^{m_{t+1}} - 1 = m_{t+1} n'_t p^{s_t} + \text{higher powers of } p.$$

Let us first consider the case when $k < t+3$. Observe that $p^{s_t} \geqslant p^{t+1} \geqslant t+3$ for any positive integer t. Then the factors of p that occur in the denominator of the last fraction in (5.34) are canceled by the factors of p that occur in the corresponding factor in the nominator. Moreover, $(k-1)k$ can have at most $k-2$ factors of p, since we exclude $p=2$. The number of factors of p in $\binom{N}{k} p^{kl}$ is then greater or equal to

$$s_t - (k-2) + kl \geqslant t+1+2+k(l-1) \geqslant t+2+2(l-1)+1 \geqslant t+2+l,$$

when $l \geqslant 1$.

Let us now consider the case when $k \geqslant t+3$. Then the number of factors of p in $\binom{N}{k} p^{kl}$ is greater or equal to

$$lk \geqslant l(t+3) \geqslant 3l+t \geqslant l+t+2l \geqslant l+t+2.$$

So far, we have proved that

$$|(b(l,I_t))^N - u + N(i_0 p^l + \cdots + i_{t+1} p^{l+t+1})|_p \leqslant 1/p^{l+t+2}.$$

Since the number of factors of p in $\binom{m_{t+1}}{j} p^{s_t} p^l$ are greater or equal to

$$js_t + l \geqslant j(t+1) + l \geqslant t+2+l$$

it follows that

$$\begin{aligned} &|(b(l,I_t))^N - a + (1+n'_t p^{s_t})^{m_{t+1}} (i_0 p^l + \cdots + i_{t+1} p^{l+t+1})|_p \\ &\qquad \leqslant |a + m_{t+1} n'_t p^{s_t} (i_0 p^l + \cdots + i_{t+1} p^{l+t+1})|_p \leqslant 1/p^{l+t+2}. \end{aligned}$$

For

$$|b(l,I_t)^n - b(l,I_t)|_p \leqslant |(i_0 p^l + \cdots + i_{t+1} p^{l+t+1}) m_{t+1} n'_t p^{s_t}|_p$$

to be less than or equal to $1/p^{l+t+2}$, it is necessary that the number of factors of p in $m_{t+1} p^{s_t}$ is greater than or equal to $t+2$.

If $t+1 < s_0$ then

$$\operatorname{ord}_p(m_{t+1}p^{s_0}) = \operatorname{ord}_p(m_{t+1}) + s_0 \geqslant \operatorname{ord}_p(m_{t+1}) + t + 2$$

so the least positive integer m_{t+1} fulfilling this must be $m_{t+1} = 1$. If $t+1 = s_0$ then

$$\operatorname{ord}_p(m_{t+1}p^{s_t}) = \operatorname{ord}_p(m_{t+1}) + s_0 = \operatorname{ord}_p(m_{t+1}) + t + 1.$$

The least positive integer m_{t+1} making this greater than or equal to $t+2$ is $m_{t+1} = p$. If $t+1 > s_0$ then

$$\operatorname{ord}_p(m_{t+1}p^{s_t}) = \operatorname{ord}_p(m_{t+1}) + t + 1,$$

so again we must choose $m_{t+1} = p$. This proves the first part of the theorem.

If $t+1 < s_0$ then

$$n^{\prod_{j=0}^{t+1} m_j} = (1 + n'_t p^{s_t})^{m_{t+1}} = (1 + n'_t p^{s_0}),$$

so $s_{t+1} = s_0$. If $t+1 = s_0$ then there is n'_{t+1} such that

$$n^{\prod_{j=0}^{t+1} m_j} = (1 + n'_t p^{s_0})^p = 1 + n'_{t+1} p^{s_0+1},$$

hence $s_{t+1} = t+1+1$. Finally if $t+1 > s_0$ then there is n'_{t+1} such that

$$n^{\prod_{j=0}^{t+1} m_j} = (1 + n'_t p^{t+1})^p = 1 + n'_{t+1} p^{t+2},$$

so $s_{t+1} = t+1+1$ also in this case. The proof of the theorem is completed. □

Notice that m in the theorem above is independent of l and the values of the elements in I_t, see also Figure 5.3. This implies that all the balls at depth t in each l-sphere with center at a belong to fuzzy cycles of the same length. At depth t there are $(p-1)p^t$ balls of radius $1/p^{l+t+1}$ in each l-sphere. Since all these balls belong to a fuzzy cycle of length m there are $(p-1)p^t/m$ fuzzy cycles of length m and order $l+t+1$ in each l-sphere. If $t < s_0$ then $m = m_0$, so there are $(p-1)p^t/m_0$ cycles of length m_0 and order $l+t+1$ in each l-sphere. If instead $t \geqslant s_0$ then $m = m_0 p^{t-s_0+1}$, so in this case there are $(p-1)p^{s_0-1}/m_0$ fuzzy cycles of length $m_0 p^{t-s_0+1}$ and order $l+t+1$ in each l-sphere of a. We have proved the following theorem.

Theorem 5.60. *Let a be a fixed point of the dynamical system f. Let l and t be integers such that $l \geqslant 1$ and $t \geqslant 0$. Then the l-sphere with center a contains*

$$\frac{(p-1)}{m_0} p^{\min\{t+1,s_0\}-1}$$

fuzzy cycles of length $m_0 p^{\max\{t+1,s_0\}-s_0}$ and order $l+t+1$.

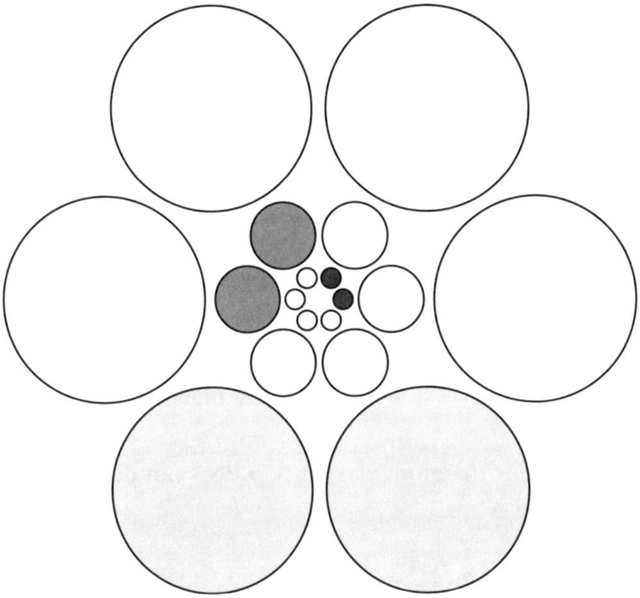

Figure 5.3. The fuzzy cycles of order $l+1$ in the l-sphere are of length m_0. In this case $m_0 = 2$. One fuzzy cycle in each sphere is indicated by different levels of gray.

So far we have studied the dynamics around fixed points. The same technique can be used to study the dynamics around cycles.

Theorem 5.61. *Let $A = (a_0, a_1, \dots, a_{r-1})$ be an r-cycle in $\mathbb{Q}_p$ of f. Let $m_0(r)$ be the order of $\overline{n^r}$ in $\mathbb{F}_p^*$ and let $s_0(r)$ be the unique number that satisfies $(n^r)^{m_0(r)} = 1 + m' p^{s_0(r)}$, $p \nmid m'$. Let $l \geqslant 1$ and $t \geqslant 0$. Then the l-sphere of A contains*

$$\frac{(p-1)}{m_0(r)} p^{\min\{t+1, s_0(r)\}-1}$$

fuzzy cycles of length $r m_0(r) p^{\max\{t+1, s_0(r)\}-s_0(r)}$ and order $l+t+1$.

Proof. Each element of A is a fixed point of $f^r(x) = x^{n^r}$. We can then copy the proof of Theorem 5.60 and multiply the length of the cycles by r. □

What are the relations between m_0 and $m_0(r)$, and s_0 and $s_0(r)$? Since $m_0(r)$ is the order of $\overline{n^r}$ in $\mathbb{F}_p^*$ and m_0 is the order of $\bar{n}$ in $\mathbb{F}_p^*$ it follows that

$$m_0(r) = \frac{m_0}{(m_0, r)}.$$

Lemma 5.62. *Let r be the length of a cycle of f in $\mathbb{Q}_p$. Then $s_0(r) = s_0$.*

Proof. The length of the longest cycle of f in $\mathbb{Q}_p$, $\hat{r}(p)$, is the order of n modulo $p^*(n)$. Remembering that $p^*(n) \mid (p-1)$ we obtain that

$$\hat{r}(p) \leqslant p^*(n) \leqslant p - 1 < p.$$

Hence, p can not divide $\hat{r}(p)$ and because $r \mid \hat{r}(p)$ we have that $p \nmid r$.

We have, since $m_0(r) = m_0/(m_0, r)$, that

$$\begin{aligned} 1 + m' p^{s_0(r)} &= (n^r)^{m_0(r)} = (n^{m_0})^{r/(m_0,r)} \\ &= (1 + n' p^{s_0})^{r/(m_0,r)} \\ &= 1 + \frac{r}{(m_0, r)} n' p^{s_0} + \text{higher powers of } p. \end{aligned}$$

We have that $p \nmid r$. It is therefore clear that p does not divide $r/(m_0, r)$. That is $s_0(r) = s_0$. □

Definition 5.63. Let $A = (a_0, a_1, \ldots, a_{r-1})$ be an r-cycle in $\mathbb{Q}_p$ of f. The number of fuzzy cycles in the set $\bigcup_{j=0}^{r-1} B_{1/p}(a_j)$ of order $j+1$ and length l is denoted by $\mathbb{N}_{\text{local}}(A, j, l, p)$. This quantity is called the *local number of fuzzy cycles*. By the *global number of fuzzy cycles* $\mathbb{N}_{\text{global}}(j, l, p)$ we denote the total number of fuzzy cycles of order j and length l in $\mathbb{Q}_p$.

Let a be a fixed point and let $\omega \in \mathbb{Z}^+$. We are now interested in counting the number of fuzzy cycles of order $\omega + 1$ in $B_{1/p}(a)$. The smallest sphere that contains balls of radius $1/p^{\omega+1}$ is the ω-sphere. It follows from Theorem 5.60 that it contains

$$\frac{(p-1)}{m_0} p^{\min\{1,s_0\}-1}$$

fuzzy cycles of length $m_0 p^{\max\{1,s_0\}-s_0}$ (at depth 0). The $(\omega - 1)$-sphere (just outside of the ω-sphere) contains

$$\frac{(p-1)}{m_0} p^{\min\{2,s_0\}-1}$$

fuzzy cycles of length $m_0 p^{\max\{2,s_0\}-s_0}$ (at depth 1), and so on until the 1-sphere that contains

$$\frac{(p-1)}{m_0} p^{\min\{\omega,s_0\}-1}$$

fuzzy cycles of length $m_0 p^{\max\{\omega,s_0\}-s_0}$ (at depth $\omega - 1$).

If $\omega \leqslant s_0$ then there are only fuzzy cycles of length m_0 and they are

$$\sum_{j=1}^{\omega} \frac{p-1}{m_0} p^{j-1} = \frac{p^{\omega} - 1}{m_0}$$

in number. If $\omega > s_0$ there are

$$\sum_{j=1}^{s_0} \frac{p-1}{m_0} p^{j-1} = \frac{p^{s_0}-1}{m_0}$$

fuzzy cycles of length m_0 and

$$\frac{p-1}{m_0} p^{s_0-1}$$

fuzzy cycles of length $m_0 p^i$, where $1 \leqslant i \leqslant \omega - s_0$.

If we generalize this in the obvious way to cycles we obtain the following theorem.

Theorem 5.64. *Let A be an r-cycle of the dynamical system f. Then*

$$\mathbb{N}_{\mathrm{local}}(A, \omega, rm_0(r), p) = \frac{p^{\min\{\omega, s_0\}} - 1}{m_0(r)},$$

and for $\omega > s_0$, $1 \leqslant i \leqslant \omega - s_0$,

$$\mathbb{N}_{\mathrm{local}}(a, \omega, m_0 p^i, p) = \frac{p-1}{m_0(r)} p^{s_0-1}.$$

Dynamics around attractors

The following theorem follows directly from Theorem 5.56.

Theorem 5.65. *If $p \mid n$ then the dynamical system generated by $f(x) = x^n$ has no fuzzy cycles except the fuzzy cycles of radius $1/p$ that correspond to the cycles of f.*

Even if the dynamical system does not have fuzzy cycles, we can still get more information about the dynamics around the cycles. We introduce a new concept, fuzzy orbits.

Definition 5.66. A set of balls $\{B_{r_0}(a_0), B_{r_1}(a_1), \ldots\}$ such that $r_i \geqslant r_{i+1}$ and $f(B_{r_i}(a_i)) \subseteq B_{r_{i+1}}(a_{i+1})$, for every $i \geqslant 0$, is called the *fuzzy orbit* of $B_{r_0}(a_0)$.

Theorem 5.67. *Let a be an attractive fixed point. Let $\{a_1, a_2, \ldots, a_{p-1}\}$ be a set of representatives of the balls of radius $1/p^{l+1}$ in the l-sphere of a. Then we have fuzzy orbits of $B_{1/p^{l+1}}(a_i)$ such that $r_j = 1/p^{l+1+kj}$, $j \geqslant 0$, where $k = \mathrm{ord}_p(n)$. Let $i \neq j$, then the fuzzy orbits of $B_{1/p^{l+1}}(a_i)$ and $B_{1/p^{l+1}}(a_j)$ never intersect, that is we can never find a ball in one of the orbits that is included in a ball of the other orbit.*

Proof. From Theorem 5.56 we know that the l-sphere of a is mapped into the $(l+k)$-sphere of a. Let $x \in B_{1/p^{l+jk+1}}(b)$ for some non-negative integer j and some b in the $(l+kj)$-sphere of a. Then

$$|f(x) - f(b)| = |x^n - b^n| = |n||x-b| \leqslant 1/p^k \cdot 1/p^{l+jk+1} = 1/p^{l+k(j+1)+1},$$

so the fuzzy orbits of $B_{1/p^{l+1}}(a_i)$ are well defined. Let x belong to the jth ball of the fuzzy orbit of $B_{1/p^{l+1}}(a_i)$ and let y belong to the jth ball of the fuzzy orbit of $B_{1/p^{l+1}}(a_h)$. Then $|x-y| = 1/p^{l+kj}$ and

$$|f(x) - f(y)| = |x^n - y^n| = |n||x-y| = 1/p^{l+k(j+1)},$$

so $f(x)$ and $f(y)$ belong to different balls in the $l+k(j+1)$-sphere of a. By induction the fuzzy orbits never intersect. □

In Figure 5.4 there is a visualization of the fuzzy orbits mentioned in the theorem above.

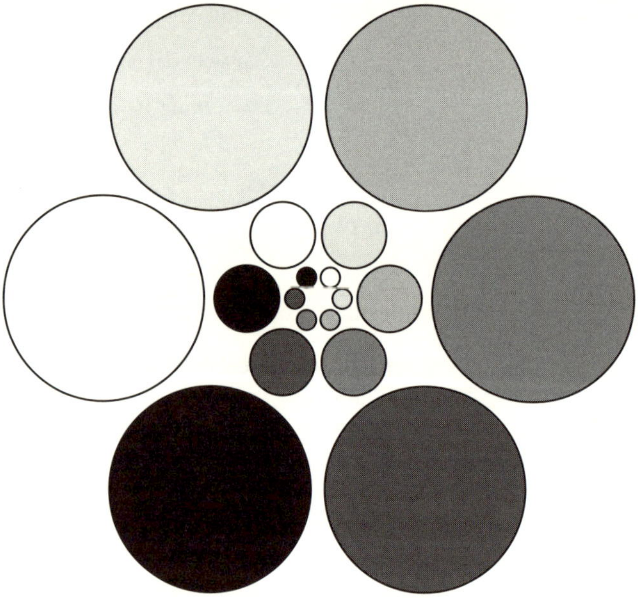

Figure 5.4. The fuzzy orbits (indicated by different levels of gray) around a fixed point in a system where $p \mid n$ but $p^2 \nmid n$.

Distribution of fuzzy cycles

Let $\omega \in \mathbb{Z}^+$ and $\rho \in \mathbb{Z}^+$. From now on we consider fuzzy cycles of order $\omega + 1$ and length ρ. We can get a fuzzy cycle of length ρ in $\mathbb{Q}_p$ only if there is $k \geqslant 0$ such that

$$\rho = rm_0(r)p^k = \frac{r}{(m_0, r)} m_0 p^k$$

where r is a length of a cycle in $\mathbb{Q}_p$. For which prime numbers p is this possible? Certainly, there must be a divisor d of ρ such that $d = m_0$. Since m_0 is the least integer such that $n^{m_0} \equiv 1 \pmod p$ it is necessary that $p < n^d$. That is, to have a chance of getting a fuzzy cycle of length ρ we must have $p < n^\rho$. We have proved the following theorem.

Theorem 5.68. *For a fixed order and a fixed length of a fuzzy cycle there is only a finite number of fields $\mathbb{Q}_p$ where it occurs.*

Let, as always, $\mathcal{P}_M$ denote the set of the first M prime numbers and let τ be a functions that counts the number of positive divisors. In Theorem 5.38 the limit

$$\lim_{M\to\infty} \frac{1}{M} \sum_{p\in\mathcal{P}_M} N(n,r,p) = \frac{1}{r}\sum_{d|r} \mu(d)\tau(n^{r/d}-1)$$

is computed. By Theorem 5.68 we have

$$\lim_{M\to\infty} \frac{1}{M} \sum_{p\in\mathcal{P}_M} \mathcal{N}_{\mathrm{global}}(\omega,\rho,p) = 0$$

since $\mathcal{N}_{\mathrm{global}}(\omega,\rho,p) = 0$ for all but finitely many prime numbers p.

Part II

The Non-Commutative Non-Archimedean Dynamics

Chapter 6

Basics of polynomial dynamics on groups

We shall study measure-preserving (in particular, ergodic) transformations on the group G (whose operation is written multiplicatively here and henceforth) in the class of all functions of the form

$$w(x_1,\dots,x_n) = g_1(x_{i_1}^{\omega_1})^{n_1} g_2(x_{i_2}^{\omega_2})^{n_2} \cdots g_k(x_{i_k}^{\omega_k})^{n_k} g_{k+1}.$$

Here $g_1,\dots,g_{k+1}$ are elements of the group G, $n_1,\dots,n_k$ are rational integers, $i_1,\dots,i_k \in \{1,2,\dots,n\}$, $\omega_1,\dots,\omega_k \in \Omega$. The image of the element $h \in G$ under the action of the operator ω is denoted by h^{ω}. Note that every operator $\omega \in \Omega$ acts on G by an endomorphism, which we denote by the same symbol ω. Thus, raising to the power $n \in \mathbb{Z}$ of the element $h \in G$ commutes with operator $\omega \in \Omega$, $(h^{\omega})^n = (h^n)^{\omega}$; so we write $h^{n\omega}$ (or $h^{\omega n}$) instead of $(h^{\omega})^n$ for short. Under these conventions, a polynomial $w(x_1,\dots,x_n)$ in variables $x_1,\dots,x_n$ over the group G with the set of operators Ω is an expression of the form

$$w(x_1,\dots,x_n) = g_1 x_{i_1}^{\omega_1 n_1} g_2 x_{i_2}^{\omega_2 n_2} \cdots g_k x_{i_k}^{\omega_k n_k} g_{k+1}. \tag{6.1}$$

Within the book, functions of the form (6.1) will be referred to as polynomial functions with operators. Note that whenever G is an 'ordinary group', that is, a group with empty set of operators, a polynomial $w(x_1,\dots,x_n)$ in variables $x_1,\dots,x_n$ over the group G can be written as

$$w(x_1,\dots,x_n) = g_1 x_{i_1}^{n_1} g_2 x_{i_2}^{n_2} \cdots g_k x_{i_k}^{n_k} g_{k+1}. \tag{6.2}$$

Sometimes it is convenient to represent polynomials in a form other than (6.1) (or (6.2)), namely in the form

$$w(x_1,\dots,x_n) = w(1,\dots,1) x_{i_1}^{h_1\omega_1 n_1} x_{i_2}^{h_2\omega_2 n_2} \cdots x_{i_k}^{h_k\omega_k n_k}, \tag{6.3}$$

where $h_1,\dots,h_k \in G$. Indeed, as $xg = gx^g$ for all $x \in G$, where $x \mapsto x^g = g^{-1}xg$ is an automorphism of G induced by a conjugation by the element $g \in G$, we can re-write (6.1) in the form (6.3) and vice versa. Note that in the case of univariate polynomials (i.e., when $n = 1$) in variable x, the polynomial can be represented in the form

$$w(x) = w(1) x^{h_1\omega_1 n_1 + \cdots + h_k\omega_k n_k}, \tag{6.4}$$

where $x^{h\omega_1 n+g\alpha m}$ stands for $x^{h\omega_1 n}x^{g\alpha m} = h^{-1}(x^{\omega})^n hg^{-1}(x^{\alpha})^m g$. Representation of the form (6.4) is convenient if, say, we consider a mapping induced by the polynomial $w(x)$ on the normal Abelian Ω-invariant subgroup $N \subset G$. In the latter case the sum $h_1\omega_1 n_1 + \dots + h_k\omega_k n_k$ can be treated as an element of the commutative ring End (N) of endomorphisms of the group N, if we put into correspondence to every $\omega \in \Omega$ an endomorphism of N induced by the operator ω, and to every $g \in G$ – an automorphism of N induced by a conjugation by g. For instance, if N is an elementary Abelian p-group, p prime, we can treat N as a vector space over $\mathbb{F}_p$ (and whence End (N) is merely an algebra of all square matrices over $\mathbb{F}_p$); so the sum $h_1\omega_1 n_1 + \dots + h_k\omega_k n_k$ can be then treated as just a sum of matrices $h_1\omega_1 n_1, \dots, h_k\omega_k n_k$, i.e., as a matrix over $\mathbb{F}_p$.

6.1 Non-commutative differential calculus

The role of this section is to develop necessary tools to study polynomial dynamics over groups (with operators). In the case of a commutative structure, e.g., a ring $\mathbb{Z}_p$ of p-adic integers, one of the key points in our study of a dynamical system $f : \mathbb{Z}_p \to \mathbb{Z}_p$ was the 'formula of small increments' that expresses the value of the function f at the point $x + h$, where h is p-adically small, via the derivative $f'(x)$: Given $f(x) \in \mathbb{Z}_p[x]$, for all $h \in \mathbb{Z}_p$,

$$f(x+h) \equiv f(x) + h \cdot f'(x) \pmod{p^{\operatorname{ord}_p h+1}}, \tag{6.5}$$

see Section 3.7. Using this formula, we actually reduced the problem to determine whether f is measure-preserving (or ergodic) to the study of action of f on the residue ring $\mathbb{Z}/p^k\mathbb{Z}$, where k is small, and to the study of the behavior of the derivative $f'(x)$ (actually to the study of the affine mapping $h \mapsto a + h \cdot f'(x)$ on the field $\mathbb{F}_p$), see e.g. Hensel's lemma 3.16 or Theorem 4.55. Our aim is to obtain an analog of the formula (6.5) for non-Abelian groups. For this purpose, we need a notion of a derivative of a polynomial over a group with operators. This notion is a further generalization of the concept of free differential calculus (i.e., derivatives of elements of a free group $F(X)$ freely generated by X) put forth by R. Fox in connection with knot theory, see [94], and of the derivative of a polynomial over a group with an empty set of operators introduced by Lausch, see [284, 286].

Let G be a group with a system of operators Ω. Then any polynomial $w(x_1, \dots, x_n)$ over G can be represented in the form (6.1), where $\omega_1, \dots, \omega_k \in \Omega$. The polynomial $w(x_1, \dots, x_n)$ is an element of the group $G[X^{\Omega}]$ of all polynomials of variables $X = \{x_1, x_2, \dots\}$ over the group G with the system of operators Ω. The group $G[X^{\Omega}]$ is a free product of the group G by the free group $F(X^{\Omega})$ freely generated by the set $\{x_i^{\omega} : i = 1, 2, \dots, \omega \in \Omega\}$. Let us consider the semigroup free product of the group $G[X^{\Omega}]$ by a free semigroup freely generated by the elements of the set Ω. We denote by $\mathbb{Z}\langle G, \Omega, X\rangle$ a semigroup ring of the above-mentioned semigroup free product over the ring of rational integers $\mathbb{Z}$. The elements of this semigroup ring can be represented as finite sums $\sum_{(i)} z_i \prod_{(j)} \omega_j w_j$, where $z_i \in \mathbb{Z}$, $\omega_j \in \Omega$, $w_j \in G[X^\Omega]$, i and j run over a finite set of subscripts. By definition, the *differentiation with respect to the variable* x_i is the map

$$\frac{\partial}{\partial x_i} : G[X^\Omega] \to \mathbb{Z}\langle G, \Omega, X\rangle,$$

which satisfies the following conditions:

1) $\frac{\partial x_j}{\partial x_i} = \delta_{ij}$ is the Kronecker delta;
2) $\frac{\partial g}{\partial x_i} = 0$ for any $g \in G$;
3) $\frac{\partial x_j^\omega}{\partial x_i} = \delta_{ij}\omega$ for any $\omega \in \Omega$;
4) $\frac{\partial uv}{\partial x_i} = \frac{\partial u}{\partial x_i} v + \frac{\partial v}{\partial x_i}$ for any $u, v \in G[X^\Omega]$.

Only the identity 4) distinguishes this differentiation from the ordinary differentiation, e.g., of polynomials over commutative rings. From this identity it follows that for $n \in \mathbb{Z}$

$$\frac{\partial x^n}{\partial x} = \begin{cases} x^{n-1} + x^{n-2} + \cdots + 1, & \text{if } n > 0; \\ 0, & \text{if } n = 0; \\ x^n + x^{n+1} + \cdots + x^{-1}, & \text{if } n < 0. \end{cases}$$

It is easy to verify that there exists a unique map that satisfies all these conditions 1)–4). Under this map the image $\frac{\partial w}{\partial x_i}$ of the polynomial $w \in G[X^\Omega]$ is called the *derivative of the polynomial* w *with respect to the variable* x_i. Furthermore, if $N \lhd G$ is an Abelian Ω-invariant normal subgroup of G, then given $g_1, g_2, \ldots \in G$, $h, h_1, h_2, \ldots \in N$, to every element $W(x_1, x_2, \ldots, x_n) = \sum_{(i)} z_i \prod_{(j)} \omega_j w_j(x_1, \ldots, x_n)$ we put into correspondence an endomorphism $W(g_1, \ldots, g_n) \in \mathrm{End}\,(N)$ induced by $W(g_1, \ldots, g_n)$ on N:

$$h^{W(g_1,\ldots,g_n)} = ((h^{z_1})^{\omega_1})^{w_1(g_1,\ldots,g_n)} \cdot ((h^{z_2})^{\omega_2})^{w_1(g_1,\ldots,g_n)} \cdots,$$

where $(\cdot)^{w_i(g_1,\ldots,g_n)}$ is a conjugation by the element $w_i(g_1, \ldots, g_n) \in G$. In the case $W = \frac{\partial w}{\partial x_i}$, this endomorphism is called the *value of the derivative* of the polynomial w at the point $(g_1, \ldots, g_n)$ and is denoted as $\frac{\partial w(g_1,\ldots,g_n)}{\partial x_i}$. The following formula, which follows directly from group laws, is now obvious:

$$w(g_1h_1, \ldots, g_nh_n) = w(g_1, \ldots, g_n) \cdot h_1^{\frac{\partial w(g_1,\ldots,g_n)}{\partial x_1}} \cdots h_n^{\frac{\partial w(g_1,\ldots,g_n)}{\partial x_n}}. \tag{6.6}$$

Example 6.1. For instance, let G be arbitrary group with empty set of operators, and let $w(x) = ax^2bx^{-1}c$ be a polynomial over G, $a, b, c \in G$. Now, if $h \in N \lhd G$, then 'pulling' the element h to the righthand position, i.e., using identities $hg = gh^g$, $(hg)^2 = g^2h^{g^2+g}, \ldots$, and $(hg)^{-1} = g^{-1}h^{-1}$, $(hg)^{-2} = g^{-2}h^{-g^{-1}-1}, \ldots$, we see that (cf. (6.4))

$$w(xh) = w(x)h^{xbx^{-1}c+bx^{-1}c-x^{-1}c}.$$

Note that $xbx^{-1}c + bx^{-1}c - x^{-1}c$ is a derivative of the polynomial $w(x)$.

In the case of polynomials of one variable x, we denote the derivative of the polynomial $w(x)$ by ∂w, for short. Thus, if $N \lhd G$ is an Abelian Ω-invariant normal subgroup of a group G with a set of operators Ω, and if $w(x)$ is a polynomial over G, then for all $g \in G$ the following equality holds:

$$w(gh) = w(g)h^{\partial w(g)}, \tag{6.7}$$

where $\partial w(g)$ is a *value of the derivative* ∂w at the point (element) $g \in G$, i.e., an endomorphism of N. Note that if, additionally, N is a minimal normal subgroup of a finite group G, then N is isomorphic to the additive group of a vector space over $\mathbb{F}_p = \mathbb{Z}/p\mathbb{Z}$. Thus, we can treat values of derivatives of polynomials as linear transformations of this vector space.

Example 6.2. In Example 6.1 let $G = \mathrm{Sym}(4)$ be a symmetric group of permutations of a set of four elements, and let $N = \mathrm{K}_4 \lhd \mathrm{Sym}(4)$ be its unique minimal normal subgroup, which is the Klein group K_4. Note that K_4 is isomorphic to the additive group of a 2-dimensional vector space over a field $\mathbb{F}_2$. The group Sym(4) is a semidirect product $\mathrm{Sym}(4) = A \leftthreetimes B \leftthreetimes \mathrm{K}_4$, where A is a cyclic group of order 2, and B is a cyclic group of order 3. Let a, b be generators of groups A, B, respectively; then $b^a = b^{-1}$. Moreover, we may assume[1] that a, b acts on K_4 by linear transformations with matrices

$$\begin{pmatrix} 1 & 0 \\ 1 & 1 \end{pmatrix} \quad \text{and} \quad \begin{pmatrix} 0 & 1 \\ 1 & 1 \end{pmatrix},$$

respectively. Let $c \in \mathrm{K}_4$, then the value of the derivative of the polynomial $w(x)$ at the point a is

$$\begin{aligned} \partial w(a) &= aba^{-1} + ba^{-1} - a^{-1} = b^{-1} + ba - a \\ &= \begin{pmatrix} 1 & 1 \\ 1 & 0 \end{pmatrix} + \begin{pmatrix} 0 & 1 \\ 1 & 1 \end{pmatrix} \cdot \begin{pmatrix} 1 & 0 \\ 1 & 1 \end{pmatrix} + \begin{pmatrix} 1 & 0 \\ 1 & 1 \end{pmatrix} = \begin{pmatrix} 1 & 0 \\ 0 & 0 \end{pmatrix}. \end{aligned}$$

If G is a finite solvable group, we can define the value of the derivative in the ring of endomorphisms of a certain chief factor of the group G similarly to the case when N is a minimal normal Ω-invariant subgroup of G. Recall that the chief factor of the group G with the system of operators Ω is, by the definition, any factor group H/K, where H and K are normal Ω-invariant subgroups in G, $H \supseteq K$, $H \neq K$, and there is no normal Ω-invariant subgroup S in G such that $H \supseteq S \supseteq K$, $H \neq S$, $S \neq K$. Thus, given a polynomial $w(x)$ over G, the action of $w(x)$ on the factor group G/K is well defined: $w(g) = (w\psi)(g)$, where $g \in G/K$, $\psi : G \to G/K$ is a canonical epimorphism. Foremost, as G is solvable and H/K is a minimal normal Ω-invariant subgroup of G/K, H/K is Abelian; thus, elementary Abelian p-group for some prime p. Therefore, the values of the derivative ∂w in the rings of endomorphism of the chief

[1] by choosing an appropriate basis of the vector space associated to K_4

factors is well defined, and can be regarded as matrices over the corresponding finite field $\mathbb{F}_p$. We denote these values as $\partial_{H/K}(g)$. Note here we may also take $g \in G$ meaning $\partial_{H/K}(g) = \partial_{H/K}(\psi(g))$. It is clear that 'small increment formulas' (6.6) and (6.7) hold in this case as well; however, they are identities in the factor group G/K rather than in the group g.

Example 6.3. Consider a group $G = \mathrm{Sym}(3) \leftthreetimes \mathrm{Q}_2$, where the symmetric group Sym(3) (of order 6) acts on the quaternion group Q_2 (of order 8) by outer automorphisms. We recall that Aut $(\mathrm{Q}_2) \cong \mathrm{Sym}(4)$, and the subgroup $\mathrm{K}_4 \subset \mathrm{Sym}(4)$ is isomorphic to the group of inner automorphisms $\mathrm{Q}_2/\mathrm{Z}(\mathrm{Q}_2)$. The center $\mathrm{Z}(\mathrm{Q}_2)$, which is of order 2, is a fully invariant subgroup in G, and $G/\mathrm{Z}(\mathrm{Q}_2) \cong \mathrm{Sym}(4)$; so $A = \mathrm{Q}_2/\mathrm{Z}(\mathrm{Q}_2)$ is a chief factor of G. As $A \cong \mathrm{K}_4$, A is isomorphic to the additive group of a 2-dimensional vector space over $\mathbb{F}_2$. We can consider a polynomial $w(x)$ from Example 6.1 as a polynomial over G, assuming that a is a transposition in Sym(3), and b is an element of order 3 in Sym(3), and $c \in \mathrm{Q}_2$. Then, identifying automorphisms induced by conjugations by a and by b with the respective 2×2 matrices over $\mathbb{F}_2$ as in Example 6.2, we conclude that the value $\partial_A w(a)$ of the derivative in the ring of endomorphisms End (A) of the chief factor A is the matrix

$$\begin{pmatrix} 1 & 0 \\ 0 & 0 \end{pmatrix} = aba^{-1} + ba^{-1} - a^{-1} = b^{-1} + ba - a = \partial_A w(a).$$

Thus, (6.7) in this case reads

$$w(ah) \cdot \mathrm{Z}(\mathrm{Q}_2) = w(a)h^{\partial_A w(a)} \cdot \mathrm{Z}(\mathrm{Q}_2),$$

for all $h \in \mathrm{Q}_2$.

It should also be pointed out that differential calculus on groups becomes noticeably simpler in one special case, namely, for finite nilpotent groups with an empty set of operators. Since all factors of the chief series of a finite nilpotent group are central (i.e., H/K lies in the center of the factor group G/K) and are prime-order groups (say, of order p), the value of the derivative of the polynomial (6.2) with respect to the ith variable at any point in the ring of endomorphisms of any principal factor is congruent modulo the corresponding p to the degree of the polynomial in ith variable:

$$\deg_i w(x_1, \dots, x_n) = \sum_{i_j = i} n_j;$$

so the 'small increment' formula (6.6) becomes especially simple:

$$w(g_1h_1, \dots, g_nh_n) = w(g_1, \dots, g_n) \cdot h_1^{\deg_1 w(x_1,\dots,x_n)} \cdots h_n^{\deg_n w(x_1,\dots,x_n)}, \qquad (6.8)$$

for all $g_1, \dots, g_n \in G$, $h_1, \dots, h_n \in A$, and for every central factor $A = H/K$ of G. Of course, (6.8) holds in G/K, and *not* necessarily in G.

6.2 Bijective polynomials over finite groups

In this section, we apply derivations on groups to determine whether a polynomial $w(x)$ over a finite solvable group G is measure-preserving; that is, whether w induces a bijective transformation $g \mapsto w(g)$ on G. Further in Section 7.3 we will see that this problem is connected to the problem whether a polynomial over a profinite group preserves the Haar measure on this group.

Let A be a minimal normal Ω-invariant subgroup of a finite solvable group G with operators Ω; then A is an elementary Abelian p-group for a suitable prime p, i.e., A is isomorphic to the additive group of a vector space over $\mathbb{F}_p = \mathbb{Z}/p\mathbb{Z}$. Thus, given a polynomial $w(x) \in G[x]$, for every $g \in G$ the derivative $\partial w(g)$ is a linear transformation on this vector space. Foremost, the polynomial $w(x)$ naturally induces a transformation on the factor group G/A: If $\varphi\colon G \to G/A$ is a canonical epimorphism, this transformation is a well-defined map $w\varphi\colon \varphi(g) \mapsto \varphi(w(g))$, $g \in G$. If this map is a bijection, we will say that w is *bijective modulo the subgroup A*. The following proposition is an immediate consequence of Proposition 2.3 combined with formula (6.7):

Proposition 6.4. *A polynomial $w(x) \in G[x]$ is bijective on G if and only if the following two conditions hold simultaneously:*

(1) *the polynomial w is bijective modulo A, and*

(2) *the derivative $\partial w(g)$ induces a non-singular linear transformation on A, for all $g \in G$.*

From here, by easy induction on the length of chief series of G we deduce the following

Theorem 6.5. *The polynomial $w(x)$ over the finite solvable group G with the set of operators Ω is bijective on G if and only if every matrix $\partial_A w(g)$ is nonsingular, for any chief factor A of the group G and any element $g \in G$.*

This theorem is a trivial generalization of the result of Lausch [284], proved by him for $\Omega = \varnothing$, to the case of a nonempty system of operators Ω. The corresponding result for nilpotent groups with $\Omega = \varnothing$ is especially simple.

Corollary 6.6. *If G is a finite nilpotent group (with an empty set of operators), then the polynomial $w(x) \in G[x]$ is bijective on G if and only if its degree is coprime with the order of G.*

Example 6.7. Let G be a symmetric group of degree 4 (with empty set of operators), and let $w(x) = ax^2bx^{-1}c$, where $a, b, c \in G$. If a, b, c are as in Example 6.2, then w is not bijective on G since $\partial_A w(g)$ is singular whenever $A = \mathrm{K}_4$ and $g = a$. However, the polynomial $v(x) = ax^2cx^{-1}b$ is bijective on G: Indeed, under notation of Example 6.2, $\partial_{\mathrm{K}_4} v(g) = b$ and $\partial_A v(g) = \partial_B v(g) = 1$, for all $g \in G$.

Chapter 7

Ergodic polynomials over groups with operators

In this chapter, we study ergodic polynomial transformations on finite (non-commutative) groups G with a set of operators Ω; that is, we study transitive transformations of form (6.1). Similarly to the commutative case, this problem inevitably leads to the ergodic theory for infinite (although profinite) groups endowed with a non-Archimedean metric. The latter theory is considered in Section 7.3.

The existence of an ergodic polynomial imposes specific constraints both on the group G and on the set of operators Ω. So at the first stage we must describe all groups G and sets of operators Ω such that the group G with the set of operators Ω has ergodic polynomials. At the second stage, we must describe these ergodic polynomials. Thus, at the first stage we must prove a group-theoretic analog of Theorem 2.7 and then develop a version of ergodic theory for groups including the non-Abelian ones. We shall see that the second stage necessarily will force us to consider ergodic (with respect to the Haar measure) transformations on profinite groups endowed with a non-Archimedean metric. Thus, the situation in the non-commutative case resembles the one for the commutative case when the problem of characterization of transitive polynomials over residue rings led us to p-adic ergodic theory on the ring of p-adic integers $\mathbb{Z}_p$.

We restrict our considerations of ergodic polynomials over groups only to the case when the groups are finite since in real-life settings we currently know only finite groups occur. However, we must note that in mathematics the study of ergodic polynomial transformations on (non-Abelian) groups has its own history started with a more than 50 year-old problem of P. Halmoš whether an automorphism of a locally compact but non-compact group can be an ergodic measure-preserving transformation, [167, p. 26]. The problem attracted notable attention and led to a related study of affine ergodic transformations on a group G (that is, ergodic transformations of the form $x \mapsto gx^{\omega}$, $g \in G, \omega \in \mathrm{Aut}(G)$), see e.g. [365] and references therein.

In the late 1960s the theory of polynomials over non-commutative algebraic structures, and especially over groups, emerged, see [286]; development of the latter naturally leads then to the study of polynomial transformations on groups with operators. Thus, results that follow can be considered as a contribution to ergodic theory for non-commutative algebraic structures.

7.1 Basic properties of groups having ergodic polynomials

Denote by $\mathcal{G}$ the class of all finite groups G with the set of operators Ω that have ergodic polynomials in one variable, that is, groups for which there exist transitive transformations of the form

$$x \mapsto w(x) = g_1 x^{\omega_1 n_1} g_2 x^{\omega_2 n_2} \cdots g_k x^{\omega_k n_k} g_{k+1}, \tag{7.1}$$

where $g_i, \ldots, g_{k+1} \in G$, $\omega_1, \ldots, \omega_k \in \Omega$, $n_1, \ldots, n_k \in \mathbb{Z}$. The class $\mathcal{G}$ obviously contains all polynomially complete groups, thus, all finite simple non-Abelian groups, see Subsection 1.2.2. In other words, any transitive transformation of a finite simple non-Abelian group can be represented by a polynomial over this group, and for applications it is important to find the explicit form of this polynomial. Note, however, that in order to solve the analogous problem for a polynomially complete universal algebra of another kind, namely, for a finite field, we use interpolation formulas which allow us to express any mapping of a finite field into itself as a polynomial over this field, see Subsection 1.3.1. As we have already stated (see the end of Subsection 2.2.3) this solution is of no practical value unless the field is of a small order. Arguments of this kind, only in the superlative degree, are also applicable to polynomials over finite simple non-Abelian groups. Indeed, to our best knowledge, currently explicit interpolation formulas are only known for one, the smallest, group of this kind, the alternating group Alt(5) of degree 5, see [32, 285]. However, transitive polynomials that were obtained this way are of length about 10^4; that is, $k \approx 10^4$ in representation (7.1) of these polynomials. This is absolutely unacceptable for any reasonable applications, especially being compared to the order of the group, which is only 60. There is no hope that in the nearest future somebody will solve the problem whether there exist short transitive polynomials over large finite simple non-Abelian groups, e.g., for $\mathrm{Alt}(n)$, $n > 5$, not speaking about expressing these polynomials explicitly.

By virtue of what has been said, it is reasonable to exclude from further consideration finite simple non-Abelian groups. But then, together with these groups, all non-solvable groups must necessarily be excluded as well. Indeed, suppose that G is a finite non-solvable group with a set of operators Ω, $w(x)$ is transitive polynomial over G, and that N is a fully invariant subgroup; that is, N is closed under action of all endomorphisms from End (G). Let $|G : N| = k$. Then it is easy to see that the kth iterate $w^k(x)$ is an ergodic polynomial over the group N considered as a group with the set of operators End (N), cf. Proposition 2.3. Furthermore, if K is a fully invariant subgroup in N, then, by Proposition 2.3, $w^k(x)$ induces a transitive polynomial transformation on the factor-group N/K. However, since the group G is non-solvable, there exist fully invariant subgroups N and K such that the factor-group N/K is isomorphic to the direct power of a finite simple non-Abelian group H, i.e., $N/K \cong H^m$. Indeed, as G is non-solvable, at least one factor G_i/G_{i+1} of composite fully invariant series $G = G_0 \rhd G_1 \rhd \cdots \rhd G_n = \{1\}$ must be non-Abelian. Recall that the series are called fully invariant whenever every G_i is a fully invariant subgroup in G; the series are composite whenever G_{i+1} is a maximal fully invariant subgroup of G that is a subgroup of G_i. So G_i/G_{i+1} is a minimal fully invariant subgroup in G_{i-1}/G_{i+1}. However, a minimal fully invariant subgroup of a finite group is isomorphic to a direct power of a simple group, either Abelian or non-Abelian. This means that if we know how to construct an ergodic polynomial $w(x)$ over the finite non-solvable group G (with some set of operators), then we could also construct an m-dimensional ergodic polynomial transformation on the finite simple non-Abelian group H (with operators). But the arguments used above show that there is no hope to solve the latter problem in the nearest future. Hence, all finite groups for which we may hope to find explicitly transitive polynomials, must not contain simple non-Abelian sections; thus, we have to restrict our considerations with solvable groups only.

Now we state some important properties of groups having transitive polynomials.

Proposition 7.1. *Let G be a finite group with a set of operators Ω, let $w(x)$ be a transitive polynomial on G, let N be an Ω-invariant normal subgroup of G, and let $|G : N| = k$. Then the following is true:*

1. *The polynomial $w^k(x)$ is transitive on the group N, which is considered as a group with a set of operators Ω.*
2. *The polynomial $(w\varphi)(x)$, where φ is a canonical epimorphism of G onto G/N, is transitive on the group G/N, which is considered as a group with a set of operators Ω.*
3. *The subgroup N is a normal Ω-invariant closure of some $g \in N$; that is, N is a minimal subgroup of G that contains all $g^{h\omega}$, where $h \in G$, $\omega \in \Omega$.*[1]
4. *If N is Abelian, then N is either a cyclic group, or N is isomorphic to the direct product of the Klein group K_4 by a cyclic group $\mathrm{C}(m)$ of odd order m, $m \in \mathbb{N}$* (i.e., the case $m = 1$ is also possible).
5. *If $N \cong \mathrm{K}_4$ then there exists either an element $a \in G$ or an operator $\alpha \in \Omega$ that acts on N as an automorphism of order 2.*

Proof. Claims 1 and 2 are just re-statements of corresponding claims of Proposition 2.3 for the case of groups with operators. In view of Claim 1, Claim 4 immediately follows from Theorem 2.4. Claim 3 is a group-theoretic version of Proposition 2.6 and can be proved along similar lines: As $w^k(x)$ is transitive on N, any $h \in N$ can be represented as $w^{ik}(1)$ for a suitable $i \in \mathbb{N}$; whence, N is a normal Ω-invariant closure of $w^k(1)$, cf. representation (6.4) of a univariate polynomial over a group. Finally, Claim 5 actually follows from the following relations that hold in the (non-commutative) ring End (K_4) of all endomorphisms of the group K_4:

$$\alpha_1 + \alpha_1 + \alpha_3 = 0; \qquad \alpha_1, \alpha_1, \alpha_3 \text{ automorphisms of order 2 of } \mathrm{K}_4, \tag{7.2}$$

$$\beta_1 + \beta_2 + 1 = 0; \qquad \beta_1, \beta_2 \text{ automorphisms of order 3 of } \mathrm{K}_4. \tag{7.3}$$

[1]Everywhere in this chapter we assume that Ω contains the identity operator Id.

Here 1 stands for an identity automorphism, and 0 for a null endomorphism of the group K_4 (i.e., $g^1 = g$, $g^0 = 1$ for all $g \in K_4$). Recall that the group K_4 is isomorphic to the additive group of the 2-dimensional vector space over the field $\mathbb{F}_2$, so $\operatorname{End}(K_4)$ is isomorphic to the algebra of all 2×2-matrices over $\mathbb{F}_2$; hence, the above mentioned identities can be verified directly.

Whenever $N \cong K_4$, from Claim 1 it follows that the polynomial $w^k(x)$ induces a transitive transformation on K_4. The latter transformation is of the form $x \mapsto ax^{\psi}$ (as K_4 is Abelian), where ψ is an integer linear combination of products of automorphisms induced on N by conjugations by elements of G and by actions of operators from Ω, see (6.4). By Note 2.5, ψ must be an automorphism of order 2. However, the group $\operatorname{Aut}(K_4)$ is isomorphic to the group $\operatorname{Sym}(3)$, a group of all permutations of 3 elements, and the group $\operatorname{Sym}(3)$ is a semidirect product (split extension) of the cyclic group of order 3 by the cyclic group of order 2. Thus, in view of the identities mentioned above, the conclusion follows. □

Claims 1 and 2 of Proposition 7.1 in combination with Proposition 2.3 can serve as a tool to determine whether a given polynomial $w(x)$ is transitive on a finite group G. The following obvious corollary holds:

Corollary 7.2. *Let G, N, φ, and k be the same as in Proposition 7.1. Then the polynomial $w(x)$ is transitive on G if and only if the polynomial $(w\varphi)(x)$ is transitive on G/N, and $w^k(x)$ is transitive on N.*

Using Corollary 7.2 we are able to determine whether a polynomial $w(x)$ is transitive on a solvable group G: We first verify whether $(w\varphi)(x)$ is transitive on the factor-group G/G', where $\varphi\colon G \to G/G'$ is a canonical epimorphism; then we verify whether $(w^k\psi)(x)$ is transitive on the factor-group G'/G'', where $\psi\colon G \to G/G''$ is a canonical epimorphism and $k = |G : G'|$, etc.

Example 7.3. The polynomial $w(x) = ax^2uvx^5b$ is transitive on the symmetric group $\operatorname{Sym}(4)$, whenever $\operatorname{Sym}(4)$ is represented as a semidirect product $A \leftthreetimes B \leftthreetimes K_4$, where A is a cyclic subgroup of order 2 with the generator a, B is a cyclic subgroup of order 3 with a generator b; $K_4 = \{1, u, v, uv\}$ is the Klein group of order 4, $b^a = b^{-1}$, $u^a = u$, $v^a = uv$, $u^b = v$, $v^b = uv$.

Indeed, $(w\varphi)(x) = ax^7b$, where $\psi\colon \operatorname{Sym}(4) \to \operatorname{Sym}(4)/K_4 = A \leftthreetimes B \cong \operatorname{Sym}(3)$ is an epimorphism. As $\#\operatorname{Sym}(3) = 6$, the polynomial $(w\varphi)(x)$ induces the same transformation on the factor group $\operatorname{Sym}(4)/K_4$ as the polynomial $\bar{w}(x) = axb$ on the group $A \leftthreetimes B$. Since every element from $A \leftthreetimes B$ has a unique representation in the form a^ib^j, where $i \in \mathbb{Z}/2\mathbb{Z}$, $j \in \mathbb{Z}/3\mathbb{Z}$, the polynomial $\bar{w}(x)$ is transitive on $A \leftthreetimes B$.

Now we calculate $w^6(h)$ for $h \in K_4$. Using derivation formulas from Section 6.1, for $s \in A \leftthreetimes B \subset \operatorname{Sym}(4)$ we obtain that $w(sh) = w(s)h^{\partial w(s)} = \bar{w}(s)\cdot(uv)^{s^5b}\cdot h^{\partial w(s)}$; whence for $i = 1, 2, \dots$ we have:

$$w^i(sh) = \bar{w}^i(s) \cdot (uv)^{\sum_{k=0}^{i-1}(\bar{w}^k(s))^5\cdot b\cdot\prod_{\ell=k+1}^{i-1}\partial w(\bar{w}^{\ell}(s))} \cdot h^{\prod_{k=0}^{i}\partial w(\bar{w}^k(s))}.$$

Note that products in this formula are not commutative; e.g.

$$(\bar{w}^k(s))^5 \cdot b \cdot \prod_{\ell=k+1}^{i-1} \partial w(\bar{w}^\ell(s))$$
$$= (\bar{w}^k(s))^5 \cdot b \cdot \partial w(\bar{w}^{k+1}(s)) \cdot \partial w(\bar{w}^{k+2}(s)) \cdots \partial w(\bar{w}^{i-1}(s))$$

in that order (we assume as usual that a product over an empty set of indices is 1). Note that we make all these calculations in the ring End (K_4) of all endomorphisms of the group K_4. As the latter group is merely a additive group of the 2-dimensional vector space over the two-element field $\mathbb{F}_2$ we may actually work with 2×2 matrices over $\mathbb{F}_2$: We choose arbitrarily a basis in this vector space, for instance, putting into correspondence to $u \in K_4$ the vector $(1, 0)$, and to $v \in K_4$ the vector $(0, 1)$, then, as $u^b = v$ and $v^b = uv$, we put into correspondence to the e.g. element b the matrix $\left(\begin{smallmatrix} 0 & 1 \\ 1 & 1 \end{smallmatrix}\right)$. Otherwise, rather then working with matrices, we can make multiplications in Aut (K_4) $= A \leftthreetimes B$ and make additions with the use of relations (7.2)–(7.3); then a and b are just automorphism of respective orders 2 and 3 in Aut (K_4) (which are induced by conjugation by $a, b \in$ Sym(4)), so relations (7.2)–(7.3) of the ring End (K_4) can be rewritten in the following form:

$$ab^2 + ab + a = 0; \tag{7.4}$$

$$b^2 + b + 1 = 0. \tag{7.5}$$

Using either of these ways, we calculate values of the derivative $\partial w(t) = (t+1)t^5 b + (t^4 + t^3 + t^2 + t + 1)b$ for relevant $t = \bar{w}^i(1)$ and finally obtain that

$$w^6(h) = (uv)^{b^2 + ab^2} h^a = vuh^a.$$

However, by Note 2.5, the transformation $h \mapsto vuh^a$ is transitive on K_4. This by Proposition 2.3 finally proves that the polynomial $w(x) = ax^2uvx^5b$ is transitive on Sym(4).

7.2 Finite solvable groups having ergodic polynomials

In this section, we characterize finite solvable groups (with operators) that have ergodic polynomials, following Anashin [19]. First we consider the multivariate case. We characterize finite solvable groups G with system of operators Ω such that there exists a transitive transformation $W = (w_1, \dots, w_n)\colon G^n \to G^n$, where $w_1, \dots, w_n$ are polynomials in n variables.

7.2.1 The multivariate case

It turns out that actually only univariate or bivariate transitive polynomial transformations may exist over finite solvable groups with operators:

Proposition 7.4. *Let G be a finite solvable group with the system of operators Ω. If the mapping $W = (w_1, \dots, w_n)\colon G^n \to G^n$ is transitive, where $w_1, \dots, w_n$ are polynomials in variables $x_1, \dots, x_n$ over the group G with operators Ω, then either $n = 1$, or $n = 2$ and $\#G = 2$.*

Proof. It suffices to show that if $n > 1$ then $n = 2$ and $\#G = 2$. Suppose that N is a minimal nontrivial normal Ω-invariant subgroup in G; then N is an elementary Abelian p-group for some prime p, see Subsection 1.2.2. Denote $m = |G : N|$ the index of N in G. If $m = 1$, then F is a transitive affine transformation of the Abelian group G^n, and by Theorem 2.4, the only possibility is $n = 2$ and G^2 is a Klein group, i.e., $\#G = 2$.

Let $m \neq 1$, i.e., let N be a proper subgroup of the group G. The restriction of the transformation of W^{nm} to the subgroup N^n is a transitive transformation of the subgroup N^n. Since N is Abelian and $n > 1$, by Claim 4 of Proposition 7.1 we conclude that $n = 2$ and $\#N = 2$. However, as N is normal, Ω-invariant and $\#N = 2$, the subgroup N must be central[2], and either $a^\omega = a$ or $a^\omega = 1$ for any $\omega \in \Omega$, $a \in N$. Therefore, if $w(x_1, \dots, x_n)$ is represented by (6.1), then by (6.3), for any $a_1, \dots, a_n \in N$, we have

$$w(a_1, \dots, a_n) = h_w a_1^{d_1(w)} \cdots a_n^{d_n(w)},$$

where $h_w = w(1, \dots, 1) = g_1 \cdots g_{k+1}$, and

$$d_i(w) = \sum_{i_s = i,\, N^{\omega_s} = N} n_s \bmod 2.$$

Now to the mapping $W = (w_1, w_2)$ we put into correspondence the 2×2 matrix $D = (d_{ij})$ over the field $\mathbb{F}_2$, where $d_{ij} = d_j(w_i)$, $i, j \in \{1, 2\}$. Then D induces the endomorphism δ of the subgroup N^2; so W can be represented as

$$W(a, b) = h \cdot (a, b)^\delta$$

for all $a, b \in N$; here $h \in G^2$ does not depend on a, b. It follows from the latter equality that for all $a, b \in N$

$$W^{2m}(a, b) = g \cdot (a, b)^{\delta^{2m}}$$

for a suitable $g \in G^2$, with g being independent of a, b. On the other hand, as was have shown above, W^{2m} is a transitive transformation of the subgroup N^2, and, hence, $g \in N^2$. Since N^2 is an elementary Abelian group of type $(2, 2)$, i.e, $N^2 \cong \mathrm{K}_4$ is a Klein group, it follows from Note 2.5 that the endomorphism δ^{2m} must be a nontrivial involution in the group of automorphisms of the group N^2. However, the algebra of all endomorphisms of the group K_4 is isomorphic to the algebra $\mathrm{L}_2(2)$ of all 2×2 matrices

[2] i.e., $N \subset \mathrm{Z}(G)$, where $\mathrm{Z}(G)$ is a center of the group G

over the field F_2; the group of all automorphisms of the group N^2 is isomorphic to the general linear group $\mathrm{GL}_2(2)$ of dimension 2 over the field $\mathbb{F}_2$; the group $\mathrm{GL}_2(2)$, in turn, is isomorphic to a symmetric group Sym(3) of degree 3, which is a split extension of the group of order 3 by the group of order 2. It is easy to show now that no even degree of any element of the group Sym(3) and, in particular, δ^{2m} can be a nontrivial involution in this group. The contradiction shows that for $m \neq 1$ only $n = 1$ is possible, and this completes the proof of the proposition. □

Now, to characterize finite solvable groups (with operators) having ergodic polynomials, we can restrict our considerations to univariate polynomials. However, we must first impose some more constraints on the system of operators.

Clearly, the existence of a transitive polynomial over a certain group G with the system of operators Ω not only restricts the possible structure of the group G, but also imposes certain constraints on Ω. A transitive polynomial may exist for the given group G with one system of operators and may not exist for the same group G with some other system of operators. The Klein group K_4, an elementary Abelian group of type $(2, 2)$, can serve as an example: If we take the whole group $\mathrm{Aut}\,(\mathrm{K}_4)$ of automorphisms of the group K_4 as Ω, then such a polynomial exists, but if we take as Ω the set of all automorphisms of order 3, then the group K_4 with this system of operators has no ergodic polynomial by Theorem 2.4. Therefore, in order to characterize all finite solvable groups with operators that have ergodic polynomials, it is reasonable to do the following. We should first try to find the description of all finite solvable groups G that admit of ergodic polynomial functions and possess the maximal system of operators Ω, i.e., a system such that any endomorphism of the group G can be induced by a certain operator from Ω, or, to put it otherwise, $\Omega = \mathrm{End}\,(G)$, where $\mathrm{End}\,(G)$ is the set of all endomorphisms of the group G. Then we should describe all ergodic polynomials over each of the finite solvable groups G with the system of operators $\Omega = \mathrm{End}\,(G)$ and, in particular, for every ergodic polynomial w to make a list $\mathrm{E}(w)$ of endomorphisms ω that occur in canonical representation (6.1) of the polynomial w. Then the final formulation of the corresponding classification theorem will be as follows: The finite solvable group G with the system of operators Ω has ergodic polynomials if and only if the group G with the system of operators $\mathrm{End}\,(G)$ has ergodic polynomials, and Ω induces on G all endomorphisms from $\mathrm{E}(w)$ for a certain ergodic polynomial w over the group G with the system of operators $\mathrm{End}\,(G)$. In other words, actually we must describe all finite solvable groups G with operators $\Omega = \mathrm{End}\,(G)$ having ergodic polynomials, and then describe all ergodic polynomials over every such group.

The corresponding classification theorem may be proved, although the proof will demand significant technical efforts and splits into a number of separated cases. Actually the proof does not exist yet since the significance of such a general theorem for applications is questionable at our view. However, to demonstrate methods of the proof, we consider further in this book several cases that look the most instructive, and also may be useful in applications to cryptography and computer science. Namely, we will describe solvable groups G having transitive polynomials in three cases, $\Omega = \varnothing$, $\Omega = \mathrm{Aut}(G)$, and $\Omega = \mathrm{End}(G)$. So denote by $\mathcal{C}_0$, $\mathcal{C}_A$, and $\mathcal{C}_E$ the class of all finite groups with the system of operators $\Omega = \varnothing$, $\Omega = \mathrm{Aut}(G)$, and $\Omega = \mathrm{End}(G)$, respectively, that have ergodic polynomials. Clearly, $\mathcal{C}_0 \subseteq \mathcal{C}_A \subseteq \mathcal{C}_E$. In description of solvable $\mathcal{C}_0$-, $\mathcal{C}_A$-, and $\mathcal{C}_E$-groups we will mainly follow the paper [19].

After we determine solvable groups from all these three classes, we describe ergodic (i.e., transitive) polynomials over some of these groups that we consider the most important in view of possible applications. The latter problem turns out to be a problem of characterization of polynomial ergodic transformations on infinite pro-2-groups endowed with a non-Archimedean metric.

We note that part of the work is already done in the paper [179] that considers the so-called *single orbit groups*. Recall that the latter are groups G having transitive affine transformations, i.e., transitive transformations of the form $x \mapsto ax^{\alpha}$, where $a \in G$, $\alpha \in \mathrm{Aut}(G)$. It turns out that all these finite groups are extensions of cyclic groups by cyclic groups: They have cyclic normal subgroups such that corresponding factor-groups are cyclic. Groups of this type are called *cyclic-by-cyclic groups*, or also *metacyclic* groups; note that the derived length of every this group is 2 whenever the group is non-Abelian. The paper [179] also describes automorphisms α that occur in transitive affine transformations of the mentioned groups.

As we will see, all three classes of solvable $\mathcal{C}_0$-, $\mathcal{C}_A$-, and $\mathcal{C}_E$-groups are wider than the class of finite single-orbit groups: There are a number of finite solvable groups that have ergodic (i.e., transitive) polynomials, and that have not transitive affine transformations.

7.2.2 The univariate case: Nilpotent groups

In this subsection, we determine all finite nilpotent groups G with operators Ω that have transitive polynomials, for the cases $\Omega = \varnothing$, $\Omega = \mathrm{Aut}(G)$, and $\Omega = \mathrm{End}(G)$, i.e., nilpotent groups from the classes $\mathcal{C}_0$, $\mathcal{C}_A$, and $\mathcal{C}_E$. The following theorem is true:

Theorem 7.5. *A finite nilpotent group lies in $\mathcal{C}_E$ if and only if it is either trivial or isomorphic to one of the following groups*:

(1) *to the cyclic group* $\mathrm{C}(m)$ *of order* m, $m = 1, 2, 3, \ldots$;

(2) *to the Klein group* K_4;

(3) *to the dihedral group* $\mathrm{D}_n = \mathrm{gp}\,(u, v \parallel u^2 = v^{2^n} = 1, v^u = v^{-1})$ *of order* 2^{n+1}, $n = 2, 3, 4, \ldots$;

(4) *to the (generalized) quaternion group* $\mathrm{Q}_n = \mathrm{gp}\,(u, v \parallel v^{2^n} = 1, v^u = v^{-1}, u^2 = v^{2^{n-1}})$ *of order* 2^{n+1}, $n = 2, 3, 4, \ldots$;

(5) *to the semidihedral group* $\mathrm{SD}_n = \mathrm{gp}\,(u, v \parallel u^2 = v^{2^n} = 1, v^u = v^{2^{n-1}-1})$ *of order* 2^{n+1}, $n = 3, 4, 5, \ldots$;

(6) *to the direct product* $H \times \mathrm{C}(m)$*, where* H *is a group of type 2–4 and* $m > 1$ *is odd.*

Out of these groups, the groups SD_n *and* $\mathrm{SD}_n \times \mathrm{C}(m)$ *with an odd* m*, and only these groups, do not lie in* $\mathcal{C}_A$*. Finally, the class* $\mathcal{C}_0$ *consists exactly of all cyclic groups* $\mathrm{C}(m)$*,* $m = 1, 2, 3, 4, \ldots$*.*

Proof. As the first derived group G' of a finite nilpotent group G is contained in the Frattini subgroup $\mathrm{Fr}(G)$ of the group G, and as G' is a fully invariant subgroup of G (see Subsection 1.2.2), the factor-group $\tilde{G} = G/G'$ must be an Abelian $\mathcal{C}_0$-group whenever $G \in \mathcal{C}_0$, by Proposition 7.1. Hence, if $w(x) \in G[x]$, then by (6.2), the polynomial $(w\varphi)(x)$ induces on $\tilde{G}$ a transformation of the form $x \mapsto gx^n$, where $g \in \tilde{G}$, $n \in \mathbb{N}_0$, φ is a canonical epimorphism of G onto $\tilde{G}$. It is clear that whenever the transformation $x \mapsto gx^n$ is transitive on $\bar{G}$, the group $\tilde{G}$ is a cyclic group generated by g. But then the group G must be also cyclic as $\ker\varphi$ lies in the Frattini subgroup $\mathrm{Fr}(G)$; see again Subsection 1.2.2. This proves the final claim of Theorem 7.5.

This argument together with Theorem 2.4 implies also that nilpotent groups of odd orders that lie in $\mathcal{C}_A$ or in $\mathcal{C}_E$ must be cyclic. As any finite nilpotent group G is a direct product of p-groups for pairwise distinct prime p that divide $\#G$ (see Subsection 1.2.2), by Proposition 2.3 it suffices to study now only the case when G is a non-Abelian 2-group that lies in $\mathcal{C}_A$ or in $\mathcal{C}_E$; in particular, $\#G = 2^{n+1}$ for some $n = 2, 3, \ldots$. Under these assumptions, Theorem 2.4 together with Proposition 7.1 imply that necessarily $\bar{G} \cong \mathrm{K}_4$.

Now we prove that necessarily G' is cyclic. Indeed, if G' is not cyclic, then combining Theorem 2.4 and Proposition 7.1 we conclude that $G'/G'' \cong \mathrm{K}_4$, as $G'' \subset \mathrm{Fr}(G')$. Thus, the group $H = G/G''$ must be of the following type: $H/H' \cong \mathrm{K}_4$, $H' \cong \mathrm{K}_4$. However, such a group H does not exist. Assuming the opposite, as H is a 2-group, whence nilpotent, the center $\mathrm{Z}(H)$ of H must be non-trivial, so there must exist $z \in \mathrm{Z}(H) \setminus \{1\}$. As $H = H' \cup aH' \cup bH' \cup abH'$ for suitable $a, b \in H \setminus \{1\}$, $a \neq b$, then at least one of elements a, b, ab must centralize H' whenever $z \in H'$, since $\mathrm{Aut}\, H' \cong \mathrm{Sym}(3)$. But then $\#\mathrm{Z}(H)$ is a multiple of 8, so $H/\mathrm{Z}(H)$ is either of order 1 or of order 2. In both cases H is Abelian in contradiction to the assumption $H' \cong \mathrm{K}_4$. Thus, $z \notin H'$; but then the same argument shows that H must be Abelian. The contradiction implies that the group H with the property $H/H' \cong \mathrm{K}_4$, $H' \cong \mathrm{K}_4$ does not exist; so G' is cyclic.

As every element from G acts on G' by conjugation, there exists a homomorphism $\rho \colon G \to \mathrm{Aut}\,(G')$. As G' is a cyclic group of order 2^n, $\mathrm{Aut}\,(G')$ is a direct product of a group of order 2 by a cyclic group of order 2^{n-1}, see e.g. [353, Theorem 9.1]. So all three cases are possible: $\rho(G)$ is a trivial group, $\rho(G)$ is a group of order 2, and $\rho(G) \cong \mathrm{K}_4$. We consider these cases separately. First we introduce some notation. As $G' \subset \mathrm{Fr}(G)$, $G/G' \cong \mathrm{K}_4$, and G is a non-Abelian 2-group, $G' = \mathrm{Fr}(G)$, and the group G is generated by two elements $a, b \in G\{1\}$, $a \neq b$. Denote c a generator of G'.

Case 1: $\rho(G)$ is a trivial group. Then both a and b centralize c; so $G' \subset Z(G)$ and G is nilpotent of class 2. It is clear then that the commutator $[a, b]$ generates G'; so we can take $c = [a, b]$. Thus, $b^{-1}ab = ac$, and hence $b^{-1}a^2b = a^2c^2$. As $a^2 \in G' \subset Z(G)$, then the latter equality implies that $c^2 = 1$. So G is a non-Abelian group of order 8; whence G is isomorphic either to a dihedral group D_4 or to a quaternion group Q_4.

Case 2: $\rho(G)$ is a group of order 2. In this case we may assume that $\rho(a) \neq 1$, $\rho(b) = 1$. Then the centralizer $C_G(G')$ of G' in G is generated by b together with G'. We claim that $C_G(G')$ is a cyclic group.

Indeed, we may assume that $n > 2$ otherwise $\#G = 8$ and $C_G(G') = G$, so $\#\rho(G) = 1$. Now take a subgroup C generated by c^4 in G. The subgroup C is fully invariant in G as a fully invariant subgroup of a fully invariant subgroup G'. Consider a factor-group $\bar{G} = G/C$ and a canonical epimorphism $\psi\colon G \to \bar{G}$. If $C_G(G')$ is not a cyclic group, then its ψ-image $\psi(C_G(G'))$ is not cyclic also. Indeed, as $b^2 \in G'$, then $b^2 = c^{2^r\ell}$ where $2 \nmid \ell$, $r \in \{0, 1, \dots, n-1\}$. If $C_G(G')$ is not cyclic then $r \neq 0$ since otherwise b^2 generates G' and whence b generates $C_G(G')$. Then $C_G(G')$ is a direct product of G' by a cyclic group of order 2 generated by $h = bc^{-2^{r-1}\ell}$. But then, $\psi(C_G(G'))$ is an Abelian group of type $(2, 4)$.

Now consider the group $\bar{G}$. Denote $\bar{a} = \psi(a)$, $\bar{b} = \psi(b)$, and $\bar{c} = \psi(c)$. We see that $\bar{G}$ is generated by two elements, $\bar{a}$ and $\bar{b}$, and that the following equalities hold: $\bar{c}^{\bar{b}} = \bar{c}$, and $\bar{b}^2 = \bar{c}^{2^r\ell}$, i.e., either $\bar{b}^2 = \bar{c}^2$ or $\bar{b}^2 = 1$, where $\bar{c} = [\bar{a}, \bar{b}]$ is an element of order 4. Note that $\bar{b}^2 \in \{1, \bar{c}^2\}$ means that $\psi(C_G(G'))$ is not a cyclic group. We will show that this leads to a contradiction.

As a induces on G' an automorphism of order 2, then $c^a = c^k$, where $k \in \mathbb{Z}/2^{n-1}\mathbb{Z}$ is an element of multiplicative order 2. That is, $k \in \{2^{n-1} - 1, 2^{n-2} - 1, 2^{n-2} + 1\}$. Hence, either $\bar{c}^{\bar{a}} = \bar{c}^{-1}$ or $\bar{c}^{\bar{a}} = \bar{c}$.

If $\bar{c}^{\bar{a}} = \bar{c}$ then $\bar{G}'$, which is generated by $\bar{c}$, lies in the center $Z(\bar{G})$. But then, as $\bar{a}^2 \in \bar{G}'$, we conclude that $[\bar{a}^2, \bar{b}] = 1$. On the other hand, $[\bar{a}^2, \bar{b}] = [\bar{a}, \bar{b}]^{\bar{a}}[\bar{a}, \bar{b}] = \bar{c}^{\bar{a}}\bar{c} = \bar{c}^2$. So $\bar{c}^2 = 1$; however, the order of $\bar{c}$ is 4. The contradiction shows that the only possibility remans: $\bar{c}^{\bar{a}} = \bar{c}^{-1}$.

However, $(\bar{a}\bar{b})^2 \in \bar{G}'$ and the pair of elements $\bar{b}$, $\bar{a}\bar{b}$ generates $\bar{G}$. Hence, as $\bar{b} \in C_{\bar{G}}(\bar{G}')$, the element $(\bar{a}\bar{b})^2$ must lie in the center of $\bar{G}$; so $(\bar{a}\bar{b})^2$ is an element of the subgroup generated by $\bar{c}^2$, which is of order 2. That is, either $(\bar{a}\bar{b})^2 = 1$ or $(\bar{a}\bar{b})^2 = \bar{c}^2$; in other words, either $(\bar{a}\bar{b})^2\bar{c} = \bar{c}$ or $(\bar{a}\bar{b})^2\bar{c} = \bar{c}^{-1}$. However, $(\bar{a}\bar{b})^2\bar{c} = \bar{a}\bar{b}\bar{a}\bar{b}\bar{c} = \bar{a}\bar{b}\bar{a}\bar{c}\bar{b} = \bar{a}\bar{b}\bar{a}[\bar{a}, \bar{b}]\bar{b} = \bar{a}^2\bar{b}^2$; so either $(\bar{a}\bar{b})^2\bar{c} = \bar{a}^2$ or $(\bar{a}\bar{b})^2\bar{c} = \bar{a}^2\bar{c}^2$, depending on $\bar{b}^2$. Thus, at least one of elements $\bar{a}^2$ and $\bar{a}^2\bar{c}^2$ must be equal to one of elements $\bar{c}$ or $\bar{c}^{-1}$. But from any of these equalities it follows that $\bar{a}^2$ is equal either to $\bar{c}$ or to $\bar{c}^{-1}$, hence implying in both cases that $\bar{c}^{\bar{a}} = \bar{c}$. From here in view of the equality $\bar{c}^{\bar{a}} = \bar{c}^{-1}$ we deduce the equality $\bar{c}^2 = 1$. However, the order of $\bar{c}$ is 4; a contradiction. So we finally conclude that $C_G(G')$ is a cyclic subgroup of G of index 2.

Now we will use a known characterization of p-groups having a cyclic subgroup of index p. We state this result for the case $p = 2$ as a lemma; for the general case, as well as for the proof, see e.g. [353, Theorem 9.4].

Lemma 7.6. *Any finite non-Abelian 2-group that has a cyclic subgroup of index 2 is isomorphic to one of the following groups*:

(1) *to the group* $\mathrm{gp}\,(u, v \parallel u^2 = v^{2^n} = 1, v^u = v^{2^{n-1}+1})$, $n = 3, 4, 5, \ldots$*;*

(2) *to the semidihedral group* SD_n, $n = 3, 4, 5, \ldots$*;*

(3) *to the dihedral group* D_n, $n = 2, 3, 4, \ldots$*;*

(4) *to the (generalized) quaternion group* Q_n, $n = 2, 3, 4, \ldots$.

Vice versa, each of the listed groups has a cyclic subgroup of index 2.

However, the group of type 1 from the statement of Lemma 7.6 does not lie in $\mathcal{C}_E$ as its factor-group by a fully invariant subgroup generated by $v^{2^{n-1}}$ is an Abelian group of type $(2, 2^{n-1})$, where $n \geq 3$, and so this factor-group is not a $\mathcal{C}_E$-group by Theorem 2.4. Finally we conclude that within the case $\#\rho(G) = 2$ only groups of type 3–5 from the statement of Theorem 7.5 may lie in $\mathcal{C}_E$.

Case 3: $\rho(G) \cong \mathrm{K}_4$. We will show that no finite 2-group G that satisfies this condition lies in $\mathcal{C}_E$. By Theorem 2.4, it suffices to prove that under this condition the subring of the ring $\mathrm{End}\,(G/G') = \mathrm{L}_2(2)$ generated by the $\tilde{\varphi}$-image $\tilde{\varphi}(\mathrm{End}\,G)$ does not contain non-identity involutions, where $\tilde{\varphi}$ is a mapping of endomorphisms induced by the canonical epimorphism $\varphi\colon G \to \tilde{G} = G/G' \cong \mathrm{K}_4$. We claim that every endomorphism of G induces on $G/G' = \{1, \tilde{a}, \tilde{b}, \tilde{a}\tilde{b}\} \cong \mathrm{K}_4$ one of the following four endomorphisms:

$\tilde{a} \mapsto 1$	$\tilde{b} \mapsto 1$	– null endomorphism
$\tilde{a} \mapsto \tilde{a}$	$\tilde{b} \mapsto \tilde{b}$	– the identity automorphism
$\tilde{a} \mapsto \tilde{a}$	$\tilde{b} \mapsto 1$	– endomorphism, not automorphism
$\tilde{a} \mapsto \tilde{1}$	$\tilde{b} \mapsto \tilde{b}$	– endomorphism, not automorphism

Here $a, b \in G$ are the same as above, $\tilde{a} = \varphi(a)$, $\tilde{b} = \varphi(b)$. In other words, our claim means that $\mathrm{End}\,(G)$ induces on $\tilde{G}$ endomorphisms that correspond respectively to the following four 2×2 matrices over $\mathbb{F}_2$, whenever we consider K_4 as a 2-dimensional vector space over $\mathbb{F}_2$ and choose an appropriate basis:

$$\begin{pmatrix} 0 & 0 \\ 0 & 0 \end{pmatrix}; \begin{pmatrix} 1 & 0 \\ 0 & 1 \end{pmatrix}; \begin{pmatrix} 1 & 0 \\ 0 & 0 \end{pmatrix}; \begin{pmatrix} 0 & 0 \\ 0 & 1 \end{pmatrix}.$$

It is clear that the subalgebra generated by these four matrices in the algebra $\mathrm{L}_2(2)$ of all 2×2 matrices over $\mathbb{F}_2$ contains no non-singular matrix whose multiplicative order is 2. This by Theorem 2.4 in view of Proposition 7.1 implies that $G \notin \mathcal{C}_E$.

To prove this claim, without loss of generality we may assume that if $c = [a,b]$ then

$$a^{-1}ca = c^{-1}, \quad b^{-1}cb = b^{2^{n-1}-1}. \tag{7.6}$$

Note that within the conditions of this case, necessarily $n > 2$. From here it can be easily deduced that the ith subgroup $L_i(G)$ from the lower central series of G is a cyclic group of order 2^{n-i+2} generated by the element $c^{2^{i-2}}$, $i = 2,3,\dots,n+2$; recall that $L_1(G) = G$, $L_2(G) = [L_1(G), G] = G'$, $L_3(G) = [L_2(G), G]$, ... It is clear that in our situation $L_{n+1}(G) = \mathrm{Z}(G)$ is a group of order 2. From (7.6) we obtain $a^{-2}b^{-1} = b^{-1}a^{-2}$; whence

$$a^2 \in \mathrm{Z}(G). \tag{7.7}$$

Further, (7.6) implies that $ac^{2^{n-1}} = b^{-2}ab^2$; as $b^2 \in G'$, from the latter equality with the use of (7.6) we deduce that

$$b^4 = c^{2^{n-1}}. \tag{7.8}$$

From here it follows that b is an element of order 8. As $b^2 = c^\ell$ for a suitable ℓ, from (7.8) it follows that $2\ell \equiv 2^{n-1} \pmod{2^n}$. Changing if necessary the system $\{a,b\}$ of generators of the group G to the system $\{a, b^{-1}\}$, we may assume that $\ell \equiv 2^{n-2} \pmod{2^n}$, i.e., that

$$b^2 = c^{2^{n-2}}. \tag{7.9}$$

With the use of relations (7.6)–(7.9), we now are able to prove our claim.

For $\varepsilon \in \mathrm{End}\,(G)$ only one of the following four possibilities may occur:

$$a^\varepsilon = ac^s; \quad a^\varepsilon = abc^s; \quad a^\varepsilon = bc^s; \quad a^\varepsilon = c^s,$$

for some $s \in \mathbb{Z}$. If $a^\varepsilon = abc^s$ then $a^{2\varepsilon} = (abc^s)^2$; from here combining (7.6)–(7.9) we deduce that $a^{2\varepsilon} = a^2c^{s2^{n-1}+2s+2^{n-2}+1}$, in contradiction to (7.7): As $\mathrm{Z}(G) = L_{n+1}(G)$ is a fully invariant subgroup of order 2 that contains a^2, then $a^{2\varepsilon}$ must be in $\mathrm{Z}(G)$, whereas $a^2c^{s2^{n-1}+2s+2^{n-2}+1}$ is in $\mathrm{Z}(G)$ for no $s \in \mathbb{Z}$ since $\mathrm{Z}(G)$ is generated by $c^{2^{n-1}}$.

If $a^\varepsilon = bc^s$, then in a similar way we obtain that $a^{2\varepsilon} = (bc^s)^2 = c^{2^{n-2}+s2^{n-1}}$, in contradiction to (7.7). By a similar argument we prove that neither $b^\varepsilon = abc^t$ nor $b^\varepsilon = ac^t$ can hold for some $t \in \mathbb{Z}$ as well. This proves our claim, thus ending considerations of the final case 3. So we proved that a finite nilpotent $\mathcal{C}_E$-group must be one of the groups listed in the statement of Theorem 7.5.

To prove the remaining assertions of the theorem, note that from the results of the paper [179] it follows that all groups of type 1–4 as well as corresponding direct products of type 6 from the statement of the theorem are single orbit groups, whence, $\mathcal{C}_A$-groups, whereas semidihedral groups (that of type 5) and hence their direct products by cyclic groups of odd orders are not single orbit groups.

We shall show that nevertheless the group $\mathrm{SD}_n = \mathrm{gp}\,(u, v \parallel u^2 = v^{2^n} = 1, v^u = v^{2^{n-1}-1})$, $n = 3,4,5,\dots$, is in $\mathcal{C}_E$; this in view of Proposition 2.3 implies that all direct products of semidihedral groups SD_n are in $\mathcal{C}_E$ as well. It suffices only to present a transitive polynomial over the group SD_n with operators $\mathrm{End}(\mathrm{SD}_n)$.

It is easy to verify that there exist endomorphisms $\alpha, \beta, \gamma \in \mathrm{End}(\mathrm{SD}_n)$ such that

$$\begin{cases} u^\alpha = u \\ v^\alpha = uv^{2^{n-1}} \end{cases} \qquad \begin{cases} u^\beta = uv^2 \\ v^\beta = v \end{cases} \qquad \begin{cases} u^\gamma = u \\ v^\gamma = 1 \end{cases}.$$

We claim that the polynomial $w(x) = uvx^\alpha x^\beta x^\gamma$ over the group SD_n with operators $\mathrm{End}(\mathrm{SD}_n)$ is transitive on this group. Direct calculations show that $w^4(v^{2t}) = v^{2(t+2^{n-2}+1)}$ for all $t = 0, 1, 2, \ldots$; that is, $w^4(h) = v^{2^{n-1}+2}h$ for all $h \in \mathrm{SD}'_n$. As the derived group SD'_n is a cyclic group of order 2^{n-1} generated by the element v^2, from Theorem 4.36 combined with Theorem 4.23 it follows that the polynomial $w^4(x)$ is transitive on SD'_n: Indeed, the latter group is isomorphic to the additive group of the residue ring $\mathbb{Z}/2^{n-1}\mathbb{Z}$, and up to this isomorphism the polynomial $w^4(z)$ induces the same transformation on SD'_n as the polynomial $f(x) = 2^{n-2} + 1 + x$ induces on the ring $\mathbb{Z}/2^{n-1}\mathbb{Z}$. However, the latter transformation is transitive on $\mathbb{Z}/2^{n-1}\mathbb{Z}$ by Theorem 4.36, so the polynomial $w^4(x)$ is transitive on SD'_n.

Further, if we consider the factor-group $\mathrm{SD}_n/\mathrm{SD}'_n \cong \mathrm{K}_4$ as the 2-dimensional vector space over $\mathbb{F}_2$, then the polynomial $w(x)$ induces on this vector space the transformation

$$(y, z) \mapsto (1, 1) + (y, z)\left(\begin{pmatrix} 1 & 0 \\ 1 & 0 \end{pmatrix} + \begin{pmatrix} 1 & 0 \\ 0 & 1 \end{pmatrix} + \begin{pmatrix} 1 & 0 \\ 0 & 0 \end{pmatrix}\right) = (1, 1) + (y, z)\begin{pmatrix} 1 & 0 \\ 1 & 1 \end{pmatrix},$$

which is obviously transitive. Thus, by Proposition 7.1, the polynomial $w(x)$ is transitive on SD_n. This finally proves Theorem 7.5. □

Note 7.7. Note that Theorem 7.5 together with results of the paper [179] imply that all $\mathcal{C}_A$-groups are single-orbit groups, whereas $\mathcal{C}_E$-groups are not: Semidihedral groups SD_n lie in $\mathcal{C}_E \setminus \mathcal{C}_A$.

7.2.3 The univariate case: Solvable groups

In this subsection, we determine all finite solvable groups G with operators Ω that have transitive polynomials, for the cases $\Omega = \varnothing$, $\Omega = \mathrm{Aut}(G)$, and $\Omega = \mathrm{End}(G)$, i.e., solvable groups from the classes $\mathcal{C}_0$, $\mathcal{C}_A$, and $\mathcal{C}_E$. It turns out that that there are not too many types of finite solvable non-nilpotent groups of this kind: Loosely speaking, these groups are either non-cyclic metacyclic groups, or extensions of (meta)cyclic groups by groups that in some sense 'look like' either a symmetric or an alternating group of degree 4. Moreover, derived lengths of all $\mathcal{C}_E$-groups are not greater than 3, although from Theorem 7.5 we know that there exist nilpotent $\mathcal{C}_E$-groups of arbitrarily large class.

In order to formulate the corresponding theorem, we introduce the following groups:

- $\mathrm{M}(m,k,s) = \mathrm{gp}\,(c,d \parallel c^m = d^k = 1, d^c = d^s)$.
  Here $m,k = 2,3,4,\ldots$, $s \not\equiv 1 \pmod k$, $s^m \equiv 1 \pmod k$, m and k are coprime; so $\mathrm{M}(m,k,s) = \mathrm{C}(m) \leftthreetimes \mathrm{C}(k)$. These groups are metacyclic, thus, metabelian, i.e., solvable of derived length exactly 2. Note that we assume that groups $\mathrm{M}(m,k,s)$ are non-abelian (otherwise $s = 1$ and the group is cyclic, $\mathrm{C}(mk)$). It is clear that all Sylow p-subgroups of these groups $\mathrm{M}(m,k,s)$ are cyclic: If p^n is the maximum power of prime p that divides mk, then either $p^n \mid m$, or $p^n \mid k$, so the Sylow p-subgroup of $\mathrm{M}(m,k,s)$ is conjugate either to a Sylow p-subgroup of the group $\mathrm{C}(m)$ or to a Sylow p-subgroup of the group $\mathrm{C}(k)$. Furthermore, these groups $\mathrm{M}(m,k,s)$ form a class of the so-called *Z-groups*, i.e., finite groups whose Sylow p-subgroups are all cyclic, for every prime $p \mid mk$, see e.g. [353]. As $\mathrm{C}(m) \leftthreetimes \mathrm{C}(k) = (\mathrm{C}(m_1) \times C) \leftthreetimes \mathrm{C}(k) = \mathrm{C}(m_1) \leftthreetimes (\mathrm{C}(k) \times C)$, where C is a direct product of all Sylow p-subgroups of $\mathrm{C}(m)$ that centralize the subgroup $\mathrm{C}(k)$, different triples m,k,s may correspond to isomorphic groups. Among all representations of a Z-group G as a semidirect products of cyclic groups of coprime orders, one is distinguished: $G = \mathrm{C}(m) \leftthreetimes \mathrm{C}(k)$ where $\mathrm{Z}(G) \cap \mathrm{C}(k) = \{1\}$; so the action of the generator of $\mathrm{C}(m)$ on $\mathrm{C}(k)$ fixes the only element from $\mathrm{C}(k)$, namely, 1. This representation will be referred to as a *canonical representation of a Z-group* and denoted by $\mathrm{Z}(m_1,k_1,s_1)$; so $\mathrm{M}(m,k,s) \cong \mathrm{Z}(m_1,k_1,s_1)$ for suitable m_1,k_1,s_1. From [353, Proposition 12.11] it follows in particular that $s_1 - 1$ is coprime to k_1. Note that $\mathrm{M}(2,3,2) = \mathrm{Z}(2,3,2) = \mathrm{Sym}(3)$ is a symmetric group of degree 3.

- $\mathrm{A}(r) = \mathrm{gp}\,(b,u,v \parallel b^{3^r} = u^2 = v^2 = 1, uv = vu, u^b = v, v^b = uv)$.
  The group $\mathrm{A}(r)$ is a split extension of the Klein group K_4 by a cyclic group of order 3^r, $r = 1,2,3,\ldots$: $\mathrm{A}(r) = \mathrm{C}(3^r) \leftthreetimes \mathrm{K}_4$. The group $\mathrm{A}(r)$ is solvable of derived length 2, i.e., a metabelian group; in particular, $\mathrm{A}(1) = \mathrm{Alt}(4)$, the alternating group of degree 4.

- $\mathrm{S}(r) = \mathrm{gp}\,(a \parallel a^2 = 1) \leftthreetimes \mathrm{A}(r)$, $r = 1,2,3,\ldots$.
  Here $b^a = b^{-1}$, $u^a = u$, $v^a = uv$. This group is a split extension of the group $\mathrm{A}(r)$ by the cyclic group $\mathrm{C}(2)$ of order 2. The derived length of $\mathrm{S}(r)$ is 3; in particular, $\mathrm{S}(1) = \mathrm{Sym}(4)$ is a symmetric group of degree 4.

- $\mathrm{AQ}(r) = \mathrm{gp}\,(b \parallel b^{3^r} = 1) \leftthreetimes \mathrm{Q}_2$, $r = 1,2,3,\ldots$.
  Here $u^b = v^{-1}$, $v^b = uv^{-1}$. The group $\mathrm{AQ}(r)$ is a split extension of the quaternion group Q_2 of order 8 by a cyclic group $\mathrm{C}(3^r)$ of order 3^r. The group $\mathrm{AQ}(r)$ is a metabelian group.

- $\mathrm{SQ}_1(r) = \mathrm{gp}\,(a \parallel a^2 = 1) \leftthreetimes \mathrm{AQ}(r)$, $r = 1,2,3,\ldots$.
  Here $b^a = b^{-1}$, $u^a = u^{-1}$, $v^a = uv$. This group is a solvable group of derived length 3.

- $\mathrm{SQ}_2(r) = \mathrm{gp}\,(a, b, u, v \parallel b^{3^r} = v^4 = 1, b^a = b^{-1}, u^a = u^{-1}, v^a = uv, u^b = v^u = v^{-1}, v^b = uv^{-1}, a^2 = u^2 = v^2), r = 1, 2, \ldots$.

  The group $\mathrm{SQ}_2(r)$ is a partial semidirect product of the group $\mathrm{AQ}(r)$ by the cyclic group $A = \mathrm{gp}\,(a \parallel a^4 = 1)$ of order 4; the amalgamated subgroups (those generated by $a^2 \in A$ and by $u^2 \in \mathrm{Q}_2 \subset \mathrm{AQ}(r)$) are cyclic groups of order 2. The group $\mathrm{SQ}_2(r)$ is a solvable group; its derived length is 3.

Neither of the above groups is nilpotent. These groups are main 'building blocks' of solvable groups with operators that have transitive polynomials: It turns out that the latter groups are (semi)direct products of the above groups as well as of nilpotent groups from Theorem 7.5.

Theorem 7.8. *A finite solvable group lies in $\mathcal{C}_E$ if and only if it is isomorphic to one of the following groups*:

(1) $\mathrm{C}(m)$, **(2)** $\mathrm{M}(m, k, s)$, **(3)** K_4, **(4)** Q_n, **(5)** D_n, **(6)** SD_n, **(7)** $\mathrm{A}(r)$, **(8)** $\mathrm{AQ}(r)$, **(9)** $\mathrm{S}(r)$, **(10)** $\mathrm{SQ}_1(r)$, **(11)** $\mathrm{SQ}_2(r)$, **(12)** *$A \leftthreetimes B$, where orders of the groups A and B are coprime, A is any group of type 3–11, B is any group of type 1–2.*

Out of these groups, the following groups lie in $\mathcal{C}_A$: All groups which are isomorphic to any group of type 1–5, 7–11 and all groups which are isomorphic to certain groups of type 12, namely, to groups of the following types 13–15:

(13) *$A \times B$, where A is any group of type 3–5, 7–11, B is any group of type 1–2;* **(14)** *A is any group of type 3–5, B is any group of type 1–2, A acts on B by an automorphism of order 2, and the centralizer of B in A is cyclic.*[3] **(15)** *A is any group of type 9–11, B is any group of type 1–2.*

Finally, out of these groups, exactly all groups which are isomorphic to any group of type 1–2, 9–11, 15, lie in $\mathcal{C}_0$.

To prove the theorem, we need several lemmas.

Lemma 7.9. *Let H and K be $\mathcal{C}_E$-groups of co-prime orders, let G be an extension* (whence, split) *of K by H. If there exists a polynomial over the group G with operators* $\mathrm{End}\,(G)$ *that is transitive on the subgroup $K \subset G$, then $G \in \mathcal{C}_E$.*

Proof. As every element $g \in G = H \leftthreetimes K$ has a unique representation of the form $g = ht$, where $h \in H$, $t \in K$, then every endomorphism $\varepsilon \in \mathrm{End}\,(H)$ can be expanded to the endomorphism $\hat{\varepsilon} \in \mathrm{End}\,(G)$ by putting $g^{\hat{\varepsilon}} = (ht)^{\hat{\varepsilon}} = h^{\varepsilon}$. Let $u(x)$ be a transitive polynomial over the group H with the set of operators $\mathrm{End}\,(H)$, represented in the form (6.1); denote by $\hat{u}(x)$ the polynomial over the group G with the set of operators $\mathrm{End}\,(G)$, obtained from $u(x)$ by substitution of $\hat{\omega}_i$ for all operators ω_i occurring in the representation (6.1) of the polynomial $u(x)$. Let $w(x)$ be a polynomial

[3] This means that if A is either a dihedral group, or a generalized quaternion group of order > 8, the centralizer is a subgroup generated by v; see representation of these groups by generators and relations in the statement of Theorem 7.5.

over the group G with the set of operators End (G) that is transitive on the subgroup K, and let $\iota \in \mathrm{Aut}\,(H)$ be an identity automorphism of H. It is clear that the polynomial $\hat{u}(x^{\hat{\iota}})w(x^{-\hat{\iota}}x)$ is a transitive polynomial over the group G with operators End (G). □

As every polynomial over the group K with empty set of operators can be considered as a polynomial over the group $G = H \leftthreetimes K$, then from Lemma 7.9 we immediately derive the following corollary:

Corollary 7.10. *Let $H \in \mathcal{C}_E$, $K \in \mathcal{C}_0$, let the orders of the groups H and K be coprime. Then the extension of K by H is in $\mathcal{C}_E$.*

Lemma 7.11. *If G is a finite solvable $\mathcal{C}_E$-group of even order, then $G = L \leftthreetimes M$, where L is a $\{2,3\}$-group, M is a $\{2,3\}'$-group, and $L, M \in \mathcal{C}_E$.*

Proof. We prove the lemma by induction on the derived length of G. For Abelian groups the statement of the lemma is obvious.

Let the lemma be true for all solvable groups whose derived length is less than t, and let G be a solvable group of derived length t. Denote by M the unique maximal fully invariant $\{2,3\}'$-subgroup of G; that is, M is a product of all fully invariant $\{2,3\}'$-subgroups of G. Denote by φ a canonical epimorphism of G onto $L = G/M$.

If the derived length of the group L is less than t, then by induction hypothesis $L = L_1 \leftthreetimes M_1$, where L_1 is a $\{2,3\}$-group and M_1 is a $\{2,3\}'$-group. Then M_1 must be trivial since M is a maximal fully invariant $\{2,3\}'$-subgroup of G. Hence, $G = L_1 \leftthreetimes M$.

If the derived length of L is t, then the $(t+1)$th derived group $L^{(t+1)}$ is trivial, whereas the tth derived group $L^{(t)}$ is not. As $L^{(t)}$ is fully invariant in L, $L^{(t)} \in \mathcal{C}_E$. As $L^{(t)}$ is Abelian, $L^{(t)} = L_1 \times M_1$, where L_1 is a $\{2,3\}$-group and M_1 is a $\{2,3\}'$-group. Then M_1 is fully invariant in $L^{(t)}$, whence, fully invariant in L; but then M_1 must be trivial as M is the unique maximal fully invariant $\{2,3\}'$-subgroup in G. Thus, $L^{(t)}$ is an Abelian $\{2,3\}$-group from $\mathcal{C}_E$.

Consider a canonical epimorphism $\psi : L \to H = L/L^{(t)}$. By induction hypothesis, $H = L_2 \leftthreetimes M_2$, where L_2 is a $\{2,3\}$-group and M_2 is a $\{2,3\}'$-group. But then the full ψ-preimage $\psi^{-1}(M_2)$ of the subgroup $M_2 \subset H$, is a split extension of $L^{(t)}$ by M_2, $\psi^{-1}(M_2) = M_2 \leftthreetimes L_1$, as orders of $L^{(t)} = L_1$ and M_2 are coprime. As L_1 is an Abelian $\{2,3\}$-group from $\mathcal{C}_E$, from Theorem 2.4 it follows that Aut L_1 is a $\{2,3\}$-group: Indeed, Aut $(\mathrm{K}_4) = \mathrm{Sym}(3)$ is a group of order 6, and the group of automorphisms of a cyclic 3-group of order 3^r is a cyclic group of order $2 \cdot 3^{r-1}$. But now, as Aut L_1 is a $\{2,3\}$-group and M_2 is a $\{2,3\}'$-group, the semidirect product $M_2 \leftthreetimes L_1$ is a direct product, $\psi^{-1}(M_2) = M_2 \leftthreetimes L_1 = M_2 \times L_1$. The subgroup $\psi^{-1}(M_2)$, which is an epimorphic preimage of a fully invariant subgroup with respect to the epimorphism ψ whose kernel $L^{(t)}$ is fully invariant, is fully invariant in L. As M_2 is fully invariant in $\psi^{-1}(M_2)$, M_2 is fully invariant in L. As M_2 is a $\{2,3\}'$-group, we conclude that M_2 must be trivial: Indeed, L has no non-trivial fully invariant $\{2,3\}'$-subgroups, as

$L = G/M$ and M is a maximal fully invariant $\{2,3\}'$-subgroup in G. Thus, H is a $\{2,3\}$-group; but then, as $L^{(t)}$ is a $\{2,3\}$-group, L must also be a $\{2,3\}$-group. From here it follows that $G = L \leftthreetimes M$. This in view of Proposition 7.1 proves Lemma 7.11. □

Lemma 7.12. *Let G be a finite solvable group of odd order. The group G lies in $\mathcal{C}_E$* (equivalently, in $\mathcal{C}_A$, in $\mathcal{C}_0$) *if and only if G is isomorphic either to a cyclic group* $\mathrm{C}(m)$*, or to a metacyclic group* $\mathrm{M}(m,k,s)$.

Proof. It is clear that $\mathrm{C}(m), \mathrm{M}(m,k,s) \in \mathcal{C}_0$: The polynomial ax is transitive on a finite cyclic group generated by a, and the polynomial cxd is transitive on a metacyclic group $M(m,k,s)$.

Now we prove that the conditions of the lemma are necessary. Let G be a finite solvable $\mathcal{C}_E$-group. Then by Proposition 7.1 all factor-groups $G^{(i)}/G^{(i+1)}$, $i = 0,1,2,\ldots$, are Abelian $\mathcal{C}_E$-groups of odd orders, thus cyclic in view of Theorem 2.4. So G is a supersolvable group. It is well known that the derived subgroup of a supersolvable group is nilpotent. Hence, as the derived subgroup is fully invariant, G' and G/G' must be cyclic in view of Proposition 7.1 and Theorem 2.4. So $G' \cong \mathrm{C}(k)$, $G/G' \cong \mathrm{C}(m)$ for suitable $k, m = 1,2,\ldots$.

If $k = 1$, then G' is trivial and therefore G is cyclic. Let $k > 1$; i.e., let G be non-Abelian. Denote d and c generators of groups G' and G/G', respectively. Denote by φ a canonical epimorphism of G onto G/G', and take an arbitrary φ-preimage $\tilde{c} \in G$ of $c \in G/G'$. Denote $\tilde{C}$ a cyclic subgroup of G generated by $\tilde{c}$; then $G = \tilde{C}G'$.

Further, $d^{\tilde{c}} = d^s$ for some $s = 1,2,\ldots$; however, $s \not\equiv 1 \pmod{k}$ since otherwise G is Abelian in contradiction to our assumption. Thus, $s-1$ and k are coprime. As $\tilde{c}^m \in G'$, then $\tilde{c}^m = d^\ell$ for a suitable rational integer ℓ; hence $d^\ell = \tilde{c}^{-1}\tilde{c}^m\tilde{c} = d^{s\ell}$ and therefore $\ell \equiv 0 \pmod{k}$ since $s-1$ and k are coprime. Thus, $\tilde{c}^m = 1$ and so $G = \tilde{C} \leftthreetimes G'$.

Now to conclude the proof of Lemma 7.12 we must only show that m and k are coprime. Assume, on the contrary, that there exists a prime p that is a factor of both m and k. Let S_1, S_2 be (unique) Sylow p-subgroups in $\tilde{C}$ and G', respectively. Denote S the subgroup of G generated by S_1 and S_2. As $S_1 \subset \tilde{C}$, S_2 is fully invariant in G', and $G = \tilde{C} \leftthreetimes G'$, then $S = S_1 \leftthreetimes S_2$, so $\#S = \#S_1\#S_2$, and therefore S is a non-cyclic fully invariant p-subgroup of G. However, from Proposition 7.1 in view of Theorem 7.5 it follows that non-cyclic fully invariant p-subgroups of $\mathcal{C}_E$-groups must have even orders; thus, the order of G is even, in contradiction to assumptions of Lemma 7.12. □

Note 7.13. Actually during the proof of Lemma 7.12 we have proved that *a supersolvable group G lies in $\mathcal{C}_E$* (equivalently, in $\mathcal{C}_A$, in $\mathcal{C}_0$) *if and only if G is isomorphic either to a cyclic group* $\mathrm{C}(m)$*, or to a metacyclic group* $\mathrm{M}(m,k,s)$, where G may be of arbitrary order, and *not* necessarily of odd order.

During the proof of Theorem 7.8 we will need some information about automorphisms of groups $M(m,k,s)$, i.e., of Z-groups. The structure of automorphism groups of Z-groups is well known, see e.g. [179, Lemma 8.6]. We formulate corresponding results as the following lemma:

Lemma 7.14. *The automorphism group of the group* $\mathrm{Z}(m,k,s) = \mathrm{C}(m) \leftthreetimes \mathrm{C}(k)$ *is isomorphic to the following group:*

$$\mathrm{Aut}\,(\mathrm{Z}(m,k,s)) \cong ((\mathbb{Z}/k\mathbb{Z})^* \leftthreetimes (\mathbb{Z}/k\mathbb{Z})^+) \times \mathrm{C}(m,k,s),$$

where $\mathrm{C}(m,k,s)$ *is a group with respect to multiplication modulo m, consisting of all* $\ell \in (\mathbb{Z}/m\mathbb{Z})^*$ *such that* $s^\ell \equiv s \pmod{k}$*; and the multiplicative group* $(\mathbb{Z}/k\mathbb{Z})^*$ *acts on the additive group* $(\mathbb{Z}/k\mathbb{Z})^+$ *of the residue ring* $\mathbb{Z}/k\mathbb{Z}$ *by multiplication. Namely, every automorphism of the group* $\mathrm{Z}(m,k,s)$ *has a unique representation of the form* $\alpha_t\beta^r\gamma_\ell$*, where* $t \in (Z/k\mathbb{Z})^*$*,* $r \in \mathbb{Z}/k\mathbb{Z}$*,* $\gamma_\ell \in \mathrm{Z}(\mathrm{Aut}\,(m,k,s)) \cong \mathrm{C}(m,k,s)$*,* ℓ *as above, and*

$$\begin{cases} c^{\alpha_t} = c \\ d^{\alpha_t} = d^t \end{cases} \qquad \begin{cases} c^{\beta} = cd \\ d^{\beta} = d \end{cases} \qquad \begin{cases} c^{\gamma_\ell} = c^\ell \\ d^{\gamma_\ell} = d \end{cases}.$$

Furthermore, let $m = p^n$*, where p is an odd prime,* $n \in \mathbb{N}$*. If the group* $\mathrm{Z}(m,k,s)$ *possesses an automorphism whose order is a power of 2, then this automorphism is of the form* $\alpha_t\beta^r$*, that is, lies in the subgroup* $(\mathbb{Z}/k\mathbb{Z})^* \leftthreetimes (\mathbb{Z}/k\mathbb{Z})^+$*.*

Proof. We prove only the last assertion of the lemma since the others are known; see their proofs in e.g. [179, Lemma 8.6]. To prove the latter assertion, it suffices to show that no γ_ℓ is of order 2.

Assume that γ_ℓ is of order 2, where $\ell \not\equiv 1 \pmod{m}$ is coprime to m, and $s^\ell \equiv s \pmod{k}$. Then $\ell^2 \equiv 1 \pmod{m}$; whence $s^{\ell^2} \equiv 1 \pmod{k}$. It is well known that the group $(\mathbb{Z}/p^n\mathbb{Z})^*$ is a cyclic group of order $(p-1)p^{n-1}$ whenever p is an odd prime, see e.g. [353, Theorem 9.1], so the only element of $\mathbb{Z}/p^n\mathbb{Z}$ whose multiplicative order is 2, is $p^n - 1$. Thus, $\ell \equiv m - 1 \pmod{m}$, so $s^{m-1} \equiv s^\ell \equiv s \pmod{k}$. Hence, $1 \equiv s^m \equiv s^2 \pmod{k}$, so the multiplicative order of s modulo k is 2, as $s \not\equiv 1 \pmod{k}$. However, $s^m \equiv 1 \pmod{k}$, so necessarily $2 \mid m$. A contradiction. □

Corollary 7.15. *Let the order of the group* $G \cong \mathrm{Z}(m,k,s)$ *be odd, and let* $\tau \in \mathrm{Aut}\,(G)$ *be an automorphism of order 2. Then there exists a representation* $G \cong \mathrm{M}(m',k',s') = \tilde{C} \leftthreetimes \tilde{D}$*, where* $\tilde{C} = \mathrm{C}(m')$*,* $\tilde{D} = \mathrm{C}(k')$*, such that* τ *acts on* $\tilde{C}$ *identically; thus* τ *acts on* $\tilde{D}$ *as an automorphism of order 2.*

Proof. As $G \cong \mathrm{Z}(m,k,s)$, then $G = C \leftthreetimes D$, where $C \cong \mathrm{C}(m)$, $D \cong \mathrm{C}(k)$ are cyclic subgroups generated by c and d, respectively. The subgroup C is a direct product of Sylow p-subgroups for all $p \mid m$. Each of these Sylow p-subgroups is cyclic, and at least one of these Sylow p-subgroups, say, C_1, acts on D non-trivially by conjugation. As every Sylow p-subgroup of D is invariant under this action, C_1

then acts non-trivially on some of these Sylow p-subgroups; say, on D_1. The subgroup G_1 of G generated by D_1 and C_1 is a characteristic subgroup of G, and is a Z-group $\mathrm{Z}(m_1, k_1, s_1)$, where $m_1 = \#C_1$, $k_1 = \#D_1$. By Lemma 7.14, τ is an automorphism of the form $\alpha_t\beta^r$; whence $c_1^\tau = c_1 d_1^r$, where c_1, d_1 are generators of C_1 and D_1, respectively. As $(\alpha_t\beta^r)^2 = \alpha_{t^2}\beta^{r(t+1)}$, then $t^2 \equiv 1 \pmod{m_1}$, $r(t+1) \equiv 0 \pmod{m_1}$. Since m_1 is a power of an odd prime, from the first of the latter congruences it follows that $t \equiv \pm 1 \pmod{m_1}$ (see the argument from the proof of Lemma 7.14). However, the assumption $t \equiv 1 \pmod{m_1}$ leads to a contradiction since the congruence $r(t+1) \equiv 0 \pmod{m_1}$ implies then that $r \equiv 0 \pmod{m_1}$ as m_1 is odd; whence $\alpha_t\beta^r$ is an identity automorphism. So $t \equiv -1 \pmod{m_1}$; then $r(t+1) \equiv 0 \pmod{m_1}$. Direct verification shows now that $c_1 d_1^{\bar{2}r}$, where $\bar{2}$ stands for the multiplicative inverse of 2 modulo m_1, is a fixed point of the automorphism τ. Furthermore, the order of the element $c_1 d_1^{\bar{2}r}$ is m_1: $(c_1 d_1^{\bar{2}r})^{m_1} = c_1^{m_1} d_1^{\bar{2}r(s_1^{m_1-1}+\cdots+s_1+1)}$, the order of c_1 is m_1, and $s_1^{m_1-1} + \cdots + s_1 + 1 \equiv 0 \pmod{k_1}$ since otherwise $(s_1 - 1)(s_1^{m_1-1} + \cdots + s_1 + 1) = s_1^{m_1} - 1 \equiv 0 \pmod{k_1}$, in contradiction to the condition $(s_1 - 1, k_1) = 1$, see the definition of a group $\mathrm{Z}(m_1, k_1, s_1)$. We conclude finally that the subgroup G_1 is a semidirect product $C_2 \leftthreetimes D_1$, where C_2 is generated by the element $c_1 d_1^{\bar{2}r}$; and τ acts identically on C_2. This way we proceed with all Sylow p-subgroups of C that do not centralize D. Now, denoting by $\tilde{C}$ a direct product of these Sylow p-subgroups, and by $\check{C}$ a direct product of all Sylow p-subgroups of C that centralize D, we see that $G = \tilde{C} \leftthreetimes (D \times \check{C})$, i.e., that $B \cong \mathrm{M}(m', k', s')$, where $m' = \#\tilde{C}$, $k' = \#\tilde{D}$, $\tilde{D} = D \times \check{C}$, and τ acts on $\tilde{C}$ identically. □

Now everything is ready for the proof of Theorem 7.8.

Proof of Theorem 7.8. We first prove that the conditions of Theorem 7.8 are necessary. Let G be a finite solvable $\mathcal{C}_E$-group. If $\#G$ is odd, then by Lemma 7.12 the group G is either cyclic or isomorphic to a metacyclic group $\mathrm{M}(m, k, s)$. If $\#G$ is even, then combining Lemma 7.11 with Lemma 7.12 we conclude that G is a split extension of a $\{2,3\}'$-group $\tilde{G}$ of type 1 or 2 by a $\{2,3\}$-group $\bar{G}$ from $\mathcal{C}_E$. If $2 \nmid \#\bar{G}$, then from Lemma 7.12 it follows that the group G is either of type 1 or of type 2.

If $3 \nmid \#\bar{G}$, then $\bar{G}$ is a 2-group from $\mathcal{C}_E$; all these groups are determined by Theorem 7.5. If $\bar{G}$ is non-cyclic then G is a group of type 12. If $\bar{G}$ is cyclic, then G is a supersolvable group, so G is either of type 1 or of type 2, by Note 7.13.

Thus we assume that $6 \mid \#\bar{G}$. We also need some extra notation: Given a finite group U and a prime p, denote by $O_p^*(U)$ the (unique) maximal fully invariant p-subgroup of U; that is, $O_p^*(U)$ is a product of all fully invariant proper p-subgroups of U, and $O_p^*(U) = \{1\}$ if U has no fully invariant proper p-subgroups. Denote $K = O_2^*(\bar{G})$ and consider two cases: K is trivial and K is non-trivial.

Case 1: $K = \{1\}$. Denote $T = O_3^*(\bar{G})$; then from Theorem 7.5 combined with Proposition 7.1 it follows that T is a cyclic group. Denote $G_1 = \bar{G}/T$, $R = O_2^*(G_1)$,

$\psi : \bar{G} \to G_1$ a canonical epimorphism, $\tilde{R} = \psi^{-1}(R)$ a preimage of the subgroup R. Then $\tilde{R} = R \leftthreetimes T$.

Note that the subgroup R cannot have proper fully invariant subgroups that centralize T: Otherwise, if R_1 is a subgroup of this kind, the preimage $\psi^{-1}(R_1)$ is a fully invariant subgroup in $\bar{G}$, and $\psi^{-1}(R_1) = R_1 \times T$, and thus $R_1 \subset O_2^*(\bar{G})$; so $O_2^*(\bar{G})$ is non-trivial, in contradiction to our assumption.

Now, as R is a 2-group from $\mathcal{C}_E$ in view of Proposition 7.1, R must be one of the groups determined by Theorem 7.5. The group R acts on T by an automorphism of order 2 and R has no fully invariant subgroups that centralize T; however, the only 2-groups from the statement of Theorem 7.5 that posses this property are a cyclic group of order 2, and the Klein group K_4.

We claim that if $\#R = 2$ then $R = G_1$. Indeed, $O_2^*(G_1/R)$ is trivial; thus if $T_1 = O_3^*(G_1/R)$ is non-trivial, then G_1 must contain a fully invariant subgroup isomorphic to $T_1 \times R$. As T_1 is a 3-group and R is a 2-group, the subgroup T_1 is then a fully invariant 3-subgroup of G_1. Whence, $O_3^*(G_1)$ is non-trivial. However, $G_1 = \bar{G}/O_3^*(\bar{G})$; a contradiction.

Thus, if $\#R = 2$ then $\bar{G} = R \leftthreetimes T$ is a group of type 2. So the group G is an extension of the $\{2,3\}'$-group of type 2 by the $\{2,3\}$-group $\bar{G} = R \leftthreetimes T$, where R and T are cyclic. Then G is a split extension of a metacyclic group of type 2 by a metacyclic group of type 2; it is easy to see that all these split extensions are supersolvable. But then G is of type 2 by Note 7.13.

Let now $R \cong \mathrm{K}_4$. If $R = G_1$ then $\bar{G}$ is a split extension of a cyclic 3-group T by the Klein group K_4. As $\mathrm{Aut}\,(\mathrm{C}(3^r))$ is a cyclic group of order $2 \cdot 3^{r-1}$, the group K_4 may act on T either trivially (then $\bar{G} = \mathrm{K}_4 \times T$), or by automorphism of order 2. In the latter case the group $\bar{G}$ is a direct product of a cyclic group of order 2 by a metacyclic group of type $M(2, 3^r, s)$ for suitable r, s. Then the group G, which is an extension of a metacyclic group by the group $\bar{G}$, is supersolvable; whence, of type 2 by Note 7.13.

We now prove that the case $R \neq G_1$ can not occur. Assuming the opposite, and combining Proposition 7.1 with Theorem 2.4, we conclude that $O_3^*(G_1/R) = T_1$ must be a non-trivial cyclic 3-group, since G_1 is a $\{2,3\}$-group. Then $G_1 \cong \mathrm{K}_4 \leftthreetimes T_1$. Hence, $T_1 = O_3^*(G_1)$; i.e., G_1 has a non-trivial maximal fully invariant 3-subgroup. However, the latter is a contradiction, as $G_1 = \bar{G}/O_3^*(\bar{G})$.

Case 2: $K \neq \{1\}$. Combining Proposition 7.1 with Theorem 7.5 we see that then K is a 2-group of either type 1, 3, 4, 5, or 6. We denote $T = O_3^*(\bar{G}/K)$, $\varphi : \bar{G} \to \bar{G}/K$, a canonical epimorphism. By Proposition 7.1, in view of Theorem 7.5 the group T is a cyclic 3-group. Then the preimage $\varphi^{-1}(T)$ is a split extension, of K by T. We consider two cases: K centralizes T (in $\varphi^{-1}(T)$) and K does not centralizes T.

In the first case $\varphi^{-1}(T) = K \times T$ is a fully invariant subgroup in $\bar{G}$ and thus both K and T are fully invariant subgroups in $\bar{G}$; hence, $\bar{G} = K \times T$ as both K and T are maximal fully invariant 2- and 3- subgroups, respectively. Thus, $\bar{G}$ is a group of type 12. As G is a split extension of a $\{2,3\}'$-group $\tilde{G}$ of type 1 or 2 by the $\{2,3\}$-group $\bar{G} = K \times T$, $G = \bar{G} \leftthreetimes \tilde{G}$, then the subgroup $T \leftthreetimes \tilde{G}$ is a fully invariant supersolvable subgroup in G; so $T \leftthreetimes \tilde{G} \in \mathcal{C}_E$ by Proposition 7.1, whence $T \leftthreetimes \tilde{G}$ is a group of type 2 by Note 7.13. Finally, G is a group of type 12.

If T does not centralizes K, then T acts on K by an automorphism of order 3^ℓ for some $\ell > 1$. However, we have already shown that K is a 2-group of either type 1, 3, 4, 5, or 6; from these groups only the Klein group K_4 and the quaternion group Q_2 of order 8 posses automorphisms whose orders are powers of 3, see e.g. [353]. So only two cases are possible, either $K \cong K_4$ or $K \cong Q_2$. Then $\ell = 1$ in both cases.

Further, either $\varphi^{-1}(T) = \bar{G}$, or $\varphi^{-1}(T)$ is a proper subgroup of $\bar{G}$. If $\varphi^{-1}(T) = \bar{G}$ then $\bar{G}$ is a group either of type 7 or of type 8. Then, the group G is of type 12.

If $\varphi^{-1}(T)$ is a proper subgroup of $\bar{G}$, then $\bar{G}/K$ is a group considered within case 1. Hence, $\bar{G}/K = R \leftthreetimes T$, where $R = O_2^*(\bar{G}/K)$, and R is either a cyclic group of order 2, or $R \cong K_4$. In both cases R acts on the cyclic 3-group T by an automorphism of order 2.

We claim that if $\#R = 2$ then $\bar{G}$ is a group of either type 9, 10, or 11; whence G is of type 12. Denote $\tilde{a}, \tilde{b}$ generators of the groups R and T, respectively. Then the group $\bar{G}$ is generated by the subgroup K and by elements $a, b \in \bar{G}$, where $a \in \varphi^{-1}(\tilde{a})$, $b \in \varphi^{-1}(\tilde{b})$. Note that $a^2 \in K$ and that the subgroup of $\bar{G}$ generated by b is isomorphic to T (whence is a cyclic 3-group).

Let $K \cong K_4$. Then b acts on K by automorphism of order 3, as $\mathrm{Aut}\,(K_4) \cong \mathrm{Sym}(3)$. As $\tilde{a}$ acts on T by automorphism of order 2, then $b^{a^2} = bw$ for a suitable $w \in K$; thus choosing if necessary new generator bw for T, we may assume that $b^{a^2} = b$. This implies that $a^2 = 1$ since $a^2 \in K \cong K_4$, and every automorphism of order 3 from $\mathrm{Aut}\,(K_4)$ has no fixed points other than 1. This proves that $\bar{G} \cong S(r)$, where r is the order of b; that is, $\bar{G}$ of type 9, whence G of type 12.

Let $K \cong Q_2$. Then b acts on K by automorphism of order 3, and a acts by automorphism of order 2 since $\mathrm{Aut}\,(Q_2) \cong \mathrm{Sym}(4)$. Moreover, as then $\bar{G}/C_{\bar{G}}(K) \cong \mathrm{Sym}(3)$, the action of a on K corresponds to a transposition from $\mathrm{Sym}(4)$. Thus, a^2 centralizes both b and K; so necessarily $a^2 \in Z(K)$. As $\#Z(K) = 2$, we conclude that $\bar{G}$ is either of type 10, if $a^2 = 1$, or of type 11, if $a^2 \neq 1$. This concludes considerations of the case when $\#R = 2$.

We argue that the rest case $R \cong K_4$ cannot occur. Indeed, in this case $\bar{G}/K = \tilde{C} \times (\tilde{A} \leftthreetimes T)$, where $\tilde{C}, \tilde{A}$ are cyclic groups of order 2, which are generated, say, by $\tilde{c}$ and $\tilde{a}$, respectively. The element $\tilde{a}$ acts on T by an automorphism of order 3. Take $c \in \varphi^{-1}(\tilde{c})$; then $c^2 \in K$. If $K \cong Q_2$, then $Z(K)$ is fully invariant in K, whence, in $\bar{G}$. So considering if necessary $\bar{G}/Z(K) \in \mathcal{C}_E$ instead of $\bar{G}$, we may assume that $K \cong K_4$. In this case necessarily $c^2 = 1$ as action of b on $K \cong K_4$ has no fixed points except 1. Furthermore, the subgroup generated by b^3 is fully invariant in $\bar{G}$; so the corresponding factor-group must be in $\mathcal{C}_E$. However, the latter factor-group is isomorphic to the direct product $H = \mathrm{Sym}(4) \times \mathrm{C}(2)$. We argue that the latter group is not in $\mathcal{C}_E$.

Let $w(x)$ be a transitive polynomial over the group H with the set of operators $\mathrm{End}\,(H)$. As $\mathrm{Sym}(4) = \mathrm{Sym}(3) \leftthreetimes \mathrm{K}_4$, then every element $y \in H$ has a unique representation of the form $y = zh$, where $z \in \mathrm{Sym}(3) \times \mathrm{C}(2)$, $h \in \mathrm{K}_4$. As K_4 is fully invariant in H, then $w(zh) = w(z)h^{\partial w(\xi)}$, where $\xi = \tau(z)$, $\tau : H \to \mathrm{Aut}\,(\mathrm{K}_4) = \mathrm{Sym}(3)$ is a canonical epimorphism with a kernel $C_H(\mathrm{K}_4) = \mathrm{K}_4 \times \mathrm{C}(2)$, and ∂w is a derivative of the polynomial x with respect to variable x. As $w(x)$ is bijective on H, then ∂w maps automorphisms (of K_4) to automorphisms, see Sections 6.1 and 6.2. Using relevant derivation formulas from Section 6.1, for every $h \in \mathrm{K}_4$ we obtain that

$$w^{12}(h) = w^{12}(1)h^{\partial w^{12}(\iota)} = w^{12}(1)h^{\partial w(\omega_0)\cdots\partial w(\omega_{11})},$$

where $\omega_j = \tau(w^j(1))$, $j = 0, 1, 2, \ldots, 11$, $\omega_0 = \iota$ is an identity automorphism. Denote $\alpha = \partial w(\omega_0)\cdots\partial w(\omega_{11})$, $u = w^{12}(1)$. By Claim 1 of Proposition 7.1, $u \in \mathrm{K}_4$, and the affine mapping $h \mapsto uh^{\alpha}$ is transitive on K_4. Then, by Theorem 2.4, the automorphism α must be a transposition in $\mathrm{Sym}(3) = \mathrm{Aut}\,(\mathrm{K}_4)$.

On the other hand, if $\varepsilon : H \to H/\mathrm{K}_4 = \mathrm{Sym}(3) \times \mathrm{C}(2)$ is a canonical epimorphism, then by Proposition 7.1 the polynomial $(w\varepsilon)(x)$ must be transitive on H/K_4 as K_4 is fully invariant in H. Thus, in the sequence $(\omega_j = \tau(w^j(1)))_{j=0}^{11}$ every element from $\mathrm{Sym}(3) = \mathrm{Aut}\,(\mathrm{K}_4)$ occurs exactly twice. However, in this case $\alpha = \partial w(\omega_0)\cdots\partial w(\omega_{11})$ lies in $\mathrm{Alt}(3)$, and whence cannot be a transposition. The contradiction proves that $H \notin \mathcal{C}_E$.

Thus we finally have proved that finite solvable $\mathcal{C}_E$-groups are groups of type 1–12. Now we are going to study the same question for $\mathcal{C}_A$-groups.

From Theorem 7.5 we already know that semidihedral groups SD_n are not in $\mathcal{C}_A$. We wonder what groups of type 12 could lie in $\mathcal{C}_A$. So let $G = A \leftthreetimes B$ be a $\mathcal{C}_A$-group of type 12, where B is a group of type 1–2 whose order is coprime to that of A, and A is a group of type 1–5, 7–11.

If A centralizes B, then $G = A \times B$ is a group of type 13. Suppose that A does not centralize B. If additionally A is a group of type 9–11, then G is of type 15.

Let now A be either a Klein group K_4, or dihedral group D_n, or A (generalized) quaternion group Q_n. Consider the case $B = \mathrm{C}(m)$ first. As in all cases the derived group A' centralizes B, then A acts on B either as a group of order 2, or as a group K_4. We argue that the latter case does not take place.

To prove this claim, it suffices to assume that $A = \mathrm{K}_4$ since $\mathrm{D}_n/\mathrm{D}_n' \cong \mathrm{Q}_n/\mathrm{Q}_n' \cong \mathrm{K}_4$. So, let $A = \{1, u, v, uv\}$, where $u^2 = v^2 = 1$, $uv = vu$. Associate $B = \mathrm{gp}\,(c \parallel c^m = 1)$ to the additive group of the residue ring $\mathbb{Z}/m\mathbb{Z}$. Then $c^u = c^i$, $c^v = c^j$, where i, j are involutions in the multiplicative group of $\mathbb{Z}/m\mathbb{Z}$, $i \not\equiv j \pmod{m}$. By Claim 2 of Proposition 7.1 in composition with Theorem 2.4, there exists $\alpha \in \mathrm{Aut}\,(G)$ that induces automorphism of order 2 on $G/B \cong \mathrm{K}_4$. Hence, only the following three cases may occur:

(1) $u^{\alpha} = vc^r$, $v^{\alpha} = uc^s$, $c^{\alpha} = c^t$;

(2) $u^\alpha = uvc^r, v^\alpha = vc^s, c^\alpha = c^t$;

(3) $u^\alpha = uc^r, v^\alpha = uvc^s, c^\alpha = c^t$.

Here $r, s, t \in \mathbb{Z}/m\mathbb{Z}$, t is coprime with m. However, each of these possibilities leads to a contradiction. For instance, consider the second one: On the one hand, $(c^u)^\alpha = (c^i)^\alpha = c^{it}$; and on the other hand, $(c^u)^\alpha = (c^\alpha)^{u^\alpha} = (c^t)^{uv} = c^{tij}$. Thus, $ti \equiv tij \pmod{m}$, whence $i \equiv j \pmod{m}$, a contradiction. Arguments for the rest possibilities are similar to the presented one, we leave details to the reader.

Thus, we have established that A acts on B as a group of order 2. In the latter case, only the following two variants may occur:

(i) $c^v = c, c^u = c^i$;

(ii) $c^u = c, c^v = c^i$,

where i is an involution in the multiplicative group of the residue ring $\mathbb{Z}/m\mathbb{Z}$, and u, v are generators of the groups D_n and Q_n in their representations by generators and relations (see the statement of Theorem 7.5), if either $A \cong \mathrm{D}_n$ or $A \cong \mathrm{Q}_n$; otherwise, u, v are elements of K_4 as above. Note that the case (i) implies that the centralizer of B in A is a cyclic subgroup of index 2. The case (ii) implies that whenever $A \cong \mathrm{D}_n$, or $A \cong \mathrm{Q}_n$ and $n > 2$, the centralizer is not cyclic, although of index 2 as well.

We assert that if either $A \cong \mathrm{D}_n$, or $A \cong \mathrm{Q}_n$ and $n > 2$, action of type (ii) can not take place. Namely, we will show that in this case no automorphism from $\mathrm{Aut}\,(G)$ acts on the factor-group $G/(A' \times B)$ as an automorphism of order 2. However, by Claim 2 of Proposition 7.1 combined with Theorem 2.4, there must exist an automorphism from $\mathrm{Aut}\,(G)$ that acts on the factor-group $G/(A' \times B) \cong \mathrm{K}_4$ as an automorphism of order 2 since otherwise $G \notin \mathcal{C}_A$: In the ring $\mathrm{End}\,(\mathrm{K}_4)$, no automorphism of order 2 from $\mathrm{Aut}\,(\mathrm{K}_4)$ can be expressed as a linear combination of automorphisms of orders other than 2, see the argument in the proof of Proposition 7.1.

Let either $A \cong \mathrm{D}_n$, or $A \cong \mathrm{Q}_n$ and $n > 2$. It is easy to verify that an automorphism $\tilde{\alpha} \in \mathrm{Aut}\,(A)$ that acts on $A/A' \cong \mathrm{K}_4$ by automorphism of order 2 must send u to uv^r and v to v^s, where both s and t are odd. Thus, if $\alpha \in \mathrm{Aut}\,(G)$ acts on $G/(A' \times B)$ as an automorphism of order 2, then, on the one hand, $(c^u)^\alpha = c^\alpha = c^t$ and $(c^u)^\alpha = c^{u^\alpha} = c^v = c^i$ for t coprime with m; so $t \equiv i \pmod{m}$. However, on the other hand, $(c^v)^\alpha = (c^i)^\alpha = c^{it}$ and $(c^v)^\alpha = c^{v^\alpha} = c^v = c^i$; so $it \equiv i \pmod{m}$. Combining the two congruences, we conclude that $i^2 \equiv i \pmod{m}$. At the same time, i is an involution in the multiplicative group of the ring $\mathbb{Z}/m\mathbb{Z}$; a contradiction. Thus we have finally proved that if A does not centralize B, and A is a group of type 3–5, B of type 1, then G is of type 14. We now consider the same problem for the case when B is of type 2; in particular, $B = C \leftthreetimes D = \mathrm{M}(m, k, s)$, where C, D are cyclic groups generated, respectively, by elements c, d.

We can assume that A centralizes C. Indeed, as B is a Z-group, we may assume that $B = \mathrm{Z}(m, k, s)$. The subgroup C is a direct product of Sylow p-subgroups for all $p \mid m$. Every this Sylow p-subgroup is cyclic, and at least one of these Sylow

p-subgroups, say, C_1, acts on D non-trivially by conjugation. As every Sylow p-subgroup of D is invariant under this action, C_1 then acts non-trivially on some of these Sylow p-subgroups; say, on D_1. The subgroup B_1 of B generated by D_1 and C_1 is a characteristic subgroup of B and a Z-group $\mathrm{Z}(m_1, k_1, s_1)$, where $m_1 = \#C_1$, $k_1 = \#D_1$. Since A is isomorphic either to K_4, or to D_n, or to Q_n, in view of Lemma 7.14 the group A acts on B_1 either trivially, or by an automorphism of order 2. If A acts on B_1 as an automorphism of order 2, then by Corollary 7.15 we conclude that the subgroup B_1 is a semidirect product $C_2 \leftthreetimes D_1$ of cyclic groups whose orders are coprime one to another, and A centralizes C_2. This way we proceed with all Sylow p-subgroups of C that do not centralize D. Now, denoting by $\tilde{C}$ a direct product of these Sylow p-subgroups, and by $\check{C}$ a direct product of all Sylow p-subgroups of C that centralize D, we see that $B = \tilde{C} \leftthreetimes (D \times \check{C})$, i.e., that $B \cong \mathrm{M}(\tilde{m}, \check{k}, \bar{s})$, where $\tilde{m} = \#\tilde{C}$, $\check{k} = \#(D \times \check{C})$, and A centralizes $\tilde{C}$.

Thus, we can assume now that A centralizes the subgroup C of $\mathrm{M}(m, k, s) = B$. However, as A does not centralize B, A must act on D non-trivially. Hence, the group G is a semidirect product $G = C \leftthreetimes (A \leftthreetimes D)$, and orders of subgroups C and $A \leftthreetimes D$ are coprime. This implies that $A \leftthreetimes D$ is a characteristic subgroup in G, whence, a $\mathcal{C}_A$-group by Proposition 7.1. However, the group $A \leftthreetimes D$ is a group of type 14, as we have already shown above. This ends considerations of the case when $G = A \leftthreetimes B$ is a $\mathcal{C}_A$-group of type 12, where B is a group of type 1–2, and A is a group of type 1–5.

We now consider the rest case of $\mathcal{C}_A$-groups; the one when $G = A \leftthreetimes B$ is a $\mathcal{C}_A$-group of type 12, where B is a group of type 1–2, and A is a group of type 7–8. We will prove that in this case A centralizes B; thus the semidirect product $G = A \leftthreetimes B$ is in fact a direct product $G = A \times B$ and so G is of type 13.

First let $A = \mathrm{A}(r)$, $B \cong \mathrm{C}(k)$ a cyclic group generated by d, where k is coprime to 6. As $\mathrm{Aut}\,(B)$ is Abelian and K_4 is a minimal normal subgroup in A, then necessarily K_4 centralizes B, and either A centralizes B, or $b \in A$ acts on B non-identically. We will show that the latter case can not occur.

As K_4 is a characteristic subgroup in G, by Proposition 7.1 there exists $\alpha \in \mathrm{Aut}\,(G)$ that acts on K_4 as an involution. Given $g \in G$, denote by $\hat{g}$ an automorphism of K_4 induced by a conjugation by g. As b acts on K_4 as an automorphism of order 3, in $\mathrm{Aut}\,(\mathrm{K}_4) = \mathrm{Sym}(3)$ the following equality then holds: $\alpha^{-1}\hat{b}\alpha = \hat{b}^2$. Thus, $b^\alpha = b^2 h$ for a suitable $h \in C_G(\mathrm{K}_4) = U \leftthreetimes (\mathrm{K}_4 \times B)$, where U is generated by b^3. We have that $d^h = d^q$, $d^b = d^t$ for suitable $t, q \in \mathbb{N}$, t, q coprime to k. Furthermore, $q \equiv t^{3\ell} \pmod{k}$ for a suitable $\ell \in \mathbb{N}_0$. Consider now the element $(d^b)^\alpha$. On the one hand, $(d^b)^\alpha = (d^t)^\alpha = (d^\alpha)^t$; whilst on the other hand, $(d^b)^\alpha = (d^\alpha)^{b^\alpha} = (d^\alpha)^{b^2 h}$. Thus, $t \equiv t^2 q \pmod{k}$, so $tq \equiv 1 \pmod{k}$; whence $t^{1+3\ell} \equiv 1 \pmod{k}$. However, at the same time $t^{3^r} \equiv 1 \pmod{k}$, as $d = d^{b^{3^r}}$. We finally conclude that necessarily $t \equiv 1 \pmod{k}$, and thus A centralizes B in this case.

If $A = \mathrm{A}(r)$, $B \cong \mathrm{M}(m, k, s) = \mathrm{C}(m) \leftthreetimes \mathrm{C}(k)$, then from the structure of automorphism groups of Z-groups (see Lemma 7.14) it follows that necessarily K_4 centralizes B. Denote by $C \cong \mathrm{C}(m)$, $D \cong \mathrm{C}(k)$ the corresponding cyclic subgroups of B, and let c, d be their respective generators. As D is a characteristic subgroup in G, the above argument (of the case when B is cyclic) proves that A centralizes D. From Lemma 7.14 it follows then that b acts on B by an automorphism β^n for some $n \in \mathbb{N}_0$, i.e., $c^b = cd^n$ (we may assume that the semidirect product $B = C \leftthreetimes D$ is the one from the canonical representation of the Z-group $B \cong \mathrm{Z}(m,k,s)$). But then $c = c^{b^{3^r}} = cd^{3^r n}$, so $3^r n \equiv 0 \pmod{k}$, and whence $n \equiv 0 \pmod{k}$ as $3 \nmid k$.

It is worth making here an important note that will be used later during the proof: *If A is of type 9–11, $B = C \leftthreetimes D \cong \mathrm{M}(m,k,s)$, where $C \cong \mathrm{C}(m)$, $D \cong \mathrm{C}(k)$, and if a semidirect product $G = A \leftthreetimes B$ is not a direct product, then A acts on B by an automorphism of order 2, the one that is induced by a conjugation by $a \in A$; moreover, we may assume that a centralizes the subgroup C of B* by just taking a proper representation of $B \cong \mathrm{M}(m',k',s')$, see Corollary 7.15. Indeed, from the structure of automorphism groups of Z-groups (see Lemma 7.14) it follows that K_4 centralizes B in this case as well, and then the above argument shows that $b \in A$ centralizes B.

Returning to the consideration of $\mathcal{C}_A$-groups, we see that the rest case $A = \mathrm{AQ}(r)$ can be easily reduced to the case $A = \mathrm{A}(n)$, which we have already considered. Indeed, the center $Z = \mathrm{Z}(\mathrm{Q}_2)$ of the quaternion group Q_2 is a fully invariant subgroup of the group $G = A \leftthreetimes B$, and Z centralizes B; the latter assertion immediately follows from the structure of automorphism groups of cyclic groups and of Z-groups. Thus, the factor group $\bar{G} = G/Z$ must lie in $\mathcal{C}_A$. However, $\bar{G} \cong \mathrm{A}(r) \leftthreetimes B$. Whence, the above argument about groups $\mathrm{A}(r) \leftthreetimes B$ proves that the subgroup $\mathrm{AQ}(r)$ centralizes B in G. This finally ends considerations of $\mathcal{C}_A$-groups.

Now we consider the remaining case, when G is a $\mathcal{C}_0$-group. By Theorem 7.5, dihedral, semidihedral and (generalized) quaternion groups are not in $\mathcal{C}_0$. By Claim 5 of Proposition 7.1, the group $\mathrm{A}(r)$ is not in $\mathcal{C}_0$ as a conjugation by the element $g \in \mathrm{A}(r)$ induces on the normal subgroup K_4 only either an identical automorphism, or an automorphisms of order 3. This in view of Claim 2 of Proposition 7.1 implies that the group $\mathrm{AQ}(r)$ is not in $\mathcal{C}_0$ either as it is a factor group of the group $\mathrm{A}(r)$. This completes considerations of the case of $\mathcal{C}_0$-groups.

Now we are going to prove that all the groups listed in the statement of Theorem 7.8 are indeed $\mathcal{C}_E$-, $\mathcal{C}_A$-, and $\mathcal{C}_0$-groups, respectively. From Theorem 7.5 we already know that semidihedral groups are in $\mathcal{C}_E$, that dihedral, (generalized) quaternion groups, and the Klein group are in $\mathcal{C}_A$, and that cyclic groups are in $\mathcal{C}_0$. It is clear that metacyclic groups of type 2 are in $\mathcal{C}_0$, as the polynomial $w(x) = cxd$ is obviously transitive on the group $\mathrm{M}(m,k,s)$ since every element of this group admits a unique representation of the form $c^i d^j$, where $i \in \mathbb{Z}/m\mathbb{Z}$, $j \in \mathbb{Z}/k\mathbb{Z}$, and $(m,k) = 1$. If we prove that the groups $\mathrm{S}(r)$, $\mathrm{SQ}_1(r)$ an $\mathrm{SQ}_2(r)$ are all in $\mathcal{C}_0$, we proof by Proposition 7.1 that the groups $\mathrm{A}(r)$ and $\mathrm{AQ}(r)$ are in $\mathcal{C}_A$, as the latter groups are normal subgroups of groups of type 9–11. In turn, this in view of Corollary 7.10 and of Proposition 2.3 will prove respectively that groups of type 12 are all in $\mathcal{C}_E$, and that groups of type 13 are $\mathcal{C}_A$-groups. By [179, Theorem 6.2], all groups of type 14 are single orbit groups, whence,

$\mathcal{C}_A$-groups. Thus, to complete the proof of Theorem 7.8 it suffices to prove that the groups of type 9–11, and 15 are $\mathcal{C}_0$-groups. For this purpose, it suffices to present a transitive polynomial for each of these groups.

In what follows, let $\ell = 8 \cdot 3^r$, $6 + t\ell \equiv 0 \pmod{m}$, $6 + t_1\ell \equiv 0 \pmod{mk}$. We assert:

1. The polynomial $w_1(x) = ax^2uvx^5b$ is transitive on either group of type 9–11.
2. The polynomial $w_2(x) = ax^2uvx^5bx^{t\ell}c$ is transitive on either group of type 15, where A is a group of type 9–11, and B is a cyclic group of order m generated by c.
3. The polynomial $w_3(x) = acx^2uvx^5bx^{t_1\ell}d$ is transitive on either group of type 15, where A is a group of type 9–11, and $B = \mathrm{M}(m,k,s)$;

To prove assertion 1, note that by Example 7.3, the polynomial $w_1(x)$ is transitive on the group $\mathrm{Sym}(4) \cong \mathrm{S}(1)$. If $r > 1$, then b^3 centralizes the subgroup K_4 of $\mathrm{S}(r)$, and following the calculations from Example 7.3, we obtain that

$$w_1^{24}(b^{3n}h) = b^{3(n+4)}(uv)^{a^3+a^2+a+1}h^{a^4} = b^{3(n+4)}h,$$

where $n \in \mathbb{N}_0$, $h \in \mathrm{K}_4$. As the transformation $w_1^{24}(b^{3n}) : b^{3n} \mapsto b^{3(n+4)}$, $n = 0, 1, 2, \ldots$, is transitive on the cyclic subgroup $\dot{B}$ generated by b^3, and as $\#(\mathrm{S}(r)/\dot{B}) = \#\,\mathrm{Sym}(4) = 24$, from Proposition 2.3 it follows that the polynomial $w_1(x)$ is transitive on the group $\mathrm{S}(r)$.

Now denote by $\mathrm{SQ}(r)$ either of groups $\mathrm{SQ}_1(r)$ and $\mathrm{SQ}_2(r)$. Denote $Z = \mathrm{Z}(\mathrm{Q}_2) = \{1, z\}$ the center of the quaternion subgroup $\mathrm{Q}_2 \subset \mathrm{SQ}(r)$ (thus, $z = u^2 = v^2$), and consider a factor group $S = \mathrm{SQ}(r)/Z \cong \mathrm{S}(r)$ and a corresponding epimorphism $\varphi : \mathrm{SQ}(r) \to S$. We already know that the polynomial $(w_1\varphi)(x)$ is transitive on S, so ti prove that the polynomial $w_1(x)$ is transitive on $\mathrm{SQ}(r)$, by Proposition 2.3 it suffices to show that the $\#S$th iterate $\tilde{w}(x) = w_1^{\#S}(x)$ of the polynomial $w_1(x)$ is transitive on Z. However, $w_1(z) = w_1(1) \cdot z^{\deg w_1} = w_1(1) \cdot z$, so $\tilde{w}(z) = \tilde{w}(1) \cdot z$; so as $\tilde{w}(1) \in Z$, we only must show that $\tilde{w}(1) = z$.

To do this, it is convenient to represent the quaternion group Q_2 as a set of all triples (α, β, γ) over $\mathbb{F}_2$ with a multiplication

$$(\alpha, \beta, \gamma) \cdot (\alpha_1, \beta_1, \gamma_1) = (\alpha + \alpha_1, \beta + \beta_1, \gamma + \gamma_1 + \alpha_1\beta + \alpha\alpha_1 + \beta\beta_1).$$

It can be verified directly that this is indeed an isomorphic representation of the quaternion group Q_2, where u corresponds to $(1, 0, 0)$, v corresponds to $(0, 1, 0)$, and $u^2 = v^2 = z$ corresponds to $(0, 0, 1)$. With the use of this representation, by direct calculations[4] in the factor group $\mathrm{SQ}(r)/\dot{B}$ we obtain that $\tilde{w}^6((\alpha, \beta, \gamma))$ and $(\alpha + \beta + 1, \beta + 1, \alpha\beta + \gamma + 1)$ are congruent modulo the subgroup $\dot{B}$ generated by b^3, i.e., lie in a common coset with respect to $\dot{B}$. $\tilde{w}^{8 \cdot 3^r}((\alpha, \beta, \gamma))$ as well) and $(\alpha, \beta, \gamma + 1)$ are congruent modulo $\dot{B}$. But the latter means that $\tilde{w}^{\#S}(1) = \tilde{w}^{\#S}((0, 0, 0)) = (0, 0, 1) = z$

[4]in a manner of these from Example 7.3

as $\tilde{w}^{\#S}(1) \in Z$ (since $\tilde{w}(x)$ is transitive on S) and $Z \cap \dot{B} = \{1\}$. This finally proves our assertion 1. In turn, in view of Proposition 2.3 this also proves our assertions 1 and 2 whenever the semidirect products $A \leftthreetimes B$ they concern are direct products.

Now we shall prove assertions 2 and 3 under the assumption that the corresponding semidirect product $G = A \leftthreetimes B$ is not a direct product. We consider two cases, when $B \cong \mathrm{C}(m)$, and when $B \cong \mathrm{M}(m, k, s)$, respectively.

Let $A = \mathrm{S}(r)$, $B = \mathrm{C}(m)$, then, as $\mathrm{Aut}\,(\mathrm{C}(m))$ is an Abelian group, in the semidirect product $A \leftthreetimes B$ the subgroup A acts on B by automorphism of order 2, which is a conjugation by the element $a \in \mathrm{S}(r)$, and all elements from $\mathrm{A}(r)$ centralize B. Thus we have that $G = \bar{A} \leftthreetimes (\mathrm{A}(r) \times C)$, where A is a cyclic group of order 2 generated by a, $C \cong \mathrm{C}(m)$ is a cyclic group generated by c.

Denote $\tilde{w}(x) = w_2(x) \cdot c^{-1}$. Note that from what we have shown above, it follows that the polynomial $\tilde{w}(x)$ is transitive on the subgroup $\bar{A} \leftthreetimes \mathrm{A}(r) \cong \mathrm{S}(r)$ since $g^{t\ell} = 1$ for all $g \in \mathrm{S}(r)$. For all $y \in \mathrm{A}(r)$, $h \in C$ we have $w_2^2(yh) = w_2^2(y) \cdot h^{\partial w_2^2(y)}$. However, as

$$\partial w_2^2(y) = (y^{6+t\ell} + y^{5+t\ell} + \cdots + y + 1) \cdot ((w_2(y))^{6+t\ell} + (w_2(y))^{5+t\ell} + \cdots + w_2(y) + 1),$$

for all $y \in \mathrm{S}(r)$ the derivative $\partial w_2^2(y)$ takes the value 1 in the ring $\mathrm{End}\,(C) \cong \mathbb{Z}/m\mathbb{Z}$ since $\mathrm{S}(r)$ acts on C by an automorphism of order 2 and m is odd. Thus, $w_2^2(h) = w_2^2(y) \cdot h$. Further, as $w_2^2(h) = \tilde{w}(\tilde{w}(h) \cdot c) \cdot c$, and values of $\partial\tilde{w}(y)$ and $\partial w_2(y)$ in the ring $\mathrm{End}\,(C)$ are equal, we have that $w_2^2(h) = \tilde{w}^2(h) \cdot c^2$; hence, $w_2^2(yh) = \tilde{w}^2(y) \cdot c^2 h$. As $\tilde{w}(x)$ transitive on $\mathrm{S}(r)$, the polynomial $\tilde{w}^2(x)$ is transitive on $\mathrm{A}(r)$ by Proposition 7.1. Foremost, the mapping $h \mapsto c^2 h$, $h \in C$ is transitive on C as $\#C$ is odd. This implies finally that the polynomial $w_2^2(x)$ is transitive on the subgroup $\mathrm{A}(r) \times C$. As this subgroup has index 2 in G, and as the polynomial $(w\varphi)(x) = a x^{7+t\ell}$ (where $\varphi : G \to G/(\mathrm{A}(r) \times C) = \bar{A}$ is an epimorphism) is transitive on the group $\bar{A}$, we conclude that the polynomial $w_2(x)$ is transitive on $\mathrm{S}(r)$ by Proposition 2.3.

A similar argument also proves our assertion 3 in the case $A \cong \mathrm{S}(r)$. Indeed, whenever $B = \mathrm{M}(m, k, s) = C \leftthreetimes D$, where C and D are cyclic groups of orders m, k generated by c and d, respectively, then, according to the note we made above during the proof of the theorem, the group $\mathrm{S}(r)$ not only acts on B by an automorphism of order 2 (which is a conjugation by a), but also a centralizes C: From here it follows that $w_3(gh) = \tilde{w}(g) \cdot chd$ for all $g \in \mathrm{A}(r)$, $h \in B$, where $\tilde{w}(x) = ax^2uvx^5bx^{t_1\ell}$. Further, as

$$w_3^2(gh) = \tilde{w}^2(g)c(ch^{\tilde{w}(g)^{6+t_1\ell}} d^{\tilde{w}(g)^{6+t_1\ell}} ch^{\tilde{w}(g)^{5+t_1\ell}} d^{\tilde{w}(g)^{5+t_1\ell}} \cdots chd),$$

and as on the subgroup B the conjugation by $\tilde{w}(g)$ coincides with the conjugation by a, we see that

$$w_3^2(gh) = \tilde{w}^2(g)c(chd(c^{\tilde{w}(g)} d^{\tilde{w}(g)} chd)^{3+2^{-1}t_1\ell})d = \tilde{w}^2(h)c^2hd^2,$$

where 2^{-1} is a multiplicative inverse of 2 modulo mk; note that $3 + 2^{-1}t_1\ell \equiv 0 \pmod{mk}$ as mk is coprime to 6. Now we finish the proof in this case by an argument similar to that from the preceding case.

To finish the proof of assertions 2 and 3, consider now the case when $G = A \leftthreetimes B$, where $A = \mathrm{SQ}(r)$. Denote $\varphi : G \to \bar{G} = G/B$, then by assertion 1, the polynomial $\bar{w}(x) = (w_1\varphi)(x)$ is transitive on $\bar{G}$. Hence, if $B = \mathrm{C}(m)$, both $w_2^{\ell}(z^j)$ and z^{j+1} lie in a common coset with respect to B since $\ell = \frac{1}{2}\#\mathrm{SQ}(r)$. Here $j \in \{0, 1\}$; we recall that $Z = \{1, z\} = \mathrm{Z}(\mathrm{Q}_2) \subset \mathrm{Z}(G)$. Thus, as $\#B$ is coprime to $\#\mathrm{SQ}(r)$, we have that $w_2^{\ell}(1) = z$. This by Proposition 2.3 concludes the proof in this case since the polynomial $w_2(x)$ is transitive on $G/Z \cong \mathrm{S}(r) \leftthreetimes B$. Proof for the case $B = \mathrm{M}(m, k, s)$ mimics the one for the case $B = \mathrm{C}(m)$, with substitution of $w_3(x)$ for $w_2(x)$. This finally ends the proof of Theorem 7.8. □

7.3 Ergodic theory for profinite groups

In this section, we develop the ergodic theory for polynomials over profinite groups: Actually we consider groups (with operators) that can be approximated by finite solvable groups. These groups can be naturally endowed with a non-Archimedean metric and a natural probabilistic measure, the normalized Haar measure. Polynomials over these groups induces continuous and measurable transformations on these groups, and we study conditions for measure-preservation or ergodicity of these transformations.

The main problem we study in this part is how to determine bijective and/or transitive polynomials over finite groups with operators. In this section we will see that this problem leads to the question how to determine measure-preservation/ergodicity of polynomial transformations on a profinite group. As a matter of fact, we will act in a manner similar to that we proceeded during the study of ergodic polynomial transformations over residue rings: In the latter case, we considered a spectrum of residue rings modulo p^k, $k = 1, 2, \ldots$, with p prime,

$$\cdots \overset{\bmod p^{k+1}}{\longrightarrow} \mathbb{Z}/p^{k+1}\mathbb{Z} \overset{\bmod p^k}{\longrightarrow} \mathbb{Z}/p^k\mathbb{Z} \overset{\bmod p^{k-1}}{\longrightarrow} \cdots \overset{\bmod p}{\longrightarrow} \mathbb{Z}/p\mathbb{Z},$$

where projection epimorphisms are reductions modulo p^k. The inverse limit of this spectrum is a ring $\mathbb{Z}_p$ of p-adic integers

$$\mathbb{Z}_p = \lim_{\substack{\longleftarrow \\ k\to\infty}} \mathbb{Z}/p^k\mathbb{Z},$$

and Theorem 4.23 states that a 1-Lipschitz transforation on $\mathbb{Z}_p$ is ergodic if and only if it is transitive modulo p^k (i.e., ergodic on $\mathbb{Z}/p^k\mathbb{Z}$) for all $k = 1, 2, \ldots$. In particular, the corresponding result for polynomials (Corollary 4.70) reads that a polynomial over $\mathbb{Z}_p$ is ergodic if and only if it is transitive modulo p^3 if $p \in \{2, 3\}$, or modulo p^2, otherwise. A practical impact of this result is that if one needs to determine whether a polynomial is transitive modulo p^k, where k is large (e.g., to use it for pseudorandom number generation, see Chapter 9) he has only to determine whether it is transitive on a much smaller set, of order p^3. This is a general effect that follows from the compatibility of polynomial mappings and from the measurable properties of $\mathbb{Z}_p$. In this section, we demonstrate that a similar effect takes place for non-commutative algebraic structures, namely, for non-Abelian groups with operators: We prove a group-theoretic analog of the mentioned result on ergodic polynomials over p-adic integers for polynomials over inverse limits of finite solvable groups. Also we develop a similar techniques to determine measure-preserving polynomials. The difference between these two cases is that measure-preserving polynomials exist over inverse limits of arbitrary finite solvable groups, whereas ergodic polynomials exist only over some special inverse limits of finite solvable groups, the ones that describes Theorem 7.8.

7.3.1 Metric and measure on a profinite group

First we recall some facts about profinite groups following [261]. Let

$$G_\infty \cdots \stackrel{\varphi_{n+1}}{\longrightarrow} G_n \stackrel{\varphi_n}{\longrightarrow} G_{n-1} \stackrel{\varphi_{n-1}}{\longrightarrow} \cdots \stackrel{\varphi_1}{\longrightarrow} G_0 \stackrel{\varphi_0}{\longrightarrow} \{1\}$$

be an inverse spectrum of groups G_n, $n = 0, 1, 2, \ldots$, and let

$$G_\infty = \varprojlim_{n\to\infty} G_n$$

be the corresponding inverse limit. That is, the group G_∞ possesses an (infinite) decreasing chain of normal subgroups $G_\infty \rhd N_n$,

$$G_\infty \rhd N_0 \rhd N_1 \rhd N_2 \rhd \cdots \rhd \{1\}$$

such that $G_\infty/N_n = G_n$, $\bigcap_{n=0}^{\infty} N_n = \{1\}$, and $\ker \varphi_n = N_{n-1}/N_n$, $n = 1, 2, \ldots$. A group G_∞ is said to be *profinite* whenever all N_n are of finite indices; that is, all G_n are finite groups, $n = 0, 1, 2, \ldots$. Given a prime p, a group G_∞ is called a *pro-p-group* whenever all G_n are p-groups, $n = 0, 1, 2, \ldots$.

A profinite group G_∞ can be endowed with a natural topology, *a profinite topology*, where $\mathcal{N} = \{N_n : n = 0, 1, 2, \ldots\}$ form a base of open neighborhoods of 1, and so all cosets with respect to all these normal subgroups N_n are a base of this topology. The group G_∞ is compact with respect to this topology. Moreover, if $\mathfrak{B}$ is the smallest σ-algebra containing the compact subsets of G_∞, then there is a unique measure μ on $\mathfrak{B}$ such that $\mu(gS) = \mu(Sg) = \mu(S)$ for $g \in G_\infty$ and $S \in \mathfrak{B}$, μ is regular, and $\mu(G_\infty) = 1$. The measure μ is the (normalized) Haar measure on G_∞; actually μ is a natural probability measure on G_∞. Now, given a measurable transformation $g \mapsto w(g)$, $g \in G_\infty$, (where, e.g., $w(x) \in G_\infty[x]$ is a polynomial over G_∞), we may speak of measure-preservation or of ergodicity of this transformation with respect to μ. Note that a polynomial transformation of G_∞ is a measurable transformation as it is a composition of multiplications, which are measurable. Foremost, the group G_∞ can be endowed with a metric d that agrees with the profinite topology on G_∞, and which is a non-Archimedean metric: If $\pi_n : G_\infty \to G_\infty/N_n$ is a canonical epimorphism, put

$$d(x, y) = 2^{-\ell} \text{ where } \ell = \min\{n : \pi_n(x) = \pi_n(y)\},$$

and $d(x, y) = 0$ if $\pi_n(x) = \pi_n(y)$ for all $n \geqslant 0$. Note that given a sequence $\mathcal{g} = (g_n \in G_n)_{n=0}^\infty$ such that $\varphi_n(g_n) = g_{n-1}$ for all $n = 1, 2, \dots$, we consider a sequence $\mathcal{g}' = (g'_n \in G_\infty)_{n=0}^\infty$ such that $\pi_n(g'_n) = g_n$, for all $n = 0, 1, 2, \dots$. The latter sequence $\mathcal{g}'$ converges with respect to metric d to some element $g \in G_\infty$, which has the following property: $\pi_n(g) = g_n$, for all $n = 0, 1, 2, \dots$. The element $g \in G_\infty$ does not depend on choice of representatives g'_n in cosets with respect to normal subgroups $\mathbb{N}_n$; so we call the element g a *limit* of the sequence $\mathcal{g} = (g_n \in G_n)_{n=0}^\infty$. Every element $g \in G_\infty$ is then a limit (in this sense) of a suitable sequence $(g_n \in G_n)_{n=0}^\infty$ such that $\varphi_n(g_n) = g_{n-1}, n = 1, 2, \dots$.

Further, if $f : G_\infty \to G_\infty$ is a compatible mapping (i.e., $f(gN) \subset f(g) \cdot N$ for every $g \in G$, $N \lhd G_\infty$), then for all $n = 0, 1, 2, \dots$ the mapping f mod $N : \pi(g) \mapsto \pi(f(g))$, $(g \in G_\infty)$, where $\pi : G_\infty \to G/N$ is a canonical epimorphism, is a well-defined mapping of G/N into G/N; so we may speak of *bijectivity and transitivity of the mapping f modulo the normal subgroup N* meaning the bijectivity (respectively, transitivity) of the mapping f mod $N : G/N \to G/N$. As usual, when we speak about mappings induces by polynomials, we do not differ polynomials and respective polynomial mappings; so in what follows we speak on measure-preserving/ergodic/transitive …etc. polynomials meaning the respective properties of the corresponding polynomial mappings. The following analog of Theorem 4.23 holds:

Theorem 7.16 ([261]). *Let $w(x) \in G_\infty[x]$ be a polynomial over the profinite group G_∞. Then, the following are equivalent*:

- *w is measure-preserving with respect to the Haar measure μ;*
- *w is bijective modulo N_n, for all $n = 0, 1, 2, \dots$;*
- *w is an isometry with respect to the metric d.*

Also, the following are equivalent:

- *w is ergodic with respect to μ;*
- *w is transitive modulo N_n, for all $n = 0, 1, 2, \dots$.*

Theorem 7.16 is a special case of [261, Theorem 1.1]; we refer the reader for proofs and more detailed information on topological, metric, and other relevant properties of profinite groups to the latter paper [261]. We note that similar statements remain true for groups with the set of operators Ω; we only must consider Ω-invariant normal subgroups rather then ordinary normal subgroups.

7.3.2 Equations, the non-commutative Hensel's lemma, and measure-preserving polynomials over profinite groups

Let $w(x)$ be a polynomial over the profinite group G_∞ from Subsection 7.3.1. We wonder how to determine whether there exists a solution of the equation $w(x) = 1$ in G, i.e., whether there exists $g \in G_\infty$ such that $w(g) = 1$; the 'root of the polynomial $w(x)$'. It is clear that such g exists if and only if the equation $w(x) = 1$ is solvable in all G_n; that is, if and only if there exist $g_n \in G_n$ such that $(w\pi_n)(g_n) = 1$ in G_n, for all $n = 0, 1, 2, \dots$. Indeed, if for every $n = 0, 1, 2, \dots$ we denote $R_n = \{g \in G_\infty : \pi_n(w(g)) = 1\}$, then R_n is closed in G_∞ with respect to the profinite topology, and as all these R_n form a nested sequence (i.e., $R_{n+1} \subset R_n$ for all $n = 0, 1, 2, \dots$) the intersection $R = \bigcap_{n=0}^{\infty} R_n$ is non-empty, see e.g. [278, Chapter 3, Section 34, I].

In notation of Subsection 7.3.1, let G_∞ be an inverse limit of finite solvable groups G_n, $n = 0, 1, 2, \dots$. We may assume that $A_n = N_n/N_{n+1}$ is a minimal normal subgroup in $G_{n+1} = G/N_{n+1}$, for all $n = 0, 1, 2, \dots$; otherwise we make correspondent refinements. Thus, every A_n is an elementary Abelian p_n-group, for a suitable prime p_n. Denote $\psi_n = \varphi_1 \circ \dots \circ \varphi_n : G_n \to G_0$ a composition of epimorphisms $\varphi_n, \dots, \varphi_1$. Then the following analog of Hensel's lemma for profinite groups holds:

Proposition 7.17. *If the equation $w(x) = 1$, where $w(x) \in G_\infty[x]$, has a solution g_0 modulo N_0* (i.e., $(w\pi_0)(g_0) = 1$ in G_0) *and if any derivative $\partial_{A_n} w(g_0')$ is a non-singular matrix over $\mathbb{F}_{p_n}$, for some (equivalently, for any) $g_0' \in \psi_n^{-1}(g_0)$, for all $n = 0, 1, 2, \dots$, then this equation has a solution $g \in G_\infty$ such that $\psi_0(g) = g_0$.*

Proof. Induction on n shows that for any $n = 0, 1, 2, \dots$ there exists a solution $g_n \in G_n$ of the equation $(w\pi_n)(x) = 1$, such that $\psi_n(g_n) = g_0$. Indeed, if $g_n \in G_n$, $(w\pi_n)(g_n) = 1$, $\psi_n(g_n) = g_0$, then $(w\pi_{n+1})(g_n') \in A_n$ for any $g_n' \in \varphi_n^{-1}(g_n)$; thus in view of (6.7), we can choose $h \in A_n$ so that $(w\pi_{n+1})(g_n' h) = 1$, and then put $g_{n+1} = g_n' h$.

It is obvious now that the sequence g_n has a limit $g \in G_\infty$, and that g is a solution we are seeking for. □

From the proof of Proposition 7.17, with the use of (6.8) we immediately deduce the following analog of Hensel's lemma for profinite pro-p-groups:

Corollary 7.18. *If in the conditions of Proposition 7.17 all groups G_n are p-groups for some prime p, and if $p \nmid \deg w$, then the equation $w(x) = 1$ has a solution in G_∞.*

This corollary has interesting connections with Part I of the book: Using it, we can solve functional equations in the group $\mathrm{Syl}_2(\infty)$ of 1-Lipschitz measure-preserving transformations on the space $\mathbb{Z}_2$ of 2-adic integers.

From Theorem 4.39 immediately follows that the latter group is an inverse limit of 2-groups (of orders 2^{2^n-1}, $n = 1, 2, \dots$). Indeed, from Theorem 4.39 it immediately follows that there are $2^{1+2+\dots+2^{n-1}} = 2^{2^n-1}$ pairwise distinct modulo 2^n 1-Lipschitz measure-preserving transformations on $\mathbb{Z}_2$. The corresponding bijective transformations on the residue ring $\mathbb{Z}/2^n\mathbb{Z}$ obviously form a group with respect to composition of transformations; actually this group is isomorphic to a Sylow 2-subgroup $\mathrm{Syl}_2(2^n)$ of the symmetric group $\mathrm{Sym}(2^n)$ of all permutations on $\mathbb{Z}/2^n\mathbb{Z}$.

Example 7.19. Given arbitrary measure-preserving transformations a, b on $\mathbb{Z}_2$, every 1-Lipschitz measure-preserving transformation g on $\mathbb{Z}_2$ can be represented as $f(a(f(b(f(x))))) = g(x)$, for a suitable 1-Lipschitz measure-preserving transformation f on $\mathbb{Z}_2$.

Indeed, we can rewrite this representation as an equation $f \circ a \circ f \circ b \circ f = g$ in indeterminate f in the group $\mathrm{Syl}_2(\infty)$, where $\circ$ stands for composition of transformations. The conclusion now follows from Corollary 7.18.

To conclude the subsection, we note that combining Theorem 7.16 and Theorem 6.5 it obviously follows a criterion for measure-preservation of polynomials over the profinite group G_∞, which is an inverse limit of finite solvable groups G_n:

Theorem 7.20. *A polynomial $w(x) \in G_\infty[x]$ is measure-preserving if and only if its is bijective modulo the subgroup N_0, and all derivatives $\partial_{A_n} w(g)$ are non-singular matrices over $\mathbb{F}_{p_n}$, for all $g \in G_{n+1}$ and all $n = 0, 1, 2, \dots$.*

Note 7.21. Theorem 7.20 remains true if G_∞ is a group with a non-empty set of operators Ω; we only must consider Ω-invariant minimal normal subgroups A_n rather than merely minimal normal subgroups.

Corollary 7.22. *If in the conditions of Theorem 7.20 all G_n are p-groups for some prime p, then the polynomial $w(x)$ is measure-preserving if and only if $p \nmid \deg w$.*

Proof. We may assume that G_0 is an (Abelian) group of order p; otherwise we make refinements to the inverse spectrum using the chief series of G_0. Foremost, we may assume that all $N_n/N_{n+1} \in \mathrm{Z}(G_n)$, by the same reason. Thus, $\partial_{A_n} w(g) = \deg w$, for all $g \in G_{n+1}$, $n = 0, 1, 2, \dots$; and $(w\pi_0)(g) = (w\pi_0)(1) \cdot g^{\deg w}$ for all $g \in G_0$. However, given $a \in G_0$, the equation $(w\pi_0)(1) \cdot x^{\deg w} = a$ in unknown x has a solution in G_0 if and only if $p \nmid \deg w$. □

In view of Example 7.19 the following assertion is obvious:

Example 7.23. Given arbitrary 1-Lipschitz measure-preserving transformations $a, b, c, d \in \mathrm{Syl}_2(\infty)$ on $\mathbb{Z}_2$, the polynomial $axbxcxd$ over $\mathrm{Syl}_2(\infty)$ induces a measure-preserving transformation on this group.

7.3.3 Ergodic polynomials over profinite groups

Contrasting to the case of measure-preserving polynomials over groups, the ergodic ones exist not over every profinite group G_∞, even if all the groups G_n forming the corresponding inverse spectrum are solvable: From Theorem 7.16 it follows that whenever a profinite group G_∞ has an ergodic polynomial, the group must be an inverse limit of finite groups having transitive polynomials; and *not* every finite solvable group has a transitive polynomial. From Theorems 7.5 and 7.8 we can see that groups listed there falls into several inverse spectra. For instance, all dihedral groups D_k, $k = 2, 3, 4, \ldots$, form an inverse spectrum

$$\cdots \xrightarrow{\varphi_{k+1}} \mathrm{D}_k \xrightarrow{\varphi_k} \mathrm{D}_{k-1} \xrightarrow{\varphi_{k-1}} \cdots \xrightarrow{\varphi_3} \mathrm{D}_2,$$

where kernels of epimorphisms φ_k are centers of corresponding dihedral groups:

$$\ker \varphi_k = \mathrm{Z}(\mathrm{D}_k) = \{1, v^{2^{k-1}}\}, \quad k = 3, 4, 5, \ldots .$$

The limit group of this inverse spectrum is a group $\mathbf{D}_\infty$, which is a split extension of the additive group $\mathbb{Z}_2^+$ of 2-adic integers by a cyclic group of order 2; the latter group acts on $\mathbb{Z}_2^+$ by taking negatives: $z \mapsto -z$, $z \in \mathbb{Z}_2$.[5] Thus, we may think of elements of the group $\mathbf{D}_\infty$ as of pairs (ε, z), where $\varepsilon \in \mathbb{F}_2 = \{0, 1\}$, $z \in \mathbb{Z}_2$. Multiplication $\cdot$ of these pairs is defined by the rule

$$(\varepsilon_1, z_1) \cdot (\varepsilon_2, z_2) = (\varepsilon_1 \oplus \varepsilon_2, (-1)^{\varepsilon_2} z_1 + z_2),$$

where $\oplus$ stands for addition modulo 2. The subgroup $Z \cong \mathbb{Z}_2^+$, as well as the subgroup $V \subset \mathrm{D}_k$, which is a cyclic subgroup of order 2^k generated by $v \subset \mathrm{D}_k$, are characteristic subgroups in D_∞ and D_k, respectively. Hence, combining Corollary 7.2 with Theorem 7.16 we conclude that a polynomial $w(x)$ over the group $\mathbf{D}_\infty$ with operators $\Omega = \mathrm{Aut}\,(\mathbf{D}_\infty)$ is ergodic if and only if it is transitive on the factor group $\mathbf{D}_\infty/Z$, and the polynomial $w^2(x)$ is ergodic on Z. However, as every automorphism of $Z \cong \mathbb{Z}_2^+$ is a multiplication by a unit from $\mathbb{Z}_2$ (and vice versa), the polynomial $w^2(x)$ induces an affine transformation $x \mapsto a + bx$ on $\mathbb{Z}_2$, for suitable $a, b \in \mathbb{Z}_2$. By Theorem 4.36, the affine transformation is ergodic on $\mathbb{Z}_2$ if and only if it is transitive modulo 4. So we finally have proved the following result:

Proposition 7.24. *A polynomial over the group* $\mathbf{D}_\infty$ *with operators* $\mathrm{Aut}\,(\mathbf{D}_\infty)$ *is ergodic if and only if it is transitive on the dihedral group* D_2 *of order 8.*

Example 7.25. The polynomial $\tilde{w}(x) = zx^{\tilde{\alpha}}$, where $z = (1, 1) \in \mathbf{D}_\infty$, and the automorphism $\tilde{\alpha}$ takes $(1, 0)$ to $(1, 1)$ and acts on the subgroup $\mathbb{Z}_2^+ \subset \mathbf{D}_\infty$ identically, is ergodic on the group $\mathbf{D}_\infty$ with operators $\mathrm{Aut}\,(\mathbf{D}_\infty)$.

[5] Note that the group $\mathbf{D}_\infty$ is *not* the infinite dihedral group D_∞; the latter group is a split extension of $\mathbb{Z}^+$ by the group of order 2.

Consider a polynomial $w(x) = uvx^{\alpha}$ over the group D_2 with operators $\operatorname{Aut}(D_2)$, where the automorphism α takes u to $u^{\alpha} = uv$ and v to $v^{\alpha} = v$. The polynomial $w(x)$ is transitive on the dihedral group D_2: Indeed, the 2-nd iterate $w^2(x) = vx^{\alpha^2}$ induces on the subgroup V generated by $v \in D_2$ a transitive transformation $v^i \mapsto v^{i+1}$, the polynomial $w(x)$ induces a transitive transformation $x \mapsto ux$ on the factor group D_2/V, so the conclusion follows in view of Corollary 7.2. This in view of Proposition 7.24 proves the ergodicity of the polynomial $\tilde{w}(x)$ on the group $\mathbf{D}_{\infty}$.

The argument that proves Proposition 7.24 after minor modification can be applied to the group $\mathbf{D}_{\infty}$ with operators $\operatorname{End}(\mathbf{D}_{\infty})$: As the subgroups Z and V are not fully invariant in respective groups, we must use first derived groups $\mathbf{D}'_{\infty}$ and D'_k instead. Note that $\mathbf{D}'_{\infty} \cong 2\mathbb{Z}_2^+$, and that D'_k is a cyclic group of order 2^{k-1} generated by v^2. Thus we obtain:

Proposition 7.26. *A polynomial over the group* $\mathbf{D}_{\infty}$ *with operators* $\operatorname{End}(\mathbf{D}_{\infty})$ *is ergodic if and only if it is transitive on the dihedral group* D_3 *of order 16.*

Combining Theorem 7.16 with Proposition 2.3, from Propositions 7.24 and7.26 we immediately deduce the following corollary:

Corollary 7.27. *A polynomial over the dihedral group* D_k *with operators* $\operatorname{Aut}(D_k)$ *(respectively,* $\operatorname{End}(D_k)$*,* $k \geq 3$*) is transitive if and only if it is transitive on the dihedral group* D_2 *of order 8 (respectively, on the dihedral group* D_3 *of order 16).*

We now can determine whether a given polynomial over a semidihedral or generalized quaternion group is transitive on these groups, although neither semidihedral groups nor generalized quaternion groups form inverse spectra. Indeed, by Corollary 7.2 a polynomial $w(x)$ over the semidihedral group SD_k with operators $\operatorname{End}(SD_k)$ is transitive on this group if and only if $w(x)$ is transitive modulo the derived group SD'_k (i.e., on the factor group $SD_k/SD'_k \cong K_4$), and the polynomial $w^4(x)$ is transitive on the subgroup SD'_k, which is a fully invariant cyclic subgroup of order 2^{k-1} generated by the element v^2. Note that $(v^2)^u = v^{2(2^{k-1}-1)} = v^{-2}$. Since $\operatorname{End}(SD'_k) \cong \mathbb{Z}/2^{k-1}\mathbb{Z}$, the polynomial $w^4(x)$ acts on $SD'_k \cong (\mathbb{Z}/2^{k-1}\mathbb{Z})^+$ as affine mapping, which is transitive on this subgroup if and only if it is transitive modulo 4, by Theorem 4.36. However, by this theorem an affine polynomial on a cyclic group of order 2^s is transitive on this group if and only if it is transitive modulo 2^{s-i}, for some (equivalently, any) $i \leq s-2$, i.e., on arbitrary proper factor group whose order is ≥ 4. Hence, the polynomial $w^4(x)$ is transitive on SD'_k if and only if the polynomial $(w^4\psi)(x)$ is transitive on the factor group SD'_k/V, where V is a cyclic subgroup generated by $v^{2^{k-1}}$, and $\psi : SD_k \to SD_k/V$ is a canonical epimorphism. However, $V = L_k(SD_k)$, the kth subgroup from the lower central series of the group SD_k; so V is fully invariant. Foremost, $SD_k/V \cong D_{k-1}$, the dihedral group of order 2^k, $SD_k/SD'_k \cong D_{k-1}/D'_{k-1} \cong K_4$, and thus $w(x)$ is transitive on SD_k/SD'_k if and only if $(w\psi)(x)$ is transitive on D_{k-1}/D'_{k-1}. So we conclude that the polynomial

$w(x)$ is transitive on SD_k if and only if the polynomial $(w\psi)(x)$ is transitive on the dihedral group D_{k-1}. However, by Corollary 7.27, the polynomial over the dihedral group D_{k-1} with operators End (D_{k-1}) is transitive if and only if it is transitive on the dihedral group of order 16. Thus, we have proved the following statement:

Corollary 7.28. *A polynomial $w(x)$ over the semidihedral group SD_k, $k \geq 4$, with operators* End (SD_k) *is transitive on this group if and only the polynomial $(w\varphi)(x)$ is transitive on the dihedral group D_3 of order 16. Here $\varphi : \mathrm{SD}_k \to \mathrm{D}_3$ is an epimorphism with a kernel $L_4(\mathrm{SD}_k)$, which is a cyclic subgroup generated by v^8.*

Note 7.29. The statement of Corollary 7.28 remains true after we replace semidihedral group SD_k by the generalized quaternion group Q_k. Foremost, if we also replace End (Q_k) by Aut (Q_k), then we may replace D_3 by D_2 without affecting validity of the statement. The proof mimics the one for semidihedral groups, and we omit it.

Example 7.30. The polynomial $w(x) = uvx^{\alpha}$, where the automorphism α takes u to $u^{\alpha} = uv$ and v to $v^{\alpha} = v$, is transitive on the generalized quaternion group Q_k with operators Aut (Q_k).

Indeed, by Note 7.29 it suffices to consider a transformation induced by this polynomial on the dihedral group D_2. By Example 7.25, the latter transformation is ergodic on $\mathbf{D}_{\infty}$; thus, it is transitive on all D_k.

It is clear now that in a similar manner one can prove the ergodicity criteria for other groups that are inverse limits of groups listed in Theorem 7.8. We will not consider all these inverse limits restricting our considerations with the some typical examples.

Cyclic groups $\mathrm{C}(p^k)$, $k = 1, 2, \ldots$, with p prime are groups of type 1 of Theorem 7.8. They form a spectrum, whose inverse limit is isomorphic to the additive group $\mathbb{Z}_p^+$ of p-adic integers. As it follows from the definition of the polynomial over a universal algebra (see Subsection 1.2.1), all polynomial transformations on this group are of the form $w(x) = g + hx$, where $g, h \in \mathbb{Z}_p$; i.e., they are affine transformations. By Theorem 4.36, the latter transformations are ergodic on $\mathbb{Z}_p^+$ if and only if they are transitive either on $\mathbb{Z}/p\mathbb{Z}$ if p is odd, or on $\mathbb{Z}/4\mathbb{Z}$, if otherwise.

Groups of type 2 of Theorem 7.8 are metacyclic groups $\mathrm{M}(m, k, s)$. They fall in different inverse spectra. For instance, let p, q be distinct primes, $p \mid q - 1$. Consider a group $\mathbf{M}(p, q, s) = \mathbb{Z}_p^+ \leftthreetimes \mathbb{Z}_q^+$, where action of $\mathbb{Z}_p^+$ on $\mathbb{Z}_q^+$ is defined as follows: Take an arbitrary pth root $s \in \mathbb{Z}_q$ of 1, $s \neq 1$. Then for every $z \in \mathbb{Z}_p$ the element $s^z \in \mathbb{Z}_q$ is well defined. Note that $s^z = 1$ for all $z \in p\mathbb{Z}_p$. Elements of the group $\mathbf{M}(p, q, s)$ can be considered as pairs (g, h), $g \in \mathbb{Z}_p$, $h \in \mathbb{Z}_q$, and multiplication $\cdot$ of these pairs is defined as

$$(g_1, h_1) \cdot (g_2, h_2) = (g_1 + g_2, s^{g_2} h_1 + h_2).$$

It is clear that the group $\mathbf{M}(p,q,s)$ is a limit group of the inverse spectrum formed by metacyclic groups of type $\mathrm{M}(p^n,q^n,s \bmod q^n)$:

$$\cdots \xrightarrow{\varphi_n} \mathrm{M}(p^n,q^n,s \bmod q^n) \xrightarrow{\varphi_{n-1}} \cdots \xrightarrow{\varphi_1} \mathrm{M}(p,q,s \bmod q).$$

If we represent elements of the group $\mathrm{M}(p^n,q^n,s \bmod q^n)$ by pairs (g,h), $g \in \mathbb{Z}/p^n\mathbb{Z}$, $h \in \mathbb{Z}/q^n\mathbb{Z}$ and define multiplication of these pairs in a way similar to that of the group $\mathbf{M}(p,q,s)$, the epimorphism φ_{n-1} is then reduction modulo p^{n-1} and q^{n-1} of respective coordinates; i.e., $\varphi_n\colon (g,h) \mapsto (g \bmod p^{n-1}, h \bmod q^{n-1})$. By Corollary 7.2, a polynomial $w(x)$ over the group $\mathrm{M}(p^n,q^n,s \bmod q^n)$ is transitive if and only if, firstly, the polynomial $w(x)$ induces a transitive transformation on the factor group $\mathrm{M}(p^n,q^n,s \bmod q^n)/Z_{q^n} \cong Z_{p^n} \cong \mathrm{C}(p^n)$, where $Z_{q^n} \cong \mathrm{C}(q^n)$ and Z_{p^n} are cyclic subgroups generated by $(0,1)$ and $(1,0)$, respectively, and, secondly, the p^nth iterate $w^{p^n}(x)$ of the polynomial $w(x)$ induces a transitive transformation on the subgroup Z_{q^n}. As both these transformations are affine transformations of the residue rings $\mathbb{Z}/p^n\mathbb{Z}$ and $\mathbb{Z}/q^n\mathbb{Z}$, respectively, sufficient and necessary conditions for their transitivity gives Theorem 4.36. So we conclude that *a polynomial over the group* $\mathbf{M}(p,q,s)$ *is ergodic if and only if it induces a transitive transformation either on the factor group* $\mathrm{M}(p,q,s \pmod q)$) *if p is odd, or on the factor-group* $\mathrm{M}(4,q^2,s \bmod q^2)$ *if $p = 2$*. Cases when p and/or q are composite can be reduced to the considered case in view of Chinese Remainder Theorem, see Subsection 1.2.3.

Example 7.31. The polynomial $w(x) = (1,0) \cdot x \cdot (0,1)$ is ergodic on the group $\mathbf{M}(p,q,s)$.

Indeed, this polynomial induces a transformation $(g,h) \mapsto (g+1,h+1)$, which is obviously transitive on the respective group.

In a similar manner we could obtain criteria of ergodicity for polynomials over inverse limits of other groups listed in Theorem 7.8. Loosely speaking, all these criteria read that *a polynomial over inverse limit of a spectrum is ergodic if and only it induces a transitive transformation on the smallest group of the spectrum.*

For instance, consider groups $\mathrm{SQ}_1(n) \leftthreetimes \mathrm{M}(p^n,q^n,s \bmod q^n)$ of type 15, $n = 1,2,\ldots$, where p,q,s as above, $p,q > 3$. These groups obviously form an inverse spectrum. During the proof of Theorem 7.8 we showed that the group $\mathrm{SQ}_1(n) \leftthreetimes \mathrm{M}(p^n,q^n,s \bmod q^n)$ can be represented as follows:

$$\mathrm{SQ}_1(n) \leftthreetimes \mathrm{M}(p^n,q^n,s \bmod q^n) = (\mathrm{C}(2) \leftthreetimes \mathrm{C}(3^n)) \times \mathrm{C}(p^n)) \leftthreetimes (\mathrm{Q}_2 \times \mathrm{C}(q^n)).$$

Thus, the limit group $\mathbf{SQ}_1 \leftthreetimes \mathbf{M}(p,q,s)$ of this inverse spectrum can be represented as $(\mathrm{C}(2) \leftthreetimes \mathbb{Z}_3^+) \times \mathbb{Z}_p^+) \leftthreetimes (\mathrm{Q}_2 \times \mathbb{Z}_q^+)$, where $\mathbf{SQ}_1 = \mathrm{C}(2) \leftthreetimes \mathbb{Z}_3 \leftthreetimes \mathrm{Q}_2$, the cyclic group $\mathrm{C}(2)$ of order 2 acts on $\mathbb{Z}_3^+$ and on $\mathbb{Z}_q^+$ by the negation $z \mapsto -z$, the group $\mathrm{C}(2) \leftthreetimes \mathbb{Z}_3^+$ acts on the quaternion group Q_2 as a symmetric group Sym(3) (so $3\mathbb{Z}_3$ centralizes Q_2)[6],

[6] Recall that $\mathrm{Aut}\,(\mathrm{Q}_2) \cong \mathrm{Sym}(3)$.

$\mathbb{Z}_p^+$ centralizes Q_2 and acts on $\mathbb{Z}_q^+$ by multiplication by s, the non-identity pth root of 1. By the argument similar to that as in the case of metacyclic groups we can prove that a polynomial over this inverse limit is ergodic if and only if it is ergodic on the group $SQ_1(1) \leftthreetimes M(p,q,s \bmod q)$.

Example 7.32. Let the group $G = \mathbf{SQ}_1 \leftthreetimes \mathbf{M}(p,q,s)$ be represented as above. Then the following polynomial $w(x)$ is ergodic: $w(x) = acx^2uvx^5bx^{24n}d$, where

- a is a generator of the subgroup $C(2)$,
- $b \in \mathbb{Z}_3^+ \subset G$ is any 3-adic integer congruent to 1 modulo 3,
- $c \in \mathbb{Z}_p^+ \subset G$ is any p-adic integer congruent to 1 modulo p,
- $d \in \mathbb{Z}_3^+$ is any q-adic integer congruent to 1 modulo q,
- n is arbitrary rational integer such that $6 + 24n \equiv 0 \pmod{pq}$; i.e., $4n \equiv -1 \pmod{pq}$.

Note that we write operation in subgroups $\mathbb{Z}_3^+, \mathbb{Z}_p^+, \mathbb{Z}_q^+ \subset G$ additively, although the operation in the group G we write in the multiplicative form.

By what was said, we only need to show that the polynomial $\bar{w}(x) = (w\varphi)(x)$ is transitive on the group $SQ_1(1) \leftthreetimes M(p,q,s \bmod q)$, where $\varphi\colon G \to SQ_1(1) \leftthreetimes M(p,q,s \bmod q)$ is an epimorphism that maps $\mathbb{Z}_3^+$, $\mathbb{Z}_p^+$, and $\mathbb{Z}_q^+$ onto $C(3) \subset SQ_1(1)$, $C(p) \subset M(p,q,s \bmod q)$, and $C(q) \subset M(p,q,s \bmod q)$, respectively. However, we have already shown this while proving sufficiency of the conditions of Theorem 7.8.

It is clear that in general an inverse limit of groups listed in Theorem 7.8 is, loosely speaking, a group that is an extension of an additive group of k-adic integers by a group combined from additive groups of m-adic integers, and/or small finite groups K_4, Q_8, $C(2)$. We do not list down all these groups here, leaving this work as an exercise to the interested reader; we only mention that *actually the corresponding dynamics can be reduced to affine actions on ℓ-adic integers $\mathbb{Z}_\ell$*, and the latter actions form as a non-autonomous dynamical system on $\mathbb{Z}_\ell$. As a matter of fact, the construction these inverse limits are based on, the semidirect products, is known under the name of skew products in ergodic theory. We will develop this approach based on actions of a dynamical system on other dynamical system in Chapter 10 to construct so-called counter-dependent pseudorandom generators, which actually are skew products of dynamical systems. However, we will consider there more complicated actions than affine ones.

Now we only illustrate how the dynamics on the group $\mathbf{D}_\infty$ can be applied to computer science. Actually we will show only how the operation of a dihedral group D_n arises in connection with computer instructions that depend on the value of a one-bit registry, a so-called "flag".[7] Consider the following instruction (or a program): If the flag value is equal to 0, then addition is carried out, and if it is 1, then subtraction is

[7] Note that usually program jumps are instructions that depend on flags. Often a flag contains a sign of a number.

carried out. This is how the $*$ operation of the non-Abelian dihedral group D_n appears: If ε, ξ are the values of the flag, a, b are n-bit words in the alphabet $\{0,1\}$, then $(\varepsilon, a) * (\xi, b) = (\varepsilon \oplus \xi, b + (-1)^{\xi} a)$, where $\oplus$ is addition modulo 2, and $+$ is addition modulo 2^n. Now, using this instruction, and endomorphisms of the group D_n, which actually can be realized as substitutions like $(1,0) \mapsto (\alpha, k)$, $(0,1) \mapsto (\beta, m)$ via look-up tables, one can evaluate a polynomial over the group D_n with a corresponding set of operators.

In connection with results of this subsection, it is natural to ask a question where this is possible to obtain a description of ergodic polynomial transformations over the considered profinite groups in *explicit form*? The reader may note that in case of p-adic ergodic transformations on $\mathbb{Z}_p$ such explicit representations were obtained. We note, however, that in the latter case we managed to do this since we obtained an explicit description of identities modulo p^k; that is, continuous transformations on $\mathbb{Z}_p$ (in particular, polynomial transformations) that are identically 0 modulo p^k, see Proposition 3.52. Using this result, we can take, say, all 16 different polynomials on the residue ring modulo 8 (see Corollary 9.16 further) and then add to these polynomials a polynomial identity modulo 8 described Proposition 3.52 and thus obtain all polynomial ergodic transformations on $\mathbb{Z}_2$ in the explicit form. Thus, to act in a way like this in the case of profinite groups, we must obtain explicitly those polynomials over 'initial', the smallest groups of corresponding inverse spectra, that are identically 1 on respective groups. Polynomial over a group G that is identically 1 everywhere on G is called a *mixed identity* of the group G. The corresponding theory of mixed identities in groups, and the related theory of *mixed varieties of groups* emerged in papers [18, 20], which were succeeded by papers [13–15]. Actually in the paper [20] there were developed techniques to characterize mixed identities of nilpotent and of metabelian groups. It might be possible that these techniques will suit to describe explicitly mixed identities of other 'initial' groups of inverse spectra considered in this subsection, thus obtaining explicit forms of ergodic polynomials over inverse limits. However, this work is not done yet; though looks as the work that can be done since adequate mathematical tools are already developed.

To conclude Part II, it is worth mentioning that methods we developed here for polynomial over groups with operators, work in a much more general setting, for polynomial dynamics over non-commutative universal algebras such as groups with multi-operators, which are merely groups with extended group signature. Although the latter groups arise in numerous applications, there is no reason at our view to develop in this book a general theory of corresponding dynamical systems; we decided to consider the concrete groups with multi-operators, e.g., rings, especially rings of p-adic integers (see Subsection 2.2.3 on the corresponding reasoning), as well as the other algebraic systems that are important for applications, the automata, see Part III. However we emphasize that our approach works in a much more general situation, for inverse limits of finite universal algebras of a very general nature; and we mention once again that the corresponding dynamical systems will inevitably be non-Archimedean.

Part III

Applications

Chapter 8

Automata, computers, combinatorics

In this chapter we apply p-adic ergodic theory to some problems from automata theory, computer science, and combinatorics.

In Section 8.1 we show that an automaton that has an m-letter input alphabet and an m-letter output alphabet, and which thus performs a transformation of words in this alphabet, can be related to a m-adic continuous map from the space $\mathbb{Z}_m$ of m-adic integers into $\mathbb{Z}_m$. The latter map reflects some important properties of the automaton, which can be studied by the use of m-adic dynamics. We prove some preliminary facts in Section 8.1 using this approach, leaving detailed development of it for further chapters.

In Section 8.2 we consider very special and important type of automata, digital computers, and demonstrate that their basic instructions, such as numerical ones (integer addition and multiplication) and bitwise logical ones (OR, the bitwise logical 'or', AND, the bitwise logical 'and', etc.) can be expanded to 2-adic functions that are continuous with respect to 2-adic distance. Thus, all compositions of these basic instructions, i.e., computer programs, can be regarded as continuous 2-adic functions as well. We develop a necessary techniques, including differential calculus, for these functions that we use further to establish results on behavior of computer programs with the use of these techniques

In Section 8.4, we apply these techniques, as well as other results from the p-adic ergodic theory, to construct huge classes of large Latin squares and mutually orthogonal Latin squares. Latin squares, which are popular combinatorial objects, are also used in various applications, such as communications, experiment design, etc.

8.1 Automata functions are continuous

We first remind some basic notions of automata theory; the reader can find these in the monographs [11, 155, 168]. We note that these monographs are mainly focused on internal states of automata, how they are changing, etc. So, this approach can be considered as more 'internal', in contrast to another, 'external' approach exhibited in [413], where major attention is paid to the question *what* transformation the automaton performs rather then to *how* it does it. Of course, these two approaches are tightly related; however, we stress that in our book we are mainly focused on transformations performed by an automaton, though we necessarily touch questions concerning internal states as well.

Actually, automata are the most general form of description of information processing, a kind of language of description of systems (so that many scientists understand a system theory merely as an automata theory). In the most general form, an *automaton* is a sextuple $\mathfrak{A} = \langle \mathcal{K}, \mathcal{N}, \mathcal{M}, f, F, u_0 \rangle$, where $\mathcal{K}$ is an input alphabet, $\mathcal{N}$ is a (nonempty) set of states, $f : \mathcal{K} \times \mathcal{N} \to \mathcal{N}$ is a state transition function (which sometimes is called also a sate update function), $\mathcal{M}$ is an output alphabet, $F : \mathcal{K} \times \mathcal{N} \to \mathcal{M}$ is an output function, $u_0 \in \mathcal{N}$ is an initial state. Thus, given an input sequence $w_0, w_1, \dots$ over the alphabet $\mathcal{K}$, the automaton transforms it into the output sequence

$$z_0 = F(w_0, u_0), z_1 = F(w_1, f(w_0, u_0)), \dots, z_j = F(w_j, f(w_{j-1}, u_{j-1})), \dots$$

over the alphabet $\mathcal{M}$, where $u_{i+1} = f(w_i, u_i) \in \mathcal{N}$, $i = 0, 1, 2, \dots$, is a corresponding sequence of states.

Note that both $\mathcal{K}$ and $\mathcal{M}$ may be empty sets; however, $\mathcal{N}$ can not. However, whenever $\mathcal{M}$ is empty (that is, whenever the automaton $\mathfrak{A}$ has no output) we always can convert it into a new automaton $\mathfrak{A}'$ with output alphabet $\mathcal{N}$ and output function $F(w, u) = u$, which is actually the same automaton as $\mathfrak{A}$, with the only difference that output of $\mathfrak{A}'$ are just states of $\mathfrak{A}$. So in the sequel we assume that every automaton $\mathfrak{A}$ always has an output, i.e., that $\mathcal{M} \neq \varnothing$.

A word of caution: In literature, there are differences in the definitions of the automaton; ours is the most general. For instance, the definition of the automaton from [11] corresponds to the case when $\mathcal{M} = \varnothing$ in our definition; whereas automata in the meaning of our definition are called transducers in [11]. Note also that sometimes automata in the sense of our definition are called Mealy machines; cf. [168]. Note also that some authors do not fix initial state letting it be arbitrary from the set $\mathcal{N}$; if initial state is fixed, they speak of *initial* automaton. In these terms, all automata in this book are initial automata; we speak of *family of automata* $\{\mathfrak{A}(u_0) : t_0 \in \mathcal{N}\}$ when we let the initial state u_0 run through the set of states $\mathcal{N}$.

For instance, the so-called *Ising automata*, which arise in connection with mathematical models of some physical phenomena related to systems whose behavior depend on spins of particles, are automata without output, see e.g. [11]; we mention also a study of Ising automata performed by J.-Y. Yao in [415, 416].

Every automaton $\mathfrak{A}$ maps the set $\mathbb{Z}_{\mathcal{K}}$ of all infinite sequences over $\mathcal{K}$ into the set $\mathbb{Z}_{\mathcal{M}}$ of all infinite sequences over $\mathcal{M}$ in a natural way: $\mathfrak{A}$ maps every input sequence $w_0, w_1, \dots$ to output sequence $F(w_0, u_0), F(w_1, f(w_0, u_0)), \dots$. Thus, to every automaton $\mathfrak{A}$ we associate the function $\Psi_{\mathfrak{A}} : \mathbb{Z}_{\mathcal{K}} \to \mathbb{Z}_{\mathcal{M}}$, which is called an *automaton function*[1] and has a special *triangular* form: Every ith term of output sequence depends only on the terms $w_0, w_1, \dots, w_i$ of input sequence. It is clear enough that every triangular function $\Psi : \mathbb{Z}_{\mathcal{K}} \to \mathbb{Z}_{\mathcal{M}}$ can be associated to some automaton $\mathfrak{A}_\Psi$;

[1]Note that sometimes automata functions are also called *determined functions*, see e.g. [413].

however, this automaton $\mathfrak{A}_\Psi$ is not unique: Different automata may evaluate the same triangular function; these automata are said to be *equivalent*. Loosely speaking, equivalent automata are machines that 'do the same thing'.

For instance, any function that corresponds to an automaton without input (that is, with $\mathcal{K} = \varnothing$) is just a constant; however, it is clear that a constant (that is, an infinite sequence over $\mathcal{M}$) can be produced by many different ways, corresponding to different automata. Note that automata without input merely generate sequences. We call these automata *generators*; these automata arise in various applications dealing with pseudorandom numbers. We study these automata intensively in Chapter 9.

Often in automata theory they study automata up to the above mentioned equivalence; that is, actually the object under study is a function rather than its representation via the automaton. Typical problems of the theory are invertibility of the automaton (that is, existence of inverse automaton function); number of states of the automaton that represents a given function; characterization of classes of functions that can be produced by all compositions of certain (simple) automata (e.g., various problems concerning completeness, pre-completeness, etc.); properties of functions that are evaluated by automata from a given class, etc. Note that in automata theory they often speak about the *serial connection of automata* (see e.g., [168]) rather then on composition of automata functions. It is clear that if $\Psi_{\mathfrak{B}}\colon \mathbb{Z}_{\mathcal{A}} \to \mathbb{Z}_{\mathcal{K}}$ is the automaton function that corresponds to the automaton $\mathfrak{B}$ with input alphabet $\mathcal{A}$ and output alphabet $\mathcal{K}$, and if $\Psi_{\mathfrak{A}}\colon \mathbb{Z}_{\mathcal{K}} \to \mathbb{Z}_{\mathcal{M}}$ is the automaton function that corresponds to the automaton $\mathfrak{A}$ with input alphabet $\mathcal{K}$ and output alphabet $\mathcal{M}$, then the automaton function that corresponds to the serial connection of automata $\mathfrak{B}$ and $\mathfrak{A}$ is the composition $\Psi_{\mathfrak{A}} \circ \Psi_{\mathfrak{B}}\colon \mathbb{Z}_{\mathcal{A}} \to \mathbb{Z}_{\mathcal{M}}$ of functions $\Psi_{\mathfrak{B}}$ and $\Psi_{\mathfrak{A}}$: $(\Psi_{\mathfrak{A}} \circ \Psi_{\mathfrak{B}})(z) = \Psi_{\mathfrak{A}}(\Psi_{\mathfrak{B}}(z))$ for every $z \in \mathbb{Z}_{\mathcal{A}}$.

We call the automaton *finite* whenever there exists an equivalent automaton with a finite number of states, and *infinite* otherwise. We stress that throughout the book, we speak about finite/infinite automata only in this meaning: Often in automata theory the automaton $\mathfrak{A}$ is called finite (or the automaton with a finite number of states, or a finite-state machine) whenever the number of its states is finite, that is $\#\mathcal{N} < \infty$; otherwise the automaton is called infinite (or the automaton with the infinite number of states). We do not use this terminology in the book!

A state $u \in \mathcal{N}$ of the automaton $\mathfrak{A}$ is called *reachable* if there exists a finite input sequence $w_0, w_1, \dots, w_i$ such that whenever the sequence is input, the ith state u_i of the automaton is u: $u_i = u$. Two states $u, v \in \mathcal{N}$ are called *equivalent* whenever there exist finite input sequences $w_0, w_1, \dots, w_i$ and $w'_0, w'_1, \dots, w'_j$ such that taking arbitrary infinite sequence $s_0, s_1, \dots$ over $\mathcal{K}$ and inputting sequences $w_0, w_1, \dots, w_i, s_0, s_1, \dots$ and $w'_0, w'_1, \dots, w'_j, s_0, s_1, \dots$, the ith and the jth states of the automaton $\mathfrak{A}$ will be, respectively, u and v, and the corresponding output sequences $z_0, z_1, \dots$ and $z'_0, z'_1, \dots$ will agree starting with the $(i+1)$th and the $(j+1)$th terms, accordingly: $z_{i+k} = z'_{j+k}$ for all $k = 1, 2, 3, \dots$.

In other words, let us vary the initial state u_0 of the automaton $\mathfrak{A}$ over the set

$\mathcal{N}$; that is, let us consider a family $\{\Psi_{\mathfrak{A}(u_0)} : u_0 \in \mathcal{N}\}$ of corresponding automata functions parametrized by the parameter u_0. Then, the states $u, v \in \mathcal{N}$ are equivalent if and only if both u and v are reachable states, and $\Psi_{\mathfrak{A}(u)}(z) = \Psi_{\mathfrak{A}(v)}(z)$ for all $z \in \mathbb{Z}_{\mathcal{K}}$. Here $\mathfrak{A}(v)$ stands for the automaton $\mathfrak{A}$ with the initial state $u_0 = v$: $\mathfrak{A}(v) = \langle \mathcal{K}, \mathcal{N}, \mathcal{M}, f, F, v\rangle$. It is obvious that a finite automaton always has equivalent states.

Often in applications it is convenient to consider automata with n inputs and m outputs over the same alphabet $\mathcal{P}$ that consists of P letters, which are usually denoted by $0, 1, 2, \dots, P-1$ and are associated to elements of the residue ring $\mathbb{Z}/P\mathbb{Z}$ modulo P under a natural correspondence. These automata obviously correspond to the case when both $\mathcal{K}$ and $\mathcal{M}$ are respective Cartesian powers of $\mathcal{P}$ in the general automaton $\mathfrak{A}$: $\mathcal{K} = \mathcal{P}^n$ and $\mathcal{M} = \mathcal{P}^m$. In this case the corresponding automaton function $\Phi = \Psi_{\mathfrak{A}}$ can be represented in the form

$$\Phi \colon \alpha_0^{\downarrow}, \alpha_1^{\downarrow}, \alpha_2^{\downarrow}, \dots \mapsto \Phi_0^{\downarrow}(\alpha_0^{\downarrow}), \Phi_1^{\downarrow}(\alpha_0^{\downarrow}, \alpha_1^{\downarrow}), \Phi_2^{\downarrow}(\alpha_0^{\downarrow}, \alpha_1^{\downarrow}, \alpha_2^{\downarrow}), \dots$$

where $\alpha_i^{\downarrow} \in \mathcal{P}^n$ is an n-letter (columnar) word over alphabet $\mathcal{P}$, and the mapping $\Phi_i^{\downarrow} \colon (\mathcal{P}^n)^{i+1} \to \mathcal{P}^m$ maps n-letter (columnar) words $\alpha_0^{\downarrow}, \dots, \alpha_i^{\downarrow}$ to an m-letter (columnar) word $\Phi_i^{\downarrow}(\alpha_0^{\downarrow}, \dots, \alpha_i^{\downarrow}) \in \mathcal{P}^m$, see Figure 8.1. That is, Φ is an m-variate triangular function; the domain of variables is $\mathbb{Z}_P$, the ring of P-adic integers. In other words, variables are infinite sequences over $\mathcal{P}$, the P-adic integers, see Section 1.7 for rigorous definitions and theoretical results on P-adic integers, P-adic arithmetics, etc.

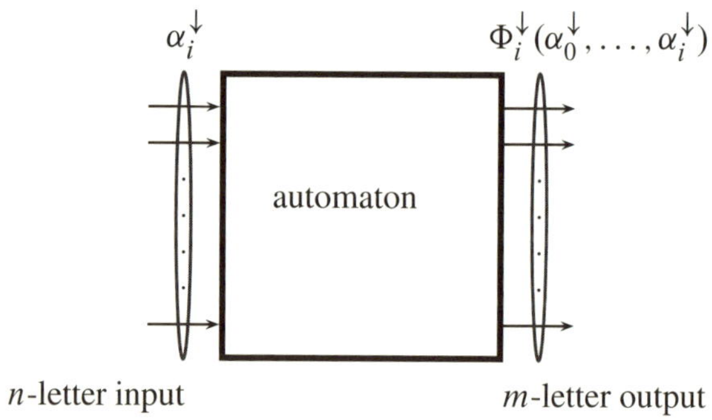

Figure 8.1. Automaton with n inputs and m outputs.

For instance, if $m = n = 1$, then the corresponding automaton evaluates a univariate triangular function Φ,

$$\chi_0, \chi_1, \chi_2, \dots \overset{\Phi}{\mapsto} \varphi_0(\chi_0), \varphi_1(\chi_0, \chi_1), \varphi_2(\chi_0, \chi_1, \chi_2), \dots$$

where $\chi_j \in \{0, 1, \dots, P-1\}$, and every $\varphi_j(\chi_0, \dots, \chi_j) \in \{0, 1, \dots, P-1\}$ is a function in variables $\chi_0, \dots, \chi_j$ of a P-valued logic. This function sends any infinite sequence over $\mathcal{P}$ to infinite sequence over $\mathcal{P}$; that is, Φ maps P-adic integers to P-adic integers. It turns out that Φ is a continuous function with respect to a P-adic metric. Although we devoted several sections in Chapter 1 and a whole Chapter 3 to p-adic numbers and p-adic analysis, here, for reader's convenience, we briefly recall some basic facts on these issues in a less formal manner.

Speaking informally, P-adic integers arise when we extend the set $\mathbb{N}_0$ of non-negative (rational) integers $0, 1, 2, 3, \ldots$, represented by their finite base-P expansions, with infinite base-P expansions; that is, with infinite sequences of symbols from $0, 1, 2, \ldots, P-1$. Addition and multiplication of these sequences can be defined via standard school-textbook algorithms for numbers represented by base-P expansions, thus converting $\mathbb{Z}_P$ into a commutative ring.

We define a distance ρ (metric) on $\mathbb{Z}_P$ in a standard way thus converting $\mathbb{Z}_P$ into a metric space: Given two infinite sequences $\mathcal{S} = s_0, s_1, \ldots$ and $\mathcal{T} = t_0, t_1, \ldots$, where $s_i, t_j \in \mathcal{P}$, we find the smallest i such that $s_i \neq t_i$; then a distance $\rho(\mathcal{S}, \mathcal{T})$ between the sequences $\mathcal{S}$ and $\mathcal{T}$ is $\rho(\mathcal{S}, \mathcal{T}) = P^{-i}$ by the definition, and the distance is 0 whenever no such i exists. The so defined distance is a metric, a P-adic metric; we refer the reader to Section 1.4 for rigorous statements. Once a metric is defined, we may speak about convergence with respect to this metric, limits, continuous functions, etc. Now we shall show that *any triangular function Φ is continuous with respect to the metric ρ.*

Indeed, let us consider a univariate triangular function $\Phi\colon \mathbb{Z}_P \to \mathbb{Z}_P$, which was mentioned above. It is obvious that given two sequences $\mathcal{S} = s_0, s_1, \ldots$ and $\mathcal{T} = t_0, s_1, \ldots$ such that $\rho(\mathcal{S}, \mathcal{T}) = P^{-i}$, then, as the function Φ is triangular, $\rho(\Phi(\mathcal{S}), \Phi(\mathcal{T})) \leq P^{-i}$ since the sequences $\Phi(\mathcal{S})$ and $\Phi(\mathcal{T}))$ agree on at least the first i terms. Hence,

$$\rho(\Phi(\mathcal{S}), \Phi(\mathcal{T})) \leq \rho(\mathcal{S}, \mathcal{T});$$

that is, Φ satisfies Lipschitz condition with a constant 1 and therefore is continuous. A similar argument shows that a multivariate function Φ also satisfies Lipschitz condition with a constant 1 with respect to a metric on n-dimensional metric space $\mathbb{Z}_P^n$. We will discuss this in more detail for the case $P = 2$, see Section 8.2.

Note, however, that we can consider any automaton $\mathfrak{A}$ with finite input and output alphabets as an automaton with n inputs and m outputs over a certain finite alphabet $\mathcal{P}$; e.g., by assuming $\mathcal{P} = \mathcal{K}$ and taking output alphabet with P^k letters, where k is large enough so that $P^k \geq \#\mathcal{M}$ (i.e., we just reserve more letter for output than are really outputted). So we summarize:

> *All automata* (that is, triangular) *functions are continuous with respect to some P-adic metric.*

This conclusion hints that P-adic theory may be useful in a study of some problems of automata theory. However, these problems must be properly re-stated beforehand, in 'analytic' terms of P-adic limits, convergence, derivatives, etc. It turns out that a number of problems can be re-stated in this manner, and P-adic analysis (also P-adic dynamics) can be applied to solve these problems. We consider some particular problems of this sort in the following sections and especially in Chapter 9. Moreover, we emphasize that to apply P-adic techniques we need the automaton function be represented explicitly in a certain meaning, as P-adic tools work with functions rather than with automata that evaluate these functions.

To illustrate this approach, we briefly discuss here a problem of invertibility of automata. The automaton $\mathfrak{A}$ is called *invertible* whenever its automaton function $\Phi = \Phi_{\mathfrak{A}}$ is invertible. The automaton is called invertible on words of length k whenever a restriction of the automaton function to input words of length k is an invertible mapping. From Theorem 4.23 it follows that an automaton with n inputs and n outputs over an alphabet $\mathcal{P} = \{0, 1, \ldots, p-1\}$, p prime, is invertible if and only if it is invertible on all words of length k for all $k = 1, 2, \ldots$; that is, if and only if the automaton function $\Phi\colon \mathbb{Z}_p^n \to \mathbb{Z}_p^n$ is measure-preserving. Now, to determine whether a given automaton is invertible one may use various techniques of Chapter 4. We conclude the section by an example that demonstrates these technique, leaving a detailed study of more specific automata for further sections in this chapter, as well as in Chapter 9.

Consider a special type of Ising automata, a Thue–Morse automaton, which generate a well-known Thue–Morse sequence. The automaton is usually defined as follows: In a general automaton $\mathfrak{A}$, assume $\mathcal{K} = \mathcal{N} = \mathcal{M} = \{0, 1\}$, $u_0 = 0$, where $f = F$, $f(0, 0) = 0$, $f(1, 0) = 1$, $f(0, 1) = 1$, and $f(1, 1) = 0$. It is obvious that f is just a XOR, addition modulo 2: $f(x, y) \equiv x + y \pmod 2$. Moreover, it is clear that the i-th symbol z_i of the output sequence is then $z_i \equiv w_i + u_i \pmod 2$; i.e., $z_i \equiv w_i + w_{i-1} + \cdots + w_0 \pmod 2$, where (w_j) is the input sequence. Thus, the corresponding automaton function Φ can be represented as $\Phi(x) = x \,\mathsf{XOR}\, 2x \,\mathsf{XOR}\, 4x \,\mathsf{XOR} \cdots \mathsf{XOR}\, 2^i x \cdots$ (read more about XOR in Section 8.2).

Example 8.1 (Thue–Morse automaton). *The Thue–Morse automaton is invertible.*

First proof: Each ith coordinate function $\delta_i(\Phi(x))$ of the automaton function $\Phi(x)$ is linear with respect to the ith variable, and the conclusion follows from Theorem 4.39.

Second proof: The automaton function $\Phi(x)$ is uniformly differentiable modulo 2, $\Phi_1'(x) \equiv 1 \pmod 2$ and $N_1(\Phi) = 1$ (see Example 8.11 further for a rigorous proof); moreover, $\Phi(x) \equiv x \pmod 2$, that is, Φ is bijective modulo 2. Now the conclusion follows from Theorem 4.45.

Of course, this result is well known and is placed here only to illustrate our methods. The following result exhibits more interesting application of p-adic techniques to automata theory:

Theorem 8.2. *Whenever the automaton function $\Psi = \Psi_{\mathfrak{A}}$ is a univariate polynomial of degree > 1 over the ring of p-adic integers $\mathbb{Z}_p$, the automaton $\mathfrak{A}$ has no equivalent states and so is infinite.*

Proof. From the definition of equivalent states it follows that whenever the equivalent states exist, there exist positive rational integers $M, N \in \mathbb{N}$ and non-negative rational integers $a \in \{0, 1, \ldots, p^N - 1\}$, $b \in \{0, 1, \ldots, p^M - 1\}$, $a \neq b$, such that

$$\frac{1}{p^N}\left(\Psi(a + p^N z) - \Psi(a) \bmod p^N\right) = \frac{1}{p^M}\left(\Psi(b + p^M z) - \Psi(b) \bmod p^M\right), \tag{8.1}$$

for all p-adic integers $z \in \mathbb{Z}_p$. Here $c \bmod p^K$ stands for the least non-negative residue of c modulo p^K: If $c = c_0 + c_1 \cdot p + c_2 \cdot p^2 + \cdots$, then $c \bmod p^K = c_0 + c_1 \cdot p + c_2 \cdot p^2 + \cdots + c_{K-1} \cdot p^{K-1}$. Indeed, loosely speaking, these a and b are p-adic representations of finite input words that send the automaton $\mathfrak{A} = \langle \mathbb{Z}/p\mathbb{Z}, \mathcal{N}, \mathbb{Z}/p\mathbb{Z}, f, F, u_0 \rangle$ to respective states $t_0, s_0 \in \mathcal{N}$, when any input sequence $z \in \mathbb{Z}_p$ to automata $\mathfrak{A}(t_0) = \langle \mathbb{Z}/p\mathbb{Z}, \mathcal{N}, \mathbb{Z}/p\mathbb{Z}, f, F, t_0 \rangle$ and $\mathfrak{A}(s_0) = \langle \mathbb{Z}/p\mathbb{Z}, \mathcal{N}, \mathbb{Z}/p\mathbb{Z}, f, F, s_0 \rangle$ results in equal outputs sequences. That is, the equivalence of states t_0 and s_0 the automaton $\mathfrak{A}$ reaches after the sequences a and b (of lengths N and M, respectively) have been input, means that output sequences (represented by p-adic integers $\Psi(a + p^N z)$ and $\Psi(b + p^M z)$) agree starting accordingly with Nth and Mth terms, for all $z \in \mathbb{Z}_p$.

As $\Psi(x)$ is a polynomial over $\mathbb{Z}_p$, by Taylor formula we have that

$$\Psi(a + p^N z) = \Psi(a) + p^N z \cdot \Psi'(a) + \cdots + p^{dN} z^d \cdot \frac{\Psi^{(d)}(a)}{d!},$$
$$\Psi(b + p^N z) = \Psi(b) + p^M z \cdot \Psi'(b) + \cdots + p^{dM} z^d \cdot \frac{\Psi^{(d)}(b)}{d!},$$

where $d = \deg \Psi(x)$. From here in view of (8.1) we conclude that

$$\begin{aligned} &\frac{1}{p^N} \cdot (\Psi(a) - \Psi(a) \bmod p^N) + z \cdot \Psi'(a) + \cdots + p^{(d-1)N} z^d \cdot \frac{\Psi^{(d)}(a)}{d!} \\ &= \frac{1}{p^M} \cdot (\Psi(b) - \Psi(b) \bmod p^M) + z \cdot \Psi'(b) + \cdots + p^{(d-1)M} z^d \cdot \frac{\Psi^{(d)}(b)}{d!}, \end{aligned} \tag{8.2}$$

for all $z \in \mathbb{Z}_p$. As both sides of (8.2) are polynomials in variable z over the integral domain $\mathbb{Z}_p$, respective coefficients of these polynomials must be pairwise equal. In particular,

$$p^{(j-1)N} \cdot \frac{\Psi^{(j)}(a)}{d!} = p^{(j-1)M} \cdot \frac{\Psi^{(j)}(b)}{d!}, \tag{8.3}$$

for all $j = 1, 2, \ldots, d$. However, as $\frac{\Psi^{(d)}(a)}{d!} = \frac{\Psi^{(d)}(b)}{d!} = \mathrm{Coef}_{x^d}(\Psi(x))$ and $d = \deg \Psi(x) > 1$, by putting $j = d$ in (8.3) we conclude that $M = N$. Now, taking $j = d - 1$ in (8.3), we see that $\mathrm{Coef}_{x^{d-1}}(\Psi(x)) + d \cdot \mathrm{Coef}_{x^d}(\Psi(x)) \cdot a = \mathrm{Coef}_{x^{d-1}}(\Psi(x)) + d \cdot \mathrm{Coef}_{x^d}(\Psi(x)) \cdot b$, i.e., that $a = b$. So the states t_0 and s_0 are equal, $t_0 = s_0$. □

Further in Subsection 11.1.2 we will show that finite automata exhibit sharp irregularities in distribution of output sequences, whereas automata whose automata functions are polynomials of degrees > 1 do not. Moreover, there in Proposition 11.15 we prove that automata functions exhibit a property that may be considered as a version of a zero-one law from probability theory.

8.2 Computers think 2-adically

In this section we consider specific, very important and very wide spread automata, digital computers. We will show that in many cases their *instructions, as well as compositions of these instructions, computer programs, can be regarded as continuous 2-adic functions.* This implies that a number of mathematical methods from 2-adic analysis and 2-adic dynamics can be exploited to develop computer programs with high performance and prescribed properties. This is a key point of the approach we apply further in Chapter 9.

A heart of a computer is the CPU, the central processing unit, a microprocessor. A contemporary microprocessor is word-oriented. That is, it works with words of zeroes and ones of a certain fixed length n (usually $n = 8, 16, 32, 64$). Each binary word z of length n can be considered as a base-2 expansion of a number $z \in \{0, 1, \dots, 2^n - 1\}$ and vise versa. We also can identify the set $\{0, 1, \dots, 2^n - 1\}$ with residues modulo 2^n; that is with elements of the residue ring $\mathbb{Z}/2^n\mathbb{Z}$ modulo 2^n.

Actually, *arithmetic* (numerical) instructions of a microprocessor are just *operations of the residue ring* $\mathbb{Z}/2^n\mathbb{Z}$: An n-bit microprocessor performing a single instruction of addition (or multiplication) of two n-bit numbers just deletes more significant digits of a sum (or of a product) of these numbers thus merely reducing the result modulo 2^n. Note that to calculate a sum of two integers (i.e., without reducing the result modulo 2^n) a 'standard' microprocessor uses not a single instruction but invokes a program (that is, a sequence of basic instructions).

The other sort of basic instructions of a microprocessor are *bitwise logical* operations, such as XOR, OR, AND, and NOT. The third type of instructions could be called a *machine* ones since they depend on an architecture of a microprocessor. But usually they include such standard instructions as left and right shifts of an n-bit word.

We now give formal definitions of these basic instructions, bitwise logical and machine: Let

$$z = \delta_0(z) + \delta_1(z) \cdot 2 + \delta_2(z) \cdot 2^2 + \delta_3(z) \cdot 2^3 + \cdots$$

be a base-2 expansion for $z \in \mathbb{N}_0 = \{0, 1, 2, \dots\}$ (that is, $\delta_j(z) \in \{0, 1\}$); then,

- $y \,\mathsf{XOR}\, z$ is a bitwise addition modulo 2: $\delta_j(y \,\mathsf{XOR}\, z) \equiv \delta_j(y) + \delta_j(z) \pmod 2$;
- $y \,\mathsf{AND}\, z$ is a bitwise multiplication modulo 2: $\delta_j(y \,\mathsf{AND}\, z) \equiv \delta_j(y) \cdot \delta_j(z) \pmod 2$;
- NOT, a bitwise logical negation: $\delta_j(\mathsf{NOT}(z)) \equiv \delta_j(z) + 1 \pmod 2$;

- y OR z is a bitwise logical 'or': $\delta_j(y \text{ OR } z) \equiv \delta_j(y) \text{ OR } \delta_j(z) \pmod 2$;
- $\lfloor \frac{z}{2} \rfloor$, the integral part of $\frac{z}{2}$, is a shift towards less significant bits;
- $2^k \cdot z$, a multiplication by kth power of 2, is a k-bit shift towards more significant bits;
- y AND z, where y is a constant, is also called a masking of z with the mask y;
- $z \bmod 2^k = z \text{ AND } (2^k - 1)$ is a reduction of z modulo 2^k; a truncation of all high order bits starting with the kth one, as $2^k - 1 = \dots 000\underbrace{11\dots11}_{k}$.

Note 8.3. All basic instructions listed above, with the exception of shift towards less significant bits, are triangular functions in the meaning of the definition from Section 8.1, for $P = 2$.

Note that in literature $\oplus$ is used along with XOR for a bitwise 'exclusive or' operator, $\vee$ along with OR, and $\wedge$ (or $\odot$) along with AND. In this book, we use only OR for bitwise logical 'or', AND for bitwise logical 'and', we use XOR for 'exclusive or' as symbols of respective operations on machine words (n-bit words, $n > 1$). And we use $\oplus$ for addition modulo 2 (i.e., for 'exclusive or') whenever we consider bits rather than binary words, e.g., when we work with Boolean functions.

We can make now the following important observation: Basic instructions of a processor are well-defined functions on the set $\mathbb{N}_0$ (of non-negative rational integers) valuated in $\mathbb{N}_0$: Actually we just represent integers from $\mathbb{N}_0$ by their base-2 expansions. Moreover, from the definitions of the mentioned basic instructions it immediately follows that actually they are defined on the set of all one-side (countably) infinite sequences of zeroes and ones, that is, on the space $\mathbb{Z}_2$ of 2-adic integers. In other words:

Basic instructions of a microprocessor are functions defined on the space of 2-adic integers and valuated in the space of 2-adic integers.

Although all necessary notions and statements of p-adic theory already are formally defined and rigorously proved, see the respective sections in Chapter 1 and the whole Chapter 3, here, for illustration and better understanding of some specific features of the 2-adic case, we (somewhat informally) discuss key issues again.

The set $\mathbb{Z}_2$ consists of all infinite binary sequences $\dots\delta_2(x)\delta_1(x)\delta_0(x) = x$, where $\delta_j(x) \in \{0, 1\}$, $j = 0, 1, 2, \dots$. Arithmetic operations (addition and multiplication) with these sequences can be defined via standard 'school-textbook' algorithms of addition and multiplication of natural numbers represented by base-2 expansions: Each term of a sequence that corresponds to the sum (respectively, to the product) of two given sequences can be calculated by these algorithms within a finite number of steps.

Thus, $\mathbb{Z}_2$ is a commutative ring with respect to the so defined addition and multiplication. The ring $\mathbb{Z}_2$ contains a subring $\mathbb{Z}$ of all rational integers: For instance, $\dots 111 = -1$, since

$$\begin{array}{rcccc} & \dots 1 & 1 & 1 & 1 \\ + & & & & \\ & \dots 0 & 0 & 0 & 1 \\ \hline & \dots 0 & 0 & 0 & 0 \end{array}$$

Moreover, the ring $\mathbb{Z}_2$ contains all rational numbers that can be represented by irreducible fractions with odd denominators. For instance, the following calculations show that $\dots 01010101 \times \dots 00011 = \dots 111$, i.e., that $\dots 01010101 = -\frac{1}{3}$ since $\dots 00011 = 3$ and $\dots 111 = -1$:

$$\begin{array}{rcccccc} & \dots 0 & 1 & 0 & 1 & 0 & 1 \\ \times & & & & & & \\ & \dots 0 & 0 & 0 & 0 & 1 & 1 \\ \hline & \dots 0 & 1 & 0 & 1 & 0 & 1 \\ + & & & & & & \\ & \dots 1 & 0 & 1 & 0 & 1 & \\ \hline & \dots 1 & 1 & 1 & 1 & 1 & 1 \end{array}$$

Sequences with only finite number of 1s correspond to non-negative rational integers in their base-2 expansions, sequences with only finite number of 0s correspond to negative rational integers, while eventually periodic sequences (that is, sequences that become periodic starting with a certain place) correspond to rational numbers represented by irreducible fractions with odd denominators: For instance, $3 = \dots 00011$, $-3 = \dots 11101$, $\frac{1}{3} = \dots 10101011$, $-\frac{1}{3} = \dots 1010101$. So the jth term $\delta_j(u)$ of the corresponding sequence $u \in \mathbb{Z}_2$ is merely the jth digit of the base-2 expansion of u whenever u is a non-negative rational integer, $u \in \mathbb{N}_0 = \{0, 1, 2, \dots\}$.

What is important, the ring $\mathbb{Z}_2$ is a metric space with respect to the metric (distance) $d_2(u, v)$ defined by the following rule: $\rho_2(u, v) = |u - v|_2 = \frac{1}{2^n}$, where n is the smallest non-negative rational integer such that $\delta_n(u) \neq \delta_n(v)$, and $d_2(u, v) = 0$ if no such n exists (i.e., if $u = v$). For instance $\rho_2(3, \frac{1}{3}) = \frac{1}{8}$:

$$\left.\begin{array}{l} \dots 1010\check{1}0101 = -\dfrac{1}{3} \\[2ex] \dots 0000\check{0}0101 = \quad 5 \end{array}\right\} \Longrightarrow \rho_2\left(-\frac{1}{3}, 5\right) = \frac{1}{2^4} = \frac{1}{16}.$$

We write then that $-\frac{1}{3} \equiv 5 \pmod{16}$; $-\frac{1}{3} \not\equiv 5 \pmod{32}$; recall the definition of mod 2^k. That is, $|u - v|_2 = 2^{-\ell}$ if and only if $u \equiv v \pmod{2^\ell}$ and $u \not\equiv v \pmod{2^{\ell+1}}$.

Further, the function $\rho_2(u, 0) = |u|_2$ is a 2-adic absolute value of a 2-adic integer u, and $\operatorname{ord}_2 u = -\log_2 |u_2|_2$ is a 2-adic valuation of u. Note that for $u \in \mathbb{N}_0$ the valuation $\operatorname{ord}_2 u$ is merely the exponent of the highest power of 2 that divides u (thus, loosely speaking, $\operatorname{ord}_2 0 = \infty$, so $|0|_2 = 0$). That is, $|u|_2 = 2^{-\ell}$ if and only if $u \equiv 0 \pmod{2^\ell}$ and $u \not\equiv 0 \pmod{2^{\ell+1}}$.

We see now that actually a reduction modulo 2^n of a 2-adic integer z is just an approximation of a 2-adic integer z by a rational integer with a precision $\frac{1}{2^n}$ with respect to the 2-adic metric. This implies:

A microprocessor actually works with approximations of 2-adic integers with respect to the 2-adic metric.

When loading a number whose base-2 expansion contains more than n significant bits into a registry of an n-bit microprocessor, the microprocessor just writes only n low order bits of the number into the registry thus reducing the number modulo 2^n. That is, a precision of the approximation is defined by a bitlength of the microprocessor. Moreover,

Every digital computer, even the simplest one, can, by its very origin, properly operate with 2-adic numbers.

Let's undertake the following 'computer experiment'. Start MS Windows XP, run the built-in Calculator. Switch to Scientific mode. Press `Dec` (that is, switch to decimals), press `1`, then `+/-`. The calculator returns `-1`, as prescribed.

Now, press `Bin`, switching the calculator to binaries. The calculator returns `...111` (64 ones), a 2-adic representation of -1, up to the highest precision the calculator can achieve, 64 bits. (Here a programmer will most likely say that the calculator just uses the *two's complement*).

Now press `Dec` again; the calculator returns `18446744073709551615`. This number is congruent to -1 modulo 2^{64}. Now press successively `/`, `3`, `=`, `Bin`, thus dividing the number by 3 and representing the result in a binary form. The calculator returns `...0101010101`, a 2-adic representation of $-1/3$, with the 2-adic precision 2^{-64}. Indeed, switching back to `Dec` the calculator returns `6148914691236517205`, a multiplicative inverse to -3 modulo 2^{64}:

$$6148914691236517205 \cdot (-3) \equiv 1 \pmod{2^{64}}.$$

This toy experiment can be performed on most calculators. However, sometimes a calculator returns an erroneous result. This usually happens when a corresponding program is written in a higher order language. Very loosely speaking, the capability of a calculator to perform 2-adic arithmetics depends on how the corresponding program is written: Programs written in assembler usually are more capable to perform 2-adic calculations than the ones written in higher-level languages. Programmers use assembler when they want to exploit CPU's resources in the most optimal way; e.g., to store negative numbers they use the two's complement rather than reserve special registry for a sign. But the usage of the two's complement of x (that is, of $\mathsf{NOT}\,x$) is just a way to represent a negative integer in a 2-adic form, as $-x = 1 + \mathsf{NOT}\,x$, see equations (8.4) further. Thus, we might conclude that a CPU is used in a more optimal way when it actually works with binary words as with 2-adic numbers. Now we are going to understand whether we can say more about relationships between basic instructions and the 2-adic metric.

Once a metric is defined, one defines notions of convergent sequences, limits, continuous functions on the metric space, and derivatives if the space is a commutative ring. Let us illustrate how it can be done in our case. We start with a notion of a limit. It reads:

Definition 8.4 (2-adic limit). A 2-adic integer z is said to be a *limit* of the sequence $\{z_i\}_{i=0}^{\infty}$ if and only if for every real $\varepsilon > 0$ there exists N such that $|z_i - z|_2 < \varepsilon$ for all $i > N$.

However, according to the definition of the 2-adic metric, $|z_i - z|_2$ can take only values $2^{-\ell}$ for a suitable $\ell = 0, 1, 2, \ldots$; so we may consider only $\varepsilon = 2^{-r}$ for $r = 0, 1, 2, \ldots$ and re-write the definition, using congruences rather than inequalities, in the following (equivalent) form:

Definition 8.5 (2-adic limit, equivalent form). A 2-adic integer z is said to be a *limit* of the sequence $\{z_i\}_{i=0}^{\infty}$ if and only if for every (sufficiently large) positive rational integer K there exists N such that $z_i \equiv z \pmod{2^K}$ for all $i > N$.

Now it is clear, for instance, that with respect to the so defined metric ρ_2 on $\mathbb{Z}_2$ the following sequence tends to $-1 = \ldots 111$,

$$1, 3, 7, 15, \ldots, 2^n - 1, \ldots \underset{\rho_2}{\longrightarrow} -1;$$

that is, $\lim^2_{n\to\infty} 2^n - 1 = -1$, where $\lim^2_{n\to\infty}$ stands for a limit with respect to the 2-adic metric. This is intuitively clear also, as

$$\begin{array}{cccccccc}
\ldots & 0 & 0 & 0 & 0 & 1 & = & 1 \\
\ldots & 0 & 0 & 0 & 1 & 1 & = & 3 \\
\ldots & 0 & 0 & 1 & 1 & 1 & = & 7 \\
\ldots & 0 & 1 & 1 & 1 & 1 & = & 15 \\
 & & & & & & \vdots & \\
\ldots & 1 & 1 & 1 & 1 & 1 & = & -1
\end{array}$$

In the same manner we can re-write the definition of a continuous function:

Definition 8.6 (2-adic continuous function). A function $f : \mathbb{Z}_2 \to \mathbb{Z}_2$ is said to be *continuous* at the point $z \in \mathbb{Z}_2$ if and only if for every (sufficiently large) positive rational integer M there exists a positive rational integer L such that $f(x) \equiv f(z) \pmod{2^M}$ whenever $x \equiv z \pmod{2^L}$.

Note 8.7. The function f is said to be *uniformly* continuous on $\mathbb{Z}_2$ if and only if f is continuous at every point $z \in \mathbb{Z}_2$, and L depends only on M, and *not* on z.

From here we immediately deduce that all triangular 2-adic (i.e., with $P = 2$, see Section 8.1) functions are uniformly continuous on $\mathbb{Z}_2$. Actually, triangular functions are 1-Lipschitz functions and vice versa; they satisfy the Lipschitz condition with a constant 1:

$$|f(a) - f(b)|_2 \le |a - b|_2.$$

In other words, triangular functions are compatible: Whenever $a \equiv b \pmod{2^\ell}$ then $f(a) \equiv f(b) \pmod{2^\ell}$; this is equivalent to 1-Lipschitz property. A similar argument shows that the same is true for multivariate triangular functions; we only mention that the 2-adic distance between two vectors over $\mathbb{Z}_2$ is a maximum of distances between respective coordinates: Whenever $\mathbf{u} = (u_1, \dots, u_n), \mathbf{v} = (v_1, \dots, v_n) \in \mathbb{Z}_2^n$ then

$$\rho_2(\mathbf{u}, \mathbf{v}) = \max\{|u_i - v_i|_2 : i = 1, 2, \dots, n\}$$

by the definition. It is easy to see that a shift towards less significant bits satisfies the Lipschitz condition with a constant 2: $\left|\lfloor \frac{a}{2} \rfloor - \lfloor \frac{b}{2} \rfloor\right|_2 \le 2 \cdot |a-b|_2$. We conclude finally:

All basic instructions of CPU are uniformly continuous 2-adic functions.

This implies that all compositions of basic instructions, that is, computer programs, are uniformly continuous 2-adic functions either. In the next section we show that a number of instructions and programs are not only uniformly continuous, but are also uniformly differentiable.

We now can expand a list of triangular functions (that is, 1-Lipschitz functions), which also are used in respective programs (e.g., in exponential and inversive pseudo-random generators, see Chapter 9), by the following ones:

$$\begin{aligned} &\text{subtraction,} && -\colon (u, v) \mapsto u - v; \\ &\text{exponentiation,} && \uparrow\colon (u, v) \mapsto u \uparrow v = (1 + 2u)^v; \\ &\text{raising to negative powers,} && u \uparrow (-n) = (1 + 2u)^{-n}; \\ &\text{division,} && //\colon (u, v) \mapsto u // v = u \cdot (v \uparrow (-1)) = \frac{u}{1 + 2v}. \end{aligned}$$

These functions are triangular (that is, 1-Lipschitz, compatible) in view of Proposition 3.65. It is worth noting here that

$$(1 + 2v)^{-1} = 1 - 2v + 4v^2 - 8v^3 + \dots + (-1)^i 2^i v^i + \cdots;$$

so while evaluating $(1 + 2v)^{-1}$ (that is, calculating a multiplicative inverse of an odd number) on a n-bit digital computer we actually use the first n terms of the series since when loading a 2-adic number into an n-bit registry a computer deletes high order bits thus reducing the number modulo 2^n. We stress again that a composition of triangular (that is, 1-Lipschitz) functions is a triangular function. The advantage of 2-adic techniques is that it can handle very complicated compositions of basic instructions, independently of how complex these compositions are; e.g., the following somewhat crazy-looking function

$$(1 + x)\ \mathsf{XOR}\ 4 \cdot \left(1 - 2 \cdot \frac{x\ \mathsf{AND}\ x^2 + x^3\ \mathsf{OR}\ x^4}{3 - 4 \cdot (5 + 6x^5)^{x^6 \mathsf{XOR} x^7}}\right)^{7 - \frac{8x^8}{9 + 10x^9}}$$

is a triangular function, and its properties can be studied by means of 2-adic analysis.

Concluding the section, we note that a look on computer instructions as on 2-adic functions immediately gives us some important identities that will be used further in some proofs and that can be applied to practical writing of programs. Namely, arithmetic and bitwise logical operations are not independent: Some of them can be expressed via the others. For instance, for all $u, v \in \mathbb{Z}_2$ the following identities hold:

$$\begin{aligned} \mathsf{NOT}\, u &= u\ \mathsf{XOR}\ (-1); \\ u + \mathsf{NOT}\, u &= -1; \\ u\ \mathsf{XOR}\ v &= u + v - 2 \cdot (u\ \mathsf{AND}\ v); \\ u\ \mathsf{OR}\ v &= u + v - (u\ \mathsf{AND}\ v); \\ u\ \mathsf{OR}\ v &= (u\ \mathsf{XOR}\ v) + (u\ \mathsf{AND}\ v). \end{aligned} \tag{8.4}$$

The proofs of identities (8.4) are just an exercise: For example, if $\alpha, \beta \in \{0, 1\}$ then $\alpha\ \mathsf{XOR}\ \beta = \alpha + \beta - 2\alpha\beta$ and $\alpha\ \mathsf{OR}\ \beta = \alpha + \beta - \alpha\beta$. Hence, as

$$\begin{aligned} u &= \delta_0(u) + \delta_1(u) \cdot 2 + \delta_2(u) \cdot 2^2 + \delta_3(u) \cdot 2^3 + \cdots \\ v &= \delta_0(v) + \delta_1(v) \cdot 2 + \delta_2(v) \cdot 2^2 + \delta_3(v) \cdot 2^3 + \cdots, \end{aligned}$$

where $\delta_i(u), \delta_i(v) \in \{0, 1\}$, $i = 0, 1, 2, \ldots$, then

$$\begin{aligned} u\ \mathsf{XOR}\ v &= \sum_{i=0}^{\infty} (\delta_i(u) \oplus \delta_i(v)) \cdot 2^i = \sum_{i=0}^{\infty} (\delta_i(u) + \delta_i(v) - 2\delta_i(u)\delta_i(v)) \cdot 2^i \\ &= \sum_{i=0}^{\infty} \delta_i(u) \cdot 2^i + \sum_{i=0}^{\infty} \delta_i(v) \cdot 2^i - 2 \cdot \sum_{i=0}^{\infty} \delta_i(u)\delta_i(v) \cdot 2^i \\ &= u + v - 2(u\ \mathsf{AND}\ v). \end{aligned}$$

The remaining identities can be proved by analogy. Identities for shifts towards more significant digits, as well as for masking and for reduction modulo 2^m can be derived from the above identities: An m-step shift of u is $2^m \cdot u$; masking of u is $u\ \mathsf{AND}\ M$, where M is an integer which base-2 expansion is a mask (i.e., a string of 0s and 1s); reduction modulo 2^m, i.e., taking the least non-negative residue of u modulo 2^m, is $u \bmod 2^m = u\ \mathsf{AND}\ (2^m - 1)$.

All these considerations (after proper modifications) remain true for arbitrary prime p, and *not* only for $p = 2$, thus leading to the notion of a p-adic integer and to p-adic analysis, see Chapter 3. We further use p-adic integers for odd p in some applications to computer science as well, see e.g. Section 8.4 and Chapter 9. Note that as a p-adic integer $z \in \mathbb{Z}_p$ has a unique representation in the p-adic canonical form $z = \delta_0(z) + \delta_1(z) \cdot p + \delta_2(z) \cdot p^2 + \cdots$, where $\delta_j(z) \in \{0, 1, \ldots, p-1\}$, further when necessary we associate a p-adic integer to the *right-infinite* string $\delta_0(z)\delta_1(z)\delta_2(z)\ldots$ and, if $\delta_j(z)$ are 0 for all $j > N$, we omit these zeros: e.g., $1011000\ldots = 1011$, and 1011 is a base-2 expansion of 13, and *not* of 11. In other words, since this moment *we write more significant digits at rightmost positions, and not at leftmost ones!*

8.3 Differentiable instructions and programs

In this section we show that basic instructions of CPU introduced in Section 8.2 are either uniformly differentiable with respect to the 2-adic metric, or are, in a definite meaning, very close to uniformly differentiable 2-adic functions. We also calculate 2-adic derivatives of basic instructions, thus obtaining a kind of 'table of derivatives', which will be used further in applications and proofs.

Although we have already stated a general definition of a derivative with respect to the p-adic distance, see Definition 3.26, in this section we give some equivalent forms of this definition for the case $p = 2$, for better exposition of essence of this extremely important notion. Actually we want to show that 2-adic differentiation is as simple as in standard real analysis; the reason that some peculiarities of 2-adic derivation look somewhat odd at the first glance, is only a matter of our habits in calculations of real derivatives, and nothing more. Moreover, in many cases (e.g., for polynomials) both 2-adic derivation and real derivation give the same result.

We start with a definition of a derivative of a univariate function. Formally it looks similar to a real case with the only difference that it uses a 2-adic absolute value rather than a real one.

Definition 8.8 (2-adic derivative). A function $f : \mathbb{Z}_2 \to \mathbb{Z}_2$ is said to be *differentiable* at the point $x \in \mathbb{Z}_2$ (and $f'(x)$ is said to be a derivative) whenever for every real $\varepsilon > 0$ there exists a real $\omega > 0$ such that

$$\left| \frac{f(x+h) - f(x)}{h} - f'(x) \right|_2 < \varepsilon \tag{8.5}$$

whenever $|h|_2 < \omega$.

We note that in a general case the derivative $f'(x)$ may not be a 2-adic *integer*, it may be a non-integral 2-adic number from $\mathbb{Q}_2$, a field of 2-adic numbers. However, in the case when f is a 1-Lipschitz (that is, triangular) function, this can not happen by Proposition 3.41. So in the sequel *we consider only triangular functions*; that is, we do not consider shifts towards less significant bits. This does *not* mean that we exclude these shifts from compositions; they may be included, we demand only that a whole composition of basic instructions, a program, must be a triangular function (that is, 1-Lipschitz, compatible).

With all this in mind, we now re-state the definition of a derivative for univariate triangular functions. Again, as 2-adic absolute value $|\cdot|_2$ can take only values $2^{-\ell}$ for a suitable $\ell = 0, 1, 2, \ldots$; we may consider only $\varepsilon = 2^{-r}$, $\omega = 2^s$ for $r, s = 0, 1, 2, \ldots$ and we may use congruences rather than inequalities, as $|z|_2 < 2^{-r}$ holds if and only if $z \equiv 0 \pmod{2^{r+1}}$. Moreover, the congruence $z \equiv 0 \pmod{2^{r+1}}$ holds if and only if $z = 2^{r+1}\tilde{z}$ for a suitable 2-adic integer $\tilde{z}$. Now, replacing inequality (8.5) by equivalent congruence and multiplying both parts of this congruence by $h = 2^\ell u$, we obtain the following equivalent definition:

Definition 8.9 (2-adic derivative, equivalent form). A (1-Lipschitz) function f defined on (and valuated in) $\mathbb{Z}_2$ is said to be differentiable at the point $x \in \mathbb{Z}_2$ (and $f'(x)$ is said to be a derivative) if for every natural number k there exists a natural number N such that the congruence

$$f(x + 2^\ell u) \equiv f(x) + 2^k u \cdot f'(x) \pmod{2^{k+\ell}}$$

holds for all $u \in \mathbb{Z}_2$ whenever $\ell \geq N$.

This definition gives rise to another important notion, a *derivative modulo* 2^k, which has no analog in real analysis. It reads:

Definition 8.10 (2-adic derivative modulo 2^k). Let k be a natural number, $k \in \mathbb{N}$. A (1-Lipschitz) function $f : \mathbb{Z}_2 \to \mathbb{Z}_2$ is said to be *differentiable modulo* 2^k at the point $x \in \mathbb{Z}_2$ (and $f_k'(x)$ is said to be a derivative modulo 2^k) if there exists a natural number N such that the congruence

$$f(x + 2^\ell u) \equiv f(x) + 2^\ell u \cdot f_k'(x) \pmod{2^{k+\ell}}$$

holds for all $u \in \mathbb{Z}_2$ whenever $\ell \geq N$.

Note that in this definition, compared to Definition 8.9, we assume that k is *fixed*; that is, the precision of approximation of a ratio of the increment of function to the increment of a variable by a derivative, see (8.5), is not worse than 2^{-k} rather than arbitrarily precise, as dictated by Definition 8.8. Definition 8.9 introduces a sort of 'derivative with a precision not worse than k digits after a point' in 2-adic analysis. The latter notion is meaningless in real analysis since there is no distinguished base to represent numbers; however, in 2-adic analysis this distinguished representation exists, namely the base-2 expansion. Now we refer the reader to a general Definition 3.27 of differentiability modulo p^k and to a discussion thereafter for more detailed introduction of this important concept; here we only mention that a derivative modulo 2^k is defined up to a summand that is congruent to zero modulo 2^k, that is, actually values of derivatives modulo p^k are residues modulo 2^k rather than an integer. Moreover, rules of derivation modulo 2^k are of the same form as in the classical case with the only difference they are congruences modulo 2^k rather than equalities; read more about this in Section 3.7.

What is really important to note is that the differentiability modulo 2^k is much looser restriction compare to ordinary differentiability. It is obvious that whenever a function is differentiable, it is differentiable modulo 2^k for all k. However, the differentiability modulo p^k for some k does not necessarily imply ordinary differentiability. A class of functions that is differentiable, say, modulo 2, is much wider than a class of differentiable functions. However, in most practical cases it is sufficient that a function is differentiable modulo 2^k for some very small k; actually, for methods we apply to computer science in Section 8.4 and Chapter 9 it is sufficient that a function is differentiable modulo 2^k for $k = 1$ or $k = 2$.

The notion of a function $f : \mathbb{Z}_2 \to \mathbb{Z}_2$ that is *uniformly* differentiable (modulo 2^k) on $\mathbb{Z}_2$ can now be introduced in a standard form: The congruence from Definition 8.9 (respectively, from Definition 8.10) must hold for all $x \in \mathbb{Z}_2$ simultaneously, that is, N must not depend on x. The smallest N with this property is defined via $N(f)$ (respectively, via $N_k(f)$). Now we introduce a short 'table of derivatives' of 2-adic analysis.

Example 8.11 (Derivatives of bitwise logical operations).

(1) *The function* $f(x) = x \text{ AND } c$ *is uniformly differentiable on* $\mathbb{Z}_2$ *for any* $c \in \mathbb{Z}$; $f'(x) = 0$ *for* $c \geq 0$, *and* $f'(x) = 1$ *for* $c < 0$.

Indeed, $f(x + 2^n s) = f(x)$, and $f(x + 2^n s) = f(x) + 2^n s$ for $n \geq l(|c|)$, where $l(|c|)$ is a bit length of a real absolute value of c (mind that for $c \geq 0$ the 2-adic representation of $-c$ starts with base-2 expansion of the number $2^{l(c)} - c$, which occupies less significant bit positions, followed by $\ldots 11$: $-1 = 111\ldots$, $-3 = 10111\ldots$, etc.).

(2) *The function* $f(x) = x \text{ XOR } c$ *is uniformly differentiable on* $\mathbb{Z}_2$ *for any* $c \in \mathbb{Z}$; $f'(x) = 1$ *for* $c \geq 0$, *and* $f'(x) = -1$ *for* $c < 0$.

This immediately follows from Claim 1 above since $u \text{ XOR } v = u + v - 2\cdot(x \text{ AND } v)$, see (8.4); thus $(x \text{ XOR } c)' = x' + c' - 2 \cdot (x \text{ AND } c)' = 1 + 2 \cdot (0,\ \text{if } c \geq 0;\ \text{or } -1,\ \text{if } c < 0)$.

(3) In a similar manner it can be shown that *functions* $(x \bmod 2^n)$, $\text{NOT}(x)$ *and* $(x \text{ OR } c)$ *for* $c \in \mathbb{Z}$ *are uniformly differentiable on* $\mathbb{Z}_2$, *and* $(x \bmod 2^n)' = 0$, $(\text{NOT } x)' = -1$, $(x \text{ OR } c)' = 1$ *for* $c \geq 0$, $(x \text{ OR } c)' = 0$ *for* $c < 0$.

(4) *The function* $f(x, y) = x \text{ XOR } y$ *is not uniformly differentiable on* $\mathbb{Z}_2^2$ (as a bivariate function)*; however, it is uniformly differentiable modulo* 2 *on* $\mathbb{Z}_2^2$, *and its partial derivatives modulo 2 are 1 everywhere on* $\mathbb{Z}_2^2$.

Indeed, as a non-zero 2-adic integer can be simultaneously considered as a limit of a sequence of positive rational integers, and as a limit of a sequence of negative rational integers, the first part of Claim 4 follows from Claim 2 above. Moreover, the second part of Claim 4 also follows from Claim 2 as $-1 \equiv 1 \pmod 2$.

Note that *some functions have zero derivatives although they are not constants* (these functions are called *pseudo-constants*); this is one of the peculiarities of 2-adic analysis.

Consider some more examples which will be used in the sequel:

Example 8.12. *The function* $f(x) = x + (x^2 \text{ OR } 5)$ *is uniformly differentiable on* $\mathbb{Z}_2$ (whence, uniformly differentiable modulo 2 and modulo 4), $N_1(f) = N_2(f) = \cdots = N(f) = 3$, and $f'(x) = 1 + 2x \cdot \frac{\partial(u \text{OR} 5)}{\partial u}\Big|_{u=x^2} = 1 + 2x$.

Indeed, it is clear that $(x + h) \text{ OR } 5 = (x \text{ OR } 5) + h$ whenever $h \equiv 0 \pmod 8$ as the base-2 expansion of 5 is $\ldots 000101$.

Example 8.13. A function

$$F(x, y) = (f(x, y), g(x, y)) = (x \text{ XOR } (2 \cdot (x \text{ AND } y)), (y + 3x^3) \text{ XOR } x)$$

is uniformly differentiable modulo 2 as a bivariate function, and $N_1(F) = 1$; namely

$$F(x + 2^n t, y + 2^m s) \equiv F(x, y) + (2^n t, 2^m s) \cdot \begin{pmatrix} 1 & x+1 \\ 0 & 1 \end{pmatrix} \pmod{2^{k+1}}$$

for all $m, n \geq 1$ (here $k = \min\{m, n\}$). The matrix $\left(\begin{smallmatrix} 1 & x+1 \\ 0 & 1 \end{smallmatrix}\right) = F_1'(x, y)$ is a Jacobi matrix modulo 2 of F (see Definition 3.27).

Here is how we calculate partial derivatives modulo 2: For instance, $\frac{\partial_1 g(x,y)}{\partial_1 x} = \frac{\partial_1 (y+3x^3)}{\partial_1 x} \cdot \frac{\partial_1 (u \text{XOR} x)}{\partial_1 u}\Big|_{u=y+3x^3} + \frac{\partial_1 x}{\partial_1 x} \cdot \frac{\partial_1 (u \text{XOR} x)}{\partial_1 x}\Big|_{u=y+3x^3} = 9x^2 \cdot 1 + 1 \cdot 1 \equiv x + 1 \pmod 2$. Note that a partial derivative modulo 2 of the function $2 \cdot (x \text{ AND } y)$ is always 0 modulo 2, due to the multiplier 2: The function x AND y is not differentiable modulo 2 as a bivariate function, however, the function $2 \cdot (x \text{ AND } y)$ is. So the Jacobian of the function F is $\det F_1' \equiv 1 \pmod 2$.

In the next section we apply techniques of 2-adic (actually, p-adic for arbitrary prime p) derivations to construct popular combinatorial objects, Latin squares. We again recall that all considerations we made above remain true for arbitrary prime p, after proper re-statements. Theoretical results we use further, were developed for a general case, see Chapters 3 and 4.

8.4 Latin squares

This section serves as the first example of how p-adic dynamics works in special applied combinatorial area, the construction of Latin squares and of mutually orthogonal Latin squares. We recall that a *Latin square of order P* is a $P \times P$ matrix containing P distinct symbols (usually denoted by $0, 1, \ldots, P-1$) such that each row and column of the matrix contains each symbol exactly once.

In algebra, Latin squares are also known as *binary quasigroups*, an algebraic system on the set $\mathcal{A} = \{0, 1, \ldots, P-1\}$ with the only binary operation $*$ defined by the Cayley table, which is a Latin square. Note that the operation $*$ is invertible with respect to each variable: given $a, b \in \mathcal{A}$, either equation $a * y = b$ and $x * a = b$ has a unique solution. However, the operation $*$ need not be associative.

In other words, a Latin square is a 2-variate mapping $f : \mathcal{A}^2 \to \mathcal{A}$, where $\mathcal{A} = \{0, 1, \ldots, P-1\}$, which is invertible (i.e., bijective) with respect to each variable.

Latins squares are used widely: For games (recall sudoku), and for more serious applications as, say, private communication networks (for password distribution), in coding theory, in some cryptographic algorithms (under a name of *multipermutations*), etc. We refer the reader to monographs [100, 287] of applied examples as well as methods to construct Latin squares. However, methods of the mentioned book may not work efficiently in some cases; thus, for these cases we need new, more effective methods.

There is no problem to construct one small Latin square; a circulant matrix serves a simple example of a Latin square. Here is a 6×6 one:

$$\begin{matrix} 0 & 1 & 2 & 3 & 4 & 5 \\ 1 & 2 & 3 & 4 & 5 & 0 \\ 2 & 3 & 4 & 5 & 0 & 1 \\ 3 & 4 & 5 & 0 & 1 & 2 \\ 4 & 5 & 0 & 1 & 2 & 3 \\ 5 & 0 & 1 & 2 & 3 & 4 \end{matrix}$$

The real problem is *how to write a software that produces a number of large Latin squares*; however, this is only a part of the problem. Another part of the problem is that in some constraint environments (e.g., in smart cards) *we can not store the whole matrix*: Given two numbers $a, b \in \{0, 1, \dots, P-1\}$ *we must calculate the* (a, b)*th entry of the matrix on-the-fly*. We apply p-adic dynamics to give a solution to this problem, in the following way.

According to Theorem 4.23 a bivariate 1-Lipschitz (that is, triangular) function $f : \mathbb{Z}_p^2 \to \mathbb{Z}_p$ is bijective modulo p^k for all $k \in \mathbb{N}$ with respect to either variable if and only if f is measure-preserving with respect to either variable. And Theorem 4.45 actually states that functions that are uniformly differentiable modulo p, are bijective modulo p^k for all $k \in \mathbb{N}$ if and only if they are bijective modulo p^k for some (in most cases, small) k. Note that polynomials with integer coefficients are uniformly differentiable functions; whence, they are uniformly differentiable modulo p. Also, polynomials are easily programmable functions as they are just compositions of additions and multiplications. Our idea is to use polynomials with integer coefficients to construct easily programmable Latin squares. Moreover, in the case $p = 2$ we can also add to numerical operations (addition and multiplication) some bitwise logical operators (e.g., XOR to construct measure-preserving functions, see Section 8.3. So the main tool we use to construct easily programmable Latin squares is the following Corollary 8.14 of Theorem 4.45.

We say that a bivariate triangular function $f : \mathbb{Z}_p^2 \to \mathbb{Z}_p$ *is a Latin square modulo* p^k whenever a reduced mapping $\bar{f} = f \bmod p^k : \mathbb{Z}/p^k\mathbb{Z} \times \mathbb{Z}/p^k\mathbb{Z} \to \mathbb{Z}/p^k\mathbb{Z}$ (that is $\bar{f}(a, b) = f(a, b) \bmod p^k$ for $a, b \in \{0, 1, \dots, p^k - 1\}$) is a Latin square on $\mathcal{A} = \mathbb{Z}/p^k\mathbb{Z} = \{0, 1, \dots, p^k - 1\}$.

Corollary 8.14. *A uniformly differentiable modulo p triangular* (i.e., 1-Lipschitz) *function* $f : \mathbb{Z}_p^2 \to \mathbb{Z}_p$ *is a Latin square modulo* p^k *for all* $k = 1, 2, \dots$ *whenever* f *is a Latin square modulo* $p^{N_1(f)}$ *and* $\frac{\partial_1 f(\mathbf{u})}{\partial_1 x_i} \not\equiv 0 \pmod{p}$ *for all* $\mathbf{u} \in (\mathbb{Z}/p^{N_1(f)}\mathbb{Z})^2$, $i = 1, 2$. *Equivalent statement: if and only if* f *is bijective modulo* $p^{N_1(f)+1}$ *with respect to either variable.*

Proof. Indeed, in view of Theorem 4.45, the function f is bijective modulo p^k with respect to either variable if and only if f is bijective modulo $p^{N_1(f)}$ with respect to either variable, and both $\frac{\partial_1 f(x,y)}{\partial_1 x}$ and $\frac{\partial_1 f(x,y)}{\partial_1 y}$ are 0 modulo p nowhere; these conditions are equivalent to the bijectivity modulo $p^{N_1(f)+1}$ of the function f with respect to either variable. □

Example 8.15 (Latin square on 2^k symbols). Take an *arbitrary* triangular function $v(x, y)$ (that is, *arbitrary* composition of numerical and bitwise logical operators, see Section 8.2) and arbitrary integer $\gamma \in \mathbb{Z}$. Then

$$f(x, y) = x + y + \gamma + 2 \cdot v(x, y)$$

is a Latin square on 2^k symbols for all $k = 1, 2, \ldots$.

Indeed, $f(x, y) \equiv x + y + \gamma \pmod 2$ is a Latin square modulo 2, and $\frac{\partial f(x,y)}{\partial x} \equiv \frac{\partial f(x,y)}{\partial x} \equiv 1 \pmod 2$.

Example 8.16 (Latin square on $2^k \cdot 3^\ell \cdots p^r$ symbols). The function $f(x, y) = x + y + 2 \cdot 3 \cdots p \cdot v(x, y)$, where $v(x, y)$ is an *arbitrary polynomial* with integer coefficients, is a Latin square on $N = 2^k \cdot 3^\ell \cdots p^r$ symbols.

Indeed, as $f(x, y)$ is a polynomial with integer coefficients, it is compatible with all congruences of the ring $\mathbb{Z}$ of rational integers. So to verify whether f is a Latin square modulo $N = 2^k \cdot 3^\ell \cdots p^r$, in view of compatibility of f it is sufficient to verify whether f is a Latin square modulo 2^k, modulo 3^ℓ, …, and modulo p^r. We use Corollary 8.14 for this purpose. The conclusion now follows, as f is a Latin square modulo $2, 3, \ldots, p$ and $\frac{\partial f(x,y)}{\partial x} \equiv \frac{\partial f(x,y)}{\partial x} \equiv 1 \pmod q$ for $q = 2, 3, \ldots, p$.

Now we expand the underlying idea of this example. Actually, given *arbitrary* Latin squares $f_2, f_3, f_5, \ldots, f_p$ on $2, 3, 5, \ldots, p$ symbols, respectively (some primes may absent), we can construct a bivariate polynomial $f(x, y)$ with integer coefficients so that $f(x, y) \equiv f_2(x, y) \pmod 2$, $f(x, y) \equiv f_3(x, y) \pmod 3$, $f(x, y) \equiv f_5(x, y) \pmod 5, \ldots, f(x, y) \equiv f_p(x, y) \pmod p$, and that $f(x, y) \bmod p^N$ is a Latin square on $N = 2^k \cdot 3^\ell \cdots p^r$ symbols, for all $k, \ell, \ldots, r \in N$.

Theorem 8.17. *Let $f_2(x, y), f_3(x, y), f_5(x, y), \ldots, f_p(x, y)$ be Latin squares on $2, 3, 5, \ldots, p$ symbols, respectively* (some primes may absent). *There exists a polynomial with rational integer coefficients $g(x, y) \in \mathbb{Z}[x, y]$ such that every function $f(x, y) \bmod p^N$, where $f(x, y) = g(x, y) + 2 \cdot 3 \cdots p \cdot v(x, y)$, is a Latin square on $N = 2^k \cdot 3^\ell \cdots p^r$ symbols, for all natural $k, \ell, \ldots, r$, and $f(x, y) \equiv f_q(x, y)$* $\pmod q$ *for all $p = 2, 3, 5, \ldots, p$. Here $v(x, y) \in \mathbb{Z}[x, y]$ is an arbitrary polynomial with rational integer coefficients.*

Sketch proof. The key idea of the proof exploits the fact that every bivariate function $f_q \colon (\mathbb{Z}/q\mathbb{Z})^2 \to \mathbb{Z}/q\mathbb{Z}$, q prime, can be represented by a polynomial with rational integer coefficients such that a derivative of this polynomial with respect to either variable defines a prescribed mapping of $\mathbb{Z}/q\mathbb{Z}$ into $\mathbb{Z}/q\mathbb{Z}$, see interpolation formula (1.9).

That is, for every $f_q(x, y)$, $q \in \{2, 3, 5, \ldots, p\}$ (some primes may absent) we construct a polynomial $g_q(x, y)$ such that $f_q(x, y) = g_q(x, y)$ for all $(x, y) \in (\mathbb{Z}/q\mathbb{Z})^2$. Then we use the Chinese Remainder Theorem 1.1 to construct a polynomial $\tilde{g}(x, y) \in \mathbb{Z}[x, y]$ such that $\tilde{g}(x, y) \equiv g_q(x, y) \pmod{q}$ for all $q \in \{2, 3, 5, \ldots, p\}$ (respective primes are absent). Then, with the use of Proposition 1.34, by adding new terms of the form $\frac{\bar{N}}{q} \cdot ((x^q - x) \cdot u_q(x, y) + (y^q - y) \cdot v_q(x, y))$ to the polynomial $\tilde{g}(x, y)$, where $\bar{N} = 2 \cdot 3 \cdot 5 \cdots p$ (respective primes in the product are absent), we construct a polynomial $g(x, y)$ such that $g(x, y) \equiv \tilde{g}(x, y) \pmod{q}$, $\frac{\partial g(x,y))}{\partial x} \not\equiv 0 \pmod{q}$ and $\frac{\partial g(x,y))}{\partial y} \not\equiv 0 \pmod{q}$ for all corresponding primes q and all $(x, y) \in \mathbb{Z}^2$. Now a combination of Theorem 4.45 with the equivalent form of the Chinese Remainder Theorem 1.30 proves Theorem 8.17. We leave details of the proof to the reader. □

Note that Theorem 8.17 not only states the existence of this polynomial $g(x, y)$ but gives also a method to construct it explicitly, as both Proposition 1.34 and Chinese Remainder Theorem 1.1 are constructive.

We must note, however, that whenever some primes in prime power decomposition of N are too large, Theorem 8.17 may be impractical since the corresponding interpolation polynomial will be of high degree and may consist of a huge number of non-zero terms. However, in most practical cases Theorem 8.17 works fine. For example, let us construct with the use of this theorem a Latin square on 10^n symbols.

We skip the first step, the construction of respective interpolation polynomials for Latin squares on 2 and 5 symbols as this procedure is clear from interpolation formula (1.9); we assume that these Latin squares are already represented by bivariate polynomials[2]: $f_2(x, y) = x + y$ and $f_5(x, y) = 1 + 3x^2 + y$. We see that $f_5(x, y) \equiv f_2(x, y) + 1 \pmod{2}$; so we only must 'tweak' constant term (note that in general case we would use Chinese Remainder Theorem 1.1 here): we put $\tilde{g}(x, y) = 6 + 3x^2 + y$ as $6 \equiv 1 \pmod{5}$ and $6 = 0 \pmod{2}$. Then, as $\frac{\partial \tilde{g}(x,y))}{\partial x} = 6x$ and $\frac{\partial \tilde{g}(x,y))}{\partial y} = 1$; we must find a tweak $g(x, y)$ for $\tilde{g}(x, y)$ to make the partial derivative $\frac{\partial g(x,y)}{\partial x}$ non-zero both modulo 2 and modulo 5 everywhere on $\mathbb{Z}/2\mathbb{Z}$ and $\mathbb{Z}/5\mathbb{Z}$, respectively; however, we must not change $\tilde{g}(x, y)$ neither modulo 2 nor modulo 5 by this tweak; that is $\tilde{g}(x, y) \equiv g(x, y) \pmod{2}$ and $\tilde{g}(x, y) \equiv g(x, y) \pmod{5}$ must hold for all $(x, y) \in \mathbb{Z}^2$. Let us tweak $\tilde{g}(x, y)$ so that, say, $\frac{\partial \tilde{g}(x,y))}{\partial x} \equiv 1 \pmod{2}$ everywhere on $\mathbb{Z}/2\mathbb{Z}$ and $\frac{\partial \tilde{g}(x,y))}{\partial x} \equiv 4 \pmod{5}$ everywhere on $\mathbb{Z}/5\mathbb{Z}$. For this purpose, according to formula from Proposition 1.34, we put $g(x, y) = 6 + 3x^2 + y + 6(x^5 - x)(x + 1) + 5(x^2 - x) = y + 6 - 11x + 2x^2 + 6x^5 + 6x^6$. That is, $f(x, y) = g(x, y) + 10 \cdot v(x, y)$, where $v(x, y)$ is arbitrary polynomial over $\mathbb{Z}$.

[2]The reader may verify by direct calculations that both $f_2(x, y)$ and $f_5(x, y)$ are Latin squares on $\mathbb{Z}/2\mathbb{Z}$ and $\mathbb{Z}/5\mathbb{Z}$, respectively.

Both $g(x, y) \bmod 10^n$ and $f(x, y) \bmod 10^n$ are Latin squares modulo 10^n for every $n = 1, 2, 3, \ldots$.

Now we will explain how p-adic dynamics may be of use to construct mutually orthogonal Latin squares. Recall that two $P \times P$ Latin squares are said to be *orthogonal* if when the squares are superimposed each of the P^2 ordered pairs of symbols appears exactly once. Here is an example of a pair of orthogonal Latin squares on 3 symbols: The Latin squares

$$\begin{matrix} 0 & 1 & 2 \\ 1 & 2 & 0 \\ 2 & 0 & 1 \end{matrix} \qquad\qquad \begin{matrix} 0 & 1 & 2 \\ 2 & 0 & 1 \\ 1 & 2 & 0 \end{matrix}$$

are orthogonal since after we superimpose them, we get a square

$$\begin{matrix} (0,0) & (1,1) & (2,2) \\ (1,2) & (2,0) & (0,1) \\ (2,1) & (0,2) & (1,0) \end{matrix}$$

where all pairs are different.

Mutually orthogonal Latin squares are used in experiment design to provide consistent testing of samples, as well as in cryptography (e.g., as block mixers for block ciphers, and as cipher combiners), etc.

For instance, consider three programs which must be tested on each of three platforms. To run all these 9 tests, we must have a sort of schedule. We can make a schedule using the just mentioned example of orthogonal Latin squares of order 3. Namely, the table of pairs of superimposed squares gives us a schedule: Columns give us days of testing, the first number in a pair is a number of platform, the second number is a number of program. As the pair (0, 2) occurs in the second column, this means that the program No 2 must be tested on the platform No 0 at the second day.

Once again, there is no problem to construct a pair of small mutually orthogonal Latin squares; a problem is to create a software that produces pairs of large Latin squares, and that does it in a somewhat 'pseudorandom' way[3]. Here we explain a corresponding method; it again utilizes Theorem 4.45. We will use the following

Corollary 8.18 (of Theorem 4.45). *Let $g, f : \mathbb{Z}_p^2 \to \mathbb{Z}_p$ be uniformly differentiable modulo p 1-Lipschitz functions, and let f and g be Latin squares modulo p^k for all $k = 1, 2, \ldots$* (cf. Corollary 8.14). *These Latin squares are orthogonal modulo p^k for all $k = 1, 2, \ldots$ if and only if the function $F(x, y) = (f(x, y), g(x, y)) : \mathbb{Z}_p^2 \to \mathbb{Z}_p^2$ preserves measure. This holds if and only if*

$$\det \begin{pmatrix} \frac{\partial_1 f(x,y)}{\partial_1 x} & \frac{\partial_1 g(x,y)}{\partial_1 x} \\ \frac{\partial_1 f(x,y)}{\partial_1 y} & \frac{\partial_1 g(x,y)}{\partial_1 y} \end{pmatrix} \not\equiv 0 \pmod p$$

for all $(x, y) \in (\mathbb{Z}/p^{N_1(F)}\mathbb{Z})^2$.

[3] Problems of this kind often arise in genetics, quantitative biology, chemistry, etc., see [100].

Proof. From the definition of orthogonal Latin squares it immediately follows that necessary and sufficient conditions for orthogonality modulo p^k is bijectivity of F modulo p^k; so the Latin squares are orthogonal modulo p^k for all $k = 1, 2, 3, \ldots$ if and only if F is measure-preserving, see Theorem 4.23. Now the conclusion follows from Theorem 4.45. □

Note that *Corollary 8.18 gives no method to construct pairs of orthogonal Latin squares on* 2^k *symbols*: From Corollaries 8.14 and 8.18 it immediately follows that for $p = 2$, no pair of functions f and g satisfy Corollary 8.18. Indeed, from Corollary 8.14 it follows that, as either of functions f and g is a Latin square modulo 2^k, every partial derivative modulo 2 of both f and g must be 1; however, this implies that a determinant from Corollary 8.18 is zero modulo 2.

However, for $p \neq 2$, Corollary 8.18 implies a method to construct large orthogonal Latin squares out of small orthogonal Latin squares. For instance, let $p = 3$, and let

$$f(x, y) \bmod 3 = \begin{pmatrix} 0 & 1 & 2 \\ 1 & 2 & 0 \\ 2 & 0 & 1 \end{pmatrix}, \qquad g(x, y) \bmod 3 = \begin{pmatrix} 0 & 1 & 2 \\ 2 & 0 & 1 \\ 1 & 2 & 0 \end{pmatrix}$$

be a pair of orthogonal Latin squares of order 3 each. Then, given arbitrary polynomials $v(x, y), w(x, y) \in \mathbb{Z}_3[x, y]$, the functions $f(x, y) = x + y + 3 \cdot v(x, y)$ and $g(x, y) = 2x + y + 3 \cdot w(x, y)$ define a pair of orthogonal Latin squares modulo 3^k, for all $k = 1, 2, \ldots$ since

$$\det \begin{pmatrix} 1 & 2 \\ 1 & 1 \end{pmatrix} \equiv 2 \pmod 3.$$

By the same reason, given a set $\mathcal{P}$ of *odd* primes and arbitrary polynomials $v(x, y)$, $w(x, y) \in \mathbb{Z}[x, y]$, the following two Latin squares are orthogonal modulo P for every P such that all prime factors of P are in $\mathcal{P}$:

$$f(x, y) = x + y + \Pi \cdot v(x, y); \; g(x, y) = -x + y + \Pi \cdot w(x, y),$$

where $\Pi = \prod_{p \in \mathcal{P}} p$. In the same fashion, Theorem 8.17 can be re-stated for pairs of orthogonal Latin squares; and a method of constructing a pair of orthogonal Latin squares on P symbols for large composite odd P can be derived from this theorem as well. Namely, given N pairs of orthogonal Latin squares on $p_1, \ldots, p_N$ symbols (p_i prime, $i = 1, 2, \ldots, N$), we construct N pairs of bivariate mappings $f_1(x, y), \ldots, f_N(x, y)$ and $g_1(x, y), \ldots, g_N(x, y)$ modulo $p_1, \ldots, p_N$, respectively, such that every pair $f_i(x, y)$ and $g_i(x, y)$ represents the ith pair of given orthogonal Latin squares on p_i symbols. For this purpose we apply interpolation formula (1.9). Then, using Chinese Remainder Theorem 1.1, we construct two bivariate polynomials $f(x, y)$ and $g(x, y)$ with rational integer coefficients such that $f(x, y) \equiv f_{p_i}(x, y) \pmod{p_i}$ and $g(x, y) \equiv g_{p_i}(x, y) \pmod{p_i}$, for all $i = 1, 2, \ldots, N$. After that, with the use of method from Proposition 1.34 we tweak the polynomials $f(x, y)$ and $g(x, y)$ so that their partial derivatives satisfy the conditions of Corollaries 8.14 and 8.18, in a manner we describe in the proof of Theorem 8.17 and in the text thereafter. We leave details to the reader.

Concluding the section, we stress that presented techniques in an obvious way can be used to construct Latin squares (and mutually orthogonal Latin squares) out of arbitrary uniformly differentiable (modulo some p^k) functions, and *not necessarily* out of polynomials; e.g., out of rational functions, analytic functions, etc., if needed.

Chapter 9

Pseudorandom numbers

As we demonstrated in Section 8.2, basic instructions of CPU are continuous with respect to the 2-adic metric; whence, so are computer programs build from these operators. These programs can be viewed as continuous 2-adic functions; whence, their behavior can be studied with the use of non-Archimedean analysis. In this chapter, we apply p-adic dynamics to construct and study pseudorandom generators.

Pseudorandom (number) generator (a PRNG for short) is an algorithm that produces a random-looking sequence of machine words, which can be also treated as a sequence of numbers in their base-2 expansions. A theory (better to say, theories) of PRNG is an important part of computer science, see e.g., [267, Chapter 3]. Actually, this Chapter 9 exhibits the non-Archimedean theory of PRNG, where a PRNG is considered as a non-Archimedean dynamical system.

We say 'theories of PRNG' rather than 'a theory' since the very definition of pseudorandomness assumes that the produced sequence must pass certain class of statistical tests, so the definition of what is a *pseudorandom sequence* (whence, what is a PRNG) depends on the choice of this class of tests. We stress that the class of tests a PRNG must pass is settled beforehand; for instance, if one takes all polynomial-time tests, he obtains a definition of pseudorandomness in the sense of the complexity theory. However, in practice they often use some standard batteries of tests, e.g. NIST, DIEHARD, or some other.

As a rule, the weakest statistical property the sequence must necessarily satisfy to be considered as pseudorandom in any reasonable meaning, is *uniform distribution*; that is, each term of the sequence must occur with the same frequency. Actually in this chapter we construct algorithms that produce uniformly distributed sequences out of a given short random string; then we study statistical properties of these sequences, other than uniform distribution.

Pseudorandom generators are widely used in numerous applications, especially in modeling, computer simulation (e.g., in quasi-Monte Carlo methods) and cryptography (e.g., in stream ciphers). The latter are ciphers that encrypt information according to the following protocol.

Let information be represented in a binary form, as a sequence of zeros and ones; so a *plaintext*, the information to be encrypted, is a sequence $\alpha_0, \alpha_1, \alpha_2, \ldots$, where $\alpha_j \in \{0, 1\}$. Let $\Gamma = \gamma_0, \gamma_1, \gamma_2, \ldots$ be another sequence of zeros and ones, which is known both to Alice and Bob, and which is known to no third party. The sequence Γ is called a *keystream*. To encrypt a plaintext, Alice just XORes it with the keystream (see Section 8.2 for the definition of XOR):

$$\begin{array}{lll} & \alpha_0, \alpha_1, \alpha_2, \dots, \alpha_i, \dots & \text{(plaintext)} \\ \mathsf{XOR} & & \text{(bitwise addition modulo 2)} \\ & \gamma_0, \gamma_1, \gamma_2, \dots, \gamma_i, \dots & \text{(keystream)} \\ \hline & \zeta_0, \zeta_1, \zeta_2, \dots, \zeta_i, \dots & \text{(encrypted text)} \end{array}$$

To decrypt, Bob acts in the opposite order:

$$\begin{array}{lll} & \zeta_0, \zeta_1, \zeta_2, \dots, \zeta_i, \dots & \text{(encrypted text)} \\ \mathsf{XOR} & & \text{(bitwise addition modulo 2)} \\ & \gamma_0, \gamma_1, \gamma_2, \dots, \gamma_i, \dots & \text{(keystream)} \\ \hline & \alpha_0, \alpha_1, \alpha_2, \dots, \alpha_i, \dots & \text{(plaintext)} \end{array}$$

Loosely speaking, Shannon's theorem yields that this encryption is secure providing the keystream Γ is picked at random for each plaintext. In real life settings we very rarely can fulfil the conditions of Shannon's theorem, and usually we use a *pseudo-random* keystream Γ rather than a random one. That is, usually in real life ciphers Γ is produced by a certain algorithm, and Γ only looks like random (that is, passes certain statistical tests). A standard reasoning at this point is that any adversary can use only a restricted number of tests to distinguish a pseudorandom keystream from a truly random one; so whenever a pseudorandom string passes all these tests, an adversary must conclude that the keystream is random and so the cipher can not be broken since otherwise a successive attack that broke the cipher actually can serve as a test that differs the keystream from a truly random. So in cryptology a stream cipher is thought of as an algorithm that takes a short random string (which is called *a key*) and stretches it into a much longer sequence, the keystream. Actually, within the scope of the book we speak about *stream cipher* meaning the latter is a PRNG which is used for encryption according to the protocol described above.

Not every PRNG is suitable for stream encryption. Stream ciphers are *cryptographically secure* PRNGs; that is, they must not only produce statistically good sequences, but also they must withstand *adversary's attacks*. We will consider mathematical problems related to some of these attacks in this chapter as well. It is worth noting here that according to postulates of modern cryptology, both the algorithm and the keystream are assumed to be known to an adversary; the only thing he does not know is a key, and in most cases an attack is aimed to *determine a key given both the algorithm and the keystream that corresponds to the unknown key*.

9.1 Pseudorandom generator is a dynamical system

Basically, the PRNG we consider in this chapter is a finite automaton $\mathfrak{A} = \langle \mathcal{N}, \mathcal{M}, f, F, u_0 \rangle$ *without input*, that is, with empty input alphabet $\mathcal{K}$, cf. the general definition of automaton in Section 8.1. Here, we recall, $\mathcal{N}$ is a finite set of states, $f: \mathcal{N} \to \mathcal{N}$ is a state transition function, $\mathcal{M}$ is a finite output alphabet, $F: \mathcal{N} \to \mathcal{M}$ is an output function (sometimes in cryptology called a *filter*), $u_0 \in \mathcal{N}$ is the initial state (which sometimes is called also a *seed*). Schematics of this typical PRNG is shown in Figure 9.1.

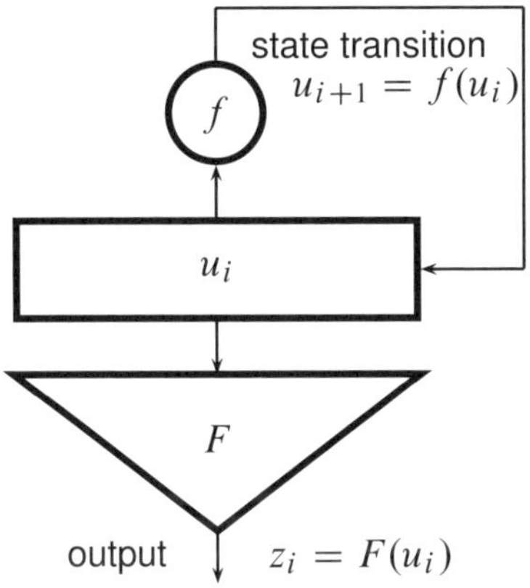

Figure 9.1. Pseudorandom generator.

Thus, this PRNG produces a sequence

$$\mathcal{Z} = \{F(u_0), F(f(u_0)), F(f^2(u_0)), \dots, F(f^j(u_0)), \dots\}$$

over the set $\mathcal{M}$, where

$$f^j(u_0) = \underbrace{f(\dots f(}_{j \text{ times}} u_0) \dots) \quad (j = 1, 2, \dots); \quad f^0(u_0) = u_0.$$

Note that the sequence depends on the initial state u_0. In cryptology, the initial state is usually a key, which is chosen from $\mathcal{N}$ at random. That is, the PRNG is considered as a mapping from $\mathcal{N}$ into the set of all (eventually) periodic sequences over $\mathcal{M}$. For better rigor of further arguments, we now state a formal definition of a generator:

Definition 9.1 (Generator). A *generator* is a family of automata $\{\mathfrak{A}(u) : u \in \mathcal{N}\}$ without input that have the same set of states $\mathcal{N}$, the same output alphabet $\mathcal{M}$, the same state transition function f, and the same output function F. The initial state of every automaton $\mathfrak{A}(u)$ is u.

The generators may be considered either as pseudorandom generators per se, or as components of more complicated automata, which are discussed in Section 10.2, the so-called *counter-dependent generators*; the latter produce sequences $\{z_0, z_1, z_2, \ldots\}$ over $\mathcal{M}$ according to the rule

$$z_0 = F_0(u_0),\ u_1 = f_0(u_0); \ldots;\ z_i = F_i(u_i),\ u_{i+1} = f_i(u_i); \ldots .$$

That is, at the $(i+1)$th step the automaton $\mathfrak{A}_i = \langle \mathcal{N}, \mathcal{M}, f_i, F_i, u_i \rangle$ is applied to the state $u_i \in \mathcal{N}$, producing a new state $u_{i+1} = f_i(u_i) \in \mathcal{N}$, and outputting a symbol $z_i = F_i(u_i) \in \mathcal{M}$. It is easy to see that actually counter-dependent generators may also be considered either as automata from Section 8.1 with input alphabet $\{0, 1, 2, \ldots\}$ or as automata without input but with a set of states $\mathbb{N}_0 \times \mathcal{N}$; however, in this chapter we consider them as *non-autonomous dynamical systems* and study in detail in Section 10.3.

For the moment we will focus on ordinary generators, that is, on PRNGs represented at Figure 8.1. Note that formally speaking the sequence of states

$$u_0, u_1 = f(u_0), u_2 = f(u_1), \ldots, u_{i+1} = f(u_i) = f^{i+1}(u_0), \ldots \tag{9.1}$$

can be considered as a trajectory of a dynamical system $\langle \mathcal{N}, f \rangle$, whereas the output sequence

$$z_0 = F(u_0), z_1 = F(u_1), \ldots, z_i = F(u_i) = F(f^i(u_0)), \ldots \tag{9.2}$$

is an observable, see Section 2.1. We will show now that this consideration is not only formal, but discloses the essence of the problem how to construct a good PRNG.

9.1.1 What pseudorandom generators are good?

A PRNG that could be considered any good obviously must meet the following conditions:

- The output sequence must be pseudorandom (i.e., must pass certain statistical tests).
- For cryptographic applications, given a segment $z_j, z_{j+1}, \ldots, z_{j+s-1}$ of the output sequence, finding the corresponding initial state (which usually is a key) must be infeasible in some properly defined sense.
- The PRNG must be suitable for software (or hardware) implementations; the performance must be sufficiently fast.

In the case the PRNG is an automaton represented by Figure 9.1, we can restate these conditions as follows:

Condition 1: The state transition function f must provide pseudorandomness; in particular, it must guarantee uniform distribution and long period of the sequence of states $\{u_i\}$.

For cryptographic purposes, it would be great if one could provide cryptographic security of this sequence as well; that is, given u_i, it must be infeasible neither to find (or to predict) u_{i+1}, nor to find u_0. Unfortunately, this is not easy to provide these properties in real life setting: PRNGs that are 'provably secure', for which there exist proofs (based on some plausible, yet still unproven conjectures) that their output sequences can not be predicted by polynomial-time algorithms, are too slow for most practical applications. In practice, one has to undertake additional efforts to make the output sequence secure: This is output functions are needed for.

> *Condition 2*: The output function F must not spoil pseudorandomness; at least, the output sequence $\{z_i\}$ must be uniformly distributed and must have a long period.
>
> Moreover, in cryptographic applications the function F must make the PRNG secure: Given z_i, F and f, it must be difficult to find u_i from the equation $z_i = F(u_i)$.

Finally, in practice, both in cryptography and computer simulations, PRNGs are implemented in software or hardware, and it is highly desirable to make these programs platform-independent to make possible to run the same algorithm on various platforms. Moreover, the performance of the corresponding programs must be sufficiently fast on all platforms. This demands the following condition:

> *Condition 3*: To make the PRNG any suitable for software/hardware implementations, and to make it platform-independent, both f and F must be (not too complicated) compositions of basic instructions from Section 8.2.

To satisfy condition 1, one may take transitive state transition function $f : \mathcal{N} \to \mathcal{N}$; the sequence of states (9.1) will have then the longest possible period (of length $\#\mathcal{N}$), and strict uniform distribution: Every element from $\mathcal{N}$ will occurs at the period exactly once, see Section 2.2.

To satisfy the first part of condition 2, one may take a balanced output function $F : \mathcal{N} \to \mathcal{M}$; see Section 2.2 for definition (in this case we assume that $\#\mathcal{N}$ is a multiple of $\#\mathcal{M}$). Whenever $\#\mathcal{N} = \#\mathcal{M}$, balanced mappings are just invertible (that is, bijective, one-to-one) mappings. Obviously, if a balanced output function is applied to a strictly uniformly distributed sequence of states, the output sequence is also strictly uniformly distributed: It is periodic with a period of length $\#\mathcal{N}$, and every element from $\mathcal{M}$ occurs at the period exactly $\frac{\#\mathcal{N}}{\#\mathcal{M}}$ times. We state this as a proposition:

Proposition 9.2. *If the state transition function f of the automaton $\mathfrak{A}$ is transitive on the state set $\mathcal{N}$, i.e., if f is a permutation with a single cycle of length $N = \#\mathcal{N}$; if, further, N is a multiple of $M = \#\mathcal{M}$, and if the output function $F : \mathcal{N} \to \mathcal{M}$ is balanced* (i.e., $\#F^{-1}(s) = \#F^{-1}(t)$ for all $s, t \in \mathcal{M}$), *then the output sequence $\mathcal{Z}$ of the automaton $\mathfrak{A}$ is purely periodic with a period length N* (i.e., maximum possible), *and each element of $\mathcal{M}$ occurs at the period the same number of times: $\frac{N}{M}$ exactly. That is, the output sequence $\mathcal{Z}$ is uniformly distributed.*

Whenever $\#\mathcal{M} \ll \#\mathcal{N}$, balanced functions may also satisfy the second part of condition 2 since the equation $z_i = F(x_i)$ has then too many solutions (namely, $\frac{\#\mathcal{N}}{\#\mathcal{M}}$), so it is infeasible to an adversary to try them all.

Finally, to satisfy condition 3, one may use only operations that are common to all platforms: These are arithmetic (numerical) operations; addition, multiplication, subtraction, division, exponentiation of integers. In this case both $\mathcal{N}$ and $\mathcal{M}$ can be associated to respective sets of rational integers $0, 1, 2, \dots, N-1$ and $0, 1, 2, \dots, M-1$; and moreover, to residue rings $\mathbb{Z}/N\mathbb{Z}$ and $\mathbb{Z}/M\mathbb{Z}$, respectively. Moreover, if one takes $N = 2^n$ and $M = 2^m$, then actually both f and F will work with n-bit to produce output sequence of m-bit words. This case is the most convenient for programming; moreover, in this case one may use along with arithmetic operations bitwise logical operations as well, and other basic instructions (see Section 8.2) to construct f and F.

9.1.2 Why p-adic ergodic theory?

Now we explain a general way to construct transitive mappings f and balanced mappings F out of arithmetic operations (in the case both N and M are composite numbers), and out of arithmetic and bitwise logical operations (in the case both N and M are powers of 2). The idea is as follows: Let, say, $N = 2^n$ and $M = 2^m$, $m \le n$, $n = kr$, $m = ks$; then using results of Chapter 4 we construct an ergodic mapping $f: \mathbb{Z}_2 \to \mathbb{Z}_2$ and a measure-preserving mapping $F: \mathbb{Z}_2^r \to \mathbb{Z}_2^s$ out of arithmetic and bitwise logical operations, as these operations are 1-Lipschitz functions defined on the space of 2-adic integers $\mathbb{Z}_2$ and valuated in $\mathbb{Z}_2$, see Section 8.2. Then, according to Theorem 4.23, taking residues of f and of F modulo 2^n and 2^k, respectively, we obtain a transitive transformation $f \bmod 2^n$ of the residue ring $\mathbb{Z}/2^n\mathbb{Z}$ and a balanced mapping $F \bmod 2^k : (\mathbb{Z}/2^k\mathbb{Z})^r \to (\mathbb{Z}/2^k\mathbb{Z})^s$. So $f \bmod 2^n$ will serve as a state transition function, whereas $F \bmod 2^k$ will serve as an output function since elements of residue ring $\mathbb{Z}/2^n\mathbb{Z}$ and of Cartesian powers $(\mathbb{Z}/2^k\mathbb{Z})^r$ and $(\mathbb{Z}/2^k\mathbb{Z})^s$ can be treated as n-bit and m-bit words, respectively. Note also that any number that is longer than a word bitlength k of a computer, is reduced modulo 2^k automatically.

The case when both N and M are composite numbers can be reduced to the case of prime powers: That is, we will construct ergodic mappings $f: \mathbb{Z}_p \to \mathbb{Z}_p$ and measure-preserving mappings $F: \mathbb{Z}_p^r \to \mathbb{Z}_p^s$ and then take $f \bmod p^n$ and $F \bmod p^k$, for all all prime factors of N and M (we assume that prime factors of N and of M form the same set). Then with the use of the Chinese Remainder Theorem 1.1 we construct mappings modulo N and M which coincides accordingly with $f \bmod p^n$ and $F \bmod p^k$ for all prime factors p of N and of M in a fashion of Section 8.4, see Theorem 8.17 and the example thereafter. We will illustrate this case by detailed examples later.

Now we make some conventions on terminology, cf. Section 2.2 and Subsection 2.1.1:

Definition 9.3. A sequence $(s_i)_{i=0}^{\infty}$ of p-adic integers is called *strictly uniformly distributed modulo* p^k whenever the sequence $(s_i \bmod p^k)_{i=0}^{\infty}$ of residues modulo p^k is strictly uniformly distributed over the residue ring $\mathbb{Z}/p^k\mathbb{Z}$.

Note 9.4. A sequence $(s_i)_{i=0}^{\infty}$ of p-adic integers is uniformly distributed (with respect to the normalized Haar measure μ on $\mathbb{Z}_p$) if and only if it is uniformly distributed modulo p^k for all $k = 1, 2, \ldots$; that is, for every $a \in \mathbb{Z}/p^k\mathbb{Z}$ relative numbers of occurrences of a in the initial segment of length ℓ in the sequence $\{s_i \bmod p^k\}$ of residues modulo p^k are asymptotically equal, i.e., $\lim_{\ell\to\infty} \frac{A(a,\ell)}{\ell} = \frac{1}{p^k}$, where $A(a,\ell) = \#\{s_i \equiv a \pmod{p^k} : i < \ell\}$ (see [276] for details). So strictly uniformly distributed sequences are uniformly distributed in a usual sense of the theory of distribution of sequences.

Note that in view of Proposition 9.2 one can vary both the state transition and the output function of a PRNG (and, for instance, make them key-dependent) *without* affecting uniform distribution of the output sequence, as the only conditions that must be satisfied to make the output uniformly distributed are ergodicity of the state transition function and measure-preservation of the output function. This idea we will exploit further, in construction of counter-dependent generators and flexible stream ciphers.

Of course, to make all these considerations practicable, we must choose these functions f and F from suitably large classes of ergodic and measure-preserving functions. In other words, we must develop certain tools to produce a number of various measure-preserving, ergodic mappings out of arithmetic (and of bitwise logical) operations. We consider these methods in the next section.

9.2 Congruential generators of the longest period

In this section we consider so-called congruential generators, a class of pseudorandom number generators which are widely used in various applications and widely studied in literature. We will show that actually the theory of these generators is a part of p-adic ergodic theory: Numerous known sporadic results of these generators can be explained in a unified way by p-adic ergodic theory represented in Chapter 4. We will show that all known results about periods of these generators can be deduced from basic theorems of p-adic ergodic theory; also, we will prove some new general results in this area.

Actually, in this section we explain how to construct a transformation on a given finite set $\mathcal{N}$ such that this transformation has a prescribed form and the longest possible period. These transformations will be compositions of arithmetic operators, and also of bitwise logical operators whenever $\#\mathcal{N}$ is a power of 2. Thus, generators based on so-called *T-functions*, which became recently of interest for modern cryptology and which are just triangular functions from Definition 3.37 when $p = 2$, are within the scope of our study as well.[1] Now we introduce the main notion of this section:

Definition 9.5. A *congruential generator* is a generator from Definition 9.1 such that $\mathcal{M} = \mathcal{N} = \mathbb{Z}/N\mathbb{Z}$, $F\colon \mathcal{M} \to \mathcal{M}$ is the identity mapping, and the state transition function $f\colon \mathbb{Z}/N\mathbb{Z} \to \mathbb{Z}/N\mathbb{Z}$ preserves all congruences of the residue ring $\mathbb{Z}/N\mathbb{Z}$: $f(a) \equiv f(b) \pmod{L}$ whenever $a \equiv b \pmod{L}$ and $L \neq 1$ is a factor of N, cf. Definition 1.18. The function f is called *recursion law* of the congruential generator.

Note 9.6. In view of the Chinese Remainder Theorem 1.30 it is obvious that the output sequence of the congruential generator has the longest possible period (of length N) if and only if every function $f \bmod p^n$ is transitive modulo p^n, where $n = \operatorname{ord}_p N$, for all prime factors of N (recall that $p^{\operatorname{ord}_p N}$ is the greatest power of p that is a factor of N, see Section 1.4).

In literature, some authors consider one more class of generators, which they call explicit congruential generators.

Definition 9.7. *Explicit congruential generators* correspond to the case when the state transition function of automaton $\mathfrak{A}$ from Definition 9.5 is a counter $f(x) = x+1 \bmod N$, whereas the output function $F\colon \mathbb{Z}/N\mathbb{Z} \to \mathbb{Z}/N\mathbb{Z}$ preserves all congruences of the residue ring $\mathbb{Z}/N\mathbb{Z}$.

Note 9.8. Obviously, the explicit congruential generator attains the longest possible period (of length N) if and only if every function $F \bmod p^n$ is bijective modulo p^n, where $n = \operatorname{ord}_p N$, for all prime factors p of N.

We stress here that according to Chapter 4 to determine whether a congruential generator (in the sense of Definition 9.5) attains the longest period (of length N), we should study *ergodicity* of the function f on space $\mathbb{Z}_p$, for all primes $p \mid N$; whenever in the case of explicit congruential generator we should study *measure-preservation* of F. This is the leading idea of the section.

In order not to misguide the reader, we note that in cryptographic literature some authors understand congruential generators in a much more general sense compare to Definition 9.5, see e.g. a paper of Krawczyk [275]. According to the latter paper, a (general) congruential generator is a number generator for which the ith element s_i of the sequence is a $\{0, 1, \dots, m-1\}$-valued number computed by the congruence

$$s_i \equiv \sum_{j=1}^{k} \alpha_j \Phi_j(s_{-n_0}, \dots, s_{-1}, s_0, \dots, s_{i-1}) \pmod{m}, \tag{9.3}$$

where $\alpha_j \in \mathbb{Z}$, $m \in \{2, 3, \dots\}$ and Φ_j, $1 \le j \le k$ is an arbitrary integer-valued function. Note that this definition can be re-stated in equivalent form: A (general)

[1] Actually, T-functions are 1-Lipschitz 2-adic functions, see Subsection 3.8.1; so the theory of T-functions is a part of p-adic theory.

congruential generator is a number generator for which the ith element s_i of the output sequence is computed by the congruence

$$s_i \equiv \Phi(s_{-n_0}, \dots, s_{-1}, s_0, \dots, s_{i-1}) \pmod{m},$$

where, as Krawczyk notes (see [275, page 531]), Φ is an *arbitrary integer-valued function that works on finite sequences of integers*.[2] This definition is too general for our purposes, and we never use it: In the sequel we refer as congruential generators only the automata from Definition 9.5, whereas automata from Definition 9.7 are referred as explicit congruential generators.

9.2.1 Types of congruential generators

Congruential generators (in the sense of Definition 9.5), as well as explicit congruential generators from Definition 9.7, were studied in a number of works, see the monographs [126, 267, 344] and references therein.[3] In this subsection, we list some known and widely used types of congruential generators. We will demonstrate that in all cases the longest possible periods are attained by these generators whenever the corresponding state transition function f is ergodic on certain subspaces of $\mathbb{Z}_p$, for some prime numbers p. This gives a unified method to calculate period length of congruential generators with the use of apparatus of Chapter 4. Further we explain how to tweak these generators to lengthen their periods if they are not the longest possible.

Linear, quadratic, and cubic congruential generators

One of the most wide-spread types of congruential generators are *linear congruential generators*[4]; they correspond to the case when $f(x) = (ax + b) \bmod N$, where a, b are rational integers and $N > 1$ is a natural number. Note that they speak about *congruential method* of generating pseudorandom numbers whenever $b = 0 \pmod{N}$; and of *mixed congruential method* otherwise, see [267]. Other congruential generators that are often used in applications are quadratic and cubic; they correspond to the cases when $f(x)$ is a polynomial with rational integer coefficients, of degree 2 or 3, respectively. Note that Corollary 4.71 yields necessary and sufficient conditions for transitivity modulo N of a polynomial of *arbitrary* degree, with rational integer coefficients; thus, Corollary 4.71 gives a criterion when a quadratic or cubic congruential generator attains the longest period.

A question when a linear congruential generator has the longest possible period (that is, of length N) was answered in 1962 by Hull and Dobell. In view of Note 9.6 and Theorem 4.23, the criterion is actually stated by Theorem 4.36. Note that

[2] The only restriction is that s_i must be evaluated in a polynomial of i time.

[3] Some more recent results are mentioned in the expository paper [396].

[4] which sometimes are also called *Lehmer* generators

the longest possible period (of length N) can be achieved only with the use of the mixed congruential method, when $b \not\equiv 0 \pmod{N}$ (actually, only when b and N are coprime, see Theorem 4.36). However, a *multiplicative* generator (with $f(x) = ax \bmod N$) is also often used in applications. In this case every ideal of the residue ring $\mathbb{Z}/N\mathbb{Z}$ is an invariant subset of the mapping $f(x) = ax$, so the longest possible period is achieved whenever f is ergodic on spheres around 0; this holds if and only if a is primitive either modulo p^2 for each prime p such that $p^2|N$, or modulo p, if $p \mid N$ and $p^2 \nmid N$, see Theorem 4.79. Usually a multiplicative generator is assumed to work only on the unit group of the residue ring $\mathbb{Z}/N\mathbb{Z}$, that is, on the multiplicative group $(\mathbb{Z}/N\mathbb{Z})^*$ of all invertible elements of the ring $\mathbb{Z}/N\mathbb{Z}$. In this case (for odd N) the generator is obviously equivalent to a linear congruential generator modulo $\varphi(N)$, the value of Euler's totient function, as the group $(\mathbb{Z}/p^k\mathbb{Z})^*$ is a cyclic group of order $(p-1)p^{k-1}$, for odd prime p; so the longest period of the generator is of length $\varphi(N)$ in this case. Note that for $N = 2^k$, $k \geq 2$, the multiplicative group $(\mathbb{Z}/2^k\mathbb{Z})^*$ is a direct product of a group of order 2 by a cyclic group of order 2^{k-2}; so the maximum length of the period of a multiplicative generator is 2^{k-2} in this case.

Power generators

Another type of congruential generators that are used in real life applications are *power generators*, with $f(x) = x^n \bmod N$. They can not achieve periods of length N since every p-adic sphere centered at 1 is an invariant subset of the transformation $x \mapsto x^n$ on $\mathbb{Z}_p$: They achieve the longest possible period when they are ergodic on p-adic spheres centered at 1; this holds if and only if n is primitive either modulo p^2 for each prime p such that $p^2|N$, or modulo p, if $p \mid N$ and $p^2 \nmid N$, see Theorem 4.14 and Theorem 4.79. Note that the maximum length of a period of the power generator can be calculated with the use of Lemma 4.76.

Inversive generators

Inversive generators are studied in numerous papers, see e.g. a survey paper [120] and references therein. When N is a prime, $f(x)$ (or $F(x)$, for explicit generators) are of the form $ax^{-1} + b$ or $(a + bx)^{-1}$; here $0^{-1} = 0$ by the definition, $a, b \in \mathbb{Z}$. These functions can not be expanded directly to residue rings modulo composite N; in the latter case domains of f and F are assumed to be restricted to the unit group $(\mathbb{Z}/N\mathbb{Z})^*$, which is a Cartesian product of unit groups $(\mathbb{Z}/p^{\mathrm{ord}_p N}\mathbb{Z})^*$, for each prime $p \mid N$. Now we can study a behavior of functions $ax^{-1} + b$ or $(b + ax)^{-1}$ on the unit group $\mathbb{Z}_p^*$ of all invertible p-adic integers to determine periods of these functions modulo N. As the unit group is a p-adic sphere of radius 1 centered at 0, and as both functions are 1-Lipschitz, the problem of maximality of the period length can be reduced to the problem of ergodicity of these functions on a p-adic sphere. We will consider corresponding examples further, see Examples 9.18 and 9.19.

There are inversive generators of another kind, which use a *generalized multiplicative inverse*. By the definition, the latter is the transformation

$$\operatorname{inv}(x)\colon x \mapsto |x|_p^{-1} \cdot \left(|x|_p^{-1} \cdot x\right)^{-1} \tag{9.4}$$

on the space $\mathbb{Z}_p$. It is known that whenever $a, b \in \mathbb{Z}$, the function $f(x) = a\cdot\operatorname{inv}(x)+b$ is transitive modulo 2^n, $n \geq 2$, if and only if $a \equiv 1 \pmod 4$ and $b \equiv 1 \pmod 2$, see [119]. We will give a short proof of this result further by p-adic ergodic theory techniques, see the text following Proposition 9.35. Here we only mention that as the function $\operatorname{inv}(x)$ is a 1-Lipschitz transformation on $\mathbb{Z}_p$, the question on transitivity of the function $a \cdot \operatorname{inv}(x) + b$ modulo 2^n is equivalent to the question on ergodicity of this function on $\mathbb{Z}_2$; the latter question can be answered with the use of methods from Chapter 4.

Exponential generators

Exponential generator is the automaton from Definition 9.5 whose state transition function f includes operation of exponentiation, $x \mapsto a^x$. Usually in literature they consider exponential generators with the recursion law $f(x) = a^x \bmod N$ (in this case a is usually assumed to be coprime with N). In cryptology, the case when N is a prime is the most often studied. Cases when N is composite are also of interest; e.g. in [144] authors consider *doubly exponential generator*, with the recursion law $f(x) \equiv a^{b^x} \bmod N$, where $N = pq$ and p, q are distinct primes.[5] These generators never achieve the longest possible period (of length N); however, in Subsection 9.2.2 we introduce a tweak that makes the period of the exponential generator the longest possible, of length N, for a given composite N, see e.g. Example 9.9 and the text thereafter. Moreover, in the next subsection we explain how p-adic ergodic theory can be applied to find period length of congruential generators modulo N whose law of recursion has a given form, even the generator of this form can never achieve the longest period N.

9.2.2 Periods of congruential generators

In this subsection, we introduce various techniques to construct congruential generators of the longest period, or to calculate lengths of periods of congruential generators mentioned above. We will illustrate the methods by examples of congruential generators from Subsection 9.2.1, reproving known results about them and obtaining new ones. We demonstrate that actually the problem is how to construct p-adic measure-preserving and/or ergodic mappings, as well as to determine whether a given mapping is measure-preserving or, respectively, ergodic. Thus, *the theory of congruential generators is essentially a part of p-adic ergodic theory.*

[5] Results of [144] where extended in [279, 312].

Techniques based on convergent p-adic series

The most general characterizations of 1-Lipschitz measure-preserving and/or ergodic transformations on $\mathbb{Z}_p$ are given in terms of Mahler expansions, that is, by representation of the transformation via convergent interpolation series, see Subsection 4.5.3. This method is the most general as every continuous transformation on $\mathbb{Z}_p$ admits Mahler expansion. In some cases, e.g. for analytic functions, we can also use representations via power series, or via falling factorial series to determine whether the function is measure-preserving or ergodic applying results of Subsection 4.6.4. Now we consider these techniques in detail. We start with an example.

As said, an exponential generator, which has the recursion law $f(x) = a^x \bmod N$, never attains the longest period, of length N. However, using Mahler expansion, we immediately can tweak generators of this kind to make lengths of their shortest periods the longest, i.e., N, just by adding a linear term to the recursion law:

Example 9.9. For every prime p and every $a \equiv 1 \pmod p$ the function $f(x) = ax + a^x$ is a 1-Lipschitz ergodic transformation on $\mathbb{Z}_p$.

Proof. Indeed, as $a = 1 + pm$ for a suitable $m \in \mathbb{Z}_p$, in view of Theorem 4.40 the function f is a 1-Lipschitz ergodic transformation on $\mathbb{Z}_p$ since $f(x) = (1 + pm)x + (1 + pm)^x = x + pmx + \sum_{i=0}^{\infty} m^i p^i \binom{x}{i} = 1 + x + 2pm\binom{x}{1} + \sum_{i=2}^{\infty} m^i p^i \binom{x}{i}$ and $i \geq \lfloor \log_p(i + 1) \rfloor + 1$ for all $i = 2, 3, 4, \ldots$. □

Now, combining Example 9.9 with Theorem 4.23 and with the Chinese Remainder Theorem 1.1, we can construct exponential generators that attain the longest period (of length N) modulo N for arbitrary composite N in an obvious way: For instance, the function $f(x) = 11x + 11^x$ is transitive modulo 10^n for all $n = 1, 2, \ldots$, as f is ergodic on $\mathbb{Z}_p$ for $p = 2$ and for $p = 5$, thus transitive modulo p^n for all $n = 1, 2, \ldots$ in view of Theorem 4.23; whence, f is transitive is transitive modulo 10^n for all $n = 1, 2, \ldots$ in view of the Chinese Remainder Theorem 1.30.

In the case $p = 2$ and $a = 1 + 2m$, the generator from Example 9.9 may have cryptographical applications, as evaluation of $f(x)$ demands not more than $n + 1$ multiplications modulo 2^n of n-bit numbers: Of course, one should use calls to the table $a^{2^j} \bmod 2^n$, $j = 1, 2, 3, \ldots, n - 1$; then $a^x = \prod_{\delta_i(x)=1} a^{2^i}$. The latter table must be precomputed, corresponding calculations involve $n-1$ multiplications modulo 2^n. Obviously, one can use m as a long-term key, with the initial state x_0 being a short-term key; i.e., one changes m from time to time, but uses new x_0 for each new message. Obviously, without a properly chosen output function this generator is not secure. The choice of output function we discuss further.

In a similar manner we can make tweaks to inversive generators modulo N to lengthen their periods to the maximum value, N. The idea is to use the mapping $\iota_p \colon x \mapsto (1 + pmx)^{-1}$ (for some $m \in \mathbb{Z}_p$) in a composition of $f(x)$ rather than the mapping $x \mapsto x^{-1}$: Although both mappings are 1-Lipschitz p-adic mappings, the first one is defined everywhere on $\mathbb{Z}_p$, whereas the domain of second one is the unit group $\mathbb{Z}_p^*$ (i.e., a p-adic sphere $S_1(0) = \bigcup_{a=1}^{p-1} a + p\mathbb{Z}_p$ of radius 1 centered at 0). Moreover, the function ι_p is a $\mathcal{C}$-function; that is, a p-adic analytic function defined by power series with p-adic integer coefficients that converges everywhere on $\mathbb{Z}_p$, see Subsection 3.10.1: $(1 + pmx)^{-1} = 1 - pmx + p^2m^2x^2 - p^3m^3x^3 + \cdots$. As the $\mathcal{C}$-function is ergodic if and only if it is transitive either modulo p^2 if $p > 3$, or modulo p^3 if $p \le 3$ (see Corollary 4.70), then the function $f(x) = x + (1 + p^3x)^{-1}$ is transitive modulo p^n for all $n = 1, 2, \ldots$ by Theorem 4.23; by the same reason, if $p > 3$, then the function $f(x) = x + (1 + p^2x)^{-1}$ is transitive modulo p^n for all $n = 1, 2, \ldots$. Now using the Chinese Remainder Theorem 1.1 we can construct inversive generator modulo N, which shortest period is of length N, modulo arbitrary composite N. For instance, taking $f(x) = (x+(1+200x)^{-1}) \bmod 10^n$, we obtain the inversive generator whose period length is a maximum, 10^n, whatever $n = 1, 2, 3, \ldots$ is taken: Again, this follows from Theorem 4.23 and the Chinese Remainder Theorem 1.30 as this transformation f is ergodic on $\mathbb{Z}_p$ for $p \in \{2, 5\}$. Moreover, the generator has the same property if we take $f(x) = (x + (1 + 100x)^{-1}) \bmod 10^n$. We need one more result concerning ergodicity of analytic functions on $\mathbb{Z}_p$ to prove this claim. The result is useful by its own:

Proposition 9.10. *Let $g\colon \mathbb{Z}_p \to \mathbb{Z}_p$ be an arbitrary 1-Lipschitz function, and let $u\colon \mathbb{Z}_p \to \mathbb{Z}_p$ be an ergodic $\mathcal{B}$-function* (e.g., an ergodic $\mathcal{C}$-function). *Then the function $f(x) = u(x) + p^2 \cdot g(x)$ is ergodic.*

Proof. If $p \notin \{2, 3\}$, the assertion trivially follows from Corollary 4.70. If $p = 2$ then, as g is 1-Lipschitz, the ith coefficient of Mahler expansion of the function $4 \cdot g(x)$ is congruent to 0 modulo $2^{2+\lfloor \log_2 i \rfloor}$ in view of Theorem 3.53, for all $i = 1, 2, \ldots$. Thus, as $2 + \lfloor \log_2 i \rfloor \ge \lfloor \log_2(i + 1) \rfloor + 1$ and the function u is ergodic, the conclusion follows from Theorem 4.40 in this case.

Finally, if $p = 3$, then in view of Corollary 4.70 it suffices to show that f is transitive modulo 27. In turn, to prove the latter claim it is sufficient to demonstrate only that $f^9(0) \not\equiv 0 \pmod{27}$, see Lemma 4.56. As g is 1-Lipschitz, easy calculation, which uses Theorem 3.62, shows that

$$f^9(x) \equiv u^9(x) + 9 \cdot \sum_{i=0}^{8} g(u^i(x)) \prod_{j=i+1}^{8} u'(u^j(x)) \pmod{27}; \tag{9.5}$$

we remind that a product over empty set is 1. However, as u is ergodic, and as $u'(0) \cdot u'(1) \cdot u'(2) \equiv 1 \pmod 3$ (see equation (4.76) and the text thereafter in the proof of Lemma 4.56), from congruence (9.5) it follows that $f^9(x) \equiv u^9(x) \not\equiv 0 \pmod{27}$. □

Note 9.11. The proof of Proposition 9.10 shows that in the case $p = 2$ the condition $u \in \mathcal{B}$ is redundant. We actually proved a stronger claim: *If $g\colon \mathbb{Z}_2 \to \mathbb{Z}_2$ is an* *arbitrary 1-Lipschitz function, and if* $u\colon \mathbb{Z}_2 \to \mathbb{Z}_2$ *is an arbitrary 1-Lipschitz ergodic function, then the function* $f(x) = u(x) + 4 \cdot g(x)$ *is ergodic.*

Example 9.12. Given a composite N, let $\check{N} = \prod_{p^2|N} p^2 \prod_{p^2\nmid N} p^{\mathrm{ord}_p N}$. Then the length of the shortest period of the inversive generator with the law of recursion $f(x) = (x + (1 + \check{N}x)^{-1}) \bmod N$ is the maximum possible, i.e., N.

For instance, the length of the shortest period of the inversive generator with the law $f(x) = (x + (1 + 100x)^{-1}) \bmod 10^n$ is 10^n, whatever $n = 2, 3, \ldots$ is taken.

With these ideas, using Proposition 9.10 in composition with Proposition 3.65 and Corollary 4.70, we immediately can construct a number of different generators of these two kinds (inversive and exponential) that have the longest periods; e.g., as the following functions $f(x)$ are ergodic on $\mathbb{Z}_p$, generators with the law $f(x) \bmod N$ have the longest possible period, N: $f(x) = 1 + x + p^2 \cdot a^{b^x}$, $a \equiv b \equiv 1 \pmod p$, (doubly exponential generator), $f(x) = 1 + x + \frac{p^2}{1+px}$ (inversive generator), $f(x) = 1 + x + p^2 \cdot (1 + px)^{\frac{1}{1+px}}$ (exponential-inversive generator) , etc.

Now we will show how one can calculate a period length of a given congruential generator with the law of recursion $f(x) \bmod N$. In view of the Chinese Remainder Theorem 1.30, it suffices to consider only prime power moduli N. For $N = p^k$, p prime, the idea is to reduce the problem of calculating the period length to the problem of finding a closed subset of $\mathbb{Z}_p$ (usually a ball or a sphere), where a certain iterate $f^i(x)$ is ergodic.

For illustration, consider an exponential generator with the law $f(x) = a^x$, where $a \equiv 1 \pmod p$; i.e., $a = 1 + pz$ for some $z \in \mathbb{Z}_p$. It is clear that f maps $\mathbb{Z}_p$ into the ball $B_{p^{-1}}(1) = 1 + p\mathbb{Z}_p$; so we can write $1 + p \cdot g(x) = (1 + pz)^x$ and then study the function $g(x)$. As $(1 + pz)^x = \sum_{i=0}^{\infty} p^i z^i \binom{x}{i}$ is the Mahler expansion for a^x, we see that $g(x) = zx + pz^2\binom{x}{2} + p^2z^3\binom{x}{3} + \cdots$. Whenever $z \not\equiv 0 \pmod p$, all p-adic spheres around 0 are invariant under action of g, so the period will be the longest possible if g is ergodic on spheres $S_{p^{-r}}(0)$ around 0. Now we can apply Theorem 4.82 and Theorem 4.79 on ergodicity on spheres. From these theorems we deduce that whenever $p \neq 2$, the derivative $g'(0)$ must be primitive modulo p^2; however, as $g'(0) \equiv z - \frac{p}{2}z^2 \pmod{p^2}$, and as $(1 - \frac{p}{2}z)^i \equiv 1 - i \cdot \frac{p}{2}z \pmod{p^2}$, the element $z - \frac{p}{2}z^2 = z \cdot (1 - \frac{p}{2}z)$ of the residue ring modulo p^n, $n \geq 2$, is primitive modulo p^2 whenever z is primitive modulo p^2 (we remind that 2 has a multiplicative inverse in $\mathbb{Z}_p$ whenever $p \neq 2$, so $\frac{p}{2} \in \mathbb{Z}_p$ in this case and least non-negative residue of $\frac{p}{2}$ modulo p^k is well defined). Now easy calculation shows that $g^{p-1}(x) \equiv xz \cdot (1 + z\frac{p}{2}) - x^2 \cdot \frac{p}{2} \pmod{p^2}$; so $g^{p-1}(x)$ is ergodic on the ball $p\mathbb{Z}_p$ in view of Proposition 9.10. Finally by Note 4.77 we conclude that g is ergodic on the sphere $S_1(0)$ of radius 1 around 0. This means, in particular, that the length of the shortest period of exponential generator with the law $f(x) = (1 + pz)^x \bmod p^k$, where $p \neq 2$ and z is primitive modulo p^2, is $(p - 1)p^{k-2}$, for all $k = 2, 3, \ldots$. Investigation of periods of exponential generator in the remaining cases, for other a, demands extra efforts; however, it is based on the same ideas, so we leave the rest of study to the reader.

In practice, congruential generators modulo 2^n are of special interest, and we consider here this case in more detail. We start with polynomial generators, which have the law of recursion of the form $f(x) \bmod 2^n$, where $f(x) \in \mathbb{Z}[x]$ is a polynomial with rational integer coefficients. From Corollary 4.71 it follows that the length of the shortest period of this generator is the longest, 2^n, $n \geq 3$, if and only if the polynomial $f(x)$ is transitive modulo 8; that is, *the polynomial generator has the longest period modulo* 2^n, $n \geq 3$, *if and only if it has the longest period modulo 8.* However, with the use of Theorem 4.40 we can obtain explicit formulas for these generators of the longest period. Moreover, we consider more general setting, when $f(x)$ is a $\mathcal{C}$-function, that is, an analytic function represented by power series with p-adic integer coefficients such that the series converges everywhere on $\mathbb{Z}_p$, see Subsection 3.10.1. The $\mathcal{C}$-functions can be represented also as falling factorial series over $\mathbb{Z}_p$; that is, in the form $f(x) = \sum_{i=0}^{\infty} e_i \cdot x^{\underline{i}}$, where $x^{\underline{0}} = 1$, $x^{\underline{1}} = x$, $x^{\underline{i}} = x(x-1)\cdots(x-i+1)$, $i = 2, 3, 4, \ldots$, and all e_i are p-adic integers.

Proposition 9.13. *The $\mathcal{C}$-function f is ergodic on $\mathbb{Z}_2$ if and only if*

$$e_0 \equiv 1 \pmod 2, \quad e_1 \equiv 1 \pmod 4, \quad e_2 \equiv 0 \pmod 2, \quad e_3 \equiv 0 \pmod 4.$$

The $\mathcal{C}$-function f is measure-preserving if and only if

$$e_1 \equiv 1 \pmod 2, \quad c_2 = 0 \pmod 2, \quad e_3 = 0 \pmod 2.$$

Proof. As f is a $\mathcal{C}$-function, all coefficients a_i of its Mahler expansion (3.32) are congruent to 0 modulo $2^{\operatorname{ord}_2(i!)}$. Now, as $\operatorname{ord}_2(i!) = i - \operatorname{wt}_2 i$ (see Lemma 3.6) is a nondecreasing function, and as $\lfloor \log_2(i+1) \rfloor + 1 \leq i - \operatorname{wt}_2 i$, $\lfloor \log_2 i \rfloor + 1 \leq i - \operatorname{wt}_2 i$ for $i > 3$, the result follows from Theorem 4.40. □

Corollary 9.14. *Let the $\mathcal{C}$-function f be represented via power series: $f(x) = \sum_{i=0}^{\infty} c_i x^i$, $c_i \in \mathbb{Z}_2$, $i = 0, 1, 2, \ldots$. Then the function f is ergodic on $\mathbb{Z}_2$ if and only if the following congruences hold simultaneously*:

$$\begin{aligned} c_3 + c_5 + c_7 + \cdots &\equiv 2c_2 \pmod 4; \\ c_4 + c_6 + c_8 + \cdots &\equiv c_1 + c_2 - 1 \pmod 4; \\ c_1 &\equiv 1 \pmod 2; \\ c_0 &\equiv 1 \pmod 2. \end{aligned}$$

The function f is measure-preserving on $\mathbb{Z}_2$ if and only if

$$\begin{aligned} c_3 + c_5 + a_7 + \cdots &\equiv 0 \pmod 2; \\ c_1 &\equiv 1 \pmod 2. \end{aligned}$$

Note 9.15. As $f \in \mathcal{C}$, $\lim^2_{i\to\infty} c_i = 0$, so infinite sums in the left-hand parts of congruences are convergent in $\mathbb{Z}_2$.

Sketch proof. As $x^i = \sum_{j=0}^{i} S(i,j)x^{\underline{j}}$ and $x^{\underline{i}} = \sum_{j=0}^{i}(-1)^{i-j}s(i,j)x^j$, where $S(i,j)$ and $s(i,j)$ are Stirling numbers of the second kind and of the first kind, respectively, we can rewrite conditions for coefficients e_i from Proposition 9.13 in terms of coefficients c_i. This demands somewhat messy calculations involving identities for Stirling numbers, so the reader is referred to e.g. [158] for useful formulas and is encouraged to complete the proof. □

We note that in the case when f is a polynomial with rational integer coefficients, the claims of Corollary 9.14 were proved in [282] with the use of another technique; the second claim for polynomial with rational integer coefficients was also proved in [370]. We will give another proof of this claim further to illustrate how to use 2-adic derivatives in order to determine whether an explicit congruential generator modulo 2^n has the longest period, see Example 9.25. We note also that Proposition 9.13 (and Corollary 9.14) is a rare case when one can give necessary and sufficient conditions for ergodicity of polynomials over $\mathbb{Z}_p$ in terms of their coefficients. Another rare case is $p = 3$; the paper [110] gives this characterization (for $p = 3$), which is, however, too lengthy to quote it here. Actually the problem is hard since it involves necessarily a characterization of transitive polynomials modulo p. The latter question can be answered currently only for small p; note that $p = 2$ and $p = 3$ are the only case when all transitive polynomial transformations modulo p can be represented by affine transformations (i.e., by polynomials of degree 1).

Proposition 9.13 shows that to provide transitivity of a polynomial generator modulo 2^n, $n \geq 3$, it is necessary and sufficient to fix only 6 bits in base-2 expansions of its coefficients, while the other bits of may vary (e.g., may be key-dependent). This guarantees transitivity of the state transition function $z \mapsto f(z) \bmod 2^n$ for each n, and hence, uniform distribution of the output sequence. This property will be used further in order to construct counter-dependent generators of the longest period, as well as flexible stream ciphers based on these generators.

As a polynomial generator has the longest period modulo 2^n, $n \geq 3$, if and only if its law of recursion is transitive modulo 8, it makes sense to list all transitive polynomial transformations on the residue ring modulo 8:

Corollary 9.16. *A $\mathcal{C}$-function f is ergodic on $\mathbb{Z}_2$ if and only if the transformation $x \mapsto f(x) \bmod 8$, $x \in \{0, 1, \dots, 7\}$, coincides with a transformation of the residue ring $\mathbb{Z}/8\mathbb{Z}$ induced by any of the following polynomials:* [6]

[6]This list of all transitive polynomial transformations on $\mathbb{Z}/8\mathbb{Z}$ was published in [282].

$x+1$	$5x+1$	$2x^2+3x+1$	$2x^2+7x+1$
$x+3$	$5x+3$	$2x^2+3x+3$	$2x^2+7x+3$
$x+5$	$5x+5$	$2x^2+3x+5$	$2x^2+7x+5$
$x+7$	$5x+7$	$2x^2+3x+7$	$2x^2+7x+7$

Proof. Follows immediately from Proposition 9.13, with the use of Proposition 3.52. □

Note 9.17. If one just reduces modulo 8 coefficients of the power series that represents *ergodic* $\mathcal{C}$-function f, he will *not* necessarily obtain a polynomial from the above list; however, the mapping $x \mapsto f(x) \bmod 8$, $x \in \{0, 1, \dots, 7\}$, induced by the function f on the residue ring $\mathbb{Z}/8\mathbb{Z}$ *will* necessarily coincide with one of transformations on $\mathbb{Z}/8\mathbb{Z}$ induced by some polynomials from the list.

Now, in order to give examples of usage of 2-adic ergodic theory in a study of periods of congruential generators modulo 2^n, we reprove some known results about inversive generators.

Example 9.18 (Inversive generator from [117]). The inversive generator with the recursion law $f(x) = (ax^{-1} + b) \bmod 2^n$, $n > 3$, $a + b \equiv 1 \pmod 2$, attains the longest possible period (that of length 2^{n-1}) if and only if $a \equiv 1 \pmod 4$ and $b \equiv 2 \pmod 4$.

Indeed, the condition $a + b \equiv 1 \pmod 2$ implies that the 2-adic ball $1 + 2\mathbb{Z}_2$ is invariant under action of f. We have then that $1 + 2 \cdot g(z) = a \cdot (1 + 2z)^{-1} + b = a + b - 2az + 4az^2 - 8az^3 + \cdots$, so $g(z) = \frac{a+b-1}{2} - az + 2az^2 - 4az^3 + \cdots$ is a $\mathcal{C}$-function of variable z. However, in view of Corollary 9.14, the function g is ergodic on $\mathbb{Z}_2$ if and only if $\frac{a+b-1}{2} = 1 \pmod 2$ (condition 4), $a = 1 \pmod 2$ (condition 3), and $0 \equiv -a - 2a - 1 \pmod 4$ (condition 2). This concludes the proof.

Example 9.19 (Inversive generator from [182]). The inversive generator with the law of recursion $f(x) = (ax^{-1} + b + cx) \bmod 2^n$, $n > 3, a + b + c \equiv 1 \pmod 2$, attains the longest possible period (that of length 2^{n-1}) if and only if $a + c \equiv 1 \pmod 4$ and $b \equiv 2 \pmod 4$.

Only minor modifications to the above proof of the Example 9.18 are needed: Actually, in this case $1 + 2 \cdot g(z) = a \cdot (1 + 2z)^{-1} + b + c \cdot (1 + 2z) = a + b + c - 2 \cdot (a - c) \cdot z + 4az^2 - 8az^3 + \cdots$; so $g(z) = \frac{a+b+c-1}{2} - (a - c) \cdot z + 2az^2 - 4az^3 + \cdots$, and the result follows.

New inversive congruential generators modulo 2^n can be constructed along this way. For instance, with the use of these ideas it is easy to find conditions when the inversive-quadratic generator with the law of recursion $f(x) = (ax^{-1} + b + cx + dx^2) \bmod 2^n$

attains the maximum possible period (that of length 2^{n-1}), as well as the ones for inversive-cubic generator with the law of recursion $f(x) = (ax^{-1} + b + cx + dx^2 + ex^3) \bmod 2^n$, etc. Also, we can use not only inversions in compositions of recursive laws, but raising to other negative powers as well. We leave all these examples as exercises for the reader.

The general method to determine whether a given transformation f of the space $\mathbb{Z}_2$ is ergodic (or measure-preserving) is as follows: We must express f via Mahler expansion and then apply Theorem 4.40. Generally speaking, this is not an easy task to find Mahler expansion for an arbitrary continuous transformation f although this expansion always exists. Nevertheless, the method works. Here we apply these techniques to prove ergodicity/measure preservation criteria for two special transformations that are used in cryptographic pseudorandom generators. Both these generators are fast: The first of them uses only additions, XOR's and multiplications by constants, the second uses additions of entries of a certain look-up table in accordance with bits of a variable, and from this view is a version of a *knapsack* generator. We recall that $\delta_i(x)$ is the value of the ith bit in a base-2 expansion of x, $i = 0, 1, 2, \ldots$.

Theorem 9.20. *The following is true*:

1° *The function* $f : \mathbb{Z}_2 \to \mathbb{Z}_2$ *of the form*

$$f(x) = a + \sum_{i=1}^{n} a_i \cdot (x \;\mathsf{XOR}\; b_i),$$

where $a, a_i, b_i \in \mathbb{Z}_2$, $i = 1, 2, 3, \ldots$, *is measure-preserving* (*respectively, ergodic*) *if and only if it is bijective* (*respectively, transitive*) *modulo 2* (*respectively, modulo 4*).

2° *The function* $f : \mathbb{Z}_2 \to \mathbb{Z}_2$ *of the form*

$$f(x) = a + \sum_{i=0}^{\infty} a_i \cdot \delta_i(x),$$

where $a, a_i \in \mathbb{Z}_2$, $i = 0, 1, 2, \ldots$, *is 1-Lipschitz and ergodic if and only if the following conditions hold simultaneously*:

$$a \equiv 1 \pmod 2;$$

$$a_0 \equiv 1 \pmod 4;$$

$$|a_i|_2 = 2^{-i},$$

for $i = 1, 2, 3, \ldots$. *The function* f *is 1-Lipschitz and measure-preserving if and only if*

$$|a_i|_2 = 2^{-i} \quad (i = 0, 1, 2, 3, \ldots).$$

Proof. Consider the Mahler expansion for the function $\delta_i(x)$, $i = 0, 1, 2, \ldots$:

$$\delta_i(x) = \sum_{j=0}^{\infty} \sigma_i(j)\binom{x}{j}. \tag{9.6}$$

To apply Theorem 4.40 we must estimate 2-adic norms of coefficients $\sigma_i(j)$ first. To do this, we need several lemmas.

Lemma 9.21. *For all $i, j = 1, 2, 3, \ldots$ the following equations hold:*

$$\sigma_i(0) = 0;$$
$$\sigma_0(j) = (-1)^{j+1}2^{j-1};$$
$$\sigma_i(j) = (-1)^{j+1}\sum_{k=1}^{\infty}(-1)^k\binom{j-1}{k2^i - 1}.$$

Proof. As $\delta_i(0) = 0$ for all $i = 0, 1, 2, \ldots$, then $\sigma_i(0) = 0$. From Mahler expansion for $\delta_i(x)$, see (9.6), by inversion formula (see Theorem 1.6) we obtain that

$$\sigma_i(j) = (-1)^j \sum_{k=0}^{\infty}(-1)^k \delta_i(k)\binom{j}{k}.$$

Hence, in view of the definition of the function $\delta_i(j)$,

$$\sigma_i(j) = (-1)^j \sum_{s=1}^{\infty}\sum_{k=(2s-1)2^i}^{s2^{i+1}-1}(-1)^k\binom{j}{k}.$$

From here, using the following well-known identity (see e.g. [158, Chapter 5]),

$$\sum_{k=m}^{n}(-1)^k\binom{a}{k} = (-1)^m\binom{a-1}{m-1} + (-1)^n\binom{a-1}{n}, \tag{9.7}$$

we conclude that

$$\sigma_i(j) = (-1)^j \sum_{s=1}^{\infty}\left(\binom{j-1}{(2s-1)2^i - 1} - \binom{j-1}{2s \cdot 2^i - 1}\right).$$

This proves the lemma since the latter identity implies that

$$\sigma_i(j) = \begin{cases} (-1)^{j+1}2^{j-1}, & \text{if } i = 0; \\ (-1)^{j+1}\sum_{k=1}^{\infty}(-1)^k\binom{j-1}{k2^i-1} & \text{otherwise.} \end{cases}$$

□

Lemma 9.22. *For all $m, t, r = 0, 1, 2, \ldots$ that satisfy simultaneously two conditions $0 \le t \le 2^m - 1$ and $m \ge r$, the following congruence holds*:

$$\binom{2^m - 1}{t} \equiv (-1)^{t - \lfloor t2^{-r} \rfloor} \binom{2^{m-r} - 1}{\lfloor t2^{-r} \rfloor} \pmod{2^{m-r+1}}.$$

In particular, for all $m, s, j \in \mathbb{N}$ that satisfy simultaneously two conditions $m > s \ge 1$ and $j \le 2^{m-s} - 1$, the following congruence holds:

$$\binom{2^m - 2}{2^s j - 1} \equiv (-1)^j 2^s j \binom{2^{m-s} - 1}{j - 1} \pmod{2^{m-s+1}}.$$

Proof. We recall that every $s \in \mathbb{Z}_2$ has a unique representation of the form $s = 2^{\operatorname{ord}_2 s} \hat{s}$, where $\hat{s}$ is the unit of $\mathbb{Z}_2$; that is, $\hat{s}$ is odd, meaning $\delta_0(\hat{s}) = 1$, and henceforth s has a multiplicative inverse $\hat{s}^{-1}$ in $\mathbb{Z}_2$, see Section 1.4. Put $M = \{i : i = 1, 2, \ldots, t;\ \operatorname{ord}_2 i \ge r\}$, and let M' be complement of M in $\{1, 2, \ldots, t\}$; then

$$\begin{aligned}\binom{2^m - 1}{t} &= \prod_{i=1}^{t} \frac{2^m - i}{i} = \prod_{i=1}^{t} \left(\frac{2^{m - \operatorname{ord}_2 i}}{\hat{\imath}} - 1 \right) \\ &\equiv (-1)^{\#M'} \prod_{i \in M} \left(\hat{s}^{-1} 2^{m - \operatorname{ord}_2 i} - 1 \right) \pmod{2^{m-r+1}}. \qquad (9.8)\end{aligned}$$

The condition $\operatorname{ord}_2 i \ge r$ for $i = 1, 2, \ldots, t$ holds if and only if $i = j2^r$ for $j = 1, 2, \ldots, \lfloor 2^{-r} t \rfloor$. This means that $\#M' = t - \lfloor 2^{-r} t \rfloor$. So, the product in the right hand part of congruence (9.8) is equal to

$$(-1)^{\#M'} \prod_{j=1}^{\lfloor 2^{-r} t \rfloor} \left(\hat{\jmath}^{-1} 2^{m - r - \operatorname{ord}_2 j} - 1 \right) = (-1)^{t - \lfloor t2^{-r} \rfloor} \binom{2^{m-r} - 1}{\lfloor t2^{-r} \rfloor}.$$

This proves the first part of the assertion of the lemma. The second part now becomes obvious since

$$\binom{2^m - 2}{2^s j - 1} = \frac{2^m - 2^s j}{2^m - 1} \binom{2^m - 1}{2^s j - 1} \equiv 2^s j \binom{2^m - 1}{2^s j - 1} \pmod{2^{m-s+1}}.$$

□

Lemma 9.23. *For $s, k = 1, 2, 3, \ldots$, the following is true*:

(1) $|\sigma_s(k)|_2 \le 2^{-\lfloor \log_2 k \rfloor + s - 1}$, *whenever $k \ne 2^s, 2^{s+1}$*;

(2) $|\sigma_s(2^s)|_2 = 1, \quad |\sigma_s(2^{s+1})|_2 = \frac{1}{2}$;

(3) $|\sigma_s(2^m - 1)|_2 \le 2^{-m+s-1}$, *whenever $m > s \ge 1$*.

Proof. Represent k as $k = 2^m + t$, where $m = \lfloor \log_2 k \rfloor$, $0 \le t < 2^m$. We may assume that $m \ge s$ since otherwise $\sigma_s(k) = 0$ in view of Lemma 9.21. Further, Lemma 9.21 implies that

$$\sigma_s(2^m + t) = (-1)^{t+1} \sum_{j=1}^{\infty} (-1)^j \binom{2^m + t - 1}{2^s j - 1}. \tag{9.9}$$

Now by the following well-known identity (see e.g. [158, Chapter 5])

$$\sum_{k=0}^{n} \binom{a}{k}\binom{b}{n-k} = \binom{a+b}{n},$$

we conclude that

$$\begin{aligned}\binom{2^m - 1 + t}{2^s j - 1} &= \sum_{k=0}^{\infty} \binom{t}{k}\binom{2^m - 1}{2^s j - k - 1} \\ &= \sum_{n=0}^{\infty} \sum_{r=0}^{2^s - 1} \binom{t}{2^s n + r}\binom{2^m - 1}{2^s(j - n - 1) + (2^s - r - 1)}.\end{aligned} \tag{9.10}$$

Here, as usual, we assume that $\binom{a}{b} = 0$ for $b < 0$. In view of Lemma 9.22, equation (9.10) implies that

$$\sum_{n=0}^{\infty} \sum_{r=0}^{2^s - 1} (-1)^{n+r+j} \binom{t}{2^s n + r}\binom{2^{m-s} - 1}{j - n - 1} \equiv \binom{2^m - 1 + t}{2^s j - 1} \pmod{2^{m-s+1}}. \tag{9.11}$$

Now (9.9) in view of (9.11) implies that

$$\begin{aligned}\sigma_s(2^m + t) &\equiv (-1)^{t+1} \sum_{n=0}^{\infty} \sum_{r=0}^{2^s - 1} (-1)^{n+r} \binom{t}{2^s n + r} \sum_{j=1}^{\infty} \binom{2^{m-s} - 1}{j - n - 1} \\ &\equiv 2^{2^{m-s}-1} (-1)^{t+1} \sum_{n=0}^{\infty} \sum_{r=0}^{2^s - 1} (-1)^{n+r} \binom{t}{2^s n + r} \pmod{2^{m-s+1}}.\end{aligned} \tag{9.12}$$

Now applying identity (9.7) and assuming that $t \neq 0$, in view of Lemma 9.21 we conclude that

$$\begin{aligned}&(-1)^{t+1} \sum_{n=0}^{\infty} \sum_{r=0}^{2^s - 1} (-1)^{n+r} \binom{t}{2^s n + r} \\ &= (-1)^{t+1} \sum_{n=0}^{\infty} (-1)^n \binom{t}{2^s n + r} \left(\binom{t-1}{2^s n - 1} - \binom{t-1}{2^s(n+1) - 1} \right)\end{aligned}$$

$$= 2(-1)^{t+1}\sum_{n=1}^{\infty}(-1)^n\binom{t-1}{2^s n-1} = 2\cdot\sigma_s(t).$$

The left hand part of this equation is equal to -1 when $t = 0$. So, taking all these arguments into account, from (9.12) we conclude that

$$\sigma_s(2^m+t) \equiv \begin{cases} 2^{2^{m-s}}\sigma_s(t) \pmod{2^{m-s+1}}, & \text{if } t \neq 0;\\ -2^{2^{m-s-1}} \pmod{2^{m-s+1}}, & \text{if } t = 0. \end{cases}$$

The latter congruence proves Claim 1 and 2 of the lemma since it easily implies that

$$\sigma_s(2^m+t) \equiv \begin{cases} 1 \pmod 2, & \text{if } m = s, t = 0;\\ 2 \pmod 4, & \text{if } m = s+1, t = 0;\\ 0 \pmod{2^{m-s+1}}, & \text{in all other cases.} \end{cases}$$

Finally, if $m > s \geq 1$, then combining together Lemmas 9.21 and 9.22, we conclude that

$$\sigma_s(2^m-1) \equiv 2^s\sum_{j=1}^{\infty}\binom{2^{m-s}-1}{j-1} \pmod{2^{m-s+1}}.$$

From here by a well-known identity $\sum_{k=1}^{n} k\binom{n}{k} = 2^{n-1}n$ (see e.g. [158, Chapter 5]), we deduce that

$$\sigma_s(2^m-1) \equiv 2^{2^{m-s}-1+s}(2^{m-s}-1) \pmod{2^{m-s+1}}.$$

This proves Claim 3 and the lemma. □

Now we are ready to prove Theorem 9.20. We start with Claim 1°. The operation XOR and, consequently, the function f are 1-Lipschitz, see Section 8.2. Further, for all $u, v \in \mathbb{Z}_2$ the following identity holds (see the proof of (8.4) in Section 8.2):

$$u \,\mathsf{XOR}\, v = \sum_{k=0}^{\infty} 2^k(\delta_k(u)+\delta_k(v)-2\delta_k(u)\delta_k(v)) = u+v-\sum_{k=0}^{\infty}2^{k+1}\delta_k(u)\delta_k(v).$$

Consequently,

$$f(x) = a + \sum_{i=1}^{n} a_i b_i + \sum_{i=1}^{n} a_i x - 2\sum_{i=1}^{n}\sum_{k=0}^{\infty} 2^k\delta_k(x)\delta_k(b_i).$$

Now, considering interpolation series for $\delta_k(x)$ and taking into account that (in view of Lemma 9.21) $\sigma_0(1) = 1$ and $\sigma_i(1) = 0$ for $i = 1, 2, 3, \ldots$, we conclude that

$$f(x) = a + \sum_{i=1}^{n} a_i b_i + x\left(\sum_{i=1}^{n} a_i - 2\sum_{i=1}^{n}\delta_0(b_i)\right) - \sum_{j=2}^{\infty}\binom{x}{j}\sum_{i=1}^{n}\sum_{k=0}^{\infty}2^{k+1}\cdot\sigma_k(j)\cdot\delta_k(b_i).$$

Lemma 9.23 immediately implies that for $k \geq 2$

$$2^{k+1}\sigma_k(j) \equiv \begin{cases} 0 \pmod{2^{\lfloor \log_2 j \rfloor + 1}}, & \text{if } j = 2^k, 2^{k+1}; \\ 0 \pmod{2^{\lfloor \log_2 j \rfloor + 2}}, & \text{otherwise.} \end{cases}$$

Now Theorem 4.40 implies that f is measure-preserving (respectively, ergodic) if and only if $\sum_{i=1}^{n} a_i \equiv 1 \pmod 2$ (respectively, if and only if $a + \sum_{i=1}^{n} a_i b_i \equiv 1 \pmod 2$ and $\sum_{i=1}^{n} a_i + 2\sum_{i=1}^{n} b_i \equiv 1 \pmod 4$). This is obviously equivalent to Claim 1° of Theorem 9.20.

To prove Claim 2° of the theorem, we first note that functions δ_i for $i > 0$ are not 1-Lipschitz. As $\sigma_i(0) = 0$ for $i > 0$ (see Lemma 9.21), we have

$$f(x) = a + \sum_{j=1}^{\infty} \binom{x}{j} \sum_{i=0}^{\infty} a_i \cdot \sigma_i(j).$$

Theorem 4.40 implies now that the function f is measure-preserving if and only if the following congruences hold simultaneously:

$$\begin{cases} \sum_{i=0}^{\infty} a_i \cdot \sigma_i(1) \equiv 1 \pmod 2; \\ \sum_{i=0}^{\infty} a_i \cdot \sigma_i(j) \equiv 0 \pmod{2^{\lfloor \log_2 j \rfloor + 1}}, \quad j = 2, 3, \ldots. \end{cases} \tag{9.13}$$

In view of Lemma 9.21, the first of conditions (9.13) is equivalent to the congruence

$$a_0 \equiv 1 \pmod 2. \tag{9.14}$$

Moreover, Lemma 9.21 implies that $\sigma_i(j) = 0$ for $i \geq \lfloor \log_2 j \rfloor$. Hence, the second of conditions (9.13) is equivalent to the following system of congruences:

$$\sum_{i=0}^{\lfloor \log_2 j \rfloor} a_i \cdot \sigma_i(j) \equiv 0 \pmod{2^{\lfloor \log_2 j \rfloor + 1}}, \quad j = 2, 3, \ldots. \tag{9.15}$$

Consider the following subsystem of system (9.15) for $j = 2^k$, $k = 1, 2, 3, \ldots$:

$$\sum_{i=0}^{k} a_i \cdot \sigma_i(2^k) \equiv 0 \pmod{2^{k+1}}, \quad k = 1, 2, 3, \ldots. \tag{9.16}$$

We claim that 2-adic integers a_i satisfy system of congruences (9.16) if and only if $a_i \equiv 2^i \pmod{2^{i+1}}$, $i = 0, 1, 2, \ldots$. We proceed with induction on i. If $i = 1$, we by Lemma 9.21 (for $k = 1$) conclude that

$$2a_0 + a_1 \cdot \sigma_1(2) \equiv 0 \pmod 4. \tag{9.17}$$

In view of Claim 2 of Lemma 9.23, the 2-adic integer $\sigma_1(2)$ has a multiplicative inverse in $\mathbb{Z}_2$, so in view of (9.14) congruence (9.17) is equivalent to the congruence

$$a_1 \equiv 2 \pmod 4.$$

Now let our claim be true for $k < n$; consider the congruence

$$\sum_{i=0}^{n} a_i \cdot \sigma_i(2^n) \equiv 0 \pmod{2^{n+1}}. \tag{9.18}$$

By induction hypothesis, $a_i = 2^i + s_i 2^{i+1}$ $(i = 0, 1, \dots, n-1)$ for suitable $s_i \in \mathbb{Z}_2$. Then, taking into account Claim 2 of Lemma 9.23, we conclude that $a_i \cdot \sigma_i(2^n) \equiv 2^{n+1} \pmod{2^{n+2}}$ for $i = 0, 1, \dots, n-2$ and $a_{n-1} \cdot \sigma_{n-1}(2^n) \equiv 2^n \pmod{2^{n+1}}$. Hence, congruence (9.18) is equivalent to the congruence $2^n + a_n \cdot \sigma_n(2^n) \equiv 0 \pmod{2^{n+1}}$. As $\sigma_n(2^n)$ is a unit of $\mathbb{Z}_2$ (in force of Claim 2 of Lemma 9.23), the latter congruence implies that $a_n \equiv 2^n \pmod{2^{n+1}}$.

From Claim 1 of Lemma 9.23 it easily follows that if $a_i \equiv 2^i \pmod{2^{i+1}}$, then a_i also satisfy each congruence of the system (9.15) for those j which are not powers of 2. This means that conditions (9.13) are equivalent to the following set of congruences:

$$a_i \equiv 2^i \pmod{2^{i+1}}, \quad i = 0, 1, 2, 3, \dots .$$

So we have proved the second part of Claim 2° of Theorem 9.20. To prove the first part of this claim, we note that since $\lfloor \log_2(i+1) \rfloor + 1 = \lfloor \log_2 i \rfloor + 1$ for $i \neq 2^k - 1$, the sufficient and necessary conditions for ergodicity of function f from Theorem 4.40 in the case under consideration can be rewritten in the following form:

$$a \equiv 1 \pmod 2; \tag{9.19}$$

$$\sum_{i=0}^{\infty} a_i \cdot \sigma_i(1) \equiv 0 \pmod 4; \tag{9.20}$$

$$\sum_{i=0}^{\infty} a_i \cdot \sigma_i(j) \equiv 0 \pmod{2^{\lfloor \log_2 j \rfloor + 1}}, \quad j = 2, 3, 4, \dots; \tag{9.21}$$

$$\sum_{i=0}^{\infty} a_i \cdot \sigma_i(2^k - 1) \equiv 0 \pmod{2^{k+1}}, \quad k = 2, 3, 4, \dots . \tag{9.22}$$

As $\sigma_i(1) = 0$ for $i \neq 0$ (see Lemma 9.21), then (9.20) is equivalent to the following condition:

$$a_0 \equiv 1 \pmod 2. \tag{9.23}$$

During the proof of the second part of Claim 2° we have established that if $a_0 \equiv 1 \pmod 2$ (and, in particular, if (9.23) is satisfied) then conditions (9.21) are equivalent to the following conditions:

$$a_i \equiv 2^i \pmod{2^{i+1}}, \quad i = 1, 2, 3, \dots . \tag{9.24}$$

Finally, combining together Claim 1 of Lemma 9.23 and Lemma 9.21, we conclude that if 2-adic integers a_i $(i = 0, 1, 2, \dots)$ satisfy conditions (9.24) and (9.23) simultaneously, then a_i also satisfy conditions (9.22). Thus, the union of conditions (9.19)–(9.22) is equivalent to the union of conditions (9.19), (9.23), and of (9.24). This proves the first part of Claim 2° and Theorem 9.20. □

Techniques based on p-adic derivations

As it was demonstrated above, the problem to determine whether a congruential generator (or, respectively, an explicit congruential generator) attains the longest period can be reduced to the problem of verifying whether given 1-Lipschitz transformations on $\mathbb{Z}_p$, for some prime p, are ergodic, or, respectively, measure-preserving. In a number of practically interesting cases these transformations are differentiable, so we can apply results of Subsections 4.6.1 and 4.6.3 to check measure-preservation and ergodicity. This method is not as general as techniques based on Mahler expansion since the class of functions it can be applied to is smaller; however, in a number of cases it is easier to calculate derivatives of compositions of functions rather than their Mahler expansions. Moreover, in the case $p = 2$ (which is one of the most interesting for applications cases) it turns out that when we limit our study to differentiable functions only, we actually do not make the class of measure-preserving functions under consideration smaller:

Proposition 9.24. *If a 1-Lipschitz function $f : \mathbb{Z}_2 \to \mathbb{Z}_2$ is measure-preserving then it is uniformly differentiable modulo 2, its derivative modulo 2 is 1 everywhere on $\mathbb{Z}_2$, and $N_1(f) = 1$.*

Proof. Indeed, by Theorem 4.44, f is measure-preserving if and only if $f(x) = c + x + 2 \cdot v(x)$, where $c \in \mathbb{Z}_2$ is a constant and $v : \mathbb{Z}_2 \to \mathbb{Z}_2$ is a 1-Lipschitz transformation. Then $f(x + 2^k h) = c + x + 2^k h + 2 \cdot v(x + 2^k h) \equiv f(x) + 2^k h \pmod{2^{k+1}}$ as $2 \cdot v(x + 2^k h) \equiv 2 \cdot v(x) \pmod{2^{k+1}}$ since v is 1-Lipschitz. Thus, f is uniformly differentiable modulo 2, $f_1'(x) \equiv 1 \pmod 2$, and $N_1(f) = 1$ by Definition 3.28. □

Thus, Proposition 9.24 implies that if a recursion law of a congruential generator is *not* differentiable modulo 2 at some point of $\mathbb{Z}_2$, then the generator is *not* transitive modulo 2^n for all sufficiently large n (actually, it is not even bijective modulo 2^n for these n). This also means that the corresponding explicit congruential generator does not achieve maximum period length on n-bit words, for all sufficiently large n.

So, to determine whether the length of the shortest period of the explicit congruential generator with the law $y_i = f(i) \bmod 2^n$, $i = 1, 2, \dots$, is equal to 2^n, we just use Theorem 4.45 which states that whenever f is uniformly differentiable modulo 2, then f is measure-preserving if and only if f is bijective modulo $2^{N_1(f)}$ and $f_1'(x) \equiv 1 \pmod 2$ for all $x \in \mathbb{Z}/2^{N_1(f)}\mathbb{Z}$. Note that to determine whether the length of the shortest period of the congruential generator with the recurrence law $f \bmod 2^n$ is equal to 2^n, we should use Theorem 4.55 which demands that the function f must be uniformly differentiable modulo 4 rather than modulo 2.

Now we consider examples of congruential generators modulo 2^n, both explicit and non-explicit, to illustrate the approach. Recall that (explicit) congruential generator modulo 2^n attains the longest period if and only if its law is (bijective) transitive modulo 2^n.

For example, we reprove results from [264] by our methods: *The following mappings of $\mathbb{Z}/2^r\mathbb{Z}$ onto $\mathbb{Z}/2^r\mathbb{Z}$ are bijective for all $r = 1, 2, \ldots$:*

$$x \mapsto (x + 2x^2) \bmod 2^r,$$
$$x \mapsto (x + (x^2 \;\mathsf{OR}\; 1)) \bmod 2^r,$$
$$x \mapsto (x \;\mathsf{XOR}\; (x^2 \;\mathsf{OR}\; 1)) \bmod 2^r.$$

Indeed, all three mappings are uniformly differentiable modulo 2, and $N_1 = 1$ for all of them. So it suffices to prove that all three mappings are bijective modulo 2, i.e., as mappings of the residue ring $\mathbb{Z}/2\mathbb{Z}$ modulo 2 onto itself (this could be checked by direct calculations), and that their derivatives modulo 2 vanish at no point of $\mathbb{Z}/2\mathbb{Z}$. The latter also holds, since the derivatives are, respectively,

$$1 + 4x \equiv 1 \pmod 2,$$
$$1 + 2x \cdot 1 \equiv 1 \pmod 2,$$
$$1 + 2x \cdot 1 \equiv 1 \pmod 2,$$

as $(x^2 \;\mathsf{OR}\; 1)' = 2x \cdot 1 \equiv 1 \pmod 2$, and $\frac{\partial_1(x \mathsf{XOR} y)}{\partial_1 x} \equiv \frac{\partial_1(x \mathsf{XOR} y)}{\partial_1 y} \equiv 1 \pmod 2$, see Example 8.11.

The following closely related variants of the previous mappings of $\mathbb{Z}/2^r\mathbb{Z}$ onto $\mathbb{Z}/2^r\mathbb{Z}$ are not bijective for all $r = 1, 2, \ldots$:

$$x \mapsto (x + x^2) \bmod 2^r,$$
$$x \mapsto (x + (x^2 \;\mathsf{AND}\; 1)) \bmod 2^r,$$
$$x \mapsto (x + (x^3 \;\mathsf{OR}\; 1)) \bmod 2^r.$$

The first two mappings are not bijective modulo 2; whereas the derivative of the third mapping is $1 + 3x^2 \cdot \frac{\partial(u \mathsf{OR} 1)}{\partial u}\Big|_{u=x^3} \equiv 1 + x \pmod 2$ (see Example 8.11), thus vanishes modulo 2 at the point 1.

Example 9.25 (see [264, 370], cf. Corollary 9.14). Let $P(x) = a_0 + a_1 x + \cdots + a_d x^d$ be a polynomial with rational integer coefficients. Then $P(x)$ is bijective modulo 2^n, $n > 1$, if and only if a_1 is odd, $(a_2 + a_4 + \cdots)$ is even, and $(a_3 + a_5 + \cdots)$ is even.

In view of Theorem 4.45 we need to verify whether the two conditions hold: First, whether P is bijective modulo 2, and second, whether $P'(z) \equiv 1 \pmod 2$ for $z \in \{0,1\}$. The first condition implies that $P(0) = a_0$ and $P(1) = a_0 + a_1 + a_2 + \cdots + a_d$ must be distinct modulo 2; hence $a_1 + a_2 + \cdots + a_d \equiv 1 \pmod 2$. The second condition implies that $P'(0) = a_1 \equiv 1 \pmod 2$, $P'(1) \equiv a_1 + a_3 + a_5 + \cdots \equiv 1 \pmod 2$. Now combining all this together we conclude that $a_2 + a_3 + \cdots + a_d \equiv 0 \pmod 2$ and $a_3 + a_5 + \cdots \equiv 0 \pmod 2$, hence $a_2 + a_4 + \cdots \equiv 0 \pmod 2$.

Note 9.26. As a bonus, we can use exactly the same proof to get exactly the same characterization of bijective modulo 2^r ($r = 1, 2, \ldots$) mappings of the form $x \mapsto P(x) = a_0 \,\mathsf{XOR}\, a_1 x \,\mathsf{XOR} \cdots \mathsf{XOR}\, a_d x^d \bmod 2^r$ since $u \,\mathsf{XOR}\, v$ is uniformly differentiable modulo 2 as a bivariate function, and its derivative modulo 2 is exactly the same as the derivative of $u + v$, and $u \,\mathsf{XOR}\, v \equiv u + v \pmod 2$.

Example 9.27 ([264]). The function $x + (x^2 \,\mathsf{OR}\, 5)$ is transitive modulo 2^n for all $n = 1, 2, \ldots$.

In [264] it is claimed that (we quote):

> ... neither the invertibility nor the cycle structure of $x + (x^2 \,\mathsf{OR}\, 5)$ could be determined by his[7] techniques.

However, this claim is not true: The proof immediately follows from our Theorem 4.55. Indeed, as the function $f(x) = x + (x^2 \,\mathsf{OR}\, 5)$ is uniformly differentiable on $\mathbb{Z}_2$, thus, f is uniformly differentiable modulo 4 (see Example 8.12), and $N_2(f) = 3$, then to prove that f is ergodic, in view of Theorem 4.55 it suffices to demonstrate only that f is transitive modulo 32; the letter can be easily done by direct calculations that complete the proof.

A bit more involved considerations show that it suffices to check transitivity of f modulo 8 rather than modulo 32, but this is of no importance at the moment since the example serves as an illustration only. More congruential generators of the longest period modulo 2^n can be constructed using this method: For instance, all the following functions f are ergodic transformations on $\mathbb{Z}_2$ (thus, transitive modulo 2^n for all $n = 1, 2, 3, \ldots$): $f(x) = x + (5x^2 \mathsf{OR} 5)$, $f(x) = x + (5^{x^2} \mathsf{OR} 5)$, $f(x) = x + (5^{-x^2} \mathsf{OR} 5)$, $f(x) = x + (5^{x^2} \,\mathsf{AND}\, (-5))$, $f(x) = 5x + (5^{x^5} \,\mathsf{AND}\, (-5))$, $f(x) = 5x + (5^{5^x} \,\mathsf{AND}\, (-5))$, etc. Corresponding proofs just mimic the proof of Example 9.27, and we leave them to the reader as exercises. We want to emphasize that the technique based on p-adic derivations can handle rather complicated compositions of both arithmetic and bitwise logical computer instructions, such as, e.g. $f(x) = x + (((1 + 4 \cdot (x^2 \,\mathsf{AND}\, (-5)))^{(1+2\cdot(x \mathsf{XOR}(-5)))^{-5}} \,\mathsf{OR}\, 5)$. The latter function f is also ergodic on $\mathbb{Z}_2$; we again leave the proof to the reader as an exercise.

[7] that is, by techniques of the paper [21], where the statement of Theorem 4.55 was proved by the first author of the book; note that the paper [21] was published nearly a decade prior to the publication of the cited paper [264]

Now we explain how to use the technique to construct various polynomial generators modulo composite N that attain the longest period. It is clear that in view of the Chinese Remainder Theorem 1.30 the problem can be reduced to the case when N is a prime power, $N = p^n$. In the latter case we first must construct a transitive polynomial modulo p and then raise it to the polynomial that is transitive modulo p^n. In view of Corollary 4.71, it is sufficient to raise a transitive polynomial over $\mathbb{F}_p$ to the transitive polynomial modulo p^3 in the case $p \in \{2, 3\}$, or, respectively, modulo p^2, if $p > 3$. Now we outline a procedure that, given a transitive transformation φ on $\mathbb{F}_p$, returns a polynomial $\tilde{f}_\varphi(x) \in \mathbb{Z}[x]$, which is transitive modulo p^n for all $n = 1, 2, 3, \ldots$, and such that $\tilde{f}_\varphi(x) \equiv \varphi(x) \pmod{p}$ for all $x \in \mathbb{F}_p$:

- *Step 1*: Consider arbitrary transitive transformation φ on $\mathbb{F}_p$ and represent φ via the corresponding interpolation polynomial $f_\varphi(x) \in \mathbb{F}_p[x]$ according to interpolation formula (1.9). Note that $f_\varphi(x)$ can be (and will be) considered as a polynomial with rational integer coefficients.

- *Step 2*: Verify whether the polynomial $f_\varphi(x)$ is transitive modulo p^3 or modulo p^2, respectively, depending on whether $p \le 3$ or $p > 3$. If yes, $f_\varphi(x)$ is the ergodic polynomial $\tilde{f}_\varphi(x) \in \mathbb{Z}[x]$ we are seeking for; otherwise go to the next step.

- *Step 3*: Note that in this case $p > 3$ since formula (1.9) gives $f_\varphi(x) = x + 1$ for $p = 2$, which is ergodic on $\mathbb{Z}_2$, and formula (1.9) gives either $f_\varphi(x) = x + 1$ or $f_\varphi(x) = x - 1$ for $p = 3$; both polynomials are ergodic on $\mathbb{Z}_3$. So it suffices to tweak the polynomial $f_\varphi(x)$ to make it transitive modulo p^2. We will do this with the use of Proposition 1.34.

  Denote $g_i = f_\varphi^i(0) \bmod p$; then the string $g_0, g_1, \ldots, g_{p-1}$ is a permutation of the string $0, 1, \ldots, p-1$. Note that $\varphi\colon g_i \mapsto g_{(i+1) \bmod p}$, $i = 0, 1, \ldots, p-1$, as $f_\varphi(x) = \varphi(x) \bmod p$ and φ is transitive on $\{0, 1, \ldots, p-1\}$. Take arbitrary $h_0, h_1, \ldots, h_{p-1} \in \{1, \ldots, p-1\}$ that satisfy the following two conditions:

$$h_0 \cdot h_1 \cdots h_{p-1} \equiv 1 \pmod{p}, \tag{9.25}$$

$$\sum_{i=0}^{p-2} h_i \cdot h_{i+1} \cdots h_{p-2} \equiv 0 \pmod{p}. \tag{9.26}$$

  It is clear that choices of $h_0, h_1, \ldots, h_{p-1}$ that satisfy this system of congruences exist: For instance, $h_1 = \cdots = h_{p-2} = 1$, $h_0 = 2$, $h_{p-1} \equiv \frac{1}{2} \pmod{p}$ is one of the possible choices as $p \ne 2$. Now take the mapping $\psi\colon \mathbb{F}_p \to \mathbb{F}_p$ such that $\psi(g_i) = h_i$, $i = 0, 1, \ldots, p-1$ and construct a polynomial $f_{\varphi,\psi}(x)$ by Proposition 1.34; thus, $f_{\varphi,\psi}(x) \equiv \varphi(x) \pmod{p}$ and $f'_{\varphi,\psi}(x) \equiv \psi(x) \pmod{p}$ for $x \in \{0, 1, \ldots, p-1\}$.[8] Consider $f_{\varphi,\psi}(x)$ as a polynomial over $\mathbb{Z}$ and ver-

[8]Note that condition (9.25) follows from Note 4.57, while condition (9.26) guarantees that the second term in (9.27) is p modulo p^2.

ify whether $f_{\varphi,\psi}(x)$ is transitive modulo p^2. If yes, $f_{\varphi,\psi}(x)$ is the polynomial $f_\varphi(x) \in \mathbb{Z}[x]$ we need; otherwise go to Step 4.

- *Step 4*: Note that by Step 3, the derivative of the polynomial $f_{\varphi,\psi}(x)$ vanishes modulo p nowhere on $\mathbb{Z}_p$, so $f_{\varphi,\psi}(x)$ is measure-preserving in view of Theorem 4.45; thus, $f_{\varphi,\psi}(x)$ is bijective modulo p^2. In view of Lemma 4.56, $f^p_{\varphi,\psi}(x) \equiv x \pmod{p^2}$ since otherwise $f_{\varphi,\psi}(x)$ would be transitive modulo p^2. Now put $\tilde{f}(x) = f_{\varphi,\psi}(x) + p$. We claim that $\tilde{f}$ is the polynomial $\tilde{f}_\varphi(x) \in \mathbb{Z}[x]$ we are seeking for.

  Indeed, $\tilde{f}(x) \equiv f_{\varphi,\psi}(x) \equiv \varphi(x) \pmod{p}$ for all $x \in \mathbb{Z}_p$, by the construction; moreover, easy induction on j shows that

$$\tilde{f}^j(x) \equiv f^j_{\varphi,\psi}(x) + p \cdot \left(1 + \sum_{i=0}^{j-2} \prod_{k=i}^{j-2} f'_{\varphi,\psi}(f^k_{\varphi,\psi}(x))\right) \pmod{p^2}. \quad (9.27)$$

  However, the latter congruence implies that $\tilde{f}^p(0) \equiv p \pmod{p^2}$ as $f^p_{\varphi,\psi}(0) \equiv 0 \pmod{p^2}$ and $f'_{\varphi,\psi}(f^k_{\varphi,\psi}(0)) \equiv h_k \pmod{p}$ for all $k = 0, 1, \dots, p-1$, see Step 3. Hence, $\tilde{f}(x)$ is transitive modulo p^2 in view of Lemma 4.56.

Note 9.28. The above procedure can be obviously modified to enumerate all polynomials that are transitive modulo p^2 (and even modulo p^3 for $p \le 3$) and thus (with the use of Proposition 3.52) to obtain a complete list of ergodic polynomials in explicit form. Note that there are exactly $(p-1)!$ pairwise distinct transitive transformations on $\mathbb{F}_p$. With the use of formula (1.9), each of these transformations can be represented by a polynomial; however, no better description of transitive polynomials on $\mathbb{F}_p$ is known.

Now we illustrate how the procedure described above works: Let us construct a polynomial generator with the recursion law $f \bmod 10^n$, such that the length of the shortest period of this generator is 10^n for all $n = 1, 2, 3, \dots$ and such that modulo 5 the generator performs a single cycle permutation $\varphi = (0, 1, 4, 3, 2)$ (i.e., $\varphi(0) = 1$, $\varphi(1) = 4, \dots, \varphi(2) = 0$).

By formula (1.9), we find interpolation polynomial $f_\varphi(x) = 1 + 3x^3$. Unfortunately, this polynomial is not bijective modulo 25, not speaking on transitivity, since its derivative $f'_\varphi(x) = 9x^2$ vanishes at 0. Consider the polynomial $f_{\varphi,\psi}(x) = 1 + 3x^3 + (x^5 - x) \cdot v(x)$, where $v(x)$ is undefined at the moment (see Proposition 1.34). We will choose $v(x)$ so that $f'_{\varphi,\psi}(x) \equiv 1 \pmod{5}$ for $x \in \{1, 3, 4\}$, $f'_{\varphi,\psi}(0) \equiv 2 \pmod{5}$, and $f'_{\varphi,\psi}(2) \equiv 3 \pmod{5}$ (see Step 3). From here, as $f'_{\varphi,\psi}(x) \equiv -x^2 - v(x) \pmod{5}$, we deduce that $v(3) \equiv 0 \pmod{5}$, and $v(x) \equiv 3 \pmod{5}$ if $x \in \mathbb{F}_5 \setminus \{3\}$. By the formula (1.9) we conclude that we may take $v(x) = -2 + x + 2x^2 - x^3 - 2x^4$; whence, $f_{\varphi,\psi}(x) = 1 + 2x - x^2 + x^3 + x^4 + x^6 + 2x^7 - x^8 - 2x^9$. Direct calculation shows that $f^5_{\varphi,\psi}(0) \equiv 20 \pmod{25}$; thus, the polynomial $f_{\varphi,\psi}(x)$ is transitive modulo 25 (whence, is ergodic on $\mathbb{Z}_5$), so Step 4 of the procedure is avoided. Now we put $f(x) = 1 + 77x + 24x^2 + 76x^3 + 76x^4 + 76x^6 + 52x^7 + 24x^8 - 52x^9$. Combining Theorem 4.36 with Proposition 9.10 we conclude that the polynomial $f(x)$ is ergodic on $\mathbb{Z}_2$; whence transitive modulo 2^n for all $n = 1, 2, \ldots$. As $f(x) \equiv f_{\varphi,\psi}(x) \pmod{25}$, by the Chinese Remainder Theorem 1.30 we finally conclude that the polynomial $f(x)$ is transitive modulo 10^n for all $n = 1, 2, \ldots$; and $f(x) \equiv \varphi(x) \pmod 5$ for all $x \in \{0, 1, 2, 3, 4\}$.

Techniques based on algebraic normal forms

In the case when we need to determine whether a given congruential generator with the recursion law $f \bmod 2^n$, where f is a 1-Lipschitz transformation of $\mathbb{Z}_2$, has the longest period, we may use one more method, that of Theorem 4.39 from Subsection 4.5.2. Compare to the two methods we presented above, the method based on Theorem 4.39 can be applied only to relatively simple compositions of arithmetic and bitwise logical instructions; however, some useful results can be obtained by this technique. We will illustrate the method by examples; some of these are of practical value. The first one presents a method to construct a family of measure-preserving (or ergodic) transformations out of a given one:

Proposition 9.29. *Let $F\colon \mathbb{Z}_2^{n+1} \to \mathbb{Z}_2$ be a 1-Lipschitz mapping such that for all $z_1, \ldots, z_n \in \mathbb{Z}_2$ the mapping $F(x, z_1, \ldots, z_n)\colon \mathbb{Z}_2 \to \mathbb{Z}_2$ is measure-preserving. Then $F(f(x), 2\cdot g_1(x), \ldots, 2\cdot g_n(x))$ is measure-preserves for all 1-Lipschitz mappings $g_1, \ldots, g_n\colon \mathbb{Z}_2 \to \mathbb{Z}_2$ and every 1-Lipschitz measure-preserving transformation $f\colon \mathbb{Z}_2 \to \mathbb{Z}_2$.*

Moreover, if f is ergodic then $f(x+4\cdot g(x))$, $f(x \,\mathsf{XOR}\, (4\cdot g(x)))$, $f(x)+4\cdot g(x)$, and $f(x) \,\mathsf{XOR}\, (4\cdot g(x))$ are ergodic for any 1-Lipschitz transformation $g\colon \mathbb{Z}_2 \to \mathbb{Z}_2$

Proof. Since the function F is 1-Lipschitz, $\delta_i(F(u_0, u_1, \ldots, u_n))$ does not depend on $\delta_j(u_k) = \chi_{j,k}$ for $j > i$, see Proposition 3.35. Consider ANF of the Boolean function $\delta_i(F(u_0, u_1, \ldots, u_n))$:

$$\delta_i(F(u_0, u_1, \ldots, u_n)) = \chi_{0,i}\Psi_i(u_0, u_1, \ldots, u_n) + \Phi_i(u_0, u_1, \ldots, u_n),$$

where Boolean functions $\Psi_i(u_0, u_1, \ldots, u_n)$ and $\Phi_i(u_0, u_1, \ldots, u_n)$ do not depend on $\chi_{0,i}$; that is, they depend only on

$$\chi_{0,0}, \ldots, \chi_{0,i-1}, \chi_{1,0}, \ldots, \chi_{1,i}, \ldots, \chi_{n,0}, \ldots, \chi_{n,i}.$$

In view of Theorem 4.39, $\Psi_i = 1$ since $F(x, z_1, \ldots, z_n)$ is measure-preserving for all $z_1, \ldots, z_n \in \mathbb{Z}_2$. Moreover, $\Phi_i(f(x), 2g_1(x), \ldots, 2g_n(x))$ does not depend on $\chi_i = \delta_i(x)$ since $\delta_j(2g(x))$ does not depend on χ_i for all $j = 1, 2, \ldots, n$. Thus, in view of Theorem 4.39, $\delta_i(f(x)) = \chi_i + \xi_i(f(x))$, where $\xi_i(f(x))$ does not depend on χ_i since f is measure-preserving. Finally,

$$\begin{aligned}\delta_i(F(f(x), 2\cdot g_1(x), \dots, 2\cdot g_n(x)))\\ &= \delta_i(f(x)) + \Phi_i(f(x), 2\cdot g_1(x), \dots, 2\cdot g_n(x))\\ &= \chi_i + \xi_i(f(x)) + \Phi_i(f(x), 2\cdot g_1(x), \dots, 2\cdot g_n(x)) = \chi_i + \Xi_i,\end{aligned}$$

where the Boolean function Ξ_i depends only on $\chi_0, \dots, \chi_{i-1}$. This proves the first assertion of Proposition 9.29 in view of Theorem 4.39.

We prove the second assertion along similar lines. For $z \in \mathbb{Z}_2$ and $i = 0, 1, 2, \dots$ let $\zeta_i = \delta_i(z)$. Thus one can represent $\delta_i(z \operatorname{XOR} 4\cdot g(z))$ and $\delta_i(z + 4\cdot g(z))$ via ANFs in Boolean variables $\zeta_0, \zeta_1, \dots, \zeta_i$. Note that $\delta_i(z \operatorname{XOR} 4\cdot g(z)) = \zeta_i + \lambda_i(z)$, where $\lambda_i(z) = 0$ for $i = 0, 1$ and $\deg \lambda_i(z) \le i - 1$ for $i > 1$, since for $i > 1$ the Boolean function $\lambda_i(z)$ depends only on $\zeta_0, \dots, \zeta_{i-2}$.

Further, we claim that $\delta_i(z + 4\cdot g(z)) = \delta_i(z) + \mu_i(z)$, where $\mu_i(z) = \mu_i^g(z)$ is 0 for $i = 0, 1$ and $\deg \mu_i(z) \le i - 1$ for $i > 1$. Indeed, $\mu_i(z) = \lambda_i(z) + \alpha_i(z)$, where the Boolean function $\alpha_i(z)$ is a carry. Yet $\alpha_i(z) = 0$ for $i = 0, 1, 2$, and $\alpha_i(z) = \zeta_{i-1}\cdot\lambda_{i-1}(z) + \zeta_{i-1}\cdot\alpha_{i-1}(z) + \lambda_{i-1}(z)\cdot\alpha_{i-1}(z)$ for $i \ge 3$, and $\alpha_i(z)$ depends only on $\zeta_0, \dots, \zeta_{i-1}$ since $\alpha_i(z)$ is a carry. However, $\deg \alpha_3(z) = 2$ and if $\deg \alpha_{i-1}(z) \le i-2$ then $\deg(\delta_{i-1}(z)\cdot\alpha_{i-1}(z)) \le i-1$, $\deg(\lambda_{i-1}(z)\cdot\alpha_{i-1}(z)) \le i-1$, and $\deg(\zeta_{i-1}\cdot\lambda_{i-1}(z)) \le i-1$ since $\alpha_{i-1}(z)$ depends only on $\zeta_0, \dots, \zeta_{i-2}$ and $\lambda_{i-1}(z)$ depends only on $\zeta_0, \dots, \zeta_{i-3}$. Thus $\deg \alpha_i(z) \le i - 1$ and hence $\deg \mu_i(z) \le i - 1$.

Now, since $f(x)$ is ergodic, $\delta_i(f(x)) = \chi_i + \xi_i(x)$, where the Boolean function ξ_i depends only on $\chi_0, \dots, \chi_{i-1}$ and, additionally, $\xi_0 = 1$, and $\deg \xi_i(x) = i$ for $i > 0$ (see Theorem 4.39); i.e. $\xi_i(x) = \chi_0\chi_1\cdots\chi_{i-1} + \vartheta_i(x)$, where $\deg \vartheta_i(x) \le i - 1$ for $i > 0$. Hence, for $* \in \{+, \operatorname{XOR}\}$ one has $\delta_i(f(x * 4\cdot g(x))) = \delta_i(x * 4\cdot g(x)) + \delta_0(x * 4\cdot g(x))\delta_1(x * 4\cdot g(x))\cdots\delta_{i-1}(x * 4\cdot g(x)) + \vartheta_i(x * 4\cdot g(x))$; thus $\delta_i(f(x * 4\cdot g(x))) = \chi_i + \chi_0\cdots\chi_{i-1} + \beta_i^*(x)$, where $\deg \beta_i^*(x) \le i - 1$ for $i > 0$, and $\delta_0(f(x * 4\cdot g(x)) = \delta_0(x * 4\cdot g(x)) + 1 = \chi_0 + 1$. Finally, $f(x * 4\cdot g(x))$ for $* \in \{+, \operatorname{XOR}\}$ is ergodic in view of Theorem 4.39.

In a similar manner it could be demonstrated that $f(x) * 4\cdot g(x)$ is ergodic for $* \in \{+, \operatorname{XOR}\}$: $\delta_i(f(x) * 4\cdot g(x)) = \delta_i(f(x))$ for $i = 0, 1$ and thus satisfy the conditions of Theorem 4.39. For $i > 1$ one has $\delta_i(f(x) \operatorname{XOR} 4\cdot g(x)) = \chi_i + \xi_i(x) + \delta_{i-2}(g(x))$; but $\delta_{i-2}(g(x))$ does not depend on χ_{i-1}, χ_i. Thus the Boolean function $\xi_i(x) + \delta_{i-2}(g(x))$ in variables $\chi_0, \dots, \chi_{i-1}$ is of odd weight, since $\xi_i(x)$ is of odd weight, thus proving that $f(x) \operatorname{XOR} 4\cdot g(x)$ is ergodic.

Now represent $g(x) = g(f^{-1}(f(x))) = h(f(x))$, where f^{-1} is the inverse mapping for f. Clearly, $f^{-1}(x)$ is well defined since the mapping $f\colon \mathbb{Z}_2 \to \mathbb{Z}_2$ is bijective; moreover $f^{-1}(x)$ is 1-Lipschitz and ergodic. Finally $\delta_i(f(x) + 4\cdot g(x)) = \delta_i(f(x)) + \mu_i'(f(x))$, where the ANF of the Boolean function $\mu_i'(x) = \mu_i^h(x)$ in Boolean variables $\chi_0, \dots, \chi_{i-1}$ does not contain a monomial $\chi_0\cdots\chi_{i-1}$ (see the claim above). This implies that the ANF of the Boolean function $\mu_i'(f(x))$ in

Boolean variables $\chi_0, \dots, \chi_{i-1}$ does not contain a monomial $\chi_0 \cdots \chi_{i-1}$ either, since $\delta_j(f(x)) = \chi_j + \xi_j(x)$ and $\xi_j(x)$ depend only on $\chi_0, \dots, \chi_{j-1}$ for $j = 2, 3, \dots$. Hence, $\delta_i(f(x) + 4 \cdot g(x)) = \chi_i + \xi_i(x) + \mu_i'(f(x))$ and the Boolean function $\xi_i(x) + \mu_i'(f(x))$ in Boolean variables $\chi_0, \dots, \chi_{i-1}$ is of odd weight. This finishes the proof in view of Theorem 4.39. □

Note 9.30. Some claims of Proposition 9.29 can be proved by other methods (cf., e.g., Note 9.11); however, we proved them applying Theorem 4.39 to illustrate the method that uses ANFs of coordinate functions.

Example 9.31 (Add-xor generator). With the use of Proposition 9.29 it is possible to construct very fast congruential generators, the so-called *add-xor generators*, that are transitive modulo 2^n. For instance, take

$$f(x) = (\dots((((x + c_0) \mathsf{XOR}\, d_0) + c_1) \mathsf{XOR}\, d_1) + \dots + c_m) \mathsf{XOR}\, d_m,$$

where $c_0 \equiv 1 \pmod 2$, and the rest of c_i, d_i are 0 modulo 4. In the general case these functions f (for arbitrary c_i, d_i) were studied in [274], where it was proved that f is ergodic if and only if it is transitive modulo 4.

With the use of Theorem 4.39 it is possible to give a short proof of the main result of [264], namely, of Theorem 3 there:

Example 9.32 (Theorem 3 of [264]). The mapping $f(x) = x + (x^2 \,\mathsf{OR}\, C)$ over n-bit words is invertible if and only if the least significant bit of C is 1. For $n \geq 3$ it is a permutation with a single cycle if and only if both the least significant bit and the third least significant bit of C are 1.

Proof. We shall prove that the function $f(x) = x + (x^2 \,\mathsf{OR}\, C)$ is measure-preserving (respectively, ergodic) if and only if the conditions on C stated above hold. Denote $c_i = \delta_i(C)$; for $x \in \mathbb{Z}_2$ and $i = 0, 1, 2, \dots$ denote $\chi_i = \delta_i(x) \in \{0, 1\}$. To calculate ANF of the Boolean function $\delta_i(x + (x^2 \,\mathsf{OR}\, C))$ in variables $\chi_0, \chi_1, \dots$, we start with the following easy claims:

- $\delta_0(x^2) = \chi_0$, $\delta_1(x^2) = 0$, $\delta_2(x^2) = \chi_0\chi_1 + \chi_1$,
- $\delta_n(x^2) = \chi_{n-1}\chi_0 + \psi_n(\chi_0, \dots, \chi_{n-2})$ for all $n \geq 3$, where ψ_n is a Boolean function in $n-1$ Boolean variables $\chi_0, \dots, \chi_{n-2}$.

The first of these claims could be easily verified by direct calculations. To prove the second one, represent $x = \bar{x}_{n-1} + 2^{n-1}s_{n-1}$ for $\bar{x}_{n-1} = x \bmod 2^{n-1}$ and calculate $x^2 = (\bar{x}_{n-1} + 2^{n-1}s_{n-1})^2 = \bar{x}_{n-1}^2 + 2^n s_{n-1}\bar{x}_{n-1} + 2^{2n-2}s_{n-1}^2 = \bar{x}_{n-1}^2 + 2^n\chi_{n-1}\chi_0 \pmod{2^{n+1}}$ for $n \geq 3$ and note that $\bar{x}_{n-1}^2$ depends only on $\chi_0, \dots, \chi_{n-2}$. This gives:

(1) $\delta_0(x^2 \,\mathsf{OR}\, C) = \chi_0 + c_0 + \chi_0 c_0$,

(2) $\delta_1(x^2 \,\mathsf{OR}\, C) = c_1$,

(3) $\delta_2(x^2 \text{ OR } C) = \chi_0\chi_1 + \chi_1 + c_2 + c_2\chi_1 + c_2\chi_0\chi_1$,

(4) $\delta_n(x^2 \text{ OR } C) = \chi_{n-1}\chi_0 + \psi_n + c_n + c_n\chi_{n-1}\chi_0 + c_n\psi_n$ for $n \geq 3$.

From here it follows that if $n \geq 3$ then $\delta_n(x^2 \text{ OR } C) = \lambda_n(\chi_0, \dots, \chi_{n-1})$ and $\deg \lambda_n \leq n-1$ since ψ_n depends only on $\chi_0, \dots, \chi_{n-2}$.

Now we successively calculate $\gamma_n = \delta_n(x + (x^2 \text{ OR } C))$ for $n = 0, 1, 2, \dots$. We have $\delta_0(x + (x^2 \text{ OR } C)) = c_0 + \chi_0 c_0$, so necessarily $c_0 = 1$ since otherwise f is not bijective modulo 2. Proceeding further with $c_0 = 1$ we obtain $\delta_1(x + (x^2 \text{ OR } C)) = c_1 + \chi_0 + \chi_1$ since χ_1 is a carry. Then $\delta_2(x + (x^2 \text{ OR } C)) = (c_1\chi_0 + c_1\chi_1 + \chi_0\chi_1) + (\chi_0\chi_1 + \chi_1 + c_2 + c_2\chi_1 + c_2\chi_0\chi_1) + \chi_2 = c_1\chi_0 + c_1\chi_1 + \chi_1 + c_2 + c_2\chi_1 + c_2\chi_0\chi_1 + \chi_2$; here $c_1\chi_0 + c_1\chi_1 + \chi_0\chi_1$ is a carry. From here in view of Theorem 4.39 we immediately deduce that $c_2 = 1$ since otherwise f is not transitive modulo 8. Now for $n \geq 3$ one has $\gamma_n = \alpha_n + \lambda_n + \chi_n$, where α_n is a carry, and $\alpha_{n+1} = \alpha_n\lambda_n + \alpha_n\chi_n + \lambda_n\chi_n$. But if $c_2 = 1$ then $\deg \alpha_3 = \deg(\mu\nu + \chi_2\mu + \chi_2\nu) = 3$, where $\mu = c_1\chi_0 + c_1\chi_1 + \chi_0\chi_1$, $\nu = (\chi_0\chi_1 + \chi_1 + c_2 + c_2\chi_1 + c_2\chi_0\chi_1) = 0$. This implies inductively in view of Claim 4 above that $\deg \alpha_{n+1} = n+1$ and that $\gamma_{n+1} = \chi_{n+1} + \xi_{n+1}(\chi_0, \dots, \chi_n)$, $\deg \xi_{n+1} = n+1$. So the conditions of Theorem 4.39 are satisfied, thus finishing the proof of Theorem 3 from [264]. □

Now we are going to study inversive generators modulo 2^n that are based on the function $\operatorname{inv}(x)$ of taking the generalized multiplicative inverse of $x \in \mathbb{Z}_2$, see equation (9.4) for the definition of $\operatorname{inv}(x)$. Before the study, we briefly discuss properties of the function $\operatorname{inv}\colon \mathbb{Z}_p \to \mathbb{Z}_p$, p prime. As proofs of claims that follow are just exercises in p-adic analysis, they are sketched or omitted.

The function $\operatorname{inv}(x)$ is defined everywhere on $\mathbb{Z}_p$: Indeed, for all $x \neq 0$, $|x|_p^{-1} \cdot x = \frac{x}{p^{\operatorname{ord}_p x}}$ is an invertible element of the ring $\mathbb{Z}_p$, see Note 1.47. As for $x = 0$, $\lim^p_{x\to 0} \operatorname{inv}(x) = 0$ since $\left|(|x|_p^{-1} \cdot x)^{-1}\right|_p = 1$ for all $x \neq 0$, and $\lim^p_{x\to 0} |x|_p = 0$; that is, $\operatorname{inv}(0) = 0$. We also write $\operatorname{inv}(x)$ in the form

$$\operatorname{inv}(x) = p^{\operatorname{ord}_p x} \cdot \left(\frac{x}{p^{\operatorname{ord}_p x}}\right)^{-1}, \quad x \in \mathbb{Z}_p \setminus \{0\}$$

assuming that $\operatorname{inv}(0) = 0$. It is easy to check that the function $\operatorname{inv}(x)$ is 1-Lipschitz, thus, uniformly continuous on $\mathbb{Z}_p$. Moreover, it is not difficult to see that $\operatorname{inv}(x)$ is differentiable (although, *not* uniformly) everywhere on $\mathbb{Z}_p$ except 0; and that the derivative $\operatorname{inv}'(x)$ is:

$$\operatorname{inv}'(x) = -\left(\frac{x}{p^{\operatorname{ord}_p x}}\right)^{-2}, \quad x \neq 0. \tag{9.28}$$

Note that $\operatorname{inv}'(x)$ is discontinuous at 0: Although both sequences $\{p^n\}_{n=0}^{\infty}$ and $\{p^n \cdot (p^2-1)\}_{n=0}^{\infty}$ tend p-adically to infinity as n goes to infinity, $\lim^p_{n\to\infty} \operatorname{inv}'(p^n) = -1$ whereas $\lim^p_{n\to\infty} \operatorname{inv}'(p^n \cdot (p^2-1)) = -(p^2-1)^{-2} \neq -1$. Moreover, the function

$\mathrm{inv}(x)$ is infinitely many times differentiable on $\mathbb{Z}_p \setminus \{0\}$, and the ith derivative of $\mathrm{inv}(x)$ is

$$\mathrm{inv}^{(i)}(x) = \frac{(-1)^i}{p^{i\,\mathrm{ord}_p x}}\left(\frac{x}{p^{\mathrm{ord}_p x}}\right)^{-i-1}$$

everywhere on $\mathbb{Z}_p$ except 0; $i = 1, 2, \dots$. However, in the case $p = 2$, the function $\mathrm{inv}(x)$ is uniformly differentiable modulo 2 on $\mathbb{Z}_2$, and $\frac{\partial_1(\mathrm{inv}(x))}{\partial_1 x} = 1$; this immediately follows from Proposition 9.24: Indeed, the function $\mathrm{inv}\colon \mathbb{Z}_p \to \mathbb{Z}_p$ is a 1-Lipschitz bijection; whence, a measure-preserving transformation of $\mathbb{Z}_p$. One more interesting property of the function $\mathrm{inv}\colon \mathbb{Z}_p \to \mathbb{Z}_p$ is that it is an automorphism of the multiplicative semigroup $\mathbb{Z}_p$; that is, $\mathrm{inv}(a \cdot b) = \mathrm{inv}(a) \cdot \mathrm{inv}(b)$ for all $a, b \in \mathbb{Z}_p$ (this follows immediately from the definition of $\mathrm{inv}(x)$, see (9.4)). In the case $p = 2$ we can obtain more information on coordinate functions $\delta_i(x)$ of the function $\mathrm{inv}(x)$:

Lemma 9.33. *Let $p = 2$. Then the ANF of the ith coordinate function $\delta_i(\mathrm{inv}(x))$ is of the form*

$$\delta_i(\mathrm{inv}(x)) = \chi_i \oplus \varphi_i(\chi_0, \dots, \chi_{i-1}),$$

where $\chi_i = \delta_i(x)$, $\varphi_0 = 0$, and the weight of every Boolean function $\varphi_i(\chi_0, \dots, \chi_{i-1})$ in Boolean variables $\chi_0, \dots, \chi_{i-1}$ is even, $i = 0, 1, 2, \dots$.

Note 9.34. Recall that the weight of the Boolean function $\varphi_i(\chi_0, \dots, \chi_{i-1})$ in Boolean variables $\chi_0, \dots, \chi_{i-1}$ is even if and only if its ANF does not contain the monomial $\chi_0 \cdots \chi_{i-1}$, see Theorem 4.39.

Proof. As $\mathrm{inv}\colon \mathbb{Z}_2 \to \mathbb{Z}_2$ is a 1-Lipschitz measure-preserving transformation on $\mathbb{Z}_2$, then in view of equation (4.25) of Subsection 4.5.2 and of Theorem 4.39, the Boolean function $\delta_i(\mathrm{inv}(x))$ depends only on Boolean variables $\chi_0, \dots, \chi_i$ and $\delta_i(\mathrm{inv}(x))$ is linear with respect to variable χ_i: $\delta_i(\mathrm{inv}(x)) = \chi_i \oplus \varphi_i(\chi_0, \dots, \chi_{i-1})$ for a suitable Boolean function $\varphi_i(\chi_0, \dots, \chi_{i-1})$ in Boolean variables $\chi_0, \dots, \chi_{i-1}$, for all $i = 0, 1, 2, \dots$ (recall that a Boolean function on empty set of variables is a constant).

Now by induction on i we prove that the weight of the Boolean function $\varphi_i(\chi_0, \dots, \chi_{i-1})$ is even, for all $i = 0, 1, 2, \dots$; that is, the number of Boolean i-dimensional vectors on which the Boolean function $\varphi_i(\chi_0, \dots, \chi_{i-1})$ takes value 1 is even. Direct calculations show that $\mathrm{inv}(x) \equiv x \pmod{2^n}$ for $n = 1, 2, 3$; so $\varphi_0 = \varphi_1 = \varphi_2 = 0$; for $n = 4$ we have $\mathrm{inv}(x) \not\equiv x \pmod{2^n}$ if and only if x is congruent 3,5,11, or 14 modulo 16, so the weight of the Boolean function $\varphi_3(\chi_0, \chi_1, \chi_2)$ is 2. Let our claim be true for Boolean functions $\varphi_0, \dots, \varphi_{i-1}$; let us prove it for the Boolean function $\varphi_i(\chi_0, \dots, \chi_{i-1})$.

For a Boolean function ψ denote by $\bar{\psi}$ its negation; that is, $\bar{\psi} = \psi \oplus 1$. Now take arbitrary $x \equiv 1 \pmod 2$ (in other words, put $\chi_0 = 1$) and consider $\delta_i(\mathrm{inv}(1 + \mathsf{NOT}(x))$. Since $x = 1+2z$, where $z = \chi_1+2\cdot\chi_2+4\cdot\chi_3+\cdots$, then $\mathrm{inv}(1+\mathsf{NOT}(x)) = (1 + 2 \cdot \mathsf{NOT}(z))^{-1} = (1 - 2 \cdot (1 + z))^{-1} = -(1 + 2z)^{-1} = 1 + \mathsf{NOT}((1 + 2z)^{-1})$ (we used the second formula from (8.4) during these conversions). It is obvious that if we denote $(1 + 2 \cdot \mathsf{NOT}(z))^{-1} = 1 + \sum_{j=1}^{\infty} 2^j \zeta_j$, then $1 + \mathsf{NOT}((1 + 2z)^{-1}) = 1 + \sum_{j=1}^{\infty} 2^j \bar{\zeta}_j$, where $\zeta_j \in \{0, 1\}$ $(j = 1, 2, \ldots)$. By this reason, the just proven equality $(1 + 2 \cdot \mathsf{NOT}(z))^{-1} = 1 + \mathsf{NOT}((1 + 2z)^{-1})$ implies that

$$\varphi_i(1, \chi_1, \ldots, \chi_{i-1}) = \varphi_i(1, \bar{\chi}_1, \ldots, \bar{\chi}_{i-1}), \tag{9.29}$$

for all $\chi_1, \ldots, \chi_{i-1} \in \{0, 1\}$, since $\zeta_i = \delta_i(\mathrm{inv}(x)) = \chi_i \oplus \varphi_i(\chi_0, \ldots, \chi_{i-1})$, $i = 1, 2, \ldots$. Further, since $\mathrm{inv}(ab) = \mathrm{inv}(a) \cdot \mathrm{inv}(b)$ for all $a, b \in \mathbb{Z}_2$, then $\mathrm{inv}(2 \cdot z) = 2 \cdot \mathrm{inv}(z)$, so $\varphi_i(0, \chi_1, \ldots, \chi_{i-1}) = \varphi_{i-1}(\chi_1, \ldots, \chi_{i-1})$; however, by induction hypothesis, the weight of the Boolean function $\varphi_{i-1}(\chi_1, \ldots, \chi_{i-1})$ in Boolean variables $\chi_1, \ldots, \chi_{i-1}$ is even. This, together with equation (9.29), completes the induction and proves the lemma. □

Now we are able to prove the following proposition that gives rise to a large new family of inversive generators modulo 2^n that involve the function inv into their compositions and whose shortest periods are of length 2^n:

Proposition 9.35. *Let f be any 1-Lipschitz transformation on $\mathbb{Z}_2$. If f is ergodic, then both compositions $f(\mathrm{inv}(x))$ and $\mathrm{inv}(f(x))$ are ergodic. Vice versa, if either of the transformations $f(\mathrm{inv}(x))$ or $\mathrm{inv}(f(x))$ is ergodic, then f is ergodic.*

Proof. For $i = 0, 1, 2, \ldots$ denote $\delta_i(x) = \chi_i$. If f is ergodic, then by Theorem 4.39,

$$\delta_i(f(x)) = \chi_i \oplus \chi_0 \cdots \chi_{i-1} \oplus \psi_i(\chi_0, \ldots, \chi_{i-1}), \tag{9.30}$$

where the ANF of the Boolean function $\psi_i(\chi_0, \ldots, \chi_{i-1})$ does not contain the monomial $\chi_0 \cdots \chi_{i-1}$, $\psi_0 = 0$, $i = 0, 1, 2, \ldots$ (we recall that the product over the empty set is 1). By Lemma 9.33,

$$\delta_i(\mathrm{inv}(x)) = \chi_i \oplus \varphi_i(\chi_0, \ldots, \chi_{i-1}), \tag{9.31}$$

where $\varphi_0 = 0$ and ANF of the Boolean function $\varphi_i(\chi_0, \ldots, \chi_{i-1})$ does not contain the monomial $\chi_0 \cdots \chi_{i-1}$, $i = 0, 1, 2, \ldots$. Whence ANF of the Boolean function $\delta_i(u(x))$, where $u(x)$ is either of functions $f(\mathrm{inv}(x))$ or $\mathrm{inv}(f(x))$, is of the form

$$\delta_i(u(x)) = \chi_i \oplus \chi_0 \cdots \chi_{i-1} \oplus \vartheta_i(\chi_0, \ldots, \chi_{i-1}), \tag{9.32}$$

where the ANF of the Boolean function $\vartheta_i(\chi_0, \ldots, \chi_{i-1})$ does not contain the monomial $\chi_0 \cdots \chi_{i-1}$, $\vartheta_0 = 1$, $i = 0, 1, 2, \ldots$. Thus, by Theorem 4.39, both $f(\mathrm{inv}(x))$ and $\mathrm{inv}(f(x))$ are ergodic.

To prove the converse statement, note that if f is not ergodic, then by Theorem 4.39, the ANF of some Boolean function $\delta_i(f(x))$ in representation (9.30) does not contain the monomial $\chi_0 \cdots \chi_{i-1}$. Thus, in view of (9.31), representation (9.32) of $\delta_i(u(x))$ does not contain the monomial $\chi_0 \cdots \chi_{i-1}$ either. Therefore $u(x)$ is not ergodic by Theorem 4.39. □

From Proposition 9.35 immediately follows the main result of [119]: *The length of the shortest period of the congruential generator with the recursion law* $(a \cdot \mathrm{inv}(x) + b) \bmod 2^n$ *is* 2^n, $n \geq 2$, *if and only if* $a \equiv 1 \pmod 4$ *and* $b \equiv 1 \pmod 2$. Indeed, by Proposition 9.35, the transformation $a \cdot \mathrm{inv}(x) + b$ is ergodic on $\mathbb{Z}_2$ if and only if the polynomial $ax + b$ is ergodic on $\mathbb{Z}_2$; by Theorem 4.36, the latter holds if and only if $ax + b$ is transitive modulo 4, or, equivalently, if and only if $a \equiv 1 \pmod 4$ and $b \equiv 1 \pmod 2$.

More complex congruential generators can be constructed with the use of Proposition 9.35: For instance, the transformation $f(x) = 3 \cdot \mathrm{inv}(x) + 3^{\mathrm{inv}(x)}$ is ergodic on $\mathbb{Z}_2$ (see Example 9.9); this transformation results in an inversive-exponential generator modulo 2^n with the shortest period of length 2^n. In a similar way we conclude that the length of the shortest period of the more complicated exponential-inversive generator with the recursion law $(\mathrm{inv}(1 + x) + 4 \cdot (1 + \mathrm{inv}(2x))^{\mathrm{inv}(x)}) \bmod 2^n$ is also 2^n (see Note 9.11); the same holds for generators with recursion laws $(\mathrm{inv}(2x^2) + \mathrm{inv}(7x) + 1) \bmod 2^n$ and $(\mathrm{inv}(2x^2 + 7x + 1)) \bmod 2^n$ (see Corollary 9.16), etc.

We conclude Subsection 9.2.2 with an open problem concerning congruential generators based on the function $\mathrm{inv}\colon \mathbb{Z}_p \to \mathbb{Z}_p$ for odd prime p. As it was said (see the text that precedes Lemma 9.33), the function $\mathrm{inv}(x)$ is infinitely many times differentiable on $\mathbb{Z}_p \setminus \{0\}$; moreover, it not difficult to see that $\mathrm{inv}(x)$ can be expressed via Taylor power series at every point of $\mathbb{Z}_p$ except 0. Unfortunately, $\mathrm{inv}(x)$ is not a $\mathcal{C}$-function (neither $\mathcal{B}$-function nor $\mathcal{A}$-function). Thus, we can not apply directly corresponding theorems from Subsection 4.6.4 on ergodicity of compositions involving the function inv. So the following (somewhat informally posed) open question reads:

Open Question 9.36. What compositions of the function inv with $\mathcal{A}$-, $\mathcal{B}$- or $\mathcal{C}$-functions are ergodic on $\mathbb{Z}_p$, for odd prime p?

Note that the answer to the analogous question on measure-preservation is rather clear: e.g., it is obvious that whenever f is 1-Lipschitz, then, as inv is measure-preserving, any composition $f(\mathrm{inv}(x))$ and $\mathrm{inv}(f(x))$ is measure-preserving if and only if f is measure-preserving.

Chapter 10

Stream ciphers

As said (see the beginning of Chapter 9), the core of a stream cipher is a cryptographically secure PRNG that generates a keystream. In most cases these PRNGs are either automata represented at Figure 9.1 or compositions of automata of this kind. Very often the state transition circuit of these automata are congruential generators of Definition 9.5. These are, for instance, the Blum–Micali generator, whose state transition circuit is an exponential generator modulo a prime p; the RSA generator, whose state transition circuit is a power generator modulo pq (p and q are primes, $p \neq q$); the BBS generator, whose state transition circuit is a quadratic generator modulo pq (p and q are primes, $p \neq q$, $p, q \equiv 3 \pmod 4$); and various generators based on T-functions. State transition circuits of the latter are congruential generators with the recursion law of the form $f \bmod 2^n$, where f is a T-function. Recall that a T-function is just a triangular function from Definition 3.37 where $p = 2$; i.e., a 2-adic 1-Lipschitz function.

We note that cryptographical security of the first three generators (Blum–Micali, RSA, and BBS) is justified by the so-called hard problems, such as a discrete logarithm problem for the Blum–Micali generator, and a problem of factorization of a composite number for RSA generator and BBS generator. As the problems of computational complexity are outside the scope of the book, we do not consider generators of this kind. These generators are studied in a number of papers and books; the monograph [375] is a good starting point.

We will focus on the last type of cryptographic generators mentioned above, on the ones based on T-functions. We will show that the theory of these generators completely follows from the 2-adic ergodic theory. Known properties of these generators are immediate consequences of corresponding theorems on measure-preservation and/or ergodicity of 2-adic 1-Lipschitz dynamical systems. We will establish also a number of new properties of these generators and introduce new types of generators, the so-called counter-dependent generators whose recursion law is a T-function that changes dynamically during encryption. This is the main goal of the chapter.

The T-functions are of growing interest for the cryptographic society. The term 'T-function' was suggested in the papers [264–266]. We note that all mathematical results of the latter three papers either are contained among or immediately and obviously follow from results on p-adic ergodic theory of the paper [21], which was published nearly a decade prior to publication of the papers [264–266]. In the paper [21], as well as in the succeeding papers [22–24] it was directly pointed out that 2-adic 1-Lipschitz functions are of great importance for cryptography, and especially for stream cipher design, and the corresponding theory emerged. To the moment, several stream ciphers based on T-functions have been developed, see [350] for details. We are not going to consider concrete cryptographic solutions in this book, we shall rather introduce and develop the underlying mathematical theory, which emerged in the mentioned works by Vladimir Anashin, succeeded by his works [25, 26, 28, 29].

10.1 How secure are congruential generators?

Cryptographic security of a PRNG implies in particular that, given an output of the PRNG, it must be infeasible to find the corresponding state of the automaton. From this point, all congruential generators of the longest period 2^n that were considered above, are not secure in the following sense: Given a residue $a \in \mathbb{Z}/2^n\mathbb{Z}$ and a 1-Lipschitz ergodic (whence, measure-preserving) transforation f on $\mathbb{Z}_2$, one can easily solve the congruence $f(x) \equiv b \pmod{2^n}$ (in unknown $x \in \mathbb{Z}/2^n\mathbb{Z}$) in n steps using the same method as in the proof of Hensel's lemma[1], with minor modification: Instead of ordinary derivatives, as in the original case of Hensel's lemma for polynomials, one should use derivatives modulo 2. Note that we can apply this method since any 1-Lipschitz measure-preserving transformation f on $\mathbb{Z}_2$ is uniformly differentiable modulo 2, and its derivative modulo 2 is 1, see Proposition 9.24.

As for congruential generators with composite $N = \#\mathcal{N}$, using Chinese Remainder Theorem 1.30, we can reduce the study of the congruential generator to the case when N is a power of a prime, i.e., when $N = p^n$. In the case when the length of the shortest period of the congruential generator is p^n (that is, a maximum possible), by Proposition 2.3 it is obvious that the length of the shortest period of the sequence $(\delta_j(f^i(u_0)))_{i=0}^{\infty}$, where $\delta_j(z)$ stands for the jth digit in the base-p expansion of z, is exactly p^{j+1}; thus, only the $(n-1)$th coordinate sequence $(\delta_{n-1}(f^i(u_0)))_{i=0}^{\infty}$ of the output sequence of the generator has the maximum period length, p^n.

This property makes no problem if we use the congruential generator in computer simulation tasks: Usually in these tasks and numerical experiments they use the sequence $(\frac{f^i(u_0) \bmod p^n}{p^n})_{i=0}^{\infty}$. However, this property is a cryptographical drawback that leads to cryptographic insecurity of the generator with the recursion law $f \bmod p^n$ whenever the function f is known to a cryptanalyst, and if p is relatively small. Indeed, to solve the congruence $z \equiv f(x) \pmod{p^n}$, and as a result to find a key, which is usually the initial state u_0, we again may use a version of the p-adic Newton's method introduced during the proof of Hensel's lemma:

First, we solve the congruence $z \equiv f(x) \pmod{p}$, thus finding the least significant digit $\delta_0(x)$ of x. Provided $\delta_j(x)$ for $j = 0, 1, \dots, k-1$ are already found, to find $\delta_k(x)$

[1] which is actually a p-adic Newton's method, see e.g. [268]

we must find a (unique) solution of the congruence $z \equiv f(\hat{x}) + p^k \cdot \check{f}_k(\hat{x}, \delta_k(x)) \pmod{p^{k+1}}$ in indeterminate $\delta_k(x)$, where $\hat{x} = \delta_0(x) + \delta_1(x) \cdot p + \cdots + \delta_{k-1}(x) \cdot p^{k-1}$ and the mapping $\check{f}_k(\cdot,\cdot)\colon \mathbb{Z}/p^k\mathbb{Z} \times \mathbb{Z}/p\mathbb{Z} \to \mathbb{Z}/p\mathbb{Z}$ is uniquely determined by f. Of course, how to express $\check{f}_k(\cdot,\cdot)$ explicitly is a separate problem, yet this is not too difficult in a number of important cases, e.g. when f is uniformly differentiable modulo p.

We may also consider the case when f is not known to a cryptanalyst: e.g., for $p = 2$ one may take $f = 1 + x + 4 \cdot g(x)$, where $g(x)$ is a 1-Lipschitz key-dependent function, which is not known to a cryptanalyst. The function f is ergodic by Proposition 9.29. This situation is a little better in comparison with a known f since a cryptanalyst can not apply the version of the 2-adic Newton's method we described above. However, the sequence formed of less significant bits of $f^i(u_0)$ is predictable in both directions, i.e. knowing k members of the sequence $\{f^i(u_0)\}$ a cryptanalyst finds $\delta_j(f^i(u_0))$ for all $j < \log_2 k$ and all $i = 0, 1, 2, \ldots$, stretching the corresponding periods in both directions.

All these considerations show that in cryptography we can not use congruential generators as stream ciphers immediately; a specially chosen output function F is needed. The simplest one is *truncation*

$$F(u) = \left\lfloor \frac{u}{p^{n-m}} \right\rfloor \bmod p^m, \tag{10.1}$$

where $m < n$. That is, we just discard less significant digits of the output sequence.[2] Thus we come to the notion of *truncated congruential generator*: The latter is the automaton $\mathfrak{A}$ of Section 9.1 such that $\mathcal{M} = \mathbb{Z}/p^n\mathbb{Z}$, $\mathcal{N} = \mathbb{Z}/p^m\mathbb{Z}$, $m < n$, $F\colon \mathcal{N} \to \mathcal{M}$ is the truncation (10.1), and the state transition function $f\colon \mathbb{Z}/p^n\mathbb{Z} \to \mathbb{Z}/p^n\mathbb{Z}$ preserves all congruences of the residue ring $\mathbb{Z}/p^n\mathbb{Z}$, cf. Definition 1.18. We can (and shall) consider f as a reduction modulo p^n of a 1-Lipschitz transformation on the space $\mathbb{Z}_p$.

Note that the function F is not compatible (see Definition 1.18), yet balanced, so the output sequence, considered as a sequence over $\mathbb{Z}/p^m\mathbb{Z}$, is purely periodic, the length of its shortest period is exactly p^n, and each element from $\mathbb{Z}/p^m\mathbb{Z}$ occurs at the period exactly p^{n-m} times. Further we are mainly focused at the case $p = 2$.

An important example of this output function F is the mapping $F(u) = \delta_j(u)$: Given $u \in \mathbb{Z}_p$, it returns the jth digit of u in the p-adic canonical expansion of u. We call the corresponding sequence $(\delta_j(f^i(u)))_{i=0}^{\infty}$ the *jth coordinate sequence*. Of course, usage of δ_j as an output function of the automaton $\mathfrak{A}$ significantly reduces performance, and the corresponding pseudorandom number generator might be not of much practical value. Nonetheless, we must study coordinate sequences to establish certain important properties of output sequences of pseudorandom generators considered further.

[2] Note that methods of [275], as it is directly pointed out there, do not apply to generators that output only parts of the numbers generated.

The truncation usually makes generators slower but more secure: General methods to predict truncated congruential generators are not known, see [77, 315]. However, these methods exist for some special types of PRNGs, e.g. for truncated linear congruential generators modulo 2^n, for linear congruential generators modulo composite N when a relatively small part of less significant bits are discarded, see [145]. To our best knowledge, there was no progress in cryptanalysis of truncated congruential generators since the time of these publications. Thus, today general truncated congruential generators seem to be rather secure with respect to the so-called 'known-plaintext attack', when the output sequence is known to a cryptanalyst.

Unfortunately, real-life applications of these generators are nonetheless not secure by another reason: Lengths of their periods are too short with respect to contemporary cryptographic limitations. Indeed, for the word bitlength $n = 32$, which is a standard for most contemporary processors, the length of the shortest period of the keystream produced by a truncated congruential generator is at most $32 \cdot 2^{32} = 2^{37}$. This figure is too small to satisfy contemporary cryptographic security restrictions: According to these, the length of the shortest period of a keystream must be at least about 2^{80}. Thus, we must make the period of a congruential generator longer and the generator more secure *leaving the output sequence uniformly distributed.* Basically, there are two approaches to the problem.

The first one is obvious: We should consider generators based on multivariate ergodic T-functions, that is, on transformations $f : \mathbb{Z}_2^n \to \mathbb{Z}_2^n$ for $n > 1$. Then the length of the shortest period of the corresponding generator modulo 2^k will be 2^{kn} in view of Theorem 4.23. Unfortunately, due to Theorem 4.51, there are no multivariate ergodic T-functions in the class of functions that are uniformly differentiable modulo 2. This implies that there are no multivariate ergodic T-functions among all natural classes of functions. e.g., among polynomials with integer coefficients, among analytic functions from class $\mathcal{C}$, etc. Thus, it is impossible to construct multivariate ergodic T-functions as a composition of additions, multiplications, exponentiations, inversions, and XORs, something else must be added into the composition. This means that we necessarily must add ORs and ANDs into the composition; the latter two operators are not uniformly differentiable modulo 2 as bivariate functions, see Section 8.3. We consider this approach in Section 10.4.

The second way to lengthen the period of the keystream is to use counter-dependent generators introduced in Section 9.1. It is obvious that whenever the counter-dependent generator consists of L congruential generators modulo 2^n each, the maximum period of the keystream it can produce is $L \cdot 2^n$: Indeed, the sequence of states of a congruential generator is then $x_{i+1} \equiv f_{i \bmod L}(x_i) \pmod{2^n}$, $i = 0, 1, 2, \ldots$. Counter-dependent generators were originally introduced in [377]. The main problem is how to guarantee the period length (and the statistical quality) of the sequence $(x_i)_{i=0}^{\infty}$. In the paper [377] length of periods were not studied, only the diversity of output sequences of counter-dependent generators. Further we use a special construct, which is called the skew product in dynamics and the wreath product in algebra, to build counter-dependent generators that produce sequences of the longest period. This construct, which is of a very general nature, will be used also to describe multivariate ergodic T-functions in Section 10.4. So we start with wreath products.

10.2 Wreath products

Seemingly wreath products originated from permutation groups and later penetrated to other mathematical theories. Here is a formal definition of the basic notion:

Definition 10.1 (Wreath product of mappings). Given a mapping $u\colon Z \to Z$, and a family[3] of mappings $\mathcal{V} = \{(v_z : X \to X) : z \in Z\}$, the *wreath product* (or, the *skew product* or, the *skew shift*) of the family $\mathcal{V}$ by the mapping u is the mapping

$$u \wr \mathcal{V} \colon (z, x) \mapsto (u(z), v_z(x))$$

of the Cartesian product $Z \times X$ into itself. We shall also denote the wreath product by $u \wr_{z \in Z} v_z$.

In other words, the wreath product is a bivariate mapping where the leftmost coordinate is a function of the variable z only, and the other coordinate is a bivariate function of z and x. The following important proposition is obvious:

Proposition 10.2. *The wreath product $u \wr \mathcal{V}$ is bijective whenever both u and all v_z are bijective.*

Some terminology notes: In automata theory (and in algebra) they used to speak of wreath products, whereas in dynamical systems theory (and in ergodic theory) the term skew product (or skew shift) is preferable. It is worth noting that semidirect products of groups we already used in Section 7.3 to construct ergodic transformations on non-commutative groups, are special case of this general construction, the wreath product.

According to Section 9.1, an ordinary PRNG corresponds to the autonomous dynamical system; whereas a counterpart of a counter-dependent PRNG in dynamics is the *non-autonomous* dynamical system. A non-autonomous dynamical system is a dynamical system driven by another dynamical system, and skew products are used to combine two dynamical systems into a new one.

In cryptology, wreath products are used in construction of *Feistel networks*. A number of cryptographic algorithms (e.g., block ciphers like DES) are based on Feistel networks.

Example 10.3 (Feistel network). The Feistel network is a composition of alternating mappings of the following two kinds: The mapping of the first kind is $\psi_f : (z, x) \mapsto (z, z \mathsf{\ XOR\ } f(x))$, where $z, x \in \mathbb{Z}/2^n\mathbb{Z}$, $f : \mathbb{Z}/2^n\mathbb{Z} \to \mathbb{Z}/2^n\mathbb{Z}$, which is obviously a

[3] whose members need not be pairwise distinct

wreath product of the mapping $u(z) = z$ with the mappings $\mathcal{V} = \{v_z(x) = z \text{ XOR } f(x) \colon z \in \mathbb{Z}/2^n\mathbb{Z}\}$. The mapping of the second kind is just a permutation $\tau \colon (z, x) \mapsto (x, z)$. The resulting mapping is the composition $\psi_{f_1} \circ \tau \circ \cdots \circ \psi_{f_k} \circ \tau \circ \psi_{f_{k+1}}$.

Another important example of wreath products are T-functions:

Example 10.4. *Any T-function is a composition of wreath products*: Let t be a T-function, that is,

$$(\chi_0, \chi_1, \chi_2, \ldots) \overset{t}{\mapsto} (\psi_0(\chi_0); \psi_1(\chi_0, \chi_1); \psi_2(\chi_0, \chi_1, \chi_2); \ldots),$$

where $\chi_0, \chi_1, \chi_2, \ldots \in \{0, 1\}$, and $\psi_0(\chi_0), \psi_1(\chi_0, \chi_1), \psi_2(\chi_0, \chi_1, \chi_2), \ldots$ are Boolean functions in respective Boolean variables. Denote $\Psi_0 = \{\psi_0\}$, $\Psi_1 = \{\psi_1(\chi_0, \cdot) \colon \chi_0 \in \{0, 1\}\}, \ldots, \Psi_i = \{\psi_i(\chi_0, \ldots, \chi_{i-1}, \cdot) \colon \chi_0, \ldots, \chi_{i-1} \in \{0, 1\}^i\}$; then

$$\begin{aligned} t_0 \colon \quad & \chi_0 && \mapsto \psi_0(\chi_0), \\ t_1 = t_0 \wr \Psi_1 \colon \quad & (\chi_0, \chi_1) && \mapsto (\psi_0(\chi_0), \psi_1(\chi_0, \chi_1)), \\ t_2 = t_1 \wr \Psi_2 \colon \quad & ((\chi_0, \chi_1), \chi_2) && \mapsto ((\psi_0(\chi_0), \psi_1(\chi_0, \chi_1)), \psi_2(\chi_0, \chi_1, \chi_2)), \\ & && \vdots \end{aligned}$$

Moreover, a similar argument immediately shows that *any triangular function is a composition of wreath products*.

Wreath products can be defined for automata. For instance, let us state a definition of the wreath product of automata with no input:

Definition 10.5 (Wreath product of automata). Let $\mathfrak{A}_j = \langle \mathcal{N}, \mathcal{M}, f_j, F_j \rangle$, $j \in \mathcal{K}$, be a family of automata without input that have the same set $\mathcal{N}$ of states, the same output alphabet $\mathcal{M}$, and the same initial state u_0. Here $\mathcal{K}$ is a non-empty (possibly, countably infinite) set of indices. Members of the family need not be necessarily pairwise distinct. Let further $\mathfrak{T}$ be an automaton with output alphabet $\mathcal{K}$, with the set of states $\mathcal{S}$, with the state transition function t, with the output function T, and with the initial state s_0. The *wreath product* $\mathfrak{T} \wr_{j \in \mathcal{K}} \mathfrak{A}_j$ of the family $\{\mathfrak{A}_j : j \in \mathcal{K}\}$ of automata by the automaton $\mathfrak{T}$ is the automaton with the set of states $\mathcal{S} \times \mathcal{N}$, with the state transition function $\breve{f}(s, u) = (t(s), f_{T(s)}(u))$, with output function $\breve{F}(s, u) = F_{T(s)}(u)$, and with the initial state (s_0, u_0).

Note that we can relate to the family $\{\mathfrak{A}_j\}$ an automaton $\mathfrak{A}$ with the input alphabet $\mathcal{K}$, with the set of states $\mathcal{N}$, with the output alphabet $\mathcal{M}$, with the state transition function $\breve{f}(j, u) = f_j(u)$, and with the output function $\breve{F}(j, u) = F_j(u)$. Then the wreath product $\mathfrak{T} \wr_{j \in \mathcal{K}} \mathfrak{A}_j$ is just a serial connection of automaton $\mathfrak{T}$ with automaton $\mathfrak{A}$, see Section 8.1. As every generator can be considered as an autonomous dynamical system (see Section 9.1), the wreath product of automata results in a non-autonomous dynamical system: To be more exact, the automaton $\mathfrak{T}$ is a controlling dynamical system (which may be autonomous or non-autonomous), whereas the automaton $\mathfrak{A}$ is a controlled (thus, non-autonomous) dynamical system.

Note also that we can in an obvious manner re-state Definition 10.5 for the case when automata $\mathfrak{A}_j$ and/or $\mathfrak{T}$ have inputs; however, actually we do not need this general case in the sequel. Further we will focus on counter-dependent generators, and for that purpose even Definition 10.5 is too general. Actually counter-dependent generators are specific wreath products of generators. Recall that according to Definition 9.1, a generator is an automaton whose initial state is a variable, and that has no input.

Definition 10.6 (Wreath product of generators). Let $\mathfrak{A}_j = \langle \mathcal{N}, \mathcal{M}, f_j, F_j \rangle$ be a family of generators with the same state set $\mathcal{N}$ and the same output alphabet $\mathcal{M}$, indexed by elements of a non-empty (possibly, countably infinite) set J; members of the family need not be necessarily pairwise distinct. Let $T : J \to J$ be an arbitrary mapping. The *wreath product* of the family $\{\mathfrak{A}_j : j \in J\}$ of generators with respect to the mapping T is the generator $T \wr_{j\in J} \mathfrak{A}_j$ that has the set of states $J \times \mathcal{N}$, the state transition function $\breve{f}(j, u) = (T(j), f_j(u))$, and the output function $\breve{F}(j, u) = F_j(u)$. We call f_j (resp., F_j) the *clock* state transition function (respectively, the clock output function).

Definition 10.6 is a formal definition of a counter-dependent generator introduced in Section 9.1. Obviously, the state transition function $\breve{f}(j, z) = (T(j), f_j(z))$ is a wreath product of the family of mappings $\{f_j : j \in J\}$ by the mapping T, see Definition 10.1.

It is worth noting here that if $J = \mathbb{N}_0$ and F_j does not depend on j, this construction gives us a number of examples of counter-dependent generators in the sense of [377, Definition 2.4], where the notion of a counter-dependent generator was originally introduced. However, we use this notion in a broader sense in comparison with that of the paper [377]: In our counter-dependent generators not only the state transition function, but also the output function depends on j. Moreover, in the paper [377] only the special case of counter-dependent generators is studied; namely, *counter-assisted generators* and their cascaded and two step modifications. The state transition function of a counter-assisted generator is of the form $f_i(x) = i \star h(x)$, where $\star$ is a binary quasigroup operation (in particular, a group operation, e.g., $+$, or XOR, or a Latin square from Section 8.4, etc.), and $h(x)$ does not depend on j. The output function of a counter-assisted generator does not depend on j either. Further in our book we study not only counter-assisted generators, but counter-dependent generators of the most general form as well.

Example 10.7. *Every generator whose recursion law is a T-function, is a composition of wreath products of linear congruential generators modulo 2.* Indeed, algebraic normal form (ANF) of any Boolean function of one Boolean variable χ is $\beta\chi \oplus \alpha$, for suitable $\alpha, \beta \in \{0, 1\}$. So the claim is just a restatement of Example 10.4.

In other words, given any T-function f, we can consider a generator $\mathfrak{T}$ with the state transition function f and with output function δ_n as a specific counter-dependent generator, a wreath product of a family consisting of linear congruential generators modulo 2 with respect to the mapping $f \bmod 2^n$. For instance, let f be a measure-preserving T-function. Then, in force of Theorem 4.39, $\delta_n(f(\chi_0 + \cdots + \chi_n \cdot 2^n)) = \chi_n \oplus \varphi_n(\chi_0, \dots, \chi_{n-1})$, where $\varphi_n(\chi_0, \dots, \chi_{n-1})$ is a Boolean function in Boolean variables $\chi_0, \dots, \chi_{n-1}$. Consider a family $\mathcal{F}$ of linear congruential generators performing the recursion $x_{j+1} = x_j + \varphi_n(\chi_0, \dots, \chi_{n-1}) \bmod 2$, $j = 0, 1, 2, \dots$, and consider a transformation $f \bmod 2^n$ of the residue ring $\mathbb{Z}/2^n\mathbb{Z}$. As every element of the ring has a unique representation of the form $\chi_0 + \cdots + \chi_{n-1} \cdot 2^{n-1}$, $\chi_0, \dots, \chi_{n-1} \in \{0, 1\}$, members of the family $\mathcal{F}$ of linear congruential generators are indexed by elements of the ring $\mathbb{Z}/2^n\mathbb{Z}$. It is clear from Definition 10.6 that the generator $\mathfrak{T}$ is a wreath product of the family $\mathcal{F}$ of linear congruential generators modulo 2 with respect to the mapping $f \bmod 2^n$: Indeed, in this case $J = \mathbb{Z}/2^n\mathbb{Z}$ and $T = f \bmod 2^n$. Note that in the general case, when f is not necessarily measure-preserving, the family $\mathcal{F}$ consists of linear congruential generators performing the recursion $x_{j+1} = x_j \cdot \theta_n(\chi_0, \dots, \chi_{n-1}) + \varphi_n(\chi_0, \dots, \chi_{n-1}) \bmod 2$, $j = 0, 1, 2, \dots$, where $\theta_n(\chi_0, \dots, \chi_{n-1})$ is a Boolean function in Boolean variables $\chi_0, \dots, \chi_{n-1} \in \{0, 1\}$.

A similar argument shows that *every generator whose recursion law is a 1-Lipschitz transformation f on $\mathbb{Z}_p$ is a composition of wreath products of congruential generators modulo p*; moreover, for odd p these congruential generators are polynomial generators modulo p, which are *not* necessarily linear. However, these polynomial generators *are linear congruential generators modulo p whenever f is uniformly differentiable modulo p*. Indeed, as in the latter case $\delta_n(f(\chi_0 + \cdots + \chi_n \cdot p^n)) \equiv \delta_n(f(\chi_0 + \cdots + \chi_{n-1} \cdot p^{n-1})) + f_1'(\chi_0 + \cdots + \chi_{n-1} \cdot p^{n-1}) \cdot \chi_n \pmod p$ for all $\chi_0, \dots, \chi_n \in \{0, 1, \dots, p-1\}$, where f_1' is a derivative modulo p, the family of congruential generators are generators performing the recursion $x_{j+1} = x_j \cdot f_1'(\chi_0 + \cdots + \chi_{n-1} \cdot p^{n-1}) + \delta_n(f(\chi_0 + \cdots + \chi_{n-1})) \bmod p$, $j = 0, 1, 2, \dots$. Note that both $f_1'(\chi_0 + \cdots + \chi_{n-1})$ and $\delta_n(f(\chi_0 + \cdots + \chi_{n-1}))$ can be expressed via polynomials over the field $\mathbb{F}_p$ in variables $\chi_0, \dots, \chi_{n-1}$.

Wreath products can be defined for families of transformations.

Definition 10.8. Let $\mathcal{U}$ be a family of transformations of the non-empty set Z; let $\mathcal{W}$ be a family of transformations of the non-empty set X. Denote $\mathcal{W}^Z$ a Cartesian power of $\mathcal{W}$. Then $\mathcal{U} \wr \mathcal{W}$ is a set of all transformations on $Z \times X$ of the form (u, w) where $u \in \mathcal{U}$ and $w \in \mathcal{W}^Z$ which act on $Z \times X$ according to the following rule:

$$(u, w)\colon (z, x) \mapsto (u(z), w_z(x)) \quad (x \in X, z \in Z),$$

where w_z is a projection of w onto coordinate z of the Cartesian product $\mathcal{W}^Z$.

In other words, as $\mathcal{W}^Z$ is a set of all mappings from Z to $\mathcal{W}$ by the definition of the Cartesian power, and as $\mathcal{W}$ is a set of mappings from X to X, every element $w \in \mathcal{W}^Z$ is a bivariate mapping, $w(*, \cdot) = w_*(\cdot)$, where the first variable (index) runs over Z, and the second runs over X; so the wreath product $\mathcal{U} \wr \mathcal{W}$ is just a union of wreath products in the sense of Definition 10.1: $\mathcal{U} \wr \mathcal{W} = \bigcup_{u \in \mathcal{U}} u \wr \mathcal{V}$, where $\mathcal{V} = \mathcal{W}^Z$.

Note that whenever both $\mathcal{U}$ and $\mathcal{W}$ are permutation groups on sets Z and X, respectively, from Proposition 10.2 it immediately follows that the wreath product $\mathcal{U} \wr \mathcal{W}$ is a permutation group on the direct product $Z \times X$. A word of caution: In permutation group theory they usually write terms of wreath products in reverse order compared to our notation; that is, the wreath product $\mathcal{U} \wr \mathcal{W}$ from our Definition 10.8 most likely would be written as $\mathcal{W} \wr \mathcal{U}$ in a paper on permutation groups.

Now we introduce a group-theoretical view on 1-Lipschitz measure-preserving transformations on $\mathbb{Z}_2$ (that is, on measure-preserving T-functions). Let $\mathrm{Sym}(2^n)$ be a symmetric group on 2^n symbols; that is, $\mathrm{Sym}(2^n)$ is a group of all permutations on the set of 2^n elements with respect to composition. The elements of the latter set can be identified with elements of the residue ring $\mathbb{Z}/2^n\mathbb{Z}$, so we can say that $\mathrm{Sym}(2^n)$ is a group of all permutations on $\mathbb{Z}/2^n\mathbb{Z}$. All compatible permutations on the residue ring $\mathbb{Z}/2^n\mathbb{Z}$ form a subgroup with respect to composition. This group is a Sylow 2-subgroup of the symmetric group $\mathrm{Sym}(2^n)$, i.e., the maximal (with respect to inclusion) 2-subgroup of the symmetric group $\mathrm{Sym}(2^n)$. It is well known (see e.g. [353]) that

$$\mathrm{Syl}_2(2^n) = \underbrace{\mathrm{Sym}(2) \wr \mathrm{Sym}(2) \wr \cdots \wr \mathrm{Sym}(2)}_{n \text{ factors}}$$

is a wreath product of symmetric groups Sym(2) on two elements; that is, of groups of order 2. In other words, *all reductions modulo* 2^n *of all measure preserving T-functions constitute the Sylow 2-subgroup* $\mathrm{Syl}_2(2^n)$ *of the symmetric group* $\mathrm{Sym}(2^n)$: This immediately follows from Example 10.4. Note that all Sylow 2-subgroups of any finite group are conjugate in this group; the meaning of the above claim is that all reductions modulo 2^n of all measure preserving T-functions lie in one Sylow 2-subgroup.

In the next section, we apply wreath products to construct counter-dependent generators of the longest period. Note that given a transitive T-function f on $\mathbb{Z}/2^n\mathbb{Z}$ (that is, a compatible transformation on the residue ring $\mathbb{Z}/2^n\mathbb{Z}$ that is a permutation consisting of the only cycle of length 2^n), we use wreath products of the family of linear congruential generators on $\mathbb{F}_2$ by the function f to construct new transitive T-function modulo 2^{n+1}, see Example 10.4. The idea of the construction we introduce in the next section is that we take a wreath product of a family of T-functions on $\mathbb{Z}/2^k\mathbb{Z}$ (rather than a family of linear congruential generators on $\mathbb{F}_2$) by a transitive permutation s on an arbitrary set (with arbitrary composite number N of elements, and not necessarily $N = 2^n$) to obtain counter-dependent generators producing sequences of n-bit words of the longest period, of length $N \cdot 2^k$. Using these wreath products, we can combine generators of different nature (e.g., linear feedback registers and generators based on T-functions) into a single counter-dependent generator and to prove that the keystream is uniformly distributed and has the longest possible period. We note that in real-life settings combining generators is a usual way to improve certain cryptographical properties of the keystream; the main problem is to *prove* that these properties are really improved. For constructs introduced further such proofs are given. Actually we find conditions the family of T-functions must satisfy to make the keystream uniformly distributed. The role of p-adic ergodic theory is then to construct involved transformations (the family of state transition functions, the family of output functions, and/or the transitive transformation s) that satisfy these conditions, and thus to provide uniform distribution of the output sequence of the corresponding counter-dependent generator.

10.3 Counter-dependent generators

A counter-dependent generator, which is by Definition 10.6 a wreath product of ordinary generators, can be used to produce a keystream in an obvious way: Choose an arbitrary key $u_0 \in \mathcal{N}$ and put

$$z_0 = F_0(x_0),\ x_1 = f_0(x_0); \dots; z_i = F_i(x_i),\ x_{i+1} = f_i(x_i); \dots. \tag{10.2}$$

That is, at the $(i+1)$th step the automaton $\mathfrak{A}_i$ is applied to the state x_i entering a new state $x_{i+1} = f_i(x_i)$ and outputting a symbol $z_i = F_i(x_i)$. The sequence (z_i) is considered as a keystream: We can treat every z_i as a number and take its base-2 expansion; then the keystream is a concatenation of these base-2 expansions.

In real-life cryptographic applications all sets J, $\mathcal{M}$ and $\mathcal{N}$ are finite; thus, the output sequence (z_i) is necessarily periodic; from the construction it immediately follows that the length of the shortest period of the sequence (z_i) can not exceed the product $\#J \cdot \#\mathcal{M}$. *The main goal of the section is to construct counter-dependent generators that produce uniformly distributed sequences of the longest possible period, i.e., of length* $\#J \cdot \#\mathcal{M}$. Note that $\#J$ is arbitrary as actually the functions f_i and F_i can be stored in memory during encryption or produced on-the-fly, and the algorithm just invokes the ith function at the ith step making calls to memory or produces this function on-the-fly sending data to the respective subroutine. However, as the functions f_i and F_i work with machine words, they are mappings of binary words to binary words. So the case when both $\#\mathcal{M}$ and $\#\mathcal{N}$ are powers of 2 is arguably the most preferable for applications to stream ciphers, and we restrict our considerations with this case only.[4]

The central result of this section is the following theorem, which is our main tool to construct further various counter-dependent generators with the longest period.

Theorem 10.9. *Let $g_0, \dots, g_{m-1}$ be a finite sequence of 1-Lipschitz measure-preserving transformations on $\mathbb{Z}_2$ such that*

(1) *the sequence $((g_{i \bmod m}(0)) \bmod 2)_{i=0}^{\infty}$ is purely periodic, and the length of its shortest period is m;*

(2) $\sum_{i=0}^{m-1} g_i(0) \equiv 1 \pmod 2$;

(3) $\sum_{j=0}^{m-1} \sum_{z=0}^{2^k-1} g_j(z) \equiv 2^k \pmod{2^{k+1}}$, *for all $k = 1, 2, \dots$.*

[4] We note however that wreath products can be used to construct generators of uniformly distributed sequences when $\#\mathcal{M}$ and $\#\mathcal{N}$ are not necessarily powers of 2, see e.g., [280].

Then the recurrence sequence $\mathcal{X}$ *defined by the recursion* $x_{i+1} = g_{i \bmod m}(x_i)$ *is strictly uniformly distributed modulo* 2^n *for all* $n = 1, 2, \ldots$*. Namely, for every* $n = 1, 2, \ldots$ *the sequence* $\mathcal{X} \bmod 2^n = (x_i \bmod 2^n)_{i=0}^{\infty}$ *is purely periodic, the length of its shortest period is* $m2^n$*, and every element from* $\mathbb{Z}/2^n\mathbb{Z}$ *occurs at the period exactly* m *times.*

Note 10.10. As, in view of Theorem 4.39, the 1-Lipschitz transformation $g_i : \mathbb{Z}_2 \to \mathbb{Z}_2$ is measure-preserving if and only if

$$\delta_k(g_i(x)) \equiv \chi_k + \varphi_k^i(\chi_0, \ldots, \chi_{k-1}) \pmod 2,$$

where $\chi_s = \delta_s(x)$, $s = 0, 1, 2, \ldots$, *condition 3 of Theorem 10.9 can be replaced by the equivalent condition*

$$\sum_{j=0}^{m-1} \operatorname{wt} \varphi_k^j \equiv 1 \pmod 2, \qquad k = 1, 2, \ldots,$$

where $\operatorname{wt} \varphi_k^j$ is the weight of the Boolean function φ_k^j (of Boolean variables $\chi_0, \ldots, \chi_{k-1}$). In turn, since the weight of every Boolean function $\varphi(\chi_0, \ldots, \chi_{k-1})$ can be expressed as $\operatorname{wt} \varphi \equiv \operatorname{Coef}_{\chi_0 \cdots \chi_{k-1}} \varphi \pmod 2$, where $\operatorname{Coef}_{\chi_0 \cdots \chi_{k-1}} \varphi$ stands for the coefficient of the monomial $\chi_0 \cdots \chi_{k-1}$ in the ANF of φ, *condition 3 of the theorem can be also replaced by either of the following two equivalent conditions:*

$$\sum_{j=0}^{m-1} \operatorname{Coef}_{\chi_0 \cdots \chi_{k-1}} \varphi_k^j \equiv 1 \pmod 2, \qquad k = 1, 2, \ldots,$$

or

$$\sum_{j=0}^{m-1} \left\lfloor \frac{\deg \varphi_k^j}{k} \right\rfloor = 1 \pmod 2, \qquad k = 1, 2, \ldots.$$

Note 10.11. For $m = 1$ Theorem 10.9 turns into the ergodicity criterion of Theorem 4.39; so Theorem 10.9 could be considered as a generalization of this criterion.

As a matter of fact, Theorem 10.9 is the immediate consequence of Lemma 10.12 that follows, see the note after the statement of the lemma. Actually the statement of the lemma gives some extra information about the structure of the sequence $\mathcal{X}$.

Lemma 10.12. *Let* $g_0, \ldots, g_{m-1}$ *be a finite sequence of 1-Lipschitz transformations of* $\mathbb{Z}_2$*, and let this sequence satisfy the following conditions*:

- $g_j(x) \equiv x + c_j \pmod 2$ *for* $j = 0, 1, \ldots, m-1$*;*
- $\sum_{j=0}^{m-1} c_j \equiv 1 \pmod 2$*;*

- *the sequence* $(c_{i \bmod m} \bmod 2)_{i=0}^{\infty}$ *is purely periodic, and* m *is the length of its shortest period;*
- $\delta_k(g_j(z)) \equiv \zeta_k + \varphi_k^j(\zeta_0, \dots, \zeta_{k-1}) \pmod 2$, $k = 1, 2, \dots$, *where* $\zeta_r = \delta_r(z)$, $r = 0, 1, 2, \dots$;
- *for each* $k = 1, 2, \dots$, *the total number of Boolean functions* $\varphi_k^j(\zeta_0, \dots, \zeta_{k-1})$ *that have odd weight, is odd.*

Then the recurrence sequence $\mathcal{X} = (x_i)_{i=0}^{\infty}$ *which is defined by the recursion* $x_{i+1} = g_{i \bmod m}(x_i)$ *is a strictly uniformly distributed sequence over* $\mathbb{Z}_2$*: Namely, the sequence* $\mathcal{X} \bmod 2^k = (x_i \bmod 2^k)_{i=0}^{\infty}$ *is purely periodic for all* $k = 1, 2, \dots$, *the length of its shortest period is* $m2^k$, *and every element from* $\mathbb{Z}/2^k\mathbb{Z}$ *occurs at the period exactly* m *times. Moreover,*

(1) $m2^{s+1}$ *is the length of some period of the sequence* $\delta_s(\mathcal{X}) = (\delta_s(x_i))_{i=0}^{\infty}$, $s = 0, 1, \dots, k-1$; [5]

(2) $\delta_s(x_{i+2^s m}) \equiv \delta_s(x_i) + 1 \pmod 2$ *for all* $s = 0, 1, \dots, k-1$, $i = 0, 1, 2, \dots$;

(3) *for each* $t = 1, 2, \dots, k$ *and each* $r = 0, 1, 2, \dots$ *the sequence*

$$x_r \bmod 2^t, x_{r+m} \bmod 2^t, x_{r+2m} \bmod 2^t, \dots$$

is a purely periodic sequence, the length of its shortest period is 2^t, *and every element from* $\mathbb{Z}/2^t\mathbb{Z}$ *occurs at the period exactly once.*

Note 10.13. In force of Theorem 4.39, the conditions of the lemma imply that all transformations g_j are measure-preserving: Actually a pair of conditions 1 and 3 of the lemma can be replaced by the single condition that all g_j are measure-preserving.

The structure of the sequence $\mathcal{X}$ from Theorem 10.9 is illustrated by Figure 10.1.

Proof of Lemma 10.12. As every g_j is bijective modulo 2^n in force of Theorem 4.39, the wreath product $\mathrm{id} \wr_{j=0}^{m-1} g_j \bmod 2^k$ of the family (g_j) by the identity transformation id on the residue ring $\mathbb{Z}/m\mathbb{Z}$ is a permutation on the direct product $\mathbb{Z}/m\mathbb{Z} \times \mathbb{Z}/2^k\mathbb{Z}$, see Proposition 10.2. Hence, the recurrence sequence $\mathcal{X} \bmod 2^k$ defined by the recursion $x_{i+1} = g_{i \bmod m}(x_i) \bmod 2^k$ is purely periodic. With this in mind, we proceed with induction on k.

If $k = 1$, we have that

$$x_{i+1} = (c_{i \bmod m} + x_i) \bmod 2.$$

Thus, $x_i \equiv x_0 + \sum_{j=0}^{i-1} c_{j \bmod m} \pmod 2$, and we must calculate the length P of the shortest period of the sequence $b_i = (\sum_{j=0}^{i-1} c_{j \bmod m}) \bmod 2$. For all i we have $0 \equiv \sum_{j=i}^{P+i-1} c_{j \bmod m} \pmod 2$; this means that the sequence $\mathcal{C} = (c_{j \bmod m} \bmod 2)_{j=0}^{\infty}$ is a linear recurrence sequence over the field $\mathbb{F}_2$, and the characteristic polynomial of this sequence is $1 + y + \dots + y^{P-1} \in \mathbb{F}_2[y]$ (see e.g. [126] for definitions). Since the latter polynomial is a factor of the polynomial $y^P - 1$, P is the length of some period

[5] that is, the sequence $\delta_s(\mathcal{X})$ *may have* periods that are shorter than $m2^{s+1}$

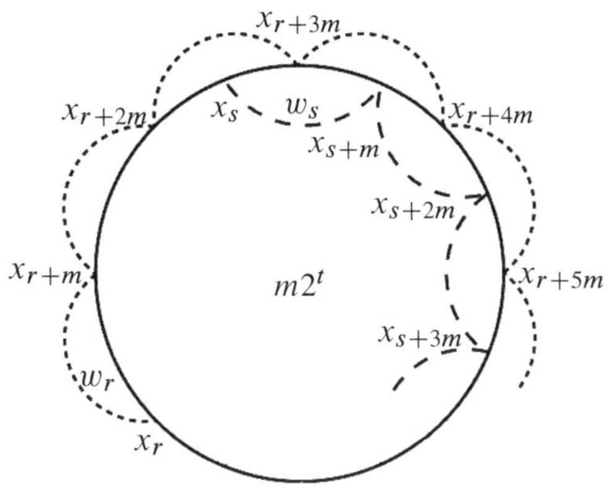

Figure 10.1. The structure of the sequence generated by the wreath product from Theorem 10.9. Every w_r, $r = 0, 1, \ldots, m-1$, is a transitive T-function of Claim 3 of Lemma 10.12: $w_r(x_{r+(\ell-1)m}) = x_{r+\ell m}$, $\ell = 1, 2, \ldots$.

of the sequence $\mathcal{C}$. Then, as m is the length of the shortest period of the sequence $\mathcal{C}$, m must be a factor of P. Yet $x_{i+m} \equiv x_0 + \sum_{j=0}^{m-1} c_{j \bmod m} \equiv x_0 + 1 \pmod 2$, and $x_{i+2m} \equiv x_0 + 2 \cdot \sum_{j=0}^{m-1} c_{j \bmod m} \equiv x_0 \pmod 2$; thus, $P = 2m$. This proves the lemma in the case $k = 1$ since $\delta_0(\mathcal{X}) = \mathcal{X} \bmod 2$ in this case.

Now let the lemma be true for $k = n$; let us prove it for $k = n + 1$. Denote $\delta_n(x_i) = \chi_n^i$, then

$$\chi_n^i \equiv \chi_n^0 + \sum_{j=0}^{i-1} \varphi_n^j(\chi_0^j, \ldots, \chi_{n-1}^j) \pmod 2. \tag{10.3}$$

Since by induction hypothesis the length of the shortest period of the sequence $\mathcal{X} \bmod 2^n$ is $m2^n$, and since all g_j are compatible transformations on $\mathbb{Z}/2^n\mathbb{Z}$, the length of the shortest period of the sequence $\mathcal{X} \bmod 2^{n+1}$ must be a multiple of $2^n m$. Thus, the only alternative can take place, either the length of the shortest period of the sequence $\mathcal{X} \bmod 2^{n+1}$ is $m2^{n+1}$, or this length is $m2^n$. We shall prove that $m2^n$ is not the case. To prove this, we only need to demonstrate that $\chi_n^{m2^n} \not\equiv \chi_n^0 \pmod 2$. In view of induction hypothesis the congruences

$$\begin{aligned} \chi_n^{m2^n+r} &\equiv \chi_n^r + \sum_{j=r}^{m2^n-1+r} \varphi_n^j(\chi_0^j, \ldots, \chi_{n-1}^j) \\ &\equiv \chi_n^r + \sum_{j=0}^{m-1} \sum_{z \in \mathbb{Z}/2^n\mathbb{Z}} \varphi_n^j(\delta_0(z), \ldots, \delta_{n-1}(z)) \equiv \chi_n^r + 1 \pmod 2, \end{aligned} \tag{10.4}$$

hold for all $r = 0, 1, 2, \ldots$, since the total number of Boolean functions $\varphi_n^0, \varphi_n^1, \ldots, \varphi_n^{m-1}$ that have odd weight is odd. This proves Claim 2 of the lemma; also, as from

(10.4) it follows that $\chi_n^{m2^n} \not\equiv \chi_n^0 \pmod 2$, the length of the shortest period of the sequence $\mathcal{X} \bmod 2^{n+1}$ is $m2^{n+1}$ in view of the note we made above.

Moreover, from (10.4) we derive that $\chi_n^{m2^{n+1}+r} \equiv \chi_n^r \pmod 2$, thus proving Claim 1 of the lemma. Finally, by Claim 3 of induction hypothesis the following string of $m2^n$ numbers

$$x_r \bmod 2^n, x_{r+m} \bmod 2^n, x_{r+2m} \bmod 2^n, \dots, x_{r+(2^n-1)m} \bmod 2^n$$

is a permutation of $0, 1, 2, \dots, 2^n - 1$. Hence, all the numbers

$$x_r, x_{r+m}, x_{r+2m}, \dots, x_{r+(2^n-1)m}$$

are pairwise distinct modulo 2^{n+1}. Thus, for each $z \in \{0, 1, \dots, 2^n - 1\}$ among the numbers

$$x_r, x_{r+m}, x_{r+2m}, \dots, x_{r+(2^{n+1}-1)m} \tag{10.5}$$

there exist exactly two numbers (say, x_u and x_v) such that $u \neq v$ and $z \equiv x_u \equiv x_v \pmod{2^n}$. Thus, $u \equiv v \pmod{m2^n}$ in view of Claim 3 of induction hypothesis. Hence necessarily $v = u + m2^n$. But then $x_u \not\equiv x_v \pmod{2^{n+1}}$ since $\delta_n(x_v) \equiv \delta_n(x_v) + 1 \pmod 2$ in view of (10.4). Thus, all 2^{n+1} numbers (10.5) are pairwise distinct modulo 2^{n+1}. This proves Claim 3 of the lemma.

As we have already proved that the sequence $\mathcal{X} \bmod 2^{n+1}$ is purely periodic, and the length of its shortest period is $m2^{n+1}$, the following finite sequence

$$x_0 \bmod 2^{n+1}, x_1 \bmod 2^{n+1}, \dots, x_{2^{n+1}-1} \bmod 2^{n+1}$$

is a period of the sequence $\mathcal{X} \bmod 2^{n+1}$. But according to already proven Claim 3, among these numbers there exist exactly m numbers that are congruent to z modulo 2^{n+1}, for every given $z \in \{0, 1, \dots, 2^{n+1} - 1\}$. This completes the proof of the lemma, and of Theorem 10.9. □

Note 10.14. Although the length P_s of the shortest period of the sequence $\delta_s(\mathcal{X})$ is a factor of $m2^{s+1}$, it is a multiple of 2^{s+1} since otherwise the length of the shortest period of the sequence $\mathcal{X} \bmod 2^{s+1}$ would be at most $m2^s$, and not $m2^{s+1}$ as Lemma 10.12 claims. Thus, $P_s \mid m2^{s+1}$ and $2^{s+1} \mid P_s$.

Note 10.15. As it follows from Claim 2 of Lemma 10.12, *the second part of the period of length $m2^{n+1}$ of the sequence $\delta_n(\mathcal{X})$ is a bitwise negation of the first part*: $\delta_n(x_{i+m2^n}) \equiv \delta_n(x_i) + 1 \pmod 2$ for all $i, n \in \mathbb{N}_0$.

We illustrate Notes 10.14 and 10.15 by an example. Consider, for instance, the sequence $\mathcal{D} = 101010\dots$, which is a purely periodic sequence, and 10 is its period of length 2. At the same time this sequence $\mathcal{D}$ can be considered as a purely periodic sequence with the period 101010, of length 6. Note that in both cases the second half of the period is a bitwise negation of the first half. This situation can never happen in the case $j = 0$: No sequence $\delta_0(\mathcal{X})$ of Lemma 10.12 coincides with this sequence $\mathcal{D}$ since the shortest period of the sequence $\mathcal{X} \bmod 2 = \delta_0(\mathcal{X})$ has the length $2m$ in view of the lemma. However, this situation can happen for senior coordinate sequences. For instance, let $\mathcal{D}_0$ be a purely periodic sequence with the period 111000; let $\mathcal{D}_1$ be a purely periodic sequence with the period 110011001100. The length of the shortest period of the sequence $\mathcal{D}_1$ is 4; however, this sequence at the same time is a sequence with the period 110011001100 of length 12, and the second half of this period is a bitwise negation of the first half. The sequence $\mathcal{D}_0 + 2 \cdot \mathcal{D}_1$ is then a purely periodic sequence with the period 331022113200. It is not difficult to demonstrate that one could construct mappings g_0, g_1, g_2 satisfying Lemma 10.12 such that $\mathcal{X} \bmod 4 = \mathcal{D}_0 + 2 \cdot \mathcal{D}_1$. A characterization of possible coordinate sequences of the sequence $\mathcal{X}$ from Theorem 10.9 is given further by Theorem 11.28.

Finally, to construct counter-dependent generators with non-identity output functions that produce uniformly distributed sequence, we can use the following obvious corollary.

Corollary 10.16. *Let a finite sequence of transformations $(g_0, \dots, g_{m-1})$ on $\mathbb{Z}_2$ satisfy the conditions of Theorem 10.9, and let $(F_0, \dots, F_{m-1})$ be an arbitrary finite sequence of balanced (and not necessarily compatible) mappings of $\mathbb{Z}/2^n\mathbb{Z}$ onto $\mathbb{Z}/2^k\mathbb{Z}$, $1 \le k \le n$. Then the sequence $\mathcal{Z} = (F_{i \bmod m}(x_i))_{i=0}^{\infty}$, where $x_{i+1} = g_{i \bmod m}(x_i) \bmod 2^n$, $i = 0, 1, 2, \dots$, is a strictly uniformly distributed sequence of elements from $\mathbb{Z}/2^k\mathbb{Z}$: It is purely periodic, it has a period of length $m2^n$, and every element from $\mathbb{Z}/2^k\mathbb{Z}$ occurs at the period exactly $m2^{n-k}$ times.*

Now we illustrate the general idea. To construct a counter-dependent generator using Theorem 10.9 together with Corollary 10.16, the following components are needed:

- The sequence $c_0, \dots, c_{m-1}, \dots$ of integers, which we call a *control sequence*.
- The sequence $h_0, \dots, h_{m-1}, \dots$ of 1-Lipschitz transformations on $\mathbb{Z}_2$, which is used to form a sequence of clock state transition functions g_i (see e.g. further Examples 10.17–10.22).
- The sequence $H_0, \dots, H_{m-1}, \dots$ of compatible mappings from $\mathbb{Z}/2^n\mathbb{Z}$ onto $\mathbb{Z}/2^k\mathbb{Z}$, $1 \le k \le n$, to produce clock output functions F_i (as, e.g., in Proposition 10.24 that follows).

Note that ergodic functions that are needed to meet the conditions of Proposition 10.24 or Example 10.20 can be constructed out of given arbitrary 1-Lipschitz transformations by Corollary 4.42 or by Proposition 9.29. A control sequence may be produced by a certain external generator (which in turn could be a counter-dependent generator or an ordinary generator), or this sequence may be just a queue the state update and output functions are called on from some look-up tables. The functions h_i and/or H_i may be either precomputed to fill these look-up tables, or these function may be produced on-the-fly in a form that is determined by the control sequence. This form may be as 'crazy-looking' as desirable; as, for instance, the following one:

$$h_i(x) = (\cdots((u_0(\delta_0(c_i)) \bigcirc_{\delta_1(c_i),\delta_2(c_i)} u_1(\delta_3(c_i))) \bigcirc_{\delta_4(c_i),\delta_5(c_i)} u_2(\delta_6(c_i)))\cdots. \tag{10.6}$$

Here $u_j(0) = x$, the variable, and $u_j(1)$ is a constant (which is determined by c_i, or is read from a precomputed look-up table, etc.), while (say) $\bigcirc_{0,0} = +$ is integer addition, $\bigcirc_{1,0} = \cdot$ is integer multiplication, $\bigcirc_{0,1} = \mathsf{XOR}$, $\bigcirc_{1,1} = \mathsf{AND}$. There is absolutely no matter what these h_i and H_i look like or how they are obtained, *the above stated results give a general method to combine all the data together to produce a uniformly distributed output sequence of the longest period.* Now we consider some examples. Actually we will only construct a state transition circuit of a counter-dependent generator according to general schematics at Figure 10.2.

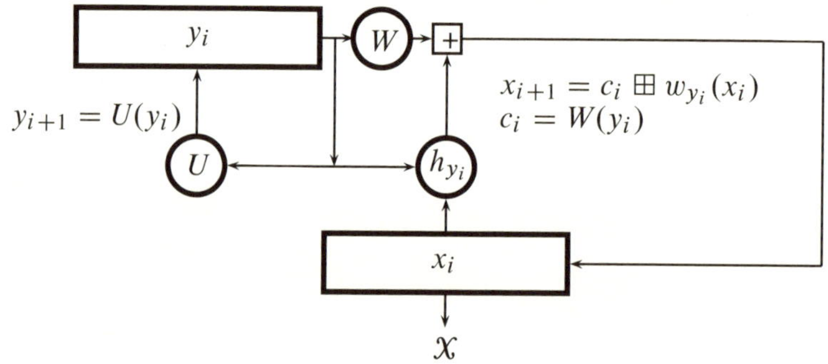

Figure 10.2. Example state transition circuit of the wreath product of automata. Here U and W are respectively the state transition function and the output function of the generator that produces the control sequence (c_i); $\boxplus$ is a binary quasigroup operation, e.g., $+$ or XOR.

Example 10.17. Let the control sequence $c_0, c_1, \dots$ be produced by the ordinary generator $\mathfrak{A} = \langle \mathbb{Z}/2^s\mathbb{Z}, \mathbb{Z}/2^s\mathbb{Z}, f, F\rangle$ of Definition 9.1, where the state transition function f is a reduction modulo 2^s of an ergodic 1-Lipschitz transformation of $\mathbb{Z}_2$, and F is a bijective output function. Then the length of the shortest period of the control sequence is $m = 2^s$, see Proposition 9.2. Now take m arbitrary ergodic 1-Lipschitz transformations $h_0, \dots, h_{m-1}$ on $\mathbb{Z}_2$, choose arbitrary odd $k \in \{0, 1, \dots, m-1\}$, and put $g_0(x) = x \mathsf{ XOR } (x+1) \mathsf{ XOR } h_0(x), \dots, g_{k-1} = x \mathsf{ XOR } (x+1) \mathsf{ XOR } h_{k-1}(x)$, $g_k = h_k, \dots, g_{m-1} = h_{m-1}$.

In other words, in this example the control sequence just defines the queue the functions g_j are called upon, thus producing the state transition sequence

$$\mathcal{X} = x_0, x_1 = g_{c_0}(x_0) \bmod 2^n, x_2 = g_{c_1}(x_1) \bmod 2^n, \dots$$

of the counter-dependent generator. Obviously, in this example the control sequence could be constructed with the use of an arbitrary permutation of $0, 1, \dots, 2^s-1$, and not necessarily as an output of the generator $\mathfrak{A}$. The proof that the sequence of mappings g_i satisfies the conditions of Theorem 10.9 is left to the reader as an exercise. Hint: use Theorem 4.39.

Example 10.18. Let $(c_0, \dots, c_{m-1})$ be an arbitrary sequence of length $m = 2^s$ of integers, i.e., $c_0, \dots, c_{m-1}$ need not be necessarily pairwise distinct. Let $(h_0, \dots, h_{m-1})$ be a finite sequence of 1-Lipschitz transformations on $\mathbb{Z}_2$. For $0 \le j \le m-1$ put $g_j(x) = c_j + x + 4 \cdot h_j(x)$. These mappings g_j satisfy the conditions of Theorem 10.9 if and only if $\sum_{j=0}^{2m-1} c_j \equiv 1 \pmod 2$.

Indeed, denote $\delta_i(x) = \chi_i \in \{0, 1\}$, then it is obvious that $\delta_0(c_i + x) \equiv \chi_0 + \delta_0(c_i) \pmod 2$ and that

$$\delta_j(c_i + x) \equiv \chi_j + \delta_0(c_i) \cdot \chi_0 \cdots \chi_{j-1} + \gamma_{ji}(\chi_0, \dots, \chi_{j-1}) \pmod 2, \qquad j > 0,$$

where $\chi_j = \delta_j(x)$, $\gamma_{ji}(\chi_0, \dots, \chi_{j-1})$ is a Boolean function of degree less than j in Boolean variables $\chi_0, \dots, \chi_{j-1}$. However, $\delta_i(4 \cdot h_j(x))$ is a Boolean function in Boolean variables $\chi_0, \dots, \chi_{j-2}$ for $j \ge 2$, and is 0 otherwise; thus

$$\delta_j(g_i(x)) \equiv \chi_j + \delta_0(c_i) \cdot \chi_0 \cdots \chi_{j-1} + \lambda_{ji}(\chi_0, \dots, \chi_{j-1}) \pmod 2,$$

where $\deg \lambda_{ji} < j$, $j = 1, 2, \dots$, and $\delta_0(g_i(x)) \equiv \chi_0 + \delta_0(c_i) \pmod 2$.

Note 10.19. From these considerations it immediately follows in view of Theorem 4.39 that *every recurrence sequence defined by recursion* $x_{i+1} = f_{i \bmod 2^m}(x_i) \bmod 2^n$, *where* f_i *are 1-Lipschitz transformations on* $\mathbb{Z}_2$ *can obtained by a truncation of* m *low order bits of the recurrence sequence defined by recursion* $z_{i+1} = G(z_i) \bmod 2^{n+m}$ *for a suitable 1-Lipschitz mapping* $G\colon \mathbb{Z}_2 \to \mathbb{Z}_2$. However, in practice it could be more convenient to produce the sequence by the recursion $x_{i+1} = f_{i \bmod 2^m}(x_i) \bmod 2^n$ than by the recursion $z_{i+1} = G(z_i) \bmod 2^{n+m}$ followed by truncation, since the mapping G may be extremely complicated although all f_i are relatively simple. Nevertheless, this note shows that *all results that are established further in the book for truncated congruential generators remain true for counter-dependent generators with recursion* $x_{i+1} = f_{i \bmod 2^m}(x_i) \bmod 2^n$.

Example 10.20. For $m > 1$ odd let $(h_0, \dots, h_{m-1})$ be a finite sequence of 1-Lipschitz ergodic transformations on $\mathbb{Z}_2$; let $(c_0, \dots, c_{m-1})$ be a finite sequence of integers such that

- $\sum_{j=0}^{m-1} c_j \equiv 0 \pmod 2$;
- the sequence $(c_{i \bmod m} \bmod 2)_{i=0}^{\infty}$ is purely periodic, and m is the length of its shortest period.

Put $g_j(x) = c_j \,\mathsf{XOR}\, h_j(x)$ (or, respectively, put $g_j(x) = c_j + h_j(x)$). Then g_j satisfy the conditions of Theorem 10.9.

The claim in the case $g_j(x) = c_j$ XOR $h_j(x)$ is obvious in view of Theorem 4.39 and Lemma 10.12; we note only that the sequence $(c_j + 1)_{j=0}^{\infty}$ satisfies the conditions of Lemma 10.12. So we only need to consider the case $g_j = c_j + h_j(x)$. The proof of the latter goes along the lines similar to those of Lemma 10.12. Namely, for $n = 1$ one has $x_{i+1} = (c_{i \bmod m} + x_i + 1) \bmod 2$, since every ergodic mapping modulo 2 is equivalent to the mapping $x \mapsto x + 1$, see Corollary 4.42; so putting substitution $c_i + 1$ for c_i returns us to the situation of Lemma 10.12 whenever $n = 1$.

Assuming the claim is true for $n = k$, prove it for $n = k + 1$. In view of Theorem 4.39, for $s > 0$ we have that

$$\delta_s(g_j(x)) \equiv \chi_s + (c_j + 1) \cdot \chi_0 \cdots \chi_{s-1} + \psi_s^j(\chi_0, \ldots, \chi_{s-1}) \pmod 2,$$

where $\deg \psi_s^j < s$ (this congruence could be easily proved by induction on s: The coefficient of the monomial $\chi_0 \cdots \chi_{s-1}$ in the ANF of the Boolean function that represents a carry to the sth position is $\delta_0(c_j)$). Thus, for $k \geq 1$ we get:

$$\begin{aligned}
\chi_k^{2^k m} &\equiv \chi_k^0 + \sum_{j=0}^{2^k m - 1} (c_{j \bmod m} + 1) \cdot \chi_0^j \cdots \chi_{k-1}^j + \sum_{j=0}^{2^k m - 1} \psi_k^j(\chi_0^j, \ldots, \chi_{k-1}^j) \\
&\equiv \chi_k^0 + \sum_{j=0}^{m-1} (c_j + 1) \sum_{z \in \mathbb{Z}/2^k\mathbb{Z}} \zeta_0 \cdots \zeta_{k-1} + \sum_{j=0}^{m-1} \sum_{z \in \mathbb{Z}/2^k\mathbb{Z}} \psi_k^j(\zeta_0, \ldots, \zeta_{k-1}) \\
&\equiv \chi_k^0 + 1 \pmod 2,
\end{aligned}$$

since all Boolean functions $\psi_k^j(\zeta_0, \ldots, \zeta_{k-1})$ are of even weight.

In connection with Example 10.20 there arises a natural question: How to construct a sequence of integers that satisfies its conditions? Here is one possible solution:

Proposition 10.21. *Let $m > 1$ be odd, and let u be a transitive transformation on $\mathbb{Z}/m\mathbb{Z}$. Take arbitrary $z \in \mathbb{Z}/m\mathbb{Z}$ and put $c_i = u^i(z) \bmod m$ if $m \equiv 1 \pmod 4$, put $c_i = (u^i(z) + 1) \bmod m$ otherwise ($i = 0, 1, 2, \ldots, m-1$). Then the sequence $\mathcal{C} = (c_{i \bmod m} \bmod 2)_{i=0}^{\infty}$ satisfies the conditions of Example 10.20; that is, $\mathcal{C}$ is a purely periodic sequence, the length of the shortest period of $\mathcal{C}$ is m, and $\sum_{j=0}^{m-1} c_j \equiv 0 \pmod 2$.*

Proof. Obviously, the sequence $\mathcal{C}$ is purely periodic. Let P be the length of the shortest period of $\mathcal{C}$. Whence, P is a factor of m. As $m = 2s + 1$, exactly s numbers of $0, 1, \ldots, m-1$ are odd. Denote r_0 (respectively, r_1) the number of even (respectively, odd) numbers at the shortest period of $\mathcal{C}$; then $\frac{m}{P} r_1 = s$, and $\frac{m}{P} r_0 = s + 1$. Thus, $\frac{m}{P}(r_0 - r_1) = 1$; hence $\frac{m}{P} = 1$, i.e., $m = P$. This completes the proof as $\sum_{i=0}^{m-1} i \equiv 0 \pmod 2$ if and only if $s \equiv 0 \pmod 2$. □

Thus, to construct a sequence (c_j) that satisfies the conditions of Example 10.20 it is sufficient to construct a transitive transformation of the residue ring $\mathbb{Z}/m\mathbb{Z}$. Of course, this can be done in a number of ways, depending on extra conditions the whole generator must meet. For instance, if one is going to use maximum of memory calls instead of computations on-the-fly, he can merely take an arbitrary array of numbers $\{0, 1, \dots, m-1\}$ in arbitrary order. On the contrary, if one needs to produce c_j on-the-fly, he could construct a corresponding generator with a compatible transitive state transition function and a bijective output function that maps $\mathbb{Z}/m\mathbb{Z}$ onto $\mathbb{Z}/m\mathbb{Z}$. This can be done with the use of p-adic ergodic theory.

Note that in the case $m = 2^s - 1$ an alternative way is to use linear feedback shift registers (LFCRs) of the maximum period length; that is, linear recurrence sequences over $\mathbb{F}_2$ of the longest period. We recall that LFCR on s cells produces a recurrence sequence over the field $\mathbb{F}_2 = \{0, 1\}$ according to the recursion $\chi_{i+s} = \sum_{j=0}^{s-1} \alpha_j \cdot \chi_{i+j}$, where $\alpha_0, \dots, \alpha_{s-1} \in \mathbb{F}_2$. The maximum length of the shortest period of this sequence is $2^s - 1$; this is the case if and only if the *characteristic polynomial* $\lambda(x) = x^s + \sum_{j=0}^{s-1} \alpha_j \cdot x^j \in \mathbb{F}_2[x]$ of the sequence is *primitive*: That is, $\lambda(x)$ is irreducible over $\mathbb{F}_2$ and $\lambda(x) \mid x^{2^s-1} - 1$ and $\lambda(x) \nmid x^d - 1$ for all $d \mid 2^s - 1$.

Outputs of LFSRs are actually sequences of non-zero s-dimensional vectors over $\mathbb{F}_2$ obtained by the recursion $\mathbf{c}_{i+1} = \mathbf{c}_i \cdot L$, where L is an $s \times s$ matrix over $\mathbb{F}_2$ with characteristic polynomial λ. Note that often sequences of this kind can be constructed with the use of XOR's and left-right shifts only, see e.g. [311]. Also, a usual way to construct these sequences (to be more exact, their conjugates) with the use of recursion $u_{i+1} = (2 \cdot u_i) \text{ XOR } (\tilde{\lambda} \cdot \delta_{s-1}(u_i))$ over the residue ring $\mathbb{Z}/2^s\mathbb{Z}$, where the base-2 expansion of $\tilde{\lambda} \in \mathbb{Z}/2^s\mathbb{Z}$ agrees with coefficients of the characteristic polynomial λ: $\tilde{\lambda} = \sum_{j=0}^{s-1} \alpha_j \cdot 2^j \in \mathbb{Z}/2^s\mathbb{Z}$. We refer the reader to [126,277,299] for extended theory of linear recurrence sequences over fields and rings.

We note that in cryptography LFCRs are very often used as sources of pseudo random sequences; actually they often produce sequences of states of corresponding PRNGs. So it is important to outline methods to construct counter-dependent generators with the use of LFCRs. Actually LFCR may serve as the generator of the control sequence in the counter-dependent generator: We can take the wreath product of LFCR with a family of T-functions to construct a counter-dependent generator of the longest period:

Example 10.22. The conditions of Example 10.20 are satisfied whenever $m = 2^s - 1$ and $c_0, \dots, c_{m-1} \in \mathbb{Z}/2^s\mathbb{Z}$ is the output sequence of a linear feedback shift register over $\mathbb{F}_2$ on s cells, of the maximum period length: Every s-bit state of the LFCR is read as a base-2 expansion of the corresponding integer.

The schematics of the corresponding counter-dependent generator is represented by Figure 10.3.

Our techniques of wreath products can also be used to reprove known results on counter-dependent generators or to make tweaks to the them to enlarge their periods.

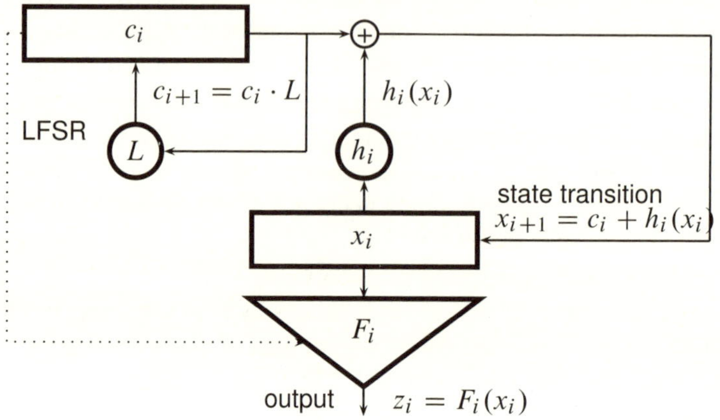

Figure 10.3. The wreath product of LFSR with a family of T-functions; a counter-dependent generator of Examples 10.20 and 10.22.

For instance, specifying mappings g_j in Example 10.20, we can strengthen Theorem 3 of the paper [265] in the following sense:

Example 10.23. Take odd $m > 1$ and consider a finite sequence $C_0, \dots, C_{m-1}$ of integers such that $\delta_0(C_j) = 1$ and $\delta_2(C_j) = 1$, $j = 0, 1, \dots, m-1$. Let the sequence $(c_j)_{j=0}^{m-1}$ satisfy the conditions of Example 10.20. Then the recurrence sequence defined by the recursion $x_{i+1} = (x_i + c_i + (x_i^2 \text{ OR } C_i)) \bmod 2^n$, $i = 0, 1, 2, \dots$, is purely periodic, the length of its shortest period $m2^n$, and each element from $\mathbb{Z}/2^n\mathbb{Z}$ occurs at the period exactly m times.

Actually, the example just represents a tweak that makes the period of the output sequence of the counter-dependent generator longer: Theorem 3 of the paper [265] gives a criterion when the sequence of pairs (y_i, x_i) defined by the recursions $y_{i+1} = (y_i + 1) \bmod m$ and $x_{i+1} = (x_i + (x_i^2 \text{ OR } C_{y_i})) \bmod 2^n$ has a period of length $m2^n$; however, the paper says nothing about periods of the sequence (x_i). The tweak represented by the example above implies that the length of the shortest period of the sequence (x_i) is $m2^n$; this can never be achieved under the conditions of Theorem 3 of [265]: For instance, the latter conditions imply that the length of the shortest period of the sequence $(x_i \pmod 2)$ is only 2, and not $2m$, as in the example above.

In a similar manner from Theorem 10.9 it could be derived that an analogous tweak works in the case m is a power of 2 (in contrast to Theorem 3 of [265], which demands that m must be odd): If $\sum_{j=0}^{2^m-1} c_j \equiv 1 \pmod 2$ and $C_j \equiv 7 \pmod 8$, then the recurrence sequence defined by the recursion $x_{i+1} = c_{i \bmod 2^m} + x_i + (x_i^2 \text{OR} C_{i \bmod 2^m})$ is strictly uniformly distributed modulo 2^n; namely, the length of its shortest period is 2^{n+m}, and each element from $\mathbb{Z}/2^n\mathbb{Z}$ occurs at the period exactly 2^m times. We leave details of the proof to the reader as an exercise, as well as further variations of the theme of wreath products with generators defined by the recursion $x_{i+1} = x_i + (x_i^2 \text{ OR } C_i)$.

10.3.1 Special output functions

All congruential generators that satisfy the conditions of Theorem 10.9 (and of Lemma 10.12) generate output sequence $\mathcal{X}$ which has a drawback: The less is j, the shorter is the period of the jth coordinate sequence $\delta_j(\mathcal{X})$, see Note 10.14. That is, although the length of the shortest period of every output sequence $\mathcal{X} \bmod 2^n$ of n-bit words is $m2^n$, only the senior coordinate sequence $\delta_{n-1}(\mathcal{X})$ may have the shortest period of length $m2^n$: Anyway, the length of the shortest period of the sequence $\delta_{n-1}(\mathcal{X})$ is $\ell 2^n$ for some $1 \le \ell \le m$, and lengths of shortest periods of the rest coordinate sequences $\delta_j(\mathcal{X})$, $j < n-1$, are shorter, $m2^{j+1}$ at most. The goal of this subsection is to demonstrate how this drawback can be cured with the use of output functions in some special way.

Denote $\tau = \tau_n$ a bit order reverse permutation on $\mathbb{Z}/2^n\mathbb{Z}$; that is,

$$\tau\left(\sum_{i=0}^{n-1} \alpha_i 2^i\right) = \sum_{i=0}^{n-1} \alpha_{n-i-1} 2^i, \quad \alpha_0, \dots, \alpha_{n-1} \in \{0, 1\}.$$

Let h_i, $i = 0, 2, \dots, m-1$, be 1-Lipschitz ergodic transformations on $\mathbb{Z}_2$. Then the composition $F_i(x)\colon x \mapsto (h_i(\tau(x))) \bmod 2^n$, $x \in \{0, 1, \dots, 2^n - 1\}$, is a bijective mapping of $\mathbb{Z}/2^n\mathbb{Z}$ onto itself. We argue that if we take F_i as an output function, then the sequence $\mathcal{Z}$ of Corollary 10.16 is free of the drawback mentioned above. To be more exact, the following proposition holds:

Proposition 10.24. *Let h_i, $i = 0, 1, 2, \dots, m-1$, be 1-Lipschitz ergodic transformations on $\mathbb{Z}_2$. Under notation of Corollary 10.16, put $F_i(x) = (h_i(\tau(x))) \bmod 2^n$. Then the length of the shortest period of each jth coordinate sequence $\delta_j(\mathcal{Z})$, $j = 0, 1, 2, \dots, n-1$, is $k_j 2^n$, where $1 \le k_j \le m$.*

In particular, the same holds if $m = 1$, i.e., when $\mathcal{Z}$ is the output sequence of the automaton $\mathfrak{A} = \langle \mathbb{Z}/2^n\mathbb{Z}, \mathbb{Z}/2^n\mathbb{Z}, f \bmod 2^n, F, u_0\rangle$ [6], where f and h are 1-Lipschitz ergodic transformations on $\mathbb{Z}_2$, $F(x) = (h(\tau(x))) \bmod 2^n$, $x \in \{0, 1, \dots, 2^n - 1\}$: The length of the shortest period of the jth coordinate sequence $\delta_j(\mathcal{Z})$ of the output sequence $\mathcal{Z}$ of the automaton $\mathfrak{A}$ is 2^n, for all $j = 0, 1, 2, \dots, n-1$.

Note 10.25. Under the conditions of Proposition 10.24, $\mathcal{Z}$ is a purely periodic sequence, the length of its shortest period is $m2^n$, and every element from $\mathbb{Z}/2^n\mathbb{Z}$ occurs at the period exactly m times (cf. Corollary 10.16 and Proposition 9.2).

To prove the proposition we need the following easy lemma:

[6] cf. Section 9.1 and Figure 9.1

Lemma 10.26. *Let $\mathcal{X} = (x_i)_{i=0}^{\infty}$ and $\mathcal{Y} = (y_i)_{i=0}^{\infty}$ be purely periodic sequences over the field $\mathbb{F}_2 = \mathbb{Z}/2\mathbb{Z}$, let lengths of their shortest periods are 2^u and 2^v respectively, and let $u > v$. Then the sequence $\mathcal{X} \mathsf{XOR} \mathcal{Y} = ((x_i + y_i) \bmod 2)_{i=0}^{\infty}$ is purely periodic, and the length of its shortest period is 2^u.*

If, additionally, $x_{i+2^{u-1}} \equiv x_i + 1 \pmod 2$ for all $i = 0, 1, 2, \ldots$, and if $\mathcal{Y}$ is a non-zero sequence, then the sequence $\mathcal{X}$ AND $\mathcal{Y} = ((x_i \cdot y_i) \bmod 2)_{i=0}^{\infty}$ is purely periodic, and the length of its shortest period is 2^u.

Proof. The first assertion of the lemma is obvious. To prove the second one assume P is the length of shortest period of the sequence $(x_i \cdot y_i)_{i=0}^{\infty}$. Then $P = 2^s$ for suitable $s \le u$. However, if $s < u$, then $x_{i+2^{u-1}} \cdot y_{i+2^{u-1}} \equiv x_i \cdot y_i \pmod 2$ for all $i = 0, 1, 2, \ldots$; thus $(x_i + 1) \cdot y_i \equiv x_i \cdot y_i \pmod 2$ and hence $y_i \equiv 0 \pmod 2$ for all $i = 0, 1, 2, \ldots$ – a contradiction. □

Proof of Proposition 10.24. In view of assertions 2 and 3 of Lemma 10.12, each subsequence $\mathcal{X}(r) = (x_{r+tm})_{t=0}^{\infty}$, $r = 0, 1, \ldots, m-1$, of the sequence $\mathcal{X} = (x_i)_{t=0}^{\infty}$ satisfies the following condition: Each coordinate sequence $\delta_j(\mathcal{X}(r))$ is a purely periodic sequence, the length of its shortest period is 2^{j+1}, and the second half of the period is a bitwise negation of the first half, i.e., $\delta_j(x_{r+(t+2^j)m}) \equiv \delta_j(x_{r+tm}) + 1 \pmod 2$ for all $t = 0, 1, 2, \ldots$. These conditions imply that this sequence is the output sequence of a suitable automaton $\mathfrak{B} = \langle \mathbb{Z}_2, \mathbb{Z}/2^n\mathbb{Z}, f, \mathrm{mod}2^n, x_r \rangle$ (cf. Section 9.1 and Figure 9.1), where the state transition function f is a 1-Lipschitz ergodic transformation on $\mathbb{Z}_2$, and the output function $\mathrm{mod}2^n$ is a reduction modulo 2^n. We omit the proof of this claim as the claim is contained in the statement of Theorem 11.26, which is proved further. However, this claim implies that the first assertion of the proposition follows from the second one, so it is sufficient to consider only the case $m = 1$.

In this case, as $h_1 = h$ is a 1-Lipschitz ergodic transformation on $\mathbb{Z}_2$, from Theorem 4.39 we deduce that

$$\delta_j(h(x)) \equiv \chi_j + \varphi_j(\chi_0, \ldots, \chi_{j-1}) \pmod 2,$$

where $\chi_k = \delta_k(x)$, and φ_j is a Boolean function of odd weight in Boolean variables $\chi_0, \ldots, \chi_{j-1}$ for $j > 0$, $\varphi_0 = 1$. Note that for $j > 0$

$$\begin{aligned} \delta_j(h(x)) &\equiv \chi_j + \chi_0 \cdot \chi_1 \cdots \chi_{j-1} + \psi_j(\chi_0, \ldots, \chi_{j-1}) \\ &\equiv \chi_j + \chi_0 \cdot \alpha_j(\chi_1, \ldots, \chi_{j-1}) + \beta_j(\chi_1, \ldots, \chi_{j-1}) \pmod 2, \end{aligned} \tag{10.7}$$

where $\psi_j, \alpha_j, \beta_j$ are Boolean functions of corresponding Boolean variables, and $\deg \psi_j < j$, so α_j is a non-zero function.

Given infinite binary sequences $\mathcal{U}, \mathcal{V}, \mathcal{W}, \ldots$ (which can be treated as 2-adic integers) and a Boolean function $\gamma(\upsilon, \nu, \omega, \ldots)$ in Boolean variables $\upsilon, \nu, \omega, \ldots$, denote $\gamma(\mathcal{U}, \mathcal{V}, \mathcal{W}, \ldots)$ a binary sequence $\mathcal{S}$ (thus, a 2-adic integer) such that

$$\delta_j(\mathcal{S}) \equiv \gamma(\delta_j(\mathcal{U}), \delta_j(\mathcal{V}), \delta_j(\mathcal{W}), \ldots) \pmod 2,$$

for all $j = 0, 1, 2, \ldots$. Loosely speaking, we just substitute, respectively, XOR and AND for $+$ and $\cdot$ in the ANF of the Boolean function γ and let variables $\upsilon, \nu, \omega, \ldots$ run through the space $\mathbb{Z}_2$ of 2-adic integers. Thus we obtain a well-defined multivariate function γ on $\mathbb{Z}_2$ valuated in $\mathbb{Z}_2$. Since there is a natural one-to-one correspondence between infinite binary sequences and 2-adic integers, the sequence $\gamma(\mathcal{U}, \mathcal{V}, \mathcal{W}, \ldots)$ is well defined. Note also that treating binary sequences as 2-adic integers we can consider base-2 expansions of infinite sequences of n-bit rational integers in the same manner we consider base-2 expansions of numbers; e.g., $\mathcal{U}+2\cdot\mathcal{V}+4\mathcal{W}$ is a sequence $\mathcal{N} = (n_0, n_1, \ldots \in \mathbb{N}_0)$ such that $n_j = \delta_j(\mathcal{U}) + 2\cdot\delta_j(\mathcal{V}) + 4\cdot\delta_j(\mathcal{W})$ for $j = 0, 1, 2, \ldots$. For instance, if $\mathcal{U} = 101\ldots$, $\mathcal{V} = 110\ldots$, and $\mathcal{W} = 010\ldots$, then $\mathcal{N} = 361\ldots$ is a sequence over $\{0, 1, \ldots, 7\} = \mathbb{Z}/8\mathbb{Z}$.

Proceeding with these conventions, denote $\mathcal{C}_j$ (respectively, $\mathcal{Z}_j$) the jth coordinate sequence of the output sequence of the automaton $\mathfrak{B}$ (respectively, of $\mathfrak{A}$). Put $\mathcal{E} = 111\ldots$. Then in view of (10.7) we get:

$$\mathcal{Z}_0 = \mathcal{C}_{n-1} \text{ XOR } \mathcal{E};$$
$$\mathcal{Z}_1 = \mathcal{C}_{n-2} \text{ XOR } \mathcal{C}_{n-1} \text{ XOR } \mathcal{B};$$
$$\mathcal{Z}_j = \mathcal{C}_{n-j-1} \text{ XOR } \mathcal{C}_{n-1} \text{ AND } \alpha_j(\mathcal{C}_{n-2}, \ldots, \mathcal{C}_{n-j}) \text{ XOR } \beta_j(\mathcal{C}_{n-2}, \ldots, \mathcal{C}_{n-j}), \quad j \ge 2,$$

where $\mathcal{B} = \beta_1\beta_1\beta_1\ldots$ is a constant binary sequence. Note that $\mathcal{C}_i$ is a purely periodic binary sequence, the length of its shortest period is 2^{i+1}, and the second half of the period is a bitwise negation of the first half, see Notes 10.14 and 10.15. Now, in view of Lemma 10.26 and conventions we made above, to complete the proof of Proposition 10.24 it suffices to show that the sequence $\alpha_j(\mathcal{C}_{n-2}, \ldots, \mathcal{C}_{n-j})$, $2 \le j \le n-1$, is a non-zero binary sequence.

Consider the sequence $\mathcal{G}_j = 2^{n-2}\cdot\mathcal{C}_{n-2} + \cdots + 2^{n-j}\cdot\mathcal{C}_{n-j}$ over $\mathbb{Z}/2^{j-1}\mathbb{Z}$. The latter sequence is just an output sequence of the generator $\mathfrak{G}_j = \langle \mathbb{Z}/2^{n-1}, \mathbb{Z}/2^{j-1}, f \bmod 2^{n-1}, T_{n-j-1}\rangle$, where T_{n-j-1} is a truncation of the first $n-j$ low order bits: $T_{n-j-1}(z) = \lfloor \frac{z}{2^{n-j}} \rfloor$, cf. (10.1). Thus, $\mathcal{G}_j$ is a purely periodic sequence, the length of its shortest period is 2^{n-1}, and each element from $\mathbb{Z}/2^{j-1}\mathbb{Z}$ occurs at the period the same number of times. However, α_j is a non-zero Boolean function (see above); thus it takes value 1 at least at one $(j-1)$-bit word. Consequently, at least one term of the sequence $\alpha_j(\mathcal{C}_{n-2}, \ldots, \mathcal{C}_{n-j})$ is 1. □

Note 10.27. As it follows from the proof of Proposition 10.24, to provide maximum period length of all coordinate sequences of the output sequence, it is sufficient only to apply the output function in such a way that the most significant bit of a state transition function substitutes for the least significant bit of argument of the output function: That is, the propositions remains true whenever τ is any permutation of bits of n-bit words such that $\delta_0(\tau(z)) = \delta_{n-1}(z)$ for $z \in \mathbb{Z}/2^n\mathbb{Z}$.

Note 10.28. There are other methods that equalize lengths of periods of coordinate sequences. For instance, using ideas of the proof of Proposition 10.24 it is not difficult to demonstrate that *if a recurrence sequence is defined by the recursion* $x_{i+1} = f(x_i)$, *where* $f : \mathbb{Z}_2 \to \mathbb{Z}_2$ *is 1-Lipschitz ergodic mapping, then the binary sequence* $(\delta_k(x_i + 2^j \cdot \delta_s(x_i)))_{i=0}^{\infty}$ *is purely periodic, and the length of its shortest period is* 2^s *whenever* $j \le k < s$. From here it could be deduced that e.g. the sequence

$$\mathcal{Z} = \left(\left(x_i + \tau\left(\left\lfloor \frac{x_i}{2^k} \right\rfloor \bmod 2^k\right)\right) \bmod 2^k\right)_{i=0}^{\infty}$$

is a purely periodic sequence over $\mathbb{Z}/2^k\mathbb{Z}$, the length of its shortest period is 2^{2k}, each element of $\mathbb{Z}/2^k\mathbb{Z}$ occurs at the period exactly 2^k times, and each coordinate sequence of $\mathcal{Z}$ is a purely periodic binary sequence such that the length of its shortest period is 2^{2k}. Note that $\mathcal{Z}$ is obtained according to a very simple rule: At the ith step take $(2k)$-bit output of a congruential generator of the maximum period length with the state transition function f, read the second half of this output as a k-bit number in reverse bit order and add this number modulo 2^k to the k-bit number that agrees with the first half of the output.

10.4 Generators based on multivariate functions

In the preceding section we introduced counter-dependent generators that produce recurrence sequences (z_i) of n-bit words (considered as elements of the residue ring $\mathbb{Z}/2^n\mathbb{Z}$) according to the recursion

$$z_i = F_i(x_i); \quad x_{i+1} \equiv f_i(x_i) \pmod{2^n}, \quad i = 0, 1, 2, \ldots,$$

where both f_i and F_i were univariate mappings. Trivially, each univariate mapping $\mathbb{Z}/2^{mn}\mathbb{Z} \to \mathbb{Z}/2^{mn}\mathbb{Z}$ of the residue ring modulo 2^{mn} can be treated as a mapping $(\mathbb{Z}/2^n\mathbb{Z})^m \to (\mathbb{Z}/2^n\mathbb{Z})^m$ of the Cartesian power $(\mathbb{Z}/2^n\mathbb{Z})^m$ of the residue ring $\mathbb{Z}/2^n\mathbb{Z}$, i.e., as an m-variate mapping. It turns out, however, that in certain practical cases it is more effective to implement a univariate mapping in its multivariate form to achieve better performance. For instance, in the paper [266] there were constructed examples of multivariate T-functions with a single cycle property (i.e., of 1-Lipschitz ergodic functions), whose program implementations are very fast (see Theorem 6 of [266] and the text thereafter).

In this section, we introduce a special method to construct multivariate 1-Lipschitz ergodic functions out of univariate ones; in fact, we merely represent univariate mappings in a multivariate form (actually the mentioned mappings from [266] have the same origin). To our best knowledge, no other methods to construct multivariate ergodic transformations on $\mathbb{Z}_p^m$ are known: We remind that according to Theorem 4.51 there are no uniformly differentiable ergodic transformations when $m > 1$.

Moreover, combining this multivariate representation with wreath products, we describe in this section how to "lift" arbitrary m-variate transitive transformation on $(\mathbb{Z}/2^n\mathbb{Z})^m$ to an m-variate transitive transformation on $(\mathbb{Z}/2^{n+K}\mathbb{Z})^m$, and how to construct counter-dependent generators based on these multivariate mappings.

Denote B a natural bijection of the mth Cartesian power $\mathbb{Z}_2^m$ of the space $\mathbb{Z}_2$ of 2-adic integers onto the space $\mathbb{Z}_2$, which is defined by the following rule:[7] For $\mathbf{x} = (x^{(0)}, \dots, x^{(m-1)}) \in \mathbb{Z}_2^m$ and all $j \in \{0, 1, 2, \dots\}$ put

$$\delta_k(B(\mathbf{x})) \equiv \delta_{(j-(j \bmod m))/m}(x^{(j \bmod m)}) \pmod 2.$$

Loosely speaking, we think of the element $(x^{(0)}, \dots, x^{(m-1)})$ of the Cartesian power $\mathbb{Z}_2^m$ as of a table of m infinite binary rows, and B puts into the correspondence to this table an infinite binary string (that is, an element of $\mathbb{Z}_2$) obtained by reading successively bits of each column, from top to bottom. Now consider a 1-Lipschitz mapping $H\colon \mathbb{Z}_2 \to \mathbb{Z}_2$ and a conjugate mapping

$$H^B(\mathbf{x}) = (h^{(0)}(\mathbf{x}), \dots, h^{(m-1)}(\mathbf{x}))$$

of $\mathbb{Z}_2^m$ into $\mathbb{Z}_2^m$; that is, $H^B(\mathbf{x}) = B^{-1}(H(B(\mathbf{x})))$, so every $h^{(k)}$ maps $\mathbb{Z}_2^m$ into $\mathbb{Z}_2$, $k = 0, 1, \dots, m-1$. Obviously, *the conjugate mapping H^B is 1-Lipschitz and ergodic whenever the mapping H is ergodic.*

For instance, consider the simplest example: Let $H(x) = 1 + x$, then

$$\delta_j(H(x)) \equiv \delta_j(x) + \prod_{s=0}^{j-1} \delta_s(x) \pmod 2, \quad j = 0, 1, 2, \dots$$

(we assume that the product over the empty set is 1); then every coordinate function $h^{(k)}\colon \mathbb{Z}_2^m \to \mathbb{Z}_2$, $k = 0, 1, \dots, m-1$ of the conjugate m variate mapping H^B is

$$\begin{aligned} &h^{(k)}(x^{(0)}, \dots, x^{(m-1)}) \\ &\quad = x^{(k)} \mathsf{XOR} \left(\left(\bigwedge_{s=0}^{k-1} x^{(s)} \right) \mathsf{AND} \left(\bigwedge_{r=0}^{m-1} ((x^{(r)} + 1) \mathsf{XOR}\, x^{(r)}) \right) \right) \\ &\quad = x^{(k)} \mathsf{XOR} \left(\left(\bigwedge_{s=0}^{k-1} x^{(s)} \right) \mathsf{AND} \left(\left(\left(\bigwedge_{r=0}^{m-1} x^{(r)} \right) + 1 \right) \mathsf{XOR} \left(\bigwedge_{r=0}^{m-1} x^{(r)} \right) \right) \right) \end{aligned}$$

for $k = 0, 1, 2, \dots, m-1$. Here $\bigwedge$ stands for AND of several variables, that is for a bitwise conjunction, or, which is the same, for a bitwise multiplication modulo 2. We assume that a bitwise conjunction $\bigwedge$ over the empty set is -1, i.e., the string of all 1s.

[7]Note that in contrast to the rest of the book, in this section we have to use superscripts to enumerate variables rather than subscripts, as subscripts are already reserved to denote the number of iteration of a PRNG.

Now we can construct various multivariate 1-Lipschitz ergodic mappings combining this representation with the ergodicity criterion of Theorem 4.39.

For instance, Theorem 4.39 implies that any univariate 1-Lipschitz ergodic transformation T of the space $\mathbb{Z}_2$ gives rise to the m-variate 1-Lipschitz ergodic transformation $T^B = (t^{(0)}, \dots, t^{(m-1)})$ of the form

$$t^{(k)}(x^{(0)}, \dots, x^{(m-1)}) = x^{(k)}$$
$$\mathsf{XOR}\left(\left(\bigwedge_{s=0}^{k-1} x^{(s)}\right)\mathsf{AND}\left(\bigwedge_{r=0}^{m-1}((x^{(r)}+1)\mathsf{XOR}x^{(r)})\right)\right)\mathsf{XOR}u^{(k)}(x^{(0)}, \dots, x^{(m-1)}),$$

where

$$\sum_{(x^{(0)},\dots,x^{(m-1)})=(0,\dots,0)}^{(2^r-1,\dots,2^r-1)} \delta_r(u^{(k)}(x^{(0)}, \dots, x^{(m-1)})) \equiv 0 \pmod 2 \tag{10.8}$$

for all $r = 0, 1, 2, \dots$. Expanding this approach, we deduce from Theorem 4.39 the following proposition:

Proposition 10.29. *Let $f_s^{(j)} \colon \mathbb{Z}_2 \to \mathbb{Z}_2$ be 1-Lipschitz ergodic transformations, let $g_s^{(j)} \colon \mathbb{Z}_2 \to \mathbb{Z}_2$ be 1-Lipschitz measure-preserving transformations, $s, j = 0, 1, \dots, m-1$. Then the mapping*

$$H^B(\mathbf{x}) = (h^{(0)}(\mathbf{x}), \dots, h^{(m-1)}(\mathbf{x}))$$

of $\mathbb{Z}_2^m$ onto $\mathbb{Z}_2^m$, where $\mathbf{x} = (x^{(0)}, \dots, x^{(m-1)})$ and

$$h^{(0)}(\mathbf{x}) = x^{(0)}\ \mathsf{XOR}\left(\bigwedge_{r=0}^{m-1} f_r^{(0)}(x^{(r)})\ \mathsf{XOR}\ x^{(r)}\right);$$

$$h^{(1)}(\mathbf{x}) = x^{(1)}\ \mathsf{XOR}\left(g_0^{(1)}(x^{(0)})\ \mathsf{AND}\left(\bigwedge_{r=0}^{m-1} f_r^{(1)}(x^{(r)})\ \mathsf{XOR}\ x^{(r)}\right)\right);$$

$$\vdots$$

$$h^{(m-1)}(\mathbf{x}) = x^{(m-1)}\ \mathsf{XOR}\left(\left(\bigwedge_{s=0}^{m-2} g_s^{(m-1)}(x^{(s)})\right)\right.$$
$$\left.\mathsf{AND}\left(\bigwedge_{r=0}^{m-1} f_r^{(m-1)}(x^{(r)})\ \mathsf{XOR}\ x^{(r)}\right)\right)$$

is ergodic. That is, for all $n = 1, 2, \dots$ the mapping $H \bmod 2^n$ is transitive on $(\mathbb{Z}/2^n\mathbb{Z})^m$.

Proof. It suffices to demonstrate that the conjugate mapping $H\colon \mathbb{Z}_2 \to \mathbb{Z}_2$ is 1-Lipschitz and ergodic. Denote $\chi_k^r = \delta_k(x^{(r)})$; we will find ANF of the Boolean function $\delta_t(h^{(s)}(\mathbf{x}))$ in Boolean variables χ_k^r. For $c \in \{0, 1, \dots, m-1\}$ put

$$F^{(c)} = \bigwedge_{r=0}^{m-1} (f_r^{(c)}(x^{(r)}) \ \mathsf{XOR}\ x^r); \quad G^{(c)} = \bigwedge_{s=0}^{c-1} g_s^c(x^s), \quad c > 0; \quad G^{(0)} = -1.$$

Now, as the functions $g_s^{(j)}$ and $f_s^{(j)}$ are 1-Lipschitz and, respectively, measure-preserving or ergodic, in view of Theorem 4.39 we obtain the following representation of Boolean functions $\delta_k(g_s^j)$ and $\delta_k(f_s^j)$ in algebraic normal forms:

$$\begin{aligned}
\delta_k(g_s^{(j)}(x^{(s)})) &= \chi_k^s \oplus \varphi_k^j(\chi_0^s, \dots, \chi_{k-1}^s);\\
\delta_0(f_s^{(j)}(x^{(s)})) &= \chi_0^s \oplus 1;\\
\delta_k(f_s^{(j)}(x^{(s)})) &= \chi_k^s \oplus \chi_0^s \cdots \chi_{k-1}^s \oplus \psi_k^j(\chi_0^s, \dots, \chi_{k-1}^s), \quad k > 0;
\end{aligned}$$

where $\deg \psi_k^j(\chi_0^s, \dots, \chi_{k-1}^s) < k$. Further, since

$$\delta_k(G^{(c)} \ \mathsf{AND}\ F^{(c)}) \equiv \prod_{s=0}^{c-1} \delta_k(g_s^{(c)}(x^{(s)})) \cdot \prod_{s=0}^{m-1} (\delta_k(f_s^{(c)}(x^{(s)}) + \delta_k(x^{(s)})) \pmod 2,$$

the above equations imply that

$$\begin{aligned}
\delta_0(G^{(0)} \ \mathsf{AND}\ F^{(0)}) &= 1;\\
\delta_0(G^{(c)} \ \mathsf{AND}\ F^{(c)}) &= \chi_0^0 \cdots \chi_0^{c-1} \oplus \Phi_0^c, \quad c > 0;\\
\delta_k(G^{(0)} \ \mathsf{AND}\ F^{(0)}) &= \chi_0^0 \cdots \chi_{k-1}^0 \cdots \chi_0^{m-1} \cdots \chi_{k-1}^{m-1} \oplus \Phi_k^0, \quad k > 0;\\
\delta_k(G^{(c)} \ \mathsf{AND}\ F^{(c)}) &= \chi_k^0 \cdots \chi_k^{c-1} \cdot \chi_0^0 \cdots \chi_{k-1}^0 \cdots \chi_0^{m-1} \cdots \chi_{k-1}^{m-1} \oplus \Phi_k^c,\\
&\qquad c > 0,\ k > 0,
\end{aligned}$$

where Φ_k^c (respectively, Φ_k^0 or Φ_0^c) are ANFs of Boolean functions in Boolean variables

$$\chi_k^0, \dots, \chi_k^{c-1}, \chi_0^0, \dots, \chi_{k-1}^0, \dots, \chi_0^{m-1}, \dots, \chi_{k-1}^{m-1}$$

(respectively, in $\chi_0^0, \dots, \chi_{k-1}^0, \dots, \chi_0^{m-1}, \dots, \chi_{k-1}^{m-1}$ or $\chi_0^0, \dots, \chi_0^{c-1}$), and $\deg \Phi_k^c < mk + c$. Finally, $\delta_k(h^{(c)}(x^{(0)}, \dots, x^{(m-1)})) = \chi_k^c \oplus \delta_k(G_k^{(c)} \ \mathsf{AND}\ F_k^{(c)})$, and the result follows in view of Theorem 4.39. □

Note 10.30. Of course, the assertion of the proposition remains true for the mappings $\hat{h}^{(s)} = h^{(s)} \ \mathsf{XOR}\ u^{(s)}$, $s = 0, 1, \dots, m-1$, where $u^{(s)}$ are arbitrary mappings that satisfy conditions (10.8), since these mappings $u^{(s)}$ add summands of degree $< mk + s$ to each Boolean function $\delta_k(h^{(s)}(x^{(0)}, \dots, x^{(m-1)}))$, see the proof of Proposition 10.29.

With this note we can deduce some consequences from Proposition 10.29.

Corollary 10.31 ([266, Theorem 6 and Lemma 1]). *The m-variate mapping H^B defined by*

$$h^{(s)}(x^{(0)}, \dots, x^{(m-1)}) = x^{(s)} \text{ XOR } (\text{AND} x^{(0)} \text{ AND} \cdots \text{ AND } x^{(s-1)})$$
$$\text{AND } ((h(x^{(0)} \text{ AND} \cdots \text{ AND } x^{(m-1)}) \text{ XOR } (x^{(0)} \text{ AND} \cdots \text{ AND } x^{(m-1)})),$$

$s = 0, 1, \dots, m-1$, is 1-Lipschitz and ergodic whenever h is a univariate 1-Lipschitz and ergodic mapping of $\mathbb{Z}_2$ onto $\mathbb{Z}_2$.

Proof. Just note that both functions, $\delta_k(\bigwedge_{t=0}^{m-1}(h(x^{(t)}) \text{ XOR } x^{(t)}))$ and

$$\delta_k\left(\left(h\left(\bigwedge_{t=0}^{m-1} x^{(t)}\right)\right) \text{XOR} \left(\bigwedge_{t=0}^{m-1} x^{(t)}\right)\right),$$

are Boolean functions of whose ANFs have degree $mk + s$. □

Corollary 10.32. *For $m > 1$ under the conditions of Proposition 10.29 the m-variate mapping H^B defined by*

$$h^{(t)}(\mathbf{x}) = x^{(t)} + \left(\left(\bigwedge_{s=0}^{t-1} g_s^{(t)}(x^{(s)})\right) \text{AND} \left(\bigwedge_{r=0}^{m-1}(f_r^t(x^r) \text{ XOR } x^r)\right)\right),$$

$t = 0, 1, \dots, m-1$, is 1-Lipschitz and ergodic.

Proof. Integer addition $+$ adds carry from the $(mk+c)$th bit to $(m(k+1)+c)$th bit of the conjugate mapping $H: \mathbb{Z}_2 \to \mathbb{Z}_2$; the carry is a Boolean function in variables

$$\chi_k^c, \chi_k^0, \dots, \chi_k^{c-1}, \chi_0^0, \dots, \chi_{k-1}^0, \dots, \chi_0^{m-1}, \dots, \chi_{k-1}^{m-1},$$

hence, integer addition just adds a Boolean function in $km + c + 1$ variables to the Boolean function $\delta_{k+1}(h^{(c)}(x^{(0)}, \dots, x^{(m-1)})$ in $(k+1)m + c$ variables. So the ANF of this extra summand is of degree at most $km + c + 1 < (k+1)m + c$, see the proof of Proposition 10.29. □

Note 10.33. The corollary remains true for the mapping $\hat{h}^{(s)} = h^{(s)} + u^{(s)}$, $s = 0, 1, \dots, m-1$, where $u^{(s)}$ are arbitrary mappings that satisfy conditions (10.8).

We recall that according to Theorem 4.44, a 1-Lipschitz univariate function $g: \mathbb{Z}_2 \to \mathbb{Z}_2$ (resp., $f: \mathbb{Z}_2 \to \mathbb{Z}_2$) is measure-preserving (resp., ergodic) if and only if it can be represented as $g(x) = d + x + 2 \cdot v(x)$ (resp., as $f(x) = 1 + x + 2 \cdot (v(x+1) - v(x))$) for suitable $d \in \mathbb{Z}_2$ and 1-Lipschitz mapping $v: \mathbb{Z}_2 \to \mathbb{Z}_2$. In other words, one can assume v to be an arbitrary (e.g., key-dependent) composition of arithmetic operations (such as addition, multiplication, subtraction, etc.) and bitwise logical operations (such as XOR, AND, OR, etc.). Thus, to obtain a cycle of length, say, 2^{256} applying the above results, one could use 8-variate mappings and work with 32-bit words, which are standard for most contemporary computers.

We note, however, that similarly to the univariate case, only senior bits of output sequence achieve maximum period length: To be more exact, if $x_i^{(j)}$ is the value of the jth variable at the ith step, $(x_{i+1}^{(0)}, \dots, x_{i+1}^{(m-1)}) = H^B(x_i^{(0)}, \dots, x_i^{(m-1)})$, then the period length of the sth coordinate sequence $(\delta_s(x_i^{(j)}))_{i=0}^{\infty}$ is 2^{ms+j+1}, for $s \in \{0, 1, \dots\}$, $j \in \{0, 1, \dots, m-1\}$. This drawback can be cured by the use of multivariate output functions in a manner of Proposition 10.24, namely:

Proposition 10.34. *Let H^B and F^B be m-variate ergodic mappings that satisfy the conditions of Proposition 10.29, and let $\tau\colon \mathbb{Z}/n\mathbb{Z} \to \mathbb{Z}/n\mathbb{Z}$ be a permutation of bits of the n-bit word $z \in \mathbb{Z}/2^n\mathbb{Z}$ such that $\delta_0(\tau(z)) = \delta_{n-1}(z)$* (e.g., τ may be a bit order reverse permutation as in Proposition 10.24, or a 1-bit cyclic shift towards senior bits, etc.). *Consider a recurrence sequence $\mathcal{Z} = (\mathbf{z}_i)_{i=0}^{\infty}$ over $(\mathbb{Z}/2^n\mathbb{Z})^m$ defined by recursions*

$$\mathbf{x}_{i+1} = H^B(\mathbf{x}_i) \bmod 2^n; \quad \mathbf{z}_i = F^B(\tau(x_i^{(m-1)}), x_i^{(0)}, \dots, x_i^{(m-2)}) \bmod 2^n,$$

where $\mathbf{x}_j = (x_j^{(0)}, \dots, x_j^{(m-1)}), \mathbf{z}_j = (z_j^{(0)}, \dots, z_j^{(m-1)}) \in (\mathbb{Z}/2^n\mathbb{Z})^m$. Then the output sequence $\mathcal{Z}$ is purely periodic, the length of its shortest period is 2^{nm}, every element from $(\mathbb{Z}/2^n\mathbb{Z})^m$ occurs at the period exactly once, and the length of the shortest period of each coordinate sequence $\delta_k(\mathcal{Z}^{(s)}) = (\delta_k(z_i^{(s)}))_{i=0}^{\infty}$ is 2^{nm}.

Proof. We just apply Proposition 10.24 to (univariate) conjugate mappings H and F; the conclusion follows in view of Note 10.27. □

Note 10.35. As it follows from Note 10.27, Proposition 10.34 remains true if one permutes variables $x^{(0)}, \dots, x^{(m-2)}$ of the function F^B in arbitrary order, or permutes bits in these variables, or applies arbitrary bijections to these variables, etc.

Now we explain how to use wreath products in order to "lift" arbitrary transitive permutation on $(\mathbb{Z}/2^n\mathbb{Z})^m$ to an ergodic transformation on $\mathbb{Z}_2^m$. From Theorem 10.9 we deduce the following proposition:

Proposition 10.36. *Let $T\colon (\mathbb{Z}/2^n\mathbb{Z})^m \to (\mathbb{Z}/2^n\mathbb{Z})^m$ be an arbitrary* (not necessarily compatible) *m-variate transitive mapping; let $H^B\colon (\mathbb{Z}_2)^m \to (\mathbb{Z}_2)^m$ be any m-variate 1-Lipschitz ergodic mapping of mentioned above* (see Proposition 10.29, Note 10.30, Corollary 10.31, Corollary 10.32, Note 10.33). *Then the m-variate mapping $W^B(\mathbf{x}) = T(\mathbf{x} \bmod 2^n) + (H^B(\mathbf{x})$ AND $((-2^n)^m))$ of $\mathbb{Z}_2^m$ onto $\mathbb{Z}_2^m$ is asymptotically 1-Lipschitz and ergodic; that is, W^B is transitive modulo 2^N for all $N \ge n$.*

Recall that a 2-adic representation of -2^n is an infinite binary string such that the first n bits of it are 0, and the rest are 1. In other words, $H^B(\mathbf{x})$ AND $((-2^n)^m)$ sends $\mathbf{x} = (x^{(0)}, \dots, x^{(m-1)})$ to $(h^{(0)}(\mathbf{x})\,\mathsf{AND}\,(-2^n), \dots, h^{(m-1)}(\mathbf{x})\,\mathsf{AND}\,(-2^n))$, thus sending to 0 the first n low order bits; whereas the mapping $\mathbf{x} \bmod 2^n = (x^{(0)} \bmod 2^n, \dots, x^{(m-1)} \bmod 2^n)$ sends to 0 all senior order bits, starting with the nth bit (we start enumerate bits with 0).

Proof of Proposition 10.36. The conjugate mapping W satisfies the conditions of Theorem 10.9 for $M = nm$ since all Boolean functions $\delta_j(h^{(s)}(\mathbf{x}))$ are of odd weight, see the proof of Proposition 10.29. □

Concluding the section we just note that it is clear now how to construct counter-dependent generators with the use of the above multivariate ergodic mappings. Take, for instance, $M > 1$ odd, and take a finite sequence[8]

$$\mathbf{c}_j = (c_j^{(0)}, \dots, c_j^{(m-1)}), \quad j = 0, 1, \dots, M-1$$

of m-dimensional vectors over $\mathbb{Z}/2^n\mathbb{Z}$ such that the sequence of its first coordinates satisfies the conditions of Example 10.20; that is, $\sum_{j=0}^{M-1} c_j^{(0)} \equiv 0 \pmod 2$, and the sequence $(c_{j \bmod M}^{(0)} \bmod 2)_{j=0}^{\infty}$ is purely periodic, and M is the length of its shortest period. Then take arbitrary m-variate ergodic mappings H_j^B and F_j^B, $j = 0, 1, \dots, M-1$ described above and consider recurrence sequences defined by recursions

$$\begin{aligned} \mathbf{x}_{i+1} &= (\mathbf{c}_{i \bmod M}\ \mathsf{XOR}\ H_{i \bmod M}^B(\mathbf{x}_i)) \bmod 2^n; \\ \mathbf{z}_i &= (\mathbf{F}_{i \bmod M}^B(\tau(x_i^{(m-1)}), x_i^{(0)}, \dots, x_i^{(m-2)})) \bmod 2^n, \end{aligned}$$

for $i = 0, 1, 2, \dots$, where τ satisfies the conditions of Proposition 10.34. Then the sequence of internal states $(\mathbf{x}_i)$ is purely periodic, the length of its shortest period is $M \cdot 2^{nm}$, and each m-dimensional vector over $\mathbb{Z}/2^n\mathbb{Z}$ occurs at the period exactly M times. The output sequence $\mathcal{Z} = (\mathbf{z}_i)$ is also purely periodic, the length of its shortest period is $M \cdot 2^{nm}$, and each m-dimensional vector over $\mathbb{Z}/2^n$ occurs at the period exactly M times. Moreover, the period length of each coordinate sequence $\delta_k(\mathcal{Z}^{(s)}) = (\delta_k(z_i^{(s)}))_{i=0}^{\infty}$ is a multiple of 2^{nm}; this length is not less than 2^{nm} and does not exceed $M \cdot 2^{nm}$. More counter-dependent generators (for $M = 2^k$ or arbitrary M) based on other examples of Section 10.3 may be constructed by analogy.

10.5 Security issues

In the preceding sections we developed techniques to construct counter-dependent generators aiming at their application to stream ciphers. These techniques guarantee in

[8] which may be stored in memory, or may be generated on the fly while implementing the corresponding generator

that the so constructed generator, which dynamically modifies itself during encryption, produces an output sequence that meets certain important cryptographic properties; namely, long period, uniform distribution and some other (e.g., high linear complexity, good distribution of overlapping n-tuples, see further Sections 11.2 and 11.3). The techniques can not guarantee per se that every such cipher will be secure – obvious degenerative cases exist. Actually in real world settings a cipher can be considered any secure after a long period of study by a number of cryptanalysts aiming at constructing specific attacks against the concrete cipher. So the goal of this section is only to give a reasoning that with the use of the mentioned techniques secure stream ciphers may be designed: First we will show that there exists an exponentially large number of mappings that can be used to construct the respective generators, and second, we will give some evidence that under some plausible assumptions the ciphers are secure against certain attacks.

10.5.1 The number of transitive compatible mappings

In this subsection, we calculate the total number of all compatible transitive mappings of $\mathbb{Z}/2^n\mathbb{Z}$ onto itself and the number of those of them that are induced by polynomials over $\mathbb{Z}$; that is, the number of transitive mappings that can be expressed as polynomials with rational integer coefficients.[9] The latter mappings form an important class; in further Section 11.1 we will show that mappings induced by polynomials of degree > 1 over $\mathbb{Z}$ exhibit some good statistical properties.

Proposition 10.37. *There are exactly 2^{2^n-n-1} compatible and transitive mappings $T\colon \mathbb{Z}/2^n\mathbb{Z} \to \mathbb{Z}/2^n\mathbb{Z}$. For $n \le 3$ all of them can be represented by polynomials over $\mathbb{Z}$. If $n > 3$, then exactly $2^{\sum_{i=0}^{\rho(n)}(n-i+\mathrm{wt}_2\, i)-6}$ of them can be represented by polynomials over $\mathbb{Z}$; and $\sum_{i=0}^{\rho(n)}(n-i+\mathrm{wt}_2\, i)-6 \sim \frac{1}{2}n^2$ as $n \to \infty$. Here $\mathrm{wt}_2\, i$ is* a binary weight of the non-negative rational integer i, and $\rho(n)$ is the biggest natural number k such that $k - \mathrm{wt}_2\, k < n$.

Proof. The first assertion is an easy consequence of Theorem 4.39: obviously, the number of Boolean functions of odd weight in i variables is exactly 2^{2^i-1}, and the result follows.

To prove the second assertion we first note that each integer-valued polynomial $f(x) \in \mathbb{Q}_p[x]$ over a field $\mathbb{Q}_p$ of p-adic numbers (that is, a polynomial, which takes values in $\mathbb{Z}_p$ at every point of $\mathbb{Z}_p$) admits a unique Mahler expansion

$$f(x) = \sum_{i=0}^{\infty} a_i \binom{x}{i} \tag{10.9}$$

[9] It is worth noticing here that some counting questions about polynomial maps in residue rings are considered in [68, 305].

where $a_0, a_1, a_2, \ldots \in \mathbb{Z}_p$, and only a finite number of $a_0, a_1, a_2, \ldots$ are non-zero, see Section 3.9. Further, the polynomial (10.9) is identically zero modulo 2^n if and only if $a_i \equiv 0 \pmod{2^n}$ for all $i = 0, 1, 2, \ldots$, see Proposition 3.52. Lastly, the polynomial (10.9) is a polynomial over $\mathbb{Z}_2$ if and only if $a_i \equiv 0 \pmod{2^{\operatorname{ord}_2 i!}}$ for all $i = 0, 1, 2, \ldots$.

Thus, each mapping of $\mathbb{Z}/2^n\mathbb{Z}$ onto $\mathbb{Z}/2^n\mathbb{Z}$ that is induced by polynomial over $\mathbb{Z}$ admits a unique representation by the polynomial (10.9) of degree not greater than $\rho(n)$, and with $a_0, a_1, a_2, \ldots \in \mathbb{Z}/2^n$ such that $a_i \equiv 0 \pmod{2^{i-\mathrm{wt}_2 i}}$ for $i = 2, 3, \ldots$ (see Lemma 3.6). By Theorem 4.40, the latter polynomial is transitive modulo 2^n if and only if $a_0 \equiv 1 \pmod 2$, $a_1 \equiv 1 \pmod 4$, and $a_i \equiv 0 \pmod{2^{\lfloor \log_2(i+1) \rfloor + 1}}$ for $i = 2, 3, \ldots$. Since $i - \mathrm{wt}_2 i < \lfloor \log_2(i+1) \rfloor + 1$ if and only if $i = 0, 1, 2, 3$, the total number of transitive permutations on $\mathbb{Z}/2^n\mathbb{Z}$ that are induced by polynomials over $\mathbb{Z}$ is exactly $2^{\eta(n)}$, where $\eta(n) = 4n - 8 + \sum_{i=4}^{\rho(n)} (n - i + \mathrm{wt}_2 i) = -6 + \sum_{i=0}^{\rho(n)} (n - i + \mathrm{wt}_2 i)$ for $n > 3$, and $\eta(1) = 1$, $\eta(2) = 2$, $\eta(3) = 16$.

Now to finish the proof of Proposition 10.37, we only have to demonstrate that $\lim_{n\to\infty} \frac{2\eta(n)}{n^2} = 1$. We start with estimating $\rho(n)$.

Represent n as $n = 2^k + t$ where $0 \le t < 2^k$. Verify that $\rho(2^{k+1} - 1) = 2^{k+1} - 1$ by direct calculations. So, $\rho(n) = n$ if $n = 2^{k+1} - 1$ (i.e., if $t = 2^k - 1$), and $\rho(n) = 2^k + s$ for certain $s \ge 0$, in the opposite case (i.e., if $t < 2^k - 1$). We claim that $s < 2^k$. Indeed, the function $k - \mathrm{wt}_2 k$, and hence the function $\rho(n)$, are nondecreasing; thus, $s \le 2^k$. However, assuming $s = 2^k$ we obtain a contradiction: On the one hand, $2^k + t = n > \rho(n) - \mathrm{wt}_2 \rho(n) = 2^k + 2^k - \mathrm{wt}_2(2^k + 2^k) = 2^{k+1} - 1$, however, $t < 2^k - 1$ on the other hand. Thus for $t < 2^k - 1$ (i.e., for $n \ne 2^{k+1} - 1$) we conclude that $\rho(n) = 2^k + s$ for some $t \le s \le 2^k - 1$ since obviously $\rho(n) \ge n$. Hence $n = 2^k + t > \rho(n) - \mathrm{wt}_2(\rho(n)) = 2^k + s - 1 - \mathrm{wt}_2 s$; consequently $s = \max\{r \in \mathbb{N} : s - \mathrm{wt}_2 s < t + 1\} = \rho(t + 1)$ by the definition of the function ρ. So we have proved the formula

$$\rho(n) = \rho(2^k + t) = \begin{cases} 2^k + t, & \text{if } t = 2^k - 1, \text{ i.e., if } n = 2^{k+1} - 1; \\ 2^k + \rho(t + 1), & \text{if } t < 2^k - 1, \text{ i.e., if } n \ne 2^{k+1} - 1. \end{cases}$$

From here an obvious recursive procedure to calculate $\rho(n)$ follows; the procedure halts not later than in k steps (we remind that $k + 1$ is the number of digits in the base-2 expansion of n). We conclude finally that $n \le \rho(n) \le n + \lfloor \log_2 n \rfloor$ since the number of digits in the base-2 expansion of n is exactly $\lfloor \log_2 n \rfloor + 1$ and $2^r - 1 = \underbrace{11\ldots1}_{r}$.

Now we successively calculate $\eta(n) = \sum_{i=0}^{n} (i + \mathrm{wt}_2 i) + \sum_{j=n+1}^{\rho(n)} (n - j + \mathrm{wt}_2 j) - 6 = \frac{n(n+1)}{2} + \sum_{i=1}^{n} \mathrm{wt}_2 i - \frac{(\rho(n)-n)(\rho(n)-n+1)}{2} + \sum_{j=1}^{\rho(n)-n} \mathrm{wt}_2(n + j) - 6$. Finally, taking into account that

$$\sum_{i=1}^{n} \mathrm{wt}_2 i \le \sum_{i=1}^{2^{\lfloor \log_2 n \rfloor + 1} - 1} \mathrm{wt}_2 i = \sum_{i=1}^{\lfloor \log_2 n \rfloor + 1} i \binom{\lfloor \log_2 n \rfloor + 1}{i}$$

$$= (\lfloor \log_2 n \rfloor + 1)2^{\lfloor \log_2 n \rfloor} \le (1 + \log_2 n)n$$

and also that $\rho(n) - n \le \log_2 n$, $\mathrm{wt}_2(a + b) \le \mathrm{wt}_2\, a + \mathrm{wt}_2\, b$, $\mathrm{wt}_2\, a \le 1 + \log_2 a$, we conclude that $\lim_{n\to\infty} \frac{2\eta(n)}{n^2} = 1$. □

Note 10.38. During the proof of Proposition 10.37 we have demonstrated that each mapping of $\mathbb{Z}/2^n\mathbb{Z}$ onto $\mathbb{Z}/2^n\mathbb{Z}$ induced by a polynomial over $\mathbb{Z}$ can be represented by a polynomial of degree not greater than $\rho(n) \le n + \log_2 n$, and this estimate is sharp. Moreover, from the final part of the proof it could be deduced that the number of transitive transformations on $\mathbb{Z}/2^n\mathbb{Z}$ that are induced by polynomials over $\mathbb{Z}$ is

$$O(2^{\frac{1}{2}n(n+1)+n(1+\log_2 n)+\frac{1}{2}(1+\log_2 n)\log_2 n+(1+\log_2\log_2 n)\log_2 n}).$$

The case $n = 2^k$ is of special interest since usually the word length of contemporary processors is a power of 2. In this case $\rho(n) = n + 1$, and for $k \ge 2$ direct calculations of $\eta(n)$ (see the proof of Proposition 10.37) imply that the number of transitive modulo 2^n mappings of $\mathbb{Z}/2^n$ onto itself that are induced by polynomials over $\mathbb{Z}$ is exactly $2^{2^{2k-1}+(k+1)2^{k-1}-4}$. For instance, in the case $n = 32$ this makes 2^{604} transitive mappings; all of them are induced by polynomials over $\mathbb{Z}$ of degree ≤ 33, i.e, can be expressed via arithmetic operations. However, for $n = 8$ this makes only 2^{44} polynomials of degree not exceeding 9. By the use of bitwise logical operations along with arithmetic operations one could significantly increase the number of transitive mappings, up to 2^{2^n-n-1}. Each of these mappings can be expressed as a polynomial over $\mathbb{Q}$, yet the bound for its degree d raises significantly either. Namely, from the proof of Proposition 10.37 it follows that $\lfloor \log_2(d + 1)\rfloor + 1 < n$ for $n > 2$, i.e., $d \le 2^{n-1} - 2$, and this bound is sharp. For $n = 8$, e.g., this makes 2^{247} transitive polynomials over $\mathbb{Q}$ of degree ≤ 126. Note that for each $1 \le d \le \rho(n)$ (resp., for each $1 < d \le 2^{n-1} - 2$) there exist an ergodic polynomial over $\mathbb{Z}$ (resp., a compatible and ergodic polynomial over $\mathbb{Q}$) of degree exactly d. The number of pairwise distinct modulo 2^n mappings induced by these polynomials may also be calculated using the ideas of the proof of Proposition 10.37. We leave these proofs and calculations to the reader.

10.5.2 Key recovery and intractability

In this subsection we are going to give some evidence that with the use of the techniques described above it might be possible to design stream ciphers such that the problem of their key recovery is intractable up to the following conjecture: Choose at random $k \le n$ Boolean functions ψ_i in n Boolean variables $\chi_0, \dots, \chi_{n-1}$ from the class of algebraic normal forms with polynomially restricted number of monomials. Define the mapping $U : \mathbb{Z}/2^n\mathbb{Z} \to \mathbb{Z}/2^k\mathbb{Z}$ by the formula

$$\begin{aligned} U(x) &= U(\chi_0, \dots, \chi_{n-1}) \\ &= \psi_0(\chi_0, \dots, \chi_{n-1}) + \psi_1(\chi_0, \dots, \chi_{n-1}) \cdot 2 + \dots + \psi_{k-1}(\chi_0, \dots, \chi_{n-1}) \cdot 2^{k-1}, \end{aligned} \tag{10.10}$$

where $\chi_j = \delta_j(x)$ for $x \in \mathbb{Z}/2^n\mathbb{Z}$. We conjecture that *this function U is one-way*, that is, one could invert it (i.e., could find an U-preimage whenever it exists) only with a negligible in n probability. Note that to find any U-preimage, i.e., to solve the equation $U(x) = y$ in unknown x one must solve a system of k Boolean equations in n variables. Recall that *to determine whether k ANFs have a common zero is an NP-complete problem*, see e.g. [147, Appendix A, Section A7.2, Problem ANT-9].

Of course, it is not sufficient to conjecture that U is one-way if we only know that the problem of whether the U-preimage exists is NP-complete; it must be hard in average to invert U. However, to our best knowledge, no polynomial-time algorithms that solve random systems of k Boolean equations in n variables for so restricted k are known. The best known results are polynomial-time algorithms that solve so-called overdefined Boolean systems of degree not more than 2, i.e., systems where the number of equations is greater than the number of unknowns and where each ANF is at most quadratic, see [44, 92].

Proceeding with the above plausible conjecture, to each Boolean function ψ_i, $i = 0, 1, 2, \ldots, k-1$ we relate a mapping $\Psi_i \colon \mathbb{Z}_2 \to \mathbb{Z}_2$ in the following way: $\Psi_i(x) = \psi_i(\delta_0(x), \ldots, \delta_{n-1}(x)) \in \{0, 1\} \subset \mathbb{Z}_2$. Now to every mapping U from (10.10) we relate a transformation on $\mathbb{Z}_2$ according to the following formula:

$$\begin{aligned} g_U(x) &= (1 + x) \operatorname{XOR} 2^{n+1} \cdot U(x) \\ &= (1 + x)\ 2^{n+1} \cdot \Psi_0(x) \operatorname{XOR} 2^{n+2} \cdot \Psi_1(x) \operatorname{XOR} \cdots \operatorname{XOR} 2^{n+k} \cdot \Psi_{k-1}(x). \end{aligned}$$

Clearly,

$$\begin{aligned} \delta_j(g_U(x)) &= \delta_j(g_U(\chi_0 + \chi_1 \cdot 2 + \chi_2 \cdot 2^2 + \cdots)) \\ &= \begin{cases} 1 \oplus \chi_0, & \text{if } j = 0; \\ \chi_j \oplus \chi_0 \cdots \chi_{j-1}, & \text{if } 0 < j \le n; \\ \chi_j \oplus \chi_0 \cdots \chi_{j-1} \oplus \psi_{j-n-1}(\chi_0, \ldots, \chi_{n-1}), & \text{if } n + 1 \le j \le n + k. \end{cases} \end{aligned}$$

By Theorem 4.39, the mapping $g_U \colon \mathbb{Z}_2 \to \mathbb{Z}_2$ is 1-Lipschitz and ergodic for every choice of Boolean functions $\psi_0, \ldots, \psi_{k-1}$.

Now for $m = 2^n$ and $i = 0, 1, 2, \ldots, m - 1$, we randomly choose mappings $U_i \colon \mathbb{Z}/2^n\mathbb{Z} \to \mathbb{Z}/2^k\mathbb{Z}$ of the above type. Put $d_0 = \cdots = d_{2^n-3} = 0$, $d_{2^n-2} = d_{2^n-1} = 1$ and consider a counter-dependent generator with the sequence of states defined by the recursion $x_{i+1} = d_{i \bmod m} \operatorname{XOR} g_{U_{i \bmod m}}(x_i)$ that generates the output sequence $F(x_0), F(x_1), \ldots$ over $\mathbb{Z}/2^k\mathbb{Z}$, where $F(x) = \lfloor \frac{x}{2^{n+1}} \rfloor \bmod 2^k$, a truncation. By Theorem 10.9, the output sequence satisfies Corollary 10.16.

We shall always take a key $x \in \{0, 1, \ldots, 2^n - 1\}$ as the initial state x_0. Let x be the only information that is not known to an attacker, let everything else, i.e., n, k, g_{U_i}, d_i, and F, as well as the first s terms of the output sequence (z_i), be known to him. As $\delta_0(x) \cdots \delta_{j-1}(x) = 1$ if and only if $x \equiv -1 \pmod{2^j}$, with probability $1 - \epsilon$ (where

ϵ is negligible if s is a polynomial in n) the attacker obtains a sequence[10]

$$z_0 = U_0(z),\ z_0 \text{XOR} z_1 = U_1(z+1),\ \ldots,\ z_{s-2}\text{XOR} z_{s-1} = U_{s-1}(z+s-1). \quad (10.11)$$

To find x, the attacker may try to solve any of these equations; however, he will find a solution with a negligible advantage since U_i is one-way. Of course, the attacker may try to express $x + i$ as a collection of ANFs of Boolean functions $\delta_0(x + i), \ldots, \delta_{n-1}(x + i)$ in variables $\chi_0 = \delta_0(x), \ldots, \chi_{n-1} = \delta_{n-1}(x)$, then substitute these ANFs for the variables into ANFs that define mappings U_i to obtain an overdefined system (10.11) in unknowns $\chi_0, \ldots, \chi_{n-1}$. However, the known formula (see e.g. [12] and fix an obvious misprint there)

$$\delta_j(x+i) \equiv \chi_j + \delta_j(i) + \sum_{r=0}^{j-1} \delta_r(i) \cdot \chi_r \prod_{t=r+1}^{j-1} (\delta_t(i) + \chi_t) \pmod 2 \qquad (10.12)$$

implies that the number of monomials in the equations of the obtained system will be, generally speaking, exponential in n; to say nothing of that the number of operations to make these substitutions and to eliminate equal terms is also exponential in n unless the degree of all ANFs that define all U_i is bounded by a constant. However, the latter is not the case according to our assumptions.

Finally, our assumption that the attacker knows all U_i seems to be too strong: It is more practical to assume that he does not know U_i in (10.11): Indeed, given clock output functions (and/or clock state transition functions) as explicit compositions of arithmetical and bitwise logical operators, 'normally' it is infeasible to represent these functions in the Boolean form (4.25): Corresponding ANFs 'as a rule' are sums of exponential in n number of monomials, cf. (10.12). Moreover, if these clock output functions F_i and/or clock state transition functions f_i are determined by a key-dependent control sequence (say, which is produced by a generator with unknown initial state), see Section 10.3, then the explicit forms of the mentioned compositions are also unknown. So in general the attacker has to find the initial state x_0 having only a segment $z_j, z_{j+1}, \ldots$ of the output sequence formed according to the rule (10.2), where both f_i and F_i are not known to him. An 'algebraic' way to do this by guessing f_i and F_i and solving corresponding systems of equations seems to be hopeless in view of the first assertion of Proposition 10.37 and the above discussion. The results of further Sections 11.2 and 11.3[11] give us reasons to conjecture that under common tests the sequence $z_j, z_{j+1}, \ldots$ behaves like a random one, so 'statistical' methods of breaking such (reasonably designed) ciphers seem to be ineffective as well.

[10] which is pseudorandom even if $U = U_0 = U_1 = \cdots$, under additional conjecture (how plausible is it?) that the function U constructed above is a pseudorandom function

[11] as well as computer experiments: output sequences of concrete generators of the type we considered here passed both DIEHARD and NIST test suites

Chapter 11

Structure of trajectories

In this chapter we study common probabilistic, cryptographic and other properties of output sequences of the generators considered in preceding sections: Linear complexity, ℓ-error linear complexity, 2-adic complexity of these sequences, their structure, distribution of k-tuples in them, etc.

11.1 Distribution in Euclidean space

In this section, we study dynamics $f: \mathbb{Z}_p \to \mathbb{Z}_p$ through its 'plots' in the Euclidean unit hypercube.

There is a well-known map m from $\mathbb{Z}_p$ onto a unit interval $[0,1] \subset \mathbb{R}$ of real numbers, which is sometimes called the *Monna map*: Given $z \in \mathbb{Z}_p$, consider a canonical p-adic expansion $z = \sum_{i=0}^{\infty} \delta_i(z) \cdot p^i$, where $\delta_i(z) \in \{0, 1, \dots, p-1\}$; then $m(z) = \sum_{i=0}^{\infty} \delta_i(z) \cdot p^{-i-1} \in [0,1]$. So, given a map $f: \mathbb{Z}_p \to \mathbb{Z}_p$, we can consider a set of all pairs $(m(z), m(f(z)))$, $z \in \mathbb{Z}_p$, which is a subset in a unit square $[0,1] \times [0,1]$, a kind of a 'graph' of the function f, see Figures 11.1, 11.2, 11.3, and 11.4. Of course, all these figures were actually obtained as sets of points $(m(z), m(f(z) \bmod p^n)$, $z \in \mathbb{Z}/p^n\mathbb{Z}$, for some n ($p = 2$ and $n = 17$, to be more exact). However, it is clear that these pictures do not depend 'visually' on n since the bigger n, the least is dependence of the position of the point $(m(z \bmod p^n), m(f(z) \bmod p^n)$ in a unit square on the nth digit in a base-p representation of the fraction $m(f(z) \bmod p^n)$ since $(m(z \bmod p^n), m(f(z) \bmod p^n) \to (m(z), m(f(z))$ as $n \to \infty$.

However, given a 1-Lipschitz transformation f on $\mathbb{Z}_p$, we can study maps of another sort: For every $n \in \mathbb{N}$ consider all points $\left(\frac{x}{p^n}, \frac{f(x) \bmod p^n}{p^n}\right)$, $x \in \{0, 1, \dots, p^n - 1\}$, as $n \to \infty$. Corresponding 'graphs' are much more informative compared to the graph obtained for the Monna map, since in the latter case more significant bits in base-p representation of $f(z)$ play the leading role: For instance, as Figures 11.1, 11.2, 11.3, and 11.4 look somewhat alike, graphs of the second type for corresponding functions are quite different visually, cf. Figures 11.10, 11.7, 11.8, and 11.5, respectively: We can observe various geometrical structures there, such as straight lines, parabolas, stripes, etc. Moreover, some of these graphs exhibit strong dependence on n, see e.g. Figures 11.9–11.12. In this section, we derive some important information about the transformation f from its graph of the second kind. This information, as

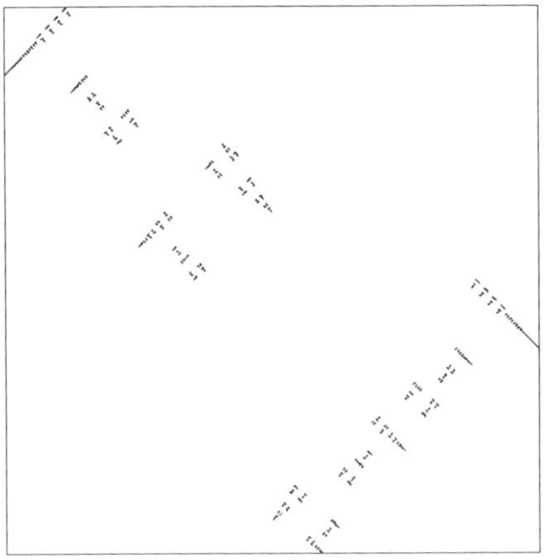

Figure 11.1. The function
$f(x) = x + x^2$ OR C,
$C = -131065$.

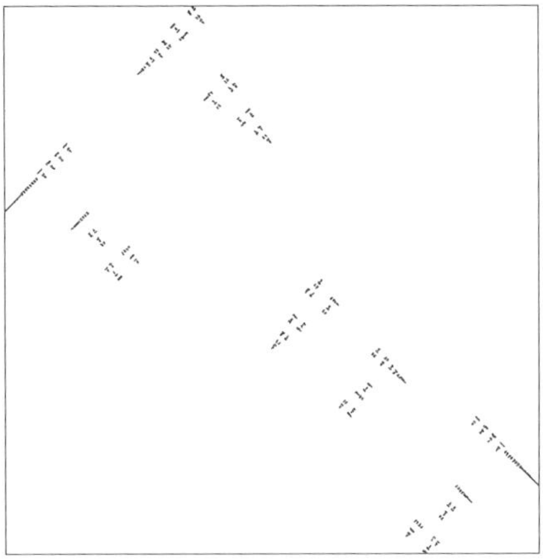

Figure 11.2. Same function,
$C = 101_2$.

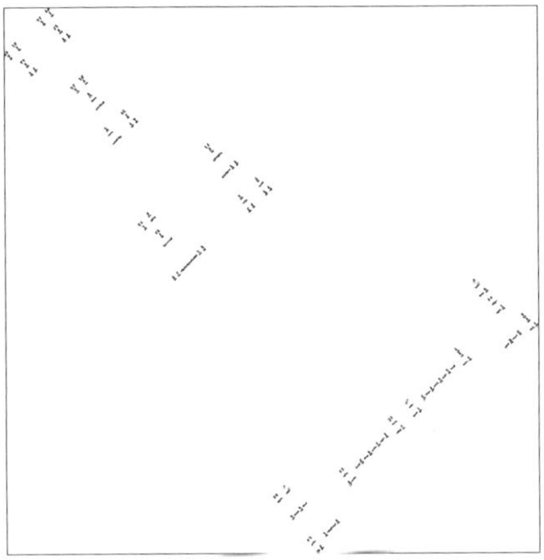

Figure 11.3. Same function,
$C = 11101010100001001_2$.

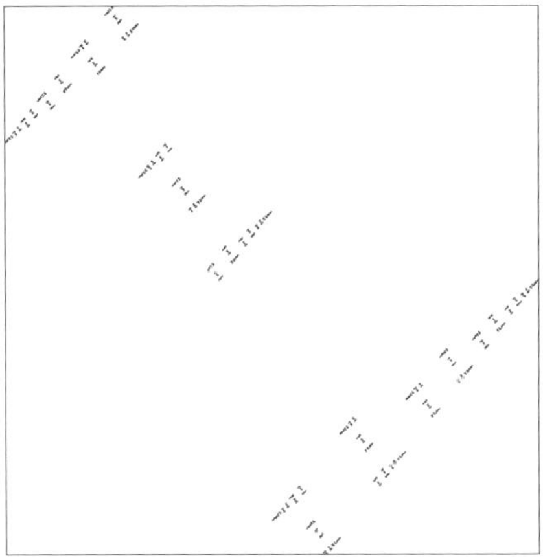

Figure 11.4. The function
$f(x) = 3 + 5x$.

we will see, is sometimes crucial whenever one is going to use f as a state transition function of pseudorandom generators, since the mentioned graph reflects a statistical quality of the produced sequence. Also, this graph says a lot about the behavior of the corresponding automaton that evaluates f.

11.1.1 Points falling on hyperplanes

In this subsection we study, loosely speaking, what do straight lines in the graphs mentioned above imply. In more precise terms, we study *linear complexity* of the sequence of iterations $x, f(x), f^2(x), \dots$. Here is a definition:

Definition 11.1. Let $\mathcal{Z} = (z_i)_{i=0}^{\infty}$ be a sequence over a commutative ring R. The *linear complexity* $\lambda_R(\mathcal{Z})$ of the sequence $\mathcal{Z}$ over R is the smallest $r \in \mathbb{N}_0$ such that there exist $c, c_0, c_1, \dots, c_{r-1} \in R$ (not all equal to 0) such that for all $i = 0, 1, 2, \dots$ holds

$$c + \sum_{j=0}^{r-1} c_j \cdot z_{i+j} = 0. \tag{11.1}$$

We say that $\lambda_R(\mathcal{Z}) = \infty$ if no such r exists.

We should notice that in this section we use the notion of linear complexity of a sequence over a ring in a somewhat broader sense than it is commonly used, see e.g. [126]. More often the linear complexity of the sequence (x_n) of elements of a commutative ring R is understood as the smallest $r > 0$ such that there exist $c_0, \dots, c_{r-1} \in R$ that satisfy simultaneously all equations $x_{n+r} = \sum_{j=0}^{r-1} c_j x_{n+j}$ for $n = 0, 1, 2, \dots$. We, in distinction to the latter, consider non-homogeneous relations (i.e., with a nonzero constant term), as well as relations where all coefficients may be zero divisors (however, not all 0 simultaneously; in the assertion of Theorem 11.5 that follows, the latter, however, is not important). If R is a field, then both notions basically do not differ one from another: If a sequence satisfies the relation $c + \sum_{i=0}^{r} c_i x_{n+i} = 0$ where $c_r \neq 0$, then it satisfies the relation $x_{n+r+1} = c_r^{-1} c_0 x_n - \sum_{j=0}^{r-1} c_r^{-1}(c_j - c_{j+1}) x_{n+j+1}$. Our definition is some more convenient for geometric interpretations.

For instance, if $R = \mathbb{Z}/p^k\mathbb{Z}$; then geometrically equation (11.1) means that all points $(\frac{z_i}{p^k}, \frac{z_{i+1}}{p^k}, \dots, \frac{z_{i+r-1}}{p^k})$, $i = 0, 1, 2, \dots$, of the unit r-dimensional Euclidean hypercube fall into parallel hyperplanes. Given a 1-Lipschitz ergodic transformation f on $\mathbb{Z}_p$, with the use of linear complexity over the residue ring $\mathbb{Z}/p^k\mathbb{Z}$ we can study distribution of r-tuples of the sequence $(f^i(x))_{i=0}^{\infty}$ modulo p^k. From Theorem 4.23, we know that *independently on what concrete transformation f is taken*, this sequence is strictly uniformly distributed as the sequence of elements from $\mathbb{Z}/p^k\mathbb{Z}$: The length of the shortest period is p^k, and every element from $\mathbb{Z}/p^k\mathbb{Z}$ occurs at the period exactly once. However, distribution of consecutive pairs of elements in this sequence (triples, etc.) varies depending on f. For example, although every linear congruential generator based on ergodic transformation $f(x) = a + bx$ of $\mathbb{Z}_p$ produces a strictly uniformly distributed sequence over $\mathbb{Z}/p^n\mathbb{Z}$ for all n, the linear complexity over $\mathbb{Z}/p^k\mathbb{Z}$ of this generator is only 2, as it immediately follows from (11.1). Hence, distribution of pairs in produced sequences is rather poor: All the points that correspond to pairs of consecutive numbers fall into a small number of parallel straight lines in a unit square, and this picture *does not depend on k*, as in Figure 11.5.

Yet another example: The already mentioned transformation $f(x) = x + x^2 \text{ OR } C$ on $\mathbb{Z}_2$ from the paper [264] is ergodic if and only if $C \equiv 5 \pmod 8$, or $C \equiv 7 \pmod 8$, see Example 9.32. However, distribution of pairs of the sequence produced

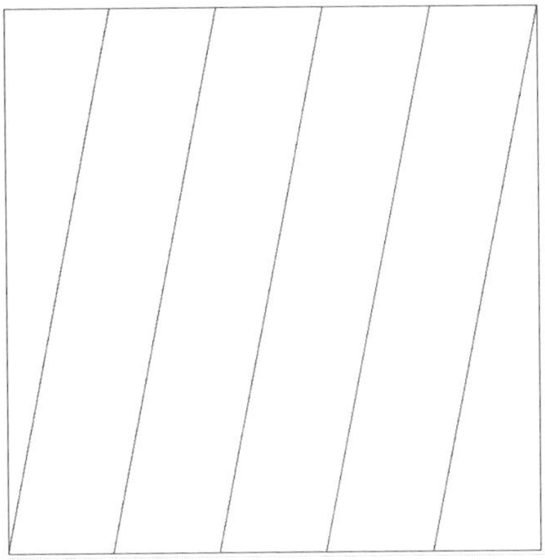

Figure 11.5. Linear congruential generator $x_{i+1} = 3 + 5x_i$, $p = 2$.

Figure 11.6. Polynomial generator of degree 8.

Figure 11.7. The generator $x_{i+1} = x_i + x_i^2 \text{ OR } C$, $C = 101$.

Figure 11.8. Same generator, $C = 11101010100001001$.

by this transformation varies from satisfactory (when there are few 1s in more significant bit positions of C, as in Figure 11.7) to poor (when there are more 1s in these positions, as in Figure 11.8).

Moreover, in some cases (e.g., when C is a negative rational integer) the distribution degenerates from satisfactory to bad whereas k unboundedly increases, see Figures 11.9–11.12; note that the limit plot (as $k \to \infty$) in this case will be the same as for the linear transformation $f(x) = x - 1$.[1]

[1]Vulnerabilities like the mentioned ones were used in [320, 321] to construct attacks against this generator.

Figure 11.9. The function $f(x) = x+((x^2)\mathrm{OR}(-131065))$, $k = 16$.

Figure 11.10. Same function, $k = 17$.

Figure 11.11. Same function, $k = 18$.

Figure 11.12. Same function, $k = 22$.

It is not easy to find an ergodic 1-Lipschitz transformation that guarantees good distribution of pairs modulo p^k. For instance, this problem is not completely solved even for quadratic generators although intensive studies were undertaken, see e.g. [118, 122] and the expository paper [120]. However, it is clear that transformations that exhibit low linear complexities over $\mathbb{Z}/p^k\mathbb{Z}$ result in low quality generators. Actually, we must judge a PRNG as bad whenever the linear complexity tends to a constant as k goes to infinity since this means that the produced pseudorandom numbers fall into relatively small numbers of hyperplanes.

The main goal of this subsection is to prove that polynomial generators of degree greater than 2 are not too bad from this view[2]: Corresponding linear complexities tend to infinity as $k \to \infty$. In other words, these generators result in sequences of p-adic numbers that have infinite linear complexities over $\mathbb{Z}_p$ (and over $\mathbb{Q}_p$). Namely, the following theorem is true (Anashin [24]):

Theorem 11.2. *Let $f(x) \in \mathbb{Q}_p[x]$ be an integer-valued 1-Lipschitz ergodic polynomial*[3] *of degree ≥ 2, and let $x_0 \in \mathbb{Z}_p$. Then the linear complexity $\lambda_{\mathbb{Z}/p^k\mathbb{Z}}(\mathcal{X}_k)$ of the sequence $\mathcal{X}_k = (f^i(x_0) \bmod p^k)_{i=0}^{\infty}$ over $\mathbb{Z}/p^k\mathbb{Z}$ tends to infinity as $k \to \infty$:*

$$\lim_{k\to\infty} \lambda_{\mathbb{Z}/p^k\mathbb{Z}}(\mathcal{X}_k) = \infty.$$

We split the proof of this theorem into several assertions that are of their own interest themselves.

Proposition 11.3. *Let $f \in \mathbb{Q}_p[x]$ be an integer-valued 1-Lipschitz ergodic polynomial of degree d over a field $\mathbb{Q}_p$ of p-adic numbers; let r be a positive rational integer such that for each $k = 0, 1, 2, \ldots$ there exist $c, c_0, \ldots, c_r \in \mathbb{Z}_p$ (not all congruent to 0 modulo p) that satisfy the following congruences:*

$$c + \sum_{i=0}^{r} c_i x_{n+i} \equiv 0 \pmod{p^k}, \qquad n = 0, 1, 2, \ldots, \tag{11.2}$$

where $x_j = f^j(x_0)$, $x_0 \in \mathbb{Z}_p$, $j = 0, 1, 2, \ldots$. Then $d = 1$.

To prove the proposition, we need the following lemma:

Lemma 11.4. *Under the assumptions of Proposition 11.3, let $c, c_0, \ldots, c_r \in \mathbb{Z}_p$ do not depend on k; that is, let there exist $c, c_0, \ldots, c_r \in \mathbb{Z}_p$ that satisfy* (11.2) *for all $k \in \mathbb{N}$ simultaneously. Then $d = 1$.*

Proof. As f is ergodic, $d \neq 0$. Assume that $d > 1$. Consider $w(x) = c + \sum_{i=0}^{r} c_i \cdot f^i(x)$. As $w(x)$ is a composition of integer-valued 1-Lipschitz polynomials over $\mathbb{Q}_p$, $w(x) \in \mathbb{Q}_p[x]$ is an integer-valued 1-Lipschitz polynomial over $\mathbb{Q}_p$. However, $\deg f^i(x) = d^i$; whence, as $d > 1$, we conclude that $w(x)$, being a sum of polynomials of pairwise distinct degrees, must be a polynomial of a nonzero degree.

On the other hand, since $x_{n+i} \equiv f^i(f^n(x_0)) \pmod{p^k}$, the assumptions of the lemma imply that $w(x_n) \equiv 0 \pmod{p^k}$ for all $n = 0, 1, 2, \ldots$. In other words, $w(z) \equiv 0 \pmod{p^k}$ for all $z \in \mathbb{Z}_p$ since x_n takes all values in $\{0, 1, \ldots, p^k - 1\}$ in view of the ergodicity of f, and $w(x)$ is 1-Lipschitz. The assumptions of the lemma now imply that $w(z) \equiv 0 \pmod{p^k}$ for all $z \in \mathbb{Z}_p$ and all $k = 1, 2, \ldots$. Consequently, $w(z) = 0$ for all $z \in \mathbb{Z}_p$ and hence the polynomial $w(x)$ must be 0 in the ring $\mathbb{Q}_p[x]$. A contradiction that proves the lemma. □

[2]cf. Figure 11.6 for distribution of pairs for a polynomial generator of degree 8

[3]these are characterized by Proposition 4.69

Proof of Proposition 11.3. By the assumption, for each $k \in \mathbb{N}$ the set $\mathcal{L}_k$ of all $\mathbf{c} = (c, c_0, \dots, c_r) \in \mathbb{Z}_p^{r+2}$ such that $|\mathbf{c}|_p = 1$ and $c, c_0, \dots, c_r$ satisfy (11.2), is not empty. Obviously, $\mathcal{L}_1 \supset \mathcal{L}_2 \supset \cdots$ since f is 1-Lipschitz.

Further, we assert that each set $\mathcal{L}_k$ is closed in the topology of the metric space $\mathbb{Z}_p^{r+2}$. Actually, if $\mathbf{c} \in \mathcal{L}_k$, $\mathbf{c}' \in \mathbb{Z}_p^{r+2}$, $|\mathbf{c} - \mathbf{c}'| \le p^{-s}$, $s \ge k$, then $\mathbf{c}' = \mathbf{c} + p^s\mathbf{z}$ for a suitable $\mathbf{z} \in \mathbb{Z}_p^{r+2}$. Hence, $|\mathbf{c}'|_p = 1$ and $\mathbf{c}'$ satisfies (11.2); consequently, $\mathbf{c}' \in \mathcal{L}_k$.

Now we apply to the sequence of nested sets $\mathcal{L}_1 \supset \mathcal{L}_2 \supset \cdots$ the p-adic analog of the classical lemma on nested closed real intervals. The analog of that lemma holds for topological spaces of much more general nature, see e.g. the corresponding theorem in [278, Chapter 3, Section 34, I]; the p-adic lemma can be easily deduced from the mentioned theorem. Thus, we conclude that the intersection of nested sets $\mathcal{L}_1 \supset \mathcal{L}_2 \supset \cdots$ is not empty. That is, there exists $\mathbf{c}'' \in \mathbb{Z}_p^{r+2}$ that satisfies the assumptions of Lemma 11.4. Yet then $d = 1$. □

Now we are able to prove the following theorem:

Theorem 11.5. *Let $f \in \mathbb{Q}_p[x]$ be an integer-valued 1-Lipschitz ergodic polynomial, let* $\deg f > 1$, *and let there exist $r \in \mathbb{N}$ such that for each $k \in \mathbb{N}$ the linear complexity over the ring $\mathbb{Z}/p^k\mathbb{Z}$ of the recurrence sequence $(x_n)_{n=0}^{\infty}$ defined by the recursion $x_{n+1} \equiv f(x_n) \pmod{p^k}$, does not exceed r. In other words, let there exist $c^{(k)}, c_0^{(k)}, \dots, c_r^{(k)} \in \mathbb{Z}_p$ such that the following congruences hold*:

$$c^{(k)} + \sum_{i=0}^{r} c_i^{(k)} x_{n+i} \equiv 0 \pmod{p^k}, \qquad n = 0, 1, 2, \dots. \tag{11.3}$$

Then $\lim^p_{k\to\infty} c^{(k)} = \lim^p_{k\to\infty} c_1^{(k)} = \cdots = \lim^p_{k\to\infty} c_r^{(k)} = 0.$

Proof. To start with, we note that from the proofs of both Lemma 11.4 and Proposition 11.3 it follows that they remain true if we let k under their assumptions range over an arbitrary infinite subset of $\mathbb{N}$ rather than the whole set $\mathbb{N}$.

Now for each $k \in \mathbb{N}$ take (and fix) $c^{(k)}, c_0^{(k)}, c_1^{(k)}, \dots, c_r^{(k)} \in \mathbb{Z}_p^{r+2}$ that satisfy (11.3). Put $\mathbf{c}_k = (c^{(k)}, c_0^{(k)}, c_1^{(k)}, \dots, c_r^{(k)}) \in \mathbb{Z}_p^{r+2}$. In view of Proposition 11.3 we have then $|\mathbf{c}_k|_p < 1$ for all $k \in \mathbb{N}$. Denote $\mathcal{N} = \{k \in \mathbb{N} : |\mathbf{c}_k|_p > p^{-k}\}$. In other words, $k \notin \mathcal{N}$ if and only if (11.3) is equivalent to the congruence $0 \equiv 0 \pmod{p^k}$. It is obvious that if $\mathcal{N}$ is finite, then the conclusion of the theorem is true. Let $\mathcal{N}$ be infinite.

For $k \in \mathcal{N}$ put $\hat{\mathbf{c}}_k = |\mathbf{c}_k|_p \mathbf{c}_k$ and denote by $\hat{\mathcal{N}}$ the set of all $m \in \mathbb{N}$ such that $p^k |\mathbf{c}_k|_p = p^m$ for a suitable $k \in \mathcal{N}$. In other words, we replace every set of congruences (11.3) with the equivalent system of congruences

$$\hat{c}^{(k)} + \sum_{i=0}^{r} \hat{c}_i^{(k)} x_{n+i} \equiv 0 \pmod{p^m}, \qquad n = 0, 1, 2, \dots,$$

where $(\hat{c}^{(k)}, \hat{c}_0^{(k)}, \hat{c}_1^{(k)}, \dots, \hat{c}_r^{(k)}) = \hat{\mathbf{c}}_k$, $p^m = p^k |\mathbf{c}_k|_p$.

If the set $\hat{\mathcal{N}}$ is finite, the conclusion of the theorem is obviously true. If $\hat{\mathcal{N}}$ is infinite, then, since $|\hat{\mathbf{c}}_k|_p = 1$, in view of Proposition 11.3 and the note at the beginning of the proof, we conclude that $\deg f = 1$. A contradiction. □

Note that Lemma 11.4 asserts that the recurrence sequence defined by the recursion $x_i = f(x_{i-1})$ has infinite linear complexity over the ring $\mathbb{Z}_p$ providing $f \in \mathbb{Q}_p[x]$ is integer-valued 1-Lipschitz ergodic polynomial of degree $d > 1$ thus proving Theorem 11.2. This assertion can be slightly strengthened.

Corollary 11.6. *If $f \in \mathbb{Q}_p[x]$ is an integer-valued 1-Lipschitz ergodic polynomial of degree $d > 1$, then the recurrence sequence (x_n) defined by the recursion $x_{n+1} = f(x_n)$ has infinite linear complexity over the field $\mathbb{Q}_p$.*

Proof. If for suitable $c, c_0, \ldots, c_r \in \mathbb{Q}_p$ that are not 0 simultaneously the equality $c + \sum_{j=0}^{r} c_j x_{n+j} = 0$ holds for all $n = 0, 1, 2, \ldots$, then the equality $hc + \sum_{j=0}^{r} hc_j x_{n+j} = 0$ where $h = 1$ if $c, c_0, \ldots, c_r \in \mathbb{Z}_p$, and $h = |(c, c_0, \ldots, c_r)|_p$ otherwise, holds either. As f is 1-Lipschitz, the conclusion now follows from Lemma 11.4. □

Note 11.7. The condition that f is a *polynomial* over the field $\mathbb{Q}_p$ is essential: For instance, let $p = 2$ and let

$$f(x) = 1 + x + 4(-1)^{1+x} = 1 + x + \sum_{j=0}^{\infty} (-1)^j 2^{j+2} \binom{x}{j}.$$

By Theorem 4.40, f is an integer-valued 1-Lipschitz ergodic function. However, it is easy to see that the recurrence sequence (x_n) over $\mathbb{Z}_2$ defined by the recursion $x_{n+1} = f(x_n)$ satisfies the relation $x_{n+2} = x_n + 2$; that is, the linear complexity over the ring $\mathbb{Z}_2$ of this sequence is 2.

11.1.2 Lacunas

In real life settings we never deal with automata that have infinite number of states. However, very often we deal with automata whose number of states is very big; a contemporary computer is an example of an automaton of this sort. In real-time, we can simulate only behavior of an automaton that has a relatively small number of states; however, judging on this behavior we want to make conclusions about the behavior of a similar (in a certain sense) automaton that has a very big number of states. In this setting we naturally come to the necessity to study the behavior of an automaton when the number of its states goes to infinity.

Any automaton $\mathfrak{A} = \langle \mathcal{K}, \mathcal{N}, \mathcal{M}, h, H, u_0 \rangle$ with the state transition function h, with the output function H, and with nonempty input alphabet $\mathcal{K}$ and nonempty output alphabet $\mathcal{M}$, can be considered as a transducer of information: It transforms sequences over $\mathcal{K}$ into sequences over $\mathcal{M}$ by means of transformation $\Psi_{\mathfrak{A}}$, see Section 8.1. For instance, every encryption device is a transducer that has some specific features: First, the transformation $f = \Psi_{\mathfrak{A}}$ must be one-to-one, otherwise decryption is not possible; and second, this transformation f must be random-looking, otherwise the cipher is not secure. Further without loss of generality we may assume that both input and output alphabets are $\{0, 1, \dots, p-1\}$, p a prime[4]; in most practical cases $p = 2$. So f is a 1-Lipschitz transformation on the space of p-adic integers $\mathbb{Z}_p$, see again Section 8.1.

Now, to study correlations between input (plain texts) and output (encrypted texts) we need to study the distribution of pairs $\left(\frac{x}{p^n}, \frac{f(x) \bmod p^n}{p^n}\right)$, $x \in \{0, 1, \dots, p^n - 1\}$, as $n \to \infty$: The more random-looking this distribution is, the better.[5] The main goal of this subsection is to demonstrate that this *distribution exhibits sharp irregularities whenever a designer uses only those computer instructions that can be represented by finite-state automata* (such as addition, multiplication by a constant, which is a rational p-adic integer, and bitwise logical operations like XOR, AND, etc.); moreover, we will show how to *avoid these irregularities using multiplication of variables*[6]. Now we give formal definitions and statements:

Definition 11.8. We say that a 1-Lipschitz function $f : \mathbb{Z}_p \to \mathbb{Z}_p$ *has lacunas* whenever there exists an open (in the standard topology of $\mathbb{R}^2$) subset O of the unit square $[0, 1]^2$ that contains no points of the form $\left(\frac{x \bmod p^n}{p^n}, \frac{f(x) \bmod p^n}{p^n}\right)$, $x \in \mathbb{Z}_p$, $n = 1, 2, 3, \dots$. We call this open subset O an f*-lacuna*. We omit 'f-' when this does not lead to misunderstanding.

Clearly, the lacunas are merely 'holes', blank spots at the graph of the function $\frac{f \bmod p^n}{p^n}$ that do not disappear as $n \to \infty$, see e.g. Figures 11.13–11.14.

On the contrary, the function f has no lacunas if and only if the set

$$\left\{\left(\frac{x \bmod p^n}{p^n}, \frac{f(x) \bmod p^n}{p^n}\right) : x \in \mathbb{Z}_p; n = 1, 2, 3, \dots\right\}$$

is everywhere dense in $[0, 1]^2$, see e.g. Figures 11.15–11.18 on page 354. It is clear that whenever an automaton is used for encryption, it is bad if the associated 1-Lipschitz function has lacunas; however, we can only say that, *may be*, the encryption is good whenever this function has no lacunas. Now we will show that all finite automata are *very bad* from this view. We first prove a lemma showing they are 'bad':

[4]Note, however, that nowhere in the proofs of Lemma 11.9 and Theorem 11.10 we assume that p is a prime number.

[5]Recall that mod p^k is a reduction modulo p^k, that is, $x \bmod p^k$ is a number from $\{0, 1, \dots, p^k - 1\}$ such that $|x - (x \bmod p^k)|_p \le p^{-k}$.

[6]It is well known that the latter operation can not be represented by a finite-state automaton, see e.g. [75, Theorem 2.2.3].

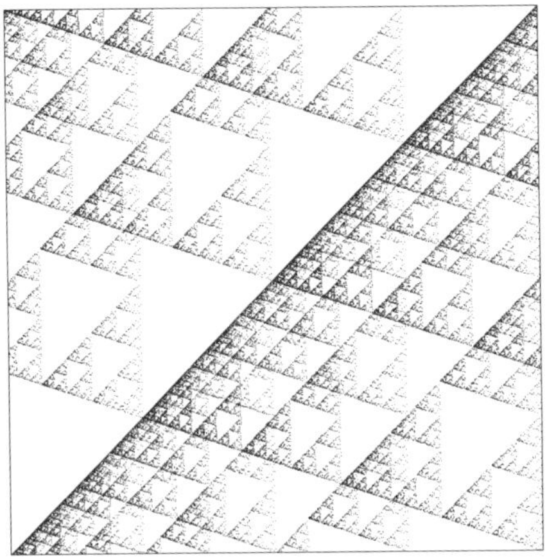

Figure 11.13. The function $f(x) = 1 + x + 4 \cdot ((7 + \frac{1}{77}x)$ OR$(3 - \frac{1}{3}x))$, $p = 2, n = 16$.

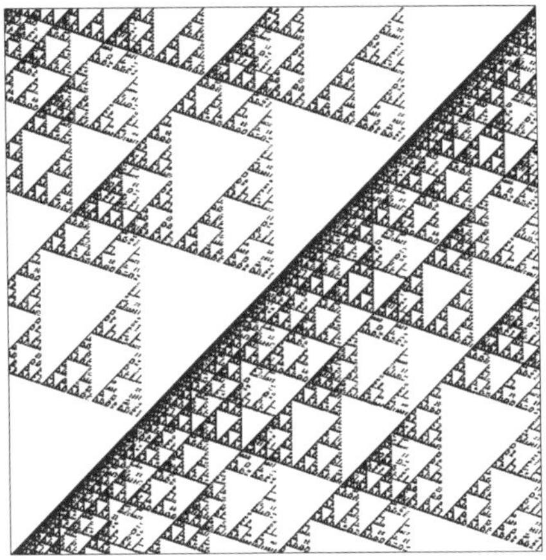

Figure 11.14. Same function, $n = 24$.

Lemma 11.9. *Whenever a 1-Lipschitz function $f : \mathbb{Z}_p \to \mathbb{Z}_p$ corresponds to a finite-state automaton, f has lacunas.*

Proof. As f is 1-Lipschitz, it is clear that given $k \in \mathbb{N}$, for all $x \in \mathbb{Z}_p$ we can represent $f(x)$ as

$$f(x) = (f(x \bmod p^k)) \bmod p^k + p^k \cdot g_z(y), \tag{11.4}$$

where $y = \frac{1}{p^k}(x - (x \bmod p^k)) \in \mathbb{Z}_p$, $z = x \bmod p^k$, and $g_z : \mathbb{Z}_p \to \mathbb{Z}_p$ is a 1-Lipschitz function.

Now, as f corresponds to a finite-state automaton $\mathfrak{A} = \langle \mathcal{K}, \mathcal{N}, \mathcal{M}, h, H, u_0 \rangle$, *the number of these functions g_z is finite*, as actually g_z is a function that corresponds to the automaton $\mathfrak{A}(z) = \langle \mathcal{K}, \mathcal{N}, \mathcal{M}, h, H, t_0 \rangle$, where $t_0 \in \mathcal{N}$ is the state of the automaton $\mathfrak{A}$ after inputting the finite sequence $z = x \bmod p^k$. That is, there exists $N \in \mathbb{N}$ such that for all $k > N$ the function $z = z(x)$ in the equality (11.4) takes values only in the same finite set, i.e., this finite number of values the function $z(x)$ takes does not depend on k. Clearly, this number does not exceed p^N, where $N = \lceil \#\mathcal{N} \rceil$; we recall that all states of the automaton $\mathfrak{A}$ are assumed to be accessible.

Now we take $n > N$, fix arbitrary $\alpha_0, \dots, \alpha_{n-1} \in \{0, 1, \dots, p-1\}$, and denote $a = \alpha_0 + \alpha_1 \cdot p + \dots + \alpha_{n-1} \cdot p^{n-1}$. There exist not more than p^N different numbers $g_z(a) \bmod p^n$, as there exist not more than p^N different functions g_z. As $n > N$, there exists a number $b \in \{0, 1, \dots, p^n - 1\}$ that differs from all these numbers $g_z(a) \bmod p^n$. We fix this number $b = \beta_0 + \beta_1 \cdot p + \dots + \beta_{n-1} \cdot p^{n-1}$; here $\beta_i \in \{0, 1, \dots, p-1\}$, $i = 0, 1, 2, \dots, n-1$.

In other words, since $\mathfrak{A}$ is a finite-state automaton, given a sufficiently long word $\alpha_{n-1} \dots \alpha_0$ over the alphabet $\{0, 1, \dots, p-1\}$, there exists a word $\beta_{n-1} \dots \beta_0$ such that no output word (of length $n + K$, $K \geq N$) of the automaton $\mathfrak{A}$ ends with $\beta_{n-1} \ldots \beta_0$ whenever the input word (of length $n + K$) of the automaton ends with $\alpha_{n-1} \ldots \alpha_0$. That is, given a number $a = \alpha_0 + \alpha_1 \cdot p + \cdots + \alpha_{n-1} \cdot p^{n-1}$ we have that if for some $x \in \mathbb{Z}_p$ and $L \geq N + n - 1$

$$\frac{x \bmod p^L}{p^L} \in I(a) = \left[\frac{p^N \cdot a}{p^{N+n-1}}, \frac{p^N - 1 + p^N \cdot a}{p^{N+n-1}}\right] = \left[\frac{a}{p^{n-1}}, \frac{a}{p^{n-1}} + \frac{p^N - 1}{p^{N+n-1}}\right],$$

then

$$\frac{f(x) \bmod p^L}{p^L} \notin I(b) = \left[\frac{b}{p^{n-1}}, \frac{b}{p^{n-1}} + \frac{p^N - 1}{p^{N+n-1}}\right].$$

As only a finite number of rational numbers of the form $\frac{x \bmod p^L}{p^L}$ (where $x \in \mathbb{Z}_p$) from the segment $I(a)$ are such that $\frac{f(x) \bmod p^L}{p^L} \in I(b)$ (may be, only those with $L < N + n - 1$), an open interval $I'(a) = \left(\frac{a}{p^{n-1}}; \frac{a}{p^{n-1}} + \frac{1}{p^{k+n-1}}\right) \subset I(a)$ contains no points of this kind. So $I'(a) \times I'(b) \subset [0, 1]^2$, where $I'(b)$ stands for an open interval $\left(\frac{b}{p^{n-1}}; \frac{b}{p^{n-1}} + \frac{1}{p^{k+n-1}}\right)$, is an f-lacuna. □

Now using Lemma 11.9 we will show that finite automata are 'very bad': Whenever the function $f : \mathbb{Z}_p \to \mathbb{Z}_p$ corresponds to a finite automaton, the graph of the function f 'consists mainly of holes'.

Theorem 11.10. *Under the conditions of Lemma 11.9, every neighborhood*[7] *of every point from the unit square* $[0, 1]^2$ *contains an* f*-lacuna.*

Proof. Take an arbitrary $m \in \mathbb{N}$ and arbitrary numbers $u, v \in \{0, 1, \ldots, p^m - 1\}$. Consider base-$p$ expansions $u = \chi_0 + \chi_1 \cdot p + \cdots + \chi_{m-1} \cdot p^{m-1}$, $v = \xi_0 + \xi_1 \cdot p + \cdots + \xi_{m-1} \cdot p^{m-1}$ of the numbers u, v and denote $\bar{u} = \chi_0\chi_1 \cdots \chi_{m-1}$, $\bar{v} = \xi_0\xi_1 \cdots \xi_{m-1}$. During the proof of Lemma 11.9 we have shown that there exists a pair of non-empty words $\bar{a} = a_{n-1} \cdots a_0$, $\bar{b} = b_{n-1} \cdots b_0$ over the alphabet $\{0, 1, \ldots, p - 1\}$ such that for all $K \geq n + N$ no output word of length K of the automaton $\mathfrak{A}$ ends with $\bar{b}$ whenever $\mathfrak{A}$ is feeded by an arbitrary input word of length K that ends with $\bar{a}$; here n, N are the same as in the proof of Lemma 11.9. Therefore, no output word of length $K \geq \ell + m + n + N$ ends with a concatenation $\bar{v}\bar{0}\bar{b}$ when the automaton $\mathfrak{A}$ is feeded by any word of length $K \geq \ell + m + n + N$ that ends with a concatenation $\bar{u}\bar{0}\bar{a}$, where $\bar{0} = 0 \ldots 0$ is a word of length $\ell > 0$.

[7] Within the context of the subsection a neighborhood of a point is understood as an open (in the topology of $\mathbb{R}^2$) subset that contains the point.

Now arguing as in the proof of Lemma 11.9, we conclude that the following open square

$$J_\ell(u) \times J_\ell(v) = \left(\frac{u}{p^{m-1}} + \frac{a}{p^{\ell+m+n-1}}; \frac{u}{p^{m-1}} + \frac{a}{p^{\ell+m+n-1}} + \frac{1}{p^{N+\ell+m+n-1}}\right)$$
$$\times \left(\frac{v}{p^{m-1}} + \frac{b}{p^{\ell+m+n-1}}; \frac{v}{p^{m-1}} + \frac{b}{p^{\ell+m+n-1}} + \frac{1}{p^{N+\ell+m+n-1}}\right)$$

is an f-lacuna. However, given a point $(x, y) \in [0,1]^2$ we can find a point $(\frac{u}{p^m}, \frac{v}{p^m}) \in [0,1]^2$ that is arbitrarily close to (x, y), and then we can take a sufficiently small lacuna of the form $J_\ell(u) \times J_\ell(v)$ by choosing ℓ sufficiently large to make the lacuna lay inside a given neighborhood of the point (x, y). □

From Theorem 11.10 it follows that whenever only instructions of the form $+$, XOR, AND, OR and NOT are used in the composition of f, the corresponding distribution in the unit square will be necessarily poor. However, this drawback can be cured in some cases if we let integer multiplication $x \cdot y$ into the composition. Namely, the following theorem is true:

Theorem 11.11. *If f is a univariate polynomial of degree ≥ 2 with rational integer coefficients, then f has no lacunas.*

Proof. As f is a polynomial, f has not more than a finite number of zeros in $\mathbb{R}$, so there exists $d \in \mathbb{N}_0$ such that for all $b \geq d$ either values $f(b)$ are all positive or they are all negative. It suffices to consider only the case when all $f(b) > 0$: Whenever we prove the theorem for this case, the conclusion for the case when all $f(b) < 0$ follows. Indeed, for every $c \in \mathbb{N}$ and every $n \in \mathbb{N}$ we have that $\frac{-c \bmod p^n}{p^n} = \frac{p^n - (c \bmod p^n)}{p^n} = 1 - \frac{c \bmod p^n}{p^n}$. Thus, a symmetry with respect to the axis $y = \frac{1}{2}$ of the unit square $[0,1]^2 \subset \mathbb{R}^2$ maps the subset

$$\mathcal{E}(f) = \left\{\left(\frac{x \bmod p^n}{p^n}, \frac{f(x) \bmod p^n}{p^n}\right) : x \in \mathbb{Z}_p, n \in \mathbb{N}\right\} \subset [0,1]^2$$

onto the subset $\mathcal{E}(-f)$ and vice versa. So f has lacunas if and only if $-f$ has lacunas.

We will show that for every sufficiently large k and every $z, u \in \{0, 1, \ldots, p^k - 1\}$ there exist $M = M(k)$ and $a \in \{0, 1, \ldots, p^M - 1\}$ such that

$$\left|\frac{a}{p^M} - \frac{u}{p^k}\right| < \frac{1}{p^k} \text{ and } \left|\frac{f(a) \bmod p^M}{p^M} - \frac{z}{p^k}\right| < \frac{1}{p^k}. \tag{11.5}$$

This will prove Theorem 11.11 as every point from $[0,1]^2$ can be approximated by points of the form $(\frac{u}{p^k}, \frac{z}{p^k})$.

The idea of the proof is as follows: We will take an arbitrary natural number $v \geq d$ whose length in a base-p expansion is less than k (so that v is not more than a k-digit number in the system with the base $\{0, 1, \ldots, p-1\}$), and then we will change zeroes in this expansion at positions starting with ℓth, $\ell > k$ to some other figures from $\{0, 1, \ldots, p-1\}$ so that the obtained natural number $a = v + p^{\ell}t$ will satisfy inequalities (11.5) for some M. To do this, we will need that $f''(v) \neq 0$. The latter condition can also be satisfied as $\deg f > 1$ and f'' is a polynomial over $\mathbb{Z}$ either; so f'' has not more than a finite number of zeros in $\mathbb{R}$.

Let $\operatorname{ord}_p(f''(v)) = s$; that is, $f''(v) = p^s \cdot \xi$ where $\xi \in \mathbb{N}$, $p \nmid \xi$. Take $r > s$ such that $p^r > v$. Now take and fix $n \in \mathbb{N}$ so that $n > \max\{\log_p f(v + p^{k+r}t) : t = 0, 1, 2, \ldots, p^k - 1\}$ and $n > 2k + 2r + 2s$. Put

$$\tilde{u} = 1 + p^{k+r+s}u, \tag{11.6}$$

$$\tilde{z} = f'(v) + p^{k+r+s}\hat{z}, \tag{11.7}$$

where $\hat{z} \in \{0, 1, \ldots, p^k - 1\}$ is such that $\lfloor \frac{\tilde{z}}{p^{k+r+s}} \rfloor \bmod p^k = z$. In other words, we choose $\hat{z}$ in such a way that the number whose base-p expansion stands in positions from $(k+r+s)$th to $(2k+r+s-1)$th in the canonical p-adic expansion of $\tilde{z}$, is equal to z. Obviously, given $f'(v)$ and z, there exists a unique $\hat{z}$ that satisfy this condition: $\hat{z} \equiv z - \lfloor \frac{f'(v)}{p^{k+r+s}} \rfloor \pmod{p^k}$; so

$$\tilde{z} \bmod p^{2k+r+s} = (f'(v) \bmod p^{k+r+s}) + p^{k+r+s} \cdot z. \tag{11.8}$$

Now for every $\zeta \in \{0, 1, \ldots, p^k - 1\}$ with the use of Taylor formula we obtain that $f(v + p^{r+k}\zeta + p^n\tilde{u}) \equiv f(v + p^{r+k}\zeta) + p^n\tilde{u} \cdot f'(v + p^{r+k}\zeta) \pmod{p^{2n}}$ and, moreover, that

$$\begin{aligned} f(v + p^{r+k}\zeta + p^n\tilde{u}) &\equiv f(v + p^{r+k}\zeta) \\ &\quad + p^n\tilde{u} \cdot (f'(v) + p^{r+k}\zeta \cdot f''(v)) \pmod{p^{n+2k+r+s}} \end{aligned} \tag{11.9}$$

as $n + 2r + 2k > n + 2k + r + s$ (since $r > s$ by the choice of r).

We claim that there exists $\zeta \in \{0, 1, \ldots, p^k - 1\}$ such that

$$\tilde{u} \cdot (f'(v) + p^{r+k}\zeta \cdot f''(v)) \equiv \tilde{z} \pmod{p^{2k+r+s}}. \tag{11.10}$$

Indeed, in view of (11.6)–(11.7) this congruence is equivalent to the congruence $(1 + p^{k+r+s}u) \cdot (f'(v) + p^{r+k}\zeta \cdot f''(v)) \equiv f'(v) + p^{k+r+s}\hat{z} \pmod{p^{2k+r+s}}$, and the latter congruence is equivalent to the congruence $f'(v) + p^{r+k}\zeta \cdot f''(v) \equiv (1 - p^{k+r+s}u) \cdot (f'(v) + p^{k+r+s}\hat{z}) \pmod{p^{2k+r+s}}$ as $(1 + p^{k+r+s}u)^{-1} \equiv 1 - p^{k+r+s}u \pmod{p^{2k+r+s}}$. That is, congruence (11.10) is equivalent to the congruence $p^{k+r}\zeta \cdot f''(v) \equiv p^{k+r+s}\hat{z} - p^{k+r+s}u \cdot f'(v) \pmod{p^{2k+r+s}}$. However, as $f''(v) = p^s\xi$, the latter congruence is equivalent to the congruence $\zeta\xi \equiv \hat{z} - u \cdot f'(v) \pmod{p^k}$.

From here we find that $\zeta \equiv \xi^{-1} \cdot (\hat{z} - u \cdot f'(v)) \pmod{p^k}$, thus proving our claim (we remind that $\xi \not\equiv 0 \pmod{p}$, so ξ has a multiplicative inverse ξ^{-1} modulo p^k).

Now we put $M = n + 2k + r + s$ and $a = v + p^{r+k}\zeta + p^n \cdot (1 + p^{k+r+s}u)$; then

$$\frac{a}{p^M} = \frac{u}{p^k} + \frac{v + p^{r+k}\zeta + p^n}{p^{n+2k+r+s}},$$

so $\left|\frac{a}{p^M} - \frac{u}{p^k}\right| < \frac{1}{p^k}$, since $v < p^r$, $\zeta < p^k$, and $n > 2r + 2s + 2k$. However, at the same time, combining (11.10), (11.7), (11.8), and (11.9), we see that

$$\frac{f(a) \bmod p^M}{p^M} = \frac{z}{p^k} + \frac{f(v + p^{r+k}\zeta)}{p^n} \cdot \frac{1}{p^{2k+r+s}} + \frac{f'(v) \bmod p^{k+r+s}}{p^{k+r+s}} \cdot \frac{1}{p^k}, \tag{11.11}$$

since $f(a) \bmod p^M = f(v + p^{r+k}\zeta) + p^n \cdot (f'(v) \bmod p^{k+r+s}) + p^{n+k+r+s}z$ (the number in the right-hand side is less than p^M due to our choice of n). Now from (11.11) it follows that $\left|\frac{f(a) \bmod p^M}{p^M} - \frac{z}{p^k}\right| < \frac{1}{p^k}$ since $0 \le f(v + p^{r+k}\zeta) \le p^n - 1$ due to our choice of n. □

Note 11.12. From the proof of Theorem 11.11 it follows that *whenever a function defined by an automaton is a polynomial of degree* > 1 *with rational integer coefficients, then, given arbitrary* k*-letter words* z *and* u (where k is large enough), *and arbitrary finite word* v' *in a* p*-letter alphabet, there exists an input word* a *that has* v' *as an initial subword and* u *as an ending subword, such that the corresponding output word of the automaton ends with the subword* z. Indeed, we may choose arbitrarily the subword v' by fixing initial less significant (i.e., rightmost) digits in the base-p expansion of $v \in \mathbb{N}$ as during the proof we impose only two restrictions on v: $v > d$ and $f''(v) \neq 0$. We can satisfy these conditions simultaneously in the case some less significant digits of v are fixed as f'' is a polynomial, and so it has not more than a finite number of zeros.

The following note is just a restatement of the above one:

Note 11.13. Under the conditions of Theorem 11.11, not only the set

$$\left\{\left(\frac{x \bmod p^n}{p^n}, \frac{f(x) \bmod p^n}{p^n}\right) : x \in \mathbb{Z}_p; n = 1, 2, 3, \ldots\right\}$$

is everywhere dense in $[0, 1]^2$, but so is every set

$$\left\{\left(\frac{x \bmod p^n}{p^n}, \frac{f(x) \bmod p^n}{p^n}\right) : x \in B_{p^{-\ell}}(v); n > k\right\},$$

for every $v \in \mathbb{Z}_p$, where $B_{p^{-\ell}}(v)$ is a ball of radius $p^{-\ell}$ centered at v.

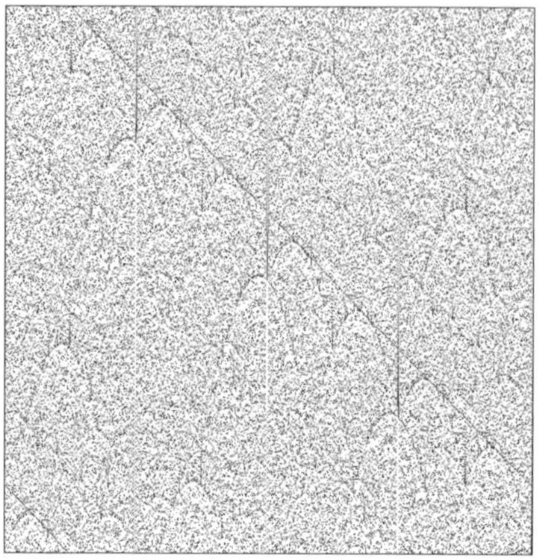

Figure 11.15. The function $f(x) = 2x^2 + 3x + 1$, $p = 2, n = 16$.

Figure 11.16. Same function, $n = 18$.

Figure 11.17. Same function, $n = 20$.

Figure 11.18. Same function, $n = 23$.

Note 11.14. In the context of quality of pseudorandom sequences produced by congruential generators, it is worth mentioning that Theorem 11.11 under suitable (and somewhat more technical) restatement holds for a wider class of functions $f : \mathbb{Z}_p \to \mathbb{Z}_p$ than polynomials over $\mathbb{Z}$. For instance, it holds for exponential generators with the recursion law $f(x) = ax + a^x$, where $a \in \mathbb{N}$, $a \neq 1$, $a \equiv 1 \pmod{p}$; see Example 9.9 about these. We omit further details[8].

The figures 11.15–11.18 illustrate Theorem 11.11: They show the behavior of points $\left(\frac{x \bmod p^n}{p^n}, \frac{f(x) \bmod p^n}{p^n}\right)$ as n increases for a quadratic polynomial f. Theorems 11.10

[8] see [31]

and 11.11 imply important practical conclusion: *To avoid lacunas in distribution of output sequence one must use multiplication of variables*, and moreover, from the results of this section it follows that *quadratic generators look as one of the best choices to produce pseudorandom numbers* for various purposes (although in cryptography extra output function is necessary). Indeed, quadratic generators satisfy Theorem 11.11 and Corollary 11.6, and program implementation of these generators is the fastest compared to other non-linear congruential generators. All quadratic generators that are transitive modulo p^n are completely characterized (see e.g. Corollary 4.71). Intensive studies of quadratic generators that produce uniform distribution of pairs $(\frac{x \bmod p^n}{p^n}, \frac{f(x) \bmod p^n}{p^n})$ in the unit square were undertaken, see e.g. [120] and references therein. Although the problem of characterization of these generators is not completely solved, large classes were described explicitly.

Now we introduce some 'measures of complexity' of 1-Lipschitz dynamics on $\mathbb{Z}_p$. Given a transformation $f\colon \mathbb{Z}_p \to \mathbb{Z}_p$, and $k, n \in \mathbb{N}$, we consider sets

$$\mathcal{P}_n^k(f) = \left\{\left(\frac{x}{p^n}, \frac{f(x) \bmod p^n}{p^n}, \dots, \frac{f^{k-1}(x) \bmod p^n}{p^n}\right) : x \in \{0, 1, \dots, p^n - 1\}\right\}$$

and

$$\mathcal{P}^k(f) = \mathrm{Cl}\left(\bigcup_{n=1}^{\infty} \mathcal{P}_n^k(f)\right),$$

where $\mathrm{Cl}(A)$ stands for a closure of a subset $A \subset [0,1]^k$ of a k-dimensional unit hypercube in a usual topology of $\mathbb{R}^k$. Thus, $\mathcal{P}^k(f)$ is a measurable subset with respect to the Lebesgue measure ζ_k on $\mathbb{R}^k$; we denote $\alpha_k(f) = \zeta_k(\mathcal{P}^k(f))$. Now, summarizing results of this subsection with Theorem 4.23 we conclude:

- $\alpha_1(f) = 1$ whenever f is a measure-preserving transformation on $\mathbb{Z}_p$;
- $\alpha_2(f) = 1$ whenever f is a polynomial of degree ≥ 2 with rational integer coefficients;
- $\alpha_2(f) = 0$ whenever f is a function that corresponds to a finite automaton.

We note that actually there are only two possibilities for the value of $\alpha_2(f)$. The following proposition may be considered as a kind of a zero-one law for 1-Lipschitz functions (whence, for automata functions).

Proposition 11.15. *For a 1-Lipschitz transformation $f\colon \mathbb{Z}_p \to \mathbb{Z}_p$, the measure $\alpha_2(f)$ can take only two values, 0 and 1.*

Proof. Indeed, let $\alpha_2(f) > 0$. Then by the definition of $\alpha_2(f)$ there exist u, v, u', v', $0 \leq u < v \leq 1$, $0 \leq u' < v' \leq 1$ such that the square $[u, v] \times [u', v'] \subset [0,1]^2$ lies completely in $\mathcal{P}^2(f)$, and every point from the real interval $(u'; v')$ is a limit (with respect to the standard Archimedean metric on $\mathbb{R}$) of some sequence of fractions $u' < \frac{f(a_m) \bmod p^m}{p^m} < v'$, where $u < \frac{a_m}{p^m} < v$, $m = 1, 2, \dots$. Thus, we can take

$n \in \mathbb{N}$ and $w = \omega_0 + \omega_1 \cdot p + \cdots + \omega_{n-1} \cdot p^{n-1}$, where $\omega_i \in \{0, 1, \ldots, p-1\}$, $i = 0, 1, \ldots, n-1$, so that the square

$$S = \left[\frac{w}{p^n}, \frac{w}{p^n} + \frac{1}{p^n}\right] \times \left[\frac{f(w) \bmod p^n}{p^n}, \frac{f(w) \bmod p^n}{p^n} + \frac{1}{p^n}\right]$$

lies completely in $\mathcal{P}^2(f)$, and every inner point (x, y) of the square S [9] is a limit as $j \to \infty$ (with respect to the standard Archimedean metric in $\mathbb{R}^2$) of a sequence of inner points

$$(r_j, t_j) = \left(\frac{z_j + p^{N_j} \cdot w}{p^{N_j+n}}, \frac{f(z_j + p^{N_j} \cdot w) \bmod p^{N_j+n}}{p^{N_j+n}}\right) \in S,$$

where $N_j \in \mathbb{N}$, $z_j \in \{0, 1, \ldots, p^{N_j} - 1\}$.

However, as f is a 1-Lipschitz transformation on $\mathbb{Z}_p$, for every $z \in \{0, 1, \ldots, p^N - 1\}$ we have that $f(z + p^N \cdot w) \equiv (f(z) \bmod p^N) + p^N \cdot \xi_N(z) \pmod{p^{N+n}}$ for a suitable $\xi_N(z) \in \{0, 1, \ldots, p^n - 1\}$; thus,

$$\frac{f(z + p^N \cdot w) \bmod p^{N+n}}{p^{N+n}} = \frac{f(z) \bmod p^N}{p^{N+n}} + \frac{\xi_N(z)}{p^n}.$$

Hence, $\xi_{N_j}(z_j) = f(w) \bmod p^n$ for all $j = 1, 2, \ldots$ as all (r_j, t_j) are inner points of S. Therefore, every inner point $(x, y) \in S$, which then can be represented as

$$(x, y) = \left(\frac{w}{p^n} + \frac{\chi}{p^n}, \frac{f(w) \bmod p^n}{p^n} + \frac{\gamma}{p^n}\right),$$

where χ and γ are real numbers, $0 < \chi < 1$, $0 < \gamma < 1$, is a limit (as $j \to \infty$) of the point sequence

$$(r_j, t_j) = \left(\frac{w}{p^n} + \frac{z_j}{p^{N_j}} \cdot \frac{1}{p^n}, \frac{f(w) \bmod p^n}{p^n} + \frac{f(z_j) \bmod p^{N_j}}{p^{N_j}} \cdot \frac{1}{p^n}\right) \in S.$$

From here it follows that every inner point $(\chi, \gamma) \in [0, 1]^2$ is a limit point of the corresponding sequence of points $\left(\frac{z_j}{p^{N_j}}, \frac{f(z_j) \bmod p^{N_j}}{p^{N_j}}\right)$ as $j \to \infty$. This means that $\mathcal{P}^2(f) = [0, 1]^2$ and thus $\alpha_2(f) = 1$. □

We can consider similar measures of complexity for sequences over $\mathbb{Z}_p$ rather than for transformations on $\mathbb{Z}_p$: Given a sequence $\mathcal{X} = (x_i \in \mathbb{Z}_p)_{i=0}^{\infty}$, we consider a set

$$\mathcal{S}_n^k(\mathcal{X}) = \left\{\left(\frac{x_i \bmod p^n}{p^n}, \frac{x_{i+1} \bmod p^n}{p^n}, \ldots, \frac{x_{k+i-1} \bmod p^n}{p^n}\right) : i = 0, 1, \ldots\right\},$$

[9] that is, (x, y) has an open neighborhood that is contained completely in S

and a set $\mathcal{S}^k(\mathcal{X}) = \mathrm{Cl}\left(\bigcup_{n=1}^{\infty} \mathcal{S}_n^k(\mathcal{X})\right)$ and then put $\gamma_k(\mathcal{S}) = \zeta_k(\mathcal{S}^k(\mathcal{X}))$. This way we can relate to, say, the output sequence of a PRNG we considered in Chapters 9 and 10, a certain real number from the unit segment $[0, 1]$. Note, for instance, that if we take a sequence $\mathcal{S} = (f^i(x))_{i=0}^{\infty}$ produced by a 1-Lipschitz ergodic transformation f on $\mathbb{Z}_2$, and a sequence $\mathcal{S}' = (\lfloor \frac{f^i(x)}{2^m} \rfloor)_{i=0}^{\infty}$ obtained from the sequence $\mathcal{S}$ by truncation of m low order bits of terms of the sequence $\mathcal{S}$, then $\gamma_k(\mathcal{S}) = \gamma_k(\mathcal{S}')$. Thus, if $\gamma_k(\mathcal{S}) < 1$, which clearly reflects that there are certain irregularities in distribution of the sequence $\mathcal{S}$ produced by a PRNG with the law of recursion $x_{i+1} = f(x_i)$, then these irregularities *cannot be cured by truncation of low order bits*; so a usual 'remedy' in cryptology to improve quality of a sequence produced by a T-function f, the truncation of lower order bits, will not work in this case. Foremost, to study a truncation of, say, a half of bits, we can consider a set

$$\mathcal{T}_{2m}^k(f) = \left\{\left(\frac{\lfloor \frac{f^i(x) \bmod 2^{2m}}{2^m} \rfloor}{2^m}, \dots, \frac{\lfloor \frac{f^{k+i-1}(x) \bmod 2^{2m}}{2^m} \rfloor}{2^m}\right) : i = 0, 1, \dots\right\},$$

a corresponding set $\mathcal{T}^k(f) = \mathrm{Cl}\left(\bigcup_{m=1}^{\infty} \mathcal{T}_{2m}^k(f)\right)$, and its measure $\beta_k(f) = \zeta_k(\mathcal{T}^k(f))$. It is clear that $\beta_k(f) = \gamma_k(\mathcal{S})$, where $\mathcal{S} = (f^i(x))_{i=0}^{\infty}$. Thus, if $\gamma_k(\mathcal{S}) < 1$, then it clearly points out that the corresponding PRNG has certain drawbacks that can not be improved by a truncation of a certain portion (a half, in this example) of output bits. So measures of corresponding sets connected to f can give a designer an important tool to make judgements about the quality of the output sequence produced by certain types of T-functions. Thus, given an *ergodic* 1-Lipschitz transformation f on $\mathbb{Z}_p$, we can consider a set

$$\mathcal{R}_n^k(f) = \left\{\left(\frac{f^i(x) \bmod p^n}{p^n}, \dots, \frac{f^{k+i-1}(x) \bmod p^n}{p^n}\right) : i = 0, 1, \dots, p^n - 1\right\},$$

which in view of Theorem 4.23 does not depend on $x \in \mathbb{Z}_p$, the corresponding set

$$\mathcal{R}^k(f) = \mathrm{Cl}\left(\bigcup_{n=1}^{\infty} \mathcal{R}_n^k(f)\right),$$

and denote $\varepsilon_k(f) = \zeta_k(\mathcal{P}^k(f))$. It is clear in view of Theorem 4.23 that when f is ergodic, $\alpha_k(f) = \varepsilon_k(f)$. Both α_k and ε_k (as well as related β_k and γ_k) reflect important properties of distribution of trajectories. For instance, it is not difficult to see that, although the following transformations $f(x) = 1+5x$, $g(x) = x+(x^2 \mathsf{OR}(-3))$, and $h(x) = 1 + 5x + 4x^2$ are ergodic on $\mathbb{Z}_2$, $\varepsilon_2(f) = \varepsilon_2(g) = 0$, whereas $\varepsilon_2(h) = 1$. Moreover, if we truncate a half of output bits, we will not improve sequences produced by T-functions f and g, as $\beta_2(f) = \varepsilon_2(f) = 0$ and $\beta_2(g) = \varepsilon_2(g) = 0$.

It would be interesting to study how the above measures are related to other measures of complexity of sequences e.g., to discrepancy[10] and to the ones considered in the next section.

[10] see [126, 276] about the latter measure and relevant results

11.2 Properties of coordinate sequences

In this section, we study properties of coordinate sequences of generators considered in Chapters 9 and 10, that is, of both ordinary congruential and counter-dependent generators. We consider only generators that produce sequences modulo 2^n of the maximum period length, that is, we restrict ourselves to the $p = 2$ only, as this case is the most important for practical applications. Note however that a number of results obtained in this section remain true after proper re-statement in the general case, when p is arbitrary prime. We follow Anashin [24–26, 28, 29].

Recall that the jth coordinate sequence $\mathcal{X}_j = \delta_j(\mathcal{X})$ is the sequence $(\delta_j(x_i))_{i=0}^{\infty}$, where $\mathcal{X} = (x_i)_{i=0}^{\infty}$ is the output sequence of the corresponding automaton. To study coordinate sequences, it is convenient to consider a generator $\mathfrak{A}'$ with the state set $\mathbb{Z}_2$, 1-Lipschitz ergodic state transition function $f: \mathbb{Z}_2 \to \mathbb{Z}_2$ and with identity output function $F(z) = z$. We also consider a generator $\mathfrak{A}'_j$ that differs from $\mathfrak{A}'$ only by the output function, which is $\delta_j(z)$ in this case. Thus, the output sequence of the generator $\mathfrak{A}'_j$ is just the jth coordinate sequence $\mathcal{X}_j$ of the generator $\mathfrak{A}'$. Recall that according to Definition 9.1, a generator is a family of automata without input that have the same set of states, same state transition and same output functions, where the initial state runs through the set of states. So when we speak of some property of a coordinate sequence of the generator we mean that this property holds for sequences obtained at all initial states; that is, the property does not depend on the choice of the initial state of the generator (i.e., holds for all automata from the family).

The jth coordinate sequence $\mathcal{X}_j$ has rather specific structure. Namely, the following theorem holds.

Theorem 11.16. *The jth coordinate sequence $\mathcal{X}_j$ is purely periodic, and 2^{j+1} is the length of its shortest period. The second half of the period is a bitwise negation of its first half; that is,*

$$\delta_j(x_{i+2^j}) \equiv \delta_j(x_i) + 1 \pmod 2 \tag{11.12}$$

for all $i = 0, 1, 2, \ldots$.

Proof. Although this theorem immediately follows from Notes 10.14 and 10.15 at $m = 1$, we give an independent proof.

Since the mapping $f: \mathbb{Z}_2 \to \mathbb{Z}_2$ is 1-Lipschitz and ergodic, the recurrence sequence defined by the recursion $x_{i+1} = f(x_i) \bmod 2^{j+1}$ is purely periodic, and 2^{j+1} is the length of its shortest period, whereas the recurrence sequence defined by the recursion $x_{i+1} = f(x_i) \bmod 2^j$ is purely periodic, and the length of its shortest period is 2^j. As $x_{i+1} \bmod 2^{j+1} = x_{i+1} \bmod 2^j + 2^j \cdot \delta_j(x_{i+1})$, the first assertion of Theorem 11.16 follows.

If $\delta_j(x_{i+1}) = \delta_j(x_{i+1+2^j})$ for some i, from the preceding equality we obtain that $x_{i+1+2^j} \equiv x_{i+1} \pmod{2^{j+1}}$; whence

$$x_{i+t+1+2^j} \equiv f^t(x_{i+1+2^j}) \equiv f^t(x_{i+1}) \equiv x_{i+t+1} \pmod{2^{j+1}}$$

for all $t = 0, 1, 2, \ldots$, as f is 1-Lipschitz. This means that the length of the shortest period of the sequence $(x_i \bmod 2^{j+1})_{i=0}^{\infty}$ does not exceed 2^j, in contradiction with ergodicity of f, see Theorem 4.23. □

Note 11.17. Theorem 11.16 can be generalized in two directions. First, to output sequences of wreath products of automata (this is already done, see Notes 10.14 and 10.15), and second, to the case p odd.

In the latter case, provided the transformation $f: \mathbb{Z}_p \to \mathbb{Z}_p$ is 1-Lipschitz and ergodic, the jth coordinate sequence $(\delta_j(f^i(z)))_{i=0}^{\infty}$ is purely periodic, and p^{j+1} is the length of its shortest period (here and further within this remark $\delta_j(z)$ stands for the value of the jth position in the base-p expansion of z). Each subsequence $(\delta_j(f^{i+p^j t}(z)))_{t=0}^{\infty}$ is a purely periodic sequence, and p is the length of its shortest period. Moreover, in the case $j > 0$, this subsequence is generated by a transitive linear congruential generator modulo p, i.e., by a polynomial $a+x$ for appropriate $a \in \{1, 2, \ldots, p-1\}$. Thus, this subsequence is strictly uniformly distributed modulo p: Every $u \in \mathbb{Z}/p\mathbb{Z}$ occurs at the period exactly once. The 0th sequence $(\delta_0(f^i(z)))_{i=0}^{\infty}$ is generated by a (generally speaking, nonlinear) polynomial congruential generator with the recursion law $x_{i+1} \equiv g(x_i) \pmod p$, where g is a transitive modulo p polynomial over a finite field $\mathbb{F}_p$ of residues modulo p.

A proof of these assertions could be extracted from the proof of Theorem 4.55 since in view of Theorem 3.53 and Proposition 3.52 a reduction modulo p^{j+1} of every 1-Lipschitz transformation on $\mathbb{Z}_p$ can be considered as a polynomial transformation induced by an integer-valued 1-Lipschitz polynomial over $\mathbb{Q}$. So the mapping $z \mapsto f(z) \bmod p^{j+1}$ can be considered as a reduction modulo p^{j+1} of a 1-Lipschitz ergodic mapping $w: \mathbb{Z}_p \to \mathbb{Z}_p$ where $w(x) \in \mathbb{Q}[x]$. As w is uniformly differentiable everywhere on $\mathbb{Z}_p$, the conditions of Theorem 4.55 are satisfied. We leave details of the proof for the reader, and for the rest of the section we consider only the case $p = 2$.

11.2.1 Linear and 2-adic complexities

In this subsection, we study two measures of complexity of coordinate sequences of sequences produced by linear congruential generators and by counter-dependent generators: The linear complexity over a field $\mathbb{F}_2$ of two elements, and the 2-adic complexity, which was introduced by Klapper and Goresky in the paper [263].

From Definition 11.1 it follows that the linear complexity $\lambda_{\mathbb{F}}(\mathcal{S})$ of the sequence $\mathcal{S} = (s_i)_{i=0}^{\infty}$ over a field $\mathbb{F}$ is the smallest $n \in \mathbb{N}$ such that every n successive members of the sequence satisfy some non-trivial linear relation of length $n+1$, i.e., there exist $a_0, a_1, \ldots, a_n \in \mathbb{F}$, not all equal to 0, such that $a_0 s_i + a_1 s_{i+1} + \cdots + a_n s_{i+n} = 0$ for all $i = 0, 1, 2, \ldots$. In this case we also say that the polynomial $a_0 + a_1 x + \cdots + a_n x^n \in F[x]$ is a *characteristic polynomial* of the sequence $\mathcal{S}$. In other words, linear complexity is just a degree of the *minimal polynomial* of the sequence $\mathcal{S}$, that is, of the characteristic polynomial of the sequence $\mathcal{S}$ that has the smallest degree among other characteristic polynomials of $\mathcal{S}$. Note that a polynomial $g(x) \in \mathbb{F}[x]$ is a characteristic polynomial of the sequence $\mathcal{S}$ if and only if the minimal polynomial of $\mathcal{S}$ is a factor of $g(x)$; see e.g. [126] or [299] for references. In this subsection, whenever $\mathbb{F} = \mathbb{F}_p = \mathbb{Z}/p\mathbb{Z}$ is a field of p elements, we denote for brevity the linear complexity over the field $\mathbb{F}_p$ by λ_p rather than by $\lambda_{\mathbb{Z}/p\mathbb{Z}}$.

Linear complexity is one of the crucial cryptographic properties: Pseudorandom generators that produce sequences of low linear complexity are not secure since having relatively short segment of output sequence and solving the corresponding system of linear equations over $\mathbb{F}$, a cryptanalyst can find $a_0, a_1, \dots, a_n$ and thus predict with probability 1 the rest terms of the sequence. Of course, high linear complexity per se does not guarantee security. However, the following theorem shows that coordinate sequences of linear congruential generators on $\mathbb{Z}/2^n\mathbb{Z}$ whose shortest periods are of length 2^n, have high linear complexities:

Theorem 11.18. *Let $\mathcal{X}$ be a recurrence sequence over $\mathbb{Z}_2$ with the recursion law $x_{i+1} = f(x_i)$, where f is a 1-Lipschitz ergodic transformation on $\mathbb{Z}_2$. Then the linear complexity $\lambda_2(\mathcal{X}_j)$ of the jth coordinate sequence $\mathcal{X}_j = \delta_j(\mathcal{X})$ is $2^j + 1$, for all $j = 0, 1, 2, \dots$.*

To prove the theorem, we need the following lemma:

Lemma 11.19. *Let p be a prime, let $\mathcal{S}$ be a purely periodic sequence over $\mathbb{Z}/p\mathbb{Z}$, and let the length of the shortest period of $\mathcal{S}$ be p^{j+1}. Then $\lambda_p(\mathcal{S}) > p^j$.*

Proof. Since p^{j+1} is the length of a period of the sequence $\mathcal{S}$, the polynomial $x^{p^{j+1}} - 1$ over the field $\mathbb{F}_p$ is a characteristic polynomial of the sequence $\mathcal{S}$. Yet $x^{p^{j+1}} - 1 = (x-1)^{p^{j+1}}$; thus, the minimal polynomial $\mu(x)$ of the sequence $\mathcal{S}$ must be of the form $(x-1)^r$, where $r \le p^{j+1}$. However, the polynomial $x^{p^j} - 1 = (x-1)^{p^j}$ is not a characteristic polynomial of the sequence $\mathcal{S}$ since otherwise the length of some period of the sequence $\mathcal{S}$ is a factor of p^j; but the sequence $\mathcal{S}$ has no periods of length less than p^{j+1}. Hence, $\deg \mu(x) = r > p^j$ since otherwise the polynomial $(x-1)^{p^j}$ is a characteristic polynomial of $\mathcal{S}$. □

Proof of Theorem 11.18. Since $x_{i+2^j} \equiv x_i + 1 \pmod 2$ for all $i = 0, 1, 2, \dots$ (see Theorem 11.16), the congruence $x_{i+1+2^j} + x_{i+2^j} + x_{i+1} + x_i \equiv 0 \pmod 2$ holds for all $i = 0, 1, 2, \dots$. Hence, the polynomial $x^{2^j+1} + x^{2^j} + x + 1 = (x+1)^{2^j+1}$ is a characteristic polynomial of the jth coordinate sequence $\mathcal{X}_j$. Now the assertion of Theorem 11.18 follows from Lemma 11.19. □

We note that expectation of the linear complexity over $\mathbb{F}_2$ of a random binary sequence of length N is $\frac{N}{2}$. Thus, from this point coordinate sequences of linear congruential generators modulo 2^n whose shortest periods are the longest possible, i.e., of lengths 2^n, could be judged as 'looking random'.

In cryptology, they often use another measure of complexity of a binary periodic sequence $\mathcal{S}$, the *ℓ-error linear complexity*. The latter is a minimum degree of the minimal polynomial of a linear recurrence sequence $\mathcal{S}'$ over $\mathbb{F}_2$ such that $\mathcal{S}'$ has a period which coincides with the period of the sequence $\mathcal{S}$ everywhere except ℓ positions (the minimum is taken over all these sequences $\mathcal{S}'$). In other words, the ℓ-error linear complexity is the length of the shortest LFSR that produces a sequence $\mathcal{S}'$ which has the same period as $\mathcal{S}$ and coincides with $\mathcal{S}$ everywhere except for not more than ℓ binary positions at the period of $\mathcal{S}$.

Obviously, a random sequence of length L coincides with a sequence that has a period of length L approximately at $\frac{L}{2}$ places. That is, the ℓ-error linear complexity makes sense only for $\ell < \frac{L}{2}$. With respect to ℓ-error liner complexity, coordinate sequences of congruential generators with the recursion law $x_{i+1} = f(x_i)$, where f is a 1-Lipschitz ergodic transformation on $\mathbb{Z}_2$, look complex enough. Namely, the following proposition holds:

Proposition 11.20. *In the conditions of Theorem 11.18, let $\ell \geq 0$ be less than the half of the length of the shortest period of the jth coordinate sequence $\mathcal{X}_j = \delta_j(\mathcal{X})$; i.e., let $0 \leq \ell < 2^j$. Then the ℓ-error linear complexity of $\mathcal{X}_j$ exceeds 2^j.*

Proof. Let $\mathcal{E} = (\varepsilon_i)_{i=0}^{\infty}$ be a linear recurrence sequence over $\mathbb{F}_2$ such that $\mathcal{E}$ has a period of length 2^{j+1}, and $\delta_j(x_i) = \varepsilon_i$ for all $i \in \{0, 1, 2, \dots, 2^{j+1} - 1\}$ with the exception of ℓ indices $i = i_1, \dots, i_\ell \in \{0, 1, \dots, 2^{j+1} - 1\}$. Let d be a degree of the minimal polynomial $\mu(x)$ of $\mathcal{E}$. Since 2^{j+1} is the length of a period of $\mathcal{E}$, $\mu(x)$ must be a multiple of the polynomial $x^{2^{j+1}} + 1 = (X + 1)^{2^{j+1}}$ over the field $\mathbb{F}_2$. Hence, $\mu(x) = (x + 1)^d$, and $d \leq 2^{j+1}$.

On the other hand, as $\ell < 2^j$, then in view of (11.12) the length of the shortest period of the sequence $\mathcal{E}$ cannot be less than 2^{j+1}. Hence, $d \geq 2^j + 1$, since otherwise $\mu(x)$ is a multiple of $(x + 1)^{2^j} = x^{2^j} + 1$, and so $\mathcal{E}$ has a period of length 2^j. □

Theorem 11.18 can be expanded to output sequences of counter-dependent generators from Theorem 10.9. Namely, the following proposition holds.

Proposition 11.21. *Let $\mathcal{X}$ be a sequence from Theorem 10.9. Then the linear complexity of the jth coordinate sequence $\mathcal{X}_j$ exceeds 2^j, for all $j = 0, 1, 2, \dots$.*

Proof. Since the sequence $\mathcal{X}_j$ has a period of length $m2^n$ (see Lemma 10.12), the polynomial $u(x) = x^{m2^{j+1}} - 1 = (x^m - 1)^{2^{j+1}}$ is a characteristic polynomial of the sequence $\mathcal{X}_j$. Thus, the minimal polynomial $\mu(x)$ of the sequence $\mathcal{X}_j$ is a factor of $u(x)$. On the other hand, $\mu(x)$ is not a factor of $w(x) = (x^m - 1)^{2^j}$ since otherwise the sequence $\mathcal{X}_j$ has a period of length $m2^j$; however, the latter is impossible since the second half of the period of length $m2^{j+1}$ of this sequence is a bitwise negation of the first half, see Note 10.15. Since both polynomials $u(x)$, $w(x)$ have the same set of roots in their splitting field, at least one of these roots must be a root of the polynomial $\mu(x)$, and the multiplicity of this root must exceed 2^j. Thus, $\deg \mu(x) > 2^j$. □

As it can be seen from the proof, Proposition 11.21 holds for $m = 1$ as well, turning into Theorem 11.16 in this case. Thus, we may say that the lower bound for $\lambda_2(\mathcal{X}_j)$ that gives Proposition 11.21 is sharp. However, this bound can be improved for special choices of m. For instance, if $m = 2^k$, then $\lambda_2(\mathcal{X}_j) = m2^j + 1$ in view of Note 10.19 and Theorem 11.18. Also, if $m = m_1 2^k$, where m_1 is odd, then the proof of Proposition 11.21 shows that $\lambda_2(\mathcal{X}_j) > 2^{j+k}$ in this case.

So it seems possible to improve significantly the bound for linear complexity that is given by Proposition 11.21 in the case $m > 1$. To do this, we have to run a bit ahead and to use Theorem 11.28 that is proved further. With the use of this theorem, the general case can be reduced to the case $m > 1$ odd. Namely, in view of Theorem 11.28, every purely periodic binary sequence with the period of length $m2^n$, $n > 1$, such that the second half of this period is a bitwise negation of its first part, can be considered as the $(n-1)$th coordinate sequence of a certain wreath product of automata that is described by Theorem 10.9. Thus, if $m = m_1 2^k$, where m_1 odd, this sequence in view of Theorem 11.28 can be considered as $(n-1+k)$th coordinate sequence of a suitable wreath product of automata mentioned in Theorem 10.9 for $m = m_1$ odd. Thus we can assume that m is odd.

Proceeding with this note and using the congruence $\delta_{n-1}(x_{i+2^{n-1}\ell}) \equiv \delta_{n-1}(x_i)+1 \pmod 2$ (see Note 10.15) we conclude that the minimal polynomial $\mu_{n-1}(x)$ of the sequence $\mathcal{X}_{n-1} = \delta_{n-1}(\mathcal{X})$ is a factor of the polynomial

$$x^{m2^{n-1}+1} + x^{m2^{n-1}} + x + 1 = (x^m + 1)^{2^{n-1}} \cdot (x+1)$$
$$= (x^{m-1} + \cdots + x + 1)^{2^{n-1}} \cdot (x+1)^{2^{n-1}+1}.$$

Thus, the root of multiplicity $> 2^{n-1}$ from the proof of Proposition 11.21 is 1 (since the polynomial $x^{m-1} + \cdots + x + 1$ is a factor of $x^m - 1$; yet $x^m - 1$ has no roots of multiplicity > 1 in its splitting field, as m is odd). Hence,

$$\mu_{n-1}(x) = v(x) \cdot (x+1)^{2^{n-1}+1}, \tag{11.13}$$

where $v(x)$ is a factor of $(x^{m-1} + \cdots + x + 1)^{2^{n-1}}$. Thus,

$$m2^{n-1} + 1 \geq \deg \mu_{n-1}(x) = \lambda_2(\delta_{n-1}(\mathcal{X})) \geq 2^{n-1} + 1. \tag{11.14}$$

We shall show now that for $n > 1$ both these bounds are sharp.

Consider a finite sequence $\mathcal{T}$ of length $m2^{n-1}$ consisting of gaps and runs (alternating blocks of 0s and 1s, respectively) of length 2^{n-1} each. Take this sequence as the first half of a period of a sequence $\mathcal{S}$, and take a bitwise negation $\hat{\mathcal{T}}$ of $\mathcal{T}$ as a second half of a period of $\mathcal{S}$ (of course $\hat{\mathcal{T}} = (\mathcal{T})\,\mathsf{XOR}\,(2^{2^{n-1}\ell} - 1)$, where we consider $\mathcal{T}$ as a canonical 2-adic representation of a suitable rational integer $\gamma_{n-1} > 0$). Obviously, $\mathcal{S}$

is a purely periodic sequence with a period of length $m2^n$, and the second half of this period is a bitwise negation of its first half. Thus, as it is shown by Theorem 11.28, the sequence $\mathcal{S}$ is the $(n-1)$th coordinate sequence of a suitable wreath product of automata described by Theorem 10.9. Yet obviously $\mathcal{S}$ is a sequence of gaps and runs of length 2^{n-1} each; thus, the length of the shortest period of the sequence $\mathcal{S}$ is 2^n. So the linear complexity $\lambda_2(\mathcal{S})$ of the sequence $\mathcal{S}$ is $2^{n-1}+1$, see the proof of Theorem 11.18.

Now we prove that the upper bound in (11.14) is also sharp. Consider a sequence $\mathcal{U}$ of gaps and runs of length 2^{n-1} each, and consider a purely periodic sequence $\mathcal{V}$ with a period of length $m2^{n-1}$; let the latter period consists of a run of length $(m-1)\cdot 2^{n-1}$ followed by a gap of length 2^{n-1}. Let $\mu_{\mathcal{U}}(x)$ and $\mu_{\mathcal{V}}(x)$ be minimal polynomials of corresponding sequences.

Since $\mathcal{U}$ is a purely periodic sequence whose shortest period is of length 2^n, and the second half of this period is a bitwise negation of its first half, the polynomial $\mu_1(x) = x^{2^{n-1}+1} + x^{2^{n-1}} + x + 1 = (x+1)^{2^{n-1}+1}$ is a characteristic polynomial of the sequence $\mathcal{U}$ (see the argument above); so $\mu_{\mathcal{U}}(x)$ is a factor of $\mu_1(x)$. However, the first 2^{n-1} overlapping (2^{n-1})-tuples considered as vectors of dimension 2^{n-1} over the field $\mathbb{F}_2$ are obviously linearly independent. Hence, $\deg \mu_{\mathcal{U}}(x) > 2^{n-1}$ (see [299, Theorem 8.51]). Finally we conclude that $\mu_{\mathcal{U}}(x) = \mu_1(x)$. A similar argument proves that $\mu_{\mathcal{V}}(x) = x^{(m-1)\cdot 2^{n-1}} + x^{(m-2)\cdot 2^{n-1}} + \cdots + x^{2^{n-1}} + 1$.

Now consider a sum $\mathcal{R}$ of these two sequences over $\mathbb{F}_2$; i.e., $\mathcal{R} = \mathcal{U}$ XOR $\mathcal{V}$. Obviously, $\mu_{\mathcal{U}}(x)$ and $\mu_{\mathcal{V}}(x)$ has no common divisor of degree > 0 since 1 is the only root of $\mu_{\mathcal{U}}(x)$, and 1 is not a root of $\mu_{\mathcal{V}}(x)$ (recall that m is odd). Thus, $\mu_{\mathcal{U}}(x)\cdot\mu_{\mathcal{V}}(x)$ is a minimal polynomial of the sequence $\mathcal{R}$ (see [299, Theorem 8.57]). Hence, $\lambda_2(\mathcal{R}) = m2^{n-1}+1$.

As m is odd, $\mathcal{R}$ is obviously a purely periodic sequence, the length of its shortest period is $m2^n$, and the second half of this period is a bitwise negation of its first half. Consequently, in force of Theorem 11.28, the sequence $\mathcal{R}$ is the $(n-1)$th coordinate sequence of a suitable wreath product of automata from Theorem 10.9.

As a bonus we have that the exact period length P of the $(n-1)$th coordinate sequence $\delta_{n-1}(\mathcal{X})$ for odd m is a multiple of 2^n: Since x^P+1 is a characteristic polynomial of the sequence $\delta_{n-1}(\mathcal{X})$, $\mu_{n-1}(x)$ is a factor of x^P+1. Yet $x^P+1 = (x^s+1)^{2^t} = (x+1)^{2^t}(x^{s-1}+\cdots+1)^{2^t}$, where $P = s2^t$, s odd, and 1 is not a root of $x^{s-1}+\cdots+1$ since s is odd. Thus, necessarily $2^t \geq 2^{n-1}+1$ in view of (11.13). Hence, $t \geq n$. So we conclude that $P = s2^n$; yet $P \leq m2^n$ since the sequence $\mathcal{X} \bmod 2^n$ is a purely periodic sequence, and the length of its shortest period is $m2^n$ in force of Theorem 10.9. Thus, $P = s2^n$, where $1 \leq s \leq m$. As it is demonstrated by sequences $\mathcal{S}$ and $\mathcal{R}$, both extreme cases $s = 1$ and $s = m$ occur.

We summarize the above considerations in the following theorem:

Theorem 11.22. *Let $\mathcal{X}_j$, $j > 0$, be the jth coordinate sequence of the sequence $\mathcal{X}$ from Theorem 10.9; so $\mathcal{X}_j$ is a purely periodic binary sequence with a period of length*

$m2^{j+1}$. Represent $m = r2^k$, where r is odd. Then the length of the shortest period of the sequence $\mathcal{X}_j$ is $s2^{k+j+1}$ for some $s \in \{1, 2, \dots, r\}$, and both extreme cases $s = 1$ and $s = r$ occur: For every sequence $s_1, s_2, \dots$ over the set $\{1, r\}$ there exists a sequence $\mathcal{X}$ from Theorem 10.9 such that the length of the shortest period of the jth coordinate sequence $\mathcal{X}_j$ is $2^{k+j+1}s_j$, for all $j = 1, 2, \dots$.

Moreover, the linear complexity $\lambda_2(\mathcal{X}_j)$ of the sequence $\mathcal{X}_j$ satisfies the following inequality:

$$2^{k+j} + 1 \le \lambda_2(\mathcal{X}_j) \le r2^{k+j} + 1.$$

Both these bounds are sharp: For every sequence $t_1, t_2, \dots$ over the set $\{1, r\}$ there exists a sequence $\mathcal{X}$ from Theorem 10.9 such that the linear complexity of the jth coordinate sequence $\mathcal{X}_j$ is $t_j 2^{k+j} + 1$, for all $j = 1, 2, \dots$.

Proof. Nearly everything is already done by the preceding argument. We only note that in view of mentioned Theorem 11.28, we can choose coordinate sequences independently one of another. That is, given purely periodic binary sequences $\mathcal{X}_1, \mathcal{X}_2, \dots$, such that every sequence $\mathcal{X}_j$, $j = 1, 2, \dots$, has a period of length $m2^{j+1}$, and the second half of this period is a bitwise negation of its first half, there exists a sequence $\mathcal{X}$ from Theorem 10.9 such that its jth coordinate sequence $\delta_j(\mathcal{X})$ coincides with the sequence $\mathcal{X}_j$, for all $j = 1, 2, \dots$. □

With the use of Theorem 11.16 it is possible to estimate two other measures of complexity of coordinate sequences. These measures were introduced in [263]; these are *2-adic complexity* and *2-adic span*. Whereas the linear complexity $\lambda_2(\mathcal{S})$ of a binary sequence $\mathcal{S}$ is the number of cells in a linear feedback shift register (LFSR) that outputs the sequence $\mathcal{S}$, the 2-adic span is the number of cells in both memory and register of a feedback with carry shift register (FCSR) that outputs $\mathcal{S}$, and the 2-adic complexity estimates the number of cells in the register of this FCSR. Actually FCSR is a generator that produces an (eventually) periodic binary sequence $s_i = (a\xi^i \bmod q) \bmod 2$, $i = 0, 1, 2, \dots$, where $a, q \in \mathbb{N}$ are some integers, q is odd, $2\xi \equiv 1 \pmod{q}$. The output can be considered as a 2-adic canonical representation of an irreducible fraction with odd denominator. By the definition, the 2-adic complexity $\mathrm{C}_2(\mathcal{S})$ of the (eventually) periodic sequence $\mathcal{S} = s_0, s_1, s_2, \dots$ over $\mathbb{Z}/2\mathbb{Z}$ is $\log_2(\mathrm{C}(u, v))$, where $\mathrm{C}(u, v) = \max\{|u|, |v|\}$ and $\frac{u}{v} \in \mathbb{Q}$ is the irreducible fraction such that its 2-adic expansion agrees with $\mathcal{S}$; that is, $\frac{u}{v} = s_0 + s_1 \cdot 2 + s_2 \cdot 2^2 + \cdots \in \mathbb{Z}_2$. The number of cells in the register of FCSR that produces $\mathcal{S}$ is then $\lceil \log_2(\mathrm{C}(u, v)) \rceil$, the least rational integer that is not smaller than $\log_2(\mathrm{C}(u, v))$. Thus, to estimate 2-adic complexity of the jth coordinate sequence $\mathcal{X}_j$ of output of a congruential generator with the recursion law $x_{i+1} = f(x_i)$, where f is a 1-Lipschitz ergodic transformation on the space $\mathbb{Z}_2$, we only need to estimate $\mathrm{C}_2(\mathcal{X}_j)$.

Theorem 11.23. *Let $\mathcal{X}_j = \zeta_0, \zeta_1, \zeta_2, \dots$ be the jth coordinate sequence of the recurrence sequence $\mathcal{X}$ defined by the recursion $x_{i+1} = f(x_i)$, where f is a 1-Lipschitz* *ergodic transformation on the space* $\mathbb{Z}_2$*; that is,* $\zeta_i = \delta_j(x_i)$, $i = 0, 1, 2, \ldots$. *Then* $\mathrm{C}_2(\mathcal{X}_j) = \log_2\left(\frac{2^{2^j}+1}{\gcd(2^{2^j}+1,\gamma+1)}\right)$, *where* $\gamma = \zeta_0 + \zeta_1 \cdot 2 + \zeta_2 \cdot 2^2 + \cdots + \zeta_{2^j-1} \cdot 2^{2^j-1}$, *and* gcd *stands for the greatest common divisor.*

Note 11.24. We note that γ is a non-negative rational integer, $0 \le \gamma \le 2^{2^j} - 1$; and that for each γ from this range there exists a 1-Lipschitz ergodic transformation f on $\mathbb{Z}_2$ such that the first half of the period of the jth coordinate sequence $\mathcal{X}_j$ of the corresponding output $\mathcal{X}$ is a base-2 expansion of γ (see further Theorem 11.26). Thus, to find all possible values of the 2-adic complexity of the jth coordinate sequence $\mathcal{X}_j$ one must decompose the jth Fermat number $2^{2^j} + 1$. It is known that the jth Fermat number is prime for $0 \le j \le 4$ and that it is composite for $5 \le j \le 23$. For each Fermat number outside this range it is not known whether it is prime or composite. The complete decomposition of jth Fermat number is not known for $j > 11$. Assuming that for some $j \ge 2$ the jth Fermat number is composite, all its factors are of the form $t2^{j+2} + 1$, see e.g. [76] for further references. So, *the following bounds for 2-adic complexity* $\mathrm{C}_2(\mathcal{X}_j)$ *of the* j*th coordinate sequence* $\mathcal{X}_j$ *hold*:

$$j + 3 \le \lceil \mathrm{C}_2(\mathcal{X}_j) \rceil \le 2^j + 1;$$

however, to prove whether the lower bound is sharp for certain $j > 11$, *or whether* $\lceil \mathrm{C}_2(\mathcal{X}_j) \rceil$ *could be actually less than* $2^j + 1$ *for* $j > 23$ *is as difficult as to decompose the* j*th Fermat number or, respectively, to determine whether the* j*th Fermat number is prime or composite.*

Proof of Theorem 11.23. We only have to express $\zeta_0 + \zeta_1 \cdot 2 + \zeta_2 \cdot 2^2 + \cdots$ as an irreducible fraction. Denote $\gamma = \zeta_0 + \zeta_1 \cdot 2 + \zeta_2 \cdot 2^2 + \cdots + \zeta_{2^j-1} \cdot 2^{2^j-1}$. Then using the identity $u + \mathsf{NOT}\, u = -1$ of (8.4), by Theorem 11.16 we conclude that $\zeta_0 + \zeta_1 \cdot 2 + \zeta_2 \cdot 2^2 + \cdots + \zeta_{2^{j+1}-1} \cdot 2^{2^{j+1}-1} = \gamma + 2^{2^j} \cdot (2^{2^j} - \gamma - 1) = \gamma'$ and hence $\zeta_0 + \zeta_1 \cdot 2 + \zeta_2 \cdot 2^2 + \cdots = \gamma' + \gamma' \cdot 2^{2^{j+1}} + \gamma' \cdot 2^{2 \cdot 2^{j+1}} + \gamma' \cdot 2^{3 \cdot 2^{j+1}} + \cdots = \frac{\gamma+1}{2^{2^j}+1} - 1$. This completes the proof in view of the definition of the 2-adic complexity of a sequence. □

Note 11.25. Similar estimates of $\mathrm{C}_2(\delta_j(\mathcal{X}))$ can be obtained for the sequence $\mathcal{X}$ produced by a wreath product of automata from Theorem 10.9. In view of Note 11.17 the argument of the proof of Theorem 11.23 shows that the representation of the binary sequence $\delta_j(\mathcal{X})$ as a 2-adic integer is $\frac{\gamma+1}{2^{2^j m}+1} - 1$, so we have only to study a fraction $\frac{\gamma+1}{2^{2^j m}+1}$, where $\gamma = \zeta_0 + \zeta_1 \cdot 2 + \zeta_2 \cdot 2^2 + \cdots + \zeta_{2^j m-1} 2^{2^j m-1}$, and m is of the statement of Theorem 10.9. Representing $m = 2^k m_1$ with $m_1 > 1$ odd, we can factorize $2^{2^j m} + 1 = (2^{2^{j+k}} + 1)(2^{2^{j+k}(m_1-1)} - 2^{2^{j+k}(m_1-2)} + \cdots - 2^{2^{j+k}} + 1)$, but the problem does not become much easier because of the first multiplier. We omit further details.

11.2.2 Structure of coordinate sequences

Both Theorems 11.18 and 11.23, as well as Proposition 11.20, show that all three measures of complexity of a sequence (linear complexity, ℓ-error linear complexity, and 2-adic complexity) are not too sensitive. For instance, consider a very simple recurrence sequence $\mathcal{X}$ of 2-adic integers that is defined by the recursion $x_{i+1} = x_i + 1$, $i = 0, 1, 2, \ldots$, $x_0 = 0$. We see that both linear and 2-adic complexities of the jth coordinate sequence $\mathcal{X}_j$ depend on j exponentially: $\lambda_2(\mathcal{X}_j) = \mathrm{C}_2(\mathcal{X}_j) = 2^j + 1$. However, in this case $\mathcal{X}_j$ is merely a sequence of gaps and runs (alternating blocks of 0s and 1s) of length 2^j each.

From the proofs of corresponding results it is easy to observe that such big figures for linear and 2-adic complexities in this example are just a consequence of a very simple law the jth coordinate sequence obeys: The second half of the period is a bitwise negation of the first half, see Theorem 11.16. Intuitively it is clear that binary sequences that satisfy this law are as complex as the first halves of their periods. So it is important to investigate what sequences of length 2^j could be outputted as the first half of the period of the jth coordinate sequence of sequences produced by 1-Lipschitz ergodic transformations on the space $\mathbb{Z}_2$ and by counter-dependent generators of the longest period. So in this subsection we study what values takes the rational integer γ from Theorem 11.23.

In other words, let $\gamma_j(f, z) \in \mathbb{N}_0$ be such a number that its base-2 expansion agrees with the first half of the period of the jth coordinate sequence produced by the 1-Lipschitz ergodic transformation f on $\mathbb{Z}_2$; i.e., let

$$\gamma_j(f, z) = \delta_j(f^0(z)) + 2 \cdot \delta_j(f^1(z)) + 4 \cdot \delta_j(f^2(z)) + \cdots + 2^{2^j-1} \cdot \delta_j(f^{2^j-1}(z)).$$

Obviously, $0 \le \gamma_j(f, z) \le 2^{2^j} - 1$. The following question arises naturally:

> Given a 1-Lipschitz ergodic mapping $f\colon \mathbb{Z}_2 \to \mathbb{Z}_2$ and a 2-adic integer $z \in \mathbb{Z}_2$, what infinite string
>
> $$\gamma_0 = \gamma_0(f, z), \gamma_1 = \gamma_1(f, z), \gamma_2 = \gamma_2(f, z), \ldots,$$
>
> where $\gamma_j \in \{0, 1, \ldots, 2^{2^j} - 1\}$ for $j = 0, 1, 2, \ldots$, can be obtained?

And the answer is *any one*. Namely, the following theorem holds:

Theorem 11.26. *Given an arbitrary sequence $\mathcal{G} = (\gamma_j)_{j=0}^{\infty}$ of non-negative rational integers that satisfies the inequalities $0 \le \gamma_j \le 2^{2^j} - 1$, $j = 0, 1, 2, \ldots$, there exist a 1-Lipschitz ergodic mapping $f\colon \mathbb{Z}_2 \to \mathbb{Z}_2$ and a 2-adic integer $z \in \mathbb{Z}_2$ such that*

$$\delta_j(f^i(z)) \equiv \delta_{i \bmod 2^j}(\gamma_j) + \left\lfloor \frac{i}{2^j} \right\rfloor \pmod 2$$

for all $i, j \in \mathbb{N}_0$.

Note 11.27. The sequence $\left(\lfloor \frac{i}{2^j} \rfloor \bmod 2\right)_{i=0}^{\infty}$ is merely a binary sequence of alternating gaps and runs (i.e., blocks of consecutive 0s or 1s, respectively) of length 2^j each.

Proof of Theorem 11.26. Put $z = z_0 = \sum_{j=0}^{\infty} \delta_0(\gamma_j)2^j$ and put

$$z_i = \sum_{j=0}^{\infty} \left(\left(\delta_{i \bmod 2^j}(\gamma_j) + \left\lfloor \frac{i}{2^j} \right\rfloor \right) \bmod 2 \right) \cdot 2^j$$

for $i = 1, 2, 3, \ldots$. Consider a sequence $\mathcal{Z} = (z_i)_{i=0}^{\infty}$ of 2-adic integers. Speaking informally, we are filling a table with countable infinite number of rows and columns in such a way that the first 2^j entries of the jth column represent γ_j in its base-2 expansion, and the other entries of this column are obtained from these by applying recursive relation (11.12) from Theorem 11.16. Then the ith row of the table can be considered as a 2-adic canonical representation of z_i, $i = 0, 1, 2, \ldots$.

We shall prove that $\mathcal{Z}$ is dense in $\mathbb{Z}_2$, and then we shall define f on $\mathcal{Z}$ in such a way that makes f 1-Lipschitz and ergodic on $\mathcal{Z}$. This will imply the assertion of the theorem.

Proceeding along this way we claim that $\mathcal{Z} \bmod 2^k = \mathbb{Z}/2^k\mathbb{Z}$ for all $k = 1, 2, \ldots$; that is, a natural ring epimorphism $\bmod 2^k : z \mapsto z \bmod 2^k$ maps $\mathcal{Z}$ onto the residue ring $\mathbb{Z}/2^k\mathbb{Z}$. Indeed, this trivially holds for $k = 1$. Assuming our claim holds for $k < m$ we shall prove it for $k = m$. Given arbitrary $t \in \{0, 1, \ldots, 2^m - 1\}$ there exists $z_i \in \mathcal{Z}$ such that $z_i \equiv t \pmod{2^{m-1}}$. If $z_i \not\equiv t \pmod{2^m}$ then $\delta_{m-1}(z_i) \equiv \delta_{m-1}(t) + 1 \pmod 2$ and thus $\delta_{m-1}(z_{i+2^{m-1}}) \equiv \delta_{m-1}(t) \pmod 2$. However, $z_{i+2^{m-1}} \equiv z_i \pmod{2^{m-1}}$. Hence $z_{i+2^{m-1}} \equiv t \pmod{2^m}$.

A similar argument shows that for each $k \in \mathbb{N}$ the sequence $(z_i \bmod 2^k)_{i=0}^{\infty}$ is purely periodic, has a period of length 2^k, and each $t \in \{0, 1, \ldots, 2^k - 1\}$ occurs at the period exactly once (in particular, all terms of $\mathcal{Z}$ are pairwise distinct 2-adic integers). Moreover, $i \equiv i' \pmod{2^k}$ if and only if $z_i \equiv z_{i'} \pmod{2^k}$. Consequently, $\mathcal{Z}$ is dense in $\mathbb{Z}_2$ since for each $t \in \mathbb{Z}_2$ and each $k \in \mathbb{N}$ there exists $z_i \in \mathcal{Z}$ such that $|z_i - t|_2 \le 2^{-k}$. Moreover, if we put $f(z_i) = z_{i+1}$ for all $i = 0, 1, 2, \ldots$ then $|f(z_i) - f(z_{i'})|_2 = |z_{i+1} - z_{i'+1}|_2 = |(i+1) - (i'+1)|_2 = |i - i'|_2 = |z_i - z_{i'}|_2$. Hence, f is well defined on $\mathcal{Z}$ and 1-Lipschitz with respect to the 2-adic metric. Thus, the continuation of f to the whole space $\mathbb{Z}_2$ is 1-Lipschitz as well. Yet f is transitive modulo 2^k for each $k \in \mathbb{N}$, so this continuation is ergodic in view of Theorem 4.23. □

Theorem 11.26 can be extended to coordinate sequences of wreath products of automata, namely, to sequences $\mathcal{X}_j = \delta_j(\mathcal{X}) = (\delta_j(x_i))_{i=0}^{\infty}$, where $\mathcal{X} = (x_i)_{i=0}^{\infty}$ is a recurrence sequence from Theorem 10.9. It turns out that, in loose terms, *each first half of the period of every coordinate sequence $\mathcal{X}_j$ ($j \ge 1$) of wreath products of automata can be chosen arbitrarily and independently of others.* Now we give a formal statement and a proof of it.

Recall that $\delta_j(\mathcal{X})$ is a purely periodic binary sequence with the period of length $2^{j+1}m$, and the second half of the period is a bitwise negation of its first half, see Lemma 10.12. Thus, we associate the sequence $\delta_j(\mathcal{X})$ to a rational number (which we denoted by the same symbol $\delta_j(\mathcal{X})$) that has canonical 2-adic representation $\delta_j(x_0) + \delta_j(x_1) \cdot 2 + \delta_j(x_2) \cdot 2^2 + \cdots$. Hence by Note 11.25,

$$\frac{2^{2^j m} - \gamma_j}{2^{2^j m} + 1} = \delta_j(\mathcal{X}), \tag{11.15}$$

where $\gamma_j = \delta_j(x_0) + \delta_j(x_1) \cdot 2 + \delta_j(x_2) \cdot 2^2 + \cdots + \delta_j(x_{2^j m-1}) \cdot 2^{2^j m-1}$, and m, x_i are from the statement of Theorem 10.9. In other words, the base-2 expansion of the number $\gamma_j \in \mathbb{N}_0$ agrees with the $2^j m$ initial terms of the sequence $(\delta_j(x_i))_{i=0}^{\infty}$, where $x_{i+1} = g_{i \bmod m}(x_i)$, and $g_0, \ldots, g_{m-1}$ is a finite sequence of 1-Lipschitz measure-preserving transformations that satisfies Theorem 10.9. Thus, $\gamma_j \in \{0, 1, \ldots, 2^{2^j m} - 1\}$, and γ_j depends on x_0 and on the sequence $g_0, \ldots, g_{m-1}$. Any purely periodic sequence with a period of length $2^{j+1}m$ such that the second half of the period is a bitwise negation of the first half, can be considered as a canonical 2-adic representation of a rational number, see (11.15) and the proof of Note 11.25. Thus, we wonder what sequences of this kind can be represented by coordinate sequences of wreath products of automata from Theorem 10.9.

In other words, to every sequence $\mathcal{X}$ from Theorem 10.9 we associate a sequence $\mathcal{G}(\mathcal{X}) = (\gamma_0, \gamma_1, \ldots)$ of non-negative rational integers γ_j such that $0 \le \gamma_j \le 2^{2^j m} - 1$ if and only if equality (11.15) holds for all $j = 0, 1, 2, \ldots$. Now we take an arbitrary sequence $\mathcal{G}$ of this type and wonder whether this sequence can be associated to some sequence $\mathcal{X}$ from Theorem 10.9. Generally speaking, the answer is *no*. Indeed, according to Theorem 10.9 the sequence $\delta_0(\mathcal{X})$ is a purely periodic sequence with the *shortest* period of length $2m$. Yet, if a purely periodic binary sequence $\mathcal{S}$ that has a period of length $2^n m$ such that the second half of this period is a bitwise negation of the its first half, i.e., the sequence $\mathcal{S}$ that can be represented in the form (11.15) as $\mathcal{S} = \frac{2^{2m} - \gamma_0}{2^{2m}+1}$ for a suitable $0 \le \gamma_0 \le 2^{2m} - 1$, then the length of the shortest period of this sequence is *not necessarily* $2^n m$; see the example that follows Note 10.15.

However, according to Note 10.14, for $j > 0$ coordinate sequences $\delta_j(\mathcal{X})$ may have periods which are shorter than $2^{j+1}m$; so it is reasonable to ask whether an arbitrary sequence $\mathcal{G} = \gamma_1, \gamma_2, \ldots$ of non-negative rational integers γ_j that satisfy the inequality $0 \le \gamma_j \le 2^{2^j m} - 1$, corresponds to some sequence $\mathcal{X}$ from Theorem 10.9 if we discard $\delta_0(\mathcal{X})$; that is, given $\mathcal{G}$, whether there exists a positive rational integer m and a sequence $\mathcal{X}$ from Theorem 10.9 such that $\delta_j(\mathcal{X})$ satisfy (11.15) for all $j > 0$. To this question, the answer is *yes*. The following theorem holds:

Theorem 11.28. *Let $m > 1$ be a rational integer, and let $\mathcal{G} = \gamma_0, \gamma_1, \ldots$ be an arbitrary sequence of non-negative rational integers $\gamma_j \in \{0, 1, \ldots, 2^{2^j m} - 1\}$, $j = 0, 1, 2, \ldots$. Then there exist a finite sequence $g_0, \ldots, g_{m-1}$ of 1-Lipschitz measure-preserving transformations on $\mathbb{Z}_2$ that satisfies the conditions of Theorem 10.9, and* *a 2-adic integer $x_0 \in \mathbb{Z}_2$, such that coordinate sequences $\delta_j(\mathcal{X})$ of the recurrence sequence $\mathcal{X} = (x_0, x_1, \ldots)$ of 2-adic integers that is defined by the recursion $x_{i+1} = g_{i \bmod m}(x_i)$, $i = 0, 1, 2, \ldots$, satisfy equality* (11.15), *for all $j = 1, 2, 3, \ldots$.*

Proof. According to Theorem 4.39, a mapping $g_i : \mathbb{Z}_2 \to \mathbb{Z}_2$ is a 1-Lipschitz measure-preserving transformation of the space $\mathbb{Z}_2$ if and only if each Boolean function $\delta_j(g_i(x))$ in Boolean variables $\chi_0 = \delta_0(x), \chi_1 = \delta_1(x), \ldots$ can be represented as

$$\delta_j(g_i(x)) = \chi_j \oplus \varphi_j^i(\chi_0, \ldots, \chi_{j-1}),$$

where $\varphi_j^i = \varphi_j^i(\chi_0, \ldots, \chi_{j-1})$ is a Boolean function in Boolean variables $\chi_0, \ldots, \chi_{j-1}$. Thus, the 1-Lipschitz measure-preserving transformation g_i is completely determined by the sequence $\varphi_0^i, \varphi_1^i, \ldots$ of corresponding Boolean functions. So, given a sequence $\mathcal{G}$, we must determine $x_0 \in \mathbb{Z}_2$ and a family $\{\varphi_j^i : i = 0, 1, \ldots, m-1; j = 0, 1, 2, \ldots\}$ of Boolean functions so that respective measure-preserving mappings g_k, $k = 0, 1, \ldots, m-1$, satisfy Theorem 10.9, and that $\delta_j(\mathcal{X})$ satisfy (11.15) for all $j = 1, 2, \ldots$, where the recurrence sequence $\mathcal{X} = (x_0, x_1, \ldots)$ is defined by the recursion $x_{i+1} = g_{i \bmod m}(x_i)$, $i = 0, 1, 2, \ldots$.

To start with, we put $x_0 = \delta_0(\gamma_0) + \delta_0(\gamma_1) \cdot 2 + \delta_0(\gamma_2) \cdot 2^2 + \cdots \in \mathbb{Z}_2$. Further we describe an inductive procedure to determine φ_j^i successively for $j = 0, 1, 2, \ldots$.

For $j = 0$ we put arbitrary $g_0(0) = \varphi_0^0, \ldots, g_{m-1}(0) = \varphi_0^{m-1} \in \{0, 1\}$ that satisfy conditions 1 and 2 of Theorem 10.9. So we define all mappings $g_i \bmod 2$, $i = 0, 1, \ldots, m-1$. Note also that the recurrence sequence $\mathcal{X}_0 = (\xi_0^0, \xi_0^1, \ldots)$ defined by recursion $\xi_0^0 = x_0 \bmod 2$, $\xi_{k+1}^0 = g_{k \bmod m}(\xi_k^0) \bmod 2$ is a purely periodic sequence over $\mathbb{Z}/2\mathbb{Z} = \{0, 1\}$ with the shortest period of length $2m$, that every element of $\mathbb{Z}/2\mathbb{Z}$ occurs at the period exactly m times, and that $\xi_{k+m}^0 \equiv \xi_k^0 + 1 \pmod 2$ (cf. the very beginning of the proof of Lemma 10.12).

Suppose that we have already find Boolean functions φ_j^i for $j = 0, 1, \ldots, n-1$, $i = 0, 1, \ldots, m-1$ so that all terms of the recurrence sequence $\mathcal{X}_{n-1} = (\xi_0^{n-1}, \xi_1^{n-1}, \ldots)$ that is defined by the recurrence $\xi_0^{n-1} = x_0 \bmod 2^n$, $\xi_{k+1}^{n-1} = g_{k \bmod m}(\xi_k^{n-1}) \bmod 2^n$, satisfy the congruence $\delta_j(\xi_{k+2^{n-1}m}^{n-1}) \equiv \delta_j(\xi_k^{n-1}) + 1 \pmod 2$, for all $j = 0, 1, \ldots, n-1$ and $k = 0, 1, 2, \ldots$. Note that then easy induction on j (which actually is already done during the proof of Claim 3 of Lemma 10.12) shows that for any k

$$\#\{\xi_{k+sm}^{n-1} : s = 0, 1, \ldots, 2^n - 1\} = 2^n. \tag{11.16}$$

Hence, $\mathcal{X}_{n-1}$ is a purely periodic sequence over the residue ring $\mathbb{Z}/2^n\mathbb{Z}$, the length of its shortest period is $2^n m$, and each element from $\mathbb{Z}/2^n\mathbb{Z}$ occurs at the period exactly m times. Now we find Boolean function φ_n^i for $i = 0, 1, \ldots, m-1$.

Given a Boolean function φ in Boolean variables $\chi_0, \ldots, \chi_s$ and a 2-adic integer $z \in \mathbb{Z}_2$, denote $\varphi(z) = \varphi(\delta_0(z), \ldots, \delta_s(z))$. Proceeding with this notation, put

$$\varphi_n^{k \bmod m}(\xi_k^{n-1}) \equiv \delta_k(\gamma_n) + \delta_{k+1}(\gamma_n) \pmod 2, \tag{11.17}$$

for $k = 0, 2, \ldots, 2^n m - 2$. Put also

$$\varphi_n^{m-1}(\xi_{2^n m-1}^{n-1}) \equiv \delta_{2^n m-1}(\gamma_n) + \delta_0(\gamma_n) + 1 \pmod 2. \tag{11.18}$$

Note that in view of (11.17) and (11.16), Boolean functions φ_n^i, $i = 0, 1, \ldots, m-2$ are well defined. Also, the Boolean function φ_n^{m-1} is well defined in view of (11.18), (11.17), and (11.16).

Consider now a recurrence sequence $\mathcal{E}_n = (\varepsilon_k)_{k=0}^{\infty}$ over $\mathbb{Z}/2\mathbb{Z}$ that is defined by the recursion $\varepsilon_0 = \delta_0(\gamma_n)$, $\varepsilon_{k+1} = \varepsilon_k + \varphi_n^{k \bmod m}(\xi_k^{n-1}) \pmod 2$. In view of (11.17) we conclude that $\varepsilon_k = \delta_k(\gamma_n)$ for $k = 0, 2, \ldots, 2^n m - 1$, and that $\varepsilon_{2^n m} \equiv \delta_0(\gamma_n) + 1 \pmod 2$, by (11.18). However, $\mathcal{X}_{n-1}$ is a purely periodic sequence over $\mathbb{Z}/2^n\mathbb{Z}$, the length of its shortest period is $2^n m$; proceeding with this we obtain successively (in view of (11.18) and (11.17)) that

$$\begin{aligned}
&\varepsilon_{2^n m} \equiv \delta_0(\gamma_n) + 1 \pmod 2, \quad \ldots, \quad \varepsilon_{2^n m + (2^n m - 1)} \equiv \delta_{2^n m-1}(\gamma_n) + 1 \pmod 2,\\
&\varepsilon_{2 \cdot 2^n m} \equiv \delta_0(\gamma_n) \pmod 2, \quad \ldots, \quad \varepsilon_{2 \cdot 2^n m + (2^n m - 1)} \equiv \delta_{2^n m-1}(\gamma_n) \pmod 2,\\
&\varepsilon_{3 \cdot 2^n m} \equiv \delta_0(\gamma_n) + 1 \pmod 2, \quad \ldots .
\end{aligned}$$

Note that in view of the definition of ε_k one has

$$\varepsilon_{2^n m} = \delta_0(\gamma_n) \oplus \sum_{k=0}^{2^n m-1} \varphi_n^{k \bmod m}(\xi_k^{n-1}).$$

However, the sum in the right hand side must be 1 modulo 2 since $\varepsilon_{2^n m} \equiv \delta_0(\gamma_n) + 1 \pmod 2$, as it was proved above. So, in view of (11.16) we conclude that

$$\sum_{k=0}^{2^n m-1} \varphi_n^{k \bmod m}(\xi_k^{n-1}) \equiv \sum_{i=0}^{m-1} \sum_{\xi \in \mathbb{Z}/2^n} \varphi_n^i(\xi) \equiv 1 \pmod 2.$$

Noticing that $\sum_{\xi \in \mathbb{Z}/2^n} \varphi_n^i(\xi)$ is just a weight of the Boolean function φ_n^i, we see that an odd number of Boolean functions from $\varphi_n^0, \ldots, \varphi_n^{m-1}$ must have odd weights (cf. conditions of Lemma 10.12).

Now putting $\xi_k^n = \xi_k^{n-1} + 2^n \cdot \varepsilon_k$ for $k = 0, 1, 2, \ldots$, we obtain a sequence $\mathcal{X}_n = (\xi_0^n, \xi_1^n, \ldots)$ over the ring $\mathbb{Z}/2^{n+1}\mathbb{Z}$. Terms of this sequence $\mathcal{X}_n$ satisfy the following relations

$$\begin{aligned}
\xi_0^n &= x_0 \bmod 2^{n+1},\\
\xi_{k+1}^n &= g_{k \bmod m}(\xi_k^n) \bmod 2^{n+1},\\
\delta_j(\xi_{k+2^n m}^n) &\equiv \delta_j(\xi_k^n) + 1 \pmod 2
\end{aligned}$$

for all $j = 0, 1, \ldots, n$ and $k = 0, 1, 2, \ldots$. The sequence $\mathcal{X}_n$ is a purely periodic sequence that has a period of length $2^{n+1}m$ (by the third of the above congruences, as the sequence $\mathcal{X}_{n-1}$ is purely periodic, and the length of its shortest period is $2^n m$, by the assumption we made above); moreover each element from $\mathbb{Z}/2^{n+1}\mathbb{Z}$ occurs at this period exactly $2^{n+1}m$ times. Finally, $\delta_n(\mathcal{X}_n) = \varepsilon_0\varepsilon_1 \ldots = \frac{2^{2^n m}-\gamma_n}{2^{2^n m}+1}$.

Using this inductive procedure for $n = 1, 2, \ldots$, we construct well-defined mappings $g_i \bmod 2^{n+1}$, $i = 0, 1, \ldots, m-1$, that are compatible bijective transformations on the residue ring $\mathbb{Z}/2^{n+1}\mathbb{Z}$. Moreover, the corresponding recurrence sequence $\mathcal{X}_n$ defined by the recursion $x_{i+1} = g_{i \bmod m}(x_i) \bmod 2^{n+1}$ satisfy (11.15) for $j = 1, \ldots, n$. The mappings g_i satisfy condition 3 of Theorem 10.9 for $k = 1, 2, \ldots, n+1$ since we have seen above that the odd number of Boolean functions from $\varphi_k^0, \ldots, \varphi_k^{m-1}$ have odd weights, for all $k = 1, 2, \ldots, n$. Finally we conclude that these mappings g_i satisfy conditions 1 and 2 of Theorem 10.9. This completes the proof in view of notices that we made at the very beginning. □

11.3 Distribution of k-tuples

In this section we study a distribution overlapping binary k-tuples in output sequences of congruential generators and of counter-dependent generators that generate sequences of the longest possible period. If $\{0, 1, 2, \ldots, 2^n - 1\} = \mathbb{Z}/2^n\mathbb{Z}$ is the output alphabet of this generator, the output sequence is strictly uniformly distributed as a sequence over $\mathbb{Z}/2^n$: That is, it is purely periodic, and each element of $\mathbb{Z}/2^n\mathbb{Z}$ occurs at the period the same number of times. However, we may consider this sequence as a binary sequence, concatenating corresponding n-bit terms of the sequence, and we ask what is a distribution of n-tuples in such binary sequence. The point is, that *strict uniform distribution of an arbitrary sequence $\mathcal{T}$ over $\mathbb{Z}/2^n\mathbb{Z}$ does not necessarily imply uniform distribution of overlapping n-tuples, if this sequence is considered as a binary sequence!*

For instance, let $\mathcal{T}$ be the following strictly uniformly distributed sequence over $\mathbb{Z}/4\mathbb{Z}$: $\mathcal{T} = 023102310231\ldots$. The length of the shortest period of this sequence is 4, and a binary representation of this sequence is $\mathcal{T} = 000111100001111000011110\ldots$; recall that according to our conventions at the very end of Section 8.2 we write more significant bits rightmost, and not leftmost; i.e., $2 = 01$, $1 = 10$, etc. Obviously, when we consider $\mathcal{T}$ as a sequence over $\mathbb{Z}/4\mathbb{Z}$, every number from $\{0, 1, 2, 3\}$ occurs in the sequence with the same frequency $\frac{1}{4}$. Yet if we consider $\mathcal{T}$ as a binary sequence, then 00, as well as 11, occur in this sequence with a frequency $\frac{3}{8}$, whereas 01, and 10, occur with a frequency $\frac{1}{8}$. Thus, the sequence $\mathcal{T}$ is uniformly distributed over $\mathbb{Z}/4\mathbb{Z}$, and it is not uniformly distributed over $\mathbb{Z}/2\mathbb{Z}$.

In this section, we show that this effect does not take place for output sequences of generators from Theorem 10.9; in particular, it is not the case for linear congruential generators with output alphabet $\{0, 1, 2, \ldots, 2^n - 1\}$ whose shortest period is the longest possible, i.e., of length 2^n, as the latter generators are special case of generators from Theorem 10.9 at $m = 1$. Namely, *if we consider any of these sequences as a* *binary sequence, the corresponding distribution of k-tuples is uniform, for all $k \le n$.* Now we state this property more formally.

Consider a (binary) *n-cycle* $C = (\varepsilon_0\varepsilon_1\dots\varepsilon_{n-1})$; that is, an oriented graph with vertices $\{a_0, a_1, \dots, a_{n-1}\}$ and with edges

$$\{(a_0, a_1), (a_1, a_2), \dots, (a_{n-2}, a_{n-1}), (a_{n-1}, a_0)\},$$

where each vertex a_j is labeled with $\varepsilon_j \in \{0, 1\}$, $j = 0, 1, \dots, n-1$. Note that then $(\varepsilon_0\varepsilon_1\dots\varepsilon_{n-1}) = (\varepsilon_{n-1}\varepsilon_0\dots\varepsilon_{n-2}) = \cdots$, etc. Clearly, every purely periodic sequence $\mathcal{S}$ over $\mathbb{Z}/2\mathbb{Z}$ with a period $\alpha_0\dots\alpha_{n-1}$ of length n can be related to a binary n-cycle $C(\mathcal{S}) = (\alpha_0\dots\alpha_{n-1})$. Conversely, to each binary n-cycle $(\alpha_0\dots\alpha_{n-1})$ we relate n purely periodic binary sequences with periods of length n: These sequences are n shifted versions of the sequence

$$\alpha_0\dots\alpha_{n-1}\alpha_0\dots\alpha_{n-1}\dots,$$

that is

$$\begin{gathered}\alpha_1\dots\alpha_{n-1}\alpha_0\alpha_1\dots\alpha_{n-1}\alpha_0\dots,\\ \alpha_2\dots\alpha_{n-1}\alpha_0\alpha_1\alpha_2\dots\alpha_{n-1}\alpha_0\alpha_1\dots,\\ \vdots \qquad \vdots \qquad \vdots\\ \alpha_{n-1}\alpha_0\alpha_1\alpha_2\dots\alpha_{n-2}\alpha_{n-1}\alpha_0\alpha_1\alpha_2\dots\alpha_{n-2}\dots.\end{gathered}$$

Further, a *k-chain in a binary n-cycle C* is a binary string $\beta_0\dots\beta_{k-1}$, $k < n$, that satisfies the following condition: There exists $j \in \{0, 1, \dots, n-1\}$ such that $\beta_i = \varepsilon_{(i+j) \bmod n}$ for $i = 0, 1, \dots, k-1$. Thus, a k-chain is just a string of length k of labels that corresponds to a chain of length k in a graph C. We call a binary n-cycle C *k-full*, if each k-chain occurs in the graph C the same number $r > 0$ of times.

Clearly, if C is k-full, then $n = 2^k r$. For instance, a well-known De Bruijn sequence is an n-full 2^n-cycle, see any book on combinatorics for De Bruijn sequence and relevant references, e.g. [165]. It is clearly that a k-full n-cycle is $(k-1)$-full: Each $(k-1)$-chain occurs in C exactly $2r$ times, etc. Thus, if an n-cycle $C(\mathcal{S})$ is k-full, then each m-tuple (where $1 \le m \le k$) occurs in the sequence $\mathcal{S}$ with the same probability (limit frequency) $\frac{1}{2^m}$. That is, the sequence $\mathcal{S}$ is *k-distributed*, see [267, Section 3.5, Definition D].

Definition 11.29. A purely periodic binary sequence $\mathcal{S}$ with the shortest period of length N is said to be *strictly k-distributed* if and only if the corresponding N-cycle $C(\mathcal{S})$ is k-full.

Thus, if a sequence $\mathcal{S}$ is strictly k-distributed, then it is strictly s-distributed, for all positive $s \le k$.

A k-distribution is a good "indicator of randomness" of an infinite sequence: The larger k, the better the sequence, i.e., "more random-looking". The best case is when a sequence is k-distributed for all $k = 1, 2, \dots$. Such sequences are called ∞-*distributed.* Obviously, a periodic sequence can not be ∞-distributed. A periodic sequence is just an infinite repetition of a finite sequence, the period. A common requirement in applications is that the length of the shortest period of a periodic sequence must be large, and the whole period is never used in practice. For instance, in cryptography normally a relatively small part of a period is used. So we are interested of "how random" a finite sequence is, namely, the period. Of course, it seems very reasonable to consider a period of length n as an n-cycle and to study the distribution of k-tuples in this n-cycle; for instance, if this n-cycle is k-full, the distribution of k-tuples is strictly uniform. However, other approaches also exist.

Donald Knuth in [267] introduced a useful "indicator of randomness" of a finite sequence over a finite alphabet A, see [267, Section 3.5, Definition Q1]. We formulate the corresponding definition only for $A = \{0, 1\}$: Knuth says that a finite binary sequence $\varepsilon_0\varepsilon_1\dots\varepsilon_{N-1}$ of length N is random, if and only if

$$\left|\frac{\nu(\beta_0\dots\beta_{k-1})}{N} - \frac{1}{2^k}\right| \le \frac{1}{\sqrt{N}} \tag{11.19}$$

for all $0 < k \le \log_2 N$, where $\nu(\beta_0\dots\beta_{k-1})$ is the number of occurrences of a binary word $\beta_0\dots\beta_{k-1}$ in a binary word $\varepsilon_0\varepsilon_1\dots\varepsilon_{N-1}$. If a finite sequence is random in this sense of Definition Q1 from the book [267], we shall say that the sequence has *property Q1*, or *satisfies Q1*, or is a *Q1-sequence*. We shall also say that an *infinite periodic sequence satisfies Q1* if and only if its shortest period satisfies Q1. Note that, contrasting to the case of strict k-distribution, which implies strict $(k-1)$-distribution, it is not enough to demonstrate only that (11.19) holds for $k = \lfloor \log_2 N \rfloor$ to prove that a finite sequence of length N satisfies Q1: For instance, the sequence 1111111100000111 satisfies (11.19) for $k = \lfloor \log_2 n \rfloor = 4$, and this sequence does not satisfy (11.19) for $k = 3$. Note that an analog of property Q1 for odd prime p could be stated in an obvious way.

Now we are able to state the following theorem:

Theorem 11.30. *Let $\mathcal{Z} = \mathcal{X} \bmod 2^n$ be a sequence over $\mathbb{Z}/2^n\mathbb{Z}$, where $\mathcal{X}$ is a sequence from Theorem 10.9.*[11] *Let $\mathcal{Z}'$ be a binary representation of $\mathcal{Z}$* (hence $\mathcal{Z}'$ is a purely periodic binary sequence whose shortest period is of length $mn2^n$). *Then the sequence $\mathcal{Z}'$ is strictly n-distributed.*

Moreover, if Z is a recurrence sequence with the recursion law $z_{i+1} = f(z_i) \bmod 2^n$, where f is a 1-Lipschitz ergodic transformation on $\mathbb{Z}_2$, then the sequence $\mathcal{Z}'$ satisfies Q1.

[11] Whence, $\mathcal{Z}$ is a purely periodic sequence with the shortest period of length $m2^n$. In particular, $\mathcal{Z}$ may be the output sequence of a congruential generator with output alphabet $\{0, 1, \dots, 2^n - 1\}$ that has the longest possible period, of length 2^n; this corresponds to the case $m = 1$.

Proof. The sequence $\mathcal{Z} = z_0 z_1 \dots$ is a recurrence sequence over $\{0, 1, \dots, 2^n - 1\}$ that satisfies the following recurrence relation:

$$z_{i+1} = f_i(z_i) \bmod 2^n, \qquad i = 0, 1, 2, \dots,$$

where f_i is a 1-Lipschitz measure-preserving transformation on $\mathbb{Z}_2$. Here and further in the proof we assume that the subscript i of f is always reduced modulo m for $m > 1$ and is empty symbol for $m = 1$, where m is from the statement of Theorem 10.9. The case $m = 1$ corresponds to a congruential generator with a state transition function $f \bmod 2^n$, where f is a 1-Lipschitz ergodic transformation on $\mathbb{Z}_2$.

Denote by $\mathcal{Z}' = \zeta_0\zeta_1 \dots$ a binary representation of the sequence $\mathcal{Z}$. Take an arbitrary binary word $\mathbf{b} = \beta_0\beta_1 \dots \beta_{n-1}$, $\beta_j \in \{0, 1\}$, and for $k \in \{0, 1, \dots, n-1\}$ denote

$$\nu_k(\mathbf{b}) = \#\{r : 0 \le r < 2^n mn;\ r \equiv k \pmod n;\ \zeta_r\zeta_{r+1} \dots \zeta_{r+n-1} = \beta_0\beta_1 \dots \beta_{n-1}\}.$$

Obviously, $\nu_0(\mathbf{b})$ is the number of occurrences of a rational integer z with base-2 expansion $\beta_0\beta_1 \dots \beta_{n-1}$ at the shortest period of the sequence $\mathcal{Z}$. Hence, $\nu_0(\mathbf{b}) = m$ since the sequence $\mathcal{Z}$ is strictly uniformly distributed modulo 2^n. Now consider $\nu_k(\mathbf{b})$ for $0 < k < n$.

Fix $k \in \{1, 2, \dots, n-1\}$ and let $r = k + tn$. As all f_i are 1-Lipschitz, the equality $\zeta_r\zeta_{r+1} \dots \zeta_{r+n-1} = \beta_0\beta_1 \dots \beta_{n-1}$ holds if and only if the following two relations hold simultaneously:

$$\zeta_{tn+k}\zeta_{tn+k+1} \dots \zeta_{tn+n-1} = \beta_0\beta_1 \dots \beta_{n-k-1}, \tag{11.20}$$

$$f_t(\overline{\zeta_{tn}\zeta_{tn+1} \dots \zeta_{tn+k-1}}) \equiv \overline{\beta_{n-k}\beta_{n-k+1} \dots \beta_{n-1}} \pmod{2^k}. \tag{11.21}$$

Here $\overline{\gamma_0\gamma_1 \dots \gamma_s} = \gamma_0 + \gamma_1 \cdot 2 + \dots + \gamma_s \cdot 2^s$ for $\gamma_0, \gamma_1, \dots, \gamma_s \in \{0, 1\}$ is a rational integer with base-2 expansion $\gamma_0\gamma_1 \dots \gamma_s$.

We consider the case $m = 1$ first; so $f_t = f$. Then, given $\mathbf{b} = \beta_0\beta_1 \dots \beta_{n-1}$, congruence (11.21) has exactly one solution $\overline{\alpha_0\alpha_1 \dots \alpha_{k-1}}$ modulo 2^k, since f is ergodic, whence, bijective modulo 2^k, by Theorem 4.23. Thus, in view of (11.20) and (11.21) we conclude that the equality $\zeta_r\zeta_{r+1} \dots \zeta_{r+n-1} = \beta_0\beta_1 \dots \beta_{n-1}$ holds if and only if

$$\zeta_s\zeta_{s+1} \dots \zeta_{s+n-1} = \alpha_0\alpha_1 \dots \alpha_{k-1}\beta_0\beta_1 \dots \beta_{n-k-1}, \tag{11.22}$$

where $s = tn$. Yet there exists exactly one $s \equiv 0 \pmod n$, $0 \le s < 2^n n$ such that (11.22) holds, since every element of $\mathbb{Z}/2^n\mathbb{Z}$ occurs at the period of $\mathcal{Z}$ exactly once. We conclude now that if $m = 1$ then $\nu_k(\mathbf{b}) = 1$ for all $k \in \{0, 1, \dots, n-1\}$; thus, $\nu(\mathbf{b}) = \sum_{j=0}^{n-1} \nu_j(\mathbf{b}) = n$ for all $\mathbf{b}$. This means that the $(2^n n)$-cycle $C(\mathcal{Z}')$ is n-full, whence, the sequence $\mathcal{Z}'$ is strictly n-distributed.

A similar argument is applied to the case $m > 1$. Namely, given $j \in \{0, 1, \dots, m-1\}$, consider those $r = k + tn < 2^n \ell n$ where $t \equiv j \pmod m$ and denote

$$\nu_k^j(\mathbf{b}) = \#\{r : 0 \le r < 2^n mn;\ r = k + tn;\ t \equiv j \pmod m;\ \zeta_r\zeta_{r+1} \dots \zeta_{r+n-1} = \mathbf{b}\}.$$

Now $\zeta_r\zeta_{r+1}\dots\zeta_{r+n-1} = \beta_0\beta_1\dots\beta_{n-1}$ holds if and only if (11.22) holds, where $\overline{\alpha_0\alpha_1\dots\alpha_{k-1}}$ is a unique solution of congruence (11.21) modulo 2^k. This solution exists since all f_j are measure-preserving, see Theorem 10.9. Yet (11.22) is equivalent to the condition

$$z_t = \overline{\alpha_0\alpha_1\dots\alpha_{k-1}\beta_0\beta_1\dots\beta_{n-k-1}},$$

where $t \in \{j, j+m, \dots, j+(2^n-1)\cdot m\}$. However, by Claim 3 of Lemma 10.12, given $\overline{\alpha_0\alpha_1\dots\alpha_{k-1}\beta_0\beta_1\dots\beta_{n-k-1}}$, there exists exactly one $t \in \{j, j+m, \dots, j+(2^n-1)\cdot m\}$ such that the latter equality holds. So we conclude that $\nu_k^j(\mathbf{b}) = 1$; whence $\nu_k(\mathbf{b}) = \sum_{j=0}^{m-1}\nu_k^j(\mathbf{b}) = m$, and finally $\nu(\mathbf{b}) = \sum_{k=0}^{n-1}\nu_k(\mathbf{b}) = nm$ for all $\mathbf{b}$. This completes the proof of the first assertion of the theorem.

To prove the second assertion, note that we return to the case $m = 1$; hence, in view of the first assertion, which is already proved, every ℓ-tuple for $1 \le \ell \le n$ occurs at the $2^n n$-cycle $C(\mathcal{Z}')$ exactly $2^{n-\ell}n$ times. Thus, every such ℓ-tuple occurs $2^{n-\ell}n-c$ times at the finite binary sequence $\hat{\mathcal{Z}} = \hat{z}_0\hat{z}_1\dots\hat{z}_{2^n-1}$, where $\hat{z}$ for $z \in \{0, 1, \dots, 2^n-1\}$ stands for an n-bit sequence that agrees with the base-2 expansion of z. Note that c depends on the ℓ-tuple, yet $0 \le c \le \ell-1$ for every ℓ-tuple. Easy algebra shows that (11.19) holds for these ℓ-tuples.

Now to prove that $\mathcal{Z}'$ satisfies Q1, we must only demonstrate that (11.19) holds for ℓ-tuples with $\ell = n+d$, where $0 < d \le \log_2 n$. We claim that such ℓ-tuple occurs in the sequence $\hat{\mathcal{Z}}$ not more than n times.

Indeed, in this case $\zeta_r\zeta_{r+1}\dots\zeta_{r+n+d-1} = \beta_0\beta_1\dots\beta_{n+d-1}$ holds if and only if along with relations (11.20) and (11.21) the following extra congruence holds:

$$f(\overline{\zeta_{tn}\zeta_{tn+1}\dots\zeta_{tn+k-1}\beta_0\beta_1\dots\beta_{d-1}}) \equiv \overline{\beta_{n-k}\beta_{n-k+1}\dots\beta_{n+d-1}} \pmod{2^{k+d}},$$

where $k = r \bmod n$. However, this extra congruence may or may not have a solution in unknowns $\zeta_{tn}, \zeta_{tn+1}, \dots, \zeta_{tn+k-1}$; this depends on $\beta_0\beta_1\dots\beta_{n+d-1}$. But if a solution exists, it is unique, given $k \in \{0, 1, \dots, n-1\}$, since f is ergodic, whence by Theorem 4.23 f is bijective modulo 2^s, for all $s = 1, 2, \dots$. This proves our claim. Now easy exercise in inequalities shows that (11.19) holds in this case, thus completing the proof of Theorem 11.30. □

Note 11.31. The first assertion of Theorem 11.30 remains true for *wreath products of truncated automata*, i.e. for the sequence $\mathcal{F}$ of Note 10.19, where $F_j(x) = \left\lfloor \frac{x}{2^{n-k}} \right\rfloor \bmod 2^k$, $j = 0, 1, \dots, m-1$, a truncation of $n-k$ low order bits. Namely, *a binary representation $\mathcal{F}'$ of the sequence $\mathcal{F}$ is a purely periodic strictly k-distributed binary sequence with a period of length $2^n mk$.*

The second assertion of Theorem 11.30 holds for arbitrary prime p. Namely, *a base-p representation of the recurrence sequence with the recursion law $z_{i+1} = f(z_i) \bmod p^n$, where f is a 1-Lipschitz ergodic transformation on the space of p-adic integers $\mathbb{Z}_p$, is a strictly n-distributed sequence* (over $\mathbb{Z}/p\mathbb{Z}$), *whose shortest period* (of length $p^n n$) *satisfies* Q1.

Moreover, the first assertion of Theorem 11.30 holds for truncated congruential generators with output function $F(x) = \lfloor \frac{x}{p^{n-k}} \rfloor \bmod p^k$. Namely, *a base-$p$ representation of the output sequence of a truncated congruential generator over $\mathbb{Z}/p^n\mathbb{Z}$ with a maximum period length, is a purely periodic strictly k-distributed sequence over $\mathbb{Z}/p\mathbb{Z}$ with a period of length $p^n k$.*

The second assertion for this generator holds whenever $2 + p^k > kp^{n-k}$; thus, *we may truncate* $\le \left(\frac{n}{2} - \log_p \frac{n}{2}\right)$ *lower order digits without affecting property* Q1.

All these claims could be proved by slight modifications of the proof of Theorem 11.30. We leave details of these proofs as exercises for the interested reader.

Chapter 12

p-adic probability theory

The development of a non-Archimedean (in particular, p-adic) mathematical physics [34, 50, 104, 137, 143, 309, 324, 351, 406–408] and especially quantum models with wave functions taking values in non-Archimedean fields (in particular, fields of p-adic numbers and their finite extensions), e.g. [7, 8, 88, 185, 193, 209, 210, 212, 214, 218,222,230,230], induced some new mathematical structures over non-Archimedean fields. In particular, probability theory with p-adic valued probabilities was developed in [195–209,211,213–215,219,220,222,223,225,226,231,233,242,244,245,251,252, 259, 260]. The main task of this probability formalism was to present the probability interpretation for p-adic valued wave functions.

12.1 Historical remarks

The first theory with p-adic probabilities was the frequency theory in which probabilities were defined as limits of relative frequencies $\nu_N = n/N$ in the p-adic topology[1]. This frequency probability theory was a natural extension of the frequency probability theory of R. von Mises [317, 318]. One of the most interesting features of the p-adic frequency theory of probability is the possibility to obtain *negative probabilities as limits of relative frequencies.* Thus negative probabilities can be obtained on the mathematical level of rigorousness as p-adic probabilities. Typically p-adic frequency negative probabilities (as well as probabilities which are larger than 1) appear in the cases of violation of the ordinary (von Mises) statistical stabilization with respect to the real metric. In fact, in this chapter we shall only consider a p-adic generalization of von Mises' principle of the statistical stabilization. The next natural step is to find a p-adic generalization of von Mises' principle of randomness. This problem will be studied in this chapter on the basis of a p-adic generalization of Martin-Löf's theory of statistical tests [297, 313].

The next step was the creation of p-adic probability formalism from theory of p-adic valued measures. It was natural to do this by following the fundamental work of A. N. Kolmogorov [270], see also [271], in which he proposed the measure-theoretical

[1] The following trivial fact is the cornerstone of this theory: the relative frequencies belong to the field of rational numbers $\mathbb{Q}$; we can study their behavior not only with respect to the real topology on $\mathbb{Q}$, but also with respect to other topologies on $\mathbb{Q}$ and, in particular, the p-adic topologies on $\mathbb{Q}$.

axiomatics of probability theory. Kolmogorov used properties of the frequency (Mises) probability (non-negativity, normalization by 1 and additivity) as the basis of his axiomatics. Then he added the technical condition of σ-additivity to incorporate probability in Lebesgue's integration theory. In [194–209] we followed A. N. Kolmogorov. p-adic frequency probability has also the properties of additivity, it is normalized by 1 and the set of possible values of this probability is the whole field of p-adic numbers $\mathbb{Q}_p$. Thus it was natural to define p-adic probability as a $\mathbb{Q}_p$-valued measure normalized by 1.

However, to find a p-adic analogue of the condition of σ-additivity was not so easy. It is the well-known fact that all σ-additive $\mathbb{Q}_p$-valued measures defined on σ-rings are discrete measures [322, 374, 399]. Therefore the creators of non-Archimedean integration theory (A. Monna and T. Springer [323]) did not try to develop abstract measure theory, but they proposed an integration formalism based on integrals of continuous functions. This integration theory has been used for creation of p-adic probability theory in the measure-theoretical framework [260]. The main disadvantage of this probability model is the strong connection with the topological structure of sample space. This is quite similar to the first attempts to create probability formalism – by Kolmogorov, Fréchet and Cramer. In such formalisms preceding the modern probability model the topological structure of sample space played the important role.

An abstract theory of non-Archimedean measures was developed by A. van Rooji [399]. The basic idea of this approach is to study measures defined on *rings* which in principle cannot be extended to measures on σ-rings. This gives the possibility for constructing non-discrete p-adic valued measures. On the other hand, the condition of continuity for measures in [399] implies the σ-additivity in all natural cases[2].

In this chapter we develop the p-adic probability formalism based on measure theory of [399]. By probabilistic reasons we use the special case of this measure theory: measures defined on *algebras* (such measures have some special properties). However, probabilistic applications stimulate also the development of the general theory of non-Archimedean measures defined on rings. We prove the formula of the change of variables for these measures and use this formula for developing the formalism of conditional expectations for p-adic valued random variables, see also [260].

We point out that the use of p-adic valued probabilistic measures gives the possibility to work on the mathematical level of rigorousness with all signed 'probabilities' (for example, with Wiegner's distribution). Such a p-adic approach to negative probabilities provides a new possibility to attack some fundamental problems of quantum physics, see e.g. [205] for the p-adic probabilistic model of Dirac's quantization of electromagnetic field with the aid of negative probabilities or [215] for the corresponding model for measurements with finite precision.

Applications of p-adic probabilities to the Einstein–Podolsky–Rosen paradox and Bell's inequality [207, 211, 213, 222, 226] are especially interesting. By applying p-

[2] Thus the σ-additivity is not a problem. The problem is to find the right domain of definition of p-adic probabilistic measures.

adic probability theory one might escape two fundamental problems of modern quantum mechanics: quantum nonlocality and "death of realism", see e.g. [242] for the detailed discussion on the mathematical level or rigorousness. In fact, so called hidden variables could peacefully coexist with locality, but under the assumption that their fluctuations are described by p-adic probability theory. In particular, this implies that relative frequencies for hidden variables do not stabilize with respect to the ordinary real metric. However, they stabilize with respect to the p-adic metric. Of course, we have the problem of the choice of the "right prime" p describing prequantum fluctuations. This problem could not be solved mathematically. Quantum physics (either theoretical or experimental) should provide the answer.

The Einstein–Podolsky–Rosen paradox and violation of Bell's inequality is a problem of great complexity. One may try to test the p-adic probabilistic model in simpler experiments. The simplest experiment (playing the fundamental role in quantum foundations) is the well-known two slit experiment, see [242] for presentation on the mathematical level or rigorousness. We proposed experimental tests for our p-adic predictions [220, 225]. Unfortunately, these tests have not yet been done.[3]

As the fields of p-adic numbers are non-Archimedean there exist infinitely large p-adic numbers (in particular, infinitely large natural numbers) in $\mathbb{Q}_p$. Thus p-adic analysis gives the possibility to use actual infinities and consider statistical ensembles with an infinite number of elements. Probabilities with respect to such ensembles are defined as the standard proportion. One of the main features of such ensemble probabilities is the appearance of negative (rational) probabilities (as well as probabilities which are larger than 1). In this approach the origin of such pathological from the real viewpoint probabilities is very clear. In particular, we shall see that a large set of negative probabilities is naturally interpreted as a set of infinitely small probabilities providing a finer structure of conventional zero probability. We shall also see that a large set of probabilities which are larger than one is naturally interpreted as a set of probabilities which differ negligibly from one. Another interesting property of p-adic ensemble probability is that the corresponding probabilistic measure is not well defined on an algebra of sets. The system of events is only a semi-algebra.

12.2 Frequency probability theory

We present a natural generalization of the von Mises frequency theory of probability. Our approach is based on the following two remarks:

(1) relative frequencies $\nu_N = n/N$ always belong to the field of rational numbers $\mathbb{Q}$;

[3]Experimenters are not extremely interested to test deviations from the conventional quantum ideology. They have been performing new tests to improve violation of Bell's inequality during the last 20 years. However, they tell that they are too busy to perform nonconventional tests. Moreover, young researchers are really afraid to do anything unconventional, since they would have problems to find job. Such unpleasant scientific situation is a sign of the deepest crises in quantum foundations.

(2) there exist topologies on $\mathbb{Q}$ which are different from the usual real topology τ_R corresponding to the real metric $\rho_R(x, y) = |x - y|$.

As in ordinary von Mises' theory, we also consider an infinite sequence

$$u = (u_1, \ldots, u_N, \ldots), \quad u_j \in L, \tag{12.1}$$

of observations. Here $L = \{\alpha_1, \ldots, \alpha_k\}$ is the label set for possible results of observations. In the simplest case $L = \{0, 1\}$, "yes/no"-observation. We restrict considerations to the case of observations with discrete label sets. Generalization to the case of continuous label sets is not trivial, cf. von Mises [318].

Denote by

$$\nu_N(\alpha_i; u) = \frac{n_N(\alpha_i; u)}{N}$$

the relative frequency of realizations of the label $\alpha_i \in L$ in the initial segment of u having the length N. Here $n_N(\alpha_i; u)$ is the number of realizations of α_i in this segment.

Von Mises formulated the following principle of the statistical stabilization of relative frequencies in a sequence of observations: for any label $\alpha \in L$, the sequence $\{\nu_N(\alpha_i; u)\}$ stabilizes when $n \to \infty$, i.e., there exists the limit $\lim_{N\to\infty} \nu_N(\alpha_i; u)$. Of course, this principle does not hold for any sequence (12.1). Von Mises selected a special class of sequences, so called collectives, which satisfy this principle.

Besides the principle of the statistical stabilization a collective should satisfy the so called principle of randomness. This principle provides the invariance of the limit of relative frequencies with respect to choices of subsequences in sequence (12.1). Von Mises considered a special class of possible choices of subsequences, so called place selections. Unfortunately, the notion of the place selection induced complicated logical problems in von Mises' frequency theory of probability. These problems have not been totally resolved. A mathematically rigorous notion of randomness corresponding to von Mises' idea of the place selection has not yet been elaborated. We can mention an attempt to define random sequences by using the notion of Kolmogorov algorithmic complexity [242, 272, 297]. However, it was not totally adequate to von Mises' approach. Another attempt to define rigorously a random sequence was performed in the measure-theoretic framework by Martin-Löf [297, 313] (who was definitely inspired by Kolmogorov during stay at Moscow State University). Martin-Löf's approach neither match with von Mises' place selection approach.

In this book we would not like to go deeply in the p-adic generalization of von Mises randomness. We shall start with "castrated frequency probability theory" which will be solely based on the generalization of the principle of the statistical stabilization. This theory will serve as the basis of the measure-theoretic formalization, in same way as it was done by Kolmogorov who took axioms of probability theory from von Mises' frequency theory (besides the condition of σ-additivity). Then we shall study the problem of p-adic randomness in the measure-theoretic framework by generalizing Martin-Löf's approach.[4]

We formulate a new topological principle of the statistical stabilization of relative frequencies:

The statistical stabilization of relative frequencies $\nu_N(\alpha_i; u)$ *can be considered not only in the real topology on the field of rational numbers* $\mathbb{Q}$, *but in any topology* τ *on* $\mathbb{Q}$.

Such a topology is said to be the *topology of statistical stabilization.* Limiting values $\mathbf{P}(\alpha_i) \equiv \mathbf{P}_u^\tau(\alpha_i)$ of frequencies $\nu_N(\alpha_i; u)$, $i = 1, \dots, k$, are said to be τ-probabilities. These probabilities belong to the completion $\mathbb{Q}_\tau$ of $\mathbb{Q}$ with respect to the topology τ. The choice of the topology τ of statistical stabilization is connected with the concrete probabilistic model.

Sequence $u =$ (12.1) for which the principle of statistical stabilization of relative frequencies for the topology τ is valid is said to be a (S, τ)-sequence. In particular, (S, τ_R)-sequences, where τ_R is the real topology, are sequences satisfying ordinary von Mises' principle of the statistical stabilization.

As was mentioned, in the frequency framework we do not try to propose any τ-analogue of von Mises' principle of randomness. We shall proceed with the remark that to define probabilities one needs only the principle of the statistical stabilization. Thus fruitful theory can be developed even for S-sequences and not only for collectives, see [242], cf. with the law of large numbers in Kolmogorov's framework.

We are mainly interested in the following situation. The real topology τ_R is not a topology of the statistical stabilization for the sequence (12.1), but another topology τ is. In this case we cannot consider (12.1) in von Mises' framework. However, we can operate with $u =$ (12.1) as a (S, τ)-sequence.

Set $U_Q = \{q \in \mathbb{Q} : 0 \leqslant q \leqslant 1\}$. These are all rational numbers in the segment $[0, 1]$. These and only these numbers can appear as relative frequencies of realizations of attributes in some sequence of observations.

We denote the closure of the set U_Q in the completion $\mathbb{Q}_\tau$ of the set of rational numbers $\mathbb{Q}$ by U_{Q_τ}. The following theorem is an evident consequence of the topological principle of the statistical stabilization:

Theorem 12.1. *The probabilities* $\mathbf{P}(\alpha_i)$ *belong to the set* U_{Q_τ} *for an arbitrary* (S, τ)-*sequence* u.

As usual, we consider the algebra F_L of all subsets of L. As in the frequency theory of von Mises we define probabilities $\mathbf{P}(A) = \sum_{\alpha_i \in A} \mathbf{P}(\alpha_i)$ for $A \in F_L$. By Theorem 12.1 the probability $\mathbf{P}(A)$ belongs to the set U_{Q_τ} for every $A \in F_L$.

Theorem 12.2. *Let the completion* $\mathbb{Q}_\tau$ *of* $\mathbb{Q}$ *with respect to the topology of the statistical stabilization* τ *be an additive topological group. Then for every* (S, τ)-*sequence*

[4] See [199] for an attempt to develop Kolmogorov algorithmic complexity in the p-adic framework.

u the probability is an additive function on F_L: $\mathbf{P}(A \cup B) = \mathbf{P}(A) + \mathbf{P}(B)$, $A, B \in F_L$, $A \cap B = \varnothing$.

Here we have used only the fact that $\lim(h_N + g_N) = \lim h_N + \lim g_N$ in an additive topological group.

Theorem 12.3. *The probability* $\mathbf{P}(L) = 1$ *for every topology of the statistical stabilization* τ *on* $\mathbb{Q}$.

We may choose the topology of the statistical stabilization τ such that $\mathbb{Q}_\tau$ is not an additive group. In this case we obtain non-additive probabilities.

Now (following Kolmogorov) we can present axiomatics corresponding to the properties of frequency probabilities. Of course, this axiomatics depends on the topology τ. Thus we have an infinite set of axiomatic theories $A(\tau)$. The simplest case (and the one most similar to the Kolmogorov axiomatics) is such that $\mathbb{Q}_\tau$ is a topological field. There, by definition, *a* τ*-probability is a* U_{Q_τ}*-valued measure with the normalization condition* $\mathbf{P}(\Omega) = 1$. Technical restrictions on $\mathbf{P}$ providing fruitful theory of integration should be chosen, compare with Kolmogorov's condition of σ-additivity.

We obtain a large class of non-Kolmogorov probabilistic models if we choose a metrizable topology τ such that the corresponding metric has the form $\rho_\tau(x, y) = |x - y|_\tau$, where $|\cdot|_\tau$ is a valuation on $\mathbb{Q}$. According to the Ostrovsky theorem, every valuation on $\mathbb{Q}$ is equivalent to the ordinary real absolute value $|\cdot|_R$ or one of the p-adic valuations $|\cdot|_p$. Therefore we may obtain only two classes of probabilistic models:

(1) the ordinary theory of probability (with the topology of the statistical stabilization τ_R);

(2) one of the p-adic valued probabilistic models (with topologies of the statistical stabilization τ_p).

We mention an interesting property of p-adic probabilities:

$$U_{\mathbb{Q}_p} = \mathbb{Q}_p,$$

see [195–209, 242, 244, 245]. To prove this, we need only to show that every $x \in \mathbb{Q}_p$ can be realized as the limit of frequencies $\nu_N = n/N$, where n, N are natural numbers, $n \leqslant N$. Thus any p-adic number x may serve as p-adic probability. In particular, every rational number can serve as p-adic probability. One can obtain such pathological probabilities (from the point of view of the usual theory of probability) as $\mathbf{P}(A) = 2$, $\mathbf{P}(A) = 100$, $\mathbf{P}(A) = 5/3$, $\mathbf{P}(A) = -1$. If $p = 1 \bmod 4$, then even the imaginary unit $i = \sqrt{-1}$ belongs to $\mathbb{Q}_p$. Thus complex quantities can be obtained as frequency probabilities; for example, $\mathbf{P}(A) = i = \sqrt{-1}$ or $\mathbf{P}(A) = 1 \pm i$.

Hence, *negative (and even complex) probabilities can be realized as p-adic frequency probabilities.*

We have presented in [63, 197, 201, 214] a large number of statistical models where frequencies oscillate with respect to the real metric ρ_R and stabilize with respect to one of the p-adic metrics ρ_p. The p plays the role of a parameter of the statistical model. The corresponding statistical simulation was carried out on computer.

Thus von Mises' principle of the statistical stabilization of frequencies can be essentially extended by considering (S, τ)-sequences for topologies τ on the set of rational numbers $\mathbb{Q}$.

As was mentioned, it would be natural to extend von Mises' second principle, namely, the principle of randomness and introduce an analogue of Mises' collective, namely, a τ-collective. However, we could not obtain any meaningful extension of the principle of randomness for p-adic topologies τ_p. It is still not clear how we can define a class of place selections which would not disturb the p-adic statistical stabilization. On the other hand, it is well known that in ordinary (real) probability theory it is possible to develop the mathematical theory of randomness by using Martin-Löf statistical recursive tests [297, 313]. We shall follow P. Martin-Löf and develop a p-adic theory of recursive statistical tests[5].

We now compare the principle of statistical stabilization and the law of large numbers. In von Mises' framework the principle of statistical stabilization is the fundamental principle preceding even probability. In Kolmogorov's framework the notion of probability is fundamental and the principle of statistical stabilization appears later in the form of the strong law of large numbers. Let $(\Omega, \mathcal{F}, \mathbf{P})$ be a Kolmogorov probability space. Here Ω is the set of elementary events, $\mathcal{F}$ is a σ-algebra of events and $\mathbf{P}$ is a probability measure on $\mathcal{F}$. Consider a sequence of random variables $\xi_1(\omega), \ldots, \xi_N(\omega), \ldots$. Assume for simplicity that these variables take values in $\{0, 1\}$. Consider relative frequencies for appearance of 1 and 0, respectively, for the first N variables:

$$\nu_N(1; \omega) = \frac{\xi_1(\omega) + \cdots + \xi_N(\omega)}{N}, \qquad \nu_N(0; \omega) = 1 - \nu_N(1; \omega).$$

The strong law of large numbers provides conditions for the existence of the limits of these frequencies for almost all $\omega \in \Omega$. In the simplest case random variables are independent and equally distributed: $\mathbf{P}(\omega : \xi_j(\omega) = 1) = P_1$ and $\mathbf{P}(\omega : \xi_j(\omega) = 0) = P_0$. Then by the strong law of large numbers:

$$\lim_{N \to \infty} \nu_N(1; \omega) = P_1, \qquad \lim_{N \to \infty} \nu_N(0; \omega) = P_0.$$

This is the measure-theoretical viewpoint on the principle of statistical stabilization of relative frequencies. In the Kolmogorov model one is not interested in randomness of the sequence produced from realizations of random variables for a fixed ω. Only existence of the limit for relative frequencies is important.

Von Mises strongly criticized the law of large numbers. He pointed out that by determining convergence of relative frequencies almost everywhere one is not able to

[5]Of course, we understood that Martin-Löf's theory does not give the fruitful notion of randomness for an individual sequence of trials.

say anything about convergence for any concrete $\omega \in \Omega$. He also remarked that people often associate with the principle of statistical stabilization the form of the law of large numbers based on convergence with respect to probability. The latter has nothing to do with statistical stabilization in a sequence of trials.

We finish the introduction to generalized frequency probability theory by a discussion on the topological principle of statistical stabilization.

A topology statistical stabilization is chosen to study asymptotic behavior of frequencies. In general its choice is the complicated problem. One may be curious about reasons of the common use of the real topology to study asymptotic behavior of various statistical data which appear in natural and social science as well as engineering. We cannot give the definite answer to this question.

It might be that the conventional real statistical stabilization of frequencies of realization of various physical and social quantities is simply a characteristic feature of the Universe, at least at the present state of its evolution.[6] In such a case it is possible to assume that the statistical stability of natural phenomena with respect to the real metric induces the same sort of the statistical stability for social phenomena.

However, we could not exclude the possibility that the total dominance of the real statistical stabilization for natural and social processes is simply an anthropological illusion. We are biological organisms and we look at physical reality only as biological organisms do. We can speculate that in the process of evolution living forms selected as observables only physical quantities which follow the law of statistical stabilization with respect to the real metric. Thus other physical variables are simply non-available for us. In such a case e.g. p-adically stable worlds could exists simultaneously with our really stable world. We can even suppose that these worlds are not independent. And we created some images of phenomena which are unstable with respect to the real metric, but stable with respect to e.g. the p-adic one.

Let us go back to the fundamental problem of quantum theory, namely, the Einstein–Podolsky–Rosen paradox. In 1933, Einstein, Podolsky and Rosen[7] pointed out that quantum mechanics is either incomplete or nonlocal. The later means that the laws of special relativity are violated for quantum observables. Observation's influence can propagate with the velocity which is higher than the velocity of light. Incompleteness of quantum mechanics means that one can go beyond the quantum model and present a deeper model with so called hidden variables. In this case quantum randomness would be reduced to classical randomness of ensembles of hidden variables. However, later

[6] One may speculate that at earlier stages of evolution of Universe physical phenomena were not statistically stable with respect to the real metric. Physical processes were based on other types of statistical stability. One could not exclude the p-adic statistical stability at the very early stage of evolution. We can speculate, following Volovich [408], see also Vladimirov, Volovich and Zelenov [407], Freund and Witten [143], Frampton et al. [137], Parisi [351], Aref'eva et al. [34], Dragovich [104], that at that stage of evolution space-time had the p-adic structure. The p-adic statistical stabilization might be a consequence of the p-adic geometry of space-time. However, at the moment these are only speculations.

[7] It seems that the idea of this argument belonged to Rosen.

Bell demonstrated theoretically[8] that if one tries to go beyond quantum mechanics he again could not escape nonlocality. Hidden variables describing components of a composite system, so called entangled particles, are coupled nonlocally. This conclusion is heavily based on the assumption that prequantum fluctuations are described by the classical probability theory which is coupled with the statistical stabilization with respect to the real metric.

Why should fluctuations of "super-microscopic" variables induce the conventional law of statistical stabilization?

The common argument is that e.g. p-adic statistical stabilization is too exotic to meet it at all in physics, even at the level of hidden variables. However, nonlocality is not less exotic than the use of local, but p-adically stable hidden variables. Such type of variables is especially natural under the assumption that prequantum space-time has the p-adic geometry. Thus quantum nonlocality might be simply an image (rather perverse) of p-adic randomness of hidden variables. However, again these are only speculations.

Finally, we recall once again that various computer simulations inducing the p-adic statistical stabilization were done in [63, 197, 201, 214]. Models considered in these works are sufficiently realistic, especially biological models. Nevertheless, no single example of the p-adic statistical stabilization in "real world" has been found. In the light of previous considerations the following reasons for the absence of p-adically stable processes can be presented:

(a) "it is too late": such stability (which is in fact the real instability) was important at the early stage of evolution of Universe;

(b) we are looking for p-adic probabilities at wrong scales of space and time: maybe to find them one should be able to go beyond quantum mechanics or even to the Planck scale;

(c) human beings belong to the form of live which evolved by taking into account only physical variables exhibiting the real statistical stabilization, one could not completely exclude the possibility that there exist other forms of life which evolved by using e.g. the p-adic statistical stabilization.

12.3 Ensemble probability

In this section we interpret p-adic integers

$$N = l_0 + l_1 p + \cdots + l_s p^s + \cdots, \quad \text{where } l_s = 0, 1, \ldots, p-1, \tag{12.2}$$

with infinite number of nonzero digits l_s as *infinitely large numbers*. Such a viewpoint provides the possibility to operate with numerous actual infinities. We can introduce

[8] His theoretical conclusion was confirmed experimentally by Aspect, Zeilinger, Weihs, et al.

probabilities on ensembles of "infinite volume" by using classical Laplace's definition of probability, but for infinite number of equally possible cases.

Everywhere below for a subset A of a set Ω the symbol $\complement A$ denotes the complement of A, that is, $\Omega \setminus A$.

12.3.1 Ensembles of infinite volumes

We shall study some special ensembles $S = S_N$ which have "p-adic volume" N, where N is a nonzero p-adic integer. If N is finite then S is the ordinary finite ensemble. But, if N is infinite then S has a special p-adic structure which is defined as follows.

Consider a sequence of ensembles M_j having volumes $m_j = l_j p^j$, $j = 0, 1, \ldots$ (consisting of m_j elements). Set

$$S = \bigcup_{j=0}^{\infty} M_j. \tag{12.3}$$

Then $|S| = N$, where $|S|$ denotes the number of elements in ensemble S. This decomposition of S will play the crucial role in our probabilistic considerations. Thus S is not just an arbitrary ensemble consisting of N elements. It is an ensemble with N elements constructed with the help of the hierarchical structure corresponding to decomposition (12.3).

One can imagine the ensemble S as being the population of a tower $T = T_S$, which has an infinite number of floors with the following distribution of population through floors: the population of the jth floor is M_j. Set

$$T_k = \bigcup_{j=0}^{k} M_j.$$

This is the population of the first $k + 1$ floors.

Let $A \subset S$. Suppose that the following limit exists:

$$n(A) = \lim_{k\to\infty} n_k(A), \quad \text{where } n_k(A) = |A \cap T_k|. \tag{12.4}$$

The quantity $n(A)$ is said to be a *p-adic volume* of the set A.

We define probability of A by the standard relation of proportion:

$$\mathbf{P}(A) \equiv \mathbf{P}_S(A) = \frac{n(A)}{N}. \tag{12.5}$$

Denote the family of all $A \subset S$ for which (12.4) exists by $\mathcal{G}_S$. In our probabilistic model such sets $A \in \mathcal{G}_S$ are called *events*. Later we shall study some properties of the family of events. First we consider the algebra of sets F which consists of all finite subsets and their complements.

Proposition 12.4. $F \subset \mathcal{G}_S$.

Proof. Let A be a finite set. Then $n(A) = |A|$ and (12.5) has the form

$$\mathbf{P}(A) = \frac{|A|}{|S|}. \tag{12.6}$$

Now let $B = \complement A$. Then $|B \cap T_k| = |T_k| - |A \cap T_k|$. Hence there exists $\lim_{k\to\infty} |B \cap T_k| = N - |A|$. This equality implies the standard formula

$$\mathbf{P}(\complement A) = 1 - \mathbf{P}(A). \tag{12.7}$$

□

In particular, we have : $\mathbf{P}(S) = 1$.

Proposition 12.5. *Let $A_1, A_2 \in \mathcal{G}_S$ and $A_1 \cap A_2 = \varnothing$. Then $A_1 \cup A_2 \in \mathcal{G}_S$ and*

$$\mathbf{P}(A_1 \cup A_2) = \mathbf{P}(A_1) + \mathbf{P}(A_2). \tag{12.8}$$

Proposition 12.6. *Let $A_1, A_2 \in \mathcal{G}_S$. The following conditions are equivalent:*

(1) $A_1 \cup A_2 \in \mathcal{G}_S$; (2) $A_1 \cap A_2 \in \mathcal{G}_S$;

(3) $A_1 \setminus A_2 \in \mathcal{G}_S$; (4) $A_2 \setminus A_1 \in \mathcal{G}_S$.

There are standard formulas:

$$\mathbf{P}(A_1 \cup A_2) = \mathbf{P}(A_1) + \mathbf{P}(A_2) - \mathbf{P}(A_1 \cap A_2); \tag{12.9}$$

$$\mathbf{P}(A_1 \setminus A_2) = \mathbf{P}(A_1) - \mathbf{P}(A_1 \cap A_2). \tag{12.10}$$

Proof. We have $n_k(A_1 \cup A_2) = n_k(A_1) + n_k(A_2) - n_k(A_1 \cap A_2)$. Therefore, if, for example, $A_1 \cap A_2 \in \mathcal{G}_S$ then there exists a limit of the right hand side. It implies $A_1 \cup A_2 \in \mathcal{G}_S$ and (12.9) holds. Other implications are proved in the same way. □

It is useful to formalize properties of the system of sets $\mathcal{G}_S$ in the abstract framework:

A system of subsets $\mathcal{G}$ of some set Ω which has the properties described by Proposition 12.5 and contains $\varnothing$ and Ω is called semi-algebra.

By definition we have:

Corollary 12.7. *The family $\mathcal{G}_S$ is a semi-algebra.*

In general $A_1, A_2 \in \mathcal{G}_S$ does not imply $A_1 \cup A_2 \in \mathcal{G}_S$. To show this, by Proposition 12.6 it suffices to find $A_1, A_2 \in \mathcal{G}_S$ such that $A_1 \cap A_2 \notin \mathcal{G}_S$ It is easy to do: let $A_1, A_2 \in \mathcal{G}_S$ such that $|A_1 \cap A_2 \cap M_l| = 1$ for a nonempty M_l (there is only one element $x \in A_1 \cap A_2$ on each nonempty floor). If N is infinite then $\lim_{k\to\infty} n_k(A_1 \cap A_2)$ does not exist. Thus:

$\mathcal{G}_S$ is not an algebra of sets.

It is closed only with respect to finite unions of sets which have empty intersections. However, $\mathcal{G}_S$ is not closed with respect to countable unions of such sets: in general the condition $(A_j \in \mathcal{G}_S,\ j = 1, 2, \ldots,\ A_i \cap A_j = \varnothing,\ i \neq j)$ does not imply that $\bigcup_{j=1}^{\infty} A_j \in \mathcal{G}_S$.

Neither the natural additional assumption

$$\sum_{j=1}^{\infty} \mathbf{P}(A_j) \text{ converges in } \mathbb{Q}_p$$

nor the stronger assumption

$$\sum_{j=1}^{\infty} |\mathbf{P}(A_j)|_p < \infty$$

imply that $A \in \mathcal{G}_S$.

Example 12.8. Let $p = 2,\ N = -1 = 1 + 2 + 2^2 + \cdots + 2^n + \cdots$. Suppose that the sets A_j have the following structure: $|A_j \cap M_{3(j-1)}| = 1,\ |A_j \cap M_{3j-1}| = 2^{3j-1} - 1$ and $A_j \cap M_i = \varnothing,\ i \neq 3(j-1),\ 3j-1$, i.e., the set A_j is located on two floors of the tower T. In particular, $A_i \cap A_j = \varnothing,\ i \neq j$. As $A_j \in F$, then $A_j \in \mathcal{G}_S$; the probability $\mathbf{P}(A_j) = -2^{3j-1}$, $j = 1, 2, \ldots$. The series $\sum_{j=1}^{\infty} |\mathbf{P}(A_j)|_2 < \infty$. We show that $A = \bigcup_{j=1}^{\infty} A_j \notin \mathcal{G}_S$. We have:

$$n_{3(j-1)}(A) = |A_j \cap T_{3(j-1)}| + \left| \bigcup_{s=1}^{j-1} A_s \cap T_{3(j-1)} \right| = 1 + \gamma,$$

where $|\gamma|_2 < 1$. Thus $|n_{3(j-1)}(A)|_2 = 1$. But $|n_{3j-1}(A)|_2 < 1$.

We present the following useful formula for computation of probabilities:

$$\mathbf{P}(A) = \sum_{j=0}^{\infty} \mathbf{P}(A \cap M_j).$$

By using the model with population living in the tower T we can say: the probability to find in the tower T an inhabitant with the property A is equal to the sum of probabilities to find an inhabitant with this property on the fixed floor.

Definition 12.9. The system

$$\mathcal{P} = (S, \mathcal{G}_S, \mathbf{P}_S) \tag{12.11}$$

is called the p-adic ensemble probability space for the ensemble S.

If N is a finite natural number then we obtain probability which was considered already by Laplace who defined probability $\mathbf{P}(A)$ as proportion between the number of cases favorable to event A to the total number of possible cases. In this case, i.e., for a natural number N, the probability space (12.11) also can be considered as the Kolmogorov probability space by assigning to each element of ensemble S the probabilistic weight $\mathbf{P}(\omega) = 1/N$. However, neither Laplace nor Kolmogorov approaches could be generalized to infinite ensembles.

We remark that any ensemble probability space $\mathcal{P}$ can be approximated by ensemble probability spaces $\mathcal{P}_k$ having ensembles of finite volumes. Set

$$n_k = l_0 + l_1 p + \cdots + l_k p^k$$

for N which has the expansion (12.2). Let l_s be the first nonzero digit in (12.2). Consider finite ensembles S_{n_k}, $|S_{n_k}| = n_k$ $(k = s, s+1, \ldots)$, and ensemble probability spaces $\mathcal{P}_{n_k} = (S_{n_k}, \mathcal{G}_{S_{n_k}}, \mathbf{P}_{S_{n_k}})$. There $\mathcal{G}_{S_{n_k}}$ coincides with the algebra $F_{S_{n_k}}$ of all subsets of the finite ensemble S_{n_k} and probability is given by ordinary proportion:

$$\mathbf{P}_{S_{n_k}}(A) = \frac{|A|}{|S_{n_k}|}, \quad A \in F_{S_{n_k}}. \tag{12.12}$$

We identify S_{n_k} with the population of the first $k+1$ floors of the tower T_S.

Proposition 12.10. *Let $A \in \mathcal{G}_S$. Then*

$$\mathbf{P}_S(A) = \lim_{k\to\infty} \mathbf{P}_{S_{n_k}}(A \cap S_{n_k}). \tag{12.13}$$

To prove (12.13), we use that $\mathbb{Q}_p$ is a topological group. This approximation depends essentially on the rule of counting. It is defined by the sequence $\{n_k\}$ which gives the approximation of the infinite ensemble S by finite ensembles $\{S_{n_k}\}$. In principle the change of this rule may change the limiting result, see [242] for the details.

Proposition 12.11 (The image of ensemble probability). *The probability* $\mathbf{P}$ *maps* $\mathcal{G}_S$ *into the ball $B_{r_S}(0)$, where $r_S = 1/|N|_p$.*

To study conditional probabilities, we have to extend the notion of the p-adic ensemble probability and consider more general ensembles.

Let S be the population of the tower T_S with an infinite number of floors M_j, $j = 0, 1, \ldots$, and the following distribution of population: there are m_j elements on the jth floor, $m_j \in \mathbb{N}$, and the series $\sum_{j=1}^{\infty} m_j$ converges in $\mathbb{Z}_p$ to a nonzero number $N = |S|$. We define the p-adic ensemble probability of a set $A \subset S$ by (12.4),

(12.5); $\mathcal{G}_S$ is the corresponding family of events. It is easy to check that Propositions 12.4–12.11 hold for this more general ensemble probability.

Let $A \in \mathcal{G}_S$ and $\mathbf{P}(A) \neq 0$. We can consider A as a new ensemble with the p-adic hierarchical structure $A = \bigcup_{j=0}^{\infty} M_{Aj}$, where $M_{Aj} = A \cap M_j$, and introduce the corresponding family of events $\mathcal{G}_A$.

Proposition 12.12 (Conditional probability). *Let $A \in \mathcal{G}_S, \mathbf{P}(A) \neq 0$ and $B \in \mathcal{G}_A$. Then $B \in \mathcal{G}_S$ and Bayes' formula*

$$\mathbf{P}_A(B) = \frac{\mathbf{P}_S(B)}{\mathbf{P}_S(A)} \tag{12.14}$$

holds true.

Proof. The tower T_A of the A has the following population structure: the population of the jth floor is given by ensemble M_{Aj}. In particular, $T_{Ak} = T_k \cap A$. Thus

$$n_{Ak}(B) = |B \cap T_{Ak}| = |B \cap T_k| = n_k(B) \tag{12.15}$$

for each $B \subset A$. Hence the existence of $n_A(B) = \lim_{k\to\infty} n_{Ak}(B)$ implies the existence of $n_S(B) = \lim_{k\to\infty} n_k(B)$. Moreover, $n_S(B) = n_A(B)$. Therefore,

$$\mathbf{P}_A(B) = \frac{n_A(B)}{n_S(A)} = \frac{n_A(B)/|S|}{n_S(A)/|S|}.$$

□

By (12.15) we obtain the following consequence:

Corollary 12.13. *Let $A, B \in \mathcal{G}_S, \mathbf{P}(A) \neq 0$, and $B \subset A$. Then $B \in \mathcal{G}_A$.*

Thus we obtain

$$\mathcal{G}_A = \{B \in \mathcal{G}_S : B \subset A\}.$$

Let $A, B, A \cap B \in \mathcal{G}_S, \mathbf{P}(A) \neq 0$. We set by definition $\mathbf{P}_A(B) = \mathbf{P}_A(A \cap B)$. Then

$$\mathbf{P}_A(B) = \frac{\mathbf{P}_S(B \cap A)}{\mathbf{P}_S(A)}. \tag{12.16}$$

If we set $\mathbf{P}_A(B) = \mathbf{P}(B|A)$ and omit the index S for the probabilities for an ensemble S, then we obtain Bayes' formula.

Remark 12.14 (Domain of applications of Bayes' formula). This question has the exact and simple mathematical answer in the p-adic ensemble probability theory. We can use Bayes' formula for events A and B if and only if$A \cap B$ is also an event, i.e., $A \cap B \in \mathcal{G}_S$.

Remark 12.15. It is important for our physical considerations that $\mathcal{G}_S$ is not an algebra of sets and $\mathbf{P}_S$ can in principle take any value $x \in B_{r_S}(0)$. The manipulations which were used to prove Bell's inequality, see [242], are not legal for the ensemble probability space $\mathcal{P} = (S, \mathcal{G}_S, \mathbf{P}_S)$. For instance, if there are three sets $Y_\varphi, Y_\theta, Y_0 \in \mathcal{G}_S$, then in principle it may happen that $Y_\varphi \cap Y_\theta,\ Y_\varphi \cap Y_0,\ Y_0 \cap Y_\theta \in \mathcal{G}_S$, but $Y_\varphi \cap Y_\theta \cap Y_0 \notin \mathcal{G}_S$. Moreover, probabilities can in principle be negative. In this case we cannot use the standard estimate for Kolmogorov probabilities.

12.3.2 The rules for working with p-adic probabilities

One of the main tools of the ordinary theory of probability is based on the order structure on the field of real numbers $\mathbb{R}$. It gives the possibility of comparing probabilities of different events; events E with probabilities $\mathbf{P}(E) \ll 1$ are considered as negligible and events E with probabilities $\mathbf{P}(E) \approx 1$ are considered as practically certain. However, the use of these relations in concrete applications is essentially based on our (real) probability intuition.

What is a large probability? What is a small probability?

Moreover, it is not easy to compare two arbitrary probabilities. For instance, do you prefer to win with the probability $\mathbf{P}(E_1) = \frac{11}{17}$ or $\mathbf{P}(E_2) = \frac{13}{19}$. Formally, because $\mathbf{P}(E_1) < \mathbf{P}(E_2)$ it would be better to choose E_2. But in practice this choice does not give many advantages. Thus ordinary probability intuition is based more on centuries of human experiment than on exact mathematical theory.

To work with p-adic probabilities, we have to develop p-adic probability intuition. However, we should paid attention to a mathematical problem which complicating direct generalization of the real scheme. It is the absence of the liner order structure on $\mathbb{Q}_p$.

In principle, we may try to proceed even without liner order. For example, we can classify ("decompose") some events with the aid of their p adic probabilities. Such an approach works well in frequency probability theory. Consider two sequences u and v of results of observations (generated by some statistical experiment) which are not S-sequences in ordinary von Mises' frequency theory, i.e., relative frequencies $\nu_N(\alpha; u)$ and $\nu_N(\alpha; v)$ do not converge in the field of real numbers. Here $\alpha \in L$ and L is the label set – the set of all possible results of observations. In the Kolmogorov framework this situation corresponds to violation of the law of large numbers. In such a situation which is pathological from the viewpoint of ordinary probability theory we could not distinguish probabilistic properties of u and v. Ordinary (real) probabilities do not exists; both these sequences are totally chaotic from the real point of view. However, if they are (S, τ_p)-sequences for some p, then can classify them by using p-adic frequency probabilities: $\mathbf{P}_u(\alpha), \mathbf{P}_v(\alpha)$. These are limits of relative frequencies in the field of p-adic numbers. Of course, it is merely the classification framework and not the probabilistic one.

Consider now the ensemble approach. Here difference of p-adic probabilities, $\mathbf{P}_S(E_1) \neq \mathbf{P}_S(E_2)$, means that events E_1 and E_2 have different p-adic volumes. The most interesting case is the case of events having infinite volumes.

However, we could do much more with the aid of p-adic probabilities by introducing a partial order structure on the ring of p-adic integers:

($\mathcal{O}$) Let

$$x = x_0 x_1 \dots x_n \dots, \qquad y = y_0 y_1 \dots y_n \dots$$

be the canonical expansions of two p-adic integers $x, y \in \mathbb{Z}_p$. We say that $x < y$ if there exists n such that $x_n < y_n$ and $x_k \leqslant y_k$ for all $k > n$.

This partial order structure on $\mathbb{Z}_p$ is the natural extension of the standard order structure on the set of natural numbers **N**. It is easy to see that $x < y$ for any $x \in \mathbb{N}$ and $y \in \mathbb{Z}_p \setminus \mathbb{N}$, i.e., any finite natural number is less than any infinite number. In general we could not compare two infinite numbers.

Example 12.16. Let $p = 2$ and let $x = -1/3 = 10101\dots1010\dots$, $z = -2/3 = 0101\dots0101\dots$ and $y = -16 = 0001\dots1111\dots$. Then $x < y$ and $z < y$, but the numbers x and z are incompatible.

It is important to remark that there exists the maximal number $N_{\max} \in \mathbb{Z}_p$. It is easy to see that

$$N_{\max} = -1 = (p-1) + (p-1)p + \dots + (p-1)p^n + \cdots .$$

Therefore the ensemble S_{-1} is the largest ensemble which can be considered in the p-adic framework.

Remark 12.17. We may assume that the volume of the ensemble increases with the increase of p, i.e., $|S_{-1}^p| < |S_{-1}^q|$, $p < q$.

Proposition 12.18. *Let $N \in \mathbb{Z}_p, N \neq 0$. Then $S_N \in \mathcal{G}_{S_{-1}}$ and*

$$\mathbf{P}_{S_{-1}}(S_N) = \frac{|S_N|}{|S_{-1}|} = -N. \tag{12.17}$$

Corollary 12.19. *Let $N \in \mathbb{Z}_p, N \neq 0$. Then $\mathcal{G}_{S_N} \subset \mathcal{G}_{S_{-1}}$ and probabilities $\mathbf{P}_{S_N}(A)$ are calculated as conditional probabilities with respect to the sub-ensemble S_N of ensemble S_{-1}:*

$$\mathbf{P}_{S_N}(A) = \mathbf{P}_{S_{-1}}(A|S_N) = \frac{\mathbf{P}_{S_{-1}}(A)}{\mathbf{P}_{S_{-1}}(S_N)}, \quad A \in \mathcal{G}_{S_N}. \tag{12.18}$$

However, $A \in \mathcal{G}_{S_{-1}}$ does not imply $A \cap S_N \in \mathcal{G}_{S_N}$.

By Corollary 12.19 we can, in fact, restrict our considerations to the case of the maximal ensemble S_{-1}. Therefore we shall study this case $S \equiv S_{-1}$.

The (partial) order $\mathcal{O}$ on the set of p-adic integers $\mathbb{Z}_p$ gives the possibility to compare p-adic volumes $n(A)$ of sets $A \in \mathcal{G}_S$. It is natural to say that probability $\mathbf{P}(B)$ is larger than probability $\mathbf{P}(A)$ if the p-adic volume $n(B)$ of B is larger than the p-adic volume $n(A)$ of A. Thus we obtain the following (partial) order on the set of probabilities:

($\tilde{\mathcal{O}}$) $\mathbf{P}(B) > \mathbf{P}(A)$ iff $n(B) > n(A)$.

We use the same symbols $>, <$ for this new order on $\mathbb{Z}_p$. We hope that the reader would not mix these two orders on $\mathbb{Z}_p$: $\mathcal{O}$-order is used to compare p-adic volumes, $\tilde{\mathcal{O}}$-order is used to compare probabilities. For example, let $p = 2$ and let $n(B) = -2$ $(= 011\ldots1\ldots), n(A) = -3$ $(= 1011\ldots1\ldots)$. Then $n(B) > n(A)$ (with respect to $\mathcal{O}$) and consequently $\mathbf{P}(B) = 2 > \mathbf{P}(A) = 3$ (with respect to $\tilde{\mathcal{O}}$).

We study some properties of probabilities.

(P1) As the order structure is only partial, it is impossible to compare probabilities of arbitrary two events A and B.

(P2) As $x \leqslant -1$ with respect to $\mathcal{O}$ for any $x \in \mathbb{Z}_p$, we have that $\mathbf{P}(A) \leqslant 1 = \mathbf{P}(S)$ for any $A \in \mathcal{G}_S$.

(P3) As $x \geqslant 0$ with respect to $\mathcal{O}$ for any $x \in \mathbb{Z}_p$, we have $\mathbf{P}(A) \geqslant 0$ for any $A \in \mathcal{G}_S$.

To illustrate further properties of p-adic probabilities, we shall use the third order structure, namely, the usual real order structure on the set $\mathbb{Z}_p \cap \mathbb{Q}$. In this case we shall say r-increase or r-decrease. This r-order on $\mathbb{Z}_p \cap \mathbb{Q}$ has no probabilistic meaning. We consider this order, because we want to use the 'real intuition' to imagine the location of rational probabilities $\mathbf{P}(A), A \in \mathcal{G}_S$, on the real line. We shall use the symbols $[a, b], \ldots, (a, b)$ for corresponding intervals of the real line. For example, let $p = 2$ and let $\mathbf{P}(B) = 2$ and $\mathbf{P}(A) = 3$. Then $\mathbf{P}(B) > \mathbf{P}(A)$, but from the viewpoint of the r-order $\mathbf{P}(B)$ is less than $\mathbf{P}(A)$.

(P4) Set $F^f = \{A \in \mathcal{G}_S : n(A) \in \mathbb{N}\}$.[9]

The restriction of the order $\mathcal{O}$ on the set of natural numbers $\mathbb{N}$ coincides with the standard (real) order on $\mathbb{N}$. Thus $n(A) < n(B),\ A, B \in F^f$, if and only if the natural number $n(A)$ is less than the natural number $n(B)$. This implies (by definition of the order $\tilde{\mathcal{O}}$ on the set of probabilities) that $\mathbf{P} : F^f \to (-\infty, 0) \cap \mathbb{Z}$ and $\mathbf{P}(A)$ is increasing if $\mathbf{P}(A)$ is r-decreasing. Therefore, for example, probabilities $\mathbf{P}(A) = -1$ or -3 are rather small with respect to probabilities $\mathbf{P}(B) = -100$ or -300.

(P5) Set $\complement F^f = \{B = \complement A : A \in F^f\}$ (in particular, $\complement F^f$ contains complements of all finite subsets of Ω). Then $\mathbf{P} : \complement F^f \to \mathbb{N}$ and $\mathbf{P}(B)$ is decreasing if $\mathbf{P}(B)$ is r-increasing. Therefore, for example, probabilities $\mathbf{P}(E) = 100$ or 200 are rather small with respect to probabilities $\mathbf{P}(C) = 1$ or 2.

[9] In particular, F^f contains all finite subsets of S. The F^f contains also some infinite subsets $A \in \mathcal{G}_S$ which have finite p-adic volumes. For example, let $|A \cap T_k| = 1 + p^k, k = 1, 2, \ldots$ $(1 + p^k$ inhabitants of the first $(k + 1)$ floors have the property $A)$. Then $n(A) = 1$ and hence $A \in F^f$.

We can use these rules for conditional probabilities. For example, let $\mathbf{P}(B) = 100$, $\mathbf{P}(B') = 200$, $\mathbf{P}(A) = 2$ and $B, B' \subset A$. Then $\mathbf{P}(B|A) = 50 > \mathbf{P}(B'|A) = 100$.

By (P4) and (P5) we can work with probabilities belonging to $F^f \cup \complement F^f$.

Now consider events $A \notin F^f \cup \complement F^f$. We can develop our intuition only by examples.

Example 12.20. Let $p = 2$. Let $|A \cap M_{2k}| = 2^{2k}$ and $A \cap M_{2k+1} = \varnothing$, $k = 0, 1, \ldots$. Then $n(A) = -1/3$ $(= 1010\ldots10\ldots)$ and $\mathbf{P}(A) = 1/3$. Let $B \subset A$ and $B \cap M_{4k} = A \cap M_{4k}$, $B \cap M_j = \varnothing$, $j \neq 4k$. Then $n(B) = -1/15$ $(= 100010001\ldots10001\ldots)$ and $\mathbf{P}(B) = 1/15$. It is evident that $-1/15 < -1/3$ in $\mathbb{Z}_2$. Hence $\mathbf{P}(B) = 1/15 < \mathbf{P}(A) = 1/3$.

The probabilistic order relation on the set $[0, 1] \cap \mathbb{Q}$ coincides with the standard real order. Moreover, it seems to be reasonable to use this relation also in the case where the numbers $n(A)$ and $n(B)$ are incompatible in $\mathbb{Z}_2$.[10]

Example 12.21. Let p and A be the same as above. Let $|C \cap M_{2k+1}| = 2^{2k+1}$, $C \cap M_{2k} = \varnothing$, $k = 0, 1, \ldots$. Then $n(C) = -2/3$ and $\mathbf{P}(C) = 2/3$. The numbers $n(A) = -1/3$ and $n(C) = -2/3$ are incompatible in $\mathbb{Z}_2$. But heuristically it seems to be evident that we can use the r-order structure on $[0, 1]$ to compare the probabilities of the events A and C. Therefore the probability of $\omega \in C$ is two times larger than the probability $\omega \in A$.

Any probability $x \in (-\infty, 0) \cap \mathbb{Z}$ is practically negligible with respect to any probability $y \in (0, 1] \cap \mathbb{Q}$. We present the following heuristic argument. Probability $\mathbf{P}(A) \in (-\infty, 0) \cap \mathbb{Z}$ is probability of an event A with finite p-adic volume in the infinitely large ensemble S. Probability $\mathbf{P}(A) \in (0, 1] \cap \mathbb{Q}$ is probability of an event A with infinite p-adic volume in the infinitely large ensemble S.

Therefore, p-adics gives the possibility to decompose zero probability into a set of probabilities, $0 \to D_0^+$; in particular, $(-\infty, 0) \cap \mathbb{Z} \subset D_0^+$.

Remark 12.22. By definition the probability $\mathbf{P}$ on a Boolean algebra $\mathcal{A}$ is non-degenerated: $\mathbf{P}(A) = 0$, $A \in \mathcal{A}$ if and only if $A = \varnothing$. The p-adic decomposition of zero probability can be considered as a step in the direction to Boolean probabilities. The set of new labels D_0^+ gives the possibility to decompose many probabilities which must be equal to probability 0 from the viewpoint of real analysis. However, probability is still non-Boolean. Numerous events $A \in \mathcal{G}_S$, $A \neq \varnothing$, have zero probability. For example, let $|A \cap T_k| = p^k, k = 1, 2, \ldots$. Then $\mathbf{P}(A) = 0$.

We can also use these rules for conditional probabilities. For example, let $\mathbf{P}(B) = 1/15 < \mathbf{P}(B') = 2/15$, $\mathbf{P}(A) = 1/5$ and $B, B' \subset A$. Then $\mathbf{P}(B|A) = 1/3 <$

[10] However, probably it is the wrong extrapolation and we must assume existence of events with incompatible probabilities.

$\mathbf{P}(B'|A) = 2/3$. Moreover, for example, let $\mathbf{P}(B) = -1 < \mathbf{P}(B') = -5$, $\mathbf{P}(A) = -100$ and $B, B' \subset A$. Then $\mathbf{P}(B|A) = 1/100 < \mathbf{P}(B'|A) = 1/20$. Thus the r-order structure on $(0, 1] \cap \mathbb{Q}$ reproduces the rule (P4).

Proposition 12.23. *If* $\mathbf{P}(B) \in \mathbb{N}$, *then* $n(\complement B) \in \{0\} \cup \mathbb{N}$; *if* $\mathbf{P}(B) \in (0, 1) \cap \mathbb{Q}$ *then* $n(\complement B) \in \mathbb{Z}_p \setminus \mathbb{N}$.

Proof. If $k = \mathbf{P}(B) \in \mathbb{N}$, then $n(B) = -k, k = 1, 2, \ldots$, and $n(\complement B) = -1 + k$. If $a = \mathbf{P}(B) \in (0, 1) \cap \mathbb{Q}$ then $n(B) = -a$ and $n(\complement B) = a - 1 \notin \mathbb{N}$. □

Thus if $\mathbf{P}(B) \in \mathbb{N}$, then the set $\complement B$ has a finite p-adic volume, $n(\complement B)$. On the other hand, if $\mathbf{P}(B) \in (0, 1) \cap \mathbb{Q}$, then the set $\complement B$ has an infinite p-adic volume, $n(\complement B)$. It is natural to assume that probability $\mathbf{P}(B) \in \mathbb{N}$ is larger than any probability $\mathbf{P}(C) \in (0, 1) \cap \mathbb{Q}$.

Therefore, p-adics also provides the possibility to decompose probability one into a set of probabilities, $1 \to D_1^-$. In particular, $\mathbb{N} \subset D_1^-$. However, probability one is still not totally decomposed. There are numerous events $A \neq \varnothing$ with $\mathbf{P}(A) = 1$. For example, let $|A \cap M_k| = p^{\lfloor (k+1)/2 \rfloor} - 1, k = 1, 2, \ldots$ (as everywhere, $\lfloor x \rfloor$ denotes the integer part of x). Then $n(A) = -1$ and $\mathbf{P}(A) = 1$. But $\complement A \neq \varnothing$.

We can also decompose all probabilities $x = \mathbf{P}(A) \in (0, 1) \cap \mathbb{Q}$. Let $A \in \mathcal{G}_S, x = \mathbf{P}(A) \in (0, 1) \cap \mathbb{Q}$, $C \in F^f, A \cap C = \varnothing$, and let $B = A \cup C$. Then $\lambda = \mathbf{P}(B) = \mathbf{P}(A) + \mathbf{P}(C) = x - k$, where $\mathbf{P}(C) = -k, k \in \mathbb{N}$. As the p-adic volume of the set C is finite (and the ensemble S is infinite) probability $\mathbf{P}(C) = -k$ is infinitely small. Thus the probability x can be decomposed into a set of probabilities D_x^+. Each probability $\lambda \in D_x^+$ is larger than probability x and probability $\epsilon = \lambda - x = -k$ is infinitely small.

Let $B \in \mathcal{G}_S, C \in F^f, B \cap C = \varnothing$, and let $A = B \cup C, x = \mathbf{P}(A) \in (0, 1) \cap \mathbb{Q}$. Then $\lambda = \mathbf{P}(B) = \mathbf{P}(A) - \mathbf{P}(C) = x + k$, where $\mathbf{P}(C) = -k, k \in \mathbb{N}$, is infinitely small probability. Thus the probability x can be split in a set of probabilities D_x^-. Each probability $\lambda \in D_x^-$ is less than probability x and probability $\epsilon = x - \lambda = -k$ is infinitely small.

Thus probability x is decomposed into a set of probabilities $D_x = D_x^- \cup D_x^+$.

We now consider probabilities with respect to an ensemble S_N for an arbitrary $N \in \mathbb{Z}_p, N \neq 0$. By using formula (12.18) we can translate to the general case results obtained for the ensemble $S = S_{-1}$. In the general case zero probability is decomposed into a set D_0^+ which contains the set $\{\lambda = \frac{k}{N} : k \in \mathbb{N}\}$; probability one is decomposed into a set D_1^- which contains the set $\{\lambda = 1 - \frac{k}{N} : k \in \mathbb{N}\}$; probability $x \in (0, 1) \cap \mathbb{Q}$ is decomposed into a set $D_x = D_x^- \cup D_x^+$. Here D_x^- contains the set $\{\lambda = x - \frac{k}{N} : k \in \mathbb{N}\}$ and D_x^+, contains the set $\{\lambda = x + \frac{k}{N} : k \in \mathbb{N}\}$.

12.3.3 Negative probabilities and p-adic ensemble probabilities

Negative probabilities appear with mystical regularity in quantum physics, see e.g. [242] for detailed review. In the conventional approach to probability such probabilities are totally meaningless. However, in p-adics an infinite series consisting of natural numbers can converge to a negative number, integer or rational. Thus infinite tower-like population can have negative "volume". Hence, some of its sub-populations (e.g. all finite) have negative probabilities of realization.

12.4 Measures

Let X be an arbitrary set and let $\mathcal{R}$ be a ring of subsets of X. The pair $(X, \mathcal{R})$ is called a *measurable space*. The ring $\mathcal{R}$ is said to be *separating* if for every two distinct elements, x and y, of X there exists an $A \in \mathcal{R}$ such that $x \in A, y \notin A$. We shall consider measurable spaces only over separating rings which cover X.

Every ring $\mathcal{R}$ can be used as a base for the zero-dimensional topology[11] which we shall call the $\mathcal{R}$-topology. This topology is Hausdorff if and only if $\mathcal{R}$ is separating.

Throughout this section, $\mathcal{R}$ is a separating covering ring of a set X.

A subcollection $\mathcal{S}$ of $\mathcal{R}$ is said to be *shrinking* if the intersection of any two elements of $\mathcal{S}$ contains an element of $\mathcal{S}$. If $\mathcal{S}$ is shrinking, and if f is a map $\mathcal{R} \to K$ or $\mathcal{R} \to \mathbb{R}$, we say that $\lim_{A\in\mathcal{S}} f(A) = 0$ if for every $\epsilon > 0$, there exists an $A_0 \in \mathcal{S}$ such that $|f(A)| \leqslant \epsilon$ for all $A \in \mathcal{S}, A \subset A_0$.

Let K be a non-Archimedean field with the valuation $|\cdot|_K$.

A *measure* on $\mathcal{R}$ is a map $\mu\colon \mathcal{R} \to K$ with the properties:

(i) μ is additive;

(ii) for all $A \in \mathcal{R}$, $\|A\|_\mu = \sup\{|\mu(B)|_K : B \in \mathcal{R}, B \subset A\} < \infty$;

(iii) if $\mathcal{S} \subset \mathcal{R}$ is shrinking and has empty intersection, then $\lim_{A\in\mathcal{S}} \mu(A) = 0$.

We call these conditions respectively *additivity*, *bounded*, *continuity*. The latter condition is equivalent to the following: $\lim_{A\in\mathcal{S}} \|A\|_\mu = 0$ for every shrinking collection $\mathcal{S}$ with empty intersection.

Condition (iii) is the replacement for σ-additivity. Clearly (iii) implies σ-additivity. Moreover, we shall see that for the most interesting cases (iii) is equivalent to σ-additivity. Of course, we could in principle restrict our attention to these cases and use the standard condition of σ-additivity. However, in that case we should use some topological restriction on the space X. This implies that we must consider some topological structure on a p-adic probability space. We do not like to do this. We shall develop the theory of p-adic probability measures in the same way as A. N. Kolmogorov (1933) developed the theory of real valued probability measures by starting with an arbitrary set algebra.

[11] A topological space $(X;\tau)$ is zero-dimensional if each point $x \in X$ has a basis of clopen (i.e., at the same time open and closed) neighborhoods.

Further, we shall briefly discuss the main properties of measures, see [242,260,322, 323,399] for the details. As always, for any set D, we denote its characteristic function by the symbol I_D. For $f: X \to K$ and $\varphi: X \to [0, \infty)$, put

$$\|f\|_{\varphi} = \sup_{x \in X} |f(x)|_K \varphi(x).$$

We set

$$N_{\mu}(x) = \inf_{U \in \mathcal{R}, x \in U} \|U\|_{\mu}$$

for $x \in X$. Then $\|A\|_{\mu} = \|I_A\|_{N_{\mu}}$ for any $A \in \mathcal{R}$. We set $\|f\|_{\mu} = \|f\|_{N_{\mu}}$.

A *step function* (or $\mathcal{R}$-step function) is a function $f: X \to K$ of the form $f(x) = \sum_{k=1}^{N} c_k \chi_{A_k}(x)$ where $c_k \in K$ and $A_k \in \mathcal{R}$, $A_k \cap A_l = \varnothing, k \neq l$. We set for such a function

$$\int_X f(x)\mu(dx) = \sum_{k=1}^{N} c_k \mu(A_k).$$

Denote the space of all step functions by the symbol $S(X)$. The integral $f \to \int_X f(x)\mu(dx)$ is the linear functional on $S(X)$ which satisfies the inequality

$$\left| \int_X f(x)\mu(dx) \right|_K \leqslant \|f\|_{\mu}. \tag{12.19}$$

A function $f: X \to K$ is called *μ-integrable* if there exists a sequence of step functions $\{f_n\}$ such that $\lim_{n \to \infty} \|f - f_n\|_{\mu} = 0$. The μ-integrable functions form a vector space $L_1(X, \mu)$ (and $S(X) \subset L_1(X, \mu)$). The integral is extended from $S(X)$ on $L_1(X, \mu)$ by continuity. The inequality (4.1) holds for $f \in L_1(X, \mu)$.

Let $\mathcal{R}_{\mu} = \{A : A \subset X, I_A \in L_1(X, \mu)\}$. This is a ring. Elements of this ring are called μ-measurable sets. By setting $\mu(A) = \int_X I_A(x)\mu(dx)$ the measure μ is extended to a measure on $\mathcal{R}_{\mu}$. This is the *maximal extension* of μ, i.e., if we repeat the previous procedure starting with the ring $\mathcal{R}_{\mu}$, we will obtain this ring again.

Set $X_{\epsilon} = \{x \in X : N_{\mu}(x) \geqslant \epsilon\}$, $X_0 = \{x \in X : N_{\mu}(x) = 0\}$, $X_+ = X \setminus X_0$. Every $A \subset X_0$ belongs to $\mathcal{R}_{\mu}$. We call such sets *μ-negligible.*

Now we construct product measures. Let $\mu_j, j = 1, 2, \ldots, n$, be measures on (separating) rings $\mathcal{R}_j$ of subsets of sets X_j. The finite unions of the sets $A_1 \times \cdots \times A_n$, $A_j \in \mathcal{R}_j$, form a (separating) ring $\mathcal{R}_1 \times \cdots \times \mathcal{R}_n$ of $X_1 \times \cdots \times X_n$. Then there exists a unique measure $\mu_1 \times \cdots \times \mu_n$ on $\mathcal{R}_1 \times \cdots \times \mathcal{R}_n$ such that $\mu_1 \times \cdots \times \mu_n(A_1 \times \cdots \times A_n) = \mu_1(A_1) \times \cdots \times \mu_n(A_n)$. We have

$$N_{\mu_1 \times \cdots \times \mu_n}(x_1, \ldots, x_n) = N_{\mu_1}(x_1) \times \cdots \times N_{\mu_n}(x_n).$$

Let X be a zero-dimensional topological space[12]. We denote the ring of *clopen* (i.e., at the same time open and closed) subsets of X by the symbol $B(X)$ (in fact, this is

[12]We consider only Hausdorff spaces.

an algebra). We denote the space of continuous bounded functions $f: X \to K$ by the symbol $C_b(X)$. We use the norm $\|f\|_\infty = \sup_{x \in \mathbb{N} X} |f(x)|_K$ on this space.

First we remark that *if X is compact and $\mathcal{R} = B(X)$ then the condition (iii) in the definition of a measure is redundant.* If X is not compact then there exist bounded additive set functions which are not continuous.

Let X be zero-dimensional $\mathbb{N}$-compact topological space, i.e., there exists a set S such that X is homeomorphic to a closed subset of $\mathbb{N}^S$. We remark that every product of $\mathbb{N}$-compact spaces is $\mathbb{N}$-compact; every closed subspace of an $\mathbb{N}$-compact space is $\mathbb{N}$-compact. Then every bounded σ-additive function $\mu: B(X) \to K$ is a measure. On the other hand, if X is a zero-dimensional space such that every bounded σ-additive function $B(X) \to K$ is a measure, then X is $\mathbb{N}$-compact.

In the theory of integration a crucial role is played by the $\mathcal{R}_\mu$-topology, i.e., the (zero-dimensional) topology that has $\mathcal{R}_\mu$ as a base. Of course, the $\mathcal{R}_\mu$-topology is stronger than the $\mathcal{R}$-topology. Every μ-negligible set is $\mathcal{R}_\mu$-clopen. The following two theorems will be important for our considerations. Proofs of these as well as following theorems can be found e.g. in [242].

Theorem 12.24. (i) *If μ is a measure on $\mathcal{R}$, then N_μ is $\mathcal{R}$-upper semicontinuous, (hence, $\mathcal{R}_\mu$-upper semicontinuous) and for every $A \in \mathcal{R}_\mu$ and $\epsilon > 0$ the set $A_\epsilon = A \cap X_\epsilon$ is $\mathcal{R}_\mu$-compact.*

(ii) *Conversely, let $\mu: \mathcal{R} \to K$ be additive. Assume that there exists an $\mathcal{R}$-upper semicontinuous $\varphi: X \to [0, \infty)$ such that $|\mu(A)|_K \leqslant \sup_{x \in A} \varphi(x)$, $A \in \mathcal{R}$, and $\{x \in A : \varphi(x) \geqslant \epsilon\}$ is $\mathcal{R}$-compact ($A \in \mathcal{R}, \epsilon > 0$). Then μ is a measure and $N_\mu \leqslant \varphi$.*

Theorem 12.25. *Let $\mu: \mathcal{R} \to K$ be a measure. A function $f: X \to K$ is μ-integrable if and only if it has the following two properties*: (1) *f is $\mathcal{R}_\mu$-continuous;* (2) *for every $\epsilon > 0$, the set $\{x : |f(x)|_K N_\mu(x) \geqslant \epsilon\}$ is $\mathcal{R}_\mu$-compact.*

Theorem 12.26. *Let $f \in L_1(X, \mu)$ and let*

$$\int_A f(x)\mu(dx) = 0 \quad \textit{for every } A \in \mathcal{R}. \tag{12.20}$$

Then supp $f \subset X_0$.

The following two lemmas are important to prove Theorem 12.29, see [242], but they are interesting by themselves as facts from non-Archimedean measure theory:

Lemma 12.27. *Let $(X_j, \mathcal{R}_j)$, $j = 1, 2$, be measurable spaces and let $f: X_1 \to X_2$ be measurable. If $\mathcal{S}$ is shrinking in $\mathcal{R}_2$ then $f^{-1}(\mathcal{S})$ is shrinking in $\mathcal{R}_1$. If $\mathcal{S}$ has empty intersection, then $f^{-1}(\mathcal{S})$ has also empty intersection.*

Lemma 12.28. *Let $(X_j, \mathcal{R}_j)$, $j = 1, 2$, be measurable spaces and let $\eta: X_1 \to X_2$ be a measurable function. Then, for every measure $\mu: \mathcal{R}_1 \to K$, the function $\mu_\eta: \mathcal{R}_2 \to$*

K defined by the equality $\mu_\eta(A) = \mu(\eta^{-1}(A))$ is a measure on $\mathcal{R}_2$ and, for every $\mathcal{R}_2$-continuous function $h\colon X_2 \to K$, the following inequality holds:

$$\|h\|_{\mu_\eta} \leqslant \|h \circ \eta\|_\mu. \tag{12.21}$$

The following theorem on the change of variables is important in our probabilistic considerations.

Theorem 12.29. *Let $(X_j, \mathcal{R}_j)$, $j = 1, 2$, be measurable spaces and let $\eta\colon X_1 \to X_2$ be a measurable function, and let $\mu\colon \mathcal{R}_1 \to K$ be a measure. If $f\colon X_2 \to K$ is an $\mathcal{R}_2$-continuous function such that the function $f \circ \eta$ belongs to $L_1(X_1, \mu)$, then $f \in L_1(X_2, \mu_\eta)$ and*

$$\int_{X_1} f(\eta(x))\mu(dx) = \int_{X_2} f(y)\mu_\eta(dy).$$

Open Question 12.30. Find a condition on function f which is weaker than continuity, but implies the formula of the change of variables.

Further we shall obtain some properties of measures which are specific for measures defined on algebras.

Throughout this section, $\mathcal{A}$ is a separating algebra of a set X. First we remark that if we start with a measure μ defined on the algebra $\mathcal{A}$ then the system $\mathcal{A}_\mu$ of μ-integrable sets is again an algebra.

Proposition 12.31. *Let $\mu\colon \mathcal{A} \to K$ be a measure. Then for each $\epsilon > 0$, the set X_ϵ is $\mathcal{A}_\mu$-compact.*

This fact is a consequence of Theorem 12.24.

Proposition 12.32. *Let $\mu\colon \mathcal{A} \to K$ be a measure. Then the algebra $B(X)$ of $\mathcal{A}_\mu$-clopen sets coincides with the algebra $\mathcal{A}_\mu$.*

As a consequence of Proposition 12.32, we obtain that $C_b(X) \subset L_1(X, \mu)$ (for the space X endowed with $\mathcal{A}_\mu$-topology) and the following inequality holds:

$$\left|\int_X f(x)\mu(dx)\right|_K \leqslant \|f\|_\infty \|X\|_\mu, \quad f \in C_b(X).$$

Let X be a zero-dimensional topological space. A measure μ defined on the algebra $B(X)$ of the clopen sets is called a *tight* measure. Thus by Proposition 12.32 every measure $\mu\colon \mathcal{A} \to K$ is extended to a tight measure on the space X endowed with the $\mathcal{A}_\mu$-topology.

Proposition 12.33. *Let $\mu\colon \mathcal{A} \to K$ be a measure and let $f \in L_1(X, \mu)$. Then f is $(\mathcal{A}_\mu, B(K))$-measurable.*

12.5 p-adic probability space

Let $\mu\colon \mathcal{A} \to \mathbb{Q}_p$ be a measure defined on a separating algebra $\mathcal{A}$ of subsets of the set Ω which satisfies the normalization condition $\mu(\Omega) = 1$. We set $\mathcal{F} = \mathcal{A}_\mu$ and denote the extension of μ on $\mathcal{F}$ by the symbol $\mathbf{P}$. A triple $(\Omega, \mathcal{F}, \mathbf{P})$ is said to be a p-adic *probability space* (Ω is a *sample space*, $\mathcal{F}$ is an algebra of *events*, $\mathbf{P}$ is a *probability*).

As in general measure theory we set

$$\Omega_\alpha = \{\omega \in \Omega : N_{\mathbf{P}}(\omega) \geqslant \alpha\},\ \alpha > 0, \quad \Omega_+ = \bigcup_{\alpha>0} \Omega_\alpha, \quad \Omega_0 = \Omega \setminus \Omega_+.$$

If a property Ξ is valid on the subset Ω_+ we say that Ξ is valid a.e. (mod $\mathbf{P}$).

Everywhere below (G, Γ) denotes a measurable space over the algebra Γ. Functions $\xi\colon \Omega \to G$ which are $(\mathcal{F}, \Gamma)$-measurable are said to be random variables.

Everywhere below Y is a zero-dimensional topological space. We consider Y as the measurable space over the algebra $B(Y)$. Every random variable $\xi\colon \Omega \to Y$ is continuous in the $\mathcal{F}$-topology. In particular, $\mathbb{Q}_p$-valued random variables are $(\mathcal{F}, B(\mathbb{Q}_p))$-measurable functions. If $\xi \in L_1(\Omega, \mathbf{P})$, we introduce an *expectation* of this random variable by setting $\mathbf{E}\xi = \int_\Omega \xi(\omega)\mathbf{P}(d\omega)$. We note that every bounded random variable $\xi\colon \Omega \to \mathbb{Q}_p$ belongs to $L_1(\Omega, \mathbf{P})$.

Let $\eta\colon \Omega \to G$ be a random variable. The measure $\mathbf{P}_\eta$ is said to be a *distribution* of the random variable. By Theorem 12.29 we have that

$$\mathbf{E}f(\eta) = \int_{\mathbb{Q}_p} f(y)\mathbf{P}_\eta(dy) \tag{12.22}$$

for every Γ-continuous function $f : G \to \mathbb{Q}_p$ such that $f \circ \eta \in L_1(\Omega, \mathbf{P})$. In particular, we have the following result.[13]

Proposition 12.34. *Let $\eta\colon \Omega \to Y$ be a random variable and let $f \in C_b(Y)$. Then the formula* (12.22) *holds.*

We shall also use the following technical result.

Proposition 12.35. *Let $\eta\colon \Omega \to Y$ be a random variable and let $\zeta \in L_1(\Omega, \mathbf{P})$, and let $f \in C_b(Y)$. Then $\xi(\omega) = \zeta(\omega) f(\eta(\omega))$ belongs to $L_1(\Omega, \mathbf{P})$ and*

$$\mathbf{E}\xi = \int_{\mathbb{Q}_p \times Y} x f(y)\mathbf{P}_z(dx dy), \quad z(\omega) = (\zeta(\omega), \eta(\omega)).$$

Proof. We have only to show that $\xi \in L_1(\Omega, \mathbf{P})$. This fact is a consequence of Theorem 12.25. □

[13] See [242] for detailed presentation, in particular, for proofs.

The random variables $\xi, \eta\colon \Omega \to G$ are called independent if

$$\mathbf{P}(\xi \in A, \eta \in B) = \mathbf{P}(\xi \in A)\mathbf{P}(\eta \in B) \quad \text{for all } A, B \in \Gamma. \tag{12.23}$$

Proposition 12.36. *Let $\xi, \eta\colon \Omega \to Y$ be independent random variables and functions $f, g \in C_b(Y)$. Then we have:*

$$\mathbf{E} f(\xi) g(\eta) = \mathbf{E} f(\xi) \mathbf{E} g(\eta). \tag{12.24}$$

Proof. If f and g are locally constant functions then (12.24) is a consequence of (12.23). Arbitrary functions $f, g \in C_b(Y)$ can be approximated by locally constant functions (with the convergence of corresponding integrals) by using the technique developed in the proof of Theorem 12.29. □

Remark 12.37. In fact, the formula (12.24) is valid for the continuous f, g such that the random variables $f(\xi), g(\eta)$ and $f(\xi)g(\eta)$ belong $L_1(\Omega, \mathbf{P})$.

Proposition 12.38. *Let ξ and η be independent random variables. Then the random vector $z = (\xi, \eta)$ has the probability distribution $\mathbf{P}_z = \mathbf{P}_\eta \times \mathbf{P}_\xi$.*

This fact is the direct consequence of (12.23).

Let ξ and η be respectively $\mathbb{Q}_p$ and G valued random variables and $\xi \in L_1(\Omega, \mathbf{P})$. A *conditional expectation* $\mathbf{E}[\xi | \eta = y]$ is defined as a function $m \in L_1(G, \mathbf{P}_\eta)$ such that

$$\int_{\{\omega \in \Omega : \eta(\omega) \in B\}} \xi(\omega) \mathbf{P}(d\omega) = \int_B m(y) \mathbf{P}_\eta(dy) \quad \text{for every } B \in \Gamma.$$

Proposition 12.39. *The conditional expectation is defined uniquely a.e.* mod $\mathbf{P}_\eta$.

As there is no analogue of the Radon–Nikodym theorem in the non-Archimedean case, it may happen that conditional expectation does not exist. Everywhere below we assume that $m(y) = \mathbf{E}[\xi | \eta = y]$ is well defined and moreover, that it belongs to the class $C_b(Y)$.

Proposition 12.40. *Let $\xi\colon \Omega \to \mathbb{Q}_p$, $\eta\colon \Omega \to Y$ be random variables, and $\xi \in L_1(\Omega, \mathbf{P})$. The equality*

$$\mathbf{E} f(\eta)\xi = \mathbf{E} f(\eta(\omega)) \mathbf{E}[\xi(\omega) | \eta = \eta(\omega)]$$

holds for every function $f \in C_b(Y)$.

We present a p-adic generalization of Martin-Löf's theory based on tests for randomness [297, 313].[14]

[14]This theory was developed by A. Khrennikov and S. Yamada [259].

We construct natural tests for randomness for p-adic valued uniform probability distribution. Each test for randomness induces a series of limit theorems. On the other hand, individual limit theorems are not good candidates for tests for randomness, because each theorem describes only behavior of a subsequence $S_{n_k}(\omega)$ of the sequence $S_n(\omega) = \xi_1(\omega) + \cdots + \xi_n(\omega)$ of independent equally distributed random variables.

As in the case of real-valued probabilities, we proved that it is possible to enumerate effectively all p-adic tests for randomness. However, opposite to Martin-Löf's theorem for real probabilities, a universal p-adic test for randomness does not exist.

We shall use the standard terminology of the book of M. Li and P. Vitànyi [297]. The abbreviation r.e. is used for *"recursive enumeration"*.

12.6 p-adic probability measures on the space of binary sequences

We set $X = \{0, 1\}$ and

$$X^n = \{x = (x_1, \dots, x_n) : x_j \in X\},$$
$$X^* = \bigcup_n X^n,$$
$$X^\infty = \{\omega = (\omega_1, \dots, \omega_n, \dots) : \omega_j \in X\}.$$

For $x \in X^n$, we set $l(x) = n$. For $x \in X^*$, $l(x) = n$, we define a cylinder U_x with basis x by

$$U_x = \{\omega \in X^\infty : \omega_1 = x_1, \dots, \omega_n = x_n\}.$$

We denote by the symbol $\mathcal{F}_{\text{cyl}}$ an algebra of subsets of X^∞ generated by all cylinders.

The map

$$j : X^\infty \to \mathbb{Z}_2, \ j(\omega) = \sum_{j=0}^{\infty} \omega_j 2^j,$$

gives a one-to-one correspondence between X^∞ and $\mathbb{Z}_2$. Thus we can identify these sets. The algebra of cylindric sets $\mathcal{F}_{\text{cyl}}$ coincides with the algebra $B(\mathbb{Z}_2)$ of all clopen, i.e., closed and open at the same time, subsets of $\mathbb{Z}_2$.

A function $\mu : \mathcal{F}_{\text{cyl}} \to \mathbb{Q}_p$ is a p-adic (valued) measure, see Section 12.4, if the following properties hold true: (i) additivity: $\mu(A \cup B) = \mu(A) + \mu(B)$, $A \cap B = \varnothing$, $A, B \in \mathcal{F}_{\text{cyl}}$; (ii) boundedness: $\|\mu\|_p = \sup\{|\mu(A)|_p : A \in \mathcal{F}_{\text{cyl}}\} < \infty$. In fact, condition of continuity (iii) is redundant as $\mathcal{F}_{\text{cyl}} = B(\mathbb{Z}_2)$ and the set $\mathbb{Z}_2$ is compact.

A function $f : X^* \to \mathbb{Q}_p$ is said to be *recursive* if and only if there is a recursive function $g : X^* \times \mathbb{N} \to \mathbb{Q}$ such that $|f(x) - g(x, k)|_p < \frac{1}{k}$. A p-adic measure $\mu : \mathcal{F}_{\text{cyl}} \to \mathbb{Q}_p$ is said to be recursive if and only if the function $f_p : X^* \to \mathbb{Q}_p$, $f_p(x) = \mu(U_x)$, is recursive.

The uniform p-adic measure μ_p $(p \neq 2)$ on X^∞ is defined by

$$\mu_p(U_x) = \frac{1}{2^{l(x)}}, \quad x \in X^*. \tag{12.25}$$

If X^* is realized as $\mathbb{Z}_2$ and $\mathcal{F}_{\mathrm{cyl}}$ as $B(\mathbb{Z}_2)$, then μ_p is the p-adic valued Haar measure (translation invariant measure) on $\mathbb{Z}_2$.

As $\left|\frac{1}{2^{l(x)}}\right|_2 = 2^{l(x)}$, the additive set function μ_2 defined by (12.25) is not bounded. Therefore *we shall consider only the case $p \neq 2$.*

Simple considerations show that the function

$$N_{\mu_p}(x) = \inf\{\|U\|_\mu : x \in U \in B(\mathbb{Z}_2)\} = 1$$

for all $x \in \mathbb{Z}_2$. This implies that $L_1(\mathbb{Z}_2, \mu_p) = C(\mathbb{Z}_2)$ (because all $B(\mathbb{Z}_2)$-step functions are continuous and each continuous function can be uniformly approximated by a sequence of $B(\mathbb{Z}_2)$-step functions). This implies that the algebra $(B(\mathbb{Z}_2))_{\mu_p} = \{A \subset \mathbb{Z}_2 : I_A \in L_1(\mathbb{Z}_2, \mu_p)\} = B(\mathbb{Z}_2)$. Thus the Haar measure μ_p cannot be extended from the algebra $B(\mathbb{Z}_2)$ to any larger algebra. In particular, the μ_p cannot be extended on the Borel σ-algebra generated by the algebra of clopen subsets $B(\mathbb{Z}_2)$.

Let a measure $\mu : \mathcal{F}_{\mathrm{cyl}} \to \mathbb{Q}_p$ be normalized: $\mu(X^\infty) = 1$. Then we can consider the p-adic probability space $\mathcal{P} = (\Omega, \mathcal{F}, \mathbf{P})$, where $\Omega = X^\infty$, $\mathcal{F} = (\mathcal{F}_{\mathrm{cyl}})_\mu$ (the set algebra which is obtained via the μ-extension of $\mathcal{F}_{\mathrm{cyl}}$), $\mathbf{P} = \bar{\mu}$ is a p-adic probability measure. As for the p-adic uniform measure μ_p (the $\mathbb{Q}_p$-valued Haar measure on $\mathbb{Z}_2$) the extension $(\mathcal{F}_{\mathrm{cyl}})_{\mu_p}$ coincides with $\mathcal{F}_{\mathrm{cyl}}$ and the extension $\bar{\mu}_p$ coincides with μ_p, the corresponding probability space is $\mathcal{P} = (\Omega, \mathcal{F}, \mathbf{P}_p)$, where $\Omega = X^\infty$, $\mathcal{F} = \mathcal{F}_{\mathrm{cyl}}$ and $\mathbf{P}_p = \mu_p$. The $\mathbf{P}_p$ is called a *uniform p-adic probability distribution.* We remark that values of $\mathbf{P}_p$ on cylinders coincide with values of the standard (real-valued) uniform probability distribution $\mathbf{P}_\infty$ on X^∞. As $\mathbb{Q} \subset \mathbb{R}$ and $\mathbb{Q} \subset \mathbb{Q}_p$, we can interpret rational numbers $\frac{1}{2^{l(x)}}$ both as real and as p-adic numbers.

In fact, we shall not use general recursive p-adic probabilities (see only definitions). We shall consider only the uniform p-adic probability distribution $\mathbf{P}_p$, $p \neq 2$ (which is, of course, recursive).

12.7 Some technical p-adic results

The results which are obtained in this section will be used to construct p-adic tests and prove limit theorems for p-adic probabilities.

As always, for any $n, k \in \mathbb{N}$, (n, k) denotes the greatest common divisor of n and k; for any $n \in \mathbb{N}$, $M_p(n)$ denotes the mod p residue of n: $n = M_p(n) \bmod p$. We set

$$\Theta_p(n) = \begin{cases} |n - M_p(n)|_p, & n \geqslant p, \\ 1, & 1 \leqslant n \leqslant p-1. \end{cases}$$

Lemma 12.41. *Let* $n, k \in \mathbb{N}$, $k \leqslant n$ *and let* $M_p(n) \geqslant M_p(k)$. *Then*

$$\left|\binom{n}{k}\right|_p = \frac{\Theta_p(n)}{\Theta_p(k)}.$$

Proof. Let $n = \alpha + ip^N$, $k = \beta + jp^l$, where $0 \leqslant \alpha, \beta \leqslant p-1$, $i, j, N, l \in \mathbb{N}$ and $(i, p) = (j, p) = 1$. We have

$$\begin{aligned}\left|\binom{n}{k}\right|_p &= \left|(ip^N) \cdot \frac{(ip^N - p)}{p} \cdot \frac{(ip^N - 2p)}{2p} \cdots \frac{(ip^N - jp^l + p)}{(jp^l - p)} \cdot \frac{1}{(jp^l)}\right|_p \\ &= \left|\frac{p^N}{p^l}\right|_p = p^{l-N}. \end{aligned} \tag{12.26}$$

To obtain (12.26), we have used that $n - k + 1 = ip^N - jp^l + (\alpha + 1 - \beta)$ and $0 < \alpha + 1 - \beta \leqslant p$; hence the last term in the nominator of $\binom{n}{k} = \frac{n \cdots (n-k+1)}{1 \cdots k}$, which is divisible by p is $(ip^N - jp^l + p)$. The cases in that $n = \alpha$ or $k = \beta$, $0 \leqslant \alpha$, $\beta \leqslant p - 1$, are considered in the same way. □

Lemma 12.42. *Let* $n, k \in \mathbb{N}$, $k \leqslant n$, *and let* $M_p(n) + 1 \leqslant M_p(k)$. *Then*

$$\left|\binom{n}{k}\right|_p = \Theta_p(n).$$

12.8 *p*-adic tests for randomness

We use the following notations. For each set $M \subset X^*$, we set $M^{(n)} = \{x \in M : l(x) = n\}$, $n = 1, 2, \ldots$. For each set $W \subset X^* \times \mathbb{N}$, we set $W_m = \{x \in X^* : (x, m) \in W\}$. Thus $W_m^{(n)} = \{x \in X^* : l(x) = n, (x, m) \in W\}$.

Everywhere in this section the *cardinality of a (finite) set A is denoted by the symbol* $\sigma(A)$. We do not use the standard symbol $|A|$, because we do not want to use expressions of the form $|\,|A|\,|_p$.

The following definition of a p-adic test for randomness is a natural generalization of Martin-Löf's definition of a test for randomness for ordinary real probabilities (in fact, in our particular case for the uniform distribution).

Definition 12.43. Let $\mathbf{P}$ be a p-adic recursive probability. A recursively enumerable (r.e.) set $V \subset X^* \times \mathbb{N}$ is called a *p-adic* **P**-*test* (p-adic test for randomness for the probability distribution $\mathbf{P}$) if it possesses the following two properties: for all $n, m \in \mathbb{N}$, we have

$$V_{m+1} \subset V_m \quad \text{and} \quad \left|\sum_{x \in V_m^{(n)}} \mathbf{P}(U_x)\right|_p \leqslant \frac{1}{p^m}. \tag{12.27}$$

The use of p-adic tests for randomness gives the possibility to formalize (in fact, to create) p-adic statistics. We are given the sample space X^* with an associated p-adic probability distribution $\mathbf{P}$. Given an element x of the sample space, we want to test hypothesis "x is a typical outcome". Practically speaking, the property of being typical is the property of belonging to reasonable majority. To ascertain whether a given element of the sample space belongs to a particular reasonable majority we use the notation of a test. As in the ordinary probability theory, a test is given by a prescription that, for every level of significance $\varepsilon = \frac{1}{p^m}$, tells us for what elements $x \in X^*$ the hypothesis "x belongs to majority M in X^*" should rejected where $\varepsilon = 1 - \mathbf{P}(M)$. The set V_m is a *critical region* on the *significance level* $\varepsilon = \frac{1}{p^m}$. If $x \in V_m$ then the hypothesis "x belongs to majority M" is rejected with the significance level ε. We say that x fails the test at the level of critical region V_m. Of course, there is a large difference between 'p-adic majority' and the ordinary 'real majority'. Populations which are very large from the point of view of ordinary real probability may be very small from the point of view of p-adic probability and vice versa.

We shall study *only the uniform p-adic probability distribution*. Everywhere below $\mathbf{P} = \mathbf{P}_p$, $p \neq 2$. Tests for randomness for this probability distribution we shall simply call p-adic test. Here condition (12.27) can be reformulated in the following way:

$$\left|\sigma(V_m^{(n)})\right|_p \leqslant \frac{1}{p^m} \tag{12.28}$$

(as $\mathbf{P}(U_x) = \frac{1}{2^n}$ for $x \in V_m^{(n)}$ and $|2^n|_p = 1$ for $p \neq 2$, (12.27) has the form $\left|\sum_{x \in V_m^{(n)}} 1\right|_p \leqslant \frac{1}{p^m}$).

Proposition 12.44. *Let V be a p-adic test. Then, for each $(x, m) \in V$, we have*[15]

$$l(x)(\log_p 2) > m \geqslant 1. \tag{12.29}$$

Proof. Set $n = l(x)$. As $x \in V_m$, we have $V_m^{(n)} \neq \varnothing$ and by (12.28) $\sigma(V_m^{(n)})$ is divisible by p^m. Thus $2^n = \sigma(X^n) \geqslant \sigma(V_m^{(n)}) \geqslant p^m$. This implies inequality (12.29). □

Proposition 12.45. *Let V be a p-adic test. Then, for each $k \geqslant m$, $n \in \mathbb{N}$,*

$$\left|\sigma(V_m^{(n)} \setminus V_k^{(n)})\right|_p \leqslant \frac{1}{p^m}.$$

Proof. As $V_k^{(n)} \subset V_m^{(n)}$, we have

$$\sigma(V_m^{(n)}) = \sigma(V_k^{(n)}) + \sigma(V_m^{(n)} \setminus V_k^{(n)}).$$

[15] Here $\log_p$ is the ordinary real log-function: $\log_p p^u = u$, $u \in \mathbb{R}$. We recall that the p-adic natural log-function was denoted by $\ln_p$.

By the strong triangle inequality we get

$$|\sigma(V_m^{(n)} \setminus V_k^{(n)})|_p \leqslant \max(|\sigma(V_m^{(n)})|_p, |\sigma(V_k^{(n)})|_p) = 1/p^m.$$

□

Condition (12.29) can be rewritten in the form

$$\lfloor l(x) \log_p 2 \rfloor \geqslant m.$$

The function $\lambda(n) = \lfloor n \log_p 2 \rfloor$, $n \in \mathbb{N}$, will play the important role in our further considerations. For any p-adic test V and $n \in \mathbb{N}$, only sets $V_m^{(n)}$, $m = 1, \ldots, \lambda(n)$, can be nonempty.

We give now a few examples of p-adic tests for randomness. All these tests are related to behavior of sums:

$$S(x) = x_1 + \cdots + x_n, \quad x \in X^*, \quad n = l(x).$$

Example 12.46. We set

$$V_m = \{x \in X^* : \Theta_p(S(x)) \geqslant p^m \Theta_p(l(x)), S(x) \neq 0 \text{ and } M_p(S(x)) \leqslant M_p(l(x))\}. \quad (12.30)$$

To show that the set $V = \{(x, m) : x \in V_m\}$ is a p-adic test, we need only to show that (12.28) holds true. We have

$$\sigma(V_m^{(n)}) = \sum_k \binom{n}{k},$$

where $0 \leqslant k \leqslant n$ and $M_p(k) \leqslant M_p(n)$, $\frac{\Theta_p(n)}{\Theta_p(k)} \leqslant \frac{1}{p^m}$. To obtain (12.28), it is sufficient to use the strong triangle inequality and Lemma 12.41.

Example 12.47. We set

$$\overline{V}_m = \left\{x \in X^* : \Theta_p(l(x)) \leqslant \frac{1}{p^m} \text{ and } M_p(S(x)) \geqslant M_p(l(x)) + 1\right\}. \quad (12.31)$$

By using Lemma 12.42 we obtain that (12.28) holds true for $\overline{V}_m$. Thus the set $\overline{V} = \{(x, m) : x \in \overline{V}_m\}$ is a p-adic test.

Example 12.48 (Finite tests). Let $n \in \mathbb{N}$ be a fixed number. Let T be some subset of X^n, $\sigma(T) = p^{-\lambda(n)}$. We set $W_m^{(n)} = T$ for $m = 1, \ldots, \lambda(n)$ and $V_j^{(n)} = \varnothing$, $j > \lambda(n)$, and $V_j^{(n)} = \varnothing$, $k \neq n$, for all $j = 1, 2, \ldots$. Then $V = \{(x, m) : x \in V_m\}$, $V_m = \bigcup_{k=1}^{\infty} V_m^{(k)}$ is a finite p-adic test.

To illustrate the statistical meaning of tests (12.30) and (12.31), it is useful to consider some subsets of them corresponding to fixed values of $M_p(n)$ and $M_p(S(x))$.

We start with test (12.30). We set

$$V_m(1,0) = \{x \in V_m : M_p(l(x)) = 1 \text{ and } M_p(S(x)) = 0\} \text{ and}$$
$$V(1,0) = \{(x,m) : x \in V_m(1,0)\}. \quad (12.32)$$

This test is connected with samples of the form

$$x = (x_1, \dots, x_{1+jp^N}), \quad j, N \in \mathbb{N}, \ (j,p) = 1. \quad (12.33)$$

Such a sample must be rejected with the level of significance $\varepsilon = \frac{1}{p^m}$ if $1 > |S(x)|_p \geqslant p^m |l(x) - 1|_p = p^{m-N}$. Thus the test $V(1,0)$ rejects all samples of the form $x = (x_1, \dots, x_{1+jp^N})$, $(j,p) = 1$, in that the sum $S(x) = x_1 + \dots + x_{1+jp^N}$ is not divisible by a sufficiently high degree of p (but divisible by p^1).

A sample x of form (12.33) with $S(x) = ip^k$, $(i,p) = 1, k \geqslant 1$, is rejected with the level of significance $\epsilon = 1/p^m$ if $k < N - m$.

For test (12.30) and $M_p(l(x)) = 1$, we can also fix $M_p(S(x)) = 1$ and obtain a new test:

$$V_m(1,1) = \{x \in V_m : M_p(l(x)) = 1 \text{ and } M_p(S(x)) = 1\} \text{ and}$$
$$V(1,1) = \{(x,m) : x \in V_m(1,1)\}.$$

A sample x of the form (12.33) must be rejected with the level of significance $\varepsilon = \frac{1}{p^m}$ if

$$1 > |S(x) - 1|_p \geqslant p^m |l(x) - 1|_p = p^{m-N}.$$

Thus the test $V(1,1)$ rejects all samples x of the form (12.33) for that $S(x) - 1$ is not divisible by a sufficiently high degree of p (but divisible by p^1).

In the same way by fixing $M_p(n) = s = 0, \dots, p-1$ we obtain tests $V_m(s,q)$, $q = 0, \dots, s$. The $V(s,q)$ rejects some samples of the form

$$x = (x_1, \dots, x_{s+jp^N}), \quad j, N \in \mathbb{N}, \ (j,p) = 1, \quad (12.34)$$

namely, samples for which $S(x) - q$ is not divisible by a sufficiently high degree of p (but divisible by p^1). A sample x of form (12.34) with $S(x) = q + ip^k$, $(i,p) = 1, k \geqslant 1$, is rejected with the level of significance $\epsilon = 1/p^m$ if $k < N - m$.

We study now test (12.31). The condition $M_p(S(x)) \geqslant M_p(l(x)) + 1 \geqslant 0$ implies that this test is used to reject (with some level of significance) some samples for that the sum $S(x)$ is not divisible by p (compare with (12.32)). We set

$$\overline{V}_m(0,1) = \{x \in \overline{V}_m : M_p(l(x)) = 0 \text{ and } M_p(S(x)) = 1\} \text{ and}$$
$$\overline{V}(0,1) = \{(x,m) : x \in \overline{V}_m(0,1)\}.$$

By this test we reject with the level of significance $\varepsilon = \frac{1}{p^m}$ all samples of the form $x = (x_1, \ldots, x_{jp^N})$, $(j, p) = 1$, for that $N < m$ and $M_p(S(x)) = 1$. We can compare the test $\overline{V}(0,1)$ with the test $V(0,0)$. The latter test is used to reject samples of the same form, but with $S(x)$ divisible by p: $S(x) = ip^k$, $(i, p) = 1, k \geqslant 1$. A sample is rejected with the level of significance $\varepsilon = \frac{1}{p^m}$ if $k < N - m$.

It is possible to introduce a p-adic test O which covers all cases of divisibility by p of $S(x)$. We start with the following simple fact:

Proposition 12.49. *Let Φ and Ψ be two p-adic tests such that $\Phi \cap \Psi = \varnothing$. Then the set $\Gamma = \Phi \cup \Psi$ is a p-adic test with critical regions $\Gamma_m = \Phi_m \cup \Psi_m$ on the significance level $\varepsilon = \frac{1}{p^m}$.*

Proof. We need only to prove that (12.28) holds true: We have

$$|\sigma(\Gamma_m^{(n)})|_p = |\sigma(\Phi_m^{(n)}) + \sigma(\Psi_m^{(n)})|_p \leqslant \max\{|\sigma(\Phi_m^{(n)})|_p, |\sigma(\Psi_m^{(n)})|_p\} \leqslant \frac{1}{p^m}.$$

□

We now turn back to tests V and $\overline{V}$ defined in Examples 3.1, 3.2. It is evident that $V_m \cap \overline{V}_m = \varnothing$ for all m. Thus sets $\Sigma_m = V_m \cup \overline{V}_m$ give critical regions (with $\varepsilon = \frac{1}{p^m}$) of a p-adic test $\Sigma = \{(x, m) : x \in \Sigma_m\}$.

12.9 Some limit theorems

As in ordinary real probability theory tests V and $\overline{V}$ of Examples 3.1, 3.2 are related to some limit theorems for p-adic probability. Let $\mathcal{P} = (\Omega, \mathcal{F}_{\text{cyl}}, \mathbf{P})$ be the probability space based on the uniform p-adic distribution $\mathbf{P}$ on the algebra $\mathcal{F}_{\text{cyl}}$ of cylindric subsets of $\Omega = X^\infty$, $p \neq 2$. For $\omega \in \Omega$, we set $S_n(\omega) = \omega_1 + \cdots + \omega_n$.

Theorem 12.50. *For each $l \in \mathbb{N}$ the probability*

$$\mathbf{P}\left(\left\{\omega \in \Omega : |S_n(\omega) - M_p(S_n(\omega))|_p = \frac{1}{p^l}, M_p(S_n(\omega)) \leqslant M_p(n)\right\}\right) \to 0$$

in $\mathbb{Q}_p$, when $|n - M_p(n)|_p \to 0$, $n \neq M_p(n)$.

Proof. By using considerations of Example 12.46 we obtain that

$$\mathbf{P}\left(\left\{\omega \in \Omega : |S_n(\omega) - M_p(S_n(\omega))|_p = \frac{1}{p^l}, M_p(S_n(\omega)) \leqslant M_p(n)\right\}\right) \leqslant p^l |n - M_p(n)|_p.$$

□

In particular, we obtain the following limit theorems:

Corollary 12.51. *For each $l \in \mathbb{N}$, the probability*

$$\mathbf{P}\left(\left\{\omega \in \Omega : S_n(\omega) \in \mathcal{S}_{\frac{1}{p^l}}(0)\right\}\right) \to 0$$

in $\mathbb{Q}_p$, when $|n|_p \to 0$.

Corollary 12.52. *For each $l \in \mathbb{N}$, the probabilities*

$$\mathbf{P}\left(\left\{\omega \in \Omega : S_n(\omega) \in \mathcal{S}_{\frac{1}{p^l}}(0)\right\}\right) \quad \text{and} \quad \mathbf{P}\left(\left\{\omega \in \Omega : S_n(\omega) \in \mathcal{S}_{\frac{1}{p^l}}(1)\right\}\right)$$

tend to zero in $\mathbb{Q}_p$, when $|n-1|_p$ tends to zero.

Formally we can interpret Corollary 12.52 in the following way. The sum $S_n(\omega)$ can be considered as the sum $S_n(\omega) = \xi_1(\omega) + \cdots + \xi_n(\omega)$ of independent equally distributed random variables $\xi_j(\omega) = 0, 1$ with probabilities $1/2$. By Corollary 12.52 the probability distribution of random variable $S_{\lim}(\omega) = \lim_{n\to 1} S_n(\omega)$ is concentrated at the points $a_0 = 0$ and a_1 of $\mathbb{Q}_p$. By symmetry reasons $\mathbf{P}_{S_{\lim}}(\{a_0\}) = \mathbf{P}_{S_{\lim}}(\{a_1\}) = 1/2$. Of course, this is just a formal statement, because Corollary 12.52 gives convergence only for spheres of $\mathbb{Q}_p$.

Theorem 12.53. *The probability*

$$\mathbf{P}(\{\omega \in \Omega : M_p(S_n(\omega)) \geq M_p(n) + 1\}) \to 0$$

when $|n - M_p(n)|_p \to 0$.

As in the case of Theorem 12.50, we can, for example, put $M_p(n) = 0$ or $M_p(n) = 1$ and obtain the following consequences of Theorem 12.53:

Corollary 12.54. *The probability*

$$\mathbf{P}(\{\omega \in \Omega : M_p(S_n(\omega)) \geq 1\}) \to 0$$

in $\mathbb{Q}_p$, when $|n|_p \to 0$.

Corollary 12.55. *The probability*

$$\mathbf{P}(\{\omega \in \Omega : M_p(S_n(\omega)) \geq 2\}) \to 0$$

when $|n-1|_p \to 0$.

We note that

$$\mathbf{P}(\{\omega \in \Omega : M_p(S_n(\omega)) \geq 1\}) = \mathbf{P}(\{\omega \in \Omega : S_n(\omega) \in \mathcal{S}_1(0)\}).$$

Thus by Corollaries 12.51 and 12.54 we obtain that

$$\mathbf{P}(\{\omega \in \Omega : S_n(\omega) \in \mathcal{U}_{\frac{1}{p^m}}(0)\}) \to 1,$$

$|n|_p \to 0$, for any $m \in \mathbb{N}$. Hence formally we obtain that the probability distribution $\mathbf{P}_{S_{\lim}}$ of $S_{\lim}(\omega) = \lim_{n \to 0} S_n(\omega)$ is concentrated at the point $a_0 = 0 \in \mathbb{Q}_p$, $\mathbf{P}_{S_{\lim}}(\{0\}) = 1$.

It seems that in the p-adic case it is more natural to use tests for randomness than limit theorems. In the opposite to ordinary real probability theory in the p-adic case we have no general limit theorems for $n \to \infty$ (in the sense of the order on $\mathbb{N}$). All limit theorems give the convergence of probabilities for some sequences $n_k \to \infty$, $k \to \infty$. For example, $|n_k|_p \to 0$, $n_k \neq 0$, implies that $n_k = jp^N$, $(j, p) = 1$, $N \to \infty$, and $|n_k - 1|_p \to 0$, $n_k \neq 1$, implies that $n_k = 1 + jp^N$, $(j, p) = 1$, $N \to \infty$, and so on.

12.10 Recursive enumeration of the set of p-adic tests

Here we shall prove that the set of all p-adic tests is recursively enumerable. The general scheme of the proof is the same as in the case of real probabilities. However, the main part of the proof (an algorithm for constructing a p-adic test on the basis of a partial recursive function) strongly differs from the standard one.

We start with the following well-known lemma (see, for example, [297]).

Lemma 12.56. *There exists a partial recursive function $f : \mathbb{N} \times \mathbb{N} \to X^* \times \mathbb{N}$ with the following properties:*

(a1) *for all $i, j \in \mathbb{N}$ such that $f(i, j) \neq \infty$, we have $f(i, k) \neq \infty$, for all $k \leqslant j$;*

(a2) *a set $A \subset X^* \times \mathbb{N}$ is r.e. iff $A = \{f(i, j) : j = 1, 2, \ldots\} \setminus \{\infty\}$, for some $i \geqslant 1$.*

Theorem 12.57. *The set of all p-adic tests is r.e.*

Proof. Through the proof we shall use the fixed partial recursive function $\varphi = \varphi_i = f(i, \cdot)$ given by Lemma 12.56. We set $A_\varphi = \varphi(\mathbb{N})$. As in the standard case, we shall construct for each φ some total recursive function $g : \mathbb{N} \to X^* \times \mathbb{N}$ such that $T = A_g = g(\mathbb{N})$ is a p-adic test and if φ is a p-adic test by itself, then $T = A_\varphi$. We construct T step by step using an algorithm which produces a p-adic test at each step. In the following algorithm we shall use sets $\mathcal{D}_m^{(n)}$ which give approximations for sets $T_m^{(n)}$ in the process of building of T (as usual $T_m = \{x \in X^* : (x, m) \in T\}$ and $T_m^{(n)} = \{x \in T_m : l(x) = n\}$). We shall also use sets $R_m^{(n)}$ which are registers for collecting elements of $(A_\varphi)_m^{(n)}$. The main difference with the standard algorithm is due to the fact that we cannot increase sets $\mathcal{D}_m^{(n)}$ at each step when φ produces a value $\varphi(j) \in (A_\varphi)_m^{(n)} = \{x : \varphi(j) = (x, m)$ for some j and $l(x) = n\}$ (because the p-adic metric is changed discontinuously: $|x|_p \leqslant \frac{1}{p^m} \Rightarrow |x+1|_p = 1, m \geqslant 1$). We collect (in $R_m^{(n)}$) elements of $(A_\varphi)_m^{(n)}$ until $\sigma(R_m^{(n)})$ becomes divisible by p^m. After this we set $\mathcal{D}_m^{(n)} = R_m^{(n)}$.

To be sure that the result of our construction will be a r.e. set, we construct parallel a function $g\colon \mathbb{N} \to X^* \times \mathbb{N}$ such that $T = g(\mathbb{N})$ and g is a total recursive function if T is an infinite set.

Algorithm

1 Put $T = \varnothing$, $\mathcal{D}_m^{(n)} = R_m^{(n)} = \varnothing$; put $j = 0, i = 0, t_m^{(n)} = 0$.
% j is the argument of φ, i is the argument of g; $t_m^{(n)} = \sigma(R_m^{(n)})$.

2 Put $j = j + 1$.

3 If $\varphi(j) = \infty$, then the computation of T is finished.

4 Find $\varphi(j) = (x, m)$ and $n = l(x)$.

5 If $m > \lfloor n \log_p 2 \rfloor$, then $T = \varnothing$ and stop.

6 Put $R_m^{(n)} = R_m^{(n)} \cup \{x\}$ and $t_m^{(n)} = t_m^{(n)} + 1$.

7 If $|t_m^{(n)}|_p > \frac{1}{p^m}$, then go to step 2.

8 If $m \geqslant 2$ and $\mathcal{D}_{m-1}^{(n)} \not\supset R_m^{(n)}$, then go to step 2.

9 Put $\mathcal{D}_m^{(n)} = R_m^{(n)}$.
% We must make step 8 before step 9 to get $T_{m-1} \supset T_m$.

10 **(a)** Enumerate elements if $\mathcal{D}_m^{(n)} = \{z_1, \dots, z_{t_m^{(n)}}\}$;

(b) for $l = 1, \dots, t_m^{(n)}$, put $g(i + l) = (z_l, m)$;

(c) put $i = i + t_m^{(n)}$
% The previous step is not related to the construction if T; here we construct the function g which gives recursive enumeration for T.

11 Put $s = m$.

12 Put $s = s + 1$.

13 If $s > \lfloor n \log_p 2 \rfloor$, go to 18.

14 If $|t_s^{(n)}|_p > \frac{1}{p^s}$, go to 18.

15 If $\mathcal{D}_{s-1}^{(n)} \not\supset R_s^{(n)}$, go to 18.

16 Put $\mathcal{D}_s^{(n)} = R_s^{(n)}$.
% We explain the meaning of steps 11–16. By step 9 the set $\mathcal{D}_m^{(n)}$ has been increased. Thus condition 8 must be reconsidered for sets $\mathcal{D}_s^{(n)}$ with $s > m$. It can be that occasionally some of sets $R_s^{(n)}$ has the number of elements which is divisible by p^s. If they pass step 15, then we increase sets $\mathcal{D}_j^{(n)}$ by 16.

17 Repeat step 10 for $m = s$.

18 Put $T = T \bigcup_{m \leqslant s \leqslant \lfloor n \log_p 2 \rfloor} \mathcal{D}_s^{(n)} \times \{s\}$ and go to step 2.

We prove now that the set T which is constructed by the algorithm is a p-adic test.

(A1) We use the parameter j to denote the step (determined by 2) of the algorithm. We have $T_m^{(n)} = \bigcup_{j=1}^{\infty} \mathcal{D}_m^{(n)}(j)$. As $\mathcal{D}_m^{(n)}(j+1) \supset \mathcal{D}_m^{(n)}(j)$ and $\sigma(\mathcal{D}_m^{(n)}(j))$ is divisible by p^m, we get that $\sigma(T_m^{(n)})$ is divisible by p^m. Thus $|\sigma(T_m^{(n)})|_p \leqslant \frac{1}{p^m}$.

(A2) By step 8 and 15 we get that $\mathcal{D}_m^{(n)} \supset \mathcal{D}_{m+1}^{(n)}$, $n, m \in \mathbb{N}$. Thus $T_m^{(n)} \supset T_{m+1}^{(n)}$, $n, m \in \mathbb{N}$.

(A3) If steps 10 and 16 are passed an infinite number of times, then g is the total recursive function and, hence, $T = A_\varphi$ is r.e. If the steps 10 and 16 are passed only a finite number of times, then the set T is finite and, hence, r.e.

We prove now that if $V = A_\varphi$ is a p-adic test, then $T = A_\varphi$.

It is evident that $T \subset A_\varphi$. We have only to prove that $V \subset T$. It is sufficient to prove that, for each n, $V_m^{(n)} \times \{m\} \subset T_m^{(n)} \times \{m\}$ for all $m \leqslant \lfloor n \log_p 2 \rfloor$.

For each n, the set $V^{(n)} = \{(x, m) \in V : l(x) = n\}$ is finite (since $m \leqslant \lfloor n \log_p 2 \rfloor$). Thus φ produces all elements of $V^{(n)}$ after a finite numbers of steps $J = J(n, \varphi)$.[16] Let $\varphi(J) = (x_J, m_J)$ (here $l(x_J) = n$ and $m_J \leqslant \lfloor n \log_p 2 \rfloor$). We have: $\mathcal{D}_1^{(n)} \supset \mathcal{D}_2^{(n)} \supset \cdots \supset \mathcal{D}_M^{(n)}$ and $|\sigma(\mathcal{D}_s^{(n)})|_p \leqslant \frac{1}{p^s}$ for $s = 1, \ldots, M = \lfloor n \log_p 2 \rfloor$. We also have: $R_s^{(n)} = V_s^{(n)}$ (because $V^{(n)} \subset \varphi(\{1, 2, \ldots, J\})$ and $\varphi(\{1, 2, \ldots, J\})_s^{(n)} = R_s^{(n)}$). Thus, for all s, $|\sigma(R_s^{(n)})|_p \leqslant \frac{1}{p^s}$. In particular, this holds for $s = m_J$. Hence, for $m = m_J$, step 7 is passed.

We prove that $\mathcal{D}_s^{(n)} = R_s^{(n)} = V_s^{(n)}$ for all $s = 1, \ldots, m_J - 1$. As $|\sigma(R_1^{(n)})|_p \leqslant \frac{1}{p}$, $R_1^{(n)}$ has passed step 7. But step 8 is trivial for $m = 1$. Thus by step 9 we get $\mathcal{D}_1^{(n)} = R_1^{(n)} = V_1^{(n)}$. For $s = 2$, we have $|\sigma(R_2^{(n)})|_p \leqslant \frac{1}{p^2}$ and step 7 is passed. As $\mathcal{D}_1^{(n)} = V_1^{(n)}$ and $R_2^{(n)} = V_2^{(n)}$, we have $\mathcal{D}_1^{(n)} \supset R_2^{(n)}$ and step 8 is passed. By step 9 we get $\mathcal{D}_2^{(n)} = R_2^{(n)} = V_2^{(n)}$. We can repeat such considerations until s takes value $m_J - 1$.

As $\mathcal{D}_{m_J-1}^{(n)} = V_{m_J-1}^{(n)} \supset V_{m_J}^{(n)} = R_{m_J}^{(n)}$, step 8 is passed for $m = m_J$ and we get $\mathcal{D}_{m_J}^{(n)} = R_{m_J}^{(n)} = V_{m_J}^{(n)}$. Thus we arrive to step 11 with $m = m_J$. For all $m_J < s \leqslant M = \lfloor n \log_p 2 \rfloor$, step 14 is passed automatically. For $s = m_J + 1$ we have $\mathcal{D}_{m_J}^{(n)} = V_{m_J}^{(n)} \supset V_{m_J+1}^{(n)} = R_{m_J+1}^{(n)}$. Hence step 15 is passed and we put $\mathcal{D}_{m_J+1}^{(n)} = R_{m_J+1}^{(n)} = V_{m_J+1}^{(n)}$. Repeating these considerations, we prove that $\mathcal{D}_s^{(n)} = R_s^{(n)} = V_s^{(n)}$ for all $s = m_J, \ldots, M$. Hence $V^{(n)} = T^{(n)}$. □

[16]Of course, some points $(x, m) \in V^{(n)}$ can appear again on some steps $J' > J$.

12.11 No p-adic universal test

A natural generalization of the definition of a universal test for randomness is the following one:

Definition 12.58. A p-adic test U is said to be *universal* if for every p-adic test V we can effectively find $c \in \mathbb{N}$ (depending upon U and V) such that $V_{m+c} \subset U_m$ for all m.

It is well known that in the ordinary real probability theory there exists a universal test for randomness (which is, of course, not unique). We shall show that in p-adic probability theory there is no universal recursive tests. We start with some technical considerations. We have to study more carefully properties of the function $\lambda(n) = \lfloor n \log_p 2 \rfloor$. As $p > 2$, we have $\log_p 2 < 1$. We set $L_k = \lfloor \frac{k}{\log_p 2} \rfloor$. If $0 < n \leqslant L_1$, then $n \log_p 2 < 1$ and $\lambda(n) = 0$; in the same way we have: if $L_{k-1} < n \leqslant L_k$, then $\lambda(n) = k - 1$, $k \geqslant 2$. We set $n_k = L_k + 1$.

Lemma 12.59. *The inequality*

$$p^{\lambda(n_k)} > 2^{n_k - 1} \tag{12.35}$$

holds true for all $k = 1, 2, \ldots$.

Proof. We have $\lambda(n_k) = k$ and $\lambda(n_k - 1) = k - 1$. By definition $\lambda(n) = \max\{l : p^l < 2^n\}$. Hence, for all n, $p^{\lambda(n)+1} > 2^n$. In particular, $p^{\lambda(n_k - 1)+1} = p^k > 2^{n_k - 1}$. Hence $p^k = p^{\lambda(n_k)} > 2^{n_k - 1}$. □

We construct now two p-adic tests W and $\widetilde{W}$ by using the following procedure. For $k = 1, 2, \ldots$ and $j = 1, \ldots, \lambda(n_k)$, we set

$$W_j^{(n_k)} = W_{\lambda(n_k)}^{(n_k)} = \{x_1, \ldots, x_{p^{\lambda(n_k)}}\}$$

and

$$\widetilde{W}_j^{(n_k)} = \widetilde{W}_{\lambda(n_k)}^{n_k} = \{x_{2^{n_k} - p^{\lambda(n_k)} + 1}, \ldots, x_{2^{n_k}}\}$$

and $W_l^{(n_k)} = \widetilde{W}_l^{(n_k)} = \varnothing$ for $n \neq n_k$ and $l = 1, 2, \ldots$. Here we have used the lexicographic enumeration of elements of X^{n_k}, $k = 1, 2, \ldots$: $x_1, x_2, \ldots, x_{2^{n_k}}$. Since $\sigma(W_j^{(n_k)}) = \sigma(\widetilde{W}_j^{(n_k)}) = p^{\lambda(n_k}$, $j = 1, 2, \ldots, \lambda(n)$, by (12.35) we obtain $W_j^{(n_k)} \cap \widetilde{W}_j^{(n_k)} \neq \varnothing$ and hence

$$X^{n_k} = W_j^{(n_k)} \cup \widetilde{W}_j^{(n_k)}, \quad j = 1, \ldots, \lambda(n_k).$$

Theorem 12.60. *A universal p-adic test does not exist.*

Proof. Let us suppose that there exists a universal p-adic test U. Thus we can effectively find $c_1, c_2 \in \mathbb{N}$ such that $W_{m+c_1} \subset U_m$ and $\widetilde{W}_{m+c_2} \subset U_m$, where W and $\widetilde{W}$ are p-adic tests constructed before this theorem. Let k be so large that $\lambda(n_k) - c_1 \geq 1$ and $\lambda(n_k) - c_2 \geq 1$. Thus $W_{1+c_1}^{(n_k)} = W_{\lambda(n_k)}^{(n_k)}$, $\widetilde{W}_{1+c_2}^{(n_k)} = \widetilde{W}_{\lambda(n_k)}^{(n_k)}$. Hence $U_1^{(n_k)} \supset W_{1+c_1}^{(n_k)} \cup \widetilde{W}_{1+c_2}^{(n_k)} = X^{n_k}$. This implies that $|\sigma(U_1^{(n_k)})|_p = |\sigma(X^{n_k})|_p = 1$. This contradicts to (12.28). □

In the p-adic case randomness of infinite sequences was studied in details in [242]. Studies performed in the latter imply that in the p-adic case (similar to Schnorr's theory of randomness [376]) the only reasonable approach to randomness of infinite sequences is to use randomness with respect to the concrete p-adic sequential test. Of course, the use of such randomness has extremely different origins in our theory and Schnorr's theory. It seems that in the p-adic case this situation is a consequence of the impossibility to define σ-additive (non-discrete) probability on the σ-algebra generated by $\mathcal{F}_{\mathrm{cyl}}$. In Schnorr's theory this situation is a consequence of the use of total recursive null-sets.

Chapter 13

p-adic valued quantization

In this chapter we present a brief review which covers an important domain of p-adic mathematical physics – quantum mechanics with p-adic valued wave functions. One of the aims of this review is to describe one of the main sources that stimulated the creation of the theory with probabilities valued in $\mathbb{Q}_p$.

13.1 Toward quantum mechanics with p-adic valued wave functions

Quantum formalism with wave functions valued in non-Archimedean fields was developed in a series of papers and books [2–4, 6–8, 87–89, 185–193, 201, 205–212, 214, 215, 218, 220, 225, 226, 230]. In this chapter we present essentials of this theory. We restrict our considerations to the fields of p-adic numbers. General quantum theory was developed for an arbitrary non-Archimedean field K, see [214].

We recall that in ordinary quantum mechanics to describe observables taking values in the field of real numbers $\mathbb{R}$, one should proceed to the complex Hilbert space. In the same way to create theory of observables taking values in a field of p-adic numbers $\mathbb{Q}_p$ for some prime number p, one should consider a Hilbert space (in fact, rather nontrivial generalization of the ordinary Hilbert space) over a quadratic extension $\mathbb{Q}_p(\sqrt{\tau})$, where $\tau \in \mathbb{Q}_p$ and equation $x^2 = \tau$ has no solution in $\mathbb{Q}_p$. The situation is essentially more complicated than in the Archimedean case, since there exist a number of nonisomorphic quadratic extensions of $\mathbb{Q}_p$. No reasonable physical interpretation of this mathematical fact has yet been provided, see [201, 214] for an attempt. However, in the purely mathematical framework quantum formalism does not depend on the choice of a quadratic extension.

The basic objects of this theory are the p-adic Hilbert space and symmetric operators acting in this space. Vectors of the p-adic Hilbert space which are normalized with respect to inner product represent quantum states.[1]

[1] In the p-adic case the norm is not determined by the inner product. Therefore normalization with respect to the norm and the inner product which coincide in the real and complex Hilbert spaces are different in the p-adic Hilbert space.

In ordinary (complex) quantum mechanics the spectrum of an operator is identified with observed values of the quantum observable which is represented by this operator – the spectral postulate of quantization. Self-adjoint operators had been chosen for the mathematical description of observables, because they have real spectra. Moreover, the spectral theorem for self-adjoint operators induces the natural probabilistic interpretation generalizing Born's rule for the square of the ψ-function. In the p-adic case spectral theory is essentially more complicated than in the complex case. It is a nontrivial task to find a class of operators (the analogue of the class of self-adjoint operators in the complex case) which would serve for the operator representation of quantum observables. The problem is the absense of a "good spectral" theorem for e.g. symmetric operators, even in the finite-dimensional case. We still keep to symmetric operators only because they have spectra belonging to $\mathbb{Q}_p$. We shall proceed in the following way:

Consider the formal differential expression $\widehat{\mathbf{H}} = H(\partial_{x_j}, x_j)$ of operators of quantum mechanics or quantum field theory. Let us realize this formal expression as a differential operator with variables x_j belonging to the field of p-adic numbers $\mathbb{Q}_p$ and study the properties of this operator in the p-adic Hilbert space. Thus we would like to perform a p-adic analogue of Schrödinger's quantization.

As was mentioned, p-adic valued quantum theory suffers of the absence of a "good spectral theorem" for symmetric operators. At the same time this theory is essentially simpler (mathematically) than the ordinary quantum mechanics, since *operators of position and momentum are bounded in the p-adic case* – as it was found by Albeverio and Khrennikov [7].

Representations of groups in Hilbert spaces are one of the cornerstones of ordinary quantum mechanics. It is very natural to develop p-adic quantum mechanics in a similar way. We construct a representation of the Weyl–Heisenberg group in the p-adic Hilbert space, the space $L_2(\mathbb{Q}_p, \nu_b)$ of L_2-functions with respect to p-adic valued Gaussian distribution ν_b (the symbol b indicates a p-adic analogue of dispersion), see [7].[2] Here the situation differs very much from that of ordinary quantum mechanics. If we denote by $\widehat{U}(\alpha)$ and $\widehat{V}(\beta)$ the groups of unitary operators corresponding to position and momentum operators, respectively, then these groups are defined only for parameters α and β belonging to balls $B_{R(b)}$ and $B_{r(b)}$, respectively, where $R(b)$ and $r(b)$ depend on dispersion b of Gaussian distribution and they are coupled by a kind of Heisenberg's uncertainty relation.

We also study the representation of the translation group in the space $L_2(\mathbb{Q}_p, \nu_b)$. Here the result also differs from that of ordinary quantum mechanics and it is more similar to the one which holds in quantum field theory where Gaussian distributions on infinite-dimensional spaces are used.

---

[2]We remark that ν_b is not a p-adic valued measure, a bounded linear functional on the space of continuous functions. It is just a distribution, a generalized function, which is primarily defined on the space of analytic test functions. An analogue of the L_2-space can be constructed by completing the space of test functions with respect to a natural norm.

Let μ be Gaussian measure on the infinite-dimensional real Hilbert space $\mathcal{H}$. It is impossible to construct a representation of translations from all of $\mathcal{H}$ in $L_2(\mathcal{H}, \mu)$, because of the well-known fact that the translation μ^h of a Gaussian measure on $\mathcal{H}$ by a vector $h \in \mathcal{H}$ can be singular with respect to μ. It is well known that μ^h is equivalent to μ if and only if h belongs to a certain proper ('Cameron–Martin') subspace. In a similar way we cannot construct in the space $L_2(\mathbb{Q}_p, \nu_b)$ a representation of translations by all elements h in $\mathbb{Q}_p$; in fact, we have to restrict the considerations to translations belonging to some ball (which is an additive subgroup in $\mathbb{Q}_p$) whose radius depends on dispersion, b. This fact is connected with nonexistence of translation invariant measures in the p-adic case (similarly as for infinite-dimensional spaces over the field of real numbers), see [4].

We study the spectrum of the position operator, see Albeverio, Cianci, Khrennikov [3]. The main problem is to find the spectrum of this operator. This problem is sufficiently complicated. It has not yet been completely solved. As $\widehat{\mathbf{x}}$ is bounded, we know that its spectrum is a proper subset of the ball of the radius $\lambda_x = \|\widehat{\mathbf{x}}\|$. At the moment we cannot give the answer to the question: Does the spectrum of $\widehat{\mathbf{x}}$ coincide with the ball of radius λ_x? We have only a particular result in this direction. It was proved that the spectrum of the p-adic position operator is concentrated in a ball of finite radius. The radius depends on dispersion b of p-adic Gaussian distribution. It should be remarked that in the p-adic case there exist non-equivalent Gaussian distributions and, consequently, non-isomorphic L_2-spaces with respect to these distributions. They induce representations of canonical commutation relations which are not unitary equivalent.

The spectral properties of the p-adic momentum operator $\widehat{\mathbf{p}}$ have also been studied, see [2].

13.2 Hilbert spaces

Let $\tau \in \mathbb{Q}_p$ and let equation $x^2 = \tau$ have no solution in $\mathbb{Q}_p$. The symbol $\mathbb{Q}_p(\sqrt{\tau})$ denotes the corresponding quadratic extension of $\mathbb{Q}_p$, see Subsection 1.8.1 for mathematical details. Its elements have the form $z = x + \sqrt{\tau} y$, where $x, y \in \mathbb{Q}_p$. The operation of conjugation is defined by $\bar{z} = x - \sqrt{\tau} y$. We remark that $z\bar{z} = x^2 - \tau y^2$ for $z \in \mathbb{Q}_p(\sqrt{\tau})$. We remark that $z\bar{z} \in Q_p$ for any $z \in \mathbb{Q}_p(\sqrt{\tau})$. The extension of the p-adic valuation from $\mathbb{Q}_p$ onto $\mathbb{Q}_p(\sqrt{\tau})$ is denoted by the same symbol $|\cdot|_p$. We have $|z|_p = \sqrt{|z\bar{z}|_p}$ for $z \in \mathbb{Q}_p(\sqrt{\tau})$. Besides quadratic extensions, we shall also operate with the field of complex p-adic numbers $\mathbb{C}_p$, Subsection 1.8.3.

We now introduce *p-adic Hilbert spaces*. We remark that various nonequivalent non-Archimedean generalizations of real and complex Hilbert spaces have been elaborated since the pioneer paper of Kalisch [181], see Bayod for fundamental study [47]. In this book we shall use the definition which was proposed by one of the authors, see, e.g., [201]. The main idea behind the latter definition is to use as the starting point of p-adic (as well as more general non-Archimedean) generalization the canonical isomorphism between the ordinary Hilbert space (real or complex) and the corresponding space of square summable sequences l^2 – "coordinate Hilbert space." We start with the definition of the p-adic analogue of the l^2-space. In general, p-adic Hilbert space is defined as an isomorphic image of coordinate Hilbert space. We proceed in the same way in the case of Hilbert space over a quadratic extension $\mathbb{Q}_p(\sqrt{\tau})$ of $\mathbb{Q}_p$.

We take a sequence of p-adic numbers $\lambda = (\lambda_n) \in Q_p^\infty$, $\lambda_n \neq 0$. We set

$$l^2(p,\lambda) = \{f = (f_n) \in \mathbb{Q}_p^\infty : \text{ the series } \sum f_n^2 \lambda_n \text{ converges in } \mathbb{Q}_p\}.$$

It is clear that $l^2(p,\lambda) = \{f = (f_n) \in \mathbb{Q}_p^\infty : \lim_{n\to\infty} |f_n|_p \sqrt{|\lambda_n|_p} = 0\}$. In the space $l^2(p,\lambda)$ we introduce the norm $\|f\|_\lambda = \max_n |f_n|_p \sqrt{|\lambda_n|_p}$. The space $l^2(p,\lambda)$ endowed with this norm is a non-Archimedean Banach space. On the space $l^2(p,\lambda)$ we also introduce the p-adic valued inner product $(\cdot,\cdot)_\lambda$ by setting $(f,g)_\lambda = \sum f_n g_n \lambda_n$. The p-adic inner product $(\cdot,\cdot)_\lambda : l^2(p,\lambda) \times l^2(p,\lambda) \to \mathbb{Q}_p$ is continuous and the *Cauchy–Bunyakovsky–Schwarz inequality* holds: $|(f,g)_\lambda|_p \leqslant \|f\|_\lambda \, \|g\|_\lambda$.

Definition 13.1. A triplet $(l^2(p,\lambda), (\cdot,\cdot)_\lambda, \|\cdot\|_\lambda)$ is called the p-adic coordinate Hilbert space.

More generally we shall define a p-adic inner product on a $\mathbb{Q}_p$-linear space E as an arbitrary non-degenerate symmetric bilinear form $(\cdot,\cdot): E \times E \to \mathbb{Q}_p$.

Remark 13.2. We cannot introduce a p-adic analogue of positive definiteness of a bilinear form. For instance, any element $\gamma \in \mathbb{Q}_p$ can be represented as $\gamma = (x,x)_\lambda$, $x \in l^2(p,\lambda)$ (this is a simple consequence of the properties of bilinear forms over $\mathbb{Q}_p$, see Dragovich [104] for further physical discussion related to this algebraic fact).

The triplets $(E_j, (\cdot,\cdot)_j, \|\cdot\|_j)$, $j = 1,2$, where E_j are non-Archimedean Banach spaces, $\|\cdot\|_j$ are norms and $(\cdot,\cdot)_j$ are inner products satisfying the Cauchy–Bunyakovsky–Schwarz inequality, are isomorphic if the spaces E_1 and E_2 are algebraically isomorphic and the algebraic isomorphism $I: E_1 \to E_2$ is isometric and unitary, i.e., $\|Ix\|_2 = \|x\|_1$, $(Ix, Iy)_2 = (x,y)_1$.

Definition 13.3. The triplet $(E, (\cdot,\cdot), \|\cdot\|)$ is the p-adic Hilbert space if it is isomorphic to the coordinate Hilbert space $(l^2(p,\lambda), (\cdot,\cdot)_\lambda, \|\cdot\|_\lambda)$ for some sequence of weights λ.

The isomorphism relation splits the family of p-adic Hilbert spaces into equivalence classes. An equivalence class is characterized by some coordinate representative $l^2(p,\lambda)$. To classify p-adic Hilbert spaces is an open mathematical problem.

Hilbert spaces over quadratic extensions $\mathbb{Q}_p(\sqrt{\tau})$ of $\mathbb{Q}_p$ can be introduced in the same way. For a given sequence $\lambda = (\lambda_n) \in \mathbb{Q}_p^\infty$, $\lambda_n \neq 0$, we set

$$l^2(p,\lambda,\sqrt{\tau}) = \{f = (f_n) \in \mathbb{Q}_p(\sqrt{\tau})^\infty : \text{ the series} \sum f_n \bar{f}_n \lambda_n \text{ converges}\};$$

$\|f\|_\lambda = \max_n |f_n|_p \sqrt{|\lambda_n|_p}$; $(f,g)_\lambda = \sum f_n \bar{g}_n \lambda_n$.

The triplet $(l^2(p, \lambda, \sqrt{\tau}), (\cdot,\cdot)_\lambda, \|\cdot\|_\lambda)$ is the coordinate Hilbert space over the quadratic extension $\mathbb{Q}_p(\sqrt{\tau})$. In general a Hilbert space over the quadratic extension $\mathbb{Q}_p(\sqrt{\tau})$, $(E, (\cdot,\cdot), \|\cdot\|)$ is by definition isomorphic to some coordinate Hilbert space. We denote a p-adic Hilbert space over $\mathbb{Q}_p(\sqrt{\tau})$ by the symbol

$$\mathcal{H}_p \equiv \mathcal{H}_p(\sqrt{\tau}).$$

The first non-Archimedean analogue of a Hilbert space was considered by Kalisch [181]. However, a class of non-Archimedean Hilbert spaces introduced in [181] is too restrictive for our applications. Kalisch introduced Hilbert spaces over a complete separable non-Archimedean field K with the valuation $|\cdot|$ which satisfies the following conditions: (K1) $|2| = 1$; (K2) every $x \in K$, $|x| = 1$, (a unit of K) possesses a square root in K. The last condition is very strong. In particular, $\mathbf{Q}_p$ and $\mathbb{Q}_p(\sqrt{\tau})$ do not satisfy this condition. The only interesting example of a non-Archimedean field which satisfies the condition (K2) is the field of complex p-adic numbers $\mathbb{C}_p$. But this field is not useful for our applications since it is an infinite-dimensional space over $\mathbb{Q}_p$ and there is no continuous involutions on $\mathbb{C}_p$.

Now let K be a non-Archimedean field which satisfies the above restrictions. Kalisch defined [181] a non-Archimedean Hilbert space as a triplet $(E, (\cdot,\cdot), \|\cdot\|)$, where E with norm $\|\cdot\|$ is a separable non-Archimedean Banach space over K, $(\cdot,\cdot)\colon E \times E \to K$ is a symmetric bilinear form which satisfies the following conditions: (K3) the Cauchy–Bunyakovsky–Schwarz inequality holds; (K4) for every $x \in E$ there exists $\alpha \in K$ such that $\|x\| = |\alpha|_K$; (K5) for every $x \in E$ there exists $x, x' \neq 0$, such that $|(x, x')|_K = \|x\|\|x'\|$. Kalisch proved that every non-Archimedean Hilbert space is isomorphic to the coordinate Hilbert space over K: $l_K^2 = c_0(K) = \{f = (f_n) \in K^\infty : \lim_{n\to\infty} f_n = 0\}$. We wish to notice that our p-adic (and complex p-adic) Hilbert spaces do not satisfy the condition (K4). An extended review on different non-Archimedean analogues of a Hilbert space is contained in the dissertation of Bayod [47]. We wish to notice that our class of p-adic Hilbert spaces does not coincide with anyone considered in [47].

Remark 13.4. It is possible to extend the formalism and to use elements of the Galois group $G(\mathbb{C}_p/\mathbb{Q}_p)$, instead of an involution. But this theory is much more complicated, see [186] for the details.

13.3 Groups of unitary isometric operators in the p-adic Hilbert space

As usual, we introduce unitary operators $\widehat{U} : \mathcal{H}_p \to \mathcal{H}_p$ as operators which preserve the inner product: $(\widehat{U}x, \widehat{U}y) = (x, y)$ for all $x, y \in \mathcal{H}_p$ and have the image $\operatorname{Im}\widehat{U} = \widehat{U}(\mathcal{H}_p) = \mathcal{H}_p$ and isometric operators as operators which preserve the norm: $\|\widehat{U}x\| = \|x\|$ and $\operatorname{Im}\widehat{U} = \mathcal{H}_p$. Denote the space of all bounded linear operators $\widehat{A}\colon \mathcal{H}_p \to \mathcal{H}_p$

by the symbol $\mathcal{L}(\mathcal{H}_p)$. It is a Banach space with respect to the operator norm $\|\widehat{A}\| = \sup_{x\neq 0} \|\widehat{A}x\|/\|x\|$. A unitary operator need not be isometric.[3] Moreover, it could be even unbounded. Denote the group of linear isometries of the p-adic Hilbert space $\mathcal{H}_p$ by the symbol $\mathrm{IS}(\mathcal{H}_p)$. Denote the group of all bounded unitary operators in $\mathcal{H}_p$ by the symbol $\mathrm{UN}(\mathcal{H}_p)$. Set $\mathrm{UI}(\mathcal{H}_p) = \mathrm{UN}(\mathcal{H}_p) \cap \mathrm{UI}(\mathcal{H}_p)$.

An operator $\widehat{A} \in \mathcal{L}(\mathcal{H}_p)$ is said to be symmetric if $(\widehat{A}x, y) = (x, \widehat{A}y)$ for all x, y. The following simple fact will be useful in our later considerations.

Theorem 13.5. *The eigenvalue α of a symmetric operator $\widehat{A}\colon \mathcal{H}_p \to \mathcal{H}_p$ corresponding to an eigenvector u with nonzero square, $(u, u) \neq 0$, belongs to $\mathbb{Q}_p$. Eigenvectors corresponding to different eigenvalues of such type are orthogonal.*

The proof is similar to the standard one for the complex Hilbert space $\mathcal{H}$.

As usual we introduce the resolvent set $\mathrm{Res}(\widehat{A})$ of an operator $\widehat{A} \in \mathcal{L}(\mathcal{H}_p)$: $\lambda \in \mathbb{Q}_p(\sqrt{\tau})$ such that the operator $(\lambda I - \widehat{A})^{-1}$ exists and the spectrum of $\widehat{A}$, $\mathrm{Spec}(\widehat{A})$, as the complement of the resolvent set.

Note that every ball B_r in $\mathbb{Q}_p$ is an additive subgroup of $\mathbb{Q}_p$. A map $\widehat{F}: B_r \to \mathcal{L}(\mathcal{H}_p)$ with the properties $\widehat{F}(t+s) = \widehat{F}(t)\widehat{F}(s)$, $\widehat{F}(0) = I, t, s \in B_r$, where I is the unit operator in $\mathcal{H}_p$, is said to be a one parameter group of operators. If we consider $\mathrm{IS}(\mathcal{H}_p), \mathrm{UN}(\mathcal{H}_p), \mathrm{UI}(\mathcal{H}_p)$ instead of $\mathcal{L}(\mathcal{H}_p)$ we obtain the definitions of parametric groups of isometric, unitary, and isometric unitary operators, respectively. If the map $F: B_r \to \mathcal{L}(\mathcal{H}_p)$ is analytic the one parameter group is called analytic.

Let a belong to R_+. Set

$$\lfloor a \rfloor_p^- = \sup\{\lambda = p^k : k = 0, \pm 1, \dots;\ \lambda < a\}. \tag{13.1}$$

For a bounded operator $\widehat{A}$, we set

$$\gamma(\widehat{A}) = \frac{1}{p^{1/(p-1)}\|\widehat{A}\|}. \tag{13.2}$$

Theorem 13.6. *Let $\widehat{A}$ be a bounded symmetric operator in $\mathcal{H}_p \equiv \mathcal{H}_p(\sqrt{\tau})$. The map*

$$t \to e^{\sqrt{\tau} t \widehat{A}}, \quad t \in B_r,\ r = \lfloor \gamma(\sqrt{\tau}\widehat{A}) \rfloor_p^-,$$

is an analytic one-parameter group of isometric unitary operators.

Thus every symmetric operator $\widehat{A} \in \mathcal{L}(\mathcal{H}_p(\sqrt{\tau}))$ generates the one-parameter operator group of isometric unitary operators $t \to \widehat{U}(t) = e^{\sqrt{\tau} t \widehat{A}}$. This theorem is a natural generalization of the standard theorem for $\mathbb{C}$-Hilbert spaces. The following result has no analogue in functional analysis over $\mathbb{C}$.

[3] We recall that the norm on the p-adic Hilbert space is not determined by the inner product. The only condition of consistency between them is the Cauchy–Bunyakovsky–Schwarz inequality.

Theorem 13.7. *Let an operator $\widehat{A}$ belong to $\mathcal{L}(\mathcal{H}_p)$. The map $\alpha \to e^{\alpha \widehat{A}}$, $\alpha \in B_r$, $r = \lfloor \gamma(\widehat{A}) \rfloor_p^-$, is an analytic one-parameter group of isometric operators.*

More details can be found in [1].

13.4 Axiomatics of quantum mechanics with p-adic valued wave functions

As was remarked, in the p-adic case canonical commutation relations can be realized by bounded operators. Naturally defined (as by Schrödinger in the complex quantum mechanics) the operators of position and momentum are bounded. Therefore we proceed by considering only bounded operators.

Postulate 1 (Quantum states). States of a quantum system are given by vectors belonging to $\mathcal{H}_p$ normalized with respect to the inner product: $(\psi, \psi) = 1$. Two vectors ψ_1 and ψ_2 represent the same state iff $\psi_1 = c\psi_2$, where $c \in \mathbb{Q}_p(\sqrt{\tau}), c\bar{c} = 1$.

Postulate 2 (Observables). Observables are represented by symmetric operators. Results of observations are given by elements of spectra.

To simplify the probabilistic interpretation, we consider a rather restricted class of quantum observables.[4] A symmetric operator has *purely discrete nondegenerate spectrum* if all its eigenvalues $\{\alpha_j\}$ are different and all eigenvectors $\{e_{\alpha_j}\}$ are such that $(e_{\alpha_j}, e_{\alpha_j}) \neq 0$, and, finally, they form a basis in the normed space $\mathcal{H}_p$. Since the operator is symmetric, all its eigenvalues belong to $\mathbb{Q}_p$ and this basis is orthogonal with respect to the inner product. Hence, this basis is consistent with two structures – the norm (convergence) and the inner product (orthogonality). The main difference from the complex case is that in general it is not possible to normalize this basis with respect to the inner product, i.e., to select eigenvectors such that $(e_{\alpha_j}, e_{\alpha_j}) = 1$.

Thus each ψ can be expanded as $\psi = \sum_j c_j e_{\alpha_j}$, where $c_j = (\psi, e_{\alpha_j})/(e_{\alpha_j}, e_{\alpha_j})$, and this series converges with respect to the norm in $\mathcal{H}_p$. In particular, for a quantum state ψ,

$$(\psi, \psi) = \sum_j c_j \bar{c}_j (e_{\alpha_j}, e_{\alpha_j}) = \sum_j (\psi, e_{\alpha_j}) \overline{(\psi, e_{\alpha_j})} / (e_{\alpha_j}, e_{\alpha_j}) = 1. \tag{13.3}$$

Postulate 3 (Probabilistic interpretation). For a quantum state $\psi \in \mathcal{H}_p$, the probability to obtain the value $A = \alpha$ for a quantum observable A represented by an operator $\widehat{A}$ having purely discrete nondegenerate spectrum is given by the p-adic generalization of Born's rule:

$$\mathbf{P}_\psi(A = \alpha) = (\psi, e_{\alpha_j}) \overline{(\psi, e_{\alpha_j})} / (e_{\alpha_j}, e_{\alpha_j}). \tag{13.4}$$

[4] As has already been noticed, the main problem is the absence of a "good spectral theorem" in the p-adic case.

Of course, the reader would be shocked by Postulate 3 – "probability" does not belong to the segment [0, 1] of the real line. However, as we have seen it is possible to proceed in the rigorous probabilistic framework even in such unusual situation, see Chapter 12. The condition of normalization (13.3) is important.

Postulate 4 (Dynamics). Time-evolution of the quantum state, $t \to \psi(t), t \in B_r$, is given by the p-adic generalization of Schrödinger's equation:

$$\frac{h}{\sqrt{\tau}}\frac{d\psi(t)}{dt} = \widehat{\mathbf{H}}\psi(t), \tag{13.5}$$

where $\widehat{\mathbf{H}}$ is a symmetric operator – quantum Hamiltonian, and $h \in \mathbb{Q}_p$ is a small parameter – a kind of p-adic Planck constant.

Thus the time-parameter belongs to a ball in $\mathbb{Q}_p$. From the physical viewpoint consideration of such an evolution is an extremely risky step. We recall that $\mathbb{Q}_p$ is not linearly ordered. It seems that p-adic causality could not be reasonably defined.[5] Another important point is that (as was remarked) the notion of positive definiteness can not be used in the p-adic Hilbert space. Hence Hamiltonian is any symmetric operator. On the other hand, it is bounded. Finally, we remark that the "p-adic Planck constant" h was introduced formally as simply a scaling parameter.

By Theorem 13.6 the Cauchy problem for Schrödinger's equation (13.5) has the unique solution $\psi(t) = \widehat{U}(t)\psi_0$, where $\widehat{U}(t)$ is the corresponding one-parameter unitary group representing the evolution operator. Opposite to ordinary quantum mechanics, the evolution operator is not well defined for an arbitrary t. But such comparing may be not justified, since we consider the p-adic time and not the real one.

13.5 Gaussian integral and spaces of square integrable functions

As was remarked, the mathematical formalism of p-adic quantization does not depend on the choice of a quadratic extension $\mathbb{Q}_p(\sqrt{\tau})$ of $\mathbb{Q}_p$. To make considerations symbolically closer to ordinary complex quantization, we shall proceed for the quadratic extension $\mathbb{Q}_p(i)$. Of course, this choice restricts essentially the class of prime numbers under consideration. By Corollary 1.16 of Subsection 1.1.2 the equation $x^2 = -1$ does not have a solution in $\mathbb{F}_p$ and, hence, in $\mathbb{Q}_p$ if $p \equiv 3 \pmod 4$, i.e., $p = 7, 11, 15, 19, \ldots$. We remind that solutions in $\mathbb{F}_p$ can be lifted to solutions in $\mathbb{Q}_p$ with the aid of Hensel's lemma 3.3.

To provide the point wise realization of elements of the p-adic analogue of the L_2-space, we shall consider analytic functions over the field of complex p-adic numbers

[5] The latter might be natural for quantum models operating on the Planck scale, see Vladimirov, Volovich, Zelenov [407]. On this scale the notion of order might become meaningless.

$\mathbb{C}_p$. In $\mathbb{C}_p$ we denote the ball of radius $s \in \mathbb{R}^+$ with center at $z = 0$ by the symbol $\mathcal{U}_s$. Thus in our standard notations it is $B_s(0; \mathbb{C}_p)$. The space of analytic functions

$$f : \mathcal{U}_s \to \mathbb{C}_p$$

we denote by the symbol $\mathcal{A}(\mathcal{U}_s)$.

In [201] the general definition of the p-adic valued Gaussian integral was proposed on the basis of distribution theory. Gaussian distribution was defined as the distribution having Laplace transform of the form $\exp\{bx^2/2\}$, where $b \in \mathbb{R}$. We recall that in the real case if $b > 0$ then Gaussian distribution is simply a countably additive measure – Gaussian measure with dispersion b. If b is negative or even complex then Gaussian distribution cannot be realized as a measure.

For our present applications to quantization we can use a simpler approach based on the definition of Gaussian distribution through the definition of its moments. Roughly speaking we know moments of Gaussian distribution over the reals. Suppose now that dispersion is a rational number $b \in \mathbb{Q}$. Then moments can be as well interpreted as elements of any $\mathbb{Q}_p$. We now can extend by continuity our definition of moments to any "dispersion" $b \in \mathbb{Q}_p$.

Let b be a p-adic number, $b \neq 0$. The p-adic Gaussian distribution ν_b is defined by its moments ($n = 0, 1, \ldots$):

$$M_{2n} = \int_{\mathbb{Q}_p} x^{2n} \nu_b(dx) \equiv (2n)! b^n / n! 2^n, \quad M_{2n+1} = \int_{\mathbb{Q}_p} x^{2n+1} \nu_b(dx) \equiv 0.$$

By linearity we define the Gaussian integral for polynomial functions. Then we can define it for some classes of analytic functions. The analytic function $f(x) = \sum_{n=0}^{\infty} c_n x^n$, $c_n \in \mathbb{C}_p$, is said to be integrable with respect to the Gaussian distribution ν_b if the series

$$\int_{\mathbb{Q}_p} f(x) \nu_b(dx) \equiv \sum_{n=0}^{\infty} c_n M_n = \sum_{n=0}^{\infty} c_{2n} M_{2n} \tag{13.6}$$

converges. It was shown in [214] that all entire analytic functions on $\mathbf{C}_p$ are integrable. In fact, we do not need analyticity on the whole of $\mathbf{C}_p$ to be able to define the Gaussian integral. The following constant,

$$\theta_b \equiv p^{\frac{1}{2(1-p)}} \sqrt{|b/2|_p},$$

will play a fundamental role. If $p \neq 2$, then $\theta_b = p^{\frac{1}{2(1-p)}} \sqrt{|b|_p}$. If $p = 2$, then $\theta_b = \sqrt{|b|_p}$.

Proposition 13.8. *Let $f(x)$ belong to the class $\mathcal{A}(\mathcal{U}_s)$. If $s > \theta_b$, then the integral* (13.6) *converges.*

Remark 13.9. There exist functions which are analytic on the ball $\mathcal{U}_{\theta_b}$ but are not integrable, see [214].

In fact, we have proved that the Gaussian distribution is a continuous linear functional on the space of analytic functions $\mathcal{A}(\mathcal{U}_s)$, i.e., this is an analytic generalized function (distribution); for the details see [201]. We shall use the symbol of the integral for duality form between the space of test functions $\mathcal{A}(\mathcal{U}_s)$ and the space of generalized functions $\mathcal{A}'(\mathcal{U}_s)$: $(\mu, f) \equiv \int f(x)\mu(dx)$ for $f \in \mathcal{A}(\mathcal{U}_s)$ and $\mu \in \mathcal{A}'(\mathcal{U}_s)$. As usual, we define a derivative of a generalized function μ by the equality $\int f(x)\mu(dx) = -\int f'(x)\mu(dx)$.

It should be remarked that the distribution ν_b is not a bounded measure on any ball of $\mathbb{Q}_p$ (it was proved for the case $p \neq 2$, in the case $p = 2$ the question is still open), see Endo and Khrennikov [123]. Thus we can not integrate continuous functions with respect to the p-adic Gaussian distribution.

We introduce *Hermite polynomials* over $\mathbb{Q}_p$ by substituting the p-adic variable, instead of the real, into ordinary Hermite polynomials over the reals: $H_{n,b}(x) = \frac{n!}{b^n}\sum_{k=0}^{\lfloor n/2 \rfloor} \frac{(-1)^k x^{n-2k} b^k}{k!(n-2k)!2^k}$. We shall use also the following representation for the Hermite polynomials: $H_{n,b}(x) = (-1)^n e^{x^2/2b} \frac{d^n}{dx^n} e^{-x^2/2b}$. This representation holds on a ball of sufficiently small radius with center at zero. As a consequence, we obtain the following equality in the space of generalized functions $\mathcal{A}'(\mathcal{U}_s), s > \theta_b$:

$$H_{n,b}(x)\nu_b(dx) = (-1)^n \frac{d^n}{dx^n}\nu_b(dx), \tag{13.7}$$

i.e., multiplication of the Gaussian distribution by a Hermite polynomial is equivalent to computation of the corresponding derivative (in the sense of distribution theory).

In the space $\mathcal{P}(\mathbb{Q}_p)$ of polynomials on $\mathbb{Q}_p$ with coefficients belonging to $\mathbb{Q}_p(i)$ we introduce the inner product $(f, g) = \int f(x)\bar{g}(x)\nu_b(dx)$. The polynomials $H_{n,b}$ verify the following orthogonal conditions with respect to this inner product:

$$\int H_{m,b}(x)H_{n,b}(x)\nu_b(dx) = \delta_{nm}\, n!/b^n.$$

Remark 13.10. In fact, such constants $\lambda_n = n!/b^n$ were one of the reasons for introducing p-adic Hilbert spaces which are isomorphic to $l^2(p, \lambda)$.

Any $f \in \mathcal{P}(\mathbb{Q}_p)$ can be written in the following way: $f(x) = \sum_{n=0}^{N} f_n H_{n,b}(x)$, $N = N(f)$, $f_n \in \mathbb{Q}_p(i)$. We introduce the norm $\|f\|^2 = \max_n |f_n|_p^2(|n!|_p/|b|_p^n)$ and we define $L_2^i(\mathbb{Q}_p, \nu_b)$ as the completion of $\mathcal{P}(\mathbb{Q}_p)$ with respect to $\|\cdot\|$. It is evident that the space $L_2^i(\mathbb{Q}_p, \nu_b)$ is the set

$$\Big\{f(x) = \sum_{n=0}^{\infty} f_n H_{n,b}(x),\ f_n \in \mathbb{Q}_p(i) : \text{the series} \sum_{n=0}^{\infty} f_n \bar{f}_n n!/b^n \text{converges}\Big\}.$$

Denote the subspace of $L_2^i(\mathbb{Q}_p, \nu_b)$ consisting of functions, which have the Hermite coefficients $f_n \in \mathbb{Q}_p$, by $L_2(\mathbb{Q}_p, \nu_b)$. This is a Hilbert space over the field $\mathbb{Q}_p$. For $f(x) \in L_2^i(\mathbb{Q}_p, \nu_b)$ we set

$$\sigma_n^2(f) \equiv \sigma_{n,b}^2(f) = |f_n|_p^2 |n!/b^n|_p, \tag{13.8}$$

where f_n are Hermite coefficients of $f(x)$ given by the following expression: $f_n = \frac{b^n}{n!} \int f(x) H_{n,b}(x) \nu_{b,p}(dx)$. Now we wish to study the relations between $L_2(\mathbb{Q}_p, \nu_b)$-functions and analytic functions. Set $\mathcal{A}_{\mathbb{Q}_p}(\mathcal{U}_r) = \{f \in \mathcal{A}(\mathcal{U}_r) : f : B_r \to \mathbb{Q}_p\}$, i.e., these are functions, which have the Taylor coefficients belonging to the field $\mathbb{Q}_p$.

Theorem 13.11. *Assume $p \neq 2$. Then $L_2(\mathbb{Q}_p, \nu_b) \subset \mathcal{A}_{\mathbb{Q}_p}(\mathcal{U}_{\theta_b})$.*

Now we consider the case $p = 2$. In general L_2-functions are not analytic on the ball $\mathcal{U}_{\theta_b}$.

Theorem 13.12. *Let $s > \theta_b$. Then $\mathcal{A}_{\mathbb{Q}_p}(\mathcal{U}_s) \subset L_2(\mathbb{Q}_p, \nu_b)$.*

Further we construct the L_2-representation of the translation group.

If $|b|_p = p^{2k+1}$ we set $s(b) = p^k$, if $|b|_p = p^{2k}$, we set $s(b) = p^{k-1}$.

Set $\widehat{T}_\beta(f)(x) = f(x + \beta), \beta \in \mathbb{Q}_p$. We shall prove that these operators are bounded for $\beta \in B_{s(b)}$. Moreover, these operators are isometries of $L_2(\mathbb{Q}_p, \nu_b)$. Using this fact we shall construct a representation of the translation group in the p-adic Hilbert space $L_2(\mathbb{Q}_p, \nu_b)$.

Lemma 13.13. *The formula*

$$\widehat{T}_\beta H_{n,b}(x) = \sum_{j=0}^{n} \binom{n}{j} \left(\frac{\beta}{b}\right)^j H_{n-j,b}(x) \tag{13.9}$$

holds for the translations of Hermite polynomials.

Theorem 13.14. *The operator $\widehat{T}_\beta$ belongs to $\mathrm{IS}(L_2(\mathbb{Q}_p, \nu_b))$ for every $\beta \in B_{s(b)}$ and the map $T : B_{s(b)} \to \mathrm{IS}(L_2(\mathbb{Q}_p, \nu_b)), \beta \to \widehat{T}_\beta$, is analytic.*

13.6 Gaussian representations of position and momentum operators

Similarly as in ordinary Schrödinger quantum mechanics let us define the coordinate and momentum operators in $L_2^i(\mathbb{Q}_p, \nu_b)$ by

$$\widehat{\mathbf{q}} f(x) = x f(x), \ \widehat{\mathbf{p}} f(x) = (-i) \left(\frac{d}{dx} - \frac{x}{2b}\right) f(x)$$

where f belongs to the $\mathbb{Q}_p(i)$-linear space $\mathcal{D}$ of linear combinations of Hermite polynomials. The coordinate and momentum operators so defined satisfy on $\mathcal{D}$ the canonical commutation relations

$$[\widehat{\mathbf{q}}, \widehat{\mathbf{p}}] = iI, \tag{13.10}$$

where I is the unit operator in $L_2^i(\mathbb{Q}_p, \nu_b)$. We shall see that these relations can be extended to the whole of $L_2^i(\mathbb{Q}_p, \nu_b)$.

Theorem 13.15. *The operators of the coordinate $\widehat{\mathbf{q}}$ and momentum $\widehat{\mathbf{p}}$ are bounded in the space $L_2^i(\mathbb{Q}_p, \nu_b)$ and*

$$\|\widehat{\mathbf{q}}\| = \sqrt{|b|_p}; \quad \|\widehat{\mathbf{p}}\| = \frac{1}{\sqrt{|b|_p}}. \tag{13.11}$$

Moreover $\widehat{\mathbf{q}}$ and $\widehat{\mathbf{p}}$ are symmetric and satisfy (13.10) *on $L_2^i(\mathbb{Q}_p, \nu_b)$.*

Proof. Let $f(x) = \sum_{n=0}^{\infty} f_n H_{n,b}(x) \in L_2^i(\mathbb{Q}_p, \nu_b)$. By the recurrence formula

$$H_{n+1,b}(x) = b^{-1}[xH_{n,b}(x) - nH_{n-1,b}(x)] \tag{13.12}$$

we have

$$\widehat{\mathbf{q}}H_{n,b}(x) = bH_{n+1,b}(x) + nH_{n-1,b}(x), \tag{13.13}$$

and $\widehat{\mathbf{q}}f(x) = \sum_{n=0}^{\infty} bf_n H_{n+1,b}(x) + \sum_{n=1}^{\infty} nf_n H_{n-1,b}(x)$. Thus by the strong triangle inequality we obtain

$$\|\widehat{\mathbf{q}}f\|^2 \leqslant \max\left[\max_n |b|_p^2 |f_n|_p^2 \frac{|(n+1)!|_p}{|b|_p^{n+1}}, \max_n |n|_p^2 |f_n|_p^2 \frac{|(n-1)!|_p}{|b|_p^{n-1}}\right]$$

$$= |b|_p \max\left[\max_n |n+1|_p |f_n|_p^2 \frac{|n!|_p}{|b|_p^n}, \max_n |n|_p |f_n|_p^2 \frac{|n!|_p}{|b|_p^n}\right] \leqslant |b|_p \|f\|^2$$

(as $|n|_p \leqslant 1$ for all $n \in \mathbf{N}$). Thus $\|\widehat{\mathbf{q}}\| \leqslant \sqrt{|b|_p}$. Now we prove that $\|\widehat{\mathbf{q}}\|^2 = |b|_p$. Let $n = p^k$, then $D_{k,b} = \|\widehat{\mathbf{q}}H_{p^k,b}\|^2 = \max\{|b|_p^2 |(p^k+1)!|_p / |b|_p^{p^k+1}, |p^k|_p^2 |(p^k - 1)!|_p / |b|_p^{p^k-1}\}$. But $|(p^k+1)!|_p = |p^k!|_p$ and $|p^{2k}(p^k-1)!|_p = p^{-k}|p^k!|_p$. Thus $D_{k,b} = |b|_p(|p^k!|_p/|b|_p^{p^k}) = |b|_p\|H_{p^k,b}\|^2$, which proves the first equality in (13.11).

Further we have $\frac{d}{dx}H_{n,b}(x) = (x/b)H_{n,b}(x) - H_{n+1,b}(x) = (n/b)H_{n-1,b}(x)$. Set $\widehat{T_x} = (d/dx - (x/2b))$. We then have $\widehat{T_x}H_{n,b}(x) = (n/2b)H_{n-1,b}(x) - (1/2)H_{n+1,b}(x)$. To compare this expression with (13.13), we rewrite it as

$$\widehat{T_x}H_{n,b}(x) = \frac{1}{2b}[-bH_{n+1,b}(x) + nH_{n-1,b}(x)]. \tag{13.14}$$

The expression in square brackets is the same as in (13.13); the sign cannot play any role in the estimates of the max type. Thus we obtain $\|\widehat{T_x}\| = (1/|b|_p)\|\widehat{\mathbf{q}}\|$, which proves the second equality in (13.11).

The symmetry of bounded operators $\widehat{\mathbf{q}}$, $\widehat{\mathbf{p}}$ is easily verified. □

13.7 One parameter groups generated by position and momentum operators

We shall compute numbers $\lfloor \gamma(\widehat{\mathbf{q}}) \rfloor_p^-$ and $\lfloor \gamma(\widehat{\mathbf{p}}) \rfloor_p^-$, see (13.1), (13.2) in Section 13.3.

If $|b|_p = p^{2k+1}$ then $\gamma(\widehat{\mathbf{q}}) = 1/(p^k p^{1/2} p^{1/(p-1)})$.
If $p \neq 3$ then $\lfloor \gamma(\widehat{\mathbf{q}}) \rfloor_p^- = 1/p^{k+1}$.
If $p = 3$ then $\lfloor \gamma(\widehat{\mathbf{q}}) \rfloor_p^- = 1/p^{k+2}$.
If $|b|_p = p^{2k}$ then $\gamma(\widehat{\mathbf{q}}) = 1/(p^k p^{1/(1-p)})$ and $\lfloor \gamma(\widehat{\mathbf{q}}) \rfloor_p^- = 1/p^{k+1}$.
Set

$$R(b) = \lfloor \gamma(\widehat{\mathbf{q}}) \rfloor_p^-.$$

If $|b|_p = p^{2k+1}$ then $\gamma(\widehat{\mathbf{p}}) = (p^{1/2}/p^{1/(p-1)})p^k$.
If $p \neq 3$ then $\lfloor \gamma(\widehat{\mathbf{p}}) \rfloor_p^- = p^k$.
If $p = 3$ then $\lfloor \gamma(\widehat{\mathbf{p}}) \rfloor_p^- = p^{k-1}$.
If $|b|_p = p^{2k}$ then $\lfloor \gamma(\widehat{\mathbf{p}}) \rfloor_p^- = p^{k-1}$.
Set

$$r(b) = \lfloor \gamma(\widehat{\mathbf{p}}) \rfloor_p^-.$$

Theorem 13.16. *The mappings $\alpha \to \widehat{U}(\alpha) = e^{i\widehat{\alpha\mathbf{q}}}$, $\alpha \in B_{R(b)}$, and $\beta \to \widehat{V}(\beta) = e^{i\widehat{\beta\mathbf{p}}}$, $\beta \in B_{r(b)}$, are analytic one parameter groups of unitary isometric operators acting on $L_2^i(\mathbb{Q}_p, \nu_b)$. They satisfy the Weyl commutation relations:*

$$\widehat{U}(\alpha)\widehat{V}(\beta) = e^{-i\alpha\beta}\widehat{V}(\beta)\widehat{U}(\alpha). \tag{13.15}$$

We set

$$\widehat{M}_\beta f(x) = e^{-\beta\widehat{\mathbf{q}}/2b} f(x) = \sum_{n=0}^{\infty} \frac{(-\beta\widehat{\mathbf{q}})^n}{n!(2b)^n} f(x), \tag{13.16}$$

for $f \in L_2(\mathbb{Q}_p, \nu_b)$. By Theorem 13.15 we easily obtain

Proposition 13.17. *The map $M : B_{r(b)} \to \mathrm{IS}(L_2(\mathbb{Q}_p, \nu_b)), \beta \to \widehat{M}_\beta$, is an analytic one parameter group (indexed by the ball $B_{r(b)}$).*

Remark 13.18. The function $x \to e^{-\beta x/2b}$ is not defined on the whole of $\mathbb{Q}_p$ and we cannot consider the operator (13.16) as an operator of point wise multiplication.

13.8 Operator calculus

It is well known that in the ordinary $L_2(\mathbb{R}, dx)$-space the unitary group $\widehat{V}(\beta) = e^{i\widehat{\beta\mathbf{p}}}$, $\beta \in \mathbb{R}$, can be realized as the translation group: $\widehat{V}(\beta)\psi(x) = \psi(x+\beta)$ for sufficiently good functions $\psi(x)$. If we consider the equivalent representation in the L_2-space with respect to the Gaussian measure $\nu_b(dx) = (e^{-x^2/2b}/\sqrt{2\pi b})dx$ on $\mathbb{R}$ we obtain

$$\widehat{V}(\beta)\psi(x) = e^{-\beta^2/4b}e^{-\beta x/2b}\psi(x+\beta), \tag{13.17}$$

or

$$\widehat{V}(\beta) = c_\beta \widehat{M}_\beta \widehat{T}_\beta, \tag{13.18}$$

where $c_\beta = e^{-\beta^2/4b}$. We shall now prove that formula (13.18) is also valid in the p-adic case.

Set $\widehat{S}(\beta) = c_\beta \widehat{M}_\beta \widehat{T}_\beta$, $\beta \in B_{r(b)}$, where the operator $\widehat{M}_\beta$ is defined by (13.16).

Theorem 13.19. *The map $\beta \to \widehat{S}_\beta$, $\beta \in B_{r(b)}$, is a one-parameter analytic group of isometric unitary operators acting in $L_2^i(\mathbb{Q}_p, \nu_b)$.*

Lemma 13.20. *The groups $\widehat{S}(\beta)$ and $\widehat{V}(\beta)$ have $\widehat{\mathbf{p}}$ as their common generator.*

As a consequence of this lemma and the analyticity of the one parameter groups $S(\beta)$ and $V(\beta)$, we easily have:

Theorem 13.21. *The representation* (13.17), (13.18) *holds for the operator group $\widehat{V}(\beta)$.*

13.9 Spectrum of p-adic position operator

In this section we study the spectrum of the p-adic position operator. This study is based on rather tricky non-Archimedean estimates. We decided to present some of them to illustrate non-Archimedean technique.

We shall use the following notation: if $|x|_p = |y|_p$, $x, y \in \mathbb{Q}_p$, we write this as $x \sim y$ in $\mathbb{Q}_p$. If x, y are natural numbers then $x \sim y$ means that x and y are divisible by exactly the same power of p.

As usual we define a position operator $\widehat{\mathbf{q}}$ in the space $L_2(\mathbb{Q}_p, \nu_b)$ as the multiplication operator $\widehat{\mathbf{q}}f(x) = xf(x)$. The norm of this operator is given by (13.11):

$$\alpha_b = \|\widehat{\mathbf{q}}\| = \sqrt{|b|_p}\,.$$

It is evident that $\mathbb{Q}_p \setminus B_{\alpha_b} \subset \mathrm{Res}(\widehat{\mathbf{q}})$. Hence $\mathrm{Spec}(\widehat{\mathbf{q}}) \subset B_{\alpha_b}$. Does the spectrum of $\widehat{\mathbf{q}}$ coincide with the ball B_{α_b}? At the moment we do not have the final solution of this problem. We have only obtained the following result.

Theorem 13.22. *Let $p \neq 2$. Then $B_{\theta_b} \subset \mathrm{Spec}\ \widehat{\mathbf{q}}$.*

Proof. We consider the equation $(\hat{\mathbf{q}}-\lambda)f(x) = 1$ in the space $L_2(\mathbb{Q}_p, \nu_b)$. The function f is analytic on the ball B_{θ_b}. Thus this equation can be considered as the equation for analytic functions on the ball $\mathcal{U}_{\theta_b}$. Of course, this equation has no solution in the space of analytic functions and consequently in the L_2-space. □

If $p \neq 2$ then $\theta_b < \alpha_b$. The balls B_{θ_b} and B_{α_b} coincide for some values of $|b|_p$, but for other values of b the set $B_{\theta_b} \setminus B_{\alpha_b}$ is not empty.

In the case $p = 2$ we have no results about the spectral set of the position operator.

Further we shall study a point spectrum of the position operator.

In quantum mechanics over the real numbers the point spectrum of the position operator $\widehat{\mathbf{q}}$ is empty, i.e., $\widehat{\mathbf{q}}$ has no eigenvalues $\lambda \in \mathbb{R}$. If we consider the standard representation of $\widehat{\mathbf{q}}$ in the space $L_2(\mathbb{R}, dx)$ then the equation

$$\widehat{\mathbf{q}} f_\lambda(x) = \lambda f_\lambda(x) \tag{13.19}$$

has no solutions in $L_2(\mathbb{R}, dx)$ for any $\lambda \in \mathbb{R}$. Here, in fact, $f_\lambda(x) = \delta(x-\lambda)$, but the δ-function does not belong to $L_2(\mathbb{R}, dx)$. If we change the representation and consider $\widehat{\mathbf{q}}$ as the multiplication operator in the space $L_2(\mathbb{R}, \nu_b)$, where ν_b is the Gaussian measure on $\mathbb{R}$ with covariance b and zero mean value, then the situation does not change. The equation (13.19) has no solution $f_\lambda(x) \in L_2(R, \nu_b)$.

Now we show that the p-adic picture does not differ from the one for reals:

Theorem 13.23. *The point spectrum of the position operator $\widehat{\mathbf{q}}$ in $L_2(\mathbb{Q}_p, \nu_b)$ is the empty set.*

The proof of this theorem is quite long. We divide it into a few steps.

Lemma 13.24. *Suppose that the equation (13.19) has a non-zero solution $f_\lambda(x)$ belonging to $L_2(\mathbb{Q}_p, \nu_b)$. Then we have the following formula for the Hermite coefficients $f_{\lambda,n}$ of the function $f_\lambda(x)$:*

$$f_{\lambda,n} = \frac{b^n}{n!} H_{n,b}(\lambda) = \sum_{k=0}^{\lfloor n/2 \rfloor} \frac{(-1)^k \lambda^{n-2k} b^k}{k!(n-2k)!2^k}, \tag{13.20}$$

where we have chosen the normalization constant $c = \int f_\lambda(x) \nu_{b,p}(dx) = 1$.

Lemma 13.25. *The point $\lambda = 0$ does not belong to the point spectrum of the position operator $\widehat{\mathbf{q}}$.*

Proof. First we recall that $|m!|_p = p^{\frac{S_m - m}{p-1}}$, where $S_m \equiv \mathrm{wt}_p\, m$ is determined by the p-adic expansion of the natural number m, see Lemma 3.6. By (13.20) we have that the coefficients $f_{0,2m+1} = 0$ and $f_{0,2m} = (-1)^m b^m / 2^m m!$, $m = 0, 1, \ldots$. We obtain, see (13.8) for definition,

$$\sigma_{2m}^2(f_0) = |2|_p^{-2m} |2m!/m!^2|_p = |2|_p^{-2m} p^{\frac{S_{2m} - 2S_m}{p-1}}.$$

We show that there exists a subsequence (m_k) such that $\sigma^2_{2m_k}(f_0)$ does not approach zero for $k \to \infty$. It suffices to choose $m_k = p^k$. Here, if $p \neq 2$ then $S_{2p^k} = 2$, i.e., $S_{2p^k} - 2S_{p^k} = 0$. Thus $\sigma^2_{2p^k}(f_0) = 1$ for all k. If $p = 2$ then $S_{2^{k+1}} = S_{2^k}$, i.e., $S_{2^{k+1}} - 2S_{2^k} = -1$. Thus, $\sigma^2_{2^{k+1}}(f_0) = 2^{2^{k+1}-1}$. □

Further, we shall prove that for small $|\lambda|_p$ the behavior of the Hermite coefficients $f_{\lambda,2p^k}$ coincides with the behavior of $f_{0,2p^k}$.

Lemma 13.26. *Let $\lambda \in B^-_{\theta_b}$, then $|f_{\lambda,2p^k}|_p = |f_{0,2p^k}|_p$.*

Proof. We rewrite the expression (13.20) for the Hermite coefficients in the form

$$f_{\lambda,2m} = (-1)^m \sum_{j=0}^{m} \frac{(-1)^j \lambda^{2j} b^{m-j}}{(m-j)!(2j)!2^{m-j}} = \sum_{j=0}^{m} a_j. \tag{13.21}$$

Here $a_0 = f_{0,2m}$. Furthermore we rewrite a_j, for $j = 1, \ldots, m$, in the form $a_j = a_0(-1)^j(\frac{2\lambda^2}{b})^j \frac{m!}{(m-j)!(2j)!}$. Further, we obtain the equality $|m!/(m-j)!(2j)!|_p = p^{\frac{j}{p-1}} p^{S(m,j)/(p-1)}$, where $S(m, j) = S_m - S_{m-j} - S_{2j}$. Finally, we have $|a_j|_p = |a_0|_p(|\lambda|_p/\theta_b)^{2j} p^{S(m,j)/(p-1)}$. We always have $S_{m-j} + S_{2j} \geqslant 1$ for $j = 1, \ldots, m$. If $m = p^k$ then $S_m = 1$. Hence, in this case $S(m, j) \leqslant 0$. Thus $|a_j|_p < |a_0|_p$ for all $j = 1, \ldots, p^k$. □

Heuristically it is evident that the term with the maximal power of λ in (13.20): $a_m = \frac{\lambda^{2m}}{2m!}$ must dominate for respectively large $|\lambda|_p$. First we study the L_2 behavior of these coefficients.

Lemma 13.27. *The function $g_\lambda(x) = \sum_{m=0}^{\infty} a_m H_{2m,b}(x)$ does not belong to the space $L_2(\mathbb{Q}_p, \nu_b)$ for all λ satisfying the inequality $\lambda \geqslant \theta_b$.*

Proof. Set $m = p^k$, then $S_{2p^k} = 2$ for $p \neq 2$ and $S_{2^{k+1}} = 1$ for $p = 2$. In any case $\sigma_{2p^k}(g_\lambda) \not\to 0$, $k \to \infty$, if $|\lambda|_p \geqslant \theta_b$. □

Lemma 13.28. *If $|\lambda|_p > \theta_b$ then $|f_{\lambda,2p^k}|_p = |a_{p^k}|_p$ for sufficiently large k.*

Proof. It is more convenient to rewrite the Hermite coefficients in the form

$$f_{\lambda,2m} = \sum_{k=0}^{m} \frac{(-1)^k \lambda^{2m-2k} b^k}{k!(2m-2k)!2^k} = \sum_{k=0}^{m} a_{m-k}.$$

We show that the term a_m strictly dominates in this sum. Let $k = 1, \ldots, m$. We have

$$|a_{m-k}|_p = \left|\frac{\lambda^{2m}}{2m!}\right|_p \left|\frac{b^k}{2^k \lambda^{2k}}\right|_p \left|\frac{2m!}{k!(2m-2k)!}\right|_p.$$

We obtain $\left|\frac{2m!}{k!(2m-2k)!}\right|_p = p^{-k/(p-1)} p^{A(m,k)/(p-1)}$ where $A(m,k) = S_{2m} - S_k - S_{2(m-k)}$. Hence we obtain $|a_{m-k}|_p = |a_m|_p \left(\frac{\theta_b}{|\lambda|_p}\right)^{2k} p^{A(m,k)/(p-1)}$.

Now set $m = p^l$. Consider the case $p \neq 2$. Here $S_{2p^l} = 2$. If $k \neq m$ then $S_k + S_{2(m-k)} \geqslant 2$ and, consequently, $A(m,k) \leqslant 0$. Now let $m = k$, then $S_{2(m-k)} = 0$ and $S_{2m} - S_m = 1$. Hence $p^{A(m,k)/(p-1)} = p^{1/(p-1)}$. Thus we have $|a_{m-k}|_p = |a_m|(\theta_b/|\lambda|_p)^{2k}$ for $k = 1, \ldots, m-1$, and $|a_0|_p = |a_m|_p(\theta_b/|\lambda|_p)^{2m} p^{1/(p-1)}$. As $|\lambda|_p > \theta_b$, both these quantities are less than $|a_m|_p$ for sufficiently large $m = p^l$. Now consider the case $p = 2$. Here $S_{2m} = S_m = 1$. If $k \neq m$, then $S_k + S_{2(m-k)} \geqslant 2$. If $k = m$, then $A(m,k) = 0$. □

Proof of Theorem 13.23. We consider the case $p \neq 2$. In the case $p = 2$ the proof is based on similar ideas.

If $|\lambda|_p \leqslant \theta_b$ then for $m = p^k$ the term $a_0 = f_{0,2m}$ dominates and $|f_{\lambda,2m}|_p = |f_{0,2m}|_p$. We need only to use Lemma 13.28. If $|\lambda|_p > \theta_b$ then for $m = p^k$ the term a_m dominates and $|f_{\lambda,2m}|_p = |a_m|_p$. Thus we need only to use Lemma 13.27. □

Corresponding study on the spectrum of the p-adic momentum operator can be found in [2].

13.10 Concluding remarks on p-adic quantization

We point to the following distinguishing features of quantum mechanics with p-adic valued wave functions:

1. Canonical commutation relations can be represented by bounded operators[6].
2. Even in the case of one degree of freedom there exist nonequivalent representations of canonical commutation relations.[7]
3. As well as in the complex case, the p-adic operators of position and momentum have empty point spectrum.
4. One parameter groups of unitary operators induced by the p-adic operators of position and momentum are defined only for parameters belonging to proper balls

[6] It is impossible in the complex case. We remark that boundedness of p-adic operators of position and momentum was discovered by Albeverio and Khrennikov, [7]. Later this was independently discovered by Kochubei [269]. Detailed analysis of the p-adic situation was performed in [184].

[7] In the complex case all finite-dimensional representations are unitary equivalent. Nonequivalent representations appear only in the case of the infinite number of degrees of freedom – quantum field theory. We remark that, since in the complex case canonical commutation relations are represented by unbounded operators, one should be careful with formulation of the statement about unitary equivalence of representations in the finite-dimensional case. The correct mathematical formulation can be performed only by using Weyl's commutation relations, instead of Heisenberg's commutation relations. We also point out that in the p-adic case even quantum field operators are bounded, as it was shown by Albeverio and Khrennikov [7].

of $\mathbb{Q}_p$. Radii of these balls are coupled by a kind of uncertainty relation, see [212] for details.

Open Question 13.29. 1. Find precisely radii of spectral balls for the p-adic operators of position and momentum.

2. Find a class of bounded symmetric operators extending essentially the class of operators with purely discrete spectra which have "good spectral decomposition". To formulate Born's probabilistic postulate for such observables.

3. Investigate spectra of Hamiltonians with nonquadratic potentials. We remark that the p-adic harmonic oscillator was studied in detail in [201] and [214]; the anharmonic oscillator was studied by Aref'eva, Dragovic, and Volovich [35]; see also [105] on adelic harmonic oscillator.

Chapter 14

m-adic modeling in cognitive science and psychology

We repeat in cognitive science and psychology the program of huge complexity which has already been realized in physics. A model of the mental space – an analog of the model of the physical space – is proposed. Associations and ideas are embedded in this space – in the same way as rigid bodies were embedded in physical space. Finally, we consider dynamical systems describing evolution of mental entities. These are analogs of physical dynamical equations. The crucial point is that, instead of to copy directly the conventional physical structures which are based on the mathematical model of the real continuum, we apply methods of recently developed non-Archimedean theoretical physics. The latter was successfully used e.g. in superstring theory, cosmology and theory of spin glasses. Thus we reject the conventional model of the space based the real continuum. Instead of real lines, we use m-adic trees to encode mental coordinates.

It is the right point to remark that in theoretical physics the main attention was paid to p-adic models. Here $p > 1$ is a prime number. In cognitive and psychological applications there is no reason to consider only trees (homogeneous) with p branches going from each vertex. The number of branches can be given by any natural number $m > 1$. Therefore mental states will be encoded by m-adic numbers, $x \in \mathbb{Q}_m$. In the simplest model which will be presented in this chapter mental states can be represented by m-adic integers, $x \in \mathbb{Z}_m$.

In general it is possible to proceed with the mental space based on an arbitrary tree, i.e., on an arbitrary ultrametric mental space, see [237]. However, we are fine with m-adic spaces. This model, although quite simplified, demonstrates main distinguishing features of the ultrametric approach. The presence of the ring structure on $\mathbb{Q}_m$ makes the model essentially more attractive from the mathematical viewpoint. Moreover, the standard systems of encoding of information in the brain based either on 1/0, firing/nonfiring, or on the frequency code produce mental spaces based on homogeneous trees, i.e., the m-adic trees. The simplest firing/nonfiring system produces the 2-adic tree. The frequency system produces general m-adic trees. We emphasize that one should sharply distinguish hierarchic neuronal trees which are objects of physical and biological realities and mental trees which are purely informational objects. Although the former produce the latter, their structures can be very different. Neuronal structures can form nonhomogeneous trees, i.e., to describe them mathematically, one could not restrict modeling to m-adic trees and numbers, general trees and ultrametric spaces should be involved.

Our dynamical model is applied for modeling of flows of unconscious and conscious information in the human brain. In a series of models, Models 1–4, we consider cognitive systems with increasing complexity of psychological behavior determined by structure of flows of associations and ideas. The most advanced model of the mental architecture, Model 4, presents a number of essential features of human psyche. Cognitive systems described by such a model can exhibit psychical behavior described by Freudian theory of inter-relation of *unconsciousness and consciousness.* This model can be used for justification of some elements of psychoanalytic treatment of patients.

Since our models are in fact models of the AI-type, one immediately recognizes that they can be used for creation of AI-systems, which we call *psycho-robots,* exhibiting important elements of human psyche. Creation of such psycho-robots may be useful improvement of domestic robots. At the moment domestic robots are merely simple working devices (e.g. vacuum cleaners or lawn mowers). However, in future one can expect demand in systems which be able not only perform simple work tasks, but would have elements of human self-developing psyche. Such AI-psyche could play an important role both in relations between psycho-robots and their owners as well as between psycho-robots. Since the presence of a huge numbers of psycho-complexes is an essential characteristic of human psychology, it would be interesting to model them in the AI-framework.

14.1 On modeling of mental quantities

14.1.1 Representation of mental states by numbers

One of the sources of the extremely successful mathematical formalization of physics was creation of the adequate mathematical model of the physical space, namely, the Cartesian product of real lines. This provides the possibility for "embedding" physical objects into a mathematical space. Coordinates of physical systems are given by points of this space. Rigid physical bodies are represented by geometric figures (cubes, balls, ...). By describing dynamics of coordinates, e.g., with the aid of differential equations, we can describe dynamics of bodies (from falling stones to Sputniks).

In a series of works [214, 221, 224, 227–229, 232, 235] there was advocated a similar approach to description of mental processes in cognitive sciences and psychology (and even information dynamics in genetics), see also [9,10,62,109,239,240,246,247]. Similar to physics, the first step should be elaboration of a mathematical model of the *mental space.* We understood well that this is a problem of huge complexity and it might take a few hundred years to create an adequate mathematical model of the mental space. We recall that it took three hundred years to create a mathematically rigorous model of the real physical space. In previous papers critical arguments were presented against the real model of space as a possible candidate for a mental space. One of the main arguments was that the real continuum is a continuous infinitely divisible space. Such a picture of space is adequate to physical space (at least in classical physics), but the *mental space is not continuous*: mind is not infinitely divisible! Another problem with the real continuum is that it is homogeneous: "all points of this space have equal rights." Opposite to such a homogeneity mental states have clearly expressed *hierarchical structure,* see for discussions [17, 48, 90, 95, 176, 214, 221, 224, 227–229, 232, 235, 386] and [9, 10, 62, 109, 237, 239, 246, 247], see also [49, 96, 146] for corresponding medical evidence. Therefore a model of the mental space that we are looking for should be (at least) *discontinuous and hierarchical.* Such a model of space was recently invented in theoretical physics. It is a non-Archimedean (ultrametric) physical space. Such a model was successfully used e.g. in superstring theory, cosmology and theory of spin glasses.[1]

The simplest class of ultrametric spaces is given by homogeneous m-adic trees (here m is a natural number giving the number of branches of a tree at each vertex). It is interesting that such trees are nicely equipped: there is a well-defined algebraic structure which gives the possibility to add, subtract, multiply branches of such a tree. Thus each m-adic tree has the structure of a ring. It is the ring of m-adic integers $\mathbb{Q}_m$. Such a tree can be endowed with a natural topology encoding the hierarchic tree structure. Thus m-adic trees are not worse equipped than the real line. However, the equipments – algebra and topology – are very different from the real ones.

We proposed in [214, 221, 224, 227–229, 232, 235] and [9, 10, 62, 109, 237, 239, 246, 247] to choose m-adic trees rings $\mathbb{Q}_m$ – as models of the *mental space* X_{mental}. Points of this space are branches of a tree, i.e., m-adic numbers. In this chapter we restrict the mental space to the ring $\mathbb{Z}_m$ of m-adic integers. Thus we plan to encode basic cognitive mental images by m-adic integers. Such a simplified model describes well "up-down" propagation of signals in the brain, see Chapter 15 for a general model based on $\mathbb{Q}_m$ and describing "up-down" as well as "down-up" propagation of signals in the brain.

We remind that elements of the ring of m-adic integers $\mathbb{Z}_m$ can be represented in the form

$$x = x_0 + x_1 m + \cdots + x_n m^n + \cdots, \quad x_j = 0, 1, \ldots, m-1. \tag{14.1}$$

Thus m-adic integers can be considered as generalized natural numbers (the latter are given by finite sums). Hence, we plan to encode mental entities by generalized natural numbers.

We recall that the idea of representing mental entities by numbers is very old. The first was definitely *Platon, then Aristotle and later Leibniz.* Taking into account the great influence of Aristotle's ideas to development of the scientific method of investi-

[1]See, e.g., [2–4, 6–8, 34, 35, 50, 87–89, 104, 137, 143, 185–193, 201, 205–212, 214, 215, 218, 220, 225, 226, 230, 309, 324, 351, 404–408].

gation of nature and spirit, we shall analyze his views in more detail. His fundamental work *"Cathegories"* has been analyzed thousands times during the last thousand years. However, studies were of merely philosophical character. The first analysis of "Cathegories" from the viewpoint of connection between number theory, topology, and cognition was performed in [238]. We present here shortly the main conclusions of [238]. We are interested in Chapter 6 of "Cathegories" entitled "Quantity". Aristotle point to the presence of two types of quantities: discrete and continuous. We cite him: "Quantity is either discrete or continuous. ... Instances of discrete quantities are number[2] and speech[3]; of continuous, lines, surfaces, solids, and, besides these, time and place. In the case of the parts of a number, there is no common boundary at which they join. For example: two fives make ten, but the two fives have no common boundary, but are separate; the parts three and seven also do not join at any boundary. Nor, to generalize, would it ever be possible in the case of number that there should be a common boundary among the parts; they are always separate. Number, therefore, is a discrete quantity. The same is true of speech. That speech is a quantity is evident: for it is measured in long and short syllables. I mean here that speech which is vocal. Moreover, it is a discrete quantity for its parts have no common boundary. There is no common boundary at which the syllables join, but each is separate and distinct from the rest. A line, on the other hand, is a continuous quantity, for it is possible to find a common boundary at which its parts join. In the case of the line, this common boundary is the point; in the case of the plane, it is the line: for the parts of the plane have also a common boundary. Similarly you can find a common boundary in the case of the parts of a solid, namely either a line or a plane. Space and time also belong to this class of quantities. Time, past, present, and future, forms a continuous whole. Space, likewise, is a continuous quantity; for the parts of a solid occupy a certain space, and these have a common boundary; it follows that the parts of space also, which are occupied by the parts of the solid, have the same common boundary as the parts of the solid. Thus, not only time, but space also, is a continuous quantity, for its parts have a common boundary."

The main message for us is that natural numbers and mental entities belong to the same class of quantities which are essentially different from time and spatial quantities. By using the language of modern mathematics one can say that these two types of quantities have different topological structures. Aristotle's "there is no common boundary at which they join" is nothing else than the definition of the *disconnected topological space*. Since this absence of common boundary takes place at any point, one can interpret Aristotle's considerations as a story about *totally disconnected* topo-

[2]We remind that at Aristotle's time numbers were identified with natural numbers. The notion of a real number representing numerically a point of the line was elaborated only at the end of the 19th century. Aristotle could not even imagine a possibility to encode lines, surfaces and bodies by some sort of numbers.

[3]In the Russian translation of "Cathegories" it was used "word", instead of "speech." It seems that "word" is more adequate to the present discussion.

logical spaces. Thus Aristotle's message to us (through two thousand years) is that mental quantities should be described by totally disconnected topological spaces, but physical and geometrical quantities by connected topological spaces. We remind that any ultrametric space and, in particular, any $\mathbb{Z}_m$, is totally disconnected. Euclidean space and Minkowski space are connected spaces. Another massage is that natural numbers are of the same geometric nature as mental quantities. Hence, it would be natural to try to establish correspondence between mental quantities and natural numbers. By taking into account development of mathematics during the last two thousand years, we can use various generalizations of natural numbers obtained via completion of $\mathbb{N}$ with respect to topologies of the totally disconnected type.

As was already remarked, we choose m-adic numbers as *mental coordinates* representing *mental states.* By using mental coordinates we are able to embed into the space mental analogs of physical rigid bodies – *associations and ideas.* They are represented, respectively, by *balls and collections of balls* in the ultrametric mental space.

14.1.2 Encoding by branches of trees

In our model mental states – represented geometrically by branches of the m-adic tree and algebraically by elements of the ring of m-adic integers $\mathbb{Z}_m$ – are *basic cognitive mental images* (in Chapter 15 we will use the ring $\mathbb{Q}_m$.)

An association connects a number of cognitive mental images. Thus an association can be represented as a subset of the mental space. The crucial point is that in our model the associative connection of cognitive mental images is fundamentally hierarchical. Therefore an association is not an arbitrary collection of cognitive mental images (not an arbitrary set of mental points), but a hierarchically coupled collection. Since in our model the mental hierarchy is encoded by the topology of the mental space, it represents the associative coupling of cognitive mental images into balls. A larger ball couples together more cognitive mental images. Thus it is a more complex association (but it is a "fuzzy-association," it is not sharp). Decreasing of a ball's radius induces decreasing of the complexity of an association which is represented by this ball. An association becomes sharper. In the limit we obtain the ball of zero radius. That is nothing else than a single mental point (the center of such a degenerated ball). This is a single cognitive mental image. This is the limiting case of an association: a cognitive mental image is "associated" with itself. We hope that such a limiting degeneration of an association into a mental image is not misleading for readers.

Ideas are identified as collections of associations, something analogous to a coherent group of individuals in the biological analogy. The identification of the fundamental structure as a mental image allows a concrete dynamical model for ideas as collections of loosely bound associations. Association is a kind of atom of cognition from which more complex ideas are built like molecules from atoms.

We mention also that a p-adic model (here $m = p$ is a prime number) of consciousness was (independently) proposed in [362]. Pitkänen's approach was not based on encoding of *mental hierarchy* by p-adic numbers. It has a deeper relation to foundations of physics, especially the quantum one.

Recently the p-adic information space was used for genetic models, see [106, 240, 248, 363]. A new exciting domain of research is use of ultrametric methods in data-analysis – from astrophysics and computer science to biology, see, e.g., [336].

14.1.3 Dynamical system approach, artificial intelligence

In the papers [214, 221, 224, 227–229, 232, 235] and [9, 10, 62, 109, 239, 246, 247], we studied merely the dynamics of mental states – mental images. We considered dynamical systems which work with mental states. There is a nonlinear relation between input and output mental states,

$$x_{n+1} = f(x_n), x_n \in X_{\text{mental}}. \tag{14.2}$$

The description of functioning of the human brain by dynamical systems (feedback processes) is a well established approach. In principle, our m-adic dynamical cognitive model could be considered as a special model of the *Dynamical System Approach* to cognitive sciences, see, e.g., [42, 121, 391, 397, 398]. Of course, the Dynamical System Approach should be considered in the very general sense: all of cognition might be accounted for with dynamical system models, see Ashby [42]. The m-adic dynamical model developed is purely information processing model. It has no direct relation to real physical processes in a single neuron or neuronal group. This model is not based on fundamental principles of the Dynamical System Approach such as

(a) *Embodiment* of mind;

(b) *Situatedness* of cognition.

We recall that the orthodox Dynamical System Approach emphasizes commonalties between behavior in neural and cognitive processes on one hand and with physiological and environmental events on the other. The most important commonality is the dimension of time shared by all these domains. In fact, sharing of time scales with physical environment implies

(c) Continuous real time evolution described by differential equations.

Postulate (c) of the Dynamical System Approach could not be accepted in the theory of m-adic dynamical systems. The ultrametric m-adic structure of the mental configuration space could not be even in principle combined with continuous evolution based on real time, $t \in \mathbb{R}$. Discreteness of time in p-adic dynamical models is one of the fundamental features of these models. Here we even could not apply the standard ideology of the Dynamical System Approach that is used to connect discrete dynamics with continuous (differential) dynamics, namely, that discrete dynamical systems are approximations of continuous dynamical systems.

Unfortunately, practically all critical arguments against the Dynamical System Approach can also be used against m-adic models describing dynamics of mental points, see again [214,221,224,227–229,232,235] and [9,10,62,109,239,246,247]. It seems that the Dynamical System Approach (neither the orthodox one based on dynamics in the real physical space nor m-adic based on dynamics in the ultrametric mental space) can not provide an adequate representation of high level mental processes. Only basic mental states are represented by points of $\mathbb{Z}_m$ and their dynamics is described by point wise iterations. Dynamics of more complex mental structures which are composed of mental states have distinguishing features which are not reduced to features of point wise dynamics.

In the present chapter we consider dynamics of mental analogs of physical rigid bodies – associations (balls in the mental metric space) and ideas (collections of balls). This model was invented in article [232] and developed in book [237]. In spite of the fact that dynamics of associations and ideas can be in principle reduced to dynamics of mental points composing those "mental bodies", those dynamics exhibit their own interesting properties which could not be seen on the level of pointwise dynamics.[4]

In Chapter 15 we will consider another system of encoding of complex mental images. Instead of the use of the set-theoretical representation of such images, e.g., by ultrametric balls, we will use probability distributions on m-adic trees.

Our approach [236] can be considered as extension of the artificial intelligence approach, Chomsky [84], Churchland [86] to simulation of psychological behavior, cf. [70–72]. Especially close relation can be found with models of AI-life, see Langton et al. [281], Yaeger [414] and Collings and Jefferson [91]. We extend modeling of AI-life to psychological processes. On the basis of the presented models, we can create AI-societies of psycho-robots interacting with real people and observe evolution of psyche of psycho-robots (and even people interacting with them). We also mention development of theory of Animats, see, e.g., Meyer [316] and Donnart and Meyer [103]. By similarity with Animats we call our psycho-robots: *Psychots.* Finally, as a motivation of our activity, we refer to the well-known declaration of Herbert Simon:

"AI can have two purposes. One is to use the power of computers to argument human thinking, just as we use motors to argument human or horse power. Robots and expert systems are major branches of this. The other is to use a computer's AI to understand how humans think, In a humanoid way. You are using AI to understand the human mind."

Our aim is precisely to understand human mind and psychology via AI-modeling.

14.1.4 Unconscious and conscious dynamics – Freudian approach

We apply our dynamical model for modeling of flows of unconscious and conscious information in the human brain.[5]

---

[4]Of course, mathematically it is evident: dynamics of balls and collections of balls have their own properties.

[5]We do not try to discuss general philosophical and cognitive problems of modeling of consciousness, see, e.g., Bechterew [49], Freud [140–142], Whitehead [411], Clark [90], Bechtel and Abrahamsen [48], Blomberg [69], Edelman [116], Churchland and Sejnovski [86], Baars [43], Pitkanen [362], Ivanitsky

In a series of models, Models 1–3, we consider cognitive systems with increasing complexity of psychological behavior determined by structure of flows of associations and ideas. Using this basic conceptual repertoire an increasingly refined cognitive model is developed starting from an animal like individual whose sexual behavior is based on instincts alone. At the first step a classification of ideas to interesting and less interesting ones is introduced and less interesting ideas are deleted. At the next level a censorship of dangerous ideas is introduced and the conflict between interesting and dangerous leads to neurotic behaviors, idées fixes, and hysteria. These aspects of the model reflect more the general structure of conscious/unconscious processing rather than properties of m-adic numbers. The basic mathematical structure for this model is the mental ultrametric space. In particular, ultrametric is used to classify ideas – to assign to each idea its measures of interest and interdiction.

Finally, we apply our approach to mathematical modeling of Freud's theory, see, for example, [140–142] of interaction between unconscious and conscious domains. One of basic features of our model is splitting the process of thinking into two separate (but at the same time closely connected) domains: *conscious* and *unconscious*, cf. [140–142].

In spite of huge diversity of viewpoints on Freud's psychoanalysis and its connections with neurophysiology, see, e.g., Macmillan [307], Gay [148], Young-Bruehl [417] as well as Green [159], Solms [387, 388] and Solms and Turnbull [389] for debates, the ideas of Sigmund Freud are still important sources of inspiration and not only in psychology, but also in cognitive sciences and even neurophysiology, neuro informatics and cybernetics as well as mental informatics and cybernetics.

We shall use the following point of view on the simultaneous work of consciousness and unconsciousness. Consciousness contains a *control center CC* that has functions of control over results of functioning of unconsciousness. *CC* formulates problems, and sends them to the unconscious domain.[6]

The process of finding a solution is hidden in the unconscious domain. In the unconscious domain there work complex dynamical systems – *thinking processors.*

Each processor is determined by a function f from the mental space into itself (describing the corresponding feedback process – psychological function). It produces iterations of mental states (points of mental space)

$$x_1 = f(x_0), \dots, x_{n+1} = f(x_n), \dots .$$

These intermediate mental states are not used by the consciousness. Consciousness (namely *CC*) controls only some exceptional moments in the work of the dynamical system in the unconscious domain – attractors and cycles. Dynamics of mental points

---

[177], Bohm and Hiley [73], Blomberg et al. [69], Khrennikov [221, 227, 237], Albeverio et al. [9], Boden [70–72], Damasio [95], Fuster [146], Macmillan [307], Gay [148], Green [159], Young-Bruehl [417].

[6] In this book we do not pay so much attention to modeling of consciousness. In our models it is more or less reduced to *CC*.

induce dynamics of mental figures, in particular, ball-associations and, hence, ideas (collections of balls).

As was pointed out, behaviors of the dynamical system in the mental space and its lifting to spaces of associations and ideas can be very different. Extremely cycling (chaotic) behavior on the level of mental states can imply nice stabilization to attractors on the level of ideas.

We also point out that systems of m-adic numbers (restricted to $m = p$ a prime number) were intensively used in theoretical physics, see, e.g., [201,214,407].

14.1.5 Neuronal hierarchy

In this chapter we do not consider in detail the neuronal basis of the m-adic mental space[7]. The neuronal basis is provided by consideration of *hierarchical neuronal trees*. Such trees provide connection of the mental space (produced by a tree) with physical space (in which neuronal trees are located). Mental processes are connected with physical and chemical processes in the brain: mental states are produced as distributed activations of neuronal pathways. We remark that mental hierarchy was discussed a lot, see already mentioned papers and [17,48,90,176,177,214,221,224,227–229, 232, 235, 386, 390] and [9, 10, 62, 109, 237, 239, 246, 247]. However, there was not so much experimental neurophysiological evidence of existence of such neuronal hierarchical structures in the brain. Therefore the recent paper of [306], that confirmed experimentally existence of neurons-directors which rule the performances of cognitive tasks (under the same context and learning conditions) is extremely important for our model. The presence of a complicated hierarchy of time scales in the brain can be considered as an indirect confirmation of the hierarchical structure of processing of information in the brain, see Geissler [149–151], Geissler and Klix and Scheidereiter [152], Geissler and Puffe [154], Geissler and Kompass [153] for the mots detailed analysis of hierarchy of time scales in the brain.

We hope that we presented sufficient motivations (from Platon and Aristotle to Luczak and Geissler) for using the ultrametric (totally disconnected) mental space. The m-adic model matches most nicely demands of ancient philosophers, ancient and modern mathematicians, physicists (working in p-adic physics, in particular, superstring theory and quantum cosmology), neurophysiologists (studying neuronal hierarchy in the brain), psychologists (studying hierarchy in psychological behavior) and cognitive scientists.

[7] See [237, 239] for corresponding models, see Section 14.5 for a brief review and Chapter 15 for general consideration.

14.2 Mental space

Each point $x \in \mathbb{Z}_m$ has the infinite number of coordinates

$$x = (\alpha_1, \ldots, \alpha_n, \ldots). \tag{14.3}$$

Each coordinate yields the finite number of values,

$$\alpha \in A_m = \{0, \ldots, m-1\}, \tag{14.4}$$

where $m > 1$ is a natural number, the base of the alphabet A_m. The standard ultrametric is introduced on this set in the following way. For two points

$$x = (\alpha_0, \alpha_1, \alpha_2, \ldots, \alpha_n, \ldots),\ y = (\beta_0, \beta_1, \beta_2, \ldots, \beta_n, \ldots) \in \mathbb{Z}_m,$$

we set

$$\rho_m(x, y) = \frac{1}{m^k} \text{ if } \alpha_j = \beta_j, j = 0, 1, \ldots, k-1, \text{ and } \alpha_k \neq \beta_k. \tag{14.5}$$

To find the distance $\rho_m(x, y)$ between two strings of digits x and y, we have to find the first position k such that strings have different digits at this position. This is a metric and even an ultrametric and it coincides with the ultrametric $\rho_m(x, y) = |x - y|_m$ given by the absolute value $|\cdot|_m$ on $\mathbb{Z}_m$. Realization of elements of $\mathbb{Z}_m$ by informational sequences (14.3) and the m-adic metric as (14.5) is more useful in cognitive applications. The number structure on $\mathbb{Z}_m$ given by (14.1) is a supplementary mathematical structure on the space informational vectors of the form (14.3).

We shall use the following mathematical model for the mental space:

(1) Set-structure: The set of mental states X_{mental} has the structure of the m-adic tree (ring): $X_{\text{mental}} = \mathbb{Z}_m$.

(2) Topology: Two mental states x and y are close if they have sufficiently long common root. This topology is described by the metric ρ_m.

In our mathematical model the *mental space is represented as the metric space* $(\mathbb{Z}_m, \rho_m)$.

14.3 Dynamical thinking in mental space

Dynamical thinking, *the evolution of a mental state,* is performed via the following procedure:

a) an initial mental state (e.g. an external sensory input or the output of the previous activity of the brain) is sent to the unconscious domain;

b) this initial mental state is iterated by some dynamical system which is given by a map from the mental metric space into itself:

$$f : \mathbb{Z}_m \to \mathbb{Z}_m;$$

c) if iterations converge (with respect to the m-adic metric) to an attractor, then this attractor will proceed to consciousness; it is the solution of the initial problem.[8] In the simplest model, see Model 1 in Section 14.8, this attractor is sent directly to consciousness – without any additional analysis. In more complicated models of the mental architecture attractor becomes the object of further detailed analysis. Some attractors are rejected as inapproriate solutions. Such rejected attractors can be either simply deleted or in a more complicated context placed in the domain of hidden forbidden wishes. These are so called psychical complexes. They induce various psychical symptoms – the objects of psychoanalysis.

By using Freud's terminology we can say that a) and the last step in c) are under Ego's control, but b) and the first step of c) are under Id's control.

Thus, our mathematical model of cognition is based on two cornerstones:

H) The first is the assumption that the coding system which is used by the brain for recording vectors of information generates a *hierarchical structure* between digits of these vectors. Thus if $x = (\alpha_1, \alpha_2, \ldots, \alpha_n, \ldots), \alpha_j = 0, 1, \ldots, m-1$, is an information vector which presents in the brain a mental state then digits α_j have different weights. The digit α_0 is the most important, α_1 dominates over $\alpha_2, \ldots, \alpha_n, \ldots$ and so on.

D) The second is the assumption that functioning of the brain is not based on the *rule of reason*. The unconsciousness is a collection of dynamical systems $f_s(x)$ (thinking processors) which produce new mental states practically automatically. The consciousness only uses and control results (attractors in spaces of ideas) of functioning of unconscious processors.

14.4 Associations and ideas

We now improve this dynamical cognitive model on hierarchic mental trees by introducing a new hierarchy. Equivalence classes of mental states given by balls are interpreted as associations, collections of associations as ideas – mathematically these are collections of balls.

For a large class of dynamical systems on m-adic trees, dynamics of ideas exhibits a new property: for each initial idea J_0 its iterations are *attracted* by some idea J_{attr}, see [232, 237].[9] The latter idea is considered by consciousness as the solution of the problem J_0. Contrary to such an attractor-like dynamics of ideas, dynamics of mental

[8] Depending on initial conditions and the dynamical law it may occur that iterations starting with some initial mental state will not approach any attractor. For example, starting with some state a dynamical system may perform the cyclic behavior in the process of thinking. In such a case a cognitive system would not find the solution of a problem under consideration.

[9] Mathematically it is about the dynamics of sets consisting of collections of balls. The ground dynamical system in $\mathbb{Z}_m$ can have a lot of cycles, but its lifting to dynamics of sets has only attracting points – for convergence with respect to specially defined distance.

states (on the tree X_{mental}) or associations need not be attractive. In particular, there can exist numerous cycles or ergodic evolution.

By using higher cognitive levels (associations and ideas) of the representation of information a cognitive system strongly improves the regularity of thinking dynamics. Finally, we note that the use of a new cognitive hierarchy (in combination with the basic hierarchy of the m-adic tree) strongly improves the information power of a cognitive system.

Special collections of mental points form new cognitive objects, *associations*. Let $s \in \{0, 1, \ldots, m-1\}$. A set

$$A_s = \{x = (\alpha_0, \ldots, \alpha_k, \ldots) \in \mathbb{Z}_m : \alpha_0 = s\}$$

is called an association of order 1. By realizing $\mathbb{Z}_m$ as the metric space we see that A_s can be represented as the ball of radius $r = \frac{1}{m}$. Any point a having α_0 as the first digit can be chosen as a center of this ball (we recall that in an ultrametric space any point a belonging to a ball can be chosen as its center): $A_s = B_{\frac{1}{m}}(a)$, where $a = (a_0, a_1, \ldots), a_0 = s$. Associations of higher orders are defined in the same way. Let $s_0, \ldots, s_{k-1} \in \{0, 1, \ldots, m-1\}$. The set

$$A_{s_0 \ldots s_k} = \{x = (\alpha_0, \ldots, \alpha_k, \ldots) \in \mathbb{Z}_m : \alpha_0 = s_0, \ldots, \alpha_{k-1} = s_{k-1}\}$$

is called an association of order k. These are balls of the radius $r = \frac{1}{m^k}$.

Denote the set of all associations by the symbol X_A. Collections of associations will be called *ideas*. Denote the set of all ideas by the symbol X_{ID}. The space X_{ID} consists of point-associations.

In this section we study the simplest dynamics of associations and ideas which are induced by corresponding dynamics of mental states,

$$x_{n+1} = f(x_n). \tag{14.6}$$

Suppose that, for each association, its image is again an association. Thus f maps balls onto balls. Then dynamics (14.6) of mental states of a cognitive system τ induces dynamics of associations

$$A_{n+1} = f(A_n). \tag{14.7}$$

We say that dynamics in the mental space X_{mental} for transformations is lifted to the space of associations X_A.

Although deterministic dynamical systems describe some essential features of processing of mental information, it is natural to suppose that real dynamics in the brain is described by *random dynamical systems*, see [5] and [256].

14.5 Neuronal realization

Let us consider a simplest model of a neuronal tree T_{neuronal} inducing a mental space X_{mental}. This model is based on the 2-adic neuronal tree given by Figure 1.1 from

Section 1.4.2. In the general case we can imagine geometrically a system of m-adic integers (which will be the mathematical basis of our cognitive models) as a homogeneous *tree* with m-branches splitting at each vertex. The distance between mental states is determined by the length of their common root: close mental states have a long common root. The corresponding geometry strongly differs from the ordinary Euclidean geometry.

Each vertex of this tree corresponds to a single neuron. In this idealized model each axon provides connections with precisely two neurons of lower level of the hierarchy in the neuronal tree. There is the root-neuron denoted by $\star$, its axon provides connections with the two neurons, 0 and 1, of the lower level. Each of these neurons sends its axon to precisely two neurons of the lower level and so on. Each branch n of this tree (a hierarchical chain of neurons) can be coded by a sequence of zeros/ones Thus this neuronal tree can be mathematically represented as the set of 2-adic numbers: $T_{\text{neuronal}} = \mathbb{Z}_2$.

Each branch of this neuronal tree is a device for producing mental states (cognitive mental images). In the simplest model we suppose that each neuron can be only in the two states: $\alpha = 1$, firing, $\alpha = 0$, non-firing. Thus each branch produces (at some moment of time) a sequence of zeros/ones: $x = \alpha_0\alpha_1 \dots \alpha_N \dots$, where $\alpha_j = 0, 1$. (In the mathematical model we can consider infinitely long sequences.) Thus the neuronal tree $T_{\text{neuronal}} = \mathbb{Z}_2$ produces the mental space $X_{\text{mental}} = \mathbb{Z}_2$. We can consider a "*body→mind field*" on the neuronal tree T_{neuronal}. This is the map

$$\varphi : T_{\text{neuronal}} \to X_{\text{mental}},$$

mathematically:

$$\varphi : \mathbb{Z}_2 \to \mathbb{Z}_2, \; \varphi(n) = x.$$

Here n is a hierarchic chain of neurons, n is an element of the neuronal tree T_{neuronal}. And x is a mental state produced by the hierarchic chain of neurons n.

In fact, the 2-adic mental space need not be based on the 2-adic morphology of the neuronal tree. In general there is no direct connection between the morphology of a the neuronal tree and the corresponding mental space. Let us consider any tree T_{neuronal} with the root $\star$ (any number of edges leaving a vertex, so an axon can provide connections with any number of neurons at the lower floor). Nevertheless, let us consider the same firing/not coding system. Each branch of the neuronal tree T_{neuronal} produces a 2-adic number. Here the mental field is a $\mathbb{Z}_2$-valued function on T_{neuronal}. This is an important property of the model: it would be not so natural to consider only homogeneous neuronal trees of the m-adic type.

The structure of the mental space is determined not by the morphology of the neuronal tree, but by the coding system for states of neurons.

Let us consider more advanced system of coding based on frequencies of spiking for neurons, e.g., [171, 172]. We assign to each neuron its frequency of spiking:

$$\alpha = k, \text{ for the frequency: } \nu = \frac{2\pi k}{m}, \; k = 0, 1, \dots, m-1. \tag{14.8}$$

Such a system of coding induces the m-adic mental space, $X_{\text{mental}} = \mathbb{Z}_m$ for any neuronal tree. Each mental function is based on its own neuronal tree: $T_{\text{neuronal}} = T_{\text{neuronal}}(f)$.

14.6 Model of cognitive psychology

We point out that the model which was considered in Section 14.5 is a model of *neuropsychology* and not at all a model of *neurophysiology.* The neuronal trees under consideration are not trees for integration-propagation of sensory stimuli forming new mental categories at each level of such a tree, see [237] and Chapter 15 for a general model.

In Section 14.5 we considered neuronal trees creating associations. As an example, let us consider a neuronal tree which is used for representation of personalities. Various hierarchical representations can be used. We choose the "sex-representation": the state of the root-neuron, $\star$, gives the sex of a person: $\alpha_0 = 1$ – female, $\alpha_0 = 0$ – male. Consider a branch of this tree. Suppose that in this representation the next neuron (after $\star$) gives the age of a person: $\alpha_1 = 1$ – young, $\alpha_1 = 0$ – not, and so on: $\alpha_2 = 1$ – blond, $\alpha_2 = 0$ – not, $\alpha_3 = 1$ – high education, $\alpha_3 = 0$ – not, ...

Of course, depending on context a cognitive system τ can use different representations of personalities. It can occur that, for another context, sex is not the most important feature and education's level plays a more important role. In such a case the τ uses another hierarchy which will be based on another neuronal tree. If the education is the most important feature then ecoding is based on a neuronal tree with the root-neuron which represents the education level.

Let us come back to the neuronal tree with the sex root-neuron and the branch of neurons responsible, respectively, for sex, age, the education level, ...

Take the ball $B_{1/2}(1) = \{x = \alpha_0\alpha_1 \dots \alpha_N \dots : \alpha_0 = 1\}$. This is the association of a woman.

Take the ball $B_{1/4}(11) = \{x = \alpha_0\alpha_1 \dots \alpha_N \dots : \alpha_0 = 1, \alpha_1 = 1\}$. This is the association of a young woman.

Take the ball $B_{1/8}(111) = \{x = \alpha_0\alpha_1 \dots \alpha_N \dots : \alpha_0 = 1, \alpha_1 = 1, \alpha_2 = 1\}$. This is the association of a young blond woman.

Take the ball $B_{1/16}(1111) = \{x = \alpha_0\alpha_1 \dots \alpha_N \dots : \alpha_0 = 1, \alpha_1 = 1, \alpha_2 = 1, \alpha_3 = 1\}$. This is the association of a young blond woman with high education.

We remark that the point $a = 1111$ can be chosen as well as center of the balls $B_{1/8}(1111)$ and $B_{1/4}(11)$. The a belongs to both these balls and, as we know from general theory of ultrametric spaces, it can be chosen as their center, see Section 1.5. We can also write centers by using the representation by natural numbers, e.g. $a = 15 = 1 + 2 + 2^2 + 2^3$. Thus $B_{1/8}(1111) \equiv B_{1/8}(15)$. From the viewpoint of encoding of mental hierarchy by 2-adic numbers, this identification of branches of the (finite) 2-adic tree with natural numbers does not have any cognitive interpretation. However, the algebraic structure on the 2-adic tree plays an extremely important role in dynamics on this tree. One could not exclude that cognitive systems might discover 2-adic algebra. They might use it to create simple dynamical systems processing mental information.

Take now the ball $B_{1/4}(01) = \{x = \alpha_0\alpha_1 \dots \alpha_N \dots : \alpha_0 = 0, \alpha_1 = 1\}$. This is the association of young man.

Take the union of two balls-associations: $W = B_{1/4}(11) \cup B_{1/4}(01) = \{x = \alpha_0\alpha_1 \dots \alpha_N \dots : \alpha_1 = 1\}$. This is an idea of young person.[10]

"Young person", W, is an idea with respect to the hierarchy based on sex. If we consider another hierarchy (based on another neuronal tree) for which the root-neuron represents not sex, but age, then it produces "young person" as the association: $B_{1/2}^{\mathrm{age}}(1) = \{y = \beta_0\beta_1 \dots \beta_N \dots : \beta_0 = 1\}$.

But $B_{1/2}^{\mathrm{age}}(1)$ is not completely the same mental object as W. The $B_{1/2}^{\mathrm{age}}(1)$ is the "unisex young person" and W "young person with sex."

We see that our considerations of changing of neuronal trees and hence mental representations is similar to the choice of coordinate systems in physics.

As was remarked at the beginning of this section, dynamics of associations and ideas need not be based on external stimuli (sensor or mental). Thus a neuronal tree can be self-activated even without signals from outside, cf. with experimental results [306].

14.7 Dynamics of associations and ideas

Dynamics of associations (14.7) automatically induces dynamics of ideas

$$J_{n+1} = f(J_n). \tag{14.9}$$

Geometrically associations are represented as bundles of branches of the m-adic tree. Ideas are represented as sets of bundles. Thus dynamics (14.6), (14.7), (14.9) are, respectively, dynamics of branches, bundles and sets of bundles on the m-adic tree. To give examples of f mapping balls onto balls, we use the standard algebraic structure on $\mathbb{Z}_m$. For example, it is clear from previous chapters that all monomial dynamical systems belong to this class.

We are interested in attractors of dynamical system (14.9) (these are ideas-solutions). To define attractors in the space of ideas X_{ID}, we have to define a convergence in this space. We must introduce a distance on the space of ideas (sets of associations). Unfortunately there is a small mathematical complication. A metric on the space of points does not induce a metric on the space of sets that provides an adequate description of the convergence of ideas. It is more useful to introduce a generalization of metric, namely so called *pseudometric*.[11] Hence dynamics of ideas is a dynamics not

[10] By using the natural number representation one can write $W = B_{1/4}(3) \cup B_{1/4}(2)$.

[11] In fact, it is possible to introduce even a metric (Hausdorff's metric) as people in general topology do. However, it seems that this metric does not give an adequate description of dynamics of associations and ideas.

in a metric space, but in more general space, so called pseudometric space.

Let (X, ρ) be a metric space. The distance between a point $a \in X$ and a subset B of X is defined as

$$\rho(a, B) = \inf_{b \in B} \rho(a, b)$$

(if B is a finite set, then $\rho(a, B) = \min_{b \in B} \rho(a, b)$).

Denote by $\mathrm{Sub}(X)$ the system of all subsets of X. *Hausdorff's* distance between two sets A and B belonging to $\mathrm{Sub}(X)$ is defined as

$$\rho(A, B) = \sup_{a \in A} \rho(a, B) = \sup_{a \in A} \inf_{b \in B} \rho(a, b). \tag{14.10}$$

If A and B are finite sets, then

$$\rho(A, B) = \max_{a \in A} \rho(a, B) = \max_{a \in A} \min_{b \in B} \rho(a, b).$$

Hausdorff's distance ρ is not a metric on the set $Y = \mathrm{Sub}(X)$. In particular, $\rho(A, B) = 0$ does not imply that $A = B$. Nevertheless, the triangle inequality

$$\rho(A, B) \leqslant \rho(A, C) + \rho(C, B), \quad A, B, C \in Y,$$

holds for Hausdorff's distance.

Let T be a set. A function $\rho\colon T \times T \to \mathbf{R}_+ = [0, +\infty)$ for which the triangle inequality holds true is called a *pseudometric* on T; (T, ρ) is called a pseudometric space. Hausdorff's distance is a pseudometric on the space Y of all subsets of the metric space X; (Y, ρ) is a pseudometric space. The strong triangle inequality $\rho(A, B) \leqslant \max[\rho(A, C), \rho(C, B)]$, $A, B, C \in Y$, holds true for Hausdorff's distance corresponding to an ultrametric ρ on X. In this case Hausdorff's distance ρ is an *ultra-pseudometric* on the set $Y = \mathrm{Sub}(X)$.

14.8 Advantages of dynamical processing of associations and ideas

As was already mentioned, the main distinguishing feature of the dynamics of associations and ideas are their regularity comparing with the dynamics of mental states. Typically the dynamics of mental states is irregular. Numerous cycles and ergodic components appear and disappear depending on m, see previous chapters. Moreover, dynamics on more complex mental spaces (larger m and more floors for mental trees) is more irregular than dynamics on simpler mental spaces. Thus cognitive systems having more complex brains would have real problems with successive processing of information, i.e., with obtaining attractors-solutions (of course, under the assumption that our model for the hierarchical dynamical processing of mental information is adequate to the functioning of the real brain).

Surprisingly such an irregularity for mental states induces the regular dynamics of associations and ideas, [232, 237]. Cycles of states disappear. They are hidden in balls-associations. Ergodic components are also unified into balls-associations.

Thus by using the associative representation of mental information the brain working as a collection of dynamical systems on hierarchical trees essentially increases the regularity of information processing. By our model primitive brains (having a few levels of mental hierarchy and rather weak networks of connections between hierarchical levels) are fine by working only with mental states. However, more complex brains *should form associations* to stabilize dynamical processing of information.

14.9 Transformation of unconscious mental flows into conscious flows

We represent a few mathematical models of the information architecture of a conscious cognitive system τ, cf., e.g., [136]. We start with a quite simple model (Model 1). This model will be developed to more complex models which describe some essential features of human cognitive behavior. The following sequence of cognitive models is related to the process of evolution of the mental architecture of cognitive systems.

Model 1:

A) The brain of τ is split into two domains: *conscious* and *unconscious*.

B) There are two control centers, namely a *conscious control center* CC and an *unconscious control center* UC.

C) The main part of the unconscious domain is a *processing domain* Π. Dynamical thinking processors π are located in Π.

In the simplest case the outputs of some group of thinking processors $\pi_{\text{un}}^1, \dots, \pi_{\text{un}}^n$ are always sent to UC and the outputs of another group $\pi_{\text{c}}^1, \dots, \pi_{\text{c}}^m$ are always sent to CC.[12] The brain of τ works in the following way.

External information is transformed by CC into some problem-idea J_0. The CC sends J_0 to a thinking processor π located in the domain Π. Starting with J_0, π produces via iteration $J_1, \dots, J_N, \dots$ an idea-attractor J.

If $\pi = \pi_{\text{un}}^i$ (one of the unconscious-output processors), then J is transmitted to the control center UC. This center sends J as an initial idea $J_0' = J$ to Π or to a physical (unconscious) performance. In the first case some π' (it can be conscious-output as well as unconscious-output processor) performs iterations $J_0', J_1', \dots, J_N', \dots$ and produces a new idea-attractor J'. In the second case J is used as a signal to some physiological system.

[12] Information produced by π_{un} cannot be directly used in the conscious domain. This information circulates in the unconscious domain. Information produced by π_{c} can be directly used in the conscious domain.

If $\pi = \pi_c^i$ (one of conscious-output processors), then J is transmitted to the control center CC. This center sends J as an initial idea $J_0' = J$ to Π (to some π') or to a physical or mental performance (speech, writing), or to memory. There is no additional analysis of an idea-attractor J which is produced in the unconscious domain. Each attractor is recognized by the control center CC as a solution of the initial problem J_0, compared with models 2-4.

Moreover, it is natural to assume that some group of thinking processors $\pi_\Pi^1, \dots, \pi_\Pi^l$ has output only inside the processing domain Π. An attractor J produced by π_Π is transmitted neither to CC nor to UC. The J is directly used as the initial condition by some processor π. Finally, we obtain the mental architecture of a brain given by Figure 14.1.

In this model CC sends all ideas obtained from the unconscious domain to realization: mental or physical performance, memory recording, transmission to Π for a new cycle of the process of thinking. If the intensity of the flow of information from the unconscious domain is rather high, then such a τ can have problems with realizations of some ideas.

The CC is the basic neurocybernetic structure of Ego. The processing domain is the basic neurocybernetic structure of Id.

Example 14.1 (Primitive love). Let τ be a 'man' described by this model and let $\pi = \pi_{\text{sex}}$ be his sexual thinking system. The image J_0 of a woman γ is sent by CC to π_{sex}. This thinking block performs iterations $J_0, J_1, \dots, J_N$ and produces an idea-attractor J. In the simplest case we have the pathway: $CC \to \pi_{\text{sex}} \to CC$ (in principle, there could be extremely complex and long pathways, for example, $CC \to \pi_{\text{sex}} \to \pi' \to \pi'' \to UC \to \pi''' \to CC$). Suppose that the idea $J_{\text{love}} =$ (love γ) is the attractor for iterations starting with the image J_0 of a woman. Then J_{love} is sent directly to realization. Thus τ has no doubts and even no craving. He performs all orders of the unconscious domain. In fact, CC can be considered as a simple control device performing the connection with the external world.[13] The τ could not have mental problems. The only problem for τ is an intensive flow of images of women. This problem can be solved if τ collects images and then chooses randomly an image for realization.

The reader may ask: Why does such a cognitive system τ need to split mental processing into conscious and unconscious domains? The main consequence of this splitting is that the τ does not observe iterations of dynamical systems performing intensive computations. Consciousness, CC, pays attention only to results (attractors) of functioning of thinking processors. As a consequence, the τ is not permanently disturbed by these iterations. It can be concentrated on processing of external information and final results of the process of thinking.

[13] In such a simple model Id totally dominates over Ego. The latter is very primitive. It just serves to wishes of Id.

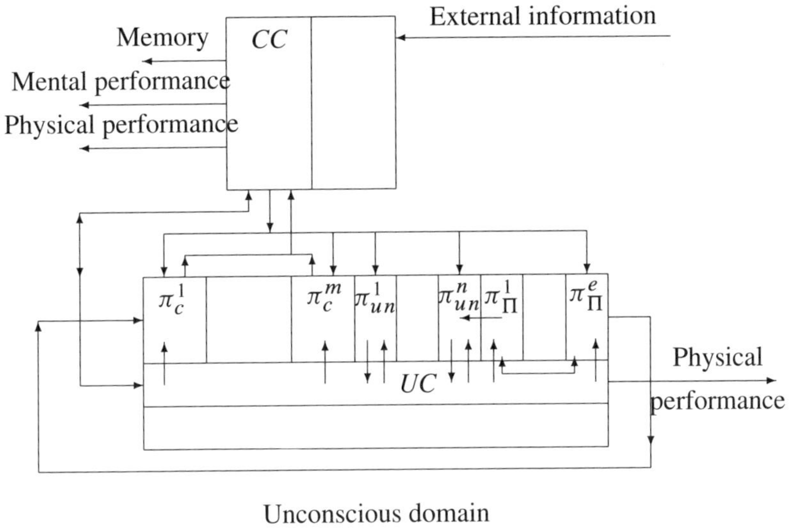

Figure 14.1. Model 1 of conscious/unconscious functioning. Besides the unconscious control center *UC* and the processing domain Π, the unconscious domain contains some other structures (empty boxes of this picture). These additional structures (in the conscious as well as unconscious domains) will be introduced in more complex models. We shall also describe the character of connections between *CC* and *UC*. In general we need not assume the specialization of processors π in Π: $(\pi_c^1, \dots, \pi_c^m) \to CC$, $(\pi_{un}^1, \dots, \pi_{un}^n) \to UC$, $(\pi_\Pi^1, \dots, \pi_\Pi^l) \to (\pi_\Pi^1, \dots, \pi_\Pi^l)$.

Model 2:

One of the possibilities to improve functioning of τ is to *create a queue of ideas J* waiting for a realization. Thus it is natural to assume that the conscious domain contains some *collector Q* in that all 'waiting ideas' are gathered.

Ideas in collector Q must be ordered for successive realizations. The same order structure can be used to delete some ideas if Q is complete. Thus all conscious ideas must be ordered.

They obtain some characteristics $I(J)$ that gives a *measure of interest* to an idea J. We may assume that I takes values in some segment $[\delta, 1]$ (in the m-adic model $\delta = 1/2$, see Remark 14.2). If $I(J) = 1$, then an idea J is extremely interesting for τ. If $I(J) = \delta$, then τ is not at all interested in J. There exists a threshold I_{rz} of the minimal interest for realization. If $I(J) < I_{rz}$, then the control center CC directly deletes J, despite the fact that J was produced in the unconscious domain as the solution of some problem J_0. If $I(J) \geq I_{rz}$, then CC sends the idea J to the collector of ideas waiting for realization, Q.[14]

[14]Thus in this model Ego has a possibility to punish uninteresting ideas which come from Id.

The τ lives in the continuously changed environment. The τ could not be concentrated on realization of only old ideas J even if they are interesting. Realizations of new ideas which are related to the present instant of time t can be more important. The time-factor must be taken into account.

Let $l(t)$, $l(0) = 1$, be some function (depending on τ) which decreases with the increasing of time t. Suppose that the interest $I(t, J)$ of an idea J in the queue Q evolves as

$$I(t, J) = l(t - t_0)\, I(J),$$

where $I(J)$ is the value of interest of J at the instant t_0 of the arrival to the collector Q. Thus the interest to J is continuously decreasing. Finally, if $I(t, J)$ becomes less than the realization threshold I_{rz} the J is deleted from Q.

Quick reactions to new circumstances can be based on an exponentially decreasing coefficient $l(t)$: $l(t) = e^{-Ct}$, where a constant $C > 0$ depends on τ.[15] If an idea J has an extremely high value of interest $I(J) \geqslant I_+$ (where I_+ is a preserving threshold), then it must be realized in any case. In our model we postulate that if $I(J) \geqslant I_+$, the interest to J is not changed with time: $I(t, J) = I(J)$.[16] We note that, of course, $I_+ \geqslant I_{\mathrm{rz}}$.

We now describe one of the possible models for finding the value $I(J)$ of interest for an idea J.

The conscious domain contains a *database D_i of ideas which are interesting for τ*. A part of this database D_i was created in the process of evolution. It is transmitted from generation to generation (perhaps even through DNA?). A part of the D_i is continuously created on the basis of τ's experience. It is the cornerstone of Ego (in coming models Ego will be essentially extended).

The conscious domain contains a special block, *comparator,* COM_c that measures the distance between two ideas, $\rho(J_1, J_2)$, and the distance between an idea J and the set D_i of interesting ideas: $\rho(J, D_i)$.

At the present level of development of neurophysiology we cannot specify a *mental distance* ρ. Moreover, such a distance may depend on a cognitive system or class of cognitive systems. The hierarchic structure of the process of thinking gives some reasons to suppose that COM_c might use the m-adic pseudometric ρ_m on the space of ideas X_{ID}. Thus the reader may assume that everywhere below ρ is generated by the m-adic metric. However, all general considerations are presented for an arbitrary metric.

We recall that the distance between a point b and a finite set A is defined as $\rho(b, A) = \min_{a \in A} \rho(b, a)$. If J is close to some interesting idea $L_0 \in D_i$, then $\rho(J, D_i)$ is small. In fact, we have $\rho(J, D_i) \leqslant \rho(J, L_0)$, $L_0 \in D_i$. If J is far from all

[15]It may be that the level of interest of J evolves in a more complex way. For example, $I(t, J) = \exp\{-C(J)t\}I(J)$. Here different ideas J have different coefficients $C(J)$ of decreasing interest.

[16]For example $I(t, J) = \exp\{-C(J)t\}I(J)$, where $C(J) = 0$ for $I(J) \geqslant I_+$. So $C(J) = \alpha\, \theta(I_+ - I(J))$, where $\alpha > 0$ is a constant (parameter of the brain) and θ is a Heavyside function: $\theta(t) = 1, t \geqslant 0$, and $\theta(t) = 0, t < 0$.

interesting ideas $L \in D_i$, then $\rho(J, D_i)$ is large. We define a measure of interest $I(J)$ as

$$I(J) = \frac{1}{1 + \rho(J, D_i)}.$$

Thus, $I(J)$ is large if $\rho(J, D_i)$ is small; $I(J)$ is small if $\rho(J, D_i)$ is large.

Remark 14.2 (The range of interest in the m-adic model). Suppose that the distance ρ is bounded from above:

$$\sup_{J_1, J_2} \rho(J_1, J_2) \leqslant C, \quad J_1, J_2 \in X_{\mathrm{ID}}.$$

Then $I(J) \geqslant \delta = \frac{1}{1+C}$. In such a case '$I(J)$ is very small' if $I(J) \approx \delta$. The function $I(J)$ takes values in the segment $[\delta, 1]$ (we remark that if $\rho(J, D_i) = 0$, then $I(J) = 1$). Let ρ be Hausdorff's pseudometric induced on the space of ideas X_{ID} by the m-adic metric ρ_m. We have $\rho(J_1, J_2) \leqslant 1$ for every pair of ideas J_1, J_2. Here $\delta = 1/2$ and $I(J)$ always belongs to $[1/2, 1]$. The sentence '$I(J)$ is very small' means that $I(J) \approx 1/2$ and, as always, '$I(J)$ is very large' means that $I(J) \approx 1$.

There should be a connection between the level $I(J)$ of interest and the strength of realization of J. Signals for mental or physical performances of J increase with increasing of $I(J)$. If, for example, the idea $J = \{$to beat this person$\}$ then the strength of the beat increases with increasing of $I(J)$.

In the process of memory recording the value of $I(J)$ also plays an important role. It is natural to suppose that in working memory the evolution of the quantity $I(t, J)$ is similar to the evolution which was considered in Q: $I(t, J) = l_{\mathrm{mem}}(t - t_0) I(J)$, where $l_{\mathrm{mem}}(0) = 1$ and $l_{\mathrm{mem}}(t)$ decreases with increasing of t. If $I(t, J)$ becomes less than the *memory preserving threshold* I_{-}^{mem}, then J is deleted from working memory.

Example 14.3 (Love with interest). Let τ be a 'man' described by Model 2. In the same way as in Example 14.1 the image J_0 of a woman γ may produce the idea $J_{\mathrm{love}} = (\text{love } \gamma)$. However, J_{love} is not sent to realization automatically. The COM_c measures $\rho(J_{\mathrm{love}}, D_i)$. Suppose that the database D_i of interesting ideas contains the idea (image) $L_{\mathrm{blond}} = (\text{blond woman})$. If γ is blond, then $\rho(J_{\mathrm{love}}, D_i)$ is small. So $I(J_{\mathrm{love}})$ is large and CC sends J_{love} to realization. However, if γ is not blond, then J_{love} is deleted (despite the unconscious demand J_{love}). Of course, the real situation is more complicated. Each τ has his canonical image L_{blond}. As $J_{\mathrm{love}} = J_{\mathrm{love};\gamma}$ depends on γ, distance $\rho(J_{\mathrm{love}}, D_i)$ can be essentially different for different women γ. Thus, for one blond woman γ, $I(J_{\mathrm{love}}) \geqslant I_{\mathrm{rz}}$, but for another blond woman γ, $I(J_{\mathrm{love}}) < I_{\mathrm{rz}}$. If there are few blond women $\gamma_1, \ldots, \gamma_l$ with $I(J_{\mathrm{love};\gamma_j}) \geqslant I_{\mathrm{rz}}$, then all ideas $J_{\mathrm{love};\gamma_j}$ are collected in Q. The queue of blond women is ordered in Q due to values $I(J_{\mathrm{love};\gamma j})$. If, for some γ, $I(J_{\mathrm{love};\gamma}) \geqslant I_{+}$, then the level of interest to $J_{\mathrm{love};\gamma}$ will not decrease with time. The level of $I(J_{\mathrm{love};\gamma})$ determines the intensity of realization of love with γ.

The mental architecture of 'brain' in Model 2 is given by Figure 14.2: A new block COM_c in the conscious domain measures the distance between an idea-attractor J which has been produced in the unconscious domain and the database D_i of interesting ideas. This distance determines the level of interest for $J : I(J) = 1/(1 + \rho(J, D_i))$. Ideas waiting for realization, $J^1, \ldots, J^s$, are collected in the special collector Q. They are ordered by values of $I(J)$: $I(J^1) \geqslant I(J^2) \geqslant \cdots \geqslant I(J^s) \geqslant I_+$. If $I(J) \geqslant I_+$ (where I_+ is the preserving threshold), then the value of interest of J does not decrease with time.

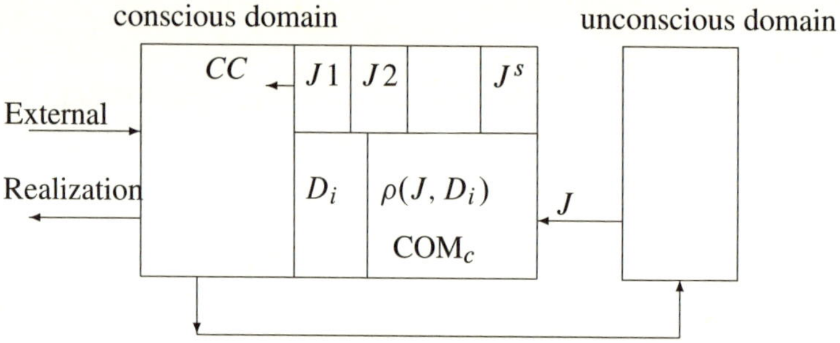

Figure 14.2. Model 2 of conscious/unconscious functioning (comparative analysis of ideas).

Model 3:

The life of τ described by Model 2 is free of contradictions. The τ is always oriented to realizations of the most interesting ideas, wishes, desires. However, environment (and, in particular, social environment) produces some constraints to realizations of some interesting ideas.

In a mathematical model we introduce a new quantity $F(J)$ which describes a measure of *interdiction* for an idea J.

It can be again assumed that $F(J)$ takes values in the segment $[\delta, 1]$. Ideas J with small $F(J)$ have low levels of interdiction. If $F(J) \approx \delta$, then J is a 'free idea'. Ideas J with large $F(J)$ have high levels of interdiction. If $F(J) \approx 1$, then J is totally forbidden.

The interdiction function is computed in the same way as the interest function. The conscious domain contains a *database* D_f *of forbidden ideas.*[17] The comparator COM_c measures not only the distance $\rho(J, D_i)$ between an idea-attractor J (which has been transmitted to the conscious domain from the unconscious domain) and the

[17] It is a cornerstone of Ego (as well as the interest-database). A part of the interdiction database belongs Super-Ego.

set of interesting idea D_i, but also the distance $\rho(J, D_i)$ between an idea-attractor J and the set of forbidden ideas D_f:

$$\rho(J, D_f) = \min_{L \in D_f} \rho(J, L).$$

If J is close to some forbidden idea L_0, then $\rho(J, D_f)$ is small. If J is far from all forbidden ideas $L \in D_f$, then $\rho(J, D_f)$ is large.

We define a *measure of interdiction* $F(J)$ as

$$F(J) = \frac{1}{1 + \rho(J, D_f)} .$$

$F(J)$ is large if $\rho(J, D_f)$ is small and $F(J)$ is small if $\rho(J, D_f)$ is large.

For the m-adic metric, $\rho(J_1, J_2) \leqslant 1$. Thus $F(J) \geqslant 1/2$. So F takes values in the segment $[1/2, 1]$. Here the sentence '$F(J)$ is very small' means that $F(J) \approx 1/2$ and '$F(J)$ is very large' means that $F(J) \approx 1$.

The control center CC must take into account not only the level of interest $I(J)$ of an idea J but also the level of interdiction $F(J)$ of an idea J. The struggle between interest $I(J)$ and interdiction $F(J)$ induces all essential features of human psychology.[18]

We consider a simple model of such a struggle. For an idea J, we define *consistency* (between interest and interdiction) as $T(J) = aI(J) - bF(J)$, where $a, b > 0$ are some weights depending on a cognitive system τ. Some τ could use more complex functionals for consistency. For example,

$$T(J) = aI^{\alpha}(J) - bI^{\beta}(J) , \tag{14.11}$$

where $\alpha, \beta > 0$, are some powers. There exists a threshold of realization T_{rz} such that if $T(J) \geqslant T_{\mathrm{rz}}$, then the idea J is sent to the collector Q for ideas waiting for realization. If $T(J) < T_{\mathrm{rz}}$, then the idea J is deleted.

It is convenient to consider a special block in the conscious domain, *analyzer*, AN_c.

This block contains the comparator COM_c which measures distances $\rho(J, D_i)$ and $\rho(J, D_f)$; a computation device which calculates measures of interest $I(J)$, interdiction $F(J)$ and consistency $T(J)$ and checks the condition $T(J) \geqslant T_{rz}$; a transmission device which sends J to Q or trash. The order in the queue Q is based on the quantity $T(J)$. It is also convenient to introduce a block SER_c, *server,* in the conscious domain which orders ideas in Q with respect to values of consistency $T(J)$.

We can again assume that there exists a threshold T_+ such that ideas J with $T(J) \geqslant T_+$ must be realized in any case. This threshold plays the important role in the process of the time evolution of consistency $T(t, J)$ of an idea J in Q: $T(t, J) = l(t - t_0)\ T(J)$, where the coefficient $l(t)$ decreases with increasing of t. Moreover,

[18] In this model the power of Ego over Id is essentially larger than in Model 2. Ego could punish both uninteresting and forbidden ideas.

$T(t, J) = T(J)$ if $T(J) \geqslant T_+$. We note that $T_+ \geqslant T_{rz}$. It can be that interest and interdiction evolve in different ways: $I(t, J) = l_i(t - t_0)I(J)$ and $F(t, J) = l_f(t - t_0)F(J)$. Here $T(t, J) = aI(t, J) - bF(t, J)$. Such a model is more realistic. A pessimist has quickly decreasing function $l_i(t)$ and slowly decreasing function $l_f(t)$. An optimist has slowly decreasing function $l_i(t)$ and quickly decreasing function $l_f(t)$.

Example 14.4 (Harmonic love). Let τ be a 'man' described by Model 3. The image J_0 of a woman γ is transformed by π_{sex} into the idea $J_{\text{love},\gamma}$. Suppose that as in Example 14.3, D_i contains L_{blond} and γ is blond. However, $J_{\text{love},\gamma}$ is not sent automatically to the queue of ideas waiting for realization. The idea $J_{\text{love},\gamma}$ must be compared with the database D_f of forbidden ideas. Suppose that the idea-image $G_{\text{tall}} =$ (tall woman) belongs to D_f. If γ is tall, then $F(J_{\text{love}})$ is quite large. The future of J_{love} depends on the value of the consistency functional $T(J)$ (the relation between coefficients a, b in (14.11) and values $I(J)$, $F(J)$). However, this process still does not induce doubts or mental problems. In this model Ego (as well as Super-Ego) controls strongly sexual drives coming from Id. It is a censor for wild sexual desires.

It seems that the consistency $T(J)$ does not determine the intensity of realization of J. An extremely interesting idea is not realized with strength that is proportional to the consistency magnitude $T(J)$. In fact, it is realized with strength that is proportional to the magnitude of interest $I(J)$. Moreover, larger interdiction also implies larger strength of realization. It seems natural to connect strength of realization with the quantity

$$S(J) = c\ I(J) + d\ F(J)\ ,$$

where $c, d > 0$ are some parameters of the brain.

We call $S(J)$ *strength* of an idea J. In particular, $S(J)$ may play an important role in memory processes. We introduce a preserving threshold S_{-}^{mem} (compare with the preserving threshold I_{-}^{mem} in model 2). The strength $S(t, J)$ of an idea J in the working memory evolves as

$$S(t, J) = l_{\text{mem}}(t - t_0)S(J),$$

where l_{mem} is a decreasing function. If $S(t, J) < S_{-}^{\text{mem}}$, then at the instance of time t the idea J is deleted from working memory.

The structure of analyzer is given by Figure 14.3.

The main disadvantage of the cognitive system τ described by Model 3 is that the analyzer AN_c permits the realization of ideas J which have at the same time very high levels of interest and interdiction (if $I(J)$ and $F(J)$ compensate each other in the consistency function). For example, let $T(J) = I(J) - F(J)$. If the realization threshold $T_{rz} = 0$ the analyzer AN_c sends to the collector Q totally forbidden ideas J (with $F(J) \approx 1$) having extremely high interest ($I(J) \approx 1$).

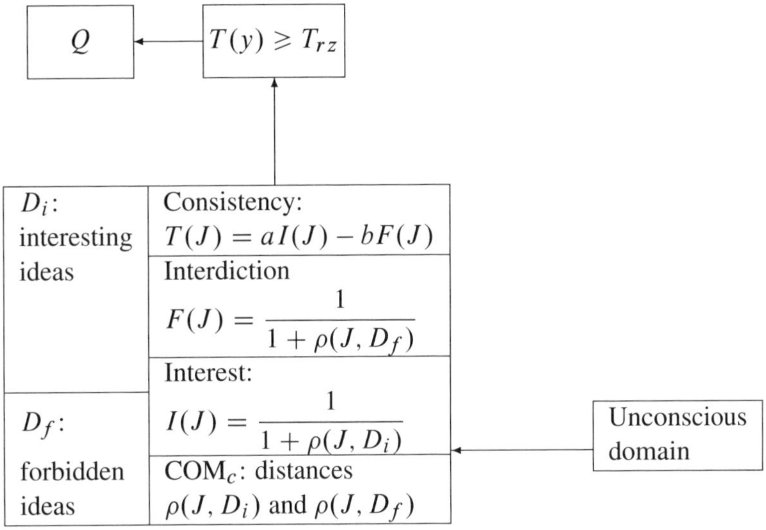

Figure 14.3. The structure of analyzer. A cognitive system τ described by Model 3 has complex cognitive behavior. However, this complexity does not imply 'mental problems'. The use of consistency functional $T(J)$ solves the contradiction between interest and interdiction.

Such a behavior ('a storm of cravings') can be dangerous (especially in a group of cognitive systems with a social structure). Therefore functioning of the analyzer AN_c must be based on more complex analysis of J which is not reduced to the calculation of $T(J)$ and testing $T(J) \geq T_{rz}$.

Model 4:

Suppose that a cognitive system described by Model 3 improves its brain by introducing two new thresholds I_{max} and F_{max}. If $I(J) \geq I_{max}$, then the idea J is extremely interesting: τ can not simply delete J. If $F(J) \geq F_{max}$, then an idea J is strongly forbidden: τ can not simply send J to Q.

If J belongs to the *'domain of doubts'*

$$O_d = \{J : I(J) \geq I_{max}\} \cap \{J : F(J) \geq F_{max}\}$$

the τ cannot take automatically (on the basis of the value of the consistency $T(J)$) the decision on realization of J.

Example 14.5 (Forbidden love). Let τ be a 'man' described by Model 4. Here the image J_0 of a woman γ contains not only the spatial image of γ, but also her social image. Suppose that the integral image J_0 is transformed by the thinking block π_{sex}

in the idea-attractor $J_{\rm love}$. Suppose that, as in all previous examples, the image $L_{\rm blond}$ belongs to D_i. Suppose that idea $G_{\rm soc}$ = (low social level) belongs to D_f. Suppose that both $\rho(J_{\rm love}, L_{\rm blond})$ and $\rho(J_{\rm love}, G_{\rm soc})$ are very small. She is blond and poor! So $I(J_{\rm love}) \geqslant I_{\max}$ (high attraction of the woman γ for the τ) and at the same time $F(J_{\rm love}) \geqslant F_{\max}$ (social restrictions are important for the τ). In such a situation the τ cannot take any decision on the idea $J_{\rm love}$: "To love or not to love?!" Here Ego demonstrates its power.

14.10 Hidden forbidden wishes, psychoanalysis

On the one hand, the creation of an additional block in analyzer $\mathrm{AN}_{\rm c}$ to perform $(I_{\max}, F_{\max})$ analysis plays the positive role. Such a τ does not realize automatically (via condition $T(J) \geqslant T_{rz}$) dangerous ideas J, despite their high attraction. On the other hand, this step in the cognitive evolution induces hard mental problems for τ. In fact, the appearance of the domain of doubts O_d in the mental space is the origin of many psychical problems and mental diseases.

Let the analyzer find that an idea-attractor belongs to the domain of doubts – forbidden wish (desire, impulse, experience).[19] The brain is not able neither to realize such an idea nor simply to delete it.The control center CC tries to perform further analysis of such a J. CC sends J to the processing domain Π as the initial problem for some processor π^1. If it produces an idea-attractor J^1 which does not belong to O_d, then the τ can continue normal cognitive functioning. However, if π^1 produces again an idea J^1 which belongs O_d, then CC must continue the struggle against this doubtful idea. In the process of such a struggle CC and some processors $\pi, \pi^1, \pi^2, \ldots$ are (at least partially) busy. An essential part of mental resources of τ is used not for reactions to external signals, but for the struggle with ideas J belonging to O_d.

Typically this is a struggle with just one *idèe fixe* J, see [141, 142].

We can explain the origin of such an idèe fixe by our cognitive model. If an idea J belonging to O_d has been produced by the processor π, then it is natural that CC will try again to use the same processor π for analyzing the idea J. As $f_\pi(J) = J$ (the J is a fixed point of the map f_π), then π starting with J will always produce the same idea J (with the trivial sequence of iterations $J, J, \ldots, J$). In general the doubtful idea J can be modified by CC (for example, on the basis of new information), $J \to J_{\rm mod}$. An idea $J_{\rm mod}$ can be considered as a perturbation of J: $\rho(J, J_{\rm mod}) < s$, where s is some constant. If $s > 0$ is relatively small (so that $J_{\rm mod}$ still belongs to the basis of attraction of J), then iterations $J_{\rm mod}, J^1_{\rm mod} = f_\pi(J_{\rm mod}), \ldots, J^N_{\rm mod} = f^N_\pi(J_{\rm mod}), \ldots$ again converge to the J.

How can CC stop this process of the permanent work with idèe fixe J?

[19] "All these experiences had involved the emergence of a wishful impulse which was in sharp contrast to the subjects other wishes and which proved incompatible with the ethical and aesthetic standards of his personality," Freud [142].

The answer to this question was given in [141, 142]: investigations of roots of hysterias and some other mental problems. By Freud idèe fixe J is shackled by CC into the unconscious domain.[20]

In our model, the unconscious domain contains (besides the processing domain Π and the unconscious control center UC) a special collector D_d for doubtful ideas, *forbidden wishes*. By Freud's model it is also a part of Ego (but an unconscious part).

After a few attempts to transform an idea J belonging to the domain of doubtful ideas O_d into some non-doubtful idea, CC sends J to D_d. We remark that the domain O_d is a mental domain (a set of ideas) and D_d is a 'hardware domain' (a set of chains of neurons used for saving of doubtful ideas).

What can we say about the further evolution of a doubtful idea J in the collector D_d? It depends on a cognitive system τ (in particular, a human individual). In the 'purely normal case' the collector D_d plays just the role of a *churchyard for doubtful ideas*. Such a D_d has no output connections and idea J will disappear after some period of time.

However, Freud demonstrated (on the basis of hundreds of cases) that advanced cognitive systems (such as human individuals) could not have 'purely normal behavior'. They could not perform the complete interment of doubtful ideas in D_d.

In our model, the collector D_d has an output connection with the unconscious control center UC. At this moment the existence of such a connection seems to be just a disadvantage in the mental architecture of τ. It seems that such a cognitive system τ was simply not able to develop a physical system for 100%-isolation of the collector D_d. However, later we shall demonstrate that the pathway

$$CC \to D_d \to UC \to CC \tag{14.12}$$

has important cognitive functions. In fact, such a connection was specially created in the process of evolution. But we start with the discussion on negative consequences of (14.12). Here we follow [141, 142].

In our mathematical model of Freud's theory of unconscious mind, an idea $J \in D_d$ is sent to UC. The unconscious control center UC sends J to one of the thinking processors ξ in Π. ξ performs iterations starting with J as an initial idea. ξ produces an idea-attractor $\tilde{J} = \lim_{N\to\infty} J_N$, $J_0 = J$. In the simplest case ξ sends the idea-attractor $\tilde{J}$ to the conscious domain. The AN_c analyzes idea $\tilde{J}$. If $\tilde{J} \notin O_d$, then AN_c

[20] "There had been a short conflict, and the end of this internal struggle was that the idea which had been appeared before consciousness as the vehicle of this irreconcilable wish fell a victim to repression, was pushed out of consciousness with all its attached memories and was forgotten. Thus the incompatibility of the wish in question with the patient's ego was the motive for the repression: the subject's ethical and other standards were the repressing forces. An acceptance of the incompatible wishful impulse or a prolongation of the conflict would have produced a high degree of unpleasure; this unpleasure was avoided by means of repression, which was thus revealed as one of devices serving to protect the mental personality," Freud [142].

sends $\tilde{J}$ to the collector Q (of ideas waiting for realization).[21] After some period of waiting $\tilde{J}$ is sent to realization.[22] By such a realization CC deletes J from the collector Q. However, CC does not delete the root of $\tilde{J}$, namely J, because J is located in the unconscious domain and CC is not able to control anything in this domain. The idea $\tilde{J}$ is nothing other than a performance of the forbidden wish J. Such unconscious transformations of forbidden wishes were studied by Freud (see [140–142], for examples).

We note that if UC sends a hidden forbidden wish J to the same thinking processor π which has already generated J for CC, then (by the same reasons as in our previous considerations) CC will again obtain the same doubtful idea J. Such a continuous reproduction of Idea fixe can take place. This is the root of obstinate doubtful wishes. This can imply mental diseases, because CC could not stop the struggle with Idea fixe even by sending it to D_d. However, UC may send J to other thinking processor $\xi \neq \pi$. Here the idea-attractor $\tilde{J}$ (which has been produced starting with J as the initial condition) differs from J. This is the real transformation of the forbidden wish. In general a new wish $\tilde{J}$ has no direct relation to the original wish J. This is nothing than a *symptom* of cognitive system τ, Freud [140, 141].

14.10.1 Hysteric reactions

In general a doubtful idea $J \in D_d$ is not only transferred into some symptom $\tilde{J}$, but it may essentially disturb functioning of the brain. Some thinking blocks π_Π are directly connected to other thinking blocks. Suppose that, for example, the following pathway is realized: $J \in D_d \to UC \to \pi_\Pi \to \pi_c \to CC$. Suppose also that ideas λ produced by π_Π play the role of parameters for the block $\pi_c : x_{n+1} = f_{\pi_c}(x_n, \lambda)$. Let CC obtain an image J_0 and send it to π_c. However, instead of the normal value of the parameter λ, the π_Π sends to π_c some abnormal value λ_{ab} induced by the hidden forbidden wish J. The π_c produces an attractor L_{ab} which may strongly differ from the attractor L_{norm} corresponding to λ_{norm}, the value of the parameter produced by the processor π_Π for the processor π_c in the absence of the hidden forbidden wish J.

In such a way we explain, for example, *hysteric reactions*. A rather innocent initial stimulus J_0 can induce via interference with a doubtful idea $J \in D_d$ inadequate performance L_{ab}. We can also explain why hidden forbidden wishes may induce *physical diseases*. Attempting to transform $J \in D_d$ into an idea which does not induce doubts and reflections, UC can send J to some thinking processor π_{phys} that is responsible for some physical activity of τ. We note that UC considers J as just a collection of mental states. This collection of mental states has different interpretations in different think-

[21] Of course, there may exist more complex pathways: $CC \to D_d \to UC \to \xi \to \xi^1 \cdots \to \xi^m \to UC \to \lambda \to \lambda^1 \to \cdots \to \lambda^k \to AN_c \to Q \to CC$, where $\xi, \dots, \xi^m, \lambda, \dots, \lambda^k$ are some thinking processors.

[22] Of course, idea $\tilde{J}$ may be simply deleted in Q if there are too many ideas in the queue and the strength $S(\tilde{J})$ of $\tilde{J}$ is not so large.

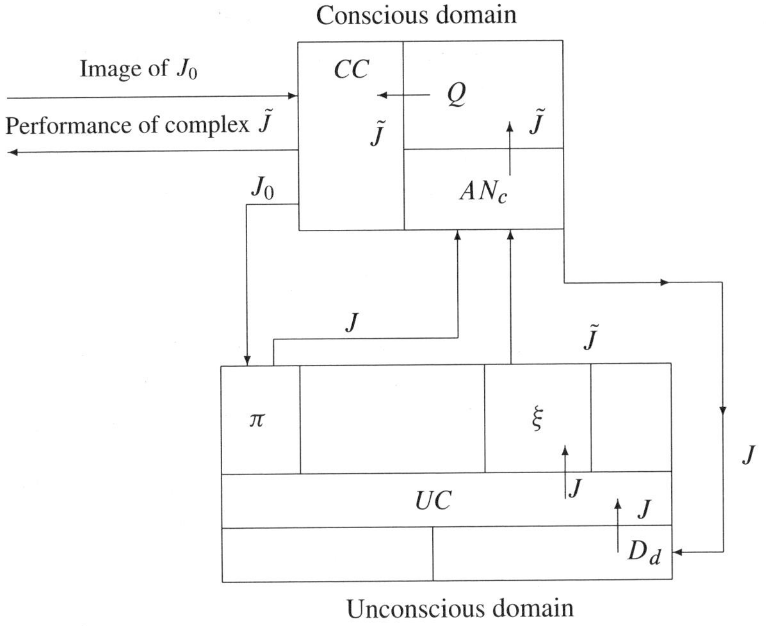

Figure 14.4. Symptom induced by a forbidden wish. Starting with an initial idea J_0 a processor π produces an attractor J; analyzer AN_c computes quantities $I(J)$, $F(J)$ (measures of interest and interdiction for the idea J); AN_c considers J as a doubtful idea: both measures of interest and interdiction are too high, $I(J) \geqslant I_{max}$, $F(J) \geqslant F_{max}$; AN_c sends J to the collector of doubtful ideas D_d; J moves from D_d to UC; UC sends it to some processor ξ; ξ produces an attractor $\tilde{J}$. Analyzer AN_c can recognize $\tilde{J}$ as an idea which could be realized (depending on the distances $\rho(\tilde{J}, D_i)$ and $\rho(\tilde{J}, D_f)$) and send $\tilde{J}$ via the collector Q to realization. This $\tilde{J}$ is a *symptom* induced by J (in fact, by the initial idea J_0).

ing systems. In particular, J can correspond in π_{phys} to some 'bad initial condition'. The corresponding attractor L_{phys} can paralyze the physical function ruled by π_{phys}.

14.10.2 Feedback control based on doubtful ideas

A cognitive system τ wants to prevent a new appearance of forbidden wishes J (collected in D_d) in the conscious domain. The brain of τ has an additional analyzer AN_d (located in the unconscious domain) that must analyze nearness of an idea attractor L produced by some processor π_c and ideas J belonging to the collector of doubtful ideas D_d.

In our model, it is supposed that each hidden forbidden wish J_{fd} in the collector D_d still remembers a thinking block π which has produced J_{fd}. This simply means that each idea J_{fd} in the D_d has the label π. Thus $J_{fd} = J_{fd}(\pi) \in D_d$ is not just a collection of mental states. There is information that these states are related to the dynamical system π. The set of doubtful ideas O which are collected in the collector D_d can be split into subsets $O(\pi)$ of forbidden wishes corresponding to different thinking systems π. $\mathrm{AN_d}$ contains a comparator COM_d that measures the distance between an idea-attractor J which has been produced by a thinking block π and the set $O(\pi)$: $\rho(J, O(\pi)) = \min_{J_{\mathrm{fd}} \in O(\pi)} \rho(J, J_{\mathrm{fd}})$. Then $\mathrm{AN_d}$ calculates the measure of interdiction

$$F_d(J) = \frac{1}{1 + \rho(J, O(\pi))}.$$

If $F_d(J)$ is large (≈ 1), then an idea J is too close to one of former hidden forbidden π-wishes. This idea should not be transmitted to the conscious domain.

Each individual τ has its own *blocking threshold* F_{bl}: if $F_d(J) < F_{\mathrm{bl}}$, then J is transmitted; if $F_d(J) \geqslant F_{\mathrm{bl}}$, then J is deleted. In the latter case J will never come to the conscious domain.[23] This threshold F_{bl} determines the degree of blocking of the thinking processor π by forbidden wishes. For some individuals (having rather small values of F_{bl}), a forbidden wish J_{fd} belonging to the set $O(\pi)$ may stop the flow of information from π to the conscious domain. The same J_{fd} may play a negligible role for individuals having rather large magnitude of F_{bl}. Therefore the blocking threshold F_{bl} is one of the important characteristics to distinguish normal and abnormal behaviors. We note that F_{bl} depends on a thinking block π: $F_{\mathrm{bl}} = F_{\mathrm{bl}}(\pi)$. Thus the same individual τ can have the normal threshold for one thinking block π, relatively large $F_{\mathrm{bl}}(\pi)$, and abnormal degree of blocking for another thinking block π', relatively small $F_{\mathrm{bl}}(\pi')$.

14.11 Neuro and mental cybernetic bases for the pleasure and reality principles

The *pleasure principle* is a psychoanalytical term coined by Sigmund Freud. Respectively, the desire for immediate gratification versus the deferral of that gratification. Quite simply, the pleasure principle drives one to seek pleasure and to avoid pain. We shall present the neurocybernetic justification of this principle. It is convenient to consider the evolution of the pleasure-function through development from Model 1 to Model 4.

We start with Model 2. In this model pleasure is identified with the interest-measure. This mental quantity takes its values in the segment [0.5, 1]. Thus we quantified pleasure. The value 0.5 corresponds to minimal pleasure and the value 1 to maximal pleasure. This quantity of interest-pleasure is the basis for ordering of ideas attractors for their realization. The brain wants most the ideas having the highest magnitude of plea-

[23] Analysis in the conscious domain could demonstrate that $T(J) \geqslant T_{\mathrm{rz}}$ and $I(J) < I_{\max}$, $F(J) < F_{\max}$. In the absence of hidden forbidden wishes J would be realized.

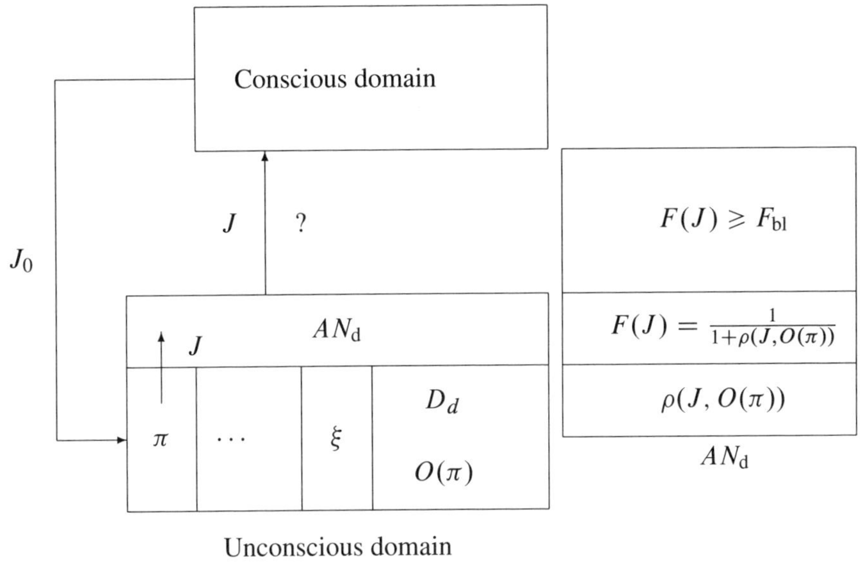

Figure 14.5. Interference of an idea-attractor with the domain of hidden forbidden wishes. Internal structure of the analyzer AN_d. Analyzer AN_d computes the distance between the idea-attractor J (produced by a thinking block π) and the domain $O(\pi)$ of hidden forbidden π-wishes. If this distance is relatively small, i.e., the measure of interdiction $F_d(J)$ is relatively large, then J does not go to the conscious domain.

sure. Such ideas have the highest priority in realization.[24] Moreover, ideas inducing not so much pleasure (e.g., pleasure $= 0.55$) might be never realized, because they might just disappear from the collector of ideas waiting for realization. Thus a brain based on Model 2 would like to maximize the pleasure-function which is defined on the space of ideas. We recall that the interest-measure increases with the decreasing of the distance from an idea-attractor to the interest-database.

We now consider Model 3. Here a brain can calculate not only the interest-function on the space of ideas, but also the interdiction function. The purpose of the latter one is to prevent such a brain from conflicts with reality. Here we chose a pleasure-reality function is identified with the consistency-function. As was mentioned, the simplest form of the consistency function is simply the difference between the functions of interest and interdiction. Thus the greatest pleasure and at the same time the greatest consistency with reality is approached in the case of the highest interest and the lowest interdiction, e.g., the interest-measure $= 1$ and the interdiction-measure $= 0.5$,

[24]The feeling of pleasure is approached at the moment of realization. The strength of this feeling is determined by the magnitude of the interest-measure.

so the the pleasure-reality function takes the value 0.5. It is a good point to remark that the pleasure-reality function (given by the measure of consistency) depends on an individual. In general it is an arbitrary linear combination of interest and interdiction:

$$\text{PLEASURE-REALITY} = a\,\text{INTEREST} + b\,\text{INTERDICTION},$$

where a and b are some coefficients.

In Model 4 the pleasure-reality function is the same as in Model 3. Finally, we consider Model 1. Here we have only dynamical systems which process external and internal stimuli and produce ideas-attractors which play the role of reactions to those stimuli. Pleasure is approached by realizations of those ideas-attractors. Here Id totally dominates.

We now analyze deeper the structure of the pleasure function. As everything in our models this function is based on the mental distance. Therefore the pleasure principle as well as the reality principle are based on the metric structure of mental space. This metric structure is based on the hierarchic encoding of mental information. Thus on the mental information level these principles are consequences of a hierarchical representation of information by cognitive systems. There would be no pleasure (nor interdiction) without the mental hierarchy.

At the neuronal level in our models the mental hierarchy is based on hierarchical neuronal trees and producing information by hierarchically ordered neuronal pathways. Thus the presence of such neuronal pathways provides the neurophysiological basis of the pleasure principle as well as the reality principle.

14.12 Consequences for psychology and neuropsychology

Our model gives the possibility to perform mathematical simulation of psychological behavior. We performed geometrization of psychology, *geometro-psychology.* By introducing a mathematical model of mental space we incorporated psychology into the same rigorous mathematical framework as it was done in physics by Newton. The crucial point is that geometries of physical and mental spaces differ very much. The presence of the rigid hierarchical structure plays the fundamental role in the m-adic mathematical model of mental space. Hierarchical representations are well accepted in psychology. Our model provides the corresponding mathematical basis.

By coupling the m-adic mental space with neuronal trees we constructed a bridge between neurophysiology and psychology. Thus our model can be considered as a contribution to neuropsychology.

We applied the m-adic mental model to mathematical modeling of Freud's psychoanalysis. We aware about diversity of views on Freud's psychoanalysis in modern psychology, see, e.g., [148,307,417] as well as [159,387–389] for debates. Our model supports the view which was presented in the journal "Neuro-psychoanalysis":

It would be possible to create an ongoing dialogue with the aim of reconciling psychoanalytic and neuroscientific perspectives on the mind. This goal is based on the assumption that these two historically divided disciplines are ultimately pursuing the same task, namely, *"attempt[ing] to make the complications of mental functioning intelligible by dissecting the function and assigning its different constituents to different component parts of the [mental] apparatus,"* Freud [140, p. 536]. Notwithstanding the fact that psychoanalysis and neuroscience have approached this important scientific task from radically different perspectives, the underlying unity of purpose has become increasingly evident in recent years as neuroscientists have begun to investigate those "complications of mental functioning" that were traditionally the preserve of psychoanalysts. This has produced an explosion of new insights into problems of vital interest to psychoanalysis, but these insights have not been reconciled with existing psychoanalytic theories and models.

We can complete this manifest by the remark that neurophysiology has an essentially higher level of the mathematical representation than traditional psychoanalysis. Therefore coupling of psychoanalysis with neurophysiology provides new perspectives in mathematization of psychoanalysis. Our m-adic model serves precisely to such a purpose. Starting with a model of the neuronal structure of the brain, hierarchical neuronal trees, we created the m-adic model of mental space. This model was then applied to mathematical modeling of psychoanalysis. The m-adic distance on mental space is the basis of forming of measures of interest and interdiction and consequently hidden forbidden wishes and, finally, symptoms and hysteria. And this m-adic distance on mental space is induced by the neuronal structures – hierarchical neuronal trees. There would be no psychical problems without mental hierarchy in the brain. Psychical problems is the price for advantages of hierarchical processing of information in the brain.

14.13 Consequences for psychoanalysis

In our model the mental cybernetic realization of resistance to reappearance of hidden forbidden wishes in the subconscious (and then conscious) domain is based on blocking thresholds. At the deeper level of the model such a resistance is a consequence of the ability of a cognitive system to measure mental distance and calculate the measure of unconscious interdiction (resistance). The unconscious comparator by calculating the distance from an idea-attractor – trying to pass from the processing domain to the conscious domain – to the database of hidden forbidden wishes prevents the appearance in the conscious domain ideas which are similar to (in particular, coincide with) former hidden forbidden wishes.

The simplest possibility to overcome this resistance and to liberate a hidden forbidden wish (complex) which generates a symptom is to increase the magnitudes of blocking thresholds. We recall that in our model an idea-attractor gets the possibility to move to the conscious domain if its measure of unconscious interdiction is less than the blocking threshold. In this way even ideas having sufficiently large measures of unconscious interdiction (so small distances to the database of hidden forbidden wishes) may come to the conscious domain.

It seems that precisely this strategy was developed at the very beginning of psychoanalysis when Dr. Breuer and Dr. Freud had been using hypnosis to overcome the resistance force and to find unconscious roots of hysteric symptoms.[25] However, already at that stage of development of psychoanalysis Freud recognized the restrictiveness of this method.[26]

We now analyze our model to find another possibility to overcome the resistance force and to find a root of symptom. We remark that the cornerstone of our model is determination of the distance between two mental states by the *first* digits in their sequential representations. The distance decreases with increasing of the length of the initial common segment in those sequences. Therefore if more digits are taken into account then smaller distances can appear. Another important remark is that *sharper associations are determined by longer initial segments of coding sequences for mental states.*

This corresponds to activating of longer neuronal branches in a hierarchical neuronal tree serving for creation of an association. Thus "sharper thinking" can produce smaller mental distances. Hence larger measures of interdiction can be obtained for ideas-attractors trying to move from the unconscious to the consciousness. Such characteristic features of our neuro and mental cybernetic models motivate the following strategy for overcoming the resistance force and liberation of complexes.

If a patient creates free associations – rather fuzzy ideas (images), then he would automatically operate with ideas(images)-attractors having lower measures of unconscious interdiction (larger distances from the database of hidden forbidden wishes). Such free and fuzzy associations which are related to complexes (hidden forbidden wishes inducing symptoms) can pass blocking thresholds and arrive to the conscious domain. Hence the psychoanalyst by asking a patient to form free associations tries to remove the processing of information in a patient's brain into the regime of operating with non sharp images. In this way the psychoanalyst would obtain some ideas coming from the unconsciousness which contain associations related to the symptom under treatment. Therefore our model supports the fundamental strategy of psychoanalytic treatment which is based on forming of free associations, Freud [140–142] and explains its origin.

[25]"The cathartic procedure, as carried out by Breuer, presupposed putting the patient into a state of deep hypnosis; for it was only in a state of hypnosis that he attained a knowledge of the pathogenic connexions which escape him in his normal state," Freud [142].

[26]"But soon I came to dislike hypnosis ...", Freud [142].

14.14 Psycho-robots

As was pointed out, Models 1–4 can be immediately realized as intelligence systems – at least on the level of software. Such a project is of the great interest. It has not been realized only because it demands a lot of programming resources which were not available for the authors of this book. We consider the possibility to design advanced psycho-robots based on Model 4.

We do not plan to present here a detailed review on other approaches to psycho-robots. To emphasize differences of our approach from other developments of psycho-robots, we present a citation from the work of Potkonjak el al. [364]:

"Man-machine communication had been recognized a long time ago as a significant issue in the implementation of automation. It influences the machine effectiveness through direct costs for operator training and through more or less comfortable working conditions. The solution for the increased effectiveness might be found in user-friendly human-machine interface. In robotics, the question of communication and its user-friendliness is becoming even more significant. It is no longer satisfactory that a communication can be called 'human-machine interface', since one must see robots as future collaborators, service workers, and probably personal helpers."

In contrast, the main aim of our modeling is not at all creation of friendly helpers to increase their effectiveness. We would like to create AI-systems which would really have essential elements of human psyche. We have shown that already psycho-robots with a rather simple AI-psyche – two emotions and two corresponding data bases – would exhibit (if one really wants to simulate human's psyche) very complicated psychological behavior. In particular, they would create various psychical complexes which would be exhibited via symptoms. We also point out to the crucial difference of our "Freudian psycho-robots" from psycho-robots created for different computer game (psycho-automata). Our aims are similar of those formulated for humanoid robots, see, e.g., Brooks [78,79]. However, we jump directly to high level psyche (without to create e.g. the visual representation of reality). The idea of Luc Steels to create a robot culture via societies of self-educating robots, see, e.g., [310], is also very attractive for us. It is clear that real humanoid psyche (including complexes and symptoms) could be created only in society of interacting Psychots and people. Moreover, such AI-socictics of Psychots can be used for modeling psychoanalytic problems and development of new methodologies of treatment of such problems.

Conclusion. *A mathematical model for hierarchical encoding of mental information was created. It was strongly motivated by development of non-Archimedean theoretical physics. Mental space (a mental analog of physical space) is realized as an m-adic tree. Processing of mental information is realized by dynamical systems on such a tree. Interplay between unconscious and conscious information flows generates interesting psychological behavior.*

Chapter 15

Neuronal hierarchy behind the ultrametric mental space

In Chapter 14 we presented the simplest model of dynamical thinking based on the m-adic mental space, $X_{\text{mental}} = \mathbb{Z}_m$ – the ring of m-adic integers.[1] In this chapter we will present in detail a neuronal hierarchical model inducing mental hierarchy. It generalizes the model which was briefly given in Chapter 14. The model of Chapter 14 described only propagation of output signals moving from the basic neuron of a hierarchic neuronal tree. Now we describe a general model which contains both input and output signals. Corresponding the mental space extends space $X_{\text{mental}} = \mathbb{Z}_m$. A new mental space is given by $X_{\text{mental}} = \mathbb{Q}_m$. *Hierarchic neural pathways* – ordered chains of neurons – are considered as fundamental units of information processing. Since a neural pathway can go through various regions of the brain, it is a nonlocal structure. Mental states produced by such a hierarchic neural pathway are not localized in physical (Euclidean) space. This model gives a new insight on the problem of localization of psychological functions, cf. Antonio R. Damasio [95]: "One held that psychological functions such as language or memory could never be traced to a particular region of brain. If one had to accept, reluctantly, that the brain did produce the mind, it did so as a whole and not as a collection of parts with special functions. The other camp held that, on the contrary, the brain did have specialized parts and those pars generate separate mind functions."

Contrary to Chapter 14, complex mental images based on ensembles of exited neural pathways are described by probability distributions on mental space (and not by special ensembles such as ultrametric balls or their collections).

[1] From the number-theoretical viewpoint $\mathbb{Z}_m$ can be considered as generalization of the system of natural numbers $\mathbb{N}$. Natural numbers are given by polynomials with respect to m with coefficients belonging to $\{0, 1, \ldots, m-1\}$, while m-adic numbers were introduced by Hensel as power series with respect to m. Thus our model of the mental space matches well with attempts of Platon and Aristotle to associate ideas with natural numbers. Moreover, the ultrametric geometry on $\mathbb{Z}_m$ matches well with Aristotle's views on discreteness of the spiritual world.

15.1 Hierarchic neural pathways

We will consider a model of the process of thinking based on the neural pathway representation of cognitive information, *Neural Pathway Model.* In our model the elementary unit of cognitive information is given not by the frequency of firing of an individual neuron, but by a *string of firings of neurons* along a pathway of neurons. Such a string of firings along a pathway we call *a mental point.* We shall use the symbol X_{mental} to denote space of mental points, *mental space.* One should sharply distinguish a neural pathway as a physical and biological structure and the string of its firings as a purely informational structure.

In our model each *psychological function* is based on a tree of neural pathways, *neuronal tree,* that is *centered* with respect to one fixed neuron. Thus basic units of processing of mental information are centered neural pathways and basic units of mental information are centered strings of firings produced by centered neural pathways.

Figure 15.1. Centered pathway.

Centring determines a *hierarchic structure* on the neuronal tree and on the corresponding space of mental points.

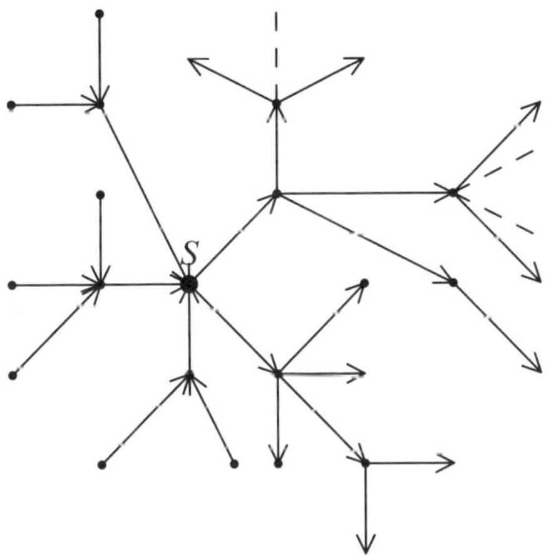

Figure 15.2. Neuronal tree.

A centering neuron S need not be considered as a kind of a *grandmother neuron.* It simply determines a system of mental coordinates (corresponding to the concrete psychological function) on the neural system of a cognitive system. Of course, the model of a psychological function based on one-neuron centrring is oversimplified. A complex psychological function is based on a few neuronal trees centered with respect to an *ensemble of neurons.*

The centering hierarchic structure on a neuronal tree induces hierarchy on each neural pathway of this tree. As in Chapter 14, this hierarchy induces a tree-structure on the space of mental points – strings of firings – which can be produced by a hierarchic neural pathway.

In Chapter 14, we considered a dynamical system, *feedback process*, in the mental space X_{mental}. Mathematically such a dynamical system was described by a map $f : X_{\text{mental}} \to X_{\text{mental}}$ that mapped strings of firings along pathways into strings of firings along pathways. In such a simplest model the process of thinking was described by a mathematical law

$$x_{n+1} = f(x_n), \tag{15.1}$$

where x belongs to the mental space X_{mental}. Dynamics of special collections of mental points – strings of firings – are of great importance, since they describe flows of associations (ultrametric balls) and ideas (collections of balls).

In this chapter we consider a *probabilistic representation* of informational flow in the brain. The new model will be also based on the m-adic mental space. However, it is not more assumed that cognitive meaning can be associated with a special symbol, or pattern of neural activation, or even result of coupling of various neural networks. Cognitive meanings are assigned to probability distributions on mental space. In particular, our feelings are feelings of probabilities.

The mental process is described as a stochastic process performing body $\to$ mind relation.[2] The evolution of a *statistical mental state* – a probability distribution of this process[3] – can be described (at least for simple psychological functions) as diffusion on an ultrametric m-adic tree: *thinking as ultrametric diffusion.* Psychological, neurophysiological, cognitive and philosophic consequences of the probabilistic thinking model will be discussed in the next section.

15.2 Model: thinking on neuronal tree

15.2.1 Mental field on the brain

Localization of psychological functions

One of the strong sides of the Neural Pathway Approach is a new viewpoint to the problem of localization of psychological functions. Since an elementary information unit of mental processing is represented by a pathway and a pathway can go through various domains of the brain and even the body, it is impossible to localize a psychological function in the Euclidean geometry of physical space. On the other hand, a psychological function can be localized in the space of all pathways. In fact, this is a

[2] We remark that in our model it is impossible to provide the mathematical description of mind $\to$ body relation. The mapping from the neuronal world to the informational world is not invertable.

[3] Such a random field can be considered as a mathematical representation of Whitehead's *field of feeling* [411].

kind of hierarchic localization – compare to A. Damasio: "What determines the contribution of a given brain unit to the operation of the system to which it belongs is not just the structure of the unit but also its place in the system. ... The mind results from the operation of each of the separate components, and from the concerted operation of the multiple systems constituted by those separate components," [95, p. 15]. In our model there is even no place for "separate components"; everything is unified from the beginning due to the pathway representation of cognitive information.

We have to distinguish the space T_{neuronal} of all pathways, chains of neurons, in the physical brain and body and the space X_{mental} of all possible mental points that can be produced by elements of T_{neuronal}. In principle, a few distinct elements of T_{neuronal}, pathways, can produce (at some instant of time) the same element of X_{mental}, a mental point. Moreover, it should be the case, since it would be very dangerous for a cognitive system to base the information representation of an important mental point on a single neural pathway. Thus mental points should be represented by ensembles of neural pathways. Such a *multiplicity* of production of a mental point by neural pathways is one of the main distinguishing feature of our model of the process of thinking, see further considerations, in Subsection 15.2.2, on the probabilistic structure of statistical mental states. In the Neural Pathway Approach it is not the end of the story about localization of a psychological function. The crucial point of our considerations is the following:

The most natural (even beautiful!) mathematical model based on m-adic geometry on the mental space X_{mental} is obtained under the assumption that each pathway contains a *Central Neuron*, say S. By choosing the definite central neuron, we obtain the hierarchic structure on the space X_{mental} produced by such a centered neural pathway.

Remark 15.1 (Centered trees as coordinate systems). Selection of a central neuron S in a neuronal tree is simply selection of center of a coordinate system in the m-adic mental space. By fixing a system of coordinates we fix a psychological function which is realized by this centered neuronal tree. It is important to point out that we do not claim that there exist a kind of the absolute central neuron or a group of neurons that 'rule' all mental processes. Our geometric model of mental processing is not similar to the model of physical processes based on Newtonian absolute space. Our model is closer to models of relativity theory. On the same neuronal tree the brain can select – depending on context – various central neurons (in reality groups of neurons) and in this way realize various psychological functions.

We now turn to localization of psychological functions. In our model complete Euclidean localization of a psychological function is impossible. However, the central neuron S of the tree of pathways representing a psychological function determines partial localization.

The simplest Neural Pathway Model is based on one fixed centered pathway. Of course, such a model is oversimplified. From neurophysiological point of view it would be more natural to suppose that a psychological function (in an advanced cognitive system) is based not on a single centered pathway, but on a system of such pathways. In the simplest case all these pathways are centered with respect to the same central neuron S.

Body → mind field

Firings of neurons along pathways of the neuronal tree produce elementary mental points involved in the realization of a psychological function. We denote a psychological function by the symbol f and a neuronal tree used by f by the symbol $T_{\text{neuronal}}(f)$. How can functioning of f be represented mathematically? Such a representation has different levels. At the basic level we should provide the description of 'body → mind' correspondence. This correspondence is described by a function

$$\varphi : T_{\text{neuronal}}(f) \to X_{\text{mental}}$$

that maps neural pathways into mental points represented by these pathways: $z \in T_{\text{neuronal}}(f) \to x = \varphi(z) \in X_{\text{mental}}$. We call the map $\varphi(z)$ *body → mind*.[4]

The psychological function f performs the evolution of the field φ. Starting with the initial field $\varphi_0(z)$, f produces a time-dependent field $\varphi(t, z)$.

We have to consider the very important problem of interpreting the evolution parameter, 'time', t; in particular, relation between physical and psychological time. We shall discuss this problem in Subsection 15.3.3. At the moment we consider the discrete time evolution, $t = t_n = 0, 1, 2, \ldots$ without further debating. By taking into account the process of wholeness of thinking[5] we describe functioning of f by an integral operator with the kernel $K(z, y)$:

$$\varphi(t_{n+1}, z) = \int_{T_{\text{neuronal}}(f)} K(z, y)\varphi(t_n, y)dy, \tag{15.2}$$

where integration is performed over the neuronal tree, $T_{\text{neuronal}}(f)$. We notice that neither the space of pathways $T_{\text{neuronal}}(f)$ nor the space of mental points X_{mental} are Euclidean. Integration is performed over the ultrametric space, the kernel $K(z, y)$, the mental field $\varphi(t_n, y)$ and the measure dy take values in $\mathbb{Q}_m$. Theory of integration of functions valued in non-Archimedean fields and defined on zero-dimensional topological spaces was developed by Mona and Springer [322, 323].

The form of the kernel $K(z, y)$ is determined by the psychological function f. We notice that mental evolution (15.2) is represented by a linear integral operator in the space of body → mind fields. In principle, we can consider more general, nonlinear models. However, the model with summation over the whole tree with a weight function $K(z, y)$ looks very natural.

[4] Of course, φ depends on the psychological function f: $\varphi = \varphi_f$.

[5] The whole neuronal tree $T_{\text{neuronal}}(f)$ is involved in the production of the mental point $x = \varphi(t_{n+1}, z)$ by the neural pathway $z \in T_{\text{neuronal}}(f)$. In general operating of z can depend on operating of any $y \in T_{\text{neuronal}}(f)$.

Thus we propose the following model of thinking:

Each psychological function f is based on a tree of neural pathways, neuronal tree $T_{\mathrm{neuronal}}(f)$. The neuronal tree has the hierarchic structure corresponding to the central neuron S of this tree. The elementary unit of information is given by a string of firings of neurons along a pathway, a branch of the tree. Various coding systems based on strings of firings through neural pathways can be proposed.

Firing/off (2-adic) coding

For each instant of time t, we assign to a neuron 1, if the neuron is firing, and 0, otherwise.

Mathematically a mental point which is produced by neurons belonging to a hierarchic pathway is represented by a sequence of zeros and ones. Each sequence is centered with respect to the position corresponding to firings of the central neuron S.

Let us consider the geometric structure of the mental space X_{mental} corresponding to firing/off coding. Here each centered pathway produces a centered sequence of zeros and ones. The most important digit in a sequence gives the state, 0 or 1, of the central neuron S.

Hierarchy on the mental space is based on the exceptional role that is played by the central neuron. This hierarchy induces 2-adic topology on the mental space.

In mathematical modeling it is convenient to consider infinitely long neural pathways and corresponding information strings, mental points (this is just a mathematical idealization). Consider the 2-adic distance ρ_2 between two mental points

$$x = (\ldots x_{-l} \ldots x_0 \ldots x_k \ldots) \text{ and } y = (\ldots y_{-l} \ldots y_0 \ldots y_k \ldots),\ x_{\pm j}, y_{\pm j} = 0, 1.$$

We use index 0 for the state x_0 of the central neuron S, negative indexes for states of neurons that produce inputs propagating to the S through the ordered neural chain, positive indexes – for states of neurons that are receivers of S-output. First suppose that all input states coincide: all $x_{-l} = y_{-l}$. Let $l \geq 0$ be the first index such that $x_l \neq y_l$. Then by definition

$$\rho_2(x, y) = \frac{1}{2^l}. \tag{15.3}$$

If two neural pathways z and w produce strings x and y having the same input part, then $\rho_2(x, y)$ goes to zero if the length l of the common output part goes to infinity.

Suppose now that input parts are different. Let l be the first index (if we go from the left hand side) such that $x_{-l} \neq y_{-l}$. Then by definition

$$\rho_2(x, y) = 2^l. \tag{15.4}$$

Larger common initial input part implies shorter distance between mental points. Consider two neural pathways starting at e.g. a sensory receptor. Suppose that states (e.g. on/off) of initial neurons in these pathways are distinct. Then the distance between corresponding mental points is large. It is important to remark that in such a situation longer pathways induce larger distance between mental points.

Thus firing/off coding in combination with hierarchy on neural pathways induces mental space $X_{\text{mental}} = \mathbb{Q}_2$. In Chapter 14 we used the mental space $X_{\text{mental}} = \mathbb{Z}_2$ (geometrically it is the unit ball of $\mathbb{Q}_2$). This space was extended to $X_{\text{mental}} = \mathbb{Q}_2$ for taking into account unput signals.

m-adic coding based on frequency of neural firing

In general m-adic coding can be induced by the frequency coding. We assign to each neuron in a pathway the frequency of firings. Frequencies of firing are a better basis for the description of processing of information by neurons than a simple on/off. This has been shown to be the fundamental element of neuronal communication in a huge number of experimental neurophysiological studies (see, e.g., [171, 172] on mathematical modeling of brain functioning in the frequency domain approach). In the mathematical model it is convenient to consider a discrete set of frequencies: $0, 1, \ldots, m-1$, where m is some natural number. Here frequency is the number of output spikes produced by a neuron during some unit of psychological time (some period Δ of physical time, see Subsection 15.3.3 for detailed consideration). Thus mathematically a mental point is represented by a sequence of numbers belonging to the set $\{0, 1, \ldots, m-1\}$. Information is not homogeneously distributed along such sequences. The presence of the central neuron S in the neuronal tree $T_{\text{neuronal}}(f)$ induces a hierarchic structure for elements of an informational sequence.

This system of frequency coding along hierarchic neural pathways should be justified by neurophysiological studies. At the moment there are no experimental technologies that provide the possibility for measuring firings of neurons along even one long pathway of individual neurons. To confirm our pathway-coding hypothesis, we have to measure simultaneously firings of neurons for a huge ensemble of neural pathways.

By repeating considerations which have been presented for firing/off coding we see that for the frequency coding mental space is given by one of the m-adic trees, $X_{\text{mental}} = \mathbb{Q}_m$.

We point out that the system of coding and not the topological structure of a neuronal tree determines the structure of the corresponding mental space, see also Section 15.3.3. Totally different neuronal trees can produce the same mental tree. For example, let us consider 2-adic, firing/off, coding. The trees a, b, c on Figure 15.3 produce the same, 2-adic, mental space.

In principle, we can consider our Neural Pathway Model as an approximation of the TNGS-model, see Edelman [116] on *Theory of Neural Groups Selection.* To combine TNGS with our model, we should consider hierarchic chains whose basis elements are not individual neurons, but groups of neurons. Such a combination of our model and Edelman's model can be called *Neuronal Group Pathway Model.* In this model states of neuronal groups are coded by natural numbers $x_j \in \{0, 1, \ldots, m-1\}$.

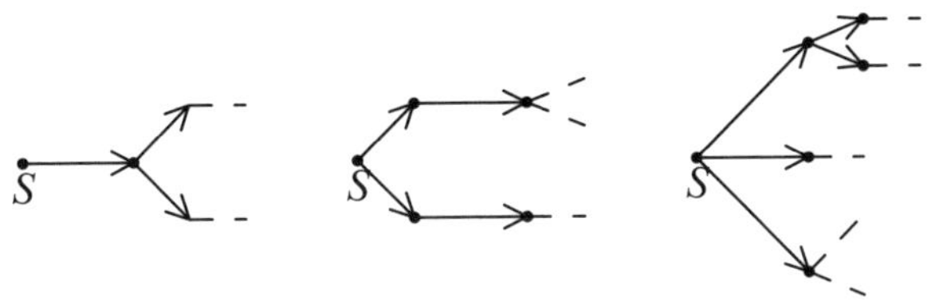

Figure 15.3. Neuronal trees: a, b, c.

Main cognitive features of the model

a) *Nonlocality (with respect to Euclidean geometry) of psychological functions.*

b) *Wholeness: integral evolution of body→mind field* φ.

c) *Sensation-thinking.* Since neural pathways go through the whole body, a part of a pathway involved in a high level psychological function can be connected to e.g. skin-sensitivity. Thus high order psychological functions also depend on various physiological stimuli.

d) *Interrelation of distinct psychological functions.* The central neuron S of a neuronal tree plays the role of center of the system of coordinates. Other neurons can also be considered as such centers. Therefore the same pathway contributes to distinct psychological functions. Thus evolution of various psychological functions is simultaneous evolution based on huge interrelation of corresponding neuronal trees, see Subsection 15.3.3.

e) *Emotion based reasoning.* Our pathway thinking model supports the fundamental conjecture of A. Damasio [95] that emotions play an important role in the process of 'reason-thinking'. Pathways going through centers creating emotions can participate in a psychological function of a high order thinking process. On the other hand, pathways going through reasoning-centers can go through some emotional center. Thus reason participate in creation of emotions and vice versa.

15.2.2 Probabilistic dynamics in the mental space

Statistical mental state

The body → mind field $\varphi(z)$ describes important features of functioning of the neural system (in particular, its part located in brain). However, we will not be concentrated on the study of dynamics, e.g. (15.2), of the body → mind field. The main reason is that $\varphi(z)$ describes merely the production of information by the neural system (in particular, its part located in brain) and not the flow of mental information by itself. The following thesis matches well with our model.

Mental Thesis. *The cognitive meaning (with respect to a psychological function f) of a mental point does not depend on a neural pathway that produces this mental point.*

The mental activity is performed not in space of neural pathways, in a neuronal tree, but in mental space. Thus mental information does not remember its neurophysiological origin. Mental Thesis is supported by experimental evidences that in some cases functions of some damaged parts of brain can be taken by other parts of brain, see, e.g., [90, 95, 116, 146]. This thesis is also supported by neurophysiological evidences that very different neural structures in brains of different species (e.g. fish and rat, [90]) can fulfill the particular psychological function.[6]

One might consider Mental Thesis as anti-materialist thesis. We would not like to be at such a position. We understand well that the relation between the brain (in fact, in our pathway model – the whole body) and mind plays the crucial role in mental activity. Mental Thesis should be considered as directed against the individual deterministic coupling between functioning physical neural pathways and the cognitive meaning of corresponding mental points. Violation of such individual determinism does not contradict to statistical determinism:

Thesis of Statistical Pathway Cognition. *The cognitive meaning (with respect to a psychological function f) of a mental point is determined by probability of production of this point in the ensemble of pathways $T_{\text{neuronal}}(f)$, the neuronal tree corresponding to f.*

Probability is considered as statistical probability, so it has nothing to do with "potentiality" or subjective probability:

Let $\mathcal{E}$ be a large ensemble of e.g. systems. States of these systems are represented by points of some space – state space. To simplify considerations, we consider the discrete state space. Probability of a point x with respect to the ensemble $\mathcal{E}$ is given by the proportion:

$$\mathbf{p}(x) = \frac{\text{the number of systems having the state } x}{\text{the total number of elements in the ensemble}}.$$

In our model systems are centered neural pathways; states are strings of firings along pathways – mental points. The ensemble $\mathcal{E}$ is a neuronal tree; the state space is the mental space. We suppose[7] that the cognitive meaning of a mental point $x \in X_{\text{mental}}$ is determined by the quantity

$$\mathbf{p}(x) = \frac{\text{the number of neural pathways that produce } x}{\text{the total number of neural pathways in the neuronal tree}}.$$

Computed in such a way $\mathbf{p}(x)$ is a possible realization of a *statistical mental state*. We remark once again that the mechanism of production of probability distributions of mental space is still unknown. It may be more complicated than considered above.

[6]Of course, we should recall that by choosing the central neuron S we chose the concrete psychological function f. Thus 'the cognitive meaning' is related to this concrete psychological function. By choosing another psychological function (a system of coordinates) we get another cognitive meaning.

[7]This is only a conjecture.

Here and in all following considerations it is assumed that a psychological function f is fixed. In fact, $\mathbf{p}(x)$ depends on f: $\mathbf{p}(x) = \mathbf{p}_f(x)$.

Mental evolution

In our probabilistic model mental processes are evolutions of statistical mental states. We have to find a mathematical model that would provide the adequate description of the evolution: $t \to \mathbf{p}(t, x)$.

Mental processes can be described as discrete time evolutions, see Subsection 15.3.3 for motivation. We consider a discrete dynamical system in the space of probability distributions:

$$\mathbf{p}(t_{n+1}, x) = L\mathbf{p}(t_n, x), \tag{15.5}$$

where L is some operator in the space of probability distributions, generator of evolution. It can be linear, but may be nonlinear. To find the form of L, experimental studies should be performed. Of course, L essentially depends on the psychological function f.

The following model of evolution can be considered:

$$\mathbf{p}(t_{n+1}, x) = \int_{X_{\text{mental}}} K(t_n, x, y)\mathbf{p}(t_n, y)dy,$$

where $K(t, x, y)$ is a time-dependent kernel of evolution. It is a real valued function. If X_{mental} (as in our model) is realized as m-adic space, then dy is Haar measure.

The statistical mental state $\mathbf{p}(t, x)$ is nothing else than the probability distribution of the body $\to$ mind field $\varphi(z)$. We can consider the neuronal tree $T_{\text{neuronal}}(f)$ as probability space $\Omega \equiv T_{\text{neuronal}}(f)$ with the uniform probability measure

$$\mathbf{P}(\omega) = \frac{1}{\text{number of elements in } \Omega},$$

for $\omega \in \Omega$. By the probabilistic tradition we use the symbol ω to denote a point of probability space. The map (body$\to$mind field) $\varphi\colon \Omega \to X_{\text{mental}}$ is a random variable and the statistical mental state

$$\mathbf{p}(x) = \mathbf{P}(\omega \in \Omega : \varphi(\omega) = x)$$

gives probability that neural pathways in the neuronal tree represent the mental point x. This is intensity of neural representation of x.[8]

Thus the evolution of the statistical mental state, $t \to \mathbf{p}(t, x)$, can be reduced to the evolution of the corresponding stochastic process $\varphi(t, \omega)$, the process of body$\to$mind correspondence.

[8] According to Kolmogorov's ideology [271] the structure, e.g., topological of probability space (in our case a neuronal tree $T_{\text{neuronal}}(f)$) does not play any role in the probabilistic formalism. This formalism depends only on the structure of configuration space in that random variables take values. In our case this is m-adic mental space $X_{\text{mental}} = \mathbb{Q}_m$.

But probability theory teaches us that it is impossible to reconstruct the stochastic process $\varphi(t, \omega)$ as a point wise map in the unique way on the basis of corresponding probability distributions. The same mental flow can be generated by various neural flows. This trivial probabilistic argument strongly supports our Mental Thesis. In our model only the body → mind field $\varphi(t, \omega)$ is well defined, but not the mind → body field. This probabilistic consideration is also a strong argument supporting nonreductionism: *Neural reduction of mental processes is impossible.*

15.3 Diffusion model of dynamics of statistical mental state

15.3.1 Markovean body → mind fields

The simplest (but nontrivial!) model of the mental evolution can be obtained by considering Markovean body → mind fields. Markov process is a stochastic process without long range statistical memory. In the Markovean case we have the following equality for the conditional probabilities:

$$\begin{aligned}\mathbf{P}(\varphi(t_{n+1}, \omega) = x_{n+1} | \varphi(t_n, \omega) = x_n, \ldots, \varphi(t_0, \omega) = x_0)\\ = \mathbf{P}(\varphi(t_{n+1}, \omega) = x_{n+1} | \varphi(t_n, \omega) = x_n),\end{aligned}$$

where $t_0 < t_1 < \cdots < t_n < t_{n+1}$. The mental position x_{n+1} at the instant of time $t = t_{n+1}$ is statistically determined by the mental position x_n at the previous instant of time $t = t_n$. Markovean statistical mental state does not remember about states at $t = t_{n-1}, \ldots, t_0$.

Such a type of mental processing is natural for primary mental activity, e.g., reactions to stimuli. Consider the following example. At $t = t_{n+1}$ one, say Ivan, reacts to the state of hunger that he had at $t = t_n$. Ivan definitely does not recall all his states of hunger for the last few days or years. His brain performs just transition $\mathbf{p}(t_n, x) \to \mathbf{p}(t_{n+1}, x)$. Here the statistical mental state $\mathbf{p}(t_n, x)$ is the state of hunger and the statistical mental state $\mathbf{p}(t_{n+1}, x)$ is the state of nourishing[9].

We now consider a simpler reaction – to pain induced by fire at the instance of time $t = t_{n+1}$. The state $\mathbf{p}(t_{n+1}, x)$, *pain,* does not depend even on the previous state $\mathbf{p}(t_n, x)$. Here the body → mind field produces a sequence of independent random variables (e.g. well-known Bernoulli process).

$$\mathbf{P}(\varphi(t_{n+1}, \omega) = x | \varphi(t_n, \omega) = y) = \mathbf{P}(\varphi(t_{n+1}, \omega) = x) \equiv \mathbf{p}(t_{n+1}, x).$$

[9] Of course, previous experiences of hunger played an important role in building the neuronal tree corresponding to hunger-nourishing function, see, for instance, Edelman [37], Chapter 3. A part of this tree architecture is even transmitted genetically.

15.3.2 Thinking as m-adic diffusion

We now consider continuous time evolution for general Markovean body $\rightarrow$ mind fields.[10]

Evolution of the statistical mental state $\mathbf{p}(t,x)$ is described by the Chapman–Kolmogorov equation:

$$\frac{\partial \mathbf{p}}{\partial t}(t,x) = L\mathbf{p}(t,x), \quad \lim_{t\downarrow 0} \mathbf{p}(t,x) = \mathbf{p}_0(x), \tag{15.6}$$

where L is the generator of Markov evolution, see [102]. If we know the initial statistical mental state $\mathbf{p}_0(x)$ and the generator of mental evolution L, then we can find the statistical mental state at any instant of time $t \geqslant 0$.

Remark 15.2 (Determinism or free will?). Our mental model combines determinism and free will. Mental determinism is a consequence of deterministic evolution equations for probability distributions. However, it is determinism of probabilities. The probabilistic representation of the statistical mental state induces feeling of free will. We remark that the situation has some similarities with quantum theory, see [73, 235, 411].

Equation (15.6) is the famous *direct Kolmogorov equation*. Besides this equation, we can consider the *inverse Kolmogorov equation*:

$$\frac{\partial \mathbf{p}}{\partial t}(t,x) = L^*\mathbf{p}(t,x), \quad \lim_{t\uparrow T} \mathbf{p}(t,x) = \mathbf{p}_T(x), \tag{15.7}$$

where L^* is the operator which is adjoint to the generator L of Markovean evolution. Starting with a statistical mental state at the instant of time $T > 0$, we can recall the statistical mental state $\mathbf{p}(t,x)$ at any instant of time $0 \leqslant t < T$. Thus the inverse Kolmogorov equation describes the process of recollection. It seems that the brain is able to solve direct and inverse Kolmogorov equations (at least for small time intervals). This gives the possibility to predict expected statistical mental states by using direct Kolmogorov equation and to recall statistical mental states from the past by using inverse Kolmogorov equation. Such ability to solve the evolution equation for the statistical mental state depends essentially on a human individual.

We remark that some mental states, stationary, $\mathbf{p}(x)$ are not changed in the process of evolution. These are stable patterns of mind. In the m-adic mental space X_{mental} simplest Markovean evolution is given by the diffusion process – Vladimirov–Volovich diffusion. This process was intensively studied in p-adic theoretical physics [407]. The corresponding evolution equation, *m-adic heat equation*, has the form

$$\frac{\partial \mathbf{p}}{\partial t}(t,x) = -\frac{1}{2}D_x^2\mathbf{p}(t,x), \quad \mathbf{p}(0,x) = \mathbf{p}_0(x), \tag{15.8}$$

[10] As we have already remarked, ultrametric topology on the mental space and continuous real time dynamics in this space are incompatible. However, such a problem does not arise for probabilistic dynamics. Here both time and the probability distribution are real valued continuous parameters.

where D_x is a kind of differential operator on the m-adic ring, Vladimirov's operator [407]. In contrast to the ordinary real derivative, Vladimirov's operator is nonlocal, i.e., $D_x\mathbf{p}(t,x)$ contains summation over all points of the mental space. This feature of m-adic diffusion is related to wholeness of mental processes.

We can easily find the fundamental solution $K(t,x)$. This is the solution for the initial mental state $\mathbf{p}_0 = \delta$. Here δ is the well-known Dirac's δ-function. In fact, δ is the probability measure concentrated at the point $x = 0$. For any Borel set A containing $x = 0, \delta(A) = 1$, and in the opposite case $\delta(A) = 0$. The dynamics of the statistical mental state $\mathbf{p}(t,x)$ for any initial distribution $\mathbf{p}_0$ can be obtained as the integral transform of the dynamics which starts with the special state, namely, δ. To simplify considerations, we suppose that all probabilities have densities with respect to Haar measure on $\mathbb{Q}_m$. In particular, at $t = 0$: $\mathbf{p}_0(x) = \delta(x)$. In such a case we have

$$\mathbf{p}(t,x) = \int_{X_{\text{mental}}} K(t, x-y)\mathbf{p}_0(y)dy. \tag{15.9}$$

It seems that such dynamics do not match with our cognitive model. It is quite unrealistic to hope that knowledge of evolution for the initial state $\mathbf{p}_0$ is which is concentrated in a *single point* of the mental space, namely, $x = 0$, determines evolution for any initial state. We remark that the possibility to represent the solution in form (15.9) is a feature of linear equations with constant coefficients. Therefore to find more realistic model for evolution of the statistical mental state we should consider either linear dynamics, but with variable coefficients, or nonlinear dynamics.

We still want to preserve linearity of evolution of probabilities. Let us consider equation (15.6), where the linear differential operator L has variable coefficients. Thus we consider general m-adic diffusion. Take Green function of this problem. It is the solution, say $G(t,x,y)$, for the initial probability distribution $\delta(x-y)$. To be more careful with notations, we should consider Dirac's δ-measure δ_y which is concentrated at the point y. To simplify considerations, we again suppose that all probabilities have densities with respect to Haar measure on $X_{\text{mental}} = \mathbb{Q}_m$. In such a case we have

$$\mathbf{p}(t,x) = \int_{X_{\text{mental}}} G(t,x,y)\mathbf{p}_0(y)dy. \tag{15.10}$$

It is a more realistic model. Knowledge of evolutions for initial states which are concentrated in all possible points of the mental space determine evolution for any initial state $\mathbf{p}_0$. It is not surprising, because the latter can be approximated by the formers.

Even a more realistic model of mental evolution is based on the m-adic diffusion equation with a mental potential V. Here $V(x,y)$ can be chosen as e.g.

$$V(x,y) = \rho_p(x,y)^\alpha, \quad \alpha \geqslant 0,$$

where $\rho_p(x,y)$ is the p-adic distance between mental points.

15.3.3 Discussion

Can consciousness be treated as a variable?

The first part of book [43] of B. J. Baars contains an interesting discussion on a possibility to treat consciousness as a variable and an importance of such a treatment in cognitive science. The m-adic attempt to quantify a mental state was partially motivated by this discussion. However, we essentially modified Baars' idea on the mental-variable. His consciousness-variable is a kind of Newtonian physical variables such as position, velocity, force; see p. 11 of [43] on similarity of the consciousness-variable to variables of Newtonian gravity. Our mental-variable, statistical mental state, is a probability distribution on $\mathbb{Q}_m$. It is more similar to variables of statistical mechanics and even quantum mechanics. Such a variable is not a local variable on the brain. It is a *distributed variable.*

Starting with a statistical mental state we can define a quantitative measure of consciousness, a kind of the consciousness-variable. First we discuss the connection between the levels of neural activity and the levels of consciousness. The idea that the direct correspondence can be established between them is the very common postulate of cognitive science. An extended discussion on this problem can be found, for example, in [43, pp. 18–19]. Our model does not support the postulate on the direct correspondence between the levels of neural activity and the levels of consciousness.

For example, let us consider the extreme case in that all possible states of the brain are activated. Such a super-activation definitely will not imply a high level of consciousness. The level of neural activity does not determine the level of consciousness.

Our conjecture is that consciousness is determined by the *variation* of a statistical mental state. One of possible numerical measures of the level of consciousness is *entropy* of a statistical mental state. It is well known that (for the discrete probability distribution) entropy approaches the maximal value for the uniform probability distribution. By our interpretation this is the lowest level of consciousness (so it may be better to use entropy with minus sign as a quantitative measure of consciousness).

Another possibility is to define a measure of consciousness as the variation of a statistical mental state (with respect to the m-adic metric on the mental space). It is a more natural measure of consciousness, since the topological structure of the mental space is taken into account. The m-adic variation is introduced on the basis of Vladimirov's differential operator D, on the m-adic tree:

$$C(\mathbf{p}) = \text{Consciousness}(\mathbf{p}) = \int_{X_{\text{mental}}} |D\mathbf{p}(x)|^2 dx, \tag{15.11}$$

where $\mathbf{p}(x)$ is the probability density of the statistical mental state of a cognitive system and dx is Haar measure. Consciousness is always nonnegative and it takes its minimal value, $C(\mathbf{p}) = 0$, for the uniform probability distribution. By our interpretation this is the lowest level of consciousness. At the moment we do not know anything about statistical mental states producing consciousness of the extremely high level, namely, solutions of the problem: $C(\mathbf{p}) \to \max$.

Are animals conscious?

Our model strongly supports the hypothesis that animals are conscious (see, e.g., [43] on the detailed neurophysiological and behavioral analysis of this problem). If consciousness is really determined by the variation of the statistical mental state, then animals are definitely able to produce such nontrivial variations by their systems of neural pathways. Moreover, in our Neural Pathway Model pathways going through the body play the important role in the creation of consciousness. Thus the role of differences in brain structures should not be overestimated, cf. [43]. On the other hand, animals have lower levels of consciousness, $C(\mathbf{p})$. We perform the following quantification. Suppose there is a threshold C_{human}. Animals can not produce statistical mental states $\mathbf{p}(x)$ such that $C(\mathbf{p})$ is larger than human's consciousness threshold C_{human}. And human beings (at least most of them) could produce (at least sometimes) statistical mental states such that consciousness is larger than this threshold.

Blindsight

Our model can explain the mystery of blindsight and similar phenomena: "The mystery of blindsight is not so much that unconscious visual knowledge remains. ... The greatest puzzle seems to be that information that is not even represented in area $V1$ is lost to consciousness when $V1$ is damaged." – [43]. However, in our Neural Pathway Model destruction of some neurons, e.g., in area $V1$, destroys (modifies) a huge ensembles of neural pathways. Of course, by our model information was never preserved in area $V1$ nor some other localized area. Information is preserved by ensembles of pathways and they are not located in some particular domain of brain.

However, we also have to explain the unique function of area $V1$ in creating of visual consciousness: "$V1$ is the only region whose loss abolishes our ability to consciously see objects, events .... But cells in $V1$ respond only to a sort of pointillist level of visual perception .... Thus it seems that area $V1$ is needed for such higher-level experiences, even though it does not contain higher-level elements! It seems like a paradox." – [43]. Yes, area $V1$ is the unique region which damage destroys the ability of conscious visualization. However, our model is, in fact, Centered Neural Pathway Model. The uniqueness of area $V1$ in conscious visualization is determined by the fact that central neurons of the neuronal trees involved into the psychological function of conscious visualization are located in area $V1$.

We continue citation of Baars [43]: "Cells that recognize objects, shapes, and textures appear only in much "higher" regions of cortex, strung in a series of specialized regions along the bottom of the temporal lobe." Yes, these cells are centers of neuronal trees corresponding to other psychological functions, e.g. object recognition.

Non-Markovean evolutions of the statistical mental state

We described Markovean models of mental evolution. Such evolutions are quite natural for primitive mental processes in which long range memory is not involved. However, advanced mental processes are definitely non-Markovean. Contrary to the Markovean case, knowledge of the statistical mental state – probability distribution – at the previous step in combination with knowledge of transition probabilities, in general, does not determine the next statistical mental state. Unfortunately, non-Markovean evolution is essentially more complicated from the mathematical viewpoint than the Markovean one.

Neural code and structure of the mental space

Suppose that the coding system of a cognitive system is based on a frequency code. There exists a time interval Δ depending on a cognitive system and a psychological function. The mental point which is produced by a centered neural pathway is a sequence with coordinates given by numbers of oscillations of neurons during the interval Δ. Thus in our model the problem of the neural code is closely related to the problem of the time-scale of the neural system. Different Δ induce different coding systems, and, consequently, different structures of mental spaces. The time interval Δ induces the natural number m that determines the m-adic structure on the mental space. It is defined as the maximal number of oscillations which can be done by neurons of the neuronal tree for some fixed psychological function during the time interval Δ. The frequency coding which is based on the 2-adic system induces the 2-adic mental space which differs crucially from the 5-adic (or 2008-adic) mental space induced by the 5-adic (or 2008-adic) system. As was demonstrated in previous chapters of this book, any change the m-adic structure crucially changes dynamics. The right choice of the time scaling parameter Δ plays an important role in the creation of an adequate mathematical model for functioning of a psychological function.

Psychological time

The time scale parameter Δ of neural coding and so called *psychological time* are closely coupled. There are strong experimental evidences, see, e.g., [325], that a moment in psychological time correlates with ≈ 100 ms of physical time for neural activity. In such a model the basic assumption is that physical time required for the transmission of information over synapses is somehow neglected in psychological time. In the model, the interval of time (≈ 100 ms) is required for the transmission of information from retina to the inferiotemporal cortex (IT) through the primary visual cortex (V1) is mapped to a moment of psychological time. It seems that by using $\Delta \approx 100$ ms we shall get the right m-adic structure of the mental space.

Unfortunately, the situation is essentially more complicated. There are experimental evidences that the temporal structure of neural functioning is not homogeneous. The interval of time required for completion of color information in V4 (≈ 60 ms) is shorter than the interval of time for the completion of shape analysis in IT (≈ 100 ms). In particular it is predicted that under certain conditions there will be a rivalry between color and form perception. This rivalry in time is one of manifestations of the complex temporal structure of the brain based on a few levels of information processing. It can be shown that at any given moment in physical time, there are neural activities in various brain regions which correspond to a range of moments in psychological time. In turn, a moment in psychological time is subserved by neural activities in different brain regions at different physical times.

Therefore it is natural to suppose that different psychological functions have different time scales and, consequently, different mental spaces. Thus one psychological function is based on the 2-adic mental space and another on the 5-adic (or 2008-adic) mental space. This is the very delicate point and we shall try to clarify it.

Consider the space T_{neuronal} of all neural pathways. The concrete psychological function f is based on some centered neuronal tree $T_{\text{neuronal}}(f)$ which is a subset of T_{neuronal}. This psychological function is based on its time scale, say $\Delta = \Delta_f$. Hence there exists the natural number m depending on Δ_f and hence on f determining the m-adic structure of the mental space for f. Thus $m = m_f$. The statistical mental state $\mathbf{p}_f(x)$ of f is defined on this m_f-adic mental space. In general, another psychological function g has its own time scale Δ_g and corresponding m_g. Its statistical mental state $\mathbf{p}_g(x)$ is defined on the m_g-adic space. If m_f is not equal to m_g, e.g. $m_f = 2$ and $m_g = 2008$, then dynamics of statistical mental states corresponding to psychological functions f and g differ crucially – even if evolutions are described by the same diffusion equation. One of them is diffusion on the 2-adic mental space and another is diffusion on the 2008-adic mental space.

Finally, we remark that psychological functions are strongly inter related on the neural pathway level. Neuronal trees $T_{\text{neuronal}}(f)$ and $T_{\text{neuronal}}(g)$ corresponding to psychological functions f and g can have large intersection. In the extreme case these trees could even coincide: $T_{\text{neuronal}}(f) = T_{\text{neuronal}}(g)$. However, the use of different time scales $\Delta_f \neq \Delta_g$ would produce totally different evolutions for corresponding statistical mental states.

Discreteness of time

Previous considerations demonstrated that the model of evolution of the statistical mental state based on continuous time $t \in \mathbb{R}$ provides only a rough approximation of mental evolution which is based on discrete psychological time. Moreover, discretization steps depend on corresponding psychological functions.

Thus the evolution of the statistical mental state $\mathbf{p}_f(t, x)$ of a psychological function f is described by discrete time dynamics, where $t_{n+1} = t_n + \Delta$.

Does consciousness benefit from long neural pathways?

Finally, we discuss one of the greatest mysteries of neuroanatomy, see, for example, [95, 116]. It seems that in the process of neural evolution cognitive systems tried to create for each psychological function as long neural pathways as possible. This mystery is explained by our neural neural pathways model. A cognitive system τ benefits by extending neural pathways for some psychological function as long as possible. For example, let the neural code be based on $m = 5$ and let a psychological function f be based on very short pathways of the length $L = 2$. Then the corresponding mental space contains $N(5, 2) = 2^5 = 32$ points. Let now $m = 5$ and $L = 10000$. Then the corresponding mental space contains a huge number of points: $N(5, 10000) = 10^{20}$. On the latter (huge) mental space there can be realized statistical mental states having essentially more complex behavior and, consequently, higher magnitude of consciousness. This argument explains spatial separation of various maps in brain, see, e.g., Edelman [116].

Why does activity of "far away" neurons play an important role?

Consider a psychological function based on a neuronal tree with one central neuron S. Suppose that interaction between mental points produced by this tree depends on the m-adic distance between these points, e.g.

$$V(x, y) = \rho_m(x, y)^\alpha, \quad \alpha > 0.$$

Then changes of states of input neurons which are located far away from the central neuron S (on neural pathways belonging to the neuronal tree) play the crucial role in variation of the magnitude of the mental potential $V(x, y)$. If states (e.g. rates of firing) of initial input neurons are different, then $\rho_m(x, y)$ is very large, see (15.4).

15.4 Postulates

Our mathematical model of probabilistic thinking on m-adic mental spaces[11] is based on the following five postulates:

1. Pr. *Statistical mental states are determined by probability distributions on mental spaces.*

Evolutions of statistical mental states are described by classical evolution equations for probability distributions of random processes, e.g. diffusion equations, on ultrametric m-adic mental spaces.

2. NeurPath. *'Quant' of mental information is given by the state of a hierarchic neural pathway.*

[11] Such spaces are produced by neuronal trees of hierarchic neural pathways.

3. NeurGr. *Each psychological function is based on a hierarchic tree of neural pathways, neuronal tree.*

4. Ult. *Mental topology is ultrametric.*

It is supposed that mental spaces, opposite to spaces used in physical models, have ultrametric topology. The presence of such geometry is equivalent to a treelike representation of mental space.

5. FrCod. *Mental encoding of information is performed by accounting frequencies of firings of neurons along hierarchic neural pathways.*

This encoding of information determines the natural number m depending on a psychological function and a cognitive system.

The first postulate, **Pr**, determines the (probabilistic) structure of advanced information processing in cognitive systems. Other postulates are related to processing of cognitive information on the primary level. In fact, **Pr** need not be rigidly connected with further postulates. Other cognitive models of probabilistic thinking can be developed. In particular, we need not base all models on the last postulate **FrCod**. Other models of mental coding can be chosen.

We briefly discuss relation of our Probabilistic Neural Pathway Model with some traditional models of cognition. As was already pointed, we do not study neural dynamics in the brain. Our model is a purely information model. We study flows of specially organized information. Of course, postulates **NeurPath** and **FrCod** provide connection with neurophysiology. However, we are not interested in investigation of functioning of neural networks producing hierarchic strings of information, mental points. The only important thing is the form of the probability distribution (statistical mental state) on the space of mental points. As was already remarked, different dynamical processes on neural level can produce the same probability distribution.

In principle, a model, for example, *connectionist* neural network model, can be proposed. It would describe "production" of information strings forming a statistical mental state. However, such a generalized connectionist model should be based on new paradigm: a hierarchic neural pathway as the basic processing unit – not single neuron! We even can not exclude a possibility that such an "underground model" can be some *AI-model*. Extended experience of computer simulations show that a complex random behavior can be simulated algorithmically – in particular, by m-adic dynamical systems, see previous chapters of this book. However, we need not presuppose existence of any deterministic "underground model".

We also mention connection with *distributed representation* models. We recall that: *A distributed representation is one in which meaning is not captured by a single symbolic unit, but rather arises from the interaction of a set of units, normally a network of some sort.* If we use just the first part of this definition, i.e., omit direct coupling with neural networks, then Probabilistic Neural Pathway Model can be considered as a model of distributed representation: mental units (points) are unified through a probability distribution – a statistical mental state.

Chapter 16

Gene expression from dynamics in the 2-adic space

In this chapter we perform geometrization of genetics by representing genetic information by points of the 4-adic "genetic space". This space can also be represented as 2-adic space. Sometimes it is more convenient. In our model the process of DNA-reproduction is described by the action of a 4-adic (or equivalently 2-adic) dynamical system. As we know, the genes contain information for production of proteins. The genetic code is a degenerate map of codons to proteins. We model this map as functioning of a polynomial dynamical system. The purely mathematical problem under consideration is to find a dynamical system reproducing the degenerate structure of the genetic code. We present one of possible solutions of this problem. A more detailed discussion on studies of the informational structure of the genetic code will be presented in Chapter 17.

As we have already seen in Chapters 14 and 15, the *m-adic mental space* can be used in cognitive science, psychology and neurophysiology, and artificial intelligence, see [237] for more details. The main distinguishing feature of encoding of information by m-adic numbers is the possibility to encode the hierarchical structure of information by the ultrametric topology on the m-adic tree. Recently it was pointed out that the same *m-adic information space* can be applied to mathematical modeling in genetics, [106–108,240,241,243,248,249,363]. The genetic version of the m-adic mental space was considered. It is an interesting attempt to embed genetics in space setup, as it was done in physics more than three hundreds years ago and as it was done in cognitive science and psychology [237].

The main problem of this model is that the presence of a hierarchic structure in genes and in genom in general has not yet been confirmed in genetics.[1] Nevertheless, it seems that one famous problem in genetics, namely, the mystery of genetic code could be explained by using the m-adic representation of information. We remind that the essence of the "m-adic encoding" of information is the possibility to take into account the presence of a hierarchic structure in the encrypted sequence. Thus in a sequence of e.g. zeros and ones, say 00011111111111111110000011110, it is important not

[1] One of the authors, Andrei Khrennikov, discussed this problem with a number of scientists of the "top-level" working in genetics, in particular, during the conference "Integrative approaches to brain complexity", 2006, Cambridge, UK, and "Quantum Bio Information", 2008, Tokyo. Opinions of experts are extremely diverse. In any event one could not exclude the possibility that genes' activity might be described by the hierarchic m-adic information space. We will use this chance in the present book.

only which symbol, 0 or 1, is located at some position, but even this position by itself. For two strings of zeros and ones the difference in the first digit is essentially more important than in the second, the second is essentially more important than the third and so on. For example, two strings

$$x = 0001111111111111111110000011110,$$
$$y = 1001111111111111111110000011110$$

are very different from the 2-adic viewpoint: $\rho_2(x, y) = 1$. However, the difference between the strings

$$x = 0001111111111111111110000011110,$$
$$z = 0001111111111111111111111100001$$

is practically negligible: $\rho_2(x, z) = 1/2^{20}$.

On the other hand, if the difference is considered with respect to the standard Hamming metric, then $\rho(x, y) = 1$, but $\rho(x, z) = 10$.

Our aim is to use m-adic mental spaces and dynamical systems in genetics.

16.1 Description of model

16.1.1 4-adic representation of nucleotides

Now we present schematically development of this model. DNA and RNA sequences are represented by 4-adic numbers. Nucleotides are mapped to digits in registers of 4-adic numbers:

CodeTCAG. Thymine – T $=$ 0, cytosine – C $=$ 1, adenine – A $=$ 2, and guanine – G $=$ 3. The U-nucleotide is represented (as well as T) by 0.

At the moment we proceed in the purely informational framework. The genetic code is considered as just a coding system. In a more advanced model one should take into account biological, chemical and physical structures. In particular, at the moment the encoding system was chosen in an arbitrary way. Roughly speaking, one may ask: “Why is T encoded by 0 and A by 2, but not vice versa?” As was mentioned, it is impossible to justify encoding without bio-chemical and bio-physical arguments. We will come back to this extremely important problem in Chapter 17. We will see that the present coding should be modified to match bio-chemistry and bio-physics. In principle, we might start directly with the code which will be used in Chapter 17. However, the m-adic genetic project is at the very beginning and it is too early to make definite conclusions. In future it might as well occur that the code of Chapter 17 should be modified as well. Therefore we will proceed in this chapter with **CodeTCAG** – to illustrate the variety of possibilities.

Hierarchic structure in DNA and RNA sequences

The *DNA and RNA sequences have the natural hierarchical structure: letters which are located at the beginning of a chain are considered as more important.* This hierarchical structure coincides with the hierarchical structure of the 4-adic tree. Such a hierarchy can also be encoded by the 4-adic metric. Algebraically DNA and RNA sequences are given by 4-adic integers. Thus we were able to enumerate DNA and RNA by numbers. Since real genetic sequences are finite, genetic reality is described by natural numbers. However, topology is induced from $\mathbb{Z}_4$. In a mathematical model it is convenient to extend the space of genetic sequences and to consider infinitely long DNA and RNA. Such genetic idealizations are represented by 4-adic generalizations of natural numbers, i.e., by $\mathbb{Z}_4$. We recall again, cf. Chapter 14, that our model is simply a modern representation of views of Platon and Aristotle as well as Leibniz.

As was pointed out, our genetic project is at the very beginning. It might occur that DNA sequences have hidden complicated hierarchic structures. At the moment the genetic community has not yet elaborated a definite viewpoint on this problem. Some scientists are sure that such a structure should exist, but it has not yet been found. Other people are rather sceptical of the existence of hidden genetic hierarchy. Of course, this problem is crucial for our modeling. However, mathematicians can not contribute so much to its solution. It is the task of experimentalists and theoreticians working in genetics. If such a hierarchic structure was found, our representation of DNA and RNA sequences by 4-adic numbers, see Subsection 16.2.1, should be modified by taking into account this yet unknown structure. It might be impossible to proceed with simply $\mathbb{Z}_4$ to represent DNA and RNA. Complex ultrametric spaces may arise.

Therefore one should not take too seriously the use of the special hierarchic structure in this chapter, namely, the structure given by the canonical sequential representation of DNA and RNA – *lexicographic order.* It might be even better to wait for clarification of the problem of the presence of a more complicated hierarchic structure in DNA which might be related to functioning of genes. However, we would like to demonstrate a possibility of the hierarchic representation in genctics, so wc proceed with lexicographic hierarchy.

16.1.2 DNA-reproduction and 4-adic dynamics

The process of DNA-reproduction is described by the action of a 4-adic dynamical system. As we know, the genes contain information for production of proteins. The genetic code is a degenerate map of codons to proteins. We model this map as functioning of a polynomial 4-adic dynamical system. Proteins are associated with cycles of such a dynamical system. By a well-known theorem of number theory this dynamics can also be represented in the 2-adic space.

16.2 Genetic space

The genetic space arises as a special case of mental spaces.

16.2.1 4-adic encoding of DNA and RNA

We will use the following mathematical model for the genetic information space. We choose the 4-adic representation for DNA and RNA based on **CodeTCAG**. An arbitrary gene in a DNA-sequence is encoded by a 4-adic integer, for example:

$$\mathrm{ATCGTA}\ldots \to 201302\ldots = 2 + 4^2 + 3 \cdot 4^3 + 2 \cdot 4^5 + \cdots .$$

As was remarked, biologically realizable sequences are finite (but very long). Thus they correspond to natural numbers. However, in a mathematical model we can use even infinitely long genetic sequences. Denote this space by the symbol X_{genetic}. This space has the following distinguishing features:

(a) *Set-structure*: The set of genetic states X_{genetic} has the structure of the 4-adic tree: $X_{\text{genetic}} = \mathbb{Z}_4$.

(b) *Topology*: Two genetic states x and y are close if they have sufficiently long common root.[2] This topology is described by the metric ρ_4.

(c) *Dynamics*: Information processing on the level of genetic states is described by 4-adic dynamical systems. In the simplest case of the discrete-time dynamics these are iteration of a map

$$f : \mathbb{Z}_4 \to \mathbb{Z}_4.$$

(d) *Hierarchical structure*: The coding system which is used in our model for recording vectors of information generates a *hierarchical structure* between digits of these vectors – between nucleotides in the gene-sequence. Thus if $x = (\alpha_0, \alpha_1, \ldots, \alpha_n, \ldots)$, $\alpha_j = 0, 1, 2, 3$, is an information vector which presents genetic information then digits α_j have different weights. The digit α_0 is the most important, α_1 dominates over $\alpha_2, \ldots, \alpha_n, \ldots$, and so on.

Transcription-map

Transcription is the process of copying a gene into RNA. This is the first step of turning a gene into protein (although not all transcriptions lead to proteins). In our coding system transcription is simply the identity map from $\mathbb{Z}_4 \to \mathbb{Z}_4$ (since the T and U nucleotides are represented by the same digit).

[2] Thus the first SNP (single nucleotide polymorphism) distinguishes two genetic states.

Encoding of proteins by codons

In the genetic code proteins are encoded by *codons* – blocks of the length 3 in the gene transcription. Each codon contains information for producing of a single amino acid. By using our 4-adic coding system we can rewrite the table of the genetic code. We collect amino acids in families with respect to a number of codons which are used to encode an amino acid:

(1) Met: 203; Trp: 033;

(2) Asn: 220, 221; Asp: 320, 321; Cys: 030, 031; Gln: 122, 123; Glu: 322, 323; His: 120, 121; Lys: 222, 223; Phe: 000, 001; Tyr: 020, 021;

(3) Ile: 200, 201, 202; Stop: 023, 032, 022;

(4) Ala: 310, 311, 312, 313; Gly: 330, 331, 332, 333; Pro: 110, 111, 112, 113; Thr: 210, 211, 212, 213; Val: 300, 301, 302, 303;

(5) Arg: 130, 131, 132, 133, 232, 233; Leu: 002, 003, 100, 101, 102, 103; Ser: 010, 011, 012, 023, 230, 231.

Codon-map

First we consider the standard left-shift

$$s_l(\alpha_0\alpha_1\alpha_2\ldots) = \alpha_1\alpha_2\ldots .$$

We also consider the following cutoff-map

$$c_3(\alpha_0\alpha_1\alpha_2\ldots) = \alpha_0\alpha_1\alpha_2.$$

Then the representation by codons of the gene-expression is given by the c_3-projections of the iterations of the left-shift:

$$x_n = c_3(s_l^{3(n-1)}(x)).$$

16.2.2 2-adic encoding

The 4-adic encoding can be easily transformed into the 2-adic encoding just by using the 2-adic representation of the genetic alphabet:

CodeTCAG (2-adic version). U = 00, A = 01, C = 10, G = 11.

We again collect amino acids in families with respect to the number of codons which are used to encode an amino acid:

(1) Met: 010011; Trp: 001111;

(2) Asn: 010100, 010110; Asp: 110100, 110110; Cys: 001100, 001110; Gln: 100101, 100111; Glu: 110101, 110111; His: 100100, 100110; Lys: 010101, 010111; Phe: 000000, 000010; Tyr: 000100, 000110;

(3) Ile: 010000, 010010, 010001; Stop: 000111, 001101, 000101;

(4) Ala: 110100, 110101, 110110, 110111; Gly: 111100, 111110, 111101, 111111; Pro: 101000, 101010, 101001, 101011; Thr: 011000, 011010, 011001, 011011; Val: 110000, 110010, 110001, 110011;

(5) Arg: 011100, 011101, 011101, 011111, 101110, 101111; Leu: 000001, 000011, 100000, 100010, 100001, 100011; Ser: 001000, 001010, 001001, 000111, 011100, 011110.

16.3 Dynamical model for degeneracy of the genetic code

We will use dynamical systems corresponding to maps:

$$\mathbb{Z}_m \to \mathbb{Z}_m, \ x \to f(x). \tag{16.1}$$

As usual, we study the behavior of iterations.

Our basic idea is associate with the genetic code some polynomial

$$f_{\text{genetic}}(x) = a_0 + a_1 x + \cdots + a_n x^n, \quad x \in \mathbb{Z}_m,$$

depending on the choice of the coding system $m = 4, 2$.

Such a polynomial encodes amino acids in the following way. The set of codons (which are considered as 2-adic numbers) is split by this polynomial into groups of cycles. Each cycle encodes one amino acid, so:

Amino acids are coded by cycles of this polynomial.

Our model cannot explain the origin of such a coding polynomial. Its origin can be a consequence of biological evolution or just purely information features of the genetic system. Since we do not know the (e.g., biological) background inducing a coding polynomial $f_{\text{genetic}}(x)$, we are not able to choose it in the unique way. In this note we propose one of the possible solutions of the problem of finding a coding polynomial.

We use Mahler polynomials. To proceed in this way, we choose the 2-adic genetic coding. The $m = 2$ is a prime number and the system of 2-adic integers $\mathbb{Z}_2$ can be extended to the field of 2-adic numbers $\mathbb{Q}_2$. In a number field division is well defined. We need this operation to define Mahler polynomials, see Section 3.9. We are looking for a map $f_{\text{genetic}} : \mathbb{Z}_2 \to \mathbb{Q}_2$ having the structure of cycles corresponding to the genetic code of amino acids.

Let the values of a function $f : \mathbb{Z}_2 \to \mathbb{Q}_2$ in the points $j = 0, 1, \ldots, n$ be known. Then its nth Mahler coefficient is defined by

$$a_n = \sum_{j=0}^{n} (-1)^{n-j} \binom{n}{j} f(j).$$

The corresponding Mahler polynomial has the form

$$F_n(x) = \sum_{k=0}^{n} a_k \binom{x}{k},$$

where the binomial polynomial

$$\binom{x}{k} = \frac{x(x-1)(x-2)\cdots(x-k+1)}{k!}.$$

The crucial is that

$$f(j) = F_n(j), \quad j = 0, 1, \ldots, n.$$

Coming back to the genetic code, we see that there are 64 different points-codons. Thus we need a Mahler polynomial of degree 63 such that

Met: $f(010011) = 010011$; Trp: $f(001111) = 001111$;
Asn: $f(010100) = 010110$, $f(010110) = 010100$;
Asp: $f(110100) = 110110$, $f(110110) = 110100, \ldots$;
Ser: $f(001000) = 001010$, $f(001010) = 001001$, $f(001001) = 000111$, $f(000111) = 011100$, $f(011100) = 011110$, $f(011110) = 001000$.

By using the 2-adic coding we can represent each codon with a 2-adic ball of radius $r = 1/64$ with center in the corresponding 2-adic word. For example, $010011 \to B_{1/64}(010011)$. This is the set of all 2-adic sequences such that the first six digits coincide with the codon word 010011. Thus the amino acid Met can be represented by the ball $B_{1/64}(010011)$ and Trp by $B_{1/64}(001111)$. But Asn by the union of two balls: $B_{1/64}(010100) \cup B_{1/64}(010110)$ and, e.g., Ser by the union of sixth balls

$$B_{1/64}(001000) \cup B_{1/64}(001010) \cup B_{1/64}(001001) \cup B_{1/64}(000111) \cup B_{1/64}(011100) \cup B_{1/64}(011110).$$

We remark that in [106], there was considered a 5-adic model to explain the origin of the gene code. In this model 5-adic balls were used to classify codons.

As we have seen in [237], consideration of fuzzy cycles is more natural, since they are stable with respect to noise (ordinary cycles can be easily disturbed by noisy perturbations).

Now we consider a model in that the "genetic polynomial" $f_{\text{genetic}}(x)$ encodes amino acids in the following way:

Amino acids are coded by fuzzy cycles of this polynomial.

However, at the moment we do not have mathematical examples of simple polynomials having the structure of fuzzy cycles corresponding to the genetic code. We shall continue the study of this problem.

Conclusion. *We have seen that the genetic code has a natural 4-adic (or 2-adic) structure. Gene expression could be coupled to a dynamical system in the genetic information space.*

Chapter 17

Genetic code on the diadic plane

In this chapter we consider another 2-adic model of the genetic code, cf. Chapter 16. It was developed by one of the authors and Sergei Kozyrev [248].

We introduce the simple parametrization for the space of codons (triples of nucleotides) by 8×8 table. This table (which we call the *diadic plane*) possesses the natural 2-adic ultrametric. We show that after this parametrization the genetic code will be a locally constant map of the simple form. The local constancy of this map will describe degeneracy of the genetic code.

The map of the genetic code defines 2-adic ultrametric on the space of amino acids. We show that hydrophobic amino acids will be clustered in two balls with respect to this ultrametric. Therefore the introduced parametrization of space of codons exhibits the hidden regularity of the genetic code.

Moreover, we show, that the construction of ultrametric introduced here is related to physical and chemical properties of amino acids. The genetic code maps the ultrametric on the diadic plane (the space of codons) onto the space of amino acids. We show that, with respect to this ultrametric, hydrophobic amino acids will be clustered in two balls. Therefore physical properties of amino acids are related to the parametrization of the genetic code considered in the present chapter.

Let us consider different variations of genetic code. We will see that the corresponding maps of the diadic plane onto the space of amino acids possess different degree of regularity (i.e. the different character of local constancy). We can conjecture that more regular maps correspond to older forms of genetic code (since it is more probable that evolution goes with symmetry breaking).

Thus parametrization of the space of codons by the diadic plane considered in the present chapter exhibits the regular structure of the genetic code.

Investigation of properties of the genetic code attracts a lot of interest, see [178, 384, 394], see Chapter 16 for references on the m-adic approach. A model to describe the degeneracy of genetic code using the quantum algebra $\mathcal{U}_q(sl(2) \oplus sl(2))$ in the limit $q \to 0$, was proposed in [138, 139]. In these papers the analogy between the genetic code and quark models of barions was discussed. Ultrametric was widely used in bioinformatics in construction of phylogenetic trees starting from nucleotide sequences, see [93]. The results presented in this chapter show that one can apply ultrametric methods to investigation of the genetic code starting from the level of single codons. In particular, one can use this observation to modify the metric used in computational genomics.

17.1 Vertebral mitochondrial and eucaryotic codes

In the present section we briefly repeat discussion on the genetic code that we started in Chapter 16. This material can be found in [410]. The genetic code is the map which gives the correspondence between codons in DNA and amino acids. Codon is a triple of nucleotides, nucleotides are of four kinds, denoted by C, A, T, G (Cytosine, Adenine, Thymine, Guanine), in total we have 64 codons. In RNA Thymine is replaced by Uracil, denoted by U.

We have 20 amino acids: alanine, threonine, glycine, proline, serine, aspartic acid, asparagine, glutamic acid, glutamine, lysine, histidine, arginine, tryptophan, tyrosine, phenylalanine, leucine, methionine, isoleucine, valine, cysteine, denoted correspondingly by Ala, Thr, Gly, Pro, Ser, Asp, Asn, Glu, Gln, Lys, His, Arg, Trp, Tyr, Phe, Leu, Met, Ile, Val, Cys, and the stop-codon Ter.

We consider the following parametrization of the set of nucleotides which is more appropriative to the present model:

CodeATGC. $\{A, T, G, C\}$ are encoded by numbers 0, 1, 2, 3.

Codons will be enumerated by triples of numbers $C_1C_2C_3$, where $C_i = 0, 1, 2, 3$. Genetic code put into correspondence to a codon $C_1C_2C_3$ an amino acid or Ter (a stop codon).

There exist several variations of genetic code. Different variants of genetic code generally coincide but can differ on few codons. Table 17.1 and 17.2 describe the vertebral mitochondrial code and the standard, or eucaryotic, code.

17.2 Parametrization of the set of codons by the diadic plane

In the present section we introduce the parametrization of the space of codons by the diadic plane.

The first step of this construction is the parametrization of the set of nucleotides by pairs of digits (x, y): 00, 01, 10, 11. These pairs can be considered as a binary representation of the numbers 0, 1, 2, 3, which enumerate the nucleotides A, T, G, C correspondingly:

$$C = x + 2y, \qquad C = 0, 1, 2, 3, \qquad x, y = 0, 1.$$

This parametrization is described by the following 2×2 table:

$$\begin{array}{|c|c|}\hline A & G \\ \hline T & C \\ \hline \end{array} = \begin{array}{|c|c|}\hline 0 & 2 \\ \hline 1 & 3 \\ \hline \end{array} = \begin{array}{|c|c|}\hline 00 & 01 \\ \hline 10 & 11 \\ \hline \end{array}. \tag{17.1}$$

000 AAA Lys	100 TAA Ter	200 GAA Glu	300 CAA Gln
001 AAT Asn	101 TAT Tyr	201 GAT Asp	301 CAT His
002 AAG Lys	102 TAG Ter	202 GAG Glu	302 CAG Gln
003 AAC Asn	103 TAC Tyr	203 GAC Asp	303 CAC His
010 ATA Met	110 TTA Leu	210 GTA Val	310 CTA Leu
011 ATT Ile	111 TTT Phe	211 GTT Val	311 CTT Leu
012 ATG Met	112 TTG Leu	212 GTG Val	312 CTG Leu
013 ATC Ile	113 TTC Phe	213 GTC Val	313 CTC Leu
020 AGA Ter	120 TGA Trp	220 GGA Gly	320 CGA Arg
021 AGT Ser	121 TGT Cys	221 GGT Gly	321 CGT Arg
022 AGG Ter	122 TGG Trp	222 GGG Gly	322 CGG Arg
023 AGC Ser	123 TGC Cys	223 GGC Gly	323 CGC Arg
030 ACA Thr	130 TCA Ser	230 GCA Ala	330 CCA Pro
031 ACT Thr	131 TCT Ser	231 GCT Ala	331 CCT Pro
032 ACG Thr	132 TCG Ser	232 GCG Ala	332 CCG Pro
033 ACC Thr	133 TCC Ser	233 GCC Ala	333 CCC Pro

Table 17.1. The vertebral mitochondrial code.

000 AAA Lys	100 TAA Ter	200 GAA Glu	300 CAA Gln
001 AAT Asn	101 TAT Tyr	201 GAT Asp	301 CAT His
002 AAG Lys	102 TAG Ter	202 GAG Glu	302 CAG Gln
003 AAC Asn	103 TAC Tyr	203 GAC Asp	303 CAC His
010 ATA Ile	110 TTA Leu	210 GTA Val	310 CTA Leu
011 ATT Ile	111 TTT Phe	211 GTT Val	311 CTT Leu
012 ATG Met	112 TTG Leu	212 GTG Val	312 CTG Leu
013 ATC Ile	113 TTC Phe	213 GTC Val	313 CTC Leu
020 AGA Arg	120 TGA Ter	220 GGA Gly	320 CGA Arg
021 AGT Ser	121 TGT Cys	221 GGT Gly	321 CGT Arg
022 AGG Arg	122 TGG Trp	222 GGG Gly	322 CGG Arg
023 AGC Ser	123 TGC Cys	223 GGC Gly	323 CGC Arg
030 ACA Thr	130 TCA Ser	230 GCA Ala	330 CCA Pro
031 ACT Thr	131 TCT Ser	231 GCT Ala	331 CCT Pro
032 ACG Thr	132 TCG Ser	232 GCG Ala	332 CCG Pro
033 ACC Thr	133 TCC Ser	233 GCC Ala	333 CCC Pro

Table 17.2. The eucaryotic code.

This parametrization of the set of nucleotides was used in [178, 394], where the Gray code model for the genetic code was considered. It was mentioned that, since the nucleotides A $=(0,0)$ and G $=(0,1)$ are purines, T $=(1,0)$ and C $=(1,1)$ are pyrimidines, the different first digits in the binary representation corresponds to the different chemical types of the nucleotides. Namely, the nucleotide (x, y) with $x=0$ is a purine, and the nucleotide (x, y) with $x=1$ is a pyrimidine.

The second digit $y=0,1$ in the considered parametrization [178] also has the physical meaning. It describes the H-bonding character (weak for $y=0$ and strong for $y=1$).

The second step of our construction is to find a parametrization of the space of codons, using the above parametrization of the set of nucleotides. To do this we take into account the importance of the nucleotides in the codon, described by the following rule [394],

$$2>1>3. \tag{17.2}$$

This means that the most important nucleotide in the codon is the second, and the less important nucleotide is the third.

The main idea of the presented in this chapter is to combine the parametrization of nucleotides by 2×2 table and the above order of nucleotides in the codon and obtain the parametrization of the space of codons by 8×8 table (the diadic plane). We call the diadic plane the square 8×8, which has the structure of the group

$$(\mathbb{Z}/8\mathbb{Z})^{+} \times (\mathbb{Z}/8\mathbb{Z})^{+}$$

(i.e., of the direct sum of two additive groups of residues modulo 8). Elements of this group we denote (x, y):

$$x=(x_0 x_1 x_2)=x_0+2x_1+4x_2, \quad y=(y_0 y_1 y_2)=y_0+2y_1+4y_2, \quad x_i, y_i=0,1$$

One can say that x and y in this formula are integer numbers from 0 to 8 in the binary representation.

Let us construct the correspondence ρ between the diadic plane and the set of codons. Using the rule (17.2), we put into correspondence to the most important (the second) nucleotide in the codon the largest scale of the 8×8 diadic plane – the pair (x_0, y_0), we correspond to the first nucleotide in the codon the pair (x_1, y_1), and the third nucleotide in the codon will determine the pair (x_2, y_2). The nucleotides define the corresponding pairs (x_i, y_i) according to the rule (17.1). We get for the codon $C_1C_2C_3$ the following representation by the pair of triples of 0 and 1, which we consider as an element of the diadic plane:

$$\rho\colon C_1C_2C_3 \mapsto (x, y)=(x_0 x_1 x_2, y_0 y_1 y_2).$$

The table 8×8 of codons on the diadic plane will take the form:

000	002	200	202	020	022	220	222
001	003	201	203	021	023	221	223
100	102	300	302	120	122	320	322
101	103	301	303	121	123	321	323
010	012	210	212	030	032	230	232
011	013	211	213	031	033	231	233
110	112	310	312	130	132	330	332
111	113	311	313	131	133	331	333

Remind that the numbers 0, 1, 2, 3 denote the nucleotides A, T, G, C correspondingly.

The diadic plane (and, correspondingly, the space of codons) possesses the 2-dimensional 2-adic ultrametric, which reflects the rules (17.1), (17.2):

$$d(C_1C_2C_3, C_1'C_2'C_3') = \max(|x - x'|_2, |y - y'|_2),$$

$$(x, y) = \rho(C_1C_2C_3), \quad (x', y') = \rho(C_1'C_2'C_3').$$

This 2-adic norm can take values 1, 1/2, 1/4.

The differences between the parametrization of the space of codons by the diadic plane introduced here and the constructions of Khrennikov [240] and Dragovich [106], respectively, are that in [240] and [106] the space of codons was parameterized by a one-dimensional parameter with 4-adic and 5-adic norms, respectively, and the rule (17.2) was not taken into account.

In the papers [178,394] the rules (17.1), (17.2) were used for the investigation of the genetic code but the combination of these rules was different from the one considered in the present chapter – instead of ultrametric parametrization the Gray code model was used.

17.3 Genetic code on the diadic plane

In the present section we discuss the vertebral mitochondrial code, which looks more regular in our approach.

The genetic code in the considered parametrization put into correspondence to elements of the diadic plane the amino acids (and the stop-codon Ter). In this way we obtain for the Vertebrate Mitochondrial Code the following table of amino acids on the diadic plane:

$\frac{\text{Lys}}{\text{Asn}}$	$\frac{\text{Glu}}{\text{Asp}}$	$\frac{\text{Ter}}{\text{Ser}}$	Gly
$\frac{\text{Ter}}{\text{Tyr}}$	$\frac{\text{Gln}}{\text{His}}$	$\frac{\text{Trp}}{\text{Cys}}$	Arg
$\frac{\text{Met}}{\text{Ile}}$	Val	Thr	Ala
$\frac{\text{Leu}}{\text{Phe}}$	Leu	Ser	Pro

Each small square of this table corresponds to the image (with respect to the genetic code) of a square 2×2 from the table of codons. For example, we have the following correspondence:

$$\begin{array}{|c|c|}\hline 000 & 002 \\ \hline 001 & 003 \\ \hline\end{array} \to \begin{array}{|c|}\hline \text{Lys} \\ \hline \text{Asn} \\ \hline\end{array}, \qquad \begin{array}{|c|c|}\hline 330 & 332 \\ \hline 331 & 333 \\ \hline\end{array} \to \begin{array}{|c|}\hline \text{Pro} \\ \hline\end{array}.$$

Some of the 2×2 squares in the table of codons map onto one amino acid (which gives degeneracy 4 for the genetic code). Some of the squares map onto two amino acids: the first line of the 2×2 square maps onto one amino acid, the second line maps onto the other amino acid, giving degeneracy 2 for the genetic code. We also have three cases of additional degeneracy. For example, the second square in the last line of the table above as well as the upper half of the first square in the last line map onto the amino acid Leucine (Leu).

2-adic balls with respect to the considered above 2-adic norm on the plane look as follows. All the table is the ball of diameter 1. A quadrant (quarter of the table), such as for example the right lower quadrant containing the amino acids Pro, Ser, Thr, Ala is a ball of diameter 1/2. A square of 4 codons (quarter of quadrant), say the square containing the amino acid Pro is a ball of diameter 1/4. Finally, any codon can be considered as a ball of zero diameter.

The major part of degeneracy of the genetic code (besides the mentioned three cases of additional degeneracy) has the clear 2-adic meaning on the diadic plane. First, the genetic code map is always locally constant on the horizontal coordinate y with the diameter of local constancy $1/4$, and is locally constant on the half of space of codons on the vertical coordinate x with the diameter of local constancy $1/4$. Second, sets with different character of local constancy are distributed on the diadic plane in the symmetric way: the lower right quadrant corresponds to local constancy with diameter $1/4$ both on x and y, the higher left quadrant corresponds to local constancy with diameter $1/4$ on y (but not on x), and the other two quadrants have similar distribution of squares with different character of local constancy.

We will say that the degeneracy of the genetic code satisfies the *principle of proximity* – similar codons are separated by small 2-adic distances on the diadic plane. Here similarity means that the corresponding codons encode the same amino acid. We will see that the principle of proximity has more general application and is also related to physical-chemical properties of the amino acids.

Let us discuss our choice of the parametrization for the genetic code. Considering the vertebral mitochondrial code, it is easy to see that we always have degeneracy of the genetic code on the third nucleotide. Moreover, this degeneracy always have the same form – it is always possible to change C by T, and to change A by G. Also, on the half of the space of codons we will have complete degeneracy of the genetic code on the last nucleotide.

Using the proximity principle we describe this degeneracy by local constancy of the map of the genetic code on small distances. Moreover, we will describe different (double and quadruple) degeneracy as a degeneracy of the map with the domain in the two-dimensional ultrametric space over one or two coordinates.

In this way we arrive to the map similar to the described above map ρ of the space of codons onto the diadic plane, where the third nucleotide in the codon corresponds to the smallest scale on the diadic 8×8 plane (where we have three scales of distance – 1, 1/2 and 1/4).

We have to put the other two scales on diadic plane into correspondence to the other two nucleotides in codon. We see that if we correspond to the first nucleotide in the codon the second (intermediate) scale on the diadic plane, then the lower right quadrant will contain four squares with degeneracy four, and the upper left quadrant will contain eight half-squares with degeneracy two. Therefore the table of amino acids on the diadic plane will have highly symmetric form. We have fixed the form of the map ρ using the local constancy and symmetry for the genetic code.

One could suggest to use for description of the genetic code the three-dimensional diadic space. Using the described above picture of degeneracy of the genetic code, we see that for the three-dimensional parametrization of the space of codons it would be natural to expect 2-times, 4-times and 8-times degeneracy of the genetic code, which will correspond to the local constancy of the map on small distances on one, two and three coordinates. But 8-times degeneracy of the genetic code does not exist. Thus the most natural object for parametrization of the genetic code should be two-dimensional, and we arrive to the diadic 8×8 plane.

Remark 17.1 (Eucaryotic and vertebrate mitochondrial code). The eucaryotic code differs from the vertebrate mitochondrial code by changing of the code for the codons AGA and AGG, for ATA, and for TGA. Compared to the eucaryotic code, the vertebrate mitochondrial code corresponds to the simpler and more regular table, since the corresponding map of nucleotides onto amino acids possess larger areas of local constancy with respect to the distance on the diadic plane. One can conjecture that the vertebrate mitochondrial code is more ancient than the eucaryotic code, since evolution with higher probability goes in the direction of breaking of symmetry.

17.4 Physical-chemical regularity of the genetic code

The map of the genetic code transfers the ultrametric on the diadic plane onto the space of amino acids. We define the distance between two amino acids as the minimum of distances between their preimages in the diadic plane. For example, distance between His and Gln is equal to 1/4, distance between Pro and Ala is equal to 1/2, distance between Asp and Ser is equal to 1. These examples show that ultrametric differs considerably from the Euclidean distance. For example, Asp and Ser, situated in the neighbor squares, have the maximal distance between them. This can be discussed as follows: ultrametric distance between the points is related to the hierarchy of balls containing these points. Codons corresponding to Asp and Ser lie in the balls which are far in the hierarchy.

A natural question arise – does this ultrametric make any physical sense? We claim that this ultrametric is related to physical properties of amino acids. Let us discuss the property of hydrophobicity. This property is related to polarity of the molecule and its charge in the solvent: hydrophobic molecules are neutral and non-polar. Hydrophobic amino acids, which are Leu, Phe, Ile, Met, Val, Cys, Trp, have high probability to be situated inside the protein (in the hydrophobic kernel), while the hydrophilic amino acids have high probability to lie on the surface of the protein and have a contact with water, see the book [134].

In the table below we put only hydrophobic amino acids and omit all the other. It is easy to see that hydrophobic amino acids are concentrated in the two balls – the lower left quadrant (Leu, Phe, Ile, Met, Val) and the third square of the second line (Cys, Trp).

—	—	—	
—	—	Trp / Cys	
Met / Ile	Val		
Leu / Phe	Leu		

We see that the property of hydrophobicity is related to 2-adic norm on the diadic plane. We say that the introduced parametrization satisfies the *proximity principle* – ultrametrically close amino acids have similar physical-chemical properties. Using the terminology of [237], one can say that proximity in ultrametric information space induces similarity of chemical properties (and, moreover, for hydrophobic amino acids, arrangement inside the protein, i.e. proximity in the physical space).

The next table contains polar amino acids. We see that their arrangement satisfies the proximity principle, in particular, all the seven amino acids in the upper left quadrant are polar.

Lys / Asn	Glu / Asp	Ter / Ser	
Ter / Tyr	Gln / His	—	Arg
—		Thr	
—		Ser	

Bibliography

[1] S. Albeverio, J. M. Bayod, C. Perez-Garcia, A. Yu. Khrennikov, and R. Cianci, *Non-Archimedean analogues of orthogonal and symmetric operators*, Izvestia Akademii Nauk 63 (1999), pp. 3–28.

[2] S. Albeverio, R. Cianci, and A. Yu. Khrennikov, *On the Fourier transform and the spectral properties of the p-adic momentum and Schrödinger operators*, J. Physics A: Math. and General 30 (1997), pp. 5767–5784.

[3] ——, *On the spectrum of the p-adic position operator*, J. Physics A: Math. and General 30 (1997), pp. 881–889.

[4] ——, *A representation of quantum field Hamiltonian in p-adic Hilbert space*, Theor. and Math. Physics 112 (1997), pp. 355–374.

[5] S. Albeverio, M. Gundlach, A. Yu. Khrennikov, and K.-O. Lindahl, *On Markovian behaviour of p-adic random dynamical systems*, Russian J. of Math. Phys. 8 (2) (2001), pp. 135–152.

[6] S. Albeverio and A. Yu. Khrennikov, *p-adic Hilbert space representation of quantum systems with an infinite number of degrees of freedom*, Int. J. of Modern Phys B 10 (1996), pp. 1665–1673.

[7] ——, *Representation of the Weyl group in spaces of square integrable functions with respect to p-adic valued Gaussian distributions*, J. Phys. A: Math. and General 29 (1996), pp. 5515–5527.

[8] ——, *A regularization of quantum field Hamiltonians with the aid of p-adic numbers*, Acta Appl. Math. 50 (1998), pp. 225–251.

[9] S. Albeverio, A. Yu. Khrennikov, and P. Kloeden, *Memory retrieval as a p-adic dynamical system*, Biosystems 49 (1999), pp. 105–115.

[10] S. Albeverio, A. Yu. Khrennikov, and B. Tirozzi, *p-adic neural networks*, Mathematical Models and Methods in Applied Sciences 9 (1999), pp. 1417–1437.

[11] J.-P. Alloche and J. Shallit, *Automatic Sequences. Theory, Applications, Generalizations*, Cambridge Univ. Press, 2003.

[12] R. C. Alperin, *p-adic binomial coefficients* mod P, The Amer. Math. Month. 92 (1985), pp. 576–578.

[13] M. G. Amaglobeli, *G-identities of nilpotent groups, I*, Algebra and Logic 40 (2001), pp. 3–21.

[14] ———, *G-identities of nilpotent groups, II*, Algebra and Logic 40 (2001), pp. 207–218.

[15] M. G. Amaglobeli and V. N. Remeslennikov, *G-identities and G-varieties*, Algebra and Logic 39 (2000), pp. 141–154.

[16] Y. Amice, *Interpolation p-adique*, Bull. Soc. Math. France 92 (1964), pp. 117–180.

[17] D. Amit, *Modeling Brain Function*, Cambridge Univ. Press, Cambridge, 1989.

[18] V. S. Anashin, *Mixed identities in groups*, Mat. Zametki 24 (1978), pp. 19–30.

[19] ———, *Solvable groups with operators and commutative rings having transitive polynomials*, Algebra and Logic 21 (1982), pp. 419–432, Translated from Russian.

[20] ———, *Mixed identities and mixed varieties of groups*, Mat. Sb. (N. S.) 129 (1986), pp. 163–174.

[21] ———, *Uniformly distributed sequences of p-adic integers*, Mathematical Notes 55 (1994), pp. 109–133.

[22] ———, *Uniformly distributed sequences over p-adic integers*, Number theoretic and algebraic methods in computer science. Proceedings of the Int. Conference (Moscow, June–July, 1993) (I. Shparlinski, A. J. van der Poorten and H. G. Zimmer, eds.), World Scientific, 1995, pp. 1–18.

[23] ———, *Uniformly distributed sequences in computer algebra, or how to constuct program generators of random numbers*, J. Math. Sci. 89 (1998), pp. 1355–1390.

[24] ———, *Uniformly distributed sequences of p-adic integers, II*, Discrete Math. Appl. 12 (2002), pp. 527–590, preprint available at `http://arXiv.org/math.NT/0209407`.

[25] ———, *Pseudorandom number generation by p-adic ergodic transformations*, available at `http://arxiv.org/abs/cs.CR/0401030`, 2004.

[26] ———, *Pseudorandom number generation by p-adic ergodic transformations: An addendum*, available at `http://arxiv.org/abs/cs.CR/0402060`, 2004.

[27] ———, *Ergodic transformations in the space of p-adic integers*, p-adic Mathematical Physics. 2nd Int. Conference (Belgrade, Serbia and Montenegro 15–21 September 2005) (Melville, New York) (Andrei Yu. Khrennikov, Zoran Rakić, and Igor V. Volovich, eds.), AIP Conference Proceedings, vol. 826, American Institute of Physics, 2006, preprint available at `http://arXiv.org/abs/math.DS/0602083`, pp. 3–24.

[28] ———, *Wreath products in stream cipher design*, Proceedings of the Int. Security and Counteracting Terrorism Conference, Moscow, 2–3 November 2005 (Moscow), Lomonosov Moscow State University, NATO-Russia Counsil, 2006, preprint available at `http://arXiv.org/abs/cs.CR/0602012`, pp. 135–161.

[29] ———, *Non-archimedean ergodic theory and pseudorandom generators*, The Computer Journal (2008), DOI: 10.1093/comjnl/bxm101.

[30] ———, *Non-archimedean theory of T-functions*, Proc. Advanced Study Institute Boolean Functions in Cryptology and Information Security, IOS Press, Amsterdam, etc., 2008, pp. 33–57.

[31] ———, *Automata: a p-adic view*, p-adic Mathematical Physics, Proc. 4th Int. Conference, Hrodna, 2009, to appear.

[32] V. S. Anashin and M. V. Larin, *Interpolation on A_5*, The 8-th all-Union Symp. on Group Theory. Abstracts. (Sumi), 1982, (Russian), pp. 6–7.

[33] T. M. Apostol, *Introduction to Analytic Number Theory*, Springer-Verlag, Berlin, New York, Heidelberg, 1976.

[34] I. Ya. Aref'eva, B. Dragovich, P. H. Frampton, and I. V. Volovich, *The wave function of the universe and p-adic gravity*, Int. J. of Modern Phys. 6 (1991), pp. 4341–4358.

[35] I. Ya. Aref'eva, B. Dragovich, and I. V. Volovich, *On the p-adic summability of the anharmonic ocillator*, Phys. Lett. B 200 (1988), pp. 512–514.

[36] L. M. Arkhipov, *Finite principal ideal rings*, Math. Notes 12 (1973), pp. 656–659.

[37] V. I. Arnold, *Ferma-Euler dynamical systems and the statistics of arithmetics of geometric progressions*, Func. An Appl. 37 (2003), pp. 1–15.

[38] ——, *Geometry and dynamics of Galois fields*, Russian Math. Surveys 59 (2003), pp. 1029–2046.

[39] ——, *Number-theoretic turbulence in Fermat-Euler arithmetics and large Young diagrams geometry statistics*, J. Math. Fluid Mech. 7 (2005), pp. 4–50.

[40] D. Arrowsmith and F. Vivaldi, *Some p-adic representations of the Smale horseshoe*, Phys. Lett. A 176 (1993), pp. 292–294.

[41] ——, *Geometry of p-adic Siegel discs*, Physica D 71 (1994), pp. 222–236.

[42] R. Ashby, *Design of a brain*, Chapman-Hall, London, 1952.

[43] B. J. Baars, *In the theater of consciousness. The workspace of mind*, Oxford University Press, Oxford, 1997.

[44] M. Bardet, J.-C. Faugère, and B. Salvy, *Complexity of Gröbner basis computation for Semi-regular Overdetermined sequences over $\mathbb{F}_2$ with solutions in $\mathbb{F}_2$*, available at `http://www.inria.fr/rrrt/rr-5049.html`, 2004.

[45] A. Batra and P. Morton, *Algebraic dynamics of polynomial maps on the algebraic closure of a finite field, I*, Rocky Mountain Journal of Mathematics 24 (1994), pp. 453–481.

[46] ——, *Algebraic dynamics of polynomial maps on the algebraic closure of a finite field, II*, Rocky Mountain Journal of Mathematics 24 (1994), pp. 905–932.

[47] J. M. Bayod, *Productos Internos en Espacios Normados non-Arquimedianos*, Univ. de Bilbao Press, Bilbao, 1976.

[48] W. Bechtel and A. Abrahamsen, *Connectionism and the mind*, Basil Blackwell, Cambridge, 1991.

[49] W. Bechterew, *Die Funktionen der Nervencentra*, Fischer, Jena, 1911.

[50] E. Beltrametti and G. Cassinelli, *Quantum mechanics and p-adic numbers*, Found. Phys. 2 (1972), pp. 1–7.

[51] S. Ben-Menahem, *p-adic iterations*, Preprint, TAUP 1627–88, Tel Aviv University, 1988.

[52] R. Benedetto, *Fatou Components in p-adic Dynamics*, Ph.D. thesis, Department of Mathematics at Brown University, May 1998.

[53] ——, *p-adic Dynamics and Sullivan's No Wandering Domains Theorem*, Composito Mathematica 122 (2000), pp. 281–298.

[54] ——, *Hyperbolic maps in p-adic dynamics*, Ergodic Theory and Dynamical Systems 21 (2001), pp. 1–11.

[55] ——, *Reduction, dynamics, and Julia sets of rational functions*, J. of Number Theory 86 (2001), pp. 175–195.

[56] ——, *Components and periodic points in non-Archimedean dynamics*, Proc. Lond. Math. Soc. 84 (2002), pp. 231–256.

[57] ——, *Examples of wandering domains in p-adic polynomial dynamics*, C. R. Math. Acad. Sci. Paris 335 (2002), pp. 615–620.

[58] ——, *Non-Archimedean holomorphic maps and the Ahlfors islands theorem*, Amer. J. Math. 125 (2003), pp. 581–622.

[59] ——, *Heights and preperiodic points of polynomials over function fields*, Int. Math. Res. Notices 62 (2005), pp. 3855–3866.

[60] ——, *Wandering domains in non-Archimedean polynomial dynamics*, Bull. London. Math. Soc 38 (2006), pp. 937–950.

[61] ——, *Periodic points of polynomials over global fields*, J. Reine Angew. Math. 608 (2007), pp. 123–153.

[62] J. Benois-Pineau, A. Yu. Khrennikov, and N. V. Kotovich, *Segmentation of images in p-adic and Euclidean metrics*, Doklady Mathematics 64 (2001), pp. 450–455.

[63] V. V. Bezgin, M. Endo, A. Yu. Khrennikov, and M. Yuoko, *Statistical biological models with p-adic stabilization*, Dokl. Akad. Nauk 334 (1994), pp. 5–8.

[64] J.-P. Bézivin, *Sur le points périodiques des applications rationelles en dynamique ultramétrique*, Acta Arith. 100 (2001), pp. 63–74.

[65] ——, *Sur les ensembles de Julia et Fatou des fonctions entiéres ultramétrique*, Ann. Inst. Fourier (Grenoble) 51 (2001), pp. 1635–1661.

[66] ——, *Fractions rationelles hyperboliques p-adiques*, Acta Arith. 112 (2004), pp. 151–175.

[67] ——, *Sur la compacité des ensembles de Julia des polynômes p-adiques*, Math. Z. 246 (2004), pp. 273–289.

[68] M. Bhargava, *The factorial function and generalizations*, Amer. Math. Monthly 107 (2000), pp. 783–799.

[69] C. Blomberg, H. Liljenström, B. I. Lindahl, and P. Arhem (eds.), *Mind and matter: Essays from biology, physics and philosophy: An introduction*, J. Theor. Biol. 171 (1994), pp. 1–230.

[70] M. A. Boden, *Artificial genius*, Discover 17 (1996), pp. 104–107.

[71] ____, *Creativity and artificial intelligence*, Artificial Intelligence 103 (1998), pp. 347–356.

[72] ____, *Mind as machine – A history of cognitive science*, Oxford University Press, Oxford, 2006.

[73] D. Bohm and B. Hiley, *The undivided universe: An ontological interpretation of quantum mechanics*, Routledge and Kegan Paul, London, 1993.

[74] D. Bosio and F. Vivaldi, *Round-off errors and p-adic numbers*, Nonlinearity 13 (2000), pp. 309–322.

[75] W. Brauer, *Automatentheorie*, B. G. Teubner, Stuttgart, 1984.

[76] R. P. Brent, *Factorization of the tenth Fermat number*, Math. Comput. 68 (1999).

[77] E. F. Brickell and A. M. Odlyzko, *Cryptanalysis: A Survey of Recent Results*, Proc. IEEE 76 (1988), pp. 578–593.

[78] R. A. Brooks, *Cambrian intelligence: The early history of the new AI*, The MIT Press, Cambridge, MA, 1999.

[79] ____, *Flesh and machines: How robots will change us*, Pantheon Books, New York, 2002.

[80] J. Bryk and C. E. Silva, *Measurable dynamics of simple p-adic polynomials*, Amer. Math. Monthly 112 (2005), pp. 212–232.

[81] P.-J. Cahen and J.-L. Chabert, *Integer-Valued Polynomials*, Math. Surv. and Monogr., vol. 48, Amer. Math. Soc., 1997.

[82] G. Call and J. Silverman, *Canonical height on varieties with morphisms*, Composito Math. 89 (1993), pp. 163–205.

[83] J. L. Chabert, A.-H. Fan, and Y. Fares, *Minimal dynamical systems on a discrete valuation domain*, Discrete and continuous dynamical systems (2008), to appear.

[84] N. Chomsky, *Formal properties of grammas*, Handbook of mathematical psychology (New-York) (R. D. Luce, R. R. Bush, and E. Galanter, eds.), Wiley, 1963, pp. 323–418.

[85] W.-S. Chou and I. E. Shparlinski, *On thye cycle structure of repeated exponentiation modulo a prime*, J. Number Theory 107 (2004), pp. 345–356.

[86] P. S. Churchland and T. Sejnovski, *The computational brain*, MITP, Cambridge, 1992.

[87] R. Cianci and A. Yu. Khrennikov, *Can p-adic numbers be useful to regularize divergent expectation values of quantum observables?*, Int. J. of Theor. Phys. 33 (1994), pp. 1217–1228.

[88] ____, *Energy levels corresponding to p-adic quantum states*, Dokl. Akad. Nauk 342 (1995), pp. 603–606.

[89] R. Cianci and A. Yu. Khrennikov, *Poisson brackets as the adelic limit of quantum commutators*, J. Physics A: Math. and General 27 (1994), pp. 7875–7882.

[90] A. Clark, *Psychological models and neural mechanisms. An examination of reductionism in psychology*, Clarendon Press, Oxford, 1980.

[91] R. J. Collings and D. R. Jefferson, *Towards simulated evolution*, Artificial life 2 (Redwood City), Addison Wesley, 1992, pp. 579–601.

[92] N. Courtois, A. Klimov, J. Patarin, and A. Shamir, *Efficient Algorithms for Solving Overdefined Systems of Multivariate Polynomial Equations*, Eurocrypt 2000, Lect. Notes Comp. Sci., vol. 1807, Springer-Verlag, 2000, pp. 392–407.

[93] N. Cristianini and M. Hahn, *Introduction to computational genomics*, Cambridge University Press, Cambridge, 2006.

[94] R. Crowell and R. Fox, *Introduction to the Knot Theory*, Ginu and Co., Boston, 1963.

[95] A. R. Damasio, *Descartes' error: emotion, reason, and the human brain*, Anton Books, New York, 1994.

[96] H. Damasio and A. R. Damasio, *Lesion analysis in neuropsychology*, Oxford Univ. Press, New York, 1989.

[97] N. De Grande-De Kimpe and A. Yu. Khrennikov, *The non-Archimedean Laplace transform*, Bull. Belgian Math. Soc. 3 (1996), pp. 225–237.

[98] N. De Grande-De Kimpe, A. Yu. Khrennikov, and L. Van Hamme, *The Fourier transform for p-adic smooth distributions*, Lecture Notes in Pure and Applied Mathematics, vol. 207, Marcel Dekker, 1999, pp. 97–112.

[99] Ch. de la Valle Poussin, *Recherches analytiques sur la théorie des nombers premiers*, Ann. Soc. Sci. Bruxelles 20 (1896), pp. 183–256, 281–297.

[100] J. Dénes and A. D. Keedwell, *Latin squares*, North-Holland, Amsterdam, 1991.

[101] D. L. Desjardins and M. E. Zieve, *On the structure of polynomial mappings modulo an odd prime power*, available at `http://arXiv.org/math.NT/0103046`, 2001.

[102] E. B. Dinkin, *Markov processes*, Fizmatgiz, Moscow, 1963.

[103] J. Y. Donnart and J. A. Meyer, *Learning reactive and planning rules in a motivationally autonomous animat*, IEEE Trans. Systems, Man., and Cybernetics. Part B: Cybernetics 26 (1996), pp. 381–395.

[104] B. Dragovich, *On signature change in p-adic space-time*, Mod. Phys. Lett. 6 (1991), pp. 2301–2307.

[105] ——, *Adelic harmonic oscillator*, Int. J. Mod. Phys. A 10 (1995), pp. 2349–2365.

[106] B. Dragovich and A. Yu. Dragovich, *A p-adic model of DNA sequence and genetic code*, available at `http://arxiv.org/abs/q-bio.GN/0607018`, 2006.

[107] ——, *A p-adic model of DNA sequence and genetic code*, P-Adic Numbers, Ultrametric Analysis, and Applications 1:1 (2009), pp. 34–41.

[108] ——, *p-adic modeling of the genome and the genetic code*, Computer Journal (2009), DOI: 10.1093/comjnl/bxm083.

[109] D. Dubischar, V. M. Gundlach, O. Steinkamp, and A. Yu. Khrennikov, *A p-adic model for the process of thinking disturbed by physiological and information noise*, J. Theor. Biology 197 (1999), pp. 451–467.

[110] F. Durand and F. Paccaut, *Minimal polynomial dynamics on the set of 3-adic integers*, Bull. London Math. Soc. 41:2 (2009), pp. 302–314.

[111] K. Schmidt, *Dynamical systems of algebraic origin*, Progress in Mathematics, vol. 128, Birkhäuser Verlag, Basel, 1995.

[112] B. Dwork, *On p-adic differential equations, 1. The Frobenius structure of differential equations*, Bull. Soc. Math. France no. 39/40 (1974), pp. 27–37.

[113] ——, *Lectures on p-adic differential equantions*, Springer-Verlag, Berlin, Heidelberg, New York, 1982.

[114] B. Dwork, G. Gerotto, and F. J. Sullivan, *An introduction to G-functions*, Annals of Math. Studies, Princeton Univ. Press, Princeton, 1994.

[115] B. Dwork and P. Robba, *On ordinary linear p-adic differential equations with algebraic functional coefficients*, Trans. Amer. Math. Soc. 231 (1977), pp. 1–46.

[116] G. M. Edelman, *The remembered present: A biological theory of consciousness*, Basic Books, New York, 1989.

[117] J. Eichenauer, J. Lehn, and A. Topuzoğlu, *A nonlinear congruential pseudorandom number generator with power of two modulus*, Math. Comp. 51 (1988), pp. 757–759.

[118] J. Eichenauer-Herrmann, *Quadratic congruential pseudorandom numbers: distribution of lagged pairs*, J. Comput. Appl. Math. 79 (1997), pp. 75–85.

[119] J. Eichenauer-Herrmann and H. Grothe, *A new inversive congruential pseudorandom number generator with power of two modulus*, ACM Trans. Modelling and Computer Simulation 2 (1992), pp. 1–11.

[120] J. Eichenauer-Herrmann, E. Herrmann, and S. Wegenkittl, *A survey of quadratic and inversive congruential pseudorandom numbers*, Monte Carlo and Quasi-Monte Carlo Methods 1996 (N.Y.) (P. Hellekalek, G. Larcher, H. Niederreiter, and P. Zinterhof, eds.), Lecture Notes in Statistics, vol. 127, Springer-Verlag, 1998, pp. 66–97.

[121] C. Eliasmith, *The third contender: A critical examination of the dynamicist theory of cognition*, Phil. Psychology 9 (1996), pp. 441–463.

[122] F. Emmerich, *Equidistribution properties of quadratic congruential pseudorandom numbers*, J. Comput. Appl. Math. 79 (1997), pp. 207–217.

[123] M. Endo and A. Yu. Khrennikov, *On the annihilators of the p-adic Gaussian distributions*, Comm. Math. Univ. Sancti Pauli 144 (1995), pp. 105–108.

[124] A. Escassut, *Analytic elements in p-adic analysis*, World Scientific, Singapore, 1995.

[125] A. Escassut and A. Yu. Khrennikov, *Nonlinear differential equations over the field of complex p-adic numbers*. *p*-adic functional analysis (W. Schikhof, C. Perez-Garcia, and J. Kakol, eds.), Lecture Notes in Pure and Applied Mathematics, vol. 192, 1997, pp. 143–151.

[126] G. Everest, A. van der Poorten, I. Shparlinski, and T. Ward, *Recurrence Sequences*, American Mathematical Society Surveys, vol. 104, American Mathematical Society, 2003.

[127] A.-H. Fan, M.-T. Li, J.-Y. Yao, and D. Zhou, *p-adic affine dynamical systems and applications*, C. R. Acad. Sci. Paris Ser. I 342 (2006), pp. 129–134.

[128] ——, *Strict ergodicity of affine p-adic dynamical systems*, Adv. Math. 214 (2007), pp. 666–700.

[129] A.-H. Fan, L. Liao, Y. F. Wang, and D. Zhou, *p-adic repellers in $\mathbb{Q}_p$ are subshifts of finite type*, C. R. Acad. Sci. Paris 344 (2007), pp. 219-224.

[130] A.-H. Fan and L.-M. Liao, *On minimal decomposition of p-adic polynomial dynamical systems*, unpublished, 2008.

[131] A.-H. Fan and Y.-F. Wang, *On p-adic Dynamical Systems*, Proc. Int. Congr. Chinese Math., vol. 2, 2007, pp. 773–799.

[132] A.-H. Fan and X.-Y. Zhang, *Some properties of Riesz product on the ring of p-adic integers*, J. Fourier Analysis and Application (2008), to appear.

[133] C. Favre and J. Rivera-Letelier, *Théorème d'équidistribution de Brolin en dynamique p-adique*, C. R. Math. Acad. Sci. Paris 339 (2004), pp. 271–276.

[134] A. V. Finkelshtein and O. B. Ptitsyn, *Physics of proteins*, Academic Press, London, 2002.

[135] S. Fishenko and E. Zenelov, *p-adic models of turbulence*, p-adic Mathematical Physics (Melville, New York) (A. Y. Khrennikov, Z. Rakic, and I. V. Volovich, eds.), AIP Conference Proceedings, no. 826, AIP, 2006, pp. 174–191.

[136] J. A. Fodor and Z. W. Pylyshyn, *Connectionism and cognitive architecture: A critical analysis*, Cognition 280 (1988), pp. 3–17.

[137] P. H. Frampton, Y. Okada, and M. R. Ubriaco, *On adelic formulas for the p-adic strings*, Phys. Lett. B 213 (1988), pp. 260–262.

[138] L. Frappat, A. Sciarrino, and P. Sorba, *Crystalizing the genetic code*, J. Biol. Phys. 27 (2001), pp. 1–38.

[139] L. Frappat, P. Sorba, and A. Sciarrino, *A crystal base for the genetic code*, Phys. Lett. A. 250 (1998), pp. 214–221.

[140] S. Freud, *The interpretation of dreams*, standard edition, 4 and 5, Vienna, 1900.

[141] ——, *New introductory lectures on psychoanalysis*, Norton, New-York, 1933.

[142] ——, *Two short accounts of psychoanalysis*, Penguin Books, New York, 1962.

[143] P. G. O. Freund and E. Witten, *Adelic string amplitudes*, Phys.Lett. B 199 (1987), pp. 191–194.

[144] J. B. Friedlander, C. Pomerance, and I. E. Shparlinski, *Period of the power generator and small values of Carmichael's function*, Math. Comp. 70 (2001), pp. 1591–1605, electronic.

[145] A. Frieze, J. Hastad, R Kannan, J. C. Lagarias, and A. Shamir, *Reconstructing truncated integer variables satisfying linear congruences*, SIAM J. Comput. 17 (1988), pp. 262–280.

[146] J. M. D. Fuster, *The prefrontal cortex: anatomy, physiology, and neuropsychology of the frontal lobe*, Lippincott-Raven, Philadelphia, 1997.

[147] M. R. Garey and D. S. Johnson, *Computers and Intractability: A Guide to the Theory of NP-completeness*, W. H. Freeman and Co., 1979.

[148] P. Gay, *Freud: A life for our time*, W. W. Norton, New York, 1988.

[149] H. G. Geissler, *The inferential basis of classification: From perceptual to memory code systems. Part 1: Theory*, Modern issues in perception (Amsterdam) (H. G. Geissler, H. F. Buffart, E. L. Leeuwenberg, and V. Sarris, eds.), North-Holland, 1983, pp. 87–105.

[150] ——, *Sources of seeming redundancy in temporally quantized information processing*, Psychophysical explorations of mental structures. Proceedings of the XXIII International Congress of Psychology of the I. U. Psychological Society (Gottingen, Germany) (H. G. Geissler, ed.), Hogrefe and Huber Publishers, 1985, pp. 193–210.

[151] ——, *New magic numbers in mental activity: On a taxonomic system for critical time periods*, Cognition, information processing and psychophysics (Hillsdale, NJ) (H. G. Geissler, S. W. Link, and J. T. Townsend, eds.), Erlbaum, 1992, pp. 293–321.

[152] H. G. Geissler, F. Klix, and U. Scheidereiter, *Visual recognition of serial structure: Evidence of a two-stage scanning model*, Formal theories of perception (Chichester) (E. L. Leeuwenberg and H. F. Buffart, eds.), John Wiley, 1978, pp. 299–314.

[153] H. G. Geissler and R. Kompass, *Psychophysical time units and the band structure of brain oscillations*, 15th Annual Meeting of the International Society for Psychophysics, 1999, pp. 7–12.

[154] H. G. Geissler and M. Puffe, *Item recognition and no end: Representation format and processing strategies*, Psychophysical judgment and the process of perception (Amsterdam) (H. G. Geissler and T. Petzold, eds.), North-Holland, 1982, pp. 270–281.

[155] A. Gill, *Introduction to the theory of finite-state machines*, McGraw-Hill Inc., 1963.

[156] D. Gorenstein, *Finite Groups*, Harper's Series in Modern Math., Harper and Row, N.Y., Evanston, London, 1968.

[157] F. Q. Gouvêa, *p-adic Numbers, An Introduction*, 2nd edition, Springer-Verlag, Berlin, Heidelberg, New York, 1997.

[158] R. L. Graham, D. E. Knuth, and O. Patashnik, *Concrete Mathematics: A Foundation for Computer Science*, 2nd edition, Addison–Wesley, Reading., Ma., 1998.

[159] V. Green, *Emotional development in psychoanalysis, attachment theory and neuroscience: Creating connections*, Routledge, New York, 2003.

[160] M. Gundlach, A. Yu. Khrennikov, and K.-O. Lindahl, *Topological transitivity for p-adic dynamical systems*, P-adic functional analysis (New York–Basel) (A. K. Katsaras, W. H. Schikhof, and L. van Hamme, eds.), Lecture Notes in Pure and Applied Mathematics, vol. 222, Marcel Dekker, 2000, pp. 127–132.

[161] ——, *On ergodic behavior of p-adic dynamical systems*, Inf. Dim. An., Quantum Prob. and Related Fields 4(4) (2001), p. 569Ů577.

[162] V. M. Gundlach, A. Yu. Khrennikov, and K.-O. Lindahl, *Ergodicity on p-adic sphere*, German Open Conference on Probability and Statistics (Hamburg), University of Hamburg Press, 2000, pp. 15–21.

[163] J. Hadamard, *Sur la distribution des zéros de la fonction $\zeta(s)$ et ses consequences arithmétiques*, Bull. Soc. Math. France 24 (1896), pp. 199–220.

[164] M. Hall Jr., *The Theory of Groups*, Macmillan, New York, 1959.

[165] ——, *Combinatorial Theory*, Blaisdell Publ. Co., 1967.

[166] R. R. Hall, *On pseudo-polynomials*, Arch. Math. 18 (1971), pp. 71–77.

[167] P. R. Halmos, *Lectures on Ergodic Theory*, Pub. Math. Soc. Japan, Kenkyusha, Tokyo, 1956.

[168] J. Hartmanis and R. E. Stearns, *Algebraic structure theory of sequential machines*, Prentice-Hall Inc., 1966.

[169] K. Hensel. *Über eine neue Begründung der Theorie der algebraischen Zahlen"*, Jahresbericht der Deutschen Mathematiker-Vereinigung, vol. 6, number 3, 1897, pp. 83–88.

[170] M. R. Herman and J.-C. Yoccoz, *Generalization of some theorem of small divisors to non-archimedean fields*, Geometric Dynamics, vol. LNM 1007, Springer-Verlag, New York, Berlin, Heidelberg, 1983, pp. 408–447.

[171] F. C. Hoppensteadt, *An introduction to the mathematics of neurons: Modeling in the frequency domain*, Cambridge Univ. Press, New York, 1997.

[172] F. C. Hoppensteadt and E. Izhikevich, *Canonical models in mathematical neuroscience*, Proc. of Int. Math. Congress (Berlin), vol. 3, 1998, pp. 593–600.

[173] L. Hsia, *A weak Néron model with applications to p-adic dynamical systems*, Compositio Mathematica 100 (1996), pp. 227–304.

[174] ——, *Closure of periodic points over a non-archimedean field*, J. London Math. Soc. 62 (2000), pp. 685–700.

[175] Z. Huang and A. Yu. Khrennikov, *p-adic valued white noise*, Quantum Probility and Related Topics 9 (1994), pp. 273–294.

[176] D. Hubel and T. S. Wiesel, *Receptive fields, binocular interaction and functional architecture in the cat's visula cortex*, J. Physiology 160 (1962), pp. 106–154.

[177] A. M. Ivanitsky, *Brain's physiology and the origin of the human's subjective world*, J. of High Nerves Activity 49 (1999), pp. 707–713.

[178] M. A. Jimenez-Montano, C. R. de la Mora-Basanez, and Th. Pöshel, *The hypercube structure of the genetic code explains conservative and non–conservative aminoacid substitutions in vivo and in vitro*, BioSystems 39 (1996), pp. 117–125.

[179] D. Jonah and B. M. Schreiber, *Transitive affine transformations on groups*, Pacific J. Math. 58 (1975), pp. 483–509.

[180] H. K. Kaiser and W. Nöbauer, *Permutation polynomials in several variables over residue class rings*, J. Austral. Math. Soc. A43 (1987), pp. 171–175.

[181] G. K. Kalisch, *On p-adic Hilbert spaces*, Ann. of Math. 48 (1947), pp. 180–192.

[182] T. Kato, L.-M. Wu, and N. Yanagihara, *On a nonlinear congruential pseudorandom number generator*, Math. Comp. 65 (1996), pp. 227–233.

[183] A. Katok and B. Hasselblatt, *Introduction to the Modern Theory of Dynamical Systems*, Cambridge University Press, 1998.

[184] H. Keller, H. Ochsenius, and W. H. Schikhof, *On the commutation relation $AB - BA = I$ for operators on nonclassical Hilbert spaces*, P-adic functional analysis (New York, Basel) (A. K. Katsaras, W. H. Schikhof, and L. van Hamme, eds.), Lecture Notes in Pure and Applied Mathematics, vol. 222, Marcel Dekker, 2000, pp. 177–190.

[185] A. Yu. Khrennikov, *Mathematical methods of non-Archimedean physics*, Uspekhi Mat. Nauk 45 (1990), pp. 79–110.

[186] ____, *Quantum mechanics over Galios extensions of number fields*, Dokl. Akad. Nauk USSR 315 (1990), pp. 860–864.

[187] ____, *Quantum mechanics over non-Archimedean number fields*, Theor. and Math. Phys. 83 (1990), pp. 406–418.

[188] ____, *Representations of Schrödinger and Bargmann-Fock in non-Archimedean quantum mechanics*, Dokl. Akad. Nauk USSR 313 (1990), pp. 325–329.

[189] ____, *Representation of the second quantization over non-Archimedean number fields*, Dokl. Akad. Nauk USSR 314 (1990), pp. 1380–1384.

[190] ____, *p-adic quantum mechanics with p-adic valued functions*, J. Math. Phys. 32 (1991), pp. 932–937.

[191] ____, *Real-non-Archimedean structure of space-time*, Theor. and Math. Phys. 86 (1991), pp. 177–190.

[192] ____, *Analysis on the p-adic superspace, 1. Generalized functions: Gaussian distribution*, J. Math. Phys. 33 (1992), pp. 1636–1642.

[193] ____, *Analysis on the p-adic superspace, 2. Differential equations*, J. Math. Phys. 33 (1992), pp. 1643–1647.

[194] ____, *Axiomatics of the p adic theory of probability*, Dokl. Akad. Nauk 326 (1992), pp. 1075–1079.

[195] ____, *p-adic probability and statistics*, Dokl. Akad. Nauk 322 (1992), pp. 1075–1079.

[196] ____, *Discrete p-adic valued probabilities*, Dokl. Akad. Nauk 333 (1993), pp. 162–164.

[197] ____, *p-adic probability theory and its applications. The principle of statistical stabization of frequencies*, Theor. and Math. Phys. 97 (1993), pp. 348–363.

[198] ____, *p-adic statistical models*, Dokl. Akad. Nauk 330 (1993), pp. 300–304.

[199] ____, *An algorithmic approach to p-adic probability theory*, Dokl. Akad. Nauk 335 (1994), pp. 35–38.

[200] ____, *Bernoulli probabilities with p-adic values*, Dokl. Akad. Nauk 338 (1994), pp. 313–316.

[201] ——, *p-adic Valued Distributions in Mathematical Physics*, Kluwer Academic Publishers, Dordrecht, 1994.

[202] ——, *A limit theorem for p-adic probabilities*, Izvestia Akad. Nauk., ser. Matem. 59 (1995), pp. 207–223.

[203] ——, *A p-adic behaviour of the standard dynamical systems*, 1995.

[204] ——, *On probablity distributions on the field of p-adic numbers*, Theory of Probab. and Appl. 40 (1995), pp. 189–192.

[205] ——, *p-adic probability description of Dirac's hypothetical world*, Int. J. Theor. Phys. 34 (1995), pp. 2423–2434.

[206] ——, *p-adic probability distribution of hidden variables*, Physica A 215 (1995), pp. 577–587.

[207] ——, *p-adic probability interpretation of Bell's inequality paradoxes*, Physics Letters A 200 (1995), pp. 119–223.

[208] ——, *Statistical interpretation of p-adic quantum theories with p-adic valued wave functions*, J. Math. Phys. 6 (1995), pp. 6625–6632.

[209] ——, *The problems of the non-Archimedean analysis generated by quantum physics*, Ann. Math. Blaise Pascal 2 (1995), pp. 181–190.

[210] ——, *p-adic quantum-classical analogue of the Heisenberg uncertainty relations*, Il Nuovo Cimento B 112 (1996), pp. 555–560.

[211] ——, *The Einstein-Podolsky-Rosen paradox and the p-adic probability theory*, Doklady Mathematics 54 (1996), pp. 790–795.

[212] ——, *Ultrametric Hilbert space representation of quantum mechanics with a finite exactness*, Found. Physics 26 (1996), pp. 1033–1054.

[213] ——, *A stochastic model with p-adic distribution of hidden variables and Kolmogorov physical probabilities*, Dokl. Akad. Nauk 356 (1997), pp. 317–320.

[214] ——, *Non-Archimedean Analysis: Quantum Paradoxes, Dynamical Systems and Biological Models*, Kluwer Academic Publishers, Dordrecht, 1997.

[215] ——, *On the physical interpretation of negative probabilities in Prugovecki's empirical theory of measurement*, Canadian J. of Physics 75 (1997), pp. 291–298.

[216] ——, *p-adic classification of fractals and chaos*, Dokl. Akad. Nauk 357 (1997), pp. 737–740.

[217] ——, *The description of brain's functioning by the p-adic dynamical systems*, 1997.

[218] ——, *The uncertainty relation for coordinate and momentum operators in the p-adic Hilbert space*, Doklady. Math. 55 (1997), pp. 283–285.

[219] ——, *Bernoulli theorem for probabilities that take p-adic values*, Doklady Mathematics 55 (1998), pp. 402–405.

[220] ——, *Description of experiments detecting p-adic statistics in quantum diffraction experiments*, Doklady Mathematics 58 (1998), pp. 478–480.

[221] ——, *Human subconscious as the p-adic dynamical system*, J. of Theor. Biology 193 (1998), pp. 179–196.

[222] ——, *Non-Kolmogorov probabilistic models with p-adic probabilities and foundations of quantum mechanics*, Stochastic Analysis and Related Topics 42 (1998), pp. 275–304.

[223] ——, *p-adic behaviour of Bernoulli probabilities*, Statistics and Probability Letters 37 (1998), pp. 375–380.

[224] ——, *p-adic dynamical systems: description of concurrent struggle in biological population with limited growth*, Dokl. Akad. Nauk 361 (1998), p. 752.

[225] ——, *p-adic probability predictions of correlations between particles in the two slit and neuron interferometry experiments*, Il Nuovo Cimento B 113 (1998), pp. 751–760.

[226] ——, *p-adic stochastic hidden variable model*, J. of Math. Physics 39 (1998), pp. 1388–1402.

[227] ——, *Classical and quantum mechanics on information spaces with applications to cognitive, psychological, social and anomalous phenomena*, Found. Phys. 29 (1999), pp. 1065–1098.

[228] ——, *Description of the operation of the human subconscious by means of p-adic dynamical systems*, Dokl. Akad. Nauk 365 (1999), pp. 458–460.

[229] ——, *Classical and quantum mechanics on p-adic trees of ideas*, Biosystems 56 (2000), pp. 95–120.

[230] ——, *Informational interpretation of p-adic physics*, Dokl. Akad. Nauk 373 (2000), pp. 174–177.

[231] ——, *Laws of large numbers in non-Archimedean probability theory*, Izvestia. Akademii Nauk, 64 (2000), pp. 211–223.

[232] ——, *p-adic discrete dynamical systems and collective behaviour of information states in cognitive models*, Discrete Dynamics in Nature and Society 5 (2000), pp. 59–69.

[233] ——, *Interpretations of probability and their p-adic extension*, Theory of Probability and its Applications 46 (2001), pp. 312–325.

[234] ——, *Small denominators in complex p-adic dynamics*, Indag. Mathem. N.S. 12(2) (2001), pp. 177–189.

[235] ——, *Classical and quantum mental models and Freud's theory of unconscious mind*, Växjö Univ. Press, Växjö, 2002.

[236] ——, *p-adic model of hierarchical intelligence*, Dokl. Akad. Nauk. 388 (2003), pp. 1–4.

[237] ——, *Information dynamics in cognitive, psychological, social, and anomalous phenomena*, Kluwer Academic Publishers, Dordreht, 2004.

[238] ——, *Modelirovanie prozessa mishleniya v p-adicheskih koordinatah (Modeling of processes of thinking in p-adic coordinates)*, Nauka, Fizmatlit, Moscow, 2004.

[239] ——, *Probabilistic pathway representation of cognitive information*, J. Theor. Biology 231 (2004), pp. 597–613.

[240] ——, *p-adic information space and gene expression*, Integrative approaches to brain complexity (Cambridge) (S. Grant, N. Heintz, and J. Noebels, eds.), Wellcome Trust Publ., 2006, p. 14.

[241] ——, *p-adic dynamical representation of gene expression*, Foundations of probability and physics 4 (G. Adenier, C. Fuchs, and A. Yu. Khrennikov, eds.), AIP Conf. Proc., vol. 889, 2007, pp. 324-331.

[242] ——, *Interpretations of Probability*, 2nd edition, Walter de Gruyter, Berlin, New York, 2008.

[243] ——, *Gene expression from polynomial dynamics in the 2-adic information space*, Chaos, Solitons & Fractals (2009), DOI: 10.1016/j.chaos.2008.12.012

[244] A. Yu. Khrennikov and Zhiyan Huang, *Generalized functionals of p-adic white noise*, Dokl. Akad. Nauk, 344 (1995), pp. 23–26.

[245] ——, *A model for white noise analysis in p-adic number fields*, Acta Mathematica Scientia (China) 16 (1996), pp. 1–14.

[246] A. Yu. Khrennikov and A. Yu. Kotovich, *Representation and compression of images with the aid of the m-adic coordinate system*, Dokl. Akad. Nauk. 387 (2002), pp. 159–163.

[247] A. Yu. Khrennikov, N. V. Kotovich, and E. L. Borzistaya, *Compression of images with the aid of representation by p-adic maps and approximation by Mahler's polynomials*, Doklady Mathematics 69 (2004), pp. 373–377.

[248] A. Yu. Khrennikov and S. V. Kozyrev, *Genetic code on the dyadic plane*, Physica A: Statistical Mechanics and its Applications 381 (2007), pp. 265–272.

[249] ——, *p-adic numbers in bioinformatics: from genetic code to PAM-matrix*, available at `http://arxiv.org/abs/0903.0137`, 2009.

[250] A. Yu. Khrennikov, K.-O. Lindahl, and M. Gundlach, *Ergodicity in the p-adic framework*, Operator Methods in Ordinary and Partial Differential Equations (Basel, Boston, Berlin), Operator Methods: Advances and Applications, vol. 132, Birkhäuser, 2002.

[251] A. Yu. Khrennikov and S. Ludkovsky, *Stochastic processes on non-Archimedean spaces with values in non-Archimedean fields*, Advanced Stud. in Contemporary Math. 5 (2002), pp. 57–91.

[252] ——, *Stochastic processes in non-Archimedean spaces with values in non-Archimedean fields*, Markov Processes and Related Fields 9 (2003), pp. 131–162.

[253] A. Yu. Khrennikov, N. Mainetti, and M. Nilsson, *Non-archimedean dynamics*, Proc. Belgian mathematical society 20 (2002), pp. 5–15.

[254] A. Yu. Khrennikov and M. Nilsson, *On the number of cycles of p-adic dynamical systems*, Journal of Number Theory 90(2) (2001), pp. 255–264.

[255] ——, *Behaviour of Hensel perturbations of p-adic monomial dynamical systems*, Analysis Mathematica 29 (2003), pp. 107–133.

[256] ——, *p-adic deterministic and random dynamics*, Kluwer Academic Publ., Dordrecht etc., 2004.

[257] A. Yu. Khrennikov, M. Nilsson, and R. Nyqvist, The asymptotic number of periodic points of discrete polynomial p-adic dynamical system, Ultrametric functional analysis, seventh international conference on p-adic analysis, Contemporary Mathematics, vol. 319, Am. Math. Soc., 2003, pp. 159–166.

[258] A. Yu. Khrennikov and P. A. Svensson, *Attracting fixed points of polynomial dynamical systems in fields of p-adic numbers*, Izvestiya Mathematics 71 (2007), pp. 753–764.

[259] A. Yu. Khrennikov and Sh. Yamada, *On the concept of a random sequence with respect to p-adic probabilities*, Theory of Probability and Applications 49 (2004).

[260] A. Yu. Khrennikov, S. Yamada, and A. van Rooij, *Measure-theoretical approach to p-adic probability theory*, Annals Math. Blaise Pascal 6 (1999), pp. 21–32.

[261] J. Kingsbery, A. Levin, A. Preygel, and C. E. Silva, *Measurable Dynamics of Maps on Profinite Groups*, Indag. Math. (N. S.) 18 (2007), pp. 561–581, preprint available at `http://arxiv.org/abs/math/0701899v1`.

[262] ——, *On measure-preserving C1 transformations of compact-open subsets of non-Archimedean local fields*, Trans. Amer. Math. Soc. 361 (2009), pp. 61–85.

[263] A. Klapper and M. Goresky, *Feedback shift registers, 2-adic span, and combiners with memory*, J. Cryptology 10 (1997), pp. 111–147.

[264] A. Klimov and A. Shamir, *A new class of invertible mappings*, Cryptographic Hardware and Embedded Systems 2002 (B. S. Kaliski Jr. et al., ed.), Lect. Notes in Comp. Sci, vol. 2523, Springer-Verlag, 2003, pp. 470–483.

[265] ——, *Cryptographic applications of T-functions*, Selected Areas in Cryptography, 2003.

[266] ——, *New Cryptographic Primitives Based on Multiword T-Functions*, Fast Software Encryption: 11th International Workshop, FSE 2004, Delhi, India, February 5–7, 2004. Revised papers (B. Roy and W. Meier, eds.), Springer-Verlag, 2004, pp. 1–15.

[267] D. Knuth, *The Art of Computer Programming*, vol. 2: Seminumerical Algorithms, 3rd edition, Addison-Wesley, 1997.

[268] N. Koblitz, *p-adic numbers, p-adic analysis, and zeta-functions*, 2nd edition, Graduate texts in math., vol. 58, Springer-Verlag, 1984.

[269] A. N. Kochubei, *p-adic commutation relations*, Phys. A: Math. Gen. 29 (1996), pp. 6375–6378.

[270] A. N. Kolmogoroff, *Grundbegriffe der Wahrscheinlichkeitsrechnung*, Springer-Verlag, Berlin, 1933.

[271] A. N. Kolmogorov, *Foundations of the Probability Theory*, Chelsea Publishing Company, New York, 1956.

[272] ——, *Three approaches to the quantitative definition of information*, Problems Inform. Transmition 1 (1965), pp. 1–7.

[273] N. Kolokotronis, *Cryptographic Properties of Nonlinear Pseudorandom Number Generators*, Designs, Codes and Cryptography 46 (2008), pp. 353–363.

[274] L. Kotomina, *Fast nonlinear congruential generators*, Diploma Thesis, Russian State University for the Humanities, Moscow, 1999, (in Russian).

[275] H. Krawczuk, *How to predict congruential generators*, J. Algorithms 13 (1992), pp. 527–545.

[276] L. Kuipers and H. Niederreiter, *Uniform Distribution of Sequences*, John Wiley & Sons, New York etc., 1974.

[277] V. L. Kurakin, A. S. Kuzmin, A. V. Mikhalev, and A. A. Nechaev, *Linear Recurring sequences over rings and modules*, J. Math. Sci. 76 (1995), pp. 2793–2915.

[278] K. Kuratowsky, *Topology*, vol. 1, Academic Press, New York, London, 1966.

[279] P. Kurlberg and C. Pomerance, *On the period of the linear congruential and power generators*, Acta Arith. 119 (2005), pp. 149–169.

[280] A. S. Kuzmin and A. A. Nechaev, *A generator of uniform pseudorandom numbers*, Probabilistic Methods in Discrete Math. Abstr. 6th Int. Conf. Petrozavodsk, June 10–16, 2004, volume 11 of Appl. and Industrial Math. Surv., 2004, pp. 969–970.

[281] C. G. Langton, C. Taylor, J. D. Farmer, and S. Rasmussen, *Artificial life-2*, Addison Wesley, Redwood City, 1992.

[282] M. V. Larin, *Transitive polynomial transformations of residue class rings*, Discrete Mathematics and Applications 12 (2002), pp. 141–154.

[283] F. Laubie, A. Movahhedi, and A. Salinier, *Systemes dynamiques non archimediens et corps des normes*, LACO, Departement de Mathematiques, Faculte des Sciences, 123 avenue Albert Thomas, 87060 Limoges, France, 2000.

[284] H. Lausch, *Zur Theorie der Polynompermutationen über endlichen Gruppen*, Arch. Math. 19 (1968), pp. 284–288.

[285] ——, *Interpolation on the alternating group A_5*, Contrib. Gen. Algebra. Proc. Klagenfurt Conf. 1978 (Klagenfurt), 1979, pp. 187–192.

[286] H. Lausch and W. Nöbauer, *Algebra of Polynomials*, North-Holland Publ. Co, American Elsevier Publ. Co, 1973.

[287] C. F. Laywine and G. L. Mullen, *Discrete mathematics using Latin squares*, John Wiley & Sons, Inc., New York, 1998.

[288] W. J. le Veque, *Topics in Number Theory*, Addison-Wesley Publishing co., Reading, Mass., 1956.

[289] H.-C. Li, *Counting periodic points of p-adic power series*, Compos. Math. 100 (1996), pp. 361–364.

[290] ——, *p-adic dynamical systems and formal groups*, Compos. Math. 104 (1996), pp. 41–54.

[291] ——, *p-adic periodic points and Sen's theorem*, J. Number Theory 56 (1996), pp. 309–318.

[292] ——, *When is a p-adic power series an endomorphism of a formal group?*, Proc. Amer. Math. Soc. 124 (1996), pp. 2325–2329.

[293] ——, *Isogenies between dynamics of formal groups*, J. Number Theory 62 (1997), pp. 284–297.

[294] ——, *p-adic power series which commute under composition*, Trans. of A.M.S. 349 (1997), pp. 1437–1446.

[295] ——, *On heights of p-adic dynamical systems*, Proc. Amer. Math. Soc. 130 (2002), pp. 379–386.

[296] ——, *p-adic dynamical systems and formal groups*, Compos. Math. 130 (2002), pp. 75–88.

[297] M. Li and P. Vitanyi, *An Introduction to Kolmogorov Complexity and its Applications*, Springer-Verlag, Berlin, Heidelberg, New York, 1997.

[298] Shujun Li, *When chaos meets computers*, available at `http://arxiv.org/abs/nlin.CD/0405038`, 2004.

[299] R. Lidl and H. Niederreiter, *Finite Fields*, Addison-Wesley Publ. Co., 1983.

[300] K.-O. Lindahl, *Dynamical systems in p-adic geometry*, Licentiate thesis, School of Mathematics and Systems Engineering, Växjö University, 2001.

[301] ——, *On Siegel's linearization theorem for fields of prime characteristic*, Nonlinearity 17 (2004), pp. 745–763.

[302] J. Lubin, *Non-archimedean dynamical systems*, Compositio Mathematica 94 (1994), pp. 321–346.

[303] ——, *Sen's theorem on iteration of power series*, Proc. Am. Math. Soc. 123 (1995), pp. 66–99.

[304] ——, *Formal flows on the non-archimedean open unit disk*, Compos. Math. 124 (2000), pp. 123–136.

[305] F. Luca and I. Shparlinski, *On the number of polynomial maps into $\mathbb{Z}_n$*, Tsukuba Journal of Mathematics 30 (2006), pp. 439–449.

[306] A. Luczak, P. Bartho, S. L. Marguet, G. Buzsaki, and K. D Hariis, *Neocortical spontaneous activity in vivo: Cellular heterogeneity and sequential structure*, Preprint of CMBN, Rutgers University, 2007.

[307] M. Macmillan, *The completed arc: Freud evaluated*, MIT Press, Cambridge MA, 1997.

[308] K. Mahler, *p-adic numbers and their functions*, Cambridge Univ. Press, 1981, 2nd edition.

[309] Yu. Manin, *New dimensions in geometry*, Lect. Notes in Math., vol. 1111, Springer-Verlag, Berlin, New York, Heidelberg, 1985.

[310] T. I. Manuel, *Creating a robot culture: An interview with Luc Steels.*, IEEE Intelligent Systems 18 (2003), pp. 59–61.

[311] G. Marsaglia, *Xorshift RNGs*, Journal of Statistical Software (electronic) 08 (2003).

[312] G. Martin and C. Pomerance, *The iterated Carmichael λ-function and the number of cycles of the power generator*, Acta Arith. 118 (2005), pp. 305–335.

[313] P. Martin-Löf, *On the concept of random sequences*, Theory of Probability Appl. (1966), pp. 177–179.

[314] B. R. McDonald, *Finite Rings with Identity*, Marcel Dekker, New York, 1974.

[315] A. Menezes, P. van Oorshot, and S. Vanstone, *Handbook of Applied Cryptography*, CRC Press, 1997.

[316] J. A. Meyer and A. Guillot, *From SAB90 to SAB94: Four years of Animat research*, Proc. of the Third International Conference of Adaptive Behavior (Cambridge), The MIT Press, 1994.

[317] R. von Mises, *Grundlagen der Wahrscheinlichkeitsrechnung*, Math. Z. 5 (1919), p. 52-99.

[318] ——, *The mathematical theory of probability and statistics*, Academic, London, 1964.

[319] M. Misiurewicz, J. G. Stevens, and D. Thomas, *Iterations of linear maps over finite fields*, Linear Algebra and its Applications 413 (2006), pp. 218–234, available at `http://www.csam.montclair.edu/~thomasd/ducci.pdf`.

[320] J. Mitra and P. Sarkar, *Time-memory trade-off attacks on multimultiplications and T-functions*, Advances in Cryptology – Asiacrypt 2004 (P. J. Lee, ed.), Lect. Notes Comp. Sci., vol. 3329, 2004, pp. 468–482.

[321] H. Molland and T. Helleseth, *A linear weakness in the Klimov-Shamir T-function*, Proc. 2005 IEEE Int. Symp. on Information Theory, 2005, pp. 1106–1110.

[322] A. Monna, *Analyse non-Archimédienne*, Springer-Verlag, Berlin, Heidelberg, New York, 1970.

[323] A. Monna and T. Springer, *Integration non-Archimédienne, 1, 2*, Indag. Math. 25 (1963), pp. 634–653.

[324] A. Monna and F. van der Blij, *Models of space and time in elementary physics*, J. Math. Anal. and Appl. 22 (1968), pp. 537–545.

[325] K. Mori, *On the relation between physical and psychological time*, Proc. Int. Conf. Toward a Science of Consciousness (Tucson, Arizona), Tucson Univ. Press, 2002, p. 102.

[326] P. Morton, *Arithmetic properties of periodic points of quadratic maps*, Acta Arithmetica 62 (1992), pp. 343–372.

[327] ——, *Characterizing Cyclic Cubic Extensions by Automorphism Polynomials*, Journal of Number Theory 49 (1992), pp. 183–208.

[328] ——, *Periodic points, multiplicities, and dynamical units*, J. Reine Angew. Math. 461 (1995), pp. 81–122.

[329] ——, *On certain algebraic curves related to polynomial maps*, Compositio Mathematica 103 (1996), pp. 319–350.

[330] ——, *Periods of Maps on Irreducible Polynomials over Finite Field*, Finite Fields and Their Applications 3 (1997), pp. 11–24.

[331] ——, *Galois Groups of Periodic Points*, Journal of Algebra 201 (1998), pp. 401–428.

[332] P. Morton and P. Patel, *The Galois theory of periodic points of polynomial maps*, Proc. London Math. Soc. 68 (1994), pp. 225–263.

[333] P. Morton and J. H. Silverman, *Rational Periodic Points of Rational Functions*, IMRN International Mathematics Research Notices 2 (1994), pp. 97–110.

[334] ——, *Periodic points, multiplicities and dynamical units*, J. Reine Angew. Math. 461 (1995), pp. 81–122.

[335] P. Morton and F. Vivaldi, *Bifurcations and discriminants for polynomial maps*, Nonlinearity 8 (1995), pp. 571–584.

[336] F. Murtagh, *On ultrametricity, data coding, and computation*, J. of Classification 21 (2004), pp. 167–184.

[337] M. Nagata, *Local Rings*, vol. 13, Interscience tracts in pure and applied mathematics, New York, London, 1962.

[338] W. Narkiewicz, *Polynomial cycles in algebraic number fields*, Coll. Math 58 (1989), pp. 151–155.

[339] ——, *Polynomial mappings*, Springer-Verlag, Berlin, 1995.

[340] ——, *Arithmetics of dynamical systems: a survey*, Tatra Mt. Math. Publ. 11 (2000), pp. 69–75.

[341] ——, *Finite polynomial orbits. A survey*, Algebraic number theory and diophantine analysis (Berlin) (F. Halter-Koch et al., ed.), Proceeding of the international conference, Graz, Austria, August 30–September 5, 1998, Walter de Gruyter, Berlin, New York, 2000, pp. 331–338.

[342] W. Narkiewicz and T. Pezda, *Finite polynomial orbits in finitely generated domains*, Monatsh. Math 124 (1997), pp. 309–316.

[343] A. A. Nechaev, *Finite rings with applications*, Handbook of Algebra, vol. 5 (M. Hazewinkel, ed.), Elsevier B.V., 2008, pp. 213–320.

[344] H. Niederreiter, *Random number generation and quasi-Monte Carlo methods*, Society for Industrial and Applied Mathematics (SIAM), Philadelphia, PA, 1992.

[345] M. Nilsson, *Cycles of monomial and perturbated monomial p-adic dynamical systems*, Ann. Math. Blaise Pascal 7(1) (2000), pp. 37–63.

[346] ——, *Distribution of cycles of monomial p-adic dynamical systems*, P-adic functional analysis (New York, Basel), Lecture Notes in Pure and Applied Mathematics, Vol. 222, Marcel Dekker, 2001, pp. 127–132.

[347] ——, *Fuzzy Cycles of p-adic Monomial Dynamical Systems*, Far. East J. Dynamical Systems 5 (2003), pp. 149–173.

[348] R. Nyqvist, *Some dynamical systems in finite field extensions of the p-adic numbers*, p-adic functional analysis (A. K. Katsaras, W. H. Schikhof, and L. Van Hamme, eds.), Lecture Notes in Pure and Applied Mathematics, vol. 222, Marcel Dekker, 2001, pp. 243–254.

[349] N. Obata, *Density of natural numbers and the Levy group*, Journal of Number Theory 30 (1988), pp. 288–297.

[350] ECRYPT Network of Excellence in Cryptology, *eSTREAM – ECRYPT Stream Cipher Project*, available at `http://www.ecrypt.eu.org/stream/`, 2005.

[351] G. Parisi, *p-adic functional integral*, Mod. Phys. Lett. 4 (1988), pp. 369–374.

[352] W. Parry and Z. Coelho, *Ergodicity of p-adic multiplications and the distribution of Fibonacci numbers*, Amer. Math. Soc. Transl. 202(2) (2001), pp. 51–70.

[353] D. Passman, *Permutation groups*, W. A. Benjamin, Inc., New York, Amsterdam, 1968.

[354] A. Peinado, F. Montoya, J. Mu noz, and A. J. Yuste, *Maximal periods of* $x^2 + c$ *in* F_q, Applied algebra, algebraic algorithms and error-correcting codes, Lecture Notes in Computer Science, vol. 2227, Springer-Verlag, 2001, pp. 219–228.

[355] J. Pettigrew, J. A. G. Roberts, and F. Vivaldi, *Complexity of regular invertible p-adic motions*, Chaos 11 (2001), pp. 849–857.

[356] T. Pezda, *Cycles of polynomial mappings in several variables*, Manuscripta Mathematica 83 (1994), pp. 279–289.

[357] ——, *Cycles of polynomials in algebraically closed fields of positive characteristic*, Colloq. Math. 67 (1994), pp. 187–195.

[358] ——, *Polynomial cycles in certain local domains*, Acta Arithmetica LXVI (1994), pp. 11–22.

[359] ——, *Cycles of polynomials in algebraically closed fields of positive characteristic, II*, Colloq. Math. 71 (1996), pp. 23–30.

[360] ——, *On cycles and orbits of polynomial mappings* $\mathbb{Z}^2 \to \mathbb{Z}^2$, Acta Math. Inform. Univ. Ostraviensis 10 (2002), pp. 95–102.

[361] ——, *Cycles of polynomial mappings in several variables over rings of integers in finite extensions of the rationals*, Acta Arith. 108 (2003), pp. 127–146.

[362] M. Pitkanen, *TGD inspired theory of consciousness with applications to biosystems*, available at `http://www.physics.helsinki.fi/~matpitka/cbookI.html`, 1998.

[363] ——, *Could genetic code be understood number theoretically?*, available at `www.helsinki.fi/~matpitka/pdfpool/genenumber.pdf`, 2006.

[364] V. Potkonjak, J. Radojicic, and S. Tzafestas, *Modeling robot "psycho-physical" state and reactions – a new option in human-robot communication. Part 1: Concept and background*, J. Intelligent and Robotic Systems 35 (2002), pp. 339–352.

[365] M. Rajagopalan and B. Schreiber, *Ergodic automorphisms and affine transformations of locally compact groups*, Pacif. J. Math. 38 (1971), pp. 167–176.

[366] J. Rivera-Letelier, *Dynamique des fractions rationnelles sur des corps locaux*, Ph.D. thesis, Université de Paris-sud, Centre d'Orsay, 2000.

[367] ——, *Dynamique des fonctions rationelles sur des corps locaux*, Astérisque 147 (2003), pp. 147–230.

[368] ——, *Expace hyperbolique p-adique et dynamique des fonctions rationelles*, Compositio Math. 138 (2003), pp. 199–231.

[369] ——, *Wild recurrent critical points*, available at `http://arXiv.org/abs/math.DS//0406417`, 2004.

[370] R. Rivest, *Permutation polynomials modulo* 2^w, Finite Fields and Appl. 7 (2001), pp. 287–292.

[371] A. M. Robert, *A Course in p-adic Analysis*, Springer-Verlag, Berlin, New York, Heidelberg, 2000.

[372] J. A. G. Roberts and F. Vivaldi, *Signature of time-reversal symmetry in polynomial automorphisms over over finite fields*, Nonlinearity 18 (2005), pp. 2171–2192.

[373] P. Ruelle, E. Thiran, D. Verstegen, and J. Weyers, *Adelic string and superstring amplitudes*, Mod. Phys. Lett. 4 (1989), pp. 1745–1753.

[374] W. H. Schikhof, *Ultrametric calculus. An introduction to p-adic analysis*, Cambridge University Press, Cambridge, 1984.

[375] B. Schneier, *Applied Cryptography*, John Wiley and Sons, 1996.

[376] C. P. Schnorr, *Zufälligkeit und Wahrscheinlichkeit*, Lect. Notes in Math. 218 (1971).

[377] A. Shamir and B. Tsaban, *Guaranteeing the diversity of number generators*, Information and Computation 171 (2001), pp. 350–363, available at `http://arXiv.org/abs/cs.CR/0112014`.

[378] A. N. Shiryayev, *Probability*, Springer-Verlag, Berlin, New York, Heidelberg, 1984.

[379] I. Shparlinski, *On some dynamical systems in finite fields and residue rings*, Discr. and Cont. Dyn. Syst. 17 (2007), pp. 901–917.

[380] J. Silverman, *Geometric and arithmetic properties of the Henon map*, Math. Zeit. 215 (1994), pp. 237–250.

[381] ——, *The field of definition of dynamical systems on* $\mathbb{P}^1$, Compos. Math. 98 (1995), pp. 269–304.

[382] ——, *Rational functions with a polynomial iterate*, J. Algebra 180 (1996), pp. 102–110.

[383] ——, *The arithmetic of dynamical systems*, Graduate Texts in Mathematics, no. 241, Springer-Verlag, New York, 2007.

[384] M. Sjöstrom and S. Wold, *A multivariate study of the relationship between the genetic code and the physical–chemical properties of amino acids*, Journal of Molecular Evolution 22 (1985), pp. 272–277.

[385] S. De Smedt and A. Yu. Khrennikov, *Dynamical systems and theory of numbers*, Comment. Math. Univ. St. Pauli 46 (1997), pp. 117–132.

[386] J. R. Smythies, *Brain mechanisms and behaviour*, Blackwell Sc. Publ., Oxford, 1970.

[387] M. Solms, *An introduction to the neuroscientific works of Freud*, The prepsychoanalytic writings of Sigmund Freud (London) (G. van der Vijver and F. Geerardyn, eds.), Karnac, 2002, pp. 25–26.

[388] ——, *Putting the psyche into neuropsychology*, vol. 19 (9), Psychologist, 2006.

[389] M. Solms and O. Turnbull, *The brain and the inner world: An introduction to the neuroscience of subjective experience*, Other Press, New York, 2003.

[390] S. M. Stringer and E. Rolls, *Invariant object recognition in the visual system with novel views of 3D objests*, Neural. Comput. 14 (2002), pp. 2585–2596.

[391] S. H. Strogatz, *Nonlinear dynamics and chaos with applications to physics, biology, chemistry, and engineering*, Addison Wesley, Reading, Mass, 1994.

[392] P.-A. Svensson, *Dynamical Systems in Unramified or Totally Ramified Extensions of the p-adic Number Field*, Ultrametric functional analysis, seventh international conference on p-adic analysis, Contemporary Mathematics, vol. 319, Am. Math. Soc., 2003, pp. 405–412.

[393] ——, *Dynamical systems in unramified or totally ramified extensions of a p-adic field*, Izvestiya Mathematics 69 (2005), pp. 1279–1287.

[394] R. Swanson, *A unifying concept for the amino acid code*, Bulletin of Mathematical Biology 46 (1984), pp. 187–203.

[395] E. Thiran, D. Verstegen, and J. Weyers, *p-adic dynamics*, J. of Stat. Phys. 54 (1989), pp. 893–913.

[396] A. Topuzoğlu and A. Winterhof, *Pseudorandom sequences*, Topics in Geometry, Coding Theory and Cryptography, Springer-Verlag, 2006, pp. 135–166.

[397] T. van Gelder, *What might cognition be, if not computation?*, J. of Philosophy 91 (1995), pp. 345–381.

[398] T. van Gelder and R. Port, *It's about time: Overview of the dynamical approach to cognition*, Mind as motion: Explorations in the dynamics of cognition (Cambridge) (T. van Gelder and R. Port, eds.), MITP, 1995, pp. 1–43.

[399] A. van Rooij, *Non-Archimedean functional analysis*, Marcel Dekker, New York, 1978.

[400] D. Verstegen, *p-adic Dynamical Systems*, Springer Proceedings in Physics 47 (1990), pp. 235–242.

[401] F. Vivaldi, *Dynamics over irreducible polynomials*, Nonlinearity 5 (1992), pp. 941–960.

[402] F. Vivaldi and S. Hatjispyros, *Galois theory of periodic orbits of rational maps*, Nonlinearity 5 (1992), pp. 961–978.

[403] F. Vivaldi and I. Vladimirov, *Pseudo-randomness of round-off errors in discretized linear maps on the plane*, Int. J. of Bifurcations and Chaos 13 (2003), pp. 3373–3393.

[404] V. S. Vladimirov and I. V. Volovich, *Superanalysis 1. Differential Calculus*, Teoret. Mat. Fiz. 59 (1984), pp. 3–27.

[405] ——, *Superanalysis 2. Integral calculus*, Teoret. Mat. Fiz. 60 (1984), pp. 169–198.

[406] ——, *p-adic quantum mechanics*, Commun. Math. Phys. 123 (1989), pp. 659–676.

[407] V. S. Vladimirov, I. V. Volovich, and E. I. Zelenov, *p-adic Analysis and Mathematical Physics*, Scientific, Singapore, 1994.

[408] I. V. Volovich, *p-adic string*, Class. Quant. Grav. 4 (1987), pp. 83–87.

[409] P. Walters, *An introduction to ergodic theory*, Springer-Verlag, Berlin, New York, Heidelberg, 2000.

[410] J. D. Watson, T. A. Baker, S. P. Bell, A. Gann, M. Levine, and R. Losick, *Molecular biology of the gene*, Benjamin Cummings, New York, 2003.

[411] A. N. Whitehead, *Process and Reality: An Essay in Cosmology*, Macmillan Publishing Company, New York, 1929.

[412] C. Woodcock and N. Smart, *p-adic chaos and random number generation*, Experimental Math. 7 (1998), pp. 333–342.

[413] S. V. Yablonsky, *Basic notions of cybernetics*, Problems of Cybernetics, Fizmatgiz, 1959, (in Russian).

[414] L. Yaeger, *Computational genetics, physiology, metabolism, neural systems, learning, vision, and behavior or Polyworld: Life in a new context*, Artificial life-3 (Redwood City), Addison Wesley, 2002, p. 102.

[415] Jia-Yan Yao, *Opacity of a finite automaton, method of calculation and the Ising chain*, Discrete Applied Mathematics 125 (2003), pp. 289–318.

[416] ——, *Some properties of Ising automata*, Theoretical Computer Science 314 (2004), pp. 251–279.

[417] E. Young-Bruehl, *Subject to Biography*, Harvard University Press, Boston, 1998.

[418] I. A. Yurov, *On p-adic functions preserving the Haar measure*, Math. Notes 63 (1998), pp. 823–836.

Notation

$\#S$	cardinality of the set S; the number of elements in S if S is a finite set
f^n	the nth iteration of the map $f\colon S \to S$
$f^{(n)}$	the nth derivative of the function f
$A \times B$	Cartesian (direct) product of A and B
$S^n = \underbrace{S \times \cdots \times S}_{n}$	the nth Cartesian (direct) power of S
$(a_i)_{i=0}^{\infty} = a_0, a_1, a_2, \ldots$	(infinite) sequence (sub- and superscripts may be omitted)
(m, n)	the greatest common divisor of m and n
$\mathrm{wt}_p\, n$	$\mathrm{wt}_p\, n = \sum_{k=0}^{m} a_k$ for a natural number $n = a_0 + a_1 p + \cdots + a_m p^m$ represented by its base-p expansion
$\delta_i^p(n) = a_i$	the ith digit in a base-p expansion of a natural number $n = a_0 + a_1 p + \cdots + a_m p^m$ (superscript may be omitted when it is clear what is p)
$\mathbb{N}$	the set of natural numbers
$\mathbb{N}_0$	the set of natural numbers and 0
$\mathbb{Z}$	the set of rational integers
$\mathbb{Q}$	the field of rational numbers
$\mathbb{R}$	the field of real numbers
$\mathbb{F}_{p^n}$	the field of p^n elements
$a \bmod n$	the least non-negative residue of a modulo n
$\mathbb{Z}/n\mathbb{Z}$	the residue ring modulo n
R^+	the additive group of the ring R
R^*	the multiplicative subgroup of the ring R; the group of units of R

$R[x_1, \dots, x_n]$	the ring of polynomials in variables $x_1, \dots, x_n$ over the ring R
$R[[x_1, \dots, x_n]]$	the ring of formal power series in variables $x_1, \dots, x_n$ over the ring R
$\Delta f(x) = f(x+1) - f(x)$	the difference
$\binom{x}{i}$	$\binom{x}{i} = \frac{x(x-1)\cdots(x-i+1)}{i!}$ for $i = 1, 2, \dots$; $\binom{x}{0} = 1$
$\left(\frac{a}{p}\right)$	Legendre symbol
$\mu(n)$	Möbius function
$\varphi(n)$	Euler's totient function
$\pi(x)$	the number of primes not exceeding x
$\lvert x \rvert$	asolute value of x
$\operatorname{ord}_p n$	p-adic valuation
$\lvert x \rvert_p$	p-adic absolute value of x
$\mathbb{Q}_p$	the field of p-adic numbers
$\lceil a \rceil$	the least rational integer that is not smaller than $a \in \mathbb{R}$
$\lfloor a \rfloor$	integral part of a, $a \in \mathbb{R}$, or $a \in \mathbb{Q}_p$
$\ln_p$	the p-adic logarithm; e.g., $\ln_p(1+x) = \sum_{i=1}^{\infty} (-1)^{i+1} \frac{x^i}{i}$
$\overset{p}{\lim}$	limit in $\mathbb{Q}_p$, limit with respect to p-adic metric (superscript may be omitted if this leads to no misunderstanding)
$\mathbb{Z}_p$	the ring of p-adic integers
$\mathbb{Q}_m$	the ring of m-adic numbers
$\mathbb{Z}_m$	the ring of m-adic integers
$B_r^-(a)$	open ball of radius r with center a
$B_r(a)$	closed ball of radius r with center a
$S_r(a)$	sphere of radius r with center a
$\overline{\mathbb{Q}}_p$	algebraic closure of $\mathbb{Q}_p$
$\mathbb{C}_p$	the field of complex p-adic numbers
$\mathbb{Q}_p(\sqrt{\tau})$	quadratic extension of $\mathbb{Q}_p$
$\mathbb{Q}_\tau$	completion of $\mathbb{Q}$ with respect to a topology τ
P	probability

Index

1-Lipschitz function, 63
α-Lipschitz function, 57
π-group, 12
π'-group, 12

A
$\mathcal{A}$-function, 87
absolute value, 19
algebraic closure, 32
algebraic normal form, ANF, 109
algebraically closed, 32
analytic function, **51**
– locally, of order r, 85
approximation by algebras, viii
arity, 6
attractive, 91
attractor, 91
attractor group, 183
automaton, **246**
– finite, 247
– generator, 247, 271
– infinite, 247
– invertible, 250
– Thue–Morse, 250
automaton function, 65, **246**

B
$\mathcal{B}$-function, 83
balance modulo p^k, 99
balanced mapping, 39
basin of attraction, 91
bijectivity modulo p^k, 99
bijectivity modulo a subgroup, 204

C
$\mathcal{C}$-function, 80
centralizer, 11
characteristic polynomial, 359
closed ball, 25
commutator (in a group), 11
compatible p-adic function, **62**
– asymptotically, 62
compatible map, **7**
complex p-adic numbers, 33
complexity of a sequence
– 2-adic, 364
– linear, 342
– ℓ-error, 361
congruence, 6
congruential generator, **276**
– add-xor, 300
– doubly exponential, 279
– explicit, 276
– exponential, 279
– inversive, 278
– knapsack, 286
– Lehmer, 277
– linear, 277
– multiplicative, 278
– power, 278
– recursion law of, 276
– truncated, 307
connected, 26
control sequence, 319
coordinate function, **63**
coordinate sequence, **307**
cyclic-by-cyclic group, 212

D
derivative, **58**
– modulo p^k, 58

– integer-valued, 60
derivative of a polynomial over a group, 201
determined function, 65, **246**
difference, Δ, 2
differentiable function, **58**
– modulo p^k, 58
– uniformly, 60
Dirichlet's theorem, 6

E

equivalent automata, **247**
equivalent states, **247**
ergodicity, xv, 93
unique, 96
Euler's totient function, 3
expectation, 176
extension (of a group by a group), **10**
– central, 12
– split, 10

F

factor(of a group), 12
Feistel network, 309
field, **15**
– primitive element of, 18
– quotient, 15
filter, 271
finite extension, 29
fixed point, 90
formal power series, **17**, 51
free product (of groups), 14
fuzzy cycles, 180
fuzzy orbit, 193

G

generator (automaton), 247, **271**
– counter-assisted, 311
– counter-dependent, 272, 311
group, **9**
– π-group, 12
– π'-group, 12
– Abelian, 11
– commutator, 11
– cyclic-by-cyclic, 212
– derived length of, 12
– dihedral, 13
– elementary Abelian, 12
– factor of, 12
– chief, 12
– free, 12
– generators and relations, 12
– Klein, 10
– metabelian, 14
– metacyclic, 212
– nilpotent, 11
– p-group, 11
– pro-p-group, 233
– profinite, 233
– quaternion (generalized), 13
– section of, 12
– semidihedral, 13
– series in, 12
– single orbit, 41, 212
– solvable, 12
– trivial, 9
– with multioperators, 14
– with operators, 14
– Z-group, 218
– canonical representation, 218
group ring, 17

H

Hensel's lemma, 52, 54
– for pro-p-groups, 235
– for profinite groups, 235

I

identity modulo p^k, 77
indifferent, 91
induced function modulo p^k, 99
integer-valued function, 15
integer-valued polynomial, 15
integral domain, 15
involution, 9

K

key, 270
keystream, 270
Krasner's lemma, 33
Krasner's theorem, 33

L

l-sphere, 185
lacuna, 348
Latin square, **262**
– modulo p^k, 263
Legendre symbol, 4
linear complexity (of a sequence), **342**, 359
linear feedback shift register, LFCR, 323
locally compact, 21
Lucas' theorem, 1

M

Mahler expansion, 75
maximum principle, 51
metacyclic group, 212
minimal polynomial, 359
mixed identity (of a group), 242
mixing, 160
Möbius function, 3
Möbius inversion formula, 169
module (over a ring), 15
Monna map, 340
monomial dynamical system, 93
multiplicative inverse
– generalized, 279

N

n-cycle, 372
– k-chain in, 372
– k-full, 372
non-Archimedean, 19
norm
of a field extension, 30
normal closure, 31
normalizer, 10

O

open ball, 25
orthogonal Latin squares, **266**
Ostrovski's theorem, 20

P

p-adic absolute value, 20, 29
p-adic expansion, 23
p-adic integers, 22
p-adic numbers, 21
p-adic valuation, 19
period, 90
periodic group, 183
periodic point, 90
plaintext, 269
polynomial completeness, 8
polynomial function on a universal algebra, 8
polynomial over a universal algebra, 7
prime number theorem, 5
primitive element (of a field), 18
primitive polynomial, 323
primitivity modulo p^k, 151
pro-p-group, 233
probability measure, 172
probability space, 173
profinite group, 233
profinite topology, 233
pseudorandom (number) generator, PRNG, 269, **271**
pseudorandom sequence, 269

Q

Q1-sequence, 373
quasigroup (binary), **262**, 311
quotient field, 15

R

r-cycle, 90
r-periodic point, 90
ramification index, 31
ramified, 31
reachable state, **247**

repeller, 91
repelling, 91
residue class field, 22, 32
ring, **14**
  – characteristic of, 15
  – commutative, 15
  – ideal of, 15
    – nilpotent, 15
    – principal, 16
    – proper, 15
  – (integral) domain, 15
  – invertible element of, 15
  – local, 15
    – residue field of, 16
  – multiplicative (sub)group of, 15
  – nilpotent element of, 15
    – nilpotency index of, 15
  – of formal power series, 17
  – of zero characteristic, 15
  – principal ideal, 16
  – radical of, 15
  – (semi)group, 17
  – unit group of, 15
  – unit of, 15
  – with identity, 15
  – zero divisor of, 15
roots of unity, 54

S

section (of a group), 12
seed, 271
semidirect product (of groups), 10
semigroup, 9
semigroup ring, 17
sequence, uniformly distributed, **38**
  – modulo p^k, 38
  – strictly, 39
    – k-distributed, 372
  – with respect to a measure, 38
serial connection (of automata), 247
series (in a group)
  – (sub)normal, 12
  – chief, 12
  – derived, 12
  – factor of, 12
    – chief, 12
  – lower central, 11
  – upper central, 11
Siegel disk, 91
single orbit group, 41, 212
skew product, **309**
skew shift, **309**
sphere, 25
stream cipher, 270
strong triangle inequality, 19, 24
subgroup, **9**
  – characteristic, 10
  – Frattini, 14
  – fully invariant, 10
  – index of, 9
  – normal, 10
  – proper, 9
  – Sylow, 11

T

T-function, **65**, 275, 305, 310
Teichmüller character, 164
topological field, 21
totally disconnected, 26
totally ramified, 31
transitive mapping, 39
transitivity modulo p^k, 99
tree, 25
triangular function, **64**, 246, 310
truncation, 307
truth set (of a Boolean function), 109
twice integer-valued function, 67

U

ultrametric, 24
ultrametric space, 24
universal algebra, **6**
  – polynomially complete, 8
  – simple, 6
unramified, 31

V

valuation group, 32
valuation ring, 32
variance, 176

W

weight
– p-adic (of a natural number), 49
– of a Boolean function, 109
wreath product, **309**

Z

Z-group, 218
– canonical representation, 218
zero-one law, 252, 355